VOLUME ONE HUNDRED AND EIGHTY EIGHT

INTERNATIONAL REVIEW OF NEUROBIOLOGY

Brain Energy Metabolism in Alzheimers Disease - Part C

INTERNATIONAL REVIEW OF NEUROBIOLOGY

VOLUME 188

SERIES EDITOR

PATRICIA JANAK

Janak Lab, Department of Neuroscience,
Dunning Hall, John Hopkins University,
Baltimore, MD, USA

PETER JENNER

Division of Pharmacology and Therapeutics,
GKT School of Biomedical Sciences
King's College, London, UK

EDITORIAL BOARD

VOLUME ONE HUNDRED AND EIGHTY EIGHT

INTERNATIONAL REVIEW OF NEUROBIOLOGY

Brain Energy Metabolism in Alzheimers Disease - Part C

Edited by

SUBASHCHANDRABOSE CHINNATHAMBI

Department of Neurochemistry, National Institute of Mental Health and Neuro Sciences Hospital (NIMHANS), Institute of National Importance, Bangalore, Karnataka, India

ELSEVIER

ACADEMIC PRESS

An imprint of Elsevier

Academic Press is an imprint of Elsevier
125 London Wall, London, EC2Y 5AS, United Kingdom
50 Hampshire Street, 5th Floor, Cambridge, MA 02139, United States

First edition 2026

ISBN: 978-0-443-34525-8
ISSN: 0074-7742

For information on all Academic Press publications
visit our website at https://www.elsevier.com/books-and-journals

Publisher: Zoe Kruze
Acquisition Editor: Mariana Kuhl
Editorial Project Manager: Palash Sharma
Production Project Manager: Abdulla Sait
Cover Designer: Bakyalakshmi S

Printed in England by CPI Group (UK) Ltd
Croydon CR0 4YY
United Kingdom

Typeset by MPS Limited, India

Contents

Contributors

Priya
Dr. B R Ambedkar Centre for Biomedical Research, University of Delhi, Delhi, India

Vignesh Pandi A
Department of Pharmacology, JSS College of Pharmacy, JSS Academy of Higher Education & Research, Ooty, The Nilgiris, Tamil Nadu, India

Ali Ariyae Motahar
Neuroscience Institute, Multiple Sclerosis Research Center, Tehran University of Medical Sciences, Tehran, Iran

Masoume Avateffazeli
National Research Institute of Tuberculosis and Lung Diseases (NRITLD), Lung Transplantation Research Center, Shahid Beheshti University of Medical Sciences, Tehran, Iran

Mercy Priyadharshini Babu Deva Irakkam
Laboratory of Stem Cells and Neuroregeneration, Department of Animal Science, Bharathidasan University, Tiruchirappalli, Tamil Nadu, India

Jyotirmoy Banerjee
Department of Biophysics, All India Institute of Medical Sciences, Delhi, India

Sugato Banerjee
Department of Pharmacology and Toxicology, National Institute of Pharmaceutical Education and Research, Kolkata, India

Gowsika Baskar
Laboratory of Stem Cells and Neuroregeneration, Department of Animal Science, School of Life Sciences, Bharathidasan University, Tiruchirappalli, Tamil Nadu, India

Sharda Bharti
Department of Biotechnology, National Institute of Technology (NIT) Raipur, Raipur, Chhattisgarh, India

Yogesh Chander
Department of Pharmaceutical Engineering and Technology, IIT(BHU), Varanasi, India

Debojyoti Chatterjee
Department of Biotechnology and Bioinformatics, JSS Academy of Higher Education and Research, Mysore, Karnataka, India

Vishal Chaurasia
Department of Pharmaceutical Engineering & Technology, Indian Institute of Technology (BHU), Varanasi, Uttar Pradesh, India

Ruchi Chawla
Department of Pharmaceutical Engineering and Technology, IIT(BHU), Varanasi, India

A. Froilanmark Chrishone
Department of Food Science and Technology, St Joseph's University, Bengaluru, Karnataka, India

Sanjib Das Modak
Department of Pharmaceutical Engineering & Technology, Indian Institute of Technology (BHU), Varanasi, Uttar Pradesh, India

Aparna Dixit
Dr. B.R. Ambedkar Center for Biomedical Research, University of Delhi, Delhi, India

Siva Sundara Kumar Durairajan
Molecular Mycology and Neurodegenerative Disease Research Laboratory, Department of Microbiology, Central University of Tamil Nadu, Tiruvarur, Tamil Nadu, India; School of Biomedical Sciences, Li Ka Shing Faculty of Medicine, The University of Hong Kong, Hong Kong, P.R. China

Dileep Francis
Department of Life Sciences, School of Biological and Forensic Sciences, Kristu Jayanti (Deemed to be University), Bengaluru, Karnataka, India

Elham Hosseini
Neuroscience Research Center, Mashhad University of Medical Sciences, Mashhad, Iran

Ashok Iyaswamy
Mr. & Mrs Ko Chi-Ming Centre for Parkinson's Disease Research, School of Chinese Medicine, Hong Kong Baptist University, Kowloon, Hong Kong, P.R. China; Department of Biochemistry, Karpagam Academy of Higher Education, Coimbatore, Tamil Nadu, India

Vedika Jain
Department of Biotechnology, National Institute of Technology (NIT) Raipur, Raipur, Chhattisgarh, India

Antony Justin
Department of Pharmacology, JSS College of Pharmacy, JSS Academy of Higher Education & Research, Ooty, The Nilgiris, Tamil Nadu, India

Mahesh Kandasamy
Laboratory of Stem Cells and Neuroregeneration, Department of Animal Science, Bharathidasan University, Tiruchirappalli, Tamil Nadu; University Grants Commission-Faculty Recharge Programme (UGC-FRP), New Delhi, India

Sanaz Khodadadi
Neuroscience Institute, Multiple Sclerosis Research Center, Tehran University of Medical Sciences, Tehran, Iran

Gopal Khodve
Department of Pharmacology and Toxicology, National Institute of Pharmaceutical Education and Research, Kolkata, India

Senthilkumar Krishnamoorthi
Mr. & Mrs Ko Chi-Ming Centre for Parkinson's Disease Research, School of Chinese Medicine, Hong Kong Baptist University, Kowloon, Hong Kong, P.R. China

Supraja M.R.
Molecular Mycology and Neurodegenerative Disease Research Laboratory, Department of Microbiology, Central University of Tamil Nadu, Tiruvarur, Tamil Nadu, India

Mohini Milmile
Department of Pharmaceutical Engineering and Technology, IIT(BHU), Varanasi, India

Ranjan Mishra
Department of Pharmaceutical Engineering and Technology, IIT(BHU), Varanasi, India

Radhika Mittal
Dr. B R Ambedkar Centre for Biomedical Research, University of Delhi, Delhi, India

Gyan Modi
Department of Pharmaceutical Engineering & Technology, Indian Institute of Technology (BHU), Varanasi, Uttar Pradesh, India

Reyhaneh Montazeri-Shatouri
Neuroscience Institute, Multiple Sclerosis Research Center, Tehran University of Medical Sciences, Tehran, Iran

Abdorreza Naser Moghadasi
Neuroscience Institute, Multiple Sclerosis Research Center, Tehran University of Medical Sciences, Tehran, Iran

Ajay Pandey
Department of Pharmaceutical Engineering and Technology, IIT(BHU), Varanasi, India

Gajanan Pawar
Department of Pharmaceutical Engineering and Technology, IIT(BHU), Varanasi, India

Pradnya Petkar
Department of Pharmaceutical Engineering and Technology, IIT(BHU), Varanasi, India

Amal P.S.
Molecular Mycology and Neurodegenerative Disease Research Laboratory, Department of Microbiology, Central University of Tamil Nadu, Tiruvarur, Tamil Nadu, India

Ignatius Raymond Peter L
Department of Life Sciences, School of Biological and Forensic Sciences, Kristu Jayanti (Deemed to be University), Bengaluru, Karnataka, India

Shaswat Raval
Department of Pharmacology and Toxicology, National Institute of Pharmaceutical Education and Research, Kolkata, India

Nasim Rezaeimanesh
Neuroscience Institute, Multiple Sclerosis Research Center, Tehran University of Medical Sciences, Tehran, Iran

Sajad Sahab Negah
Neuroscience Institute, Multiple Sclerosis Research Center, Tehran University of Medical Sciences, Tehran, Iran

Abhay Kumar Singh
Molecular Mycology and Neurodegenerative Disease Research Laboratory, Department of Microbiology, Central University of Tamil Nadu, Tiruvarur, Tamil Nadu, India

Gireesh Kumar Singh
Department of Pharmacy, School of Health Science, Central University of South Bihar, Gaya, Bihar, India

Shweta Singh
Department of Pharmaceutical Engineering and Technology, IIT(BHU), Varanasi, India

Manisha Tiwari
Dr. B.R. Ambedkar Center for Biomedical Research, University of Delhi, Delhi, India

Gopika Venu
Department of Pharmacology, JSS College of Pharmacy, JSS Academy of Higher Education & Research, Ooty, The Nilgiris, Tamil Nadu, India

Nitin Yadav
Dr. B.R. Ambedkar Center for Biomedical Research, University of Delhi, Delhi, India

Fateme Yousefi
School of Medicine, Kermanshah University of Medical Sciences, Kermanshah, Iran

CHAPTER ONE

Metabolic therapeutic targets in Alzheimer's disease

Ignatius Raymond Peter L[a], Debojyoti Chatterjee[b], A. Froilanmark Chrishone[c], and Dileep Francis[a,*]

[a]Department of Life Sciences, School of Biological and Forensic Sciences, Kristu Jayanti (Deemed to be University), Bengaluru, Karnataka, India
[b]Department of Biotechnology and Bioinformatics, JSS Academy of Higher Education and Research, Mysore, Karnataka, India
[c]Department of Food Science and Technology, St Joseph's University, Bengaluru, Karnataka, India
*Corresponding author. e-mail address: dileep@kristujayanti.com; dilipfrans@gmail.com

Contents

International Review of Neurobiology, Volume 188
ISSN 0074-7742, https://doi.org/10.1016/bs.irn.2026.05.010

Abstract

Alzheimer's disease (AD) is increasingly recognized as a disorder marked not only by amyloid-β and tau pathology, but also by profound disturbances in brain energy metabolism that arise early in disease progression. Accumulating evidence indicates that impairments in glucose utilization, insulin signaling, and mitochondrial function precede neurodegeneration and contribute directly to synaptic failure and cognitive decline. This chapter presents a comprehensive overview of Alzheimer's disease through the lens of metabolic dysfunction, highlighting disrupted neuronal bioenergetics as a central and unifying feature of pathogenesis. We examine key metabolic pathways implicated in AD, including cerebral glucose hypometabolism, brain insulin resistance, impaired glycolysis, mitochondrial oxidative phosphorylation deficits, oxidative stress, and altered mitochondrial dynamics. The chapter further discusses therapeutic strategies aimed at restoring metabolic homeostasis, such as insulin sensitization, enhancement of glucose transport, activation of mitochondrial biogenesis, modulation of the electron transport chain, and the use of mitochondria-targeted antioxidants. In parallel, alternative energy approaches—including ketone metabolism, fatty acid oxidation, and pentose phosphate pathway activation—are explored as promising avenues to bypass glucose-dependent energy deficits and reinforce neuronal resilience. Emerging directions in metabolic therapeutics are also highlighted, including combination treatment strategies, NAD^+–sirtuin and AMPK signaling, and the expanding role of the gut microbiome–brain metabolism axis. By integrating insights from experimental models, neuroimaging studies, and clinical trials, this chapter underscores the potential of metabolic interventions to enable early, disease-modifying strategies for Alzheimer's disease.

1. Introduction

Alzheimer disease (AD) is a progressive neurodegenerative disease that involves the cerebral cortex and hippocampus mostly. AD is the most prevalent type of dementia, representing the case of 55 million dementia cases across the globe, with the number rising significantly as people get older (WHO, 2023). The disease course has a preclinical phase of about 20 years before a clinical period of 8 years that is marked by progressive memory loss, cognitive deficiency and eventually proceeds to death.

There are several molecular pathologies which are in the development and progression of AD. Plaques of beta-amyloid (Aβ) peptides and hyperphosphorylated tau (τ) protein in the form of neurofibrillary tangles are the classical signs and bring about oxidative damage and neuroinflammation. Besides, metabolic dysfunction is becoming a growing implicated factor in neurodegeneration, where cognitive impairment is commonly anteceded by neuronal energy metabolism impairment (Fig. 1).

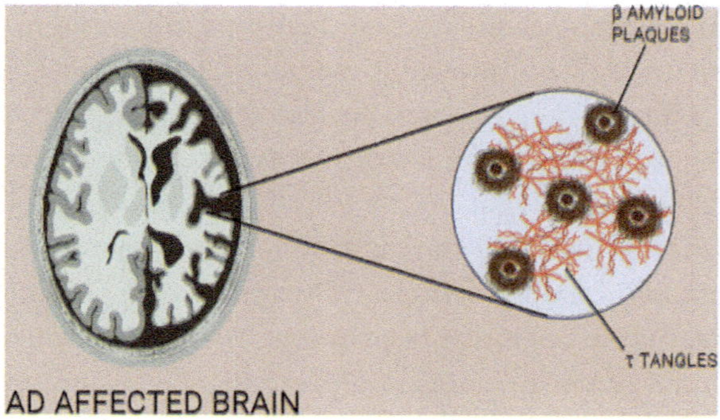

Fig. 1 Alzheimer disease pathology in brain tissue.

It has been hypothesized that the connection between glucose metabolism dysfunction of the brain and AD can be termed as Type 3 diabetes. Neuronal insulin resistance causes a problem in the absorption and use of glucose thus leaving one in an energy deficit state which reduces the normal operation of the mitochondria. It is an energy failure that leads to a series of pathological processes such as an overproduction of reactive oxygen species (ROS), dysfunction of synapses, and neurodegeneration (Meng et al., 2024).

The targeting of these metabolic pathways (especially brain insulin resistance and mitochondrial dysfunction) is a promising potential treatment and prevention option in the treatment of AD. In this chapter the metabolic foundations of AD pathogenesis are analyzed and the therapeutic approaches aimed at the restoration of neuron energy homeostasis and delaying the development of the disease are discussed.

2. The metabolic hypothesis of AD

2.1 Neurodegeneration and metabolic syndrome

Systemic metabolic dysfunction is often a precursor to cognitive impairment of AD. Metabolic syndrome is a group of interconnected diseases such as glucose unresponsiveness, insulin resistance, lipid disorders, and obesity (Fahed et al., 2022; Grundy, 2016; Lann & LeRoith, 2007). It is increasingly pointing to the idea that AD has similar metabolic characteristics to peripheral insulin resistance, which may be an independent cerebral phenomenon or a result of systemic metabolic dysfunction (Kim & Feldman, 2015).

Preclinical research has shown that, neuroinflammation in white matter of AD rodent models is caused by metabolic syndrome (Ivanova et al., 2023). This white matter disease can also be an early neurological sign further leading to the cognitive impairments of mild cognitive impairment (MCI) and AD, which might be the cause of the mechanistic relationship between neurodegeneration and metabolic syndrome. These results are supported with clinical observations: AD patients with comorbid metabolic syndrome have worse cognitive impairment, increased vascular endothelial dysfunction, and intense brain insulin resistance as compared to AD patients who do not have a metabolic syndrome (Fukui et al., 2016).

2.2 Early biomarkers of brain glucose hypometabolism

Fluorodeoxyglucose positron emission tomography (FDG-PET) is a quantitative evaluation of cerebral glucose metabolism, which is an indirect measure of synaptic activity and neuronal energy consumption. Notably, glucose hypometabolism may be identified as the primary functional change in brain networks, which may be observed years before neurodegeneration leads to irreversible structural changes, so it can be considered a useful early biomarker relative to downstream markers of neurodegeneration (Alhasan et al., 2025).

The localized distribution of glucose hypometabolism in diabetes and insulin resistance is closely similar to those observed in AD, which implies that insulin resistance may also play a direct role in the modification of glucose metabolism, which is among the major preclinical characteristics of AD. A low intake of FDG in the brain affected areas defines this hypometabolism, which is the inability of neurons to use glucose (Soltani et al., 2025).

2.3 Diabetes type 3 and brain energy crisis

Type 3 diabetes has become a term that underlines the important role of insulin to the homeostasis of brain energy (Steen et al., 2005). The brain is a highly metabolically active organ; it utilizes a substantial amount of energy in the body accounting to 20 % yet it only makes up 2 % of the body mass. This energy need is fulfilled by mitochondrial oxidative phosphorylation (OXPHOS), whereby glucose is the major substrate. Facilitated by glycolysis and OXPHOS, glucose is processed to produce adenosine triphosphate (ATP), which is the universal cellular energy currency (Kapogiannis & Avgerinos, 2020).

Several studies show that glucose metabolism changes are the early signs of cognitive impairment and neurodegeneration in AD. Resistance to insulin in the cells of brain makes them insensitive to insulin, and this results in a functional glucose deprivation although glucose is present. The neurons cannot use glucose efficiently, and they therefore fall into an energy-deprived condition with low levels of ATP synthesis.

Some of the consequences of this metabolic crisis to start with are, the build-up of unused glucose in the tissue of the neurons which demonstrates the lack of metabolism of the energy and not the lack of substrate. Second, compensatory metabolic adaptations may be attempted by neurons but these are usually inadequate or maladaptive. Third, and most importantly, the energy deficit is accompanied by mitochondrial dysfunction that is manifested by the decreased OXPHOS efficiency and the excess production of ROS. This energy crisis and oxidative stress disorder allow conditions that have been conducive to Aβ accumulation, tau hyperphosphorylation, synaptic dysfunction, and progressive neuronal death (De Felice et al., 2008).

3. Glucose metabolism as a therapeutic target in Alzheimer's disease

3.1 Rationale for targeting glucose metabolism

Impairment of cerebral glucose metabolism is one of the earliest and most consistently replicated metabolic abnormalities in Alzheimer's disease (AD). Notably, this decline appears well before measurable cognitive deficits or the accumulation of hallmark pathological features such as amyloid-β (Aβ) plaques and neurofibrillary tau tangles. Extensive neuroimaging studies using fluorodeoxyglucose positron emission tomography (FDG-PET) have documented regional hypometabolism in the posterior cingulate cortex, hippocampus, and temporoparietal regions—areas that support memory formation, learning, and executive processing (Mosconi et al., 2008; Mosconi, 2013).

This reduction in glucose utilization aligns more closely with synaptic dysfunction than with neuronal loss, reinforcing the view that metabolic decline plays a primary, causal role in AD progression rather than reflecting a downstream degenerative consequence. At a molecular level, disruptions across insulin signaling, glucose transporter expression, and glycolytic enzyme activity collectively impair mitochondrial function and ATP production, promoting oxidative stress and accelerating synaptic degeneration (Butterfield & Halliwell, 2019).

As a result, therapeutic strategies that aim to restore glucose metabolism have become an essential area of research, with potential not only to relieve symptoms but to modify disease trajectory.

3.2 Insulin sensitization and restoration of neuronal metabolic signaling

The characterization of AD as a form of "Type 3 diabetes" has gained traction due to growing evidence of brain-specific insulin resistance (De la Monte & Wands, 2008).

Under normal conditions, insulin binding activates the PI3K/AKT pathway, supporting glucose uptake, glycogen synthesis, mitochondrial activity, and cell survival. In AD, however, this pathway becomes compromised through phenomenon like downregulation of insulin receptors, impaired receptor phosphorylation, Aβ interference by binding to insulin receptors which leads to disrupted signaling (Talbot et al., 2012).

One of the most promising strategies to counteract these defects is intranasal insulin therapy, which delivers insulin directly to the central nervous system through the nasal-olfactory tract while avoiding systemic side effects. Clinical trials have demonstrated improvements in memory and executive function in individuals with mild cognitive impairment and early AD, likely due to enhanced AKT (protein kinase B) activation which leads to enhanced glucose metabolism and the suppression of glycogen synthase kinase-3β (GSK-3β)—a major contributor to tau hyperphosphorylation (Craft et al., 2012; Steen et al., 2005).

Additional insulin-sensitizing approaches include the repurposing of antidiabetic drugs such as Metformin, via AMPK activation which enhances mitochondrial function and glucose metabolism. In AD models metformin also induced neurogenesis synaptic plasticity and reduced Aβ and Tau pathology.

Another focus targets PPAR-γ agonists such as pioglitazone and rosiglitazone which increase insulin receptor expression thereby promoting insulin binding and increasing glucose uptake by the neurons. They also reduce neuroinflammation, contributing to improved metabolic stability (Heneka et al., 2015; Rotermund et al., 2018).

Together, these interventions highlight the therapeutic value of restoring insulin signaling as a means to preserve neuronal health.

3.3 Enhancement of cerebral glucose transport

Apart from insulin resistance, AD is further characterized by reduced expression of key glucose transporters at the blood–brain barrier and within neurons. GLUT1, found on endothelial cells and astrocytes, regulates glucose entry into the brain, while GLUT3 is the principal neuronal glucose transporter. Both are markedly diminished in AD, correlating with reduced regional glucose uptake and cognitive decline severity (Jurcovicova, 2014; Liu et al., 2008). Loss of GLUT1 restricts glucose passage across the blood–brain barrier, creating an energy deficit that cannot be compensated even under normal or elevated systemic glucose levels. Reduced GLUT3 expression further limits glucose availability to metabolically demanding neuronal populations, exacerbating synaptic dysfunction and mitochondrial impairment.

Therapeutic strategies aimed at restoring transporter expression include the activation of pathways that promote the translocation of GLUT, upregulation of GLUT expression through gene therapy and enhancement of vascular endothelial growth factor (VEG-F) signaling to support blood-brain barrier health and heightened density of GLUT1 (Winkler et al., 2015; Zhao & Townsend, 2009).

These findings underscore that glucose delivery itself is a rate-limiting step in sustaining neuronal energy balance.

3.4 Modulating the functionality of the glycolytic pathway

Once glucose enters the neuron, efficient glycolysis is essential for energy generation. In glycolysis, glycogen is to be broken down into usable glucose in order to meet the energy requirements of the cell. Under AD conditions, this pathway is disrupted due to decreased activity of several key enzymes, including hexokinase, phosphofructokinase-1, and pyruvate kinase (Bigl et al., 2000).

Proteomic evidence suggests that oxidative modifications (e.g., carbonylation and nitration) impair enzyme stability and function, restricting pyruvate formation and downstream mitochondrial oxidative metabolism (Butterfield et al., 2010).

Therapeutic strategies to enhance glycolytic efficiency focus on metabolic and redox regulation specifically targeting AMPK activators to boost the glycolytic flux thereby promoting mitochondrial phosphorylation efficiency resulting in increased ATP production and improved clearance of damaged and toxic proteins through the process of autophagy (Hardie et al., 2012). Certain neutraceuticals like resveratrol and α-lipoic acid have shown to protect glycolytic enzymes from degradation due to oxidative stress and also help to restore the redox equilibrium (Timmers et al., 2012).

Monitoring glycolytic status has also evolved as a valuable clinical and research tool. FDG-PET remains the gold standard for assessing cerebral glucose metabolism, while biomarkers such as lactate–pyruvate ratios, NAD^+/NADH balance, and glycolytic enzyme levels in cerebrospinal fluid provide additional insights into metabolic restoration and therapy response (Butterfield & Halliwell, 2019; Mosconi et al., 2008).

4. Mitochondrial-targeted therapy

The mitochondria are the most important organelles in the cell that are the source of cellular energy production and are double-membrane structures. In addition to the production of ATP by OXPHOS, calcium homeostasis, Lipid metabolism, and the cell death through apoptotic pathways are regulated by the mitochondria. Under physiological conditions, mitochondria produce regulated levels of ROS which act as signaling molecules. Nonetheless, excessive production of ROS results in oxidative damages to cellular components such as lipids, proteins and DNA due to mitochondrial dysfunction.

Mitochondrial maladaptation in AD can take place in various ways: impaired bioenergetics, disrupted mitochondrial dynamics (fission and fusion), defective mitophagy (clearance of damaged mitochondria), and age dependent, diminished quality control of mitochondria. These changes facilitate the aberrant clearance of amyloid precursor protein (APP), which leads to the toxic burden of Aβ and hyperphosphorylated tau (Wang et al., 2025).

Since mitochondrial dysfunction is at the core of AD pathogenesis, interventions based on mitochondrial biology therapies are promising ones. There are four major strategies: (1) improving mitochondrial biogenesis which leads to an increase of mitochondrial mass, (2) improving the functioning of the electron transport chain, (3) the activation of mitochondrial antioxidant systems, and (4) restoring equilibrium in mitochondrial dynamics.

4.1 Improving mitochondrial biogenesis

Mitochondrial biogenesis is the cellular process of mitochondrial production and is necessary to sustain the high metabolic needs of neurons. The process is strictly controlled and can be stimulated by physiological stressors such as exercise and caloric deprivation. Since the mitochondrial dysfunction is an AD characteristic, the promotion of mitochondrial biogenesis is a rational approach towards treatment.

4.1.1 The PGC-1α/SIRT1/AMPK pathway

The high-expression of peroxisome proliferator-activated receptor-γ coactivator-1α (PGC-1α) in the central nervous system is a master regulator of mitochondrial biogenesis. Research indicates that the PGC-1α concentration in the brains of AD patients is drastically lower (Zehnder et al., 2021), which argues in favor of the involvement of the malfunction of mitochondrial biogenesis in the disease pathology.

In addition to acting as a biogenesis protein, PGC-1α is anti-amyloidogenic. High levels of PGC-1α inhibit the BACE1 (β-site amyloid precursor protein cleaving enzyme 1) transcription as the rate-limiting enzyme in the production of Aβ out of APP (amyloid precursor protein) (Wang et al., 2022). This forms an inverse correlation between the levels of PGC-1α and Aβ plaque burden. Notably, SIRT1 (silent information regulator 1) controls the PGC-1α activity by deacetylation, and SIRT1 may be stimulated by AMPK (AMP-activated protein kinase) (Price et al., 2012) (Fig. 2).

The network of this interrelation between PGC-1α/SIRT1/AMPK provides several therapeutic targets.

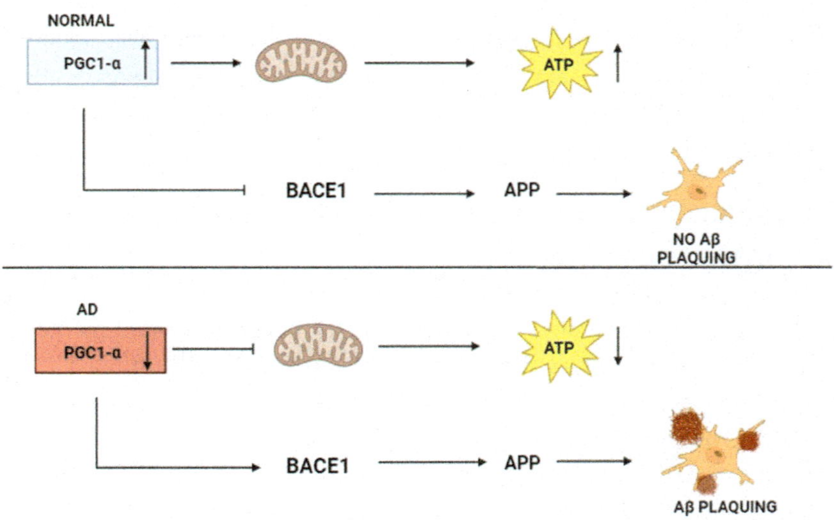

Fig. 2 Correlational effect of PGC-1α on ATP and Aβ plaquing: The normal regulation of PGC-1α promotes normal mitochondrial function leading to sufficient ATP production, simultaneously attenuating the activity of BACE-1 resulting to no Aβ plaquing. On the contrary, the downregulation of PGC-1α leads to mitochondrial dysfunction and BACE-1 activity promoting Aβ plaquing.

4.1.2 Therapeutic compounds

Resveratrol: This polyphenol is plant-derived and proves to be neuroprotective in neurodegenerative diseases (Caruana et al., 2016). Resveratrol activates SIRT1 and in turn deacetylates and activates PGC-1α. This stimulation has two effects, mitochondrial biogenesis and inhibition of BACE1, that result in bioenergetic support and anti-amyloidogenic outcome (Lagouge et al., 2006). Nevertheless, translating to clinical use is hampered by low bioavailability which requires high dosages or novel delivery systems.

Sildenafil: Sildenafil was originally used in treating erectile dysfunction and pulmonary hypertension, also activates SIRT1 and results in PGC-1α deacetylation and activation. Experimental studies show that sildenafil decreases neuroinflammatory markers and stimulates neurogenesis (Sanders, 2020). It has a good safety history and brain penetrance capacity, which makes it a potential candidate in AD treatment, although it requires clinical trials in AD populations.

Nicotinamide Riboside (NR): NR is a precursor of NAD^+ (nicotinamide adenine dinucleleotide) and facilitates the activation of PGC-1α by NAD^+ dependent SIRT1 deacetylase. Research indicates that NR supplementation alleviates Aβ toxicity in the neuron tissue (Kim & Feldman, 2015; She et al., 2025) and promotes the functionality of the mitochondrial system by replenishment NAD^+. Since the decrease in the NAD^+ levels with age occurs, NR supplement may be especially helpful in at-risk elderly populations with AD.

4.2 Enhancing electron transport chain function

4.2.1 Mitochondrial bioenergetics

Mitochondrial electron transport chain (ETC) (occurring in the inner mitochondrial membrane) links the transfer of electrons to proton pumping, to generate an electrochemical gradient to be utilized in the production of ATP. The ETC includes four protein complexes which include Complex I (NADH dehydrogenase), Complex II (succinate dehydrogenase), Complex III (cytochrome bc1 complex) and Complex IV (cytochrome c oxidase). The proton translocation between the mitochondrial matrix and the intermembrane space is caused by the flow of electrons into these complexes by redox reactions facilitated by NADH and FADH2 (products of the TCA cycle).

Proton-motive force resulting thereof is a combination of electrical potential (mitochondrial membrane potential, $\Delta\psi m$) and a pH gradient

(ΔpH). In physiological conditions, $\Delta\psi m$ contributes to about 70% of the proton-motive force. This electrochemical gradient is utilized in the process of producing ATP by ATP synthase (Complex V) through the phosphorylation of ADP (adenosine diphosphate).

4.2.2 ROS generation and control

The ETC, though necessary in producing energy, is also the major source of cellular ROS (reactive oxygen species). Electrons in complexes I and III may react with oxygen prematurely, producing superoxide (O_2^{-}) and hydrogen peroxide (H_2O_2). In the natural state, these ROS are neutralized by antioxidant enzymes such as superoxide dismutase, glutathione peroxidase, as well as catalase. Nonetheless, in disease conditions, this balance is lost, which causes oxidative damage (Mandal et al., 2025).

4.2.3 Therapeutic interventions

Coenzyme Q10 (CoQ10): CoQ10 plays the role of an electron carrier in the ETC and a lipid soluble antioxidant. It receives the electrons of Complexes I and II and passes them into Complex III hence playing an important role in the flow of electrons. Research proves the CoQ10 depletion in neurodegenerative disorders. The CoQ10 or its synthetic counterpart idebenone supplementation demonstrates potential in improving mitochondrial dysfunction (Manzar et al., 2020). Nevertheless, bioavailability is deplorable, which impairs the efficacy of CoQ10; new formulations with ubiquinol (reduced CoQ10) and liposomal preparations have the potential to enhance target tissue delivery.

Metformin: Metformin is a medication derived from *Galega officinalis* which is prescribed extensively to treat type 2 diabetes. Ironically, metformin has a therapeutic effect by the mechanism of slight Complex I inhibition. Metformin partially inhibits the complex I, avoiding complete oxidative stress, in AD where Complex I dysfunction causes excessive ROS production, which is sufficient to sustain adequate ATP production (Mandal et al., 2025). There is clinical and epidemiological evidence that metformin use has a lower risk of dementia, but supporting trials are required.

NAD^+ Precursors: NAD^+ is a required electron carrier in the ETC, which accepts electrons in the glycolysis pathway and the TCA cycle to become NADH. NAD^+ precursors such as nicotinamide riboside (NR), nicotinamide mononucleotide (NMN), nicotinamide (NAM), and niacin are beneficial in enhancing mitochondrial functions in a variety of ways.

These drugs enhance cognitive ability, decrease oxidative stress and alleviate mitochondrial dysfunction in preclinical AD models (Campbell, 2022). In human populations, clinical trials are being conducted to determine efficacy and optimum dosage.

4.3 Activating mitochondrial antioxidant systems

4.3.1 Oxidative stress in the AD pathogenesis

Mitochondrial dysfunction leads to oxidative stress which causes and accelerates cognitive impairment in AD. ROS also induce direct destruction of neuronal parts and also facilitate Aβ production and aggregation by excess APP processing. Also, oxidative stress inhibits the expression of SIRT1, which is a neuroprotective factor that is critical in neurogenesis and anti-inflammatory processes (Alzarea et al., 2025).

Considering that ROS are majorly produced due to electron leakage at ETC, antioxidants at the source of ROS, that is, in mitochondria, could be more effective than generic antioxidants, which are used throughout the cell.

4.3.2 Mitochondria-targeted antioxidants

MitoQ (Mitoquinone): MitoQ is an analog of CoQ10 to which a lipophilic triphenylphosphonium cation has been conjugated. This confers a positive charge and contributes to the accumulation of MitoQ in the mitochondria due to the negative membrane potential. The ubiquinone moiety is then reduced to its active form by Complex II and neutralizes ROS, forming a self-regenerative antioxidant system in the mitochondria (McManus et al., 2011) . Neuroprotective effects have been shown in preclinical studies, but there is limited clinical information on AD.

SkQ1: This is a mitochondria-targeted antioxidant that is composed of plastoquinone (a natural antioxidant in plants) to which lipophilic cation has been conjugated to assist its delivery to mitochondria. Research on OXYS rats (an accelerated aging oxidative stress model) shows that SkQ1 supplementation delays senescence and Aβ and tau pathology by inhibiting oxidative stress. SkQ1 also upregulates the mTOR (mammalian target of rapamycin), a kinase that is related to neurogenesis and autophage-mediated clearance of toxic proteins (Muraleva et al., 2019).

Alpha-Lipoic Acid (LA): LA is an organosulfur compound, which serves as a cofactor for mitochondrial enzymes and as an antioxidant. LA is an endogenous product of young, healthy individuals, but the amount of production reduces as a person gets older. LA administration increases the activity of antioxidant enzymes such as superoxide dismutase, catalase, and glutathione peroxidase, which decreases the consequences of mitochondrial malfunction (Pei et al., 2023). The capability of LA to penetrate the blood-brain barrier and its known safety profile makes it preferred for usage in a clinical setting, however, large-scale studies in AD are necessary (Fig. 3).

4.4 Regulation of mitochondrial dynamics

4.4.1 Fission and fusion balance

The mitochondria are dynamic organelles that constantly divide (fission) and merge (fusion) according to the needs of the cellular energy. Fission processes that are mediated by dynamin-related protein 1 (Drp1) and its receptors (MFF, Fis1, Mid49, Mid51) produce smaller mitochondria and promote removal of damaged components by mitophagy. Outer membrane fusion mediated by mitofusins 1 and 2 (Mfn1, Mfn2) and inner membrane fusion mediated by optic atrophy 1 (Opa1) make the mitochondria unified which buffers the metabolic stress and shares the energy requirement load (Fig. 4).

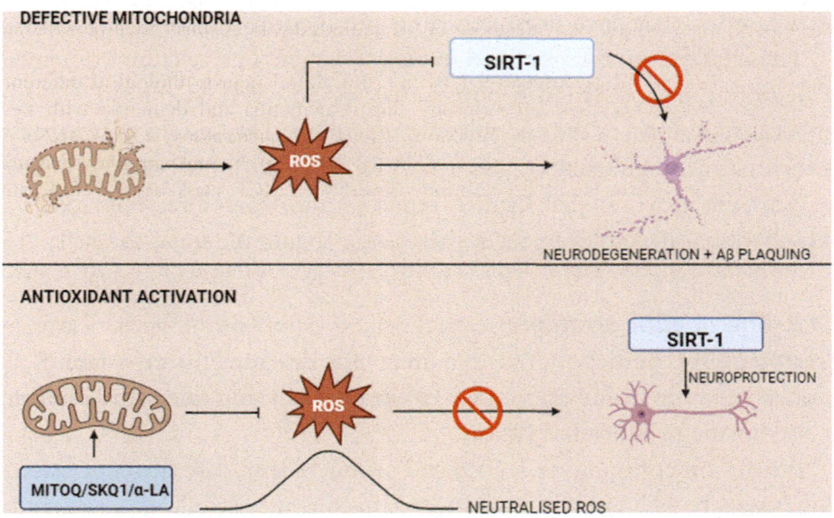

Fig. 3 Effect of ROS on neurons before and after activation of mitochondrial antioxidants: The upper panel depicts a defective mitochondrion releasing ROS which inhibits SIRT-1's neuroprotective activities leading to neurodegeneration. The lower panel shows the activity of antioxidants, whereby the ROS is neutralized leading to uninhibited neuroprotective effects of SIRT-1.

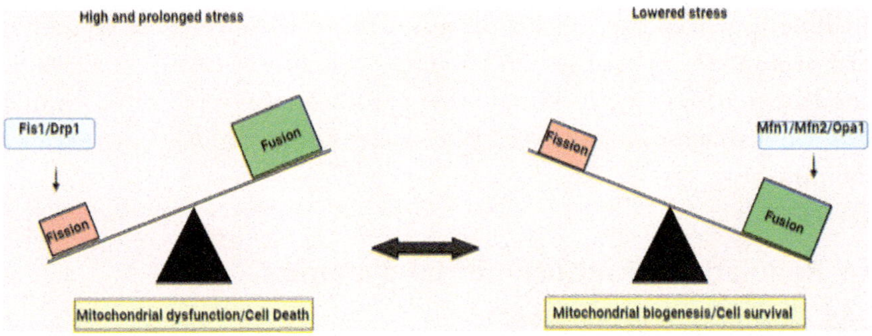

Fig. 4 Mitochondrial quality control through fission and fusion: Under high and prolonged stress conditions, mitochondrial dysfunction occurs mediated by factors like Fis1 and Drp1 where excessive fission happens to meet the energy requirements of the cell. During lowered stress conditions mitochondrial fusion occurs, mediated by Mfn1,Mfn2 and Opa1to promote mitochondrial biogenesis and neuronal homeostasis.

4.4.2 Dynamics dysfunction in AD

In AD, this fine balance is lost and fissions are more excessive and fusion is impaired. The analysis of neuroblastoma cells subjected to Aβ shows more expression of fission proteins (Drp1, Fis1) and less expression of fusion proteins (Mfn1, Mfn2). This imbalance leads to fragmented dysfunctional mitochondria that have disrupted ATP production (Hayat et al., 2025).

These effects are not confined to bioenergetics. The calcium homeostasis of neurons relies on SERCA pumps that are powered by the ATP to preserve calcium stores in the endoplasmic reticulum. This system is damaged by mitochondrial fragmentation and low ATP production. Moreover, neurons are post-mitotic cells that cannot replicate and, therefore, rely solely on mitochondrial dynamics to meet the energy requirements of the cell.

4.4.3 Therapeutic strategies

Treatments of mitochondrial dynamics are designed to re-establish the balance between fission and fusion either by inhibiting uncontrolled fission or by enhancing impaired fusion.

Pro-fusion compounds: Icariin and probucol stimulate the expression of Mfn2 which facilitates mitochondrial fusion and stabilization of mitochondrial function (Sita et al., 2020). Nilotinib is a tyrosine kinase inhibitor that elevates the levels of Mfn1 in astroglia of familial AD mouse models, improving fusion and mitochondrial bioenergetics (Eshraghi et al., 2021).

Anti-fission compounds: Mdivi-1 (mitochondrial division inhibitor-1) is a direct inhibitor of Drp1 that inhibits mitochondrial fragmentation and

Aβ plaque formation and increases mitochondrial biogenesis and respiration (Hayat et al., 2025). Ginkgolide, a compound extracted from *Ginkgo biloba* inhibits fission by adding a phosphate to serine moiety at 637 position of Drp1. Other natural products such as cordycepin, luteolin, quercetin and silibinin have a similar Drp1-inhibitory effect (Hayat et al., 2025).

5. Activating alternate energy pathways

The excessive dependence on glucose metabolism by the brain leaves the brain vulnerable when there is impairment of glucose utilization as is the case in AD. Hypoglycemia also leads to sensory and cognitive impairment which highlights the dependence of the brain on glucose. Low cerebral metabolic rate of glucose (CMRglu) has been associated with cognitive deficiency in AD and has been considered as an effective early marker of disease, which can be identified early in the disease before significant structural alterations have occurred in the brain (Costantini et al., 2008).

Since in AD pathogenesis, impaired glucose metabolism is key, the therapeutic approaches based on the diversification of neuronal fuel sources can potentially bypass this metabolic bottleneck. Three promising directions include-the metabolism of ketone bodies, the improvement of fat oxidation, and upregulation of pentose phosphate pathway.

5.1 Ketones as alternative fuel

The Ketone bodies β-hydroxybutyrate (BHB), acetoacetate (ACA), and acetone are the other energy substrates that are used in the absence of carbohydrates. Ketones are some of the most favored fuels in the development of the brain, possibly due to their easy penetration of the blood-brain barrier and the fast rate at which neurons use them. Ketones metabolism may be stimulated in adults by dietary regimens or exogenous supplementation.

The ketones get around the impaired glucose metabolism and get into the TCA cycle, producing ATP and acetylcholine precursors. BHB metabolism has some bioenergetic benefits: (1) BHB oxidation to ACA provides cellular redox potential because it produces NADH; (2) ACA to acetoacetyl-CoA produces succinate, which is a substrate of Complex II, which bypasses Complex I inhibition; and (3) BHB metabolism elevates acetyl-CoA levels in the mitochondria, aiding the TCA cycle (Costantini et al., 2008) (Fig. 5).

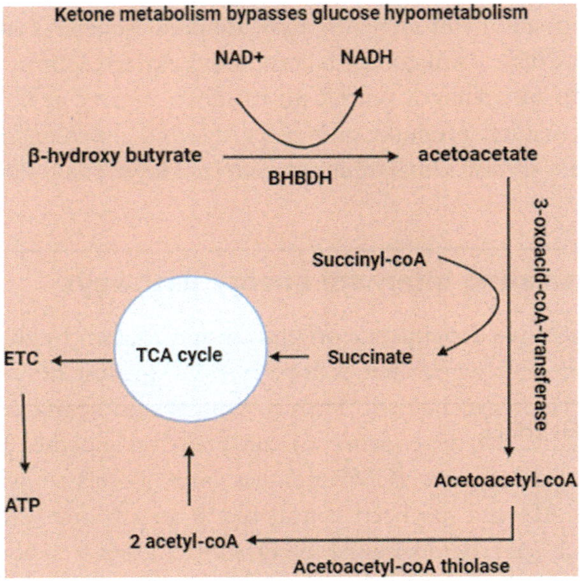

Fig. 5 Efficient bypassing of glucose hypometabolism by ketone metabolism: Ketone metabolism bypasses glucose hypometabolism by activation of alternate pathways that converge to finally produce ATP to meet the energy needs of the cell.

5.1.1 Therapeutic implementation

Ketogenic Diet: It is a dietary plan that is a high-fat, extremely-low-carbohydrate diet that causes physiological ketosis due to a metabolic switch to fat oxidation. Fatty acids undergo hepatic β-oxidation to produce ketone bodies that bypass the blood-brain barrier to supply neurons with an alternative source of energy. Neuroprotective action of ketogenic diets is preclinically indicated in models of AD (Lange, 2019).

Nevertheless, there are problems with clinical implementation. Ketogenic diets are also restrictive and cannot be sustained over a long period making the patient less compliant. Also, patients who have specific metabolic disorders (e.g., familial hyperlipidemia) are not able to accept high-fat diets. Close screening and follow-ups of patients are critical.

AC-1202 (Medium-Chain Triglyceride Supplement): This ketogenic nutrient is safe and does not need dietary restriction; it contains MCTs, which increase the plasma ketone levels. The MCTs are quickly metabolized in the liver and oxidized giving rise to ketones which diffuse to circulation and penetrate blood-brain barriers. AC-1202 has benefits relative to the ketogenic diets: more rapid absorption, simpler to deliver, and no chance of ketoacidosis (which may be caused by extended elevated concentrations of ketones) (Costantini et al., 2008).

5.2 Improvement of fatty acid oxidation

5.2.1 The lipotoxicity paradox

Fatty acids are hydrocarbon chains which terminates with carboxylic acid groups. High concentrations of free fatty acids and their metabolic intermediates (acyl-CoA, acyl-carnitines) have neurotoxic effects including destabilization of mitochondrial coupling and bioenergetics, which ultimately leads to cognitive impairments (Mi et al., 2023). Nonetheless, the therapeutic objective will not be to elevate fatty acids but instead to promote their oxidation and clearance to avoid lipotoxic build up.

The breakdown of fatty acids takes place in the form of mitochondrial β-oxidation and peroxisomal oxidation in an alternate way. Peroxisomal oxidation in the neural tissue through astrocytes is essential to ensure the well being of neurons and is less preferred to mitochondrial oxidation due to the production of less ROS. This pathway of peroxisomal degradation is disrupted in patients of AD (Mi et al., 2023).

5.2.2 PPAR-targeted interventions

The peroxisome proliferator-activated receptors (PPARs) are lipid sensitive nuclear receptors that control the metabolic processes of lipids in the peripheral tissues and the brain. There are two PPAR subtypes with potential in the treatment of AD:

PPARγ: PPARγ agonists, especially the thiazolidinediones (TZDs) have anti-inflammatory effects, such as blocking pro-inflammatory transcriptional factors such as NFκBs and STATs. Such effects can retard neurodegeneration caused by neuroinflammation. Nevertheless, TZDs have been tested in AD in clinical trials with inconsistent results and this may be because of insufficient brain penetrance, or because of patient heterogeneity.

PPARα: This subtype particularly controls the oxidation of fatty acids via up-regulation of expression of oxidative enzymes such as carnitine palmitoyltransferase and acyl-CoA oxidase in astrocytes. They also promote the production of α-secretase (anti-amyloidogenic activity) and inhibit the production of β-secretase (pro-amyloidogenic activity). A potent candidate for PPARα agonism, Wy-14643, has shown to ameliorate AD related pathologies in mice model (Cao et al., 2024) (Fig. 6).

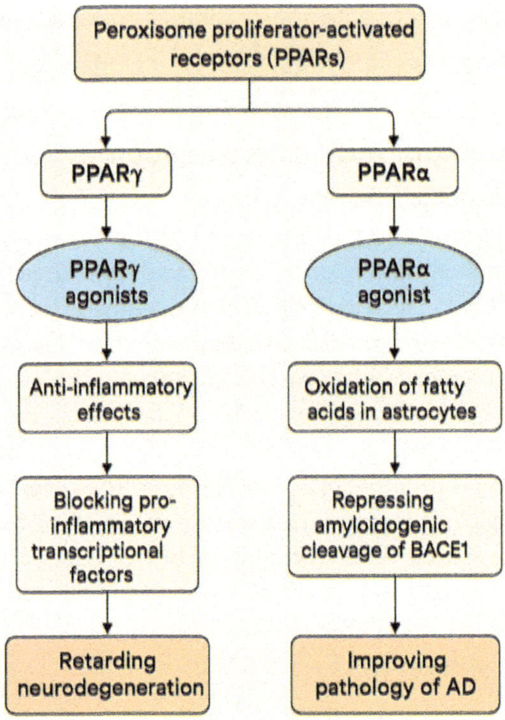

Fig. 6 PPAR agonism in anti-inflammation, enhanced protein clearance and fatty acid oxidation: PPAR agonists bring about anti-inflammatory effects and fatty acid oxidation repressing amyloidogenic cleavage of BACE-1, thereby preventing plaquing and neurodegeneration.

5.3 Activation of pentose phosphate pathway

5.3.1 Redox homeostasis

Pentose phosphate pathway (PPP) does not generate ATP but is imperative in the preservation of cell redox balance. The PPP generates NADPH that maintains the low glutathione (GSH), the primary antioxidant of the cell. When subjected to oxidative stress, GSH gets oxidized to GSSG; glutathione reductase, with the help of NADPH then reduces GSH to create a nuclear chain of antioxidants.

Both enzymatic activity and NADPH/NADP$^+$ ratio regulate the PPP. The important enzymes are glucose-6-phosphate dehydrogenase (G6PD) which catalyzes the rate limiting oxidative step and transketolase which mediates non oxidative phase. G6PD is inhibited by high NADPH/NADP$^+$ ratios that provide feedback regulation.

5.3.2 PPP dysfunction in AD

The excessive ROS produced by mitochondrion dysfunction in AD causes the presence of oxidative stress. At the same time, the G6PD and transketolase levels go down, decreasing the production of NADPH and GSH at the moment when the antioxidant capability is the most required (Yan et al., 2020). This dual impairment of increased oxidative stress and reduced antioxidant capacity causes neuronal impairment.

5.3.3 Therapeutic strategies

Hollow Manganese Prussian White Nanocapsules (HMPWCs): These nanoparticles exhibit several positive performances such as neuronal anti-inflammation, decreased tau hyperphosphorylation, and direct ROS scavenging (Gao et al., 2020). Although nanotechnology-based strategies promise well in preclinical models, there are regulatory and manufacturing issues before clinical translation.

G6PD Overexpression: G6PD genetic overexpression in mouse models of AD (APP/PS1) enhances the neuromuscular coordination and metabolism and decreases the oxidation of DNA and Aβ plaque loads (Correas et al., 2024). These results indicate that improvement of PPP flux may be of therapeutic value, but gene therapy strategies need widespread development. Alternatively, more amenable therapeutic pathway could be found through the finding of small molecules that stimulate G6PD.

Metabolic Redirection: Metabolic redirection means to redirect the glucose flux through the glycolytic pathway to the PPP, thereby increasing the production of NADPH, reducing GSSG to GSH, and increasing antioxidant ability. The latter method necessitates a fine balance to prevent exhausting glycolytic ATP generation, which can be achieved by accompanying the use of ketogenic support to balance the total energy condition.

6. Current state and future directions in metabolic therapeutics for Alzheimer's disease

The metabolic hypothesis of Alzheimer's disease has gained substantial traction over recent decades, yet disease-modifying treatments addressing the full spectrum of cerebral metabolic dysfunction remain elusive. The field now faces a critical translational challenge: converting promising preclinical discoveries into effective therapeutic strategies for human populations. This transition requires moving beyond single-target approaches toward multifaceted interventions that reflect the interconnected nature of metabolic pathology in AD.

6.1 Combination and multitarget therapeutic strategies

6.1.1 The rationale for combination approaches

The consistent failure of single-target therapies to halt AD progression has redirected attention toward combination strategies addressing multiple pathophysiological mechanisms simultaneously. This paradigm shift acknowledges a fundamental reality: metabolic dysfunction in AD operates through interconnected networks rather than isolated defects (Cummings et al., 2023). Network medicine approaches, which model protein interactions and metabolic pathways as integrated systems, provide frameworks for identifying combinations that target disease at multiple nodes. This systems-level perspective shifts focus from individual molecular targets toward restoration of entire metabolic networks (Cummings et al., 2024).

6.1.2 Insulin sensitizers and metabolic modulators

Recent clinical evidence demonstrates the potential of combining complementary metabolic interventions. A four-week randomized trial in individuals with mild cognitive impairment and early AD examined empagliflozin (a sodium-glucose cotransporter-2 inhibitor) and intranasal insulin, revealing distinct but synergistic effects (Craft et al., 2025). Empagliflozin improved high-density lipoprotein cholesterol and metabolic parameters even in non-diabetic patients, while intranasal insulin enhanced cognitive performance on sensitive memory tests and reduced cerebrospinal fluid levels of neurogranin and vascular markers associated with disease progression.

These agents work through complementary mechanisms. For instance empagliflozin enhances glucose and sodium homeostasis resulting in improved systemic insulin sensitivity and vascular health, while intranasal insulin directly addresses brain insulin resistance. Both medications modulated multiple immune and inflammatory proteins in cerebrospinal fluid and blood, activating protective immune responses while suppressing harmful inflammation. Also, intranasal insulin influenced proteins in the olfactory pathway, a recently identified route connecting the brain's glymphatic waste clearance system and peripheral immune networks (Craft et al., 2025).

6.1.3 Immunometabolic pathway targeting

Immunometabolic pathways, including insulin signaling cascades, are dysregulated at early stages of AD, simultaneously with amyloid accumulation (Craft & Claxton, 2023). Agents targeting these pathways may work synergistically with anti-amyloid therapies to slow disease progression. Metabolic modulators that enhance insulin sensitivity and function including glucagon-like peptide-1 (GLP-1) receptor agonists, SGLT2 inhibitors, and insulin itself improve immune function and reduce chronic inflammation. Additional effects demonstrated in preclinical and clinical studies include enhanced amyloid-β clearance, reduced tau pathology, improved vascular function and lipid metabolism, decreased toxicity in the synapse, and better cognitive activity (Craft & Claxton, 2023).

The promise of combination therapy extends beyond simple drug pairing. A substantial number of metabolic disorder medications have been extensively characterized with respect to mechanisms of action and safety, making them readily available for repurposing in combination protocols. Determining optimal regimens, appropriate treatment timing, and effective combinations with disease-modifying therapies represents the next frontier in AD therapeutics (Craft & Claxton, 2023).

6.2 Clinical development pipeline: From preclinical models to human trials

The translational pipeline for metabolic interventions in AD involves agents at various developmental stages, from early preclinical studies to late-phase clinical trials. This diversity reflects both growing recognition of metabolism as a therapeutic target and the complexity of validating these approaches in human populations.

6.2.1 GLP-1 receptor agonists

GLP-1 receptor agonists have progressed furthest in clinical development. A six-month randomized, placebo-controlled, double-blind trial showed that liraglutide prevented decline in brain glucose metabolism in AD patients (Gejl et al., 2016). Based on these results, larger Phase 3 trials of semaglutide (marketed as Ozempic and Wegovy for diabetes and weight management) are underway. Observational data showing reduced dementia risk in diabetic patients taking semaglutide prompted trials specifically targeting AD populations (Alzheimer's Society, 2025).

The mechanisms underlying these effects extend beyond glucose regulation. GLP-1 receptor agonists activate AMP-activated protein kinase

(AMPK) signaling through the calcium/calmodulin-dependent protein kinase kinase 2 (CaMKK2)-AMPK pathway, subsequently reducing BACE1-mediated APP cleavage and amyloid-β generation (Zhang et al., 2025). These drugs also increase AMPK activity in microglia, inhibiting neuroinflammation and promoting amyloid-β phagocytosis.

6.2.2 Metformin

Metformin, the first-line treatment for type 2 diabetes, is currently in a Phase 2/3 trial involving about 370 individuals with amnesia (Cummings et al., 2023). Participants are randomized to drug or placebo for 24 months, with the Free and Cued Selective Reminding Test as the primary outcome measure. Beyond glucose-lowering effects, metformin inhibits proinflammatory responses including nuclear factor-κB (NF-κB) and YKL-40 in vitro (Craft & Claxton, 2023).

However, metformin's clinical track record remains inconsistent. A randomized placebo-controlled trial of 80 non-diabetic participants with mild cognitive impairment receiving metformin or placebo daily for 12 months found no treatment-related differences in the Alzheimer's Disease Assessment Scale-Cognitive subscale (ADAS-Cog), cerebrospinal fluid amyloid-β42, or cerebral glucose metabolism assessed by FDG-PET (Craft & Claxton, 2023). This boundary between theoretical promise and clinical reality characterizes many metabolic interventions and underscores the need for more sophisticated trial designs and patient sampling strategies.

6.2.3 Intranasal insulin

Intranasal insulin has shown variable results across studies, with differential effects based on apolipoprotein E (APOE) genotype and dosing regimen. While the mechanistic basis for these genotype-specific responses remains unclear, they suggest that patient stratification using genetic markers might improve outcomes in future trials (Burns & Swerdlow, 2013). Recent positive results using validated delivery devices provided promising results for this approach (Craft et al., 2025).

6.2.4 Thiazolidinediones

Thiazolidinediones, particularly rosiglitazone, showed initial promise in small studies but failed in multiple large Phase 3 trials. Neither primary outcome measures nor APOE ε4 genotype-stratified analysis demonstrated efficacy (Burns & Swerdlow, 2013). The consistent negative results across well-executed trials effectively closed this therapeutic avenue, though the lessons regarding trial design, patient selection, and the challenges of translating PPARγ modulation into clinical benefit remain valuable for the field.

6.3 Novel therapeutic targets in AD

The metabolic landscape of AD extends beyond traditional insulin signaling pathways. Several emerging targets offer opportunities for therapeutic intervention, each with distinct mechanisms and translational potential.

6.3.1 NAD$^+$ metabolism and sirtuin modulation

Nicotinamide adenine dinucleotide (NAD$^+$) metabolism and the NAD$^+$ dependent sirtuin enzyme family represent one of the most promising frontiers in AD therapeutics. Sirtuins regulate diverse cellular processes including energy metabolism, aging, DNA repair, and inflammation through their deacetylase activity (Imai & Guarente, 2016). The seven mammalian sirtuins (SIRT1–7) distribute across cellular compartments with varying enzymatic activities, though SIRT1, SIRT2, and SIRT3 exhibit prominent deacetylase functions.

6.3.2 SIRT1 and neuroprotection

SIRT1 has undergone extensive investigation in AD. In vitro SIRT1 overexpression, mediated by NAD$^+$ or resveratrol, reduces oligomerized amyloid-β peptides in a concentration-dependent manner and ameliorates oxidative stress (Donmez & Guarente, 2010). These effects involve switching APP processing toward the non-amyloidogenic pathway through increased α-secretase activation. SIRT1 overexpression also prevents microglia-dependent amyloid-β toxicity through NF-κB signaling inhibition (Donmez & Guarente, 2010).

6.3.3 SIRT3 and mitochondrial protection

Beyond SIRT1, mitochondrial SIRT3 inhibits oxidative stress and mitochondrial damage, suppressing activation of the NLRP3 inflammasome, a key mediator of neuroinflammation (Grabowska et al., 2022). Age-related decline in sirtuin activity has been documented, and activation of SIRT1, SIRT2, SIRT3, and SIRT6 shows beneficial effects in preventing AD pathology (Grabowska et al., 2022).

6.3.4 NAD$^+$ supplementation strategies

NAD$^+$ supplementation offers a route to restore sirtuin function. NAD$^+$ supplementation normalizes key AD features and DNA damage responses in mouse models (Hou et al., 2018). Clinical translation has begun with oral nicotinamide mononucleotide administration is well-tolerated and increases nicotinamide levels in healthy individuals (Irie et al., 2020).

6.3.5 Isoform-specific targeting

The catalytic AMPKα subunit exists as two isoforms: AMPKα1 and AMPKα2. Emerging evidence suggests these exhibit subtype-specific effects in AD, including distinct impacts on amyloid-β pathology, tau pathology, metabolic disorders, inflammation, mitochondrial dysfunction, inflammasome activation, and pyroptosis (Li et al., 2024). This isoform specificity provides more precise targets for drug development.

Phytochemicals acting as AMPK agonists offer novel multi-mechanistic therapeutic strategies, exerting effects through inhibiting amyloid-β aggregation, preventing tau hyperphosphorylation, suppressing inflammatory responses and glial activation. They also promote autophagy, and reduce oxidative stress. Several AMPK-related pathways are involved, including AMPK/CaMKKβ/mTOR, AMPK/SIRT1/PGC-1α, AMPK/NF-κB/NLRP3, and PERK/eIF2α pathways (Li et al., 2024).

Despite controversy, AMPK remains a viable therapeutic target. Determining the detailed molecular mechanisms underlying AMPK interactions during disease progression—particularly distinguishing early versus late-stage effects—will provide insights into disease etiology and guide therapeutic development (Ma, 2019). Understanding how AMPK signaling changes with normal aging will help distinguish pathological from age-related alterations.

6.3.6 The microbiome-metabolism axis

The gut microbiome has emerged as an unexpected but powerful modulator of brain metabolism in AD. The microbiota-gut-brain axis represents a bidirectional communication system involving neural, immune, endocrine, and metabolic pathways (Kowalski & Mulak, 2019). Disturbances along this axis contribute significantly to neurodegenerative pathogenesis.

6.3.7 Mechanisms of microbiome-mediated neurodegeneration

Alterations in gut microbiota composition enhance intestinal barrier permeability and immune activation, leading to systemic inflammation that impairs the blood-brain barrier and promotes neuroinflammation, neural injury, and neurodegeneration (Kowalski & Mulak, 2019). Amyloid-β, recognized as an antimicrobial peptide participating in innate immunity, reveals harmful properties in dysregulated states. Bacterial amyloids, through molecular mimicry, elicit cross-seeding of protein misfolding and induce microglial priming (Kowalski & Mulak, 2019).

6.3.8 Microbiome as a biomarker

Recent studies demonstrate that gut microbiome composition serves as an indicator of preclinical AD. Individuals with preclinical AD, indicated by altered brain amyloid and tau proteins, possess different gut microbiome compositions than healthy individuals (Ferreiro et al., 2023). Gut microbial taxa correlate with amyloid and tau markers of preclinical disease but not with signatures of neurodegeneration. Specific bacterial species associated with preclinical AD improved machine learning predictions of disease status when included in predictive models (Ferreiro et al., 2023).

6.3.9 Microbial metabolites and brain function

Metabolites derived from gut microbiota play central roles in brain-gut communication. Short-chain fatty acids (SCFAs), produced by bacterial fermentation of dietary fiber, affect brain health through multiple mechanisms. Changes in conversion of primary bile acids to secondary bile acids by gut microbiota influence brain function, with elevated secondary bile acid levels associated with cognitive decline (National Institute on Aging, 2024). Notably, bacterially-produced secondary bile acids are detected in brain tissue, confirming blood-to-brain transport (National Institute on Aging, 2024).

6.3.10 Evidence from microbiota transfer studies

Fecal microbiota transplantation studies provide causal evidence for the microbiome's role. Transplanting gut microbiota from AD patients into microbiota-depleted young adult rats resulted in impairments in behaviors associated with adult hippocampal neurogenesis, an essential process for memory functions and mood regulation (Grabrucker et al., 2023). Impairment severity correlated with clinical cognitive scores in donor patients, directly linking microbiome alterations to cognitive symptoms. Discrete changes in rat fecal and hippocampal metabolomes were also evident (Grabrucker et al., 2023).

6.3.11 Microbiome regulation of brain glucose metabolism

Gut microbiota regulates cerebral glucose metabolism through direct communication via microbe-derived metabolites (Liang et al., 2024). FDG-PET studies revealed that mice with altered gut microbial compositions showed decreased brain glucose uptake and memory impairment, with reduced glucose uptake mediated by microbiota alterations (Liang et al., 2024). This connection between microbiome dysbiosis and brain glucose metabolism correlates microbial alterations to core metabolic deficits in AD.

6.3.12 Therapeutic interventions and research initiatives

Microbiota-based therapeutic strategies are under active investigation. Probiotic supplementation with specific strains such as Bifidobacterium breve improved cognitive function and prevented brain atrophy in older patients with suspected mild cognitive impairment in a 24-week randomized, double-blind, placebo-controlled trial (Liang et al., 2024). *Akkermansia muciniphila* prevented AD-like symptoms in animal models (Seo & Holtzman, 2024).

The National Institute on Aging is funding the Alzheimer's Gut Microbiome Project, led by Rima Kaddurah-Daouk and involving more than 40 laboratories, to understand the microbiome's dynamic role across different AD stages (National Institute on Aging, 2024). This large-scale effort reflects recognition that the microbiome represents more than a disease correlate, but it is an active participant in pathogenesis and a potential therapeutic target.

The integration of microbiome manipulation with traditional metabolic interventions offers a systems-level approach to restoring brain health. Diet, exercise, and lifestyle modifications are recognized as powerful modulators of metabolism and immune function and are being investigated as preventative or therapeutic strategies precisely because they influence both peripheral metabolism and gut microbiota simultaneously (Craft & Claxton, 2023).

7. Conclusion

A paradigm shift away from the pure amyloid-centric conceptions would be the metabolic hypothesis of the Alzheimer disease, which emphasizes the pathogenesis of neurodegeneration in the disruptions of brain energy and mitochondrial dysfunction. There exists strong evidence showing that, before the onset of the classical AD hallmarks like Aβ plaques and tau tangles, impaired glucose metabolism, insulin resistance, and mitochondrial pathology are present.

Such a metabolic view provides a wide range of therapeutic possibilities. An increase in sensitivity to glucose, enhancing the expression of glucose transporters, mitochondrial biogenesis using PGC-1α/SIRT1/AMPK activation, enhancing the functionality of the electron transport chain using targeted cofactors and mild modulators, restoring the balance between mitochondrial dynamics, and stimulating mitochondrial antioxidant pathways all are promising in preclinical models. In addition to mitochondria, the differentiation of neuronal fuel sources by ketogenic treatments, an increased rate of fatty acid oxidation by activating PPARs, and the activation of the pentose phosphate pathway are complementary measures to ensure neuronal energy homeostasis.

The way to clinical success lies in overcoming the major obstacles like finding sufficient brain penetration, finding the most appropriate patient groups and the moment of intervention, and understanding that AD is more than likely a heterogeneous disorder that has to be treated by different combinatorial regimens and not with single-fits-all measures. However, the metabolic approach provides an opportunity of having disease-modifying interventions especially when implemented at the onset of at-risk populations before the occurrence of widespread neurodegeneration.

With the increasing knowledge about brain metabolism in health and disease, and with improvements in technology of measuring metabolic profiles and delivering drugs to specific areas of the brain, it is possible that metabolic treatment will soon turn AD into something that can be prevented or treated as opposed to something that has to progressively worsen. Future studies that combine fundamental processes and clinical research would be necessary to achieve this therapeutic potential in the amelioration of Alzheimer disease.

References

Alhasan, M. S., Alhasan, A. S., Milburn, J., Khalil, M., Almaghraby, A., Alharthi, O., ... Azzam, A. Y. (2025). FDG-PET brain glucose hypometabolism predicts Alzheimer's disease progression pathways in cognitively normal adults: A longitudinal competing risks modeling. *Metabolism Open*, 100400. https://doi.org/10.1016/j.metop.2025.100400.

Alzarea, E. A., Al-Kuraishy, H. M., Al-Gareeb, A. I., Alexiou, A., Papadakis, M., Beshay, O. N., & Batiha, G. E. S. (2025). The conceivable role of metabolic syndrome in the pathogenesis of Alzheimer's disease: Cellular and subcellular alterations in underpinning a tale of two. *Neuromolecular Medicine, 27*(1), 35. https://doi.org/10.1007/s12017-025-08832-6.

Alzheimer's Society. (2025, March 19). Researching new drugs for Alzheimer's disease. https://www.alzheimers.org.uk/what-we-do/researchers/news/researching-new-drugs-alzheimers-disease.

Bigl, M., Apelt, J., Luschekina, E. A., Lange-Dohna, C., Rossner, S., Schliebs, R., et al. (2000). Expression of beta-secretase mRNA in transgenic Tg2576 mouse brain and its modulation by metabolic conditions. *Journal of Neural Transmission, 107*(6), 617–632. https://doi.org/10.1016/S0304-3940(00)01452-X.

Burns, J. M., & Swerdlow, R. H. (2013). Insulin: An emerging treatment for Alzheimer's disease dementia? *Drugs, 73*(15), 1605–1616. https://doi.org/10.1007/s40265-013-0106-5.

Butterfield, D. A., & Halliwell, B. (2019). Oxidative stress, dysfunctional glucose metabolism and Alzheimer disease. *Nature Reviews. Neuroscience, 20*(3), 148–160. https://doi.org/10.1038/s41583-019-0132-6.

Butterfield, D. A., Hardas, S. S., & Lange, M. L. (2010). Oxidatively modified glyceraldehyde-3-phosphate dehydrogenase (GAPDH) and Alzheimer's disease. *Journal of Alzheimer's Disease, 20*(3), 673–681. https://doi.org/10.3233/JAD-2010-1375.

Campbell, J. M. (2022). Supplementation with NAD+ and its precursors to prevent cognitive decline across disease contexts. *Nutrients, 14*(15), 3231. https://doi.org/10.3390/nu14153231.

Cao, Y., Zhao, L. W., Chen, Z. X., & Li, S. H. (2024). New insights in lipid metabolism: Potential therapeutic targets for the treatment of Alzheimer's disease. *Frontiers in Neuroscience, 18*, 1430465.

Caruana, M., Cauchi, R., & Vassallo, N. (2016). Putative role of red wine polyphenols against brain pathology in Alzheimer's and Parkinson's disease. *Frontiers in Nutrition, 3*, 31. https://doi.org/10.3389/fnut.2016.00031.

Correas, A. G., Olaso-Gonzalez, G., Roca, M., Blanco-Gandía, M. C., Nascimento, C., Lahoz, A., ... & Viña, J. (2024). Glucose 6 phosphate dehydrogenase overexpression rescues the loss of cognition in the double transgenic APP/PS1 mouse model of Alzheimer's disease. *Redox Biology, 75*, 103242.

Costantini, L. C., Barr, L. J., Vogel, J. L., & Henderson, S. T. (2008). Hypometabolism as a therapeutic target in Alzheimer's disease. *BMC Neuroscience, 9*(2), S16. https://doi.org/10.1186/1471-2202-9-S2-S16.

Craft, S., Baker, L. D., Montine, T. J., Minoshima, S., Watson, G. S., Claxton, A., et al. (2012). Intranasal insulin therapy for Alzheimer disease and amnestic mild cognitive impairment. *Archives of Neurology, 69*(1), 29–38.

Craft, S., Baker, L. D., Shiling, M. A., Callaghan, M., Claxton, A., Cholerton, B., ... Montine, T. J. (2025). Effects of intranasal insulin and empagliflozin on cognition and brain health in early Alzheimer's disease: A randomized clinical trial. *Alzheimer's & Dementia*. https://doi.org/10.1002/alz.14440.

Craft, S., & Claxton, A. (2023). Targeting immunometabolic pathways for combination therapy in Alzheimer's disease. *Alzheimer's & Dementia, 19*(11), 5324–5338. https://doi.org/10.1002/alz.13432.

Cummings, J., Apostolova, L., Ballard, C., Dougall, N., Chen, C., Lee, G., & Salloway, S. (2024). Alzheimer's disease: Combination therapies and clinical trials for combination therapy development. *CNS Drugs, 38*(8), 613–624. https://doi.org/10.1007/s40263-024-01103-1.

Cummings, J., Apostolova, L., Rabinovici, G. D., Atri, A., Aisen, P., Greenberg, S., ... Salloway, S. (2023). Alzheimer's disease: Novel targets and investigational drugs for disease modification. *Drugs, 83*(15), 1387–1408. https://doi.org/10.1007/s40265-023-01938-w.

De Felice, Wu, D., Lambert, M. P., Fernandez, S. J., Velasco, P. T., ... Klein, W. L. (2008). Alzheimer's disease-type neuronal tau hyperphosphorylation induced by Aβ oligomers. *Neurobiology of aging, 29*(9), 1334–1347.

de la Monte, S. M., & Wands, J. R. (2008). Alzheimer's disease is type 3 diabetes—evidence reviewed. *Journal of Diabetes Science and Technology, 2*(6), 1101–1113. https://doi.org/10.1177/193229680800200619.

Donmez, G., & Guarente, L. (2010). Aging and disease: Connections to sirtuins. *Aging Cell, 9*(2), 285–290. https://doi.org/10.1111/j.1474-9726.2010.00548.x.

Eshraghi, M., Adlimoghaddam, A., Mahmoodzadeh, A., Sharifzad, F., Yasavoli-Sharahi, H., Lorzadeh, S., et al. (2021). Ghavami Alzheimer's disease pathogenesis: Role of autophagy and mitophagy focusing in microglia. *International Journal of Molecular Sciences, 22*(7), 3330. https://doi.org/10.3390/ijms22073330.

Fahed, G., Aoun, L., Bou Zerdan, M., Allam, S., Bou Zerdan, M., Bouferraa, Y., & Assi, H. I. (2022). Metabolic syndrome: Updates on pathophysiology and management in 2021. *International journal of molecular sciences, 23*(2), 786.

Ferreiro, A. L., Choi, J., Ryou, J., Newcomer, E. P., Thompson, R., Bollinger, R. M., ... Dantas, G. (2023). Gut microbiome composition may be an indicator of preclinical Alzheimer's disease. *Science Translational Medicine, 15*(700), eabo2984. https://doi.org/10.1126/scitranslmed.abo2984.

Fukui, Y., Hishikawa, N., Shang, J., Sato, K., Nakano, Y., Morihara, R., ... Abe, K. (2016). Peripheral arterial endothelial dysfunction of neurodegenerative diseases. *Journal of the Neurological Sciences, 366*, 94–99. https://doi.org/10.1016/j.jns.2016.04.042.

Gao, Y., Yu, G., Xing, K., Gorin, D., Kotelevtsev, Y., Tong, W., & Mao, Z. (2020). Finely tuned Prussian blue-based nanoparticles and their application in disease treatment. *Journal of Materials Chemistry B, 8*(32), 7121–7134.

Gejl, M., Gjedde, A., Egefjord, L., Møller, A., Hansen, S. B., Vang, K., ... Rungby, J. (2016). Alzheimer's disease, 6-month treatment with GLP-1 analog prevents decline of brain glucose metabolism: Randomized, placebo-controlled, double-blind clinical trial. *Frontiers in Aging Neuroscience, 8*, 108. https://doi.org/10.3389/fnagi.2016.00108.

Grabowska, W., Sikora, E., & Bielak-Zmijewska, A. (2022). Sirtuins promote brain homeostasis, preventing Alzheimer's disease through targeting neuroinflammation. *Frontiers in Physiology, 13*, 962769. https://doi.org/10.3389/fphys.2022.962769.

Grabrucker, S., Marizzoni, M., Silajdžić, E., Lopizzo, N., Mombelli, E., Nicolas, S., ... Frisoni, G. B. (2023). Microbiota from Alzheimer's patients induce deficits in cognition and hippocampal neurogenesis. *Brain, 146*(12), 4916–4934. https://doi.org/10.1093/brain/awad303.

Grundy, S. M. (2016). Metabolic syndrome update. *Trends in Cardiovascular Medicine, 26*(4), 364–373. https://doi.org/10.1016/j.tcm.2015.10.004.

Hardie, D. G., Ross, F. A., & Hawley, S. A. (2012). AMPK: A nutrient and energy sensor that maintains energy homeostasis. *Nature Reviews. Molecular Cell Biology, 13*(4), 251–262. https://doi.org/10.1038/nrm3311.

Hayat, M., Syed, R. A., Qaiser, H., Uzair, M., Al-Regaiey, K., Khallaf, R., ... Bashir, S. (2025). Decoding molecular mechanisms: brain aging and Alzheimer's disease. *Neural Regeneration Research, 20*(8), 2279–2299.

Heneka, M. T., Reyes-Irisarri, E., Hull, M., & Kummer, M. P. (2015). Impact and therapeutic potential of PPARs in Alzheimer's disease. *Current Neuropharmacology, 9*(4), 643–650. https://doi.org/10.2174/157015911798376325.

Hou, Y., Lautrup, S., Cordonnier, S., Wang, Y., Croteau, D. L., Zavala, E., ... Bohr, V. A. (2018). NAD+ supplementation normalizes key Alzheimer's features and DNA damage responses in a new AD mouse model with introduced DNA repair deficiency. *Proceedings of the National Academy of Sciences, 115*(8), E1876–E1885. https://doi.org/10.1073/pnas.1718819115.

Imai, S.-I., & Guarente, L. (2016). It takes two to tango: NAD+ and sirtuins in aging/longevity control. *NPJ Aging and Mechanisms of Disease, 2*, 16017. https://doi.org/10.1038/npjamd.2016.17.

Irie, J., Inagaki, E., Fujita, M., Nakaya, H., Mitsuishi, M., Yamaguchi, S., ... Itoh, H. (2020). Effect of oral administration of nicotinamide mononucleotide on clinical parameters and nicotinamide metabolite levels in healthy Japanese men. *Endocrine Journal, 67*(2), 153–160. https://doi.org/10.1507/endocrj.EJ19-0313.

Ivanova, O., Martínez-Nicolás, I., García-Piñuela, E., & Meilán, J. J. G. (2023). Defying syntactic preservation in Alzheimer's disease: What type of impairment predicts syntactic change in dementia (if it does) and why? *Frontiers in Language Sciences, 2*, 1199107. https://doi.org/10.3389/flang.2023.1199107.

Jurcovicova, J. (2014). Glucose transport in brain—effect of inflammation. *Endocrine Regulations, 48*(1), 35–48.

Kapogiannis, D., & Avgerinos, K. I. (2020). Brain glucose and ketone utilization in brain aging and neurodegenerative diseases. *International Review of Neurobiology, 154*, 79–110. https://doi.org/10.1016/bs.irn.2020.03.015.

Kim, B., & Feldman, E. L. (2015). Insulin resistance as a key link for the increased risk of cognitive impairment in the metabolic syndrome. *Experimental & Molecular Medicine, 47*(3), e149.

Kowalski, K., & Mulak, A. (2019). Brain-gut-microbiota axis in Alzheimer's disease. *Journal of Neurogastroenterology and Motility, 25*(1), 48–60. https://doi.org/10.5056/jnm18087.

Lagouge, M., Argmann, C., Gerhart-Hines, Z., Meziane, H., Lerin, C., Daussin, F., ... Auwerx, J. (2006). Resveratrol improves mitochondrial function and protects against metabolic disease by activating SIRT1 and PGC-1α. *Cell, 127*(6), 1109–1122. https://doi.org/10.1016/j.cell.2006.11.013.

Lange, K. W. (2019). Mediterranean diet and Alzheimer's disease. *Movement and Nutrition in Health and Disease, 3*. https://doi.org/10.5283/mnhd.13.

Lann, D., & LeRoith, D. (2007). Insulin resistance as the underlying cause for the metabolic syndrome. *Medical Clinics of North America, 91*(6), 1063–1077.

Li, Y., Wei, T., Qu, M., Mou, L., Jiang, T., Chen, L., & Yu, J. (2024). The role of AMPKα subunit in Alzheimer's disease: In-depth analysis and future prospects. *Ageing Research Reviews, 99*, 102385. https://doi.org/10.1016/j.arr.2024.102385.

Liang, Y., Zhao, Q., Zhang, Y., Li, X., Wang, M., Liang, K., ... Qiu, C. (2024). The link between gut microbiome and Alzheimer's disease: From the perspective of new revised criteria for diagnosis and staging of Alzheimer's disease. *Alzheimer's & Dementia, 20*(9), 6464–6487. https://doi.org/10.1002/alz.14057.

Liu, Y., Liu, F., Grundke-Iqbal, I., Iqbal, K., & Gong, C.-X. (2008). Deficient glucose transporters in Alzheimer's disease brain. *Journal of Alzheimer's Disease, 14*(4), 491–498.

Ma, T. (2019). Therapeutic potential of AMP-activated protein kinase in Alzheimer's disease. *Journal of Alzheimer's Disease, 68*(4), 1301–1310. https://doi.org/10.3233/JAD-181039.

Mandal, P., Trushina, E., Arnold, M., Kaddurah-Daouk, R., & Baloni, P. (2025). Mitochondrial bioenergetic signatures differentiate asymptomatic from symptomatic Alzheimer's disease. bioRxiv, 2025–11. https://doi.org/10.1101/2025.11.04.686626.

Manzar, H., Abdulhussein, D., Yap, T. E., & Cordeiro, M. F. (2020). Cellular consequences of coenzyme Q10 deficiency in neurodegeneration of the retina and brain. *International Journal of Molecular Sciences, 21*(23), 9299. https://doi.org/10.3390/ijms21239299.

McManus, M. J., Murphy, M. P., & Franklin, J. L. (2011). The mitochondria-targeted antioxidant MitoQ prevents loss of spatial memory retention and early neuropathology in a transgenic mouse model of Alzheimer's disease. *Journal of Neuroscience, 31*(44), 15703–15715. https://doi.org/10.1523/JNEUROSCI.0552-11.2011.

Meng, X., Song, Q., Liu, Z., Liu, X., Wang, Y., & Liu, J. (2024). Neurotoxic β-amyloid oligomers cause mitochondrial dysfunction—The trigger for PANoptosis in neurons. *Frontiers in Aging Neuroscience, 16*, 1400544. https://doi.org/10.3389/fnagi.2024.1400544.

Mi, Y., Qi, G., Vitali, F., Shang, Y., Raikes, A. C., Wang, T., Yin, F., et al. (2023). Loss of fatty acid degradation by astrocytic mitochondria triggers neuroinflammation and neurodegeneration. *Nature Metabolism, 5*(3), 445–465. https://doi.org/10.1038/s42255-023-00756-4.

Mosconi, L. (2013). Glucose metabolism in normal aging and Alzheimer's disease: Methodological and physiological considerations for PET studies. *Clinical and Translational Imaging, 1*, 217–233. https://doi.org/10.1007/s40336-013-0026-y.

Mosconi, L., De Santi, S., Brys, M., Tsui, W. H., Pirraglia, E., Glodzik, L., et al. (2008). Hypometabolism and cognitive decline in Alzheimer's disease. *European Journal of Nuclear Medicine and Molecular Imaging, 35*, 132–145.

Muraleva, N. A., Kozhevnikova, O. S., Fursova, A. Z., & Kolosova, N. G. (2019). Suppression of AMD-like pathology by mitochondria-targeted antioxidant SkQ1 is associated with a decrease in the accumulation of amyloid β and in mTOR activity. *Antioxidants, 8*(6), 177. https://doi.org/10.3390/antiox8060177.

National Institute on Aging. (2024). Beyond the brain: The gut microbiome and Alzheimer's disease. https://www.nia.nih.gov/news/beyond-brain-gut-microbiome-and-alzheimers-disease.

Pei, X., Hu, F., Hu, Z., Luo, F., Li, X., Xing, S., ... Long, D. (2023). Neuroprotective effect of α-lipoic acid against Aβ25–35-induced damage in BV2 cells. *Molecules (Basel, Switzerland), 28*(3), 1168. https://doi.org/10.3390/molecules28031168.

Price, N. L., Gomes, A. P., Ling, A. J., Duarte, F. V., Martin-Montalvo, A., North, B. J., ... Sinclair, D. A. (2012). SIRT1 is required for AMPK activation and the beneficial effects of resveratrol on mitochondrial function. *Cell Metabolism, 15*(5), 675–690. https://doi.org/10.1016/j.cmet.2012.04.003.

Rotermund, C., Machetanz, G., & Fitzgerald, J. C. (2018). The therapeutic potential of metformin in neurodegenerative diseases. *Frontiers in Endocrinology, 9*, 400. https://doi.org/10.3389/fendo.2018.00400.

Sanders, O. (2020). Sildenafil for the treatment of Alzheimer's disease: A systematic review. *Journal of Alzheimer's Disease Reports, 4*(1), 91–106. https://doi.org/10.3233/ADR-200166.

Seo, D. O., & Holtzman, D. M. (2024). Current understanding of the Alzheimer's disease-associated microbiome and therapeutic strategies. *Experimental & Molecular Medicine, 56*(1), 86–94. https://doi.org/10.1038/s12276-023-01146-2.

She, J., Zhang, H., Xu, H., Li, Y. Y., Wu, J. C., Han, R., ... Qin, Z. H. (2025). Nicotinamide riboside restores nicotinamide adenine dinucleotide levels and alleviates brain injury by inhibiting oxidative stress and neuroinflammation in a mouse model of intracerebral hemorrhage. *Molecular Neurobiology, 62*(2), 1321–1336. https://doi.org/10.1007/s12035-024-04335-w.

Sita, G., Hrelia, P., Graziosi, A., & Morroni, F. (2020). Back to the fusion: mitofusin-2 in Alzheimer's disease. *Journal of Clinical Medicine, 9*(1), 126. https://doi.org/10.3390/jcm9010126.

Soltani, S., Dolatshahi, M., Soltani, S., Khazaei, K., Rahmani, M., & Raji, C. A. (2025). Relationships between brain glucose metabolism patterns and impaired glycemic status: A systematic review of FDG-PET studies with a focus on Alzheimer's disease. *Human Brain Mapping, 46*(4), e70180. https://doi.org/10.1002/hbm.70180.

Steen, E., Terry, B. M., J. Rivera, E., Cannon, J. L., Neely, T. R., Tavares, R., ... de La Monte, S. M. (2005). Impaired insulin and insulin-like growth factor expression and signaling mechanisms in Alzheimer's disease–is this type 3 diabetes? *Journal of Alzheimer's Disease, 7*(1), 63–80. https://doi.org/10.3233/JAD-2005-7107.

Talbot, K., Wang, H.-Y., Kazi, H., Han, L.-Y., Bakshi, K. P., Stucky, A., et al. (2012). Demonstrated brain insulin resistance in Alzheimer's disease patients. *Journal of Clinical Investigation, 122*(4), 1316–1338.

Timmers, S., Konings, E., Bilet, L., Houtkooper, R. H., van de Weijer, T., Goossens, G. H., et al. (2012). Calorie restriction-like effects of resveratrol in humans. *Cell Metabolism, 14*(5), 612–622. https://doi.org/10.1016/j.cmet.2011.10.002.

Wang, S., Liao, Z., Zhang, Q., Han, X., Liu, C., & Wang, J. (2025). Mitochondrial dysfunction in Alzheimer's disease: A key frontier for future targeted therapies. *Frontiers in Immunology, 15*, 1484373. https://doi.org/10.3389/fimmu.2024.1484373.

Wang, J., Liu, W. J., Shi, H. Z., Zhai, H. R., Qian, J. J., & Zhang, W. N. (2022). A role for PGC-1a in the control of abnormal mitochondrial dynamics in Alzheimer's disease. *Cells, 11*(18), 2849. https://doi.org/10.3390/cells11182849.

Winkler, E. A., Nishida, Y., Sagare, A. P., Rege, S. V., Bell, R. D., Perlmutter, D., et al. (2015). GLUT1 reductions exacerbate Alzheimer's disease vascular and metabolic dysfunction. *Journal of Cerebral Blood Flow & Metabolism, 35*(8), 1247–1258. https://doi.org/10.1038/nn.3966.

Yan, X., Hu, Y., Wang, B., Wang, S., & Zhang, X. (2020). Metabolic dysregulation contributes to the progression of Alzheimer's disease. *Frontiers in neuroscience, 14*, 530219.

Zehnder, T., Petrelli, F., Romanos, J., Figueiredo, E. C. D. O., Lewis, T. L., Déglon, N., ... Bezzi, P. (2021). Mitochondrial biogenesis in developing astrocytes regulates astrocyte maturation and synapse formation. *Cell Reports, 35*(2), https://doi.org/10.1016/j.celrep.2021.108952.

https://www.who.int/news-room/fact-sheets/detail/dementia.

Zhang, T., Zhang, Y., Ren, Y., Li, Y., Liang, J., Deng, X., ... Xu, Y. (2025). Activation of AMPK by GLP-1R agonists mitigates Alzheimer-related phenotypes in transgenic mice. *Nature Aging, 5*(2), 310–329. https://doi.org/10.1038/s43587-025-00869-3.

Zhao, Z., & Townsend, R. R. (2009). Blood-brain barrier dysfunction in Alzheimer's disease: Mechanisms and therapeutic approaches. *Journal of Clinical Investigation, 119*(2), 287–295.

CHAPTER TWO

Brain energy metabolism changes and neuropsychological impairment in Parkinson disease vs Alzheimer disease

Sanaz Khodadadi[a], Nasim Rezaeimanesh[a], Ali Ariyae Motahar[a], Fateme Yousefi[b], Masoume Avateffazeli[c], Reyhaneh Montazeri-Shatouri[a], Elham Hosseini[d], Sajad Sahab Negah[a], and Abdorreza Naser Moghadasi[a,*]

[a]Neuroscience Institute, Multiple Sclerosis Research Center, Tehran University of Medical Sciences, Tehran, Iran
[b]School of Medicine, Kermanshah University of Medical Sciences, Kermanshah, Iran
[c]National Research Institute of Tuberculosis and Lung Diseases (NRITLD), Lung Transplantation Research Center, Shahid Beheshti University of Medical Sciences, Tehran, Iran
[d]Neuroscience Research Center, Mashhad University of Medical Sciences, Mashhad, Iran
*Corresponding author. e-mail address: abdorrezamoghadasi@gmail.com

Contents

Abstract

Alzheimer's disease (AD) and Parkinson's disease (PD) are the most common neurodegenerative disorders, posing a significant public health risk. Although they are separate diseases, they have similar neuropsychological characteristics. Despite making up only a small portion of total body mass, the brain requires a disproportionately large amount of energy to maintain neuronal activity, synapse

International Review of Neurobiology, Volume 188
ISSN 0074-7742, https://doi.org/10.1016/bs.irn.2026.06.001

function, and cellular homeostasis. Disruption of energy metabolism is thus a major contributor to neurodegeneration. Energy metabolism has a wide-ranging impact on brain function, including cognitive and psychological processes, and gets increasingly compromised in neurodegenerative disorders. This chapter aims to offer a thorough overview of the link between altered brain energy metabolism and neuropsychological impairment in AD and PD.

1. Introduction

The two most prevalent neurodegenerative diseases, Alzheimer's disease (AD) and Parkinson's disease (PD), impact a wide range of people globally and pose an increasing public health concern (Pereira et al., 2024; Xiaopeng et al., 2025; Xie et al., 2014). AD is pathologically defined by a notable accumulation of amyloid-β (Aβ) plaques and neurofibrillary tangles (NFTs) in the brain, which primarily results in cognitive loss and amnesia (Zhang et al., 2024). In contrast, PD is characterized by the degeneration of dopaminergic neurons in the midbrain's substantia nigra pars compacta (SNpc) and is associated with Lewy bodies, which are aggregates of insoluble α-synuclein (α-syn). The main motor symptoms of PD include stiffness, bradykinesia, and resting tremor (Tolleson & Fang, 2013; Ferrer, 2009). In addition to the key symptoms used to identify these two diseases, research indicates that neuropsychiatric symptoms such as cognitive impairment (CI), attention problems, executive dysfunction, and language deficits are common. Furthermore, common neuropsychiatric symptoms such as depression, anxiety, apathy, agitation, irritability, sleep disruptions, disinhibition, and hallucinations are frequently experienced (Erkkinen et al., 2018; Aarsland et al., 2009; Lanctôt et al., 2017). Understanding the biological processes underlying these symptoms can bring new insights and lead to novel therapeutic options.

There have been numerous research studies on brain energy metabolism disruptions and their association with neurodegenerative diseases. Despite the adult brain making up only around 2 % of total body mass, it consumes a disproportionate portion of circulating glucose and other carbon substrates that are metabolized by neurons, glia, and other central nervous system (CNS) cells via glycolysis and mitochondrial oxidative phosphorylation (OXPHOS) (Bano et al., 2023; Cunnane et al., 2020; Goyal et al., 2023). Disruption in any component of the energy synthesis pathways, including the tricarboxylic acid (TCA) cycle and OXPHOS can limit adenosine triphosphate (ATP) synthesis and cause neuronal malfunction and degeneration (Saxena, 2012). In AD, a breakdown in any stage of the glucose production or consumption cycle might be one of the leading causes of neuronal injury. In PD, mitochondria's failure to make ATP can have a significant impact on neuronal damage (Beard et al., 2022; Moon & Paek, 2015; Croteau et al., 2018).

The objective of this chapter is to gain a thorough understanding of the association between brain energy metabolism disruption and neuropsychological impairment in AD and PD. The combination of neuroenergetic markers with molecular and clinical signs may thus improve diagnostic accuracy and allow the development of metabolic-based treatment approaches.

2. Parkinson's disease (PD): Metabolic & cognitive profile

2.1 Primary energy metabolism alterations

2.1.1 Mitochondrial dysfunction

Neurons possess a complicated network of mitochondria, which have various functions, including serving as intracellular powerhouses by producing ATP, epigenetic signaling, reactive oxygen species (ROS) generation, regulation of cell death, and calcium (Ca^{2+}) buffering (Moon & Paek, 2015; Henrich et al., 2023).

Neurons have a high demand for bioenergy and are highly dependent on mitochondria for ATP production. Neurons can use both glycolysis and mitochondrial OXPHOS pathways to meet their energy needs. Although the glycolytic pathway is fast, it is inefficient and produces much less ATP than mitochondrial mechanisms. Therefore, neurons appear to rely heavily on ATP produced in mitochondria to meet their high energy demands (Henrich et al., 2023).

Abnormalities in mitochondrial morphology and function in dopaminergic neurons are recognized as key pathogenic mechanisms in a subset of patients with PD (Moon & Paek, 2015). Mitochondrial dysfunction may disproportionately affect neurons that are vulnerable in PD. The best-known example of this phenomenon is dopaminergic neurons located in the SNpc, neurons that are continuously active and possess extensive axonal arbors with one to two million sites of transmitter release in each axon in humans (Henrich et al., 2023).

Mitochondrial dysfunction is primarily defined by increased formation of ROS, release of cytochrome C, reduction in mitochondrial complex I enzyme activity, activation of caspase-3, and reduced ATP levels (Moon & Paek, 2015). A systemic deficiency in mitochondrial complex I has been linked to the pathogenesis of idiopathic PD and can lead to increased oxidative stress (OS) in substantia nigra neurons. OS can damage neuronal integrity and lead to neuronal degeneration. Increased OS and ROS not only directly damage cells, however, they also cause neuronal death by activating intracellular signaling pathways (Moon & Paek, 2015).

2.1.2 Glucose metabolism impairment

Glucose is the primary and essential source of energy for brain neurons. Nearly all of the brain's energy requirements are met by glucose, which is continuously delivered to the brain under normal physiological conditions, as it contributes to the ATP synthesis, OS modulation, production of neurotransmitters, regulation of neural function, and the development and maintenance of brain structure. Glucose metabolism is higher in functionally active brain areas and during periods of brain activity (Dai et al., 2023).

A systemic impairment of glucose metabolism has been recorded in PD and has been identified as one of its non-motor symptoms. Glucose tolerance abnormality is identified in 50–80% of PD patients. Diabetes and hyperglycemia are linked with a greater risk of PD (Dai et al., 2023). Moreover, glucose metabolism impairment is commonly detected in the brains of individuals with PD, characterized by glucose hypometabolism in occipital, posterior temporoparietal, and occasionally frontal regions, along with glucose hypermetabolism in the sensorimotor cortex, putamen, and cerebellum (Dai et al., 2023). Different PD subtypes are linked with impaired glucose metabolism in numerous areas of the brain. Compared to individuals with the tremor-dominant form of PD, a more marked reduction in glucose metabolism, accompanied by a drop in dopamine uptake in the ventral striatum, has been detected in patients with the akinetic-rigid subtype (Eggers et al., 2014).

The connection between glucose metabolism disorders and PD is very complex, and their cause-and-effect relationship is still unclear However, glucose metabolism disruption at both systemic and regional levels is shown across all stages of PD, involving various metabolic processes such as glucose transport and uptake, the pentose phosphate pathway (PPP), the TCA cycle, glycolysis, OXPHOS, and gluconeogenesis (Dai et al., 2023; Chen et al., 2025).

Positron emission tomography (PET) scans have highlighted that glucose consumption is reduced in areas of the brain affected by PD (Camandola & Mattson, 2017). In addition, it seems that downregulation of 6-phosphogluconate dehydrogenase and glucose-6-phosphate dehydrogenase and occurs in the cerebellum and putamen of individuals with PD (Camandola & Mattson, 2017). On the other hand, lack of energy in neurons, in fact reduction in ATP levels, has been reported in individuals with PD. Disturbance in any component of the energy generating pathways, including glycolysis, the TCA cycle, and oxidative phosphorylation, can reduce ATP production, leading to neuronal dysfunction and degeneration, thereby contributing to the PD pathogenesis (Saxena, 2012).

2.1.3 Lactate & oxidative stress dynamics

Besides glucose, its metabolic intermediates can also be a source of energy in the brain, with lactate being one of the major contributors (Dai et al., 2023). In the CNS, astrocytes take up glucose from surrounding capillaries into the cytoplasm via glucose transporter 1 (GLUT1), which is then converted to pyruvate through glycolysis. When cells are exposed to hypoxia, the enzyme lactate dehydrogenase (LDH) converts pyruvate to lactate. Lactate generated in astrocytes is transferred into the intercellular space between astrocytes and neurons by monocarboxylate transporter 1 (MCT1) and monocarboxylate transporter 4 (MCT4), where it is absorbed by neurons via monocarboxylate transporter 2 (MCT2). This shuttle that leads to cell-to-cell lactate transfer is called the Astrocyte–Neuron Lactate Shuttle (ANLS). Lactate received by neurons is converted to pyruvate using lactate dehydrogenase B (LDHB), which enters the Krebs cycle to provide the energy needed for synaptic transmission and neuronal excitability (Yang et al., 2024).

Studies reveal that lactate performs dual functions in different circumstances. On one hand, it plays a role in regulating whole-brain metabolism; on the other hand, when neurons take up an elevated amount of lactate, it promotes the production of ROS, increases mitochondrial energy metabolism, and leads to OS in neurons (Yang et al., 2024).

Existing evidence suggests that neuroinflammation is regulated in part by lactate metabolism in microglia. Under normal physiological conditions, microglia receive most of their energy from mitochondrial OXPHOS, but in neurodegenerative disorders, the metabolism of proinflammatory microglia shifts to aerobic glycolysis, which results in high production of lactate. This lactate could lead to the production of pro-inflammatory cytokines such as interleukin (IL)-6, tumor necrosis factor-α, and interleukin-1β (IL-1β) (Yang et al., 2024).

Studies document that lactate levels are abnormally increased in PD patients and animal models and elevated CSF level of lactate are reported in late-onset PD patients (Yang et al., 2024; Veloz Castillo et al., 2021). In PD, increased lactate resulting from elevated expression of hexokinase 2 and LDHA leads to apoptosis of dopaminergic neurons (Yang et al., 2024).

2.2 Associated neuropsychological impairment

2.2.1 Executive dysfunction and attentional deficit

Executive dysfunction is the most prevalent form of cognitive impairment (CI) in PD and can manifest early on (Tong et al., 2025; Dirnberger & Jahanshahi, 2013), marked by deficits in the higher-order cognitive functions necessary for conduct that is goal-directed behavior, including planning, organization, cognitive flexibility, and inhibitory control (Zhang et al., 2024; Aarsland et al., 2021). Patients struggle with cognitive flexibility, which is the capacity to change between tasks or rules, as well as the inability to maintain or switch attention (Zhang et al., 2024; Aarsland et al., 2021). There are several neuropsychological tests for assessing executive function in PD, including Verbal Fluency Test, Wisconsin Card Sorting Test (WCST), Stroop Test, Trail Making Test Part B(TMT-B), and the Tower of London test (Zhang et al., 2024; Tong et al., 2025).

Specifically, executive dysfunction in PD is driven by the disruption of the frontostriatal circuit, where dopamine depletion in the dorsal caudate nucleus (specifically the dorsolateral head) disrupts its regulatory loop with the dorsolateral prefrontal cortex (DLPFC) (Tong et al., 2025; Gratwicke et al., 2015). The loss of noradrenergic neurons in the locus coeruleus adds to this deficiency, which impairs attentional switching and cognitive flexibility normally facilitated by projections to the medial prefrontal cortex (Tong et al., 2025).

Attentional deficits in PD are primarily characterized by an inability to direct focus voluntarily to specific stimuli or to switch attention between tasks (González-Usigli et al., 2023). The scope of these deficits encompasses several specific domains: selective attention (the ability to focus while inhibiting distractors), divided attention (multitasking), and vigilant attention (maintaining a state of arousal) (Gratwicke et al., 2015; González-Usigli et al., 2023). This impairment is both early and common and is characterized by fluctuating clarity, bradyphrenia, slowed and error-prone performance on attention-demanding tasks, and frequent co-occurrence with non-motor symptoms such as hallucinations, sleep issues, and autonomic dysfunction, leaving patients increasingly vulnerable to environmental distractions (Gratwicke et al., 2015; Degirmenci et al., 2023).

The physiological and neural mechanisms of attention deficit in PD are primarily driven by the early degeneration of the locus coeruleus, which leads to a reduction of norepinephrine in cortical projections and disrupts attentional switching and cognitive flexibility (Tong et al., 2025). This neurochemical impairment is further compounded by dopaminergic depletion in the dorsal caudate nucleus, which hinders the frontostriatal circuit's ability to maintain attention control. Structurally, attention deficits are specifically linked to gray matter atrophy in the middle temporal gyrus, superior frontal gyrus, and the insula, in addition to localized volume loss in the central body of the corpus callosum (Tong et al., 2025; Gratwicke et al., 2015; Fornari et al., 2021). At the system level, these changes cause a functional decoupling between the salience, central executive, and default mode networks, resulting in an inability to coordinate complex, goal-oriented tasks (Tong et al., 2025).

2.2.2 Visuospatial processing deficit

Visuospatial processing deficits are prominent early cognitive symptoms in PD, characterized by an inability to interpret visual information regarding object location and identity (Ghazi-Saidi, 2020). While classically categorized as a non-motor symptom, these deficits contribute significantly to physical disability through their influence on locomotion and gait (Kawashima et al., 2021). Under the "dual syndrome hypothesis," unlike the executive-dominant subtype (Parkinson's disease executive-dominant subtype; PD-EL), the Parkinson's disease visuospatial, attention, and memory-dominant subtype (PD-VAM) is primarily linked to non-dopaminergic dysfunction, specifically involving cholinergic and noradrenergic systems (Kehagia et al., 2013; Wang et al., 2025).

Visuospatial processing deficits in PD reflect disrupted dorsal (occipitoparietal) and ventral (occipitotemporal) streams, cortical thinning in parietal and temporal regions as well as the hippocampus and precuneus, failure of the oculomotor circuit supporting visuospatial working memory, and reduced white matter integrity in the optic radiations and fronto-occipital connections that impair spatial attention (Gratwicke et al., 2015; Ghazi-Saidi, 2020). Deficits in visuospatial perception, specifically the inability to switch attention or judge line orientation, are strongly correlated with axial symptoms, including postural instability and an increased risk of falls (Brandão et al., 2020). Visual hallucinations in PD are closely linked with severe visuospatial and visuoperceptual deficits, as both symptoms frequently co-occur (Ruppert et al., 2020)and represent core features of the progression toward dementia. This correlation is underpinned by a reduction in gray matter volume within parieto-occipital association areas, with marked visuospatial impairment serving as a significant clinical predictor for the emergence of vivid hallucinations (González-Usigli et al., 2023; Ghazi-Saidi, 2020; Marsh, 2000).

2.2.3 Cognitive impairment vs dementia

The shift from **mild cognitive impairment (MCI)** to **dementia in PD (PDD)** is a common and often predictable trajectory in the natural history of the disease. It is estimated that **nearly most patients** diagnosed with PD **will** experience CI (Aarsland et al., 2021; Trošt et al., 2019). In the PD-MCI stage, pathology is regionally constrained; as patients progress toward PDD, the pathological burden shifts from localized dopaminergic loss to a synergistic proteinopathy in which α-syn Lewy pathology spreads from the brainstem and subcortical nuclei into limbic and neocortical association areas (Aarsland et al., 2021; Brandão et al., 2020). Roughly one third of PDD cases exhibit coexisting severe Alzheimer-type amyloid-β and tau pathology that accelerates cognitive decline and increases mortality (Aarsland et al., 2021; Smith et al., 2019).

This process represents a critical shift from localized network dysfunction and active compensation to widespread structural and metabolic failure. Magnetic resonance (MR) imaging shows cortical thinning in fronto-temporo-parietal areas, while fluorodeoxyglucose positron emission tomography (FDG-PET) reveals posterior hypometabolism that is notably counterbalanced by compensatory hypermetabolism in the cerebellum (Ghazi-Saidi, 2020; Ruppert et al., 2020; Trošt et al., 2019). Metabolically, this early phase is defined by reduced N-acetylaspartate (NAA) in the frontal gray matter, signaling initial neuronal energy stress, but typically normal myo-inositol levels (Chaudhary et al., 2021; Huang et al., 2024).

2.3 Mechanistic links

2.3.1 Dopamine depletion → Cortico-striatal circuit failure

PD is fundamentally defined by the gradual degradation of **dopamine (DA)**-producing neurons in the SNpc, leading to profound DA depletion in the striatum (Aarsland et al., 2021; González-Usigli et al., 2023). This deficit initiates a mechanistic cascade that disrupts cortico-striatal circuits, particularly the fronto-striatal networks, which serve as metabolic powerhouses for high-energy cognitive and motor processes (González-Usigli et al., 2023; Watanabe et al., 2025). The associative striatum, especially the caudate nucleus, bears the brunt of early DA loss, severing connections to the dorsolateral prefrontal cortex (DLPFC) essential for executive function (Tong et al., 2025; Gratwicke et al., 2015).

Beyond dopamine depletion in the cortico-striatal circuit, impairments in noradrenergic, cholinergic, and serotonergic systems contribute significantly to cognitive deficits (Tong et al., 2025; Halliday et al., 2014; McGregor & Nelson, 2019). Noradrenergic loss from the locus coeruleus leads to attention deficits (Aarsland et al., 2021; Gratwicke et al., 2015), while cholinergic degeneration in the pedunculopontine nucleus and basal forebrain disrupts visual function and memory (Gratwicke et al., 2015). Serotonergic dysfunction in the raphe nuclei and cortex exacerbates mood, impulse control, and working memory issues (Tong et al., 2025; Brandão et al., 2020), as depicted in Fig. 1, linking these pathways (norepinephrine, acetylcholine, serotonin) to the cortex, striatum, hippocampus, and amygdala via the substantia nigra. These interconnected failures amplify overall cognitive impairment.

2.3.2 Mitochondrial damage → Neuronal bioenergetic crisis

Mitochondrial dysfunction in PD is an intricate and central pathophysiological process defined by a profound mismatch between energy supply and demand (Watanabe et al., 2025). This dysfunction serves as a primary driver of the selective degeneration of dopaminergic neurons in the SNpc, which have an exceptionally high metabolic demand due to their massive axonal networks (Watanabe et al., 2025; Pokotylo et al., 2025).

The hallmark of mitochondrial impairment in PD is bioenergetic failure, largely driven by a deficiency in Complex I of the mitochondrial electron transport chain (ETC) (Nicoletti et al., 2021; Li et al., 2021). While historically labeled a "Complex I disease," recent imaging mass cytometry indicates a generalized deficiency across all ETC complexes (I–V) in the PD midbrain (Li et al., 2021). These defects significantly reduce ATP production and respiratory capacity (Henrich et al., 2023; Nicoletti et al., 2021). To compensate, neurons experience an alteration to anaerobic glycolysis, known as the "neuronal Warburg effect" (Pokotylo et al., 2025). Although this shift temporarily meets energy needs, it is ultimately unsustainable and results in the formation of advanced glycation end products and lactic acidosis, which worsen

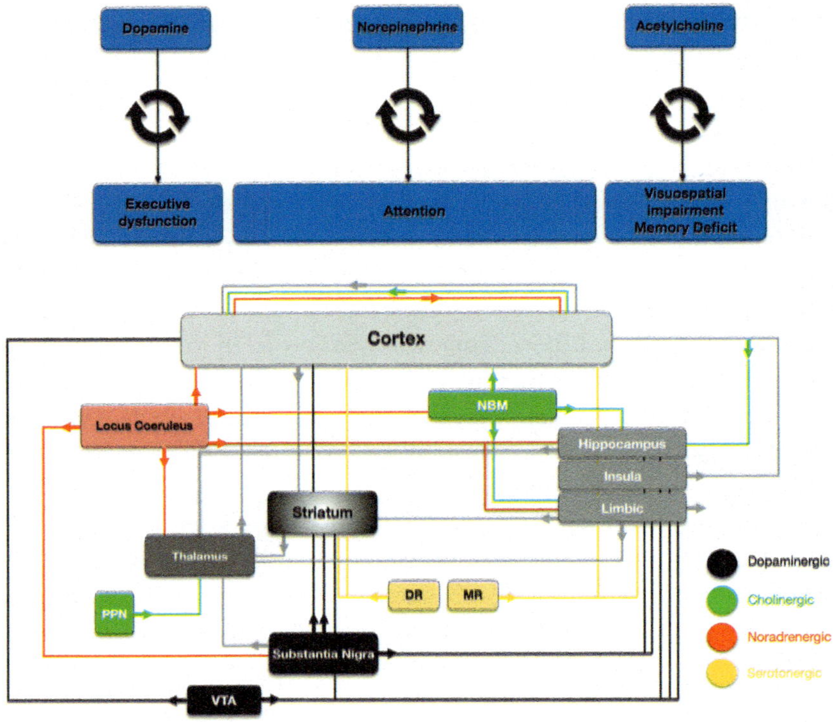

Fig. 1 The provided diagram illustrates the complex interplay between neurochemical depletion and the resulting cognitive-motor impairments in PD.**SNpc:** Substantia Nigra pars compacta; **VTA:** Ventral Tegmental Area; **NBM:** Nucleus Basalis of Meynert; **PPN:** Pedunculopontine Nucleus; **DR / MR:** Dorsal Raphe / Median Raphe.

neurotoxicity (Pokotylo et al., 2025). Failure in the ETC leads to the leakage of high-energy electrons, resulting in the excessive production of ROS and OS (Gao et al., 2022). This oxidative surge primarily targets mitochondria, resulting in mutations and deletions of mitochondrial DNA (mtDNA) (Nicoletti et al., 2021; Moradi Vastegani et al., 2023).

Mitochondrial health in PD is further compromised by the breakdown of protective quality control mechanisms, primarily regulated by the PTEN-induced kinase 1 (PINK1)/Parkin pathway (Gao et al., 2022; Huang et al., 2021). Disruption of this pathway prevents the selective removal of damaged organelles through mitophagy, causing a toxic accumulation of dysfunctional mitochondria (Nicoletti et al., 2021; Moradi Vastegani et al., 2023). Mitochondria in PD exhibit abnormal dynamics, characterized by impaired fission, fusion, and axonal transport (Nicoletti et al., 2021; Moradi Vastegani et al., 2023). Production of new mitochondria is hindered by the Parkin interacting substrate (PARIS)/peroxisome proliferator-activated receptor gamma coactivator-1 alpha (PGC-1α) pathway; without functional Parkin, the repressor PARIS accumulates and shuts down mitochondrial biogenesis (Gao et al., 2022; Moradi Vastegani et al., 2023).

Mitochondria play a critical role in Ca^{2+} buffering, particularly at mitochondria-associated endoplasmic reticulum membranes (MAMs). Disrupted signaling at MAMs causes mitochondrial calcium excess, resulting in the opening of the mitochondrial permeability transition pore (mPTP). When mPTP opens, cytochrome c and other apoptogenic substances are released into the cytosol, which starts the apoptotic cell death cascade (Gao et al., 2022; Moradi Vastegani et al., 2023; You et al., 2024).

A crucial pathogenic feedback loop exists between mitochondria and α-syn pathology. Mitochondrial stress and ROS overproduction increase the misfolding and accumulation of α-syn (Nicoletti et al., 2021). Pathological α-syn oligomers bind to proteins of the external mitochondrial membrane, such as translocase of the outer mitochondrial membrane 20 (TOM20), inhibiting the import of essential mitochondrial proteins, and interact with ATP synthase to induce membrane depolarization (Henrich et al., 2023).

Damaged mitochondria serve as a source of damage-associated molecular patterns (DAMPs). The escape of circulating cell-free mtDNA (ccf-mtDNA) and other proteins into the cytoplasm or extracellular space activates the NOD-like receptor family pyrin domain-containing 3 (NLRP3) inflammasome and the cyclic GMP-AMP synthase (cGAS)–stimulator of interferon genes (STING) pathway in microglia (Borsche et al., 2021). This triggers chronic neuroinflammation, which further damages the remaining neuronal population.

3. Alzheimer's disease: Metabolic & cognitive profile

3.1 Primary energy metabolism alterations

3.1.1 Early hippocampal & parietal hypometabolism

The early emergence of cerebral hypometabolism strongly supports the hypothesis that alterations in energy metabolism constitute an initial event in the pathogenesis of AD, rather than a later consequence of amyloid or tau pathology. This paradigm shift reframes the disorder as a disorder of bioenergetic homeostasis. These metabolic changes are characterized by reduced glucose uptake and expression of glucose transporters, attenuated astrocytic aerobic glycolysis and lactate release, impaired mitochondrial function, and concomitantly increased oxidative stress (Beard et al., 2022).

Imaging and pathology studies have shown a hypometabolic pattern in the frontal, posterior cingulate, and hippocampus cortices—regions crucial for memory and spatial processing—while the cerebellum, thalamus, and basal ganglia are comparatively spared (Goyal et al., 2023; An et al., 2018; Yuan et al., 2024; Nguyen et al., 2020) However, studies reveal that relatively the primary motor and visual cortices are relatively preserved (Beard et al., 2022).

MRI findings consistently show that the hippocampus undergoes the earliest and most prominent atrophy among affected regions in AD (Villain et al., 2010). Despite this structural vulnerability, glucose hypometabolism is more prominent in the posterior association cortex and posterior cingulate gyrus, indicating that early metabolic abnormalities do not always correlate directly with the regions showing the greatest structural injury. A related hypothesis is that disruption of the cingulum bundle leads to posterior cingulate hypometabolism due to disconnection from the hippocampus (Villain et al., 2008).

In the five-familial AD (5XFAD) transgenic mouse model, early hippocampal dysfunction is marked by increased spontaneous excitatory signaling accompanied by reduced TCA cycle activity, reduced glucose utilization, and mitochondrial morphological abnormalities that drive hippocampal cell loss and subsequent AD pathology. Astrocytic deficits in glutamine synthesis further reduce neuronal γ-aminobutyric acid (GABA) production, exacerbating local excitatory–inhibitory imbalance. In contrast, the cortex shows hypermetabolism through elevated oxidative glucose metabolism and elevated (U-^13 C) glucose metabolism before detectable amyloid deposition. This contrast suggests early cortical compensation during disease progression, while the hippocampus remains vulnerable to hypometabolic decline (Villain et al., 2010; Andersen et al., 2021; Sun et al., 2020).

3.1.2 Insulin resistance & brain glucose hypometabolism

The progressive decline in regional cerebral glucose metabolism observed in AD represents a central pathophysiological process (Bano et al., 2023; Cunnane et al., 2020; Goyal et al., 2023).

In the CNS, insulin and insulin-like growth factor 1 (IGF-1) receptors are expressed in all cell types. Insulin has multiple functions within the CNS, including supporting synaptic plasticity, neural survival, and modulating neuronal receptors including GABA, N-methyl-D-aspartate (NMDA), and α-amino-3-hydroxyl-5-methyl-4-isoxazole-propionate (AMPA). High insulin-receptor densities in the hippocampus, frontal cortex, and entorhinal cortex highlight its critical role in cognition and spatial learning (Lim et al., 2016; Sędzikowska & Szablewski, 2021).

The metabolic dysregulation of AD is significantly influenced by insulin signaling dysfunction (Yuan et al., 2024). Dysregulated insulin signaling affects amyloid precursor protein (AβPP) processing and Aβ clearance, resulting in intracellular accumulation of neurotoxic peptides, synaptic toxicity, and neuronal stress. These changes lead to memory decline and executive function, hippocampal atrophy, and cognitive decline. Moreover, Aβ competes for insulin-receptor binding, induces receptor resistance, suppresses autophosphorylation, and reduces receptor expression and activity. It has been demonstrated that insulin resistance (IR) is involved in AD pathogenesis through several mechanisms: defective insulin-receptor signaling, insulin receptor substrate 1 (IRS-1) inhibition, and downstream metabolic collapse. This signaling failure establishes a vicious cycle: defective insulin signaling promotes Aβ accumulation, while Aβ exacerbates IR. In parallel, insulin and IGF-1 also modulate tau phosphorylation through kinase activation, linking metabolic signaling defects to neurodegenerative processes (Sędzikowska & Szablewski, 2021; Chen & Zhong, 2013). In conjunction with reduced enzyme activity, decreased antioxidant defenses, and impaired OXPHOS, these mechanisms collectively exacerbate cerebral glucose hypometabolism in AD (Yuan et al., 2024; Nguyen et al., 2020).

Brain glucose hypometabolism is one of the earliest alterations that can be detected up to a decade before symptom manifestation, and the extent of these metabolic changes highly correlates with clinical severity of AD (Beard et al., 2022; Croteau et al., 2018). In AD, widespread reductions in both the uptake rate constants (KGlu) and cerebral metabolic rate of glucose (CMRGlu) occur across the brain and major cortical regions, sparing the cerebellum, occipital cortex, and white matter. The cingulate cortex shows significant reductions in both KGlu (approximately 7 %) and CMRGlu in patients with MCI. Following adjustment for cerebral atrophy, significant reductions in CMRGlu in frontal, temporal, parietal, and subcortical regions have been demonstrated. These findings suggest impaired cerebral glucose uptake and utilization rather than a systemic reduction in substrate availability (Croteau et al., 2018).

3.1.3 Astrocytic glycogen dysregulation & lactate shuttle failure

Astrocytes are regulators of brain energy metabolism, preserving neuronal survival, memory processing, neuroprotection, oxidative-stress resistance, and overall homeostasis. They mediate neuroenergetic coupling via glutamate uptake, which stimulates astrocytic aerobic glycolysis, enhancing glucose uptake and lactate release through the ANLS to supply neurons with available metabolic fuel (Beard et al., 2022) Astrocytic dysfunction, ANLS failure, and glycogen dysregulation are increasingly recognized as central features of AD pathology (Beard et al., 2022; Yuan et al., 2024; Nguyen et al., 2020; Chen & Zhong, 2013).

The ANLS highlights lactate's role in neuronal metabolism and neurotransmission. In AD models, astrocytic lactate production through aerobic glycolysis is reduced, leading to ANLS disruption and subsequent neuronal energy deficits (Aldana, 2019; Boland et al., 2018; Le Douce et al., 2020; Pan et al., 2022). Lactate was regarded as a waste product but it is now recognized as a key metabolic substrate that can be converted to glucose or pyruvate, as well as a signaling molecule that modulates transcription and cellular activity, and its reduced availability can therefore lead to neuronal energy deficits (Chen & Zhong, 2013; Brooks, 2009).

On the other hand, studies report elevated CSF lactate. Elevated lactate and reduced pyruvate in AD brains indicate a shift toward anaerobic glycolysis as a compensatory energy source (Yuan et al., 2024; Bonomi et al., 2021; Xiang et al., 2021; Yang et al., 2024).

Furthermore, as the primary source of glycogen in the central nervous system, astrocytic glycogen is essential for sustaining both the energy requirements of neurons and more general metabolic functions (Falkowska et al., 2015; McKenna et al., 2019). Beyond glycogen roles as an energy reserve, it actively contributes to synaptic plasticity, learning, and memory, and provides protection against metabolic stress conditions such as hypoglycemia or neural injury (Muddapu et al., 2020) In the early stages of AD, dysregulated astrocytic glycogen metabolism is a characteristic feature; tau and Aβ plaques further disrupt glycogen storage and mobilization, impairing neuronal and glial energy homeostasis and inducing pro-inflammatory states that speed up neurodegeneration (Dorsey et al., 2024; Jensen et al., 2020; Wu et al., 2019).

3.2 Associated neuropsychological impairments

3.2.1 Episodic memory loss

Impairments in episodic memory are considered primary manifestations of AD (Hodges, 2000). During the early stages of the disease, these disruptions can lead to problems such as difficulty remembering the location of personal items, appointments, or phone calls (Gold & Budson, 2008). These problems in daily life can be mistaken for events that happened because of getting older (Gallagher & Koh, 2011). clinically, AD is diagnosed according to criteria and when important signs and symptoms, such as severity in memory loss and cognitive and behavioral deficits, occur (McKhann et al., 2011). Extensive brain areas support the episodic memory system, consisting of widespread neocortical association areas and medial temporal lobe structures, including the para hippocampal cortex and hippocampus (Dickerson & Eichenbaum, 2010). The hippocampus is considered one of the regions that are extensively damaged in AD patients (Gold & Budson, 2008). The pathology of AD, involving synaptic and marked neuronal loss as well as aggregation of Aβ plaques and tau tangles, affects the entorhinal cortex-hippocampal circuitry, which plays a key role in memory (Spires-Jones & Hyman, 2014). Therefore, dysfunctions in episodic memory are one of the most important clinical problems that occur in the mild and moderate stages of AD (Gold & Budson, 2008).

3.2.2 Language deficits (Aphasia) & semantic memory decline

Language deficits appear in early stages of Alzheimer's disease, and when the disease progresses language function gets worse and leads to significant disability (Verma & Howard, 2012). Some disorders are placed in typical language deficits category, including dysfunction in naming, auditory and written understanding, speech, and semantic paraphasia, although the ability for repeating and articulation remains relatively unimpaired (Appell et al., 1982; Bayles et al., 1992; Croot et al., 2000). Language impairments occur in both early and later stages of AD, in the early stages because of semantic dysfunction, but in the later stages, they result from severe semantic deficits and other cortical processes (Verma & Howard, 2012). Therefore, the language functions, whose disruption in the initial stages has a clear connection with semantic memory (Verma & Howard, 2012). One of the frequent impairments in Alzheimer's patients is a semantic deficit (Joyal et al., 2020). Semantic knowledge is related to general knowledge about the world (Joyal et al., 2020). This knowledge is used by semantic processing or semantic cognition for different purposes, including recognizing objects in the environment and doing cognitive tasks, such as identifying the existence of links between concepts (Hoffman, 2018; Lambon Ralph et al., 2009; Ralph et al., 2017). Semantic knowledge is related to temporal and parietal neocortical areas, whereas it does not depend on medial temporal lobe structures, unlike episodic memory (Salmon, 2011). Decline in semantic memory leads to difficulty remembering facts, and in addition, naming and verbal tests are disrupted in patients with AD (Salmon, 2011).

3.2.3 Visuospatial & executive dysfunction

Some of the impairments that occur in patients with AD include deficits in memory, visual–spatial skills, and executive functions (van der Flier et al., 2011). When AD progresses, cognitive functions, like language, praxis, visuospatial functioning, and executive functioning, become affected, eventually leading to global cognitive decline (van der Flier et al., 2011). Several Alzheimer's patients show significant deficits in visuospatial skills (Huang et al., 2023). Visuospatial perception involves the obtaining and analysis of spatial information from visual stimuli, which consists of different mechanisms, including the segregation of visual information at the retina to its systematic, hierarchical, and modular integration in the primary and associative visual cortical areas. This information is processed in the striate and extrastriate regions, and the higher associative brain areas (Mandal et al., 2012). VSP abilities enable people to understand the location of objects in space, and VSP dysfunction can disrupt daily activities in patients with Alzheimer's disease (Mandal et al., 2012). Executive functions enable people to respond and harmonize with their environment, which include various cognitive processes, such as verbal reasoning, problem-solving, planning, sustaining attention, resistance to interference, multitasking, cognitive flexibility, and the ability to cope with novelty (Guarino et al., 2019). The prefrontal cortex plays a key role in the performance of these functions (Funahashi & Andreau, 2013). Therefore, healthy executive functions are necessary for coordination and adaptation of people to the world, and impairment in them leads to serious problems and dysfunction in everyday activities (Cahn-Weiner et al., 2002).

3.3 Mechanistic links

3.3.1 Aβ/Tau → Mitochondrial impairment & oxidative stress

The pathology of AD is related to the aggregation of amyloid-β and tau protein (Bhatia & Sharma, 2021). Mitochondrial dysfunction occurs in AD due to the accumulation of a AβPP in the mitochondrial import channels and the interaction of *Aβ* with several mitochondrial proteins (Pavlov et al., 2011). In addition, tau protein can cause mitochondrial impairment in several ways. Firstly, damage to axonal transport occurs because of the overexpression and aggregation of hyperphosphorylated tau, and this damage leads to abnormal mitochondrial distribution. Secondly, hyperphosphorylated tau affects mitochondrial fission and fusion proteins; as a result, mitochondrial dynamics are impaired. On the other hand, mitochondrial deficits result in the acceleration of tau pathology (Cheng & Bai, 2018).

Therefore, mitochondrial impairment has a significant role in the pathophysiology of AD (Cheng & Bai, 2018). Mitochondria have several important roles, including providing ATP, synaptic maintenance, intracellular Ca2+ signaling regulation, free radical production and scavenging, and activation of caspases. Mitochondria dysfunction leads to an immediate increase in reactive oxygen species (ROS) production, a byproduct of the electron transport chain, that promotes oxidative damage to DNA, RNA, proteins, and lipids (Calvo-Rodriguez et al., 2024). A state in which increased levels of cellular ROS are generated is oxidative stress (Cioffi et al., 2019). The brain is more sensitive to oxidative stress compared to other organs. In AD several events like mitochondrial impairment, high metal levels, inflammation, and β-amyloid (Aβ) peptides lead to the oxidation of neuronal components (lipids, proteins, and nucleic acids). Moreover, oxidative stress leads to increased Aβ deposition, tau hyperphosphorylation, and synaptic and neuronal loss in AD. Therefore, in the pathology of AD oxidative stress has a important role (Chen & Zhong, 2014).

3.3.2 Glucose transporters (GLUTs) downregulation

The neurons in the brain cannot produce or store glucose, and glucose must be transported from the blood-brain barrier, and glucose transporters (GLUTs) facilitate this transport (Liu et al., 2008). GLUT1 and 3 are predominant in both expression levels and functional contribution to the brain (Albaik et al., 2024). The expression of GLUT1 occurs in the endothelium of the blood-brain barrier and its role is to transport glucose from blood to the extracellular space of the brain (Liu et al., 2008). The main neuronal GLUT is GLUT3, which transports glucose from the extracellular space into neurons (Dwyer et al., 2002). It has been discovered that levels of GLUT1 and GLUT3 proteins have decreased in brain samples from AD patients, especially in the cerebral cortex, with a high decrease in GLUT3 (Szablewski, 2025). The progression of pathology and symptoms of AD is both related to decreased levels of GLUT3 (An et al., 2018). The reduction in levels of GLUT1 and GLUT3 is related to hyperphosphorylated tau and O-GlcNAcylation (Liu et al., 2008). In addition, decline in the levels of both GLUTs may depend on downregulation of hypoxia-inducible factor-1 (HIF-1), which regulates their expression (Szablewski, 2025). Moreover, an increase in amyloid-β levels can also cause a reduction in GLUT1 and GLUT3 level (Kyrtata et al., 2021). The reduction in both GLUTs results in a reduced supply of glucose in the brain, which causes impairment of the processes that depend on ATP

(Iadecola, 2015). Due to GLUTs deficiency, glucose hypometabolism may lead to abnormal tau phosphorylation and/or neurofibrillary degeneration through downregulation of the hexosamine biosynthesis pathway (Szablewski, 2025).

3.3.3 Acetylcholine deficit & synaptic energy failure

One of the most important pathologies in AD is cholinergic dysfunction (Hampel et al., 2018). Cholinergic loss in AD is linked to deficits in the neurotransmitter acetylcholine (Wenk, 2006). It is shown that levels of acetylcholine synthesis are reduced by approximately one half in patients with AD compared with healthy people (Sims et al., 1983). Acetylcholine is an important neurotransmitter in the brain, involved in activity throughout the cortex, basal ganglia, and basal forebrain (Mesulam, 2013). Acetylcholine is synthesized from choline and acetyl co-enzyme A acetyl coenzyme A (acetyl-CoA), and reaction catalyzed by enzyme called choline acetyltransferase (ChAT). Choline is obtained from the diet, whereas mitochondria synthesize acetyl-CoA by using pyruvate from glycolysis. Then, the synthesized acetylcholine (ACh) is transported through ACh transporter into synaptic vesicles via an ATP consuming process and stored in vesicles (Guo et al., 2017). Therefore, synthesis, packaging and storage of neurotransmitters are extremely dependent on mitochondria (Guo et al., 2017). It has been found that Mitochondrial dysfunction occurs in AD patients (Guo et al., 2017). Thus, the 'bioenergetic failure' of mitochondria can lead to Acetylcholine deficit (Wong et al., 2020). On the other hand, Synaptic transmission has a very intensive energy demand and synaptic activity consumes the majority of brain ATP (Shulman et al., 2004). Mitochondria provide neuronal ATP through oxidative phosphorylation (OXPHOS) (Guo et al., 2017). Consequently, Synaptic function is related to mitochondria (Cai & Tammineni, 2017). Therefore, Synaptic Energy Failure can result because of mitochondrial dysfunction in AD patients.

4. Comparative analysis: PD vs. AD

4.1 Metabolic divergences

4.1.1 Regional specificity: Striatum (PD) vs. temporoparietal (AD)

There are both similar and distinct patterns in cerebral perfusion between AD and PD, which are discussed in several studies (Chen et al., 2022). Overall, the evidence supports a dual pattern in AD and PD, where broad frontal–posterior hypoperfusion is shared but disease-specific regional signatures remain critical for interpretation. Although absolute cerebral blood flow (CBF) variability can blur pathology-relevant effects, normalized arterial spin labeling still demonstrates a common distribution of compromised perfusion, with divergent deficits that align with cognitive impairment in AD and motor severity in PD. This hemodynamic framework is reinforced by overlapping glucose hypometabolism in association cortices and the posterior cingulate; however, PD shows greater visual cortex metabolic loss and relative preservation of medial temporal metabolism compared with AD. Taken together, these regional divergences are consistent with different pathological and neurochemical anchors—striatal dopaminergic and α-syn–related mitochondrial vulnerability in PD versus temporoparietal and cingulate cortex involvement in AD—while co-pathology across conditions plausibly underlies their convergent features. Thus, a comparative regional–metabolic view suggests that shared network-level dysfunction coexists with distinct epicenters of degeneration, emphasizing the need to interpret perfusion and metabolic data through both common and disease-specific topographies (Batra et al., 2024).

While perfusion in the sensorimotor cortex and basal ganglia areas remained stable, both the AD and PD groups showed decreased normalized CBF, primarily in the temporal and frontal gyrus regions. The PD group maintained perfusion in the right inferior parietal lobule (IPL) and right precuneus, but exhibited decreased perfusion in the right putamen and right supplementary motor area (SMA) when compared with the AD group. Significant correlations were seen in the AD group between normalized CBF values in the right precuneus and memory function scores, as well as between normalized CBF values in the right IPL and general cognitive function ratings. The degree of motor symptoms in the PDD group was significantly correlated with normalized CBF in the right putamen (Chen et al., 2021).

4.1.2 Key pathways: Mitochondrial complex I (PD) vs. insulin signaling (AD)

There are various mechanisms for the pathophysiology of both AD and PD. Neuronal metabolism and homeostasis are disrupted in PD and AD disease (Al-Kuraishy et al., 2024; Alshahrani et al., 2025). Evidence indicates that PD can be predominantly caused by mitochondrial dysfunction, whereas AD is more strongly associated with alterations of insulin signaling pathways.

Early PD is characterized by two key preclinical alterations: a shift in brain energy substrate utilization/production and mitochondrial dysfunction, both occurring before α-syn inclusions and neuronal loss, supporting a central role in disease initiation. Mitochondrial dysfunction leads to impairments in OXPHOS that diminish ATP synthesis and increase ROS, resulting in a bioenergetic crisis that increases neuronal susceptibility to degeneration by weakening neuronal capacity to maintain ion gradients, synaptic transmission, and stress-response programs. Although reduced ATP is a central outcome of mitochondrial failure, the transition to apoptosis reflects a broader cascade, where oxidative injury, disrupted redox signaling, and metal ion imbalance act alongside bioenergetic decline to trigger cell death (Toomey et al., 2022; Du et al., 2025). Dopaminergic neurons exhibit heightened susceptibility to mitochondrial dysfunction, attributable to their polarized and enlarged morphology and their high energetic demands, and mutations in mitochondrial regulatory proteins further exacerbate bioenergetic stress, initiating molecular cascades that promote neuronal degeneration and Parkinson's disease progression (Blagov et al.). As PD progresses, mitochondrial impairment broadens, with deficits in mitochondrial metabolism, fatty-acid oxidation, and respiration across brain regions (Toomey et al., 2022).

PD-linked genetic mutations converge on mitochondrial dysfunction as an early and central pathogenic mechanism, disrupting bioenergetic homeostasis across both autosomal dominant and recessive forms of the disease (Malpartida et al., 2021; Lill, 2016; Devi et al., 2008; Papkovskaia et al., 2012). Mutations affecting α-synuclein, DJ-1, PINK1, and other mitochondrial regulatory proteins amplify oxidative stress, promoting ROS-mediated damage to mtDNA and mitochondrial proteins and leading to impaired respiration, reduced ATP production, and membrane depolarization (Simon et al., 2020; Stojkovska et al., 2018; Tansey et al., 2022; Celardo et al., 2014). Moreover, neuroinflammatory mechanisms—via activation of the NLRP3 inflammasome and increased IL-1β, IL-18, and gasdermin D—further reinforce oxidative stress and trigger stress-responsive cell-death programs, including ferroptosis, parthanatos, and pyroptosis, the latter being linked to PD progression (Naren et al., 2023; Colca & Finck, 2022).

It has been demonstrated that IR develops several years before clinical manifestation of AD and involved in AD pathogenesis through several mechanisms: defective insulin-receptor signaling, IRS-1 inhibition, and downstream metabolic collapse. There is a vicious cycle: defective insulin signaling promotes Aβ accumulation, while Aβ exacerbates IR (Yuan et al., 2025; Nguyen et al., 2020) As accumulation of AβPP lead to synaptic toxicity and neuronal stress. These changes lead to memory decline and executive function, hippocampal atrophy, and cognitive decline (Lim et al., 2016). Insulin signaling modulates neuronal GABA, NMDA, and AMPA receptors, catecholamine release and reuptake, and GLUT3-mediated uptake. In astrocytes, insulin enhances GLUT1-mediated uptake, promotes glycogen storage, and regulates inflammatory responses (Falkowska et al., 2015; McKenna et al., 2019). There are evidence that blocking insulin signaling specifically in neurons is adequate to elicit Alzheimer's symptoms in mice (Soto et al., 2019). Moreover, Tau expression is regulated by insulin/IGF-I signaling and AMPK activity which are disrupted in early stages of AD, linking tangle formation to cellular energy impairment (Chen & Zhong, 2013; Pedros et al., 2016; Joo et al., 2022). At the molecular level, dysregulation of AMP-activated protein kinase (AMPK) provides a mechanistic link between energy stress and neurodegeneration by impairing mitochondrial biogenesis via PGC-1α signaling.

4.1.3 Early presentation: Motor/executive (PD) vs. memory (AD)

PD is associated with motor and non-motor signs and symptoms. Main diagnosis signs are bradykinesia, rigidity, and tremor. Other hand non-motor presentations are autonomic dysfunction like orthostatic hypotension. Other presentations are constipation, urinary and erectile dysfunction, and excess salivation (Shin et al., 2022). Sleep and circadian system disturbances are new finding in Parkinson's patient. They have change in rapid eye movement (REM) cycle thereupon REM behavior disorder and constantly sleepy exhausted, metabolic abnormality and nightmare dreams (Munhoz et al., 2024). Patients frequently exhibit unilateral resting tremor, typically affecting distal hand muscles, with a characteristic pill-rolling motion. Rigidity often begins unilaterally and, in combination with tremor, impairs gait and posture, leading to musculoskeletal pain. On examination, cogwheel rigidity and increased resistance to passive movement are observed, with progressive gait impairment.

One of the most frequent early cognitive presentations in PD is executive dysfunction and CI, involving motor, sensory, and balance systems, thereby increasing fall risk. Even before formal PD diagnosis, patients often show slowed gait, shortened step length, visuospatial impairment, and altered executive function with attentional disruption (Pirtošek & Trošt, 2023).

The initial and most common presentation of AD is memory impairment, beginning with episodic short-term memory loss and progressing to deficits in problem-solving and judgment. Patients struggle with multitasking and abstract thinking and show impairments in instrumental activities such as driving and cooking, reflecting executive dysfunction. AD patients often exhibit disproportionate impairment of recent memory compared with remote memory. Impairments in memory, language, executive function, and visuospatial abilities constitute the core cognitive domains affected in AD. Memory deficits are particularly evident in anterograde episodic memory, with impairments in both visual and verbal episodic memory tasks, alongside executive, visuospatial, and linguistic dysfunction (Porsteinsson et al., 2021).

4.1.4 Dementia profiles: Fluctuating attention in PD dementia vs. amnestic AD dementia (ADD)

Dementia in PD is associated with deficits in attention, language abnormalities, verbal fluency impairment, naming difficulties, and reduced comprehension. Patients frequently exhibit slowed attentional processing and marked fluctuations in attention (Walker et al., 2019). In contrast, Alzheimer's disease initially presents with episodic memory loss due to hippocampal damage. As the disease progresses, patients develop anomia and disorientation. Alzheimer's disease is also associated with misbinding errors involving features such as color, orientation, shape, and spatial location (Schumacher et al., 2019; Zvěřová, 2019).

4.2 Overlapping energetic pathologies

4.2.1 Shared mitochondrial dysfunction

Mitochondria are essential organelles responsible for ATP production via OXPHOS and for maintaining calcium homeostasis critical for synaptic signaling. Evidence indicates that mitochondrial dysfunction contributes to the pathogenesis of both AD and PD. In AD, mitochondrial failure promotes extracellular deposition of Aβ peptides from AβPP cleavage and intracellular accumulation of NFTs composed of hyperphosphorylated tau. Aβ directly interacts with mitochondria, inhibiting complex IV and α-ketoglutarate dehydrogenase, and binds to **amyloid-β–binding alcohol dehydrogenase (ABAD)** in the mitochondrial matrix, thereby promoting free radical generation and apoptosis. In PD, mitochondrial integrity in dopaminergic neurons of the substantia nigra critically depends on the PINK1–Parkin pathway. Upon mitochondrial depolarization,

PINK1 accumulates on the outer mitochondrial membrane, leading to Parkin-mediated ubiquitination of outer membrane proteins and targeting of damaged mitochondria for mitophagy. Mutations in PINK1 or Parkin (PARK2) disrupt this surveillance mechanism, resulting in accumulation of dysfunctional mitochondria and neuronal degeneration (Klemmensen et al., 2024).

4.2.2 Neuroinflammation-induced metabolic suppression

AD and PD are progressive neuronal damage and CI caused by a multifactorial etiology involving chronic neuroinflammation, OS, mitochondrial dysfunction, and metabolic dysregulation. Studies observe the link between these pathological mechanisms of neurodegeneration and neuroinflammation through metabolic impairments in the brain. Microglia and astrocytes are important cells in modulating synaptic function and responding to injury or stress in the CNS. However, in chronic pathological conditions, such as the accumulation of proteins (e.g., Aβ in AD or α-syn in PD), these cells are activated persistently. Microglia release pro-inflammatory mediators, including cytokines (e.g., TNF-α, IL-1β), prostaglandins, nitric oxide (NO), and ROS, and astrocytes, following these inflammatory signals, adopt a reactive phenotype that leads to neuroinflammation and contributes to neuronal damage. **Intracellular signaling pathways—**nuclear factor kappa-light-chain-enhancer of activated B cells (NF-κB) and the Janus kinase/signal transducer and activator of transcription (JAK/STAT) pathway—have roles in mediating the inflammatory response in glial cells. The NF-κB pathway, often activated in neurodegenerative diseases, acts as a transcriptional regulator in some of the pro-inflammatory genes. Mitochondria and NADPH oxidase release ROS that are potent activators of NF-κB, leading to an increase in inflammatory enzymes such as cyclooxygenase-2 (COX-2), nitric oxide synthase (iNOS), and matrix metalloproteinases (e.g., MMP-9). This not only promotes inflammation but also increases oxidative damage. Cytokines similarly activate the JAK/STAT pathway in astrocytes and microglia, contributing to the expression of pro-inflammatory genes. Persistent activation of this signaling disrupts homeostatic mechanisms and leads to the inflammatory environment within the CNS. Studies suggest an association between systemic metabolic disorders (obesity, type 2 diabetes mellitus, and insulin resistance) and neurodegeneration. They increase the risk of cognitive decline and neurodegeneration. Chronic hyperglycemia has a role in the formation of advanced glycation end-products (AGEs), which activate inflammatory pathways in glial cells. Additionally, insulin resistance disturbs cerebral glucose metabolism, a vital energy source for neurons that increases ROS production. Obesity is associated with systemic inflammation and mediated by the release of pro-inflammatory cytokines, including TNF-α and IL-6 from adipose tissue. They can cross the blood–brain barrier and activate microglia in neuroinflammation (Zhang et al., 2023).

4.2.3 Role of APOE ε4 in both disorders

Apolipoprotein E (APOE) is a polymorphic gene in humans with three major isoforms—APOE2, APOE3, and APOE4—encoded by the E2, E3, and E4 alleles that have six genotypes (APOE2/2, APOE3/3, APOE4/4, APOE2/3, APOE2/4, and APOE3/4). Astrocytes, microglia, oligodendrocytes, pericytes, the choroid plexus, and, under certain conditions, neurons express the APOE protein that supports synaptic plasticity, neural signaling, lipid transport, immune regulation, and neuronal repair. APOE4 is an isoform of APOE, associated with CI and neurodegenerative disease, especially AD, PD, and dementia with Lewy bodies (DLB). APOE4 increases tau-related pathology, impairs cholesterol and lipid transport, disrupts synaptic function, and has a role in neuroinflammation, as well as mitochondrial dysfunction, reduced vascular integrity, impaired glucose metabolism, and autophagy deficits that impact cellular homeostasis. APOE4 may increase the accumulation of α-syn in Lewy bodies. Preclinical models have shown that APOE4 expression increases α-syn burden, accelerates memory impairment, and promotes lipid droplet accumulation compared with APOE2 or APOE3, which is linked with an increased risk of cognitive decline, including PD-MCI, PDD, and DLB. Overall, APOE4 represents a major genetic risk factor for AD and an emerging modifier of CI in PD and DLB (Emrani et al., 2020).

5. Discussion

In this chapter, we have examined the relationship between changes in brain energy metabolism and neuropsychological disorders in AD and PD. Various aspects of this relationship and their causes have been examined in detail. Therefore, we can now pay attention to the therapeutic and diagnostic opportunities that this relationship provides and determine our future path for further research.

Although these two diseases are separate from each other and have different causes, symptoms, and treatments, both are considered part of neurodegenerative diseases (Breijyeh & Karaman, 2020; Kashif et al., 2023; Marogianni et al., 2020), and this also means that despite the differences, common clinical manifestations can be seen in these two, the most prominent of which is neuropsychological disorders.

AD is known as a hallmark disease in the category of dementia diseases (Bondi et al., 2017) with memory disorders (Vyhnálek et al., 2019). Of course, the spectrum of neuropsychological disorders in Alzheimer's disease is much more than memory disorders. These disorders include a wide range. From attention disorders to problems in visual perception and visuomotor coordination (Giovagnoli et al., 2008). These patients also present with symptoms such as restlessness, anxiety, depression, hallucinations, and delusions (Kim et al., 2021). Neuropsychological disorders are also common in PD, and the symptoms of this disease are not limited to motor disorders. Symptoms such as cognitive rigidity, mild cognitive impairment, and memory disorders fall into this category (Kehagia et al., 2010). Anxiety and depression, as well as apathy, are other symptoms of this disease (Mueller et al., 2018). These symptoms can be extremely debilitating and cause many problems for both patients and their families. Therefore, their comprehensive understanding is very important and can provide researchers with new therapeutic options.

In this chapter, we have discussed the relationship of these symptoms to changes in energy metabolism in the brain. We now understand that energy metabolism plays a significant role in neurodegeneration. Both the generation and use of energy are necessary for brain functions. This energy is crucial for maintaining neuronal activity, appropriate synaptic function, and the regular activity of cells (Camandola & Mattson, 2017).

It is clear that energy metabolism can affect all aspects of brain function, including cognitive and psychological functions. Energy metabolism is a fundamental driver of brain function and plays a fundamental role everywhere. This can also provide diagnostic and therapeutic opportunities in this group of patients. The brain requires continuous use of energy from ATP to carry out its activities and supplies this energy mainly through glucose. In neurodegenerative diseases, this cycle is disrupted and progressively deteriorates. Investigating the factors involved in this process, such as mitochondrial function and restoration of OXPHOS and glycolysis, is considered a therapeutic opportunity (Cunnane et al., 2020).

This research has now found broad aspects. We now know that insulin resistance is directly related to mitochondrial dysfunction in AD patients. Therefore, increasing insulin sensitivity can also be considered a potential therapeutic method (Cunnane et al., 2020). On the other hand, one of the methods that can replace sugar for energy production in the brain is the use of ketones. Therefore, therapeutic methods based on ketone production can be considered one of the ways to improve the conditions of neurodegenerative patients (Norwitz et al., 2020). These strategies are all based on our understanding of the role of energy in neurodegenerative disorders. Since improving energy metabolism can improve the conditions of neuronal, synaptic and cellular functioning in the brain, it affects all aspects of its function, and of course, one of the most important of them is neuropsychological disorders. What has been said is based on general methods for improving energy metabolism in the brain, which is of course valuable. But perhaps the next important step is related to more specific pathways that specifically address the role of energy metabolism in neuropsychological problems in these diseases. This is a topic that goes beyond the general issues discussed in this chapter. To achieve this, research must specifically address this issue. In addition, new questions should be explored: Can common methods for treating neuropsychological disorders, such as pharmacological treatments and psychological and cognitive therapies, also affect energy metabolism in the brain of these patients? To answer these questions, imaging methods should be used to examine these changes in energy metabolism. If studies show these changes, then improving these specific methods could be considered an important goal in treating neuropsychological disorders in these patients.

References

Aarsland, D., Batzu, L., Halliday, G. M., Geurtsen, G. J., Ballard, C., Ray Chaudhuri, K., et al. (2021). Parkinson disease-associated cognitive impairment. *Nature reviews Disease primers*, *7*(1), 47.

Aarsland, D., Marsh, L., & Schrag, A. (2009). Neuropsychiatric symptoms in Parkinson's disease. *Movement Disorders: Official Journal of the Movement Disorder Society*, *24*(15), 2175–2186.

Albaik, M., Sheikh Saleh, D., Kauther, D., Mohammed, H., Alfarra, S., Alghamdi, A., et al. (2024). Bridging the gap: Glucose transporters, Alzheimer's, and future therapeutic prospects. *Frontiers in Cell and Developmental Biology*, *12*, 1344039.

Aldana, B. I. (2019). Microglia-specific metabolic changes in neurodegeneration. *Journal of Molecular Biology*, *431*(9), 1830–1842.

Al-Kuraishy, H. M., Fahad, E. H., Al-Windy, S., El-Sherbeni, S. A., Negm, W. A., & Batiha, G. E.-S. (2024). The effects of cholesterol and statins on Parkinson's neuropathology: A narrative review. *Inflammopharmacology*, *32*(2), 917–925.

Alshahrani, S. M., Al-Kuraishy, H. M., Al-Gareeb, A. I., Albuhadily, A. K., Shokr, M. M., Kadasah, S. F., et al. (2025). LRP1 at the crossroads of Parkinson's and Alzheimer's: Divergent roles in α-synuclein and amyloid pathology. *European Journal of Pharmacology*, 177830.

An, Y., Varma, V. R., Varma, S., Casanova, R., Dammer, E., Pletnikova, O., et al. (2018). Evidence for brain glucose dysregulation in Alzheimer's disease. *Alzheimer's & Dementia: The Journal of the Alzheimers Association, 14*(3), 318–329.

Andersen, J. V., Skotte, N. H., Christensen, S. K., Polli, F. S., Shabani, M., Markussen, K. H., et al. (2021). Hippocampal disruptions of synaptic and astrocyte metabolism are primary events of early amyloid pathology in the 5xFAD mouse model of Alzheimer's disease. *Cell Death & Disease, 12*(11), 954.

Appell, J., Kertesz, A., & Fisman, M. (1982). A study of language functioning in Alzheimer patients. *Brain and Language, 17*(1), 73–91.

Bano, D., Ehninger, D., & Bagetta, G. (2023). Decoding metabolic signatures in Alzheimer's disease: A mitochondrial perspective. *Cell Death Discovery, 9*(1), 432.

Batra, R., Krumsiek, J., Wang, X., Allen, M., Blach, C., Kastenmüller, G., et al. (2024). Comparative brain metabolomics reveals shared and distinct metabolic alterations in Alzheimer's disease and progressive supranuclear palsy. *Alzheimer's & Dementia, 20*(12), 8294–8307.

Bayles, K. A., Tomoeda, C. K., & Trosset, M. W. (1992). Relation of linguistic communication abilities of Alzheimer's patients to stage of disease. *Brain and Language, 42*(4), 454–472.

Beard, E., Lengacher, S., Dias, S., Magistretti, P. J., & Finsterwald, C. (2022). Astrocytes as key regulators of brain energy metabolism: New therapeutic perspectives. *Frontiers in Physiology, 12*, 825816.

Bhatia, V., & Sharma, S. (2021). Role of mitochondrial dysfunction, oxidative stress and autophagy in progression of Alzheimer's disease. *Journal of the Neurological Sciences, 421*, 117253.

Boland, B., Yu, W. H., Corti, O., Mollereau, B., Henriques, A., Bezard, E., et al. (2018). Promoting the clearance of neurotoxic proteins in neurodegenerative disorders of ageing. *Nature Reviews. Drug Discovery, 17*(9), 660–688.

Bondi, M. W., Edmonds, E. C., & Salmon, D. P. (2017). Alzheimer's disease: Past, present, and future. *Journal of the International Neuropsychological Society, 23*(9-10), 818–831.

Bonomi, C. G., De Lucia, V., Mascolo, A. P., Assogna, M., Motta, C., Scaricamazza, E., et al. (2021). Brain energy metabolism and neurodegeneration: Hints from CSF lactate levels in dementias. *Neurobiology of Aging, 105*, 333–339.

Borsche, M., Pereira, S. L., Klein, C., & Grünewald, A. (2021). Mitochondria and Parkinson's disease: Clinical, molecular, and translational aspects. *Journal of Parkinson's Disease, 11*(1), 45–60.

Brandão, P. R. P., Munhoz, R. P., Grippe, T. C., Cardoso, F. E. C., e Castro, B. M. D. A., Titze-de-Almeida, R., et al. (2020). Cognitive impairment in Parkinson's disease: A clinical and pathophysiological overview. *Journal of the Neurological Sciences, 419*, 117177.

Breijyeh, Z., & Karaman, R. (2020). Comprehensive review on Alzheimer's disease: Causes and treatment. *Molecules (Basel, Switzerland), 25*(24), 5789.

Brooks, G. A. (2009). Cell–cell and intracellular lactate shuttles. *The Journal of Physiology, 587*(23), 5591–5600.

Cahn-Weiner, D. A., Boyle, P. A., & Malloy, P. F. (2002). Tests of executive function predict instrumental activities of daily living in community-dwelling older individuals. *Applied Neuropsychology, 9*(3), 187–191.

Cai, Q., & Tammineni, P. (2017). Mitochondrial aspects of synaptic dysfunction in Alzheimer's disease. *Journal of Alzheimer's disease, 57*(4), 1087–1103.

Calvo-Rodriguez, M., Kharitonova, E. K., Snyder, A. C., Hou, S. S., Sanchez-Mico, M. V., Das, S., et al. (2024). Real-time imaging of mitochondrial redox reveals increased mitochondrial oxidative stress associated with amyloid β aggregates in vivo in a mouse model of Alzheimer's disease. *Molecular Neurodegeneration, 19*(1), 6.

Camandola, S., & Mattson, M. P. (2017). Brain metabolism in health, aging, and neurodegeneration. *The EMBO Journal, 36*(11), 1474–1492.

Celardo, I., Martins, L. M., & Gandhi, S. (2014). Unravelling mitochondrial pathways to Parkinson's disease. *British Journal of Pharmacology, 171*(8), 1943–1957.

Chaudhary, S., Kumaran, S. S., Goyal, V., Kalaivani, M., Kaloiya, G. S., Sagar, R., et al. (2021). Frontal lobe metabolic alterations characterizing Parkinson's disease cognitive impairment. *Neurological Sciences, 42*(3), 1053–1064.

Chen, A. P., Chen, L., Kim, T. A., & Xiong, Q. (2021). Integrating the roles of midbrain dopamine circuits in behavior and neuropsychiatric disease. *Biomedicines, 9*(6), 647.

Chen, H., Xu, Y., Chen, L., Shang, S., Luo, X., Wang, X., et al. (2022). The convergent and divergent patterns in brain perfusion between Alzheimer's disease and Parkinson's disease with dementia: An ASL MRI study. *Frontiers in neuroscience, 16*, 892374.

Chen, L., Wang, C., Qin, L., & Zhang, H. (2025). Parkinson's disease and glucose metabolism impairment. *Translational Neurodegeneration, 14*(1), 10.

Chen, Z., & Zhong, C. (2013). Decoding Alzheimer's disease from perturbed cerebral glucose metabolism: implications for diagnostic and therapeutic strategies. *Progress in Neurobiology, 108*, 21–43.

Chen, Z., & Zhong, C. (2014). Oxidative stress in Alzheimer's disease. *Neuroscience Bulletin, 30*(2), 271–281.

Cheng, Y., & Bai, F. (2018). The association of tau with mitochondrial dysfunction in Alzheimer's disease. *Frontiers in neuroscience, 12*, 163.

Cioffi, F., Adam, R. H. I., & Broersen, K. (2019). Molecular mechanisms and genetics of oxidative stress in Alzheimer's disease. *Journal of Alzheimer's Disease, 72*(4), 981–1017.

Colca, J. R., & Finck, B. N. (2022). Metabolic Mechanisms connecting Alzheimer's and Parkinson's diseases: Potential avenues for novel therapeutic approaches. *Frontiers in Molecular Biosciences, 9*, 929328.

Croot, K., Hodges, J. R., Xuereb, J., & Patterson, K. (2000). Phonological and articulatory impairment in Alzheimer's disease: a case series. *Brain and Language, 75*(2), 277–309.

Croteau, E., Castellano, C. A., Fortier, M., Bocti, C., Fulop, T., Paquet, N., et al. (2018). A cross-sectional comparison of brain glucose and ketone metabolism in cognitively healthy older adults, mild cognitive impairment and early Alzheimer's disease. *Experimental Gerontology, 107*, 18–26.

Cunnane, S. C., Trushina, E., Morland, C., Prigione, A., Casadesus, G., Andrews, Z. B., et al. (2020). Brain energy rescue: an emerging therapeutic concept for neurodegenerative disorders of ageing. *Nature Reviews. Drug Discovery, 19*(9), 609–633.

Dai, C., Tan, C., Zhao, L., Liang, Y., Liu, G., Liu, H., et al. (2023). Glucose metabolism impairment in Parkinson's disease. *Brain Research Bulletin, 199*, 110672.

Degirmenci, Y., Angelopoulou, E., Georgakopoulou, V. E., & Bougea, A. (2023). Cognitive impairment in Parkinson's disease: an updated overview focusing on emerging pharmaceutical treatment approaches. *Medicina, 59*(10), 1756.

Devi, L., Raghavendran, V., Prabhu, B. M., Avadhani, N. G., & Anandatheerthavarada, H. K. (2008). Mitochondrial import and accumulation of α-synuclein impair complex I in human dopaminergic neuronal cultures and Parkinson disease brain. *Journal of Biological Chemistry, 283*(14), 9089–9100.

Dickerson, B. C., & Eichenbaum, H. (2010). The episodic memory system: neurocircuitry and disorders. *Neuropsychopharmacology: Official Publication of the American College of Neuropsychopharmacology, 35*(1), 86–104.

Dirnberger, G., & Jahanshahi, M. (2013). Executive dysfunction in P arkinson's disease: A review. *Journal of neuropsychology, 7*(2), 193–224.

Dorsey, E. R., De Miranda, B. R., Horsager, J., & Borghammer, P. (2024). The body, the brain, the environment, and Parkinson's disease. *Journal of Parkinson's Disease, 14*(3), 363–381.

Du, Y., Fan, S.-S., Wu, H., He, J., He, Y., Meng, X.-Y., et al. (2025). Convergent and Divergent Mitochondrial Pathways as Causal Drivers and Therapeutic Targets in Neurological Disorders. *Current Issues in Molecular Biology, 47*(8), 636.

Dwyer, D. S., Vannucci, S. J., & Simpson, I. A. (2002). Expression, regulation, and functional role of glucose transporters (GLUTs) in brain. *International Review of Neurobiology, 51*, 159–188.

Eggers, C., Schwartz, F., Pedrosa, D. J., Kracht, L., & Timmermann, L. (2014). Parkinson's disease subtypes show a specific link between dopaminergic and glucose metabolism in the striatum. *PLoS One, 9*(5), e96629.

Emrani, S., Arain, H. A., DeMarshall, C., & Nuriel, T. (2020). APOE4 is associated with cognitive and pathological heterogeneity in patients with Alzheimer's disease: a systematic review. *Alzheimer's Research & Therapy, 12*(1), 141.

Erkkinen, M. G., Kim, M. O., & Geschwind, M. D. (2018). Clinical Neurology and Epidemiology of the Major Neurodegenerative Diseases. *Cold Spring Harb Perspect Biol, 10*(4).

Falkowska, A., Gutowska, I., Goschorska, M., Nowacki, P., Chlubek, D., & Baranowska-Bosiacka, I. (2015). Energy metabolism of the brain, including the cooperation between astrocytes and neurons, especially in the context of glycogen metabolism. *International journal of molecular sciences, 16*(11), 25959–25981.

Ferrer, I. (2009). Early involvement of the cerebral cortex in Parkinson's disease: convergence of multiple metabolic defects. *Progress in Neurobiology, 88*(2), 89–103.

Fornari, L. H. T., da Silva Júnior, N., Carpenedo, C. M., Hilbig, A., & de Mello Rieder, C. R. (2021). Striatal dopamine correlates to memory and attention in Parkinson's disease. *American Journal of Nuclear Medicine and Molecular Imaging, 11*(1), 10.

Funahashi, S., & Andreau, J. M. (2013). Prefrontal cortex and neural mechanisms of executive function. *Journal of Physiology-Paris, 107*(6), 471–482.

Gallagher, M., & Koh, M. T. (2011). Episodic memory on the path to Alzheimer's disease. *Current Opinion in Neurobiology, 21*(6), 929–934.

Gao, X.-Y., Yang, T., Gu, Y., & Sun, X.-H. (2022). Mitochondrial dysfunction in Parkinson's disease: from mechanistic insights to therapy. *Frontiers in aging neuroscience, 14*, 885500.

Ghazi-Saidi, L. (2020). Visuospatial and executive deficits in Parkinson's disease: A review. *Acta Scientific Neurology (ISSN: 2582-1121), 3*(4).

Giovagnoli, A. R., Erbetta, A., Reati, F., & Bugiani, O. (2008). Differential neuropsychological patterns of frontal variant frontotemporal dementia and Alzheimer's disease in a study of diagnostic concordance. *Neuropsychologia, 46*(5), 1495–1504.

Gold, C. A., & Budson, A. E. (2008). Memory loss in Alzheimer's disease: implications for development of therapeutics. *Expert Review of Neurotherapeutics, 8*(12), 1879–1891.

González-Usigli, H. A., Ortiz, G. G., Charles-Niño, C., Mireles-Ramírez, M. A., Pacheco-Moisés, F. P., Torres-Mendoza, B. Md. G., et al. (2023). Neurocognitive psychiatric and neuropsychological alterations in Parkinson's disease: a basic and clinical approach. *Brain Sciences, 13*(3), 508.

Goyal, M. S., Blazey, T., Metcalf, N. V., McAvoy, M. P., Strain, J. F., Rahmani, M., et al. (2023). Brain aerobic glycolysis and resilience in Alzheimer disease. *Proceedings of the National Academy of Sciences of the United States of America, 120*(7), e2212256120.

Gratwicke, J., Jahanshahi, M., & Foltynie, T. (2015). Parkinson's disease dementia: a neural networks perspective. *Brain, 138*(6), 1454–1476.

Guarino, A., Favieri, F., Boncompagni, I., Agostini, F., Cantone, M., & Casagrande, M. (2019). Executive functions in Alzheimer disease: A systematic review. *Frontiers in aging neuroscience, 10*, 437.

Guo, L., Tian, J., & Du, H. (2017). Mitochondrial dysfunction and synaptic transmission failure in Alzheimer's disease. *Journal of Alzheimer's Disease, 57*(4), 1071–1086.

Halliday, G. M., Leverenz, J. B., Schneider, J. S., & Adler, C. H. (2014). The neurobiological basis of cognitive impairment in Parkinson's disease. *Movement disorders, 29*(5), 634–650.

Hampel, H., Mesulam, M.-M., Cuello, A. C., Farlow, M. R., Giacobini, E., Grossberg, G. T., et al. (2018). The cholinergic system in the pathophysiology and treatment of Alzheimer's disease. *Brain, 141*(7), 1917–1933.

Henrich, M. T., Oertel, W. H., Surmeier, D. J., & Geibl, F. F. (2023). Mitochondrial dysfunction in Parkinson's disease–a key disease hallmark with therapeutic potential. *Molecular Neurodegeneration, 18*(1), 83.

Hodges, J. R. (2000). Memory in the dementias. *The Oxford handbook of memory*, 441–459.

Hoffman, P. (2018). An individual differences approach to semantic cognition: Divergent effects of age on representation, retrieval and selection. *Scientific reports, 8*(1), 8145.

Huang, M., Yu, H., Lyu, X., Pu, W., Yin, J., & Gao, B. (2024). Region-specific Cerebral Metabolic Alterations in Parkinson's Disease Patients With/without Mild Cognitive Impairment. *Neuroscience, 551*, 254–261.

Huang, Y., Xu, J., Zhang, X., Liu, Y., & Yu, E. (2023). Research progress on vestibular dysfunction and visual–spatial cognition in patients with Alzheimer's disease. *Frontiers in Aging Neuroscience, 15*, 1153918.

Huang, Y., Xu, W., & Zhou, R. (2021). NLRP3 inflammasome activation and cell death. *Cellular & Molecular Immunology, 18*(9), 2114–2127.

Iadecola, C. (2015). Sugar and Alzheimer's disease: a bittersweet truth. *Nature Neuroscience, 18*(4), 477–478.

Jensen, N. J., Wodschow, H. Z., Nilsson, M., & Rungby, J. (2020). Effects of ketone bodies on brain metabolism and function in neurodegenerative diseases. *International journal of molecular sciences, 21*(22), 8767.

Joo, I. L., Lam, W. W., Oakden, W., Hill, M. E., Koletar, M. M., Morrone, C. D., et al. (2022). Early alterations in brain glucose metabolism and vascular function in a transgenic rat model of Alzheimer's disease. *Progress in Neurobiology, 217*, 102327.

Joyal, M., Groleau, C., Bouchard, C., Wilson, M. A., & Fecteau, S. (2020). Semantic processing in healthy aging and Alzheimer's disease: a systematic review of the N400 differences. *Brain sciences, 10*(11), 770.

Kashif, M., Sivaprakasam, P., Vijendra, P., Waseem, M., & Pandurangan, A. K. (2023). A recent update on pathophysiology and therapeutic interventions of Alzheimer's disease. *Current Pharmaceutical Design, 29*(43), 3428–3441.

Kawashima, S., Shimizu, Y., Ueki, Y., & Matsukawa, N. (2021). Impairment of the visuospatial working memory in the patients with Parkinson's Disease: an fMRI study. *BMC Neurology, 21*(1), 335.

Kehagia, A. A., Barker, R. A., & Robbins, T. W. (2013). Cognitive impairment in Parkinson's disease: the dual syndrome hypothesis. *Neurodegenerative diseases, 11*(2), 79–92.

Kehagia, A. A., Barker, R. A., & Robbins, T. W. (2010). Neuropsychological and clinical heterogeneity of cognitive impairment and dementia in patients with Parkinson's disease. *The Lancet Neurology, 9*(12), 1200–1213.

Kim, B., Noh, G. O., & Kim, K. (2021). Behavioural and psychological symptoms of dementia in patients with Alzheimer's disease and family caregiver burden: a path analysis. *BMC Geriatrics, 21*(1), 160.

Klemmensen, M. M., Borrowman, S. H., Pearce, C., Pyles, B., & Chandra, B. (2024). Mitochondrial dysfunction in neurodegenerative disorders. *Neurotherapeutics: the Journal of the American Society for Experimental NeuroTherapeutics, 21*(1), e00292.

Kyrtata, N., Emsley, H. C., Sparasci, O., Parkes, L. M., & Dickie, B. R. (2021). A systematic review of glucose transport alterations in Alzheimer's disease. *Frontiers in neuroscience, 15*, 626636.

Lambon Ralph, M. A., Pobric, G., & Jefferies, E. (2009). Conceptual knowledge is underpinned by the temporal pole bilaterally: convergent evidence from rTMS. *Cerebral Cortex, 19*(4), 832–838.

Lanctôt, K. L., Amatniek, J., Ancoli-Israel, S., Arnold, S. E., Ballard, C., Cohen-Mansfield, J., et al. (2017). Neuropsychiatric signs and symptoms of Alzheimer's disease: New treatment paradigms. *Alzheimer's & Dementia: Translational Research & Clinical Interventions, 3*(3), 440–449.

Le Douce, J., Maugard, M., Veran, J., Matos, M., Jégo, P., Vigneron, P.-A., et al. (2020). Impairment of glycolysis-derived l-serine production in astrocytes contributes to cognitive deficits in Alzheimer's disease. *Cell Metabolism, 31*(3), 503–517.e8.

Li, J.-L., Lin, T.-Y., Chen, P.-L., Guo, T.-N., Huang, S.-Y., Chen, C.-H., et al. (2021). Mitochondrial function and Parkinson's disease: from the perspective of the electron transport chain. *Frontiers in molecular neuroscience, 14*, 797833.

Lill, C. M. (2016). Genetics of Parkinson's disease. *Molecular and Cellular Probes, 30*(6), 386–396.

Lim, Y., Cho, H., & Kim, E. K. (2016). Brain metabolism as a modulator of autophagy in neurodegeneration. *Brain Research, 1649*(Pt B), 158–165.

Liu, Y., Liu, F., Iqbal, K., Grundke-Iqbal, I., & Gong, C.-X. (2008). Decreased glucose transporters correlate to abnormal hyperphosphorylation of tau in Alzheimer disease. *FEBS Letters, 582*(2), 359–364.

Malpartida, A. B., Williamson, M., Narendra, D. P., Wade-Martins, R., & Ryan, B. J. (2021). Mitochondrial dysfunction and mitophagy in Parkinson's disease: from mechanism to therapy. *Trends in Biochemical Sciences, 46*(4), 329–343.

Mandal, P. K., Joshi, J., & Saharan, S. (2012). Visuospatial perception: an emerging biomarker for Alzheimer's disease. *Journal of Alzheimer's Disease, 31*(s3), S117–S135.

Marogianni, C., Sokratous, M., Dardiotis, E., Hadjigeorgiou, G. M., Bogdanos, D., & Xiromerisiou, G. (2020). Neurodegeneration and inflammation-an interesting interplay in Parkinson's disease. *International journal of molecular sciences, 21*(22), 8421.

Marsh, L. (2000). Neuropsychiatric aspects of Parkinson's disease. *Psychosomatics, 41*(1), 15–23.

McGregor, M. M., & Nelson, A. B. (2019). Circuit mechanisms of Parkinson's disease. *Neuron, 101*(6), 1042–1056.

McKenna, M. C., Schuck, P. F., & Ferreira, G. C. (2019). Fundamentals of CNS energy metabolism and alterations in lysosomal storage diseases. *Journal of Neurochemistry, 148*(5), 590–599.

McKhann, G. M., Knopman, D. S., Chertkow, H., Hyman, B. T., Jack Jr, C. R., Kawas, C. H., et al. (2011). The diagnosis of dementia due to Alzheimer's disease: recommendations from the National Institute on Aging-Alzheimer's Association workgroups on diagnostic guidelines for Alzheimer's disease. *Alzheimer's & dementia,* 7(3), 263–269.

Mesulam, M. M. (2013). Cholinergic circuitry of the human nucleus basalis and its fate in Alzheimer's disease. *Journal of Comparative Neurology, 521*(18), 4124–4144.

Moon, H. E., & Paek, S. H. (2015). Mitochondrial dysfunction in Parkinson's disease. *Experimental neurobiology, 24*(2), 103.

Moradi Vastegani, S., Nasrolahi, A., Ghaderi, S., Belali, R., Rashno, M., Farzaneh, M., et al. (2023). Mitochondrial dysfunction and Parkinson's disease: pathogenesis and therapeutic strategies. *Neurochemical Research, 48*(8), 2285–2308.

Muddapu, V., Dharshini, S., Chakravarthy, V., & Gromiha, M. (2020). Neurodegenerative diseases—is metabolic deficiency the root cause? *Front Neurosci, 14*, 213.
Mueller, C., Rajkumar, A. P., Wan, Y. M., Velayudhan, L., Ffytche, D., Chaudhuri, K. R., et al. (2018). Assessment and management of neuropsychiatric symptoms in Parkinson's disease. *CNS Drugs, 32*(7), 621–635.
Munhoz, R. P., Tumas, V., Pedroso, J. L., & Silveira-Moriyama, L. (2024). The clinical diagnosis of Parkinson's disease. *Arquivos de Neuro-psiquiatria, 82*(06), 001–010.
Naren, P., Cholkar, A., Kamble, S., Khan, S. S., Srivastava, S., Madan, J., et al. (2023). Pathological and Therapeutic Advances in Parkinson's Disease: Mitochondria in the Interplay. *Journal of Alzheimer's disease: JAD, 94*(s1), s399–s428.
Nguyen, T. T., Ta, Q. T. H., Nguyen, T. K. O., Nguyen, T. T. D., & Van Giau, V. (2020). Type 3 diabetes and its role implications in Alzheimer's disease. *International journal of molecular sciences, 21*(9), 3165.
Nicoletti, V., Palermo, G., Del Prete, E., Mancuso, M., & Ceravolo, R. (2021). Understanding the multiple role of mitochondria in Parkinson's disease and related disorders: lesson from genetics and protein–interaction network. *Frontiers in Cell and Developmental Biology, 9*, 636506.
Norwitz, N. G., Jaramillo, J. G., Clarke, K., & Soto, A. (2020). Ketotherapeutics for neurodegenerative diseases. *International Review of Neurobiology, 155*, 141–168.
Pan, R. Y., Zhang, J., Wang, J., Wang, Y., Li, Z., Liao, Y., et al. (2022). Intermittent fasting protects against Alzheimer's disease in mice by altering metabolism through remodeling of the gut microbiota. *Nature aging, 2*(11), 1024–1039.
Papkovskaia, T. D., Chau, K.-Y., Inesta-Vaquera, F., Papkovsky, D. B., Healy, D. G., Nishio, K., et al. (2012). G2019S leucine-rich repeat kinase 2 causes uncoupling protein-mediated mitochondrial depolarization. *Human Molecular Genetics, 21*(19), 4201–4213.
Pavlov, P. F., Wiehager, B., Sakai, J., Frykman, S., Behbahani, H., Winblad, B., et al. (2011). Mitochondrial γ-secretase participates in the metabolism of mitochondria-associated amyloid precursor protein. *The FASEB Journal, 25*(1), 78–88.
Pedros, I., Patraca, I., Martinez, N., Petrov, D., Sureda, F. X., Auladell, C., et al. (2016). Molecular links between early energy metabolism alterations and Alzheimer's disease. *Front Biosci (Landmark Ed), 21*, 8–19.
Pereira, G. M., Teixeira-dos-Santos, D., Soares, N. M., Marconi, G. A., Friedrich, D. C., Saffie Awad, P., et al. (2024). A systematic review and meta-analysis of the prevalence of Parkinson's disease in lower to upper-middle-income countries. *npj Parkinson's Disease, 10*(1), 181.
Pirtošek, Z. & Trošt, M. (2023). 1.2 Clinical presentation and diagnosis of Parkinson's disease. Where are We? Continuous Dopaminergic Stimulation for Parkinson's Disease [Internet].
Pokotylo, M., Brüggemann, N., & Prasuhn, J. (2025). Metabolic Dysregulation in Parkinson's Disease: Non-Oxidative Phosphorylation and Its Role in Brain Energy Metabolism. *Aging and Disease, 16*(5), 2721.
Porsteinsson, A. P., Isaacson, R. S., Knox, S., Sabbagh, M. N., & Rubino, I. (2021). Diagnosis of early Alzheimer's disease: clinical practice in 2021. *The journal of prevention of Alzheimer's disease, 8*(3), 371–386.
Ralph, M. A. L., Jefferies, E., Patterson, K., & Rogers, T. T. (2017). The neural and computational bases of semantic cognition. *Nature Reviews. Neuroscience, 18*(1), 42–55.
Ruppert, M. C., Greuel, A., Tahmasian, M., Schwartz, F., Stürmer, S., Maier, F., et al. (2020). Network degeneration in Parkinson's disease: multimodal imaging of nigro-striato-cortical dysfunction. *Brain, 143*(3), 944–959.
Salmon, D. P. (2011). Neuropsychological features of mild cognitive impairment and preclinical Alzheimer's disease. *Behavioral neurobiology of aging*, 187–212.

Saxena, U. (2012). *Bioenergetics failure in neurodegenerative diseases: back to the future.* Taylor & Francis, 351–354.

Schumacher, J., Cromarty, R., Gallagher, P., Firbank, M. J., Thomas, A. J., Kaiser, M., et al. (2019). Structural correlates of attention dysfunction in Lewy body dementia and Alzheimer's disease: an ex-Gaussian analysis. *Journal of Neurology, 266*(7), 1716–1726.

Sędzikowska, A., & Szablewski, L. (2021). Insulin and Insulin Resistance in Alzheimer's Disease. *International journal of molecular sciences, 22*(18).

Shin, H. W., Hong, S. W., & Youn, Y. C. (2022). Clinical Aspects of the Differential Diagnosis of Parkinson's Disease and Parkinsonism. *J Clin Neurol, 18*(3), 259–270.

Shulman, R. G., Rothman, D. L., Behar, K. L., & Hyder, F. (2004). Energetic basis of brain activity: implications for neuroimaging. *Trends in Neurosciences, 27*(8), 489–495.

Simon, D. K., Tanner, C. M., & Brundin, P. (2020). Parkinson Disease Epidemiology, Pathology, Genetics, and Pathophysiology. *Clinics in Geriatric Medicine, 36*(1), 1–12.

Sims, N., Bowen, D., Allen, S., Smith, C., Neary, D., Thomas, D., et al. (1983). Presynaptic cholinergic dysfunction in patients with dementia. *Journal of Neurochemistry, 40*(2), 503–509.

Smith, C., Malek, N., Grosset, K., Cullen, B., Gentleman, S., & Grosset, D. G. (2019). Neuropathology of dementia in patients with Parkinson's disease: a systematic review of autopsy studies. *Journal of Neurology, Neurosurgery & Psychiatry, 90*(11), 1234–1243.

Soto, M., Cai, W., Konishi, M., & Kahn, C. R. (2019). Insulin signaling in the hippocampus and amygdala regulates metabolism and neurobehavior. *Proc Natl Acad Sci U S A, 116*(13), 6379–6384.

Spires-Jones, T. L., & Hyman, B. T. (2014). The intersection of amyloid beta and tau at synapses in Alzheimer's disease. *Neuron, 82*(4), 756–771.

Stojkovska, I., Krainc, D., & Mazzulli, J. R. (2018). Molecular mechanisms of α-synuclein and GBA1 in Parkinson's disease. *Cell and Tissue Research, 373*(1), 51–60.

Sun, Y., Wang, Y., Chen, S.-T., Chen, Y.-J., Shen, J., Yao, W.-B., et al. (2020). Modulation of the astrocyte-neuron lactate shuttle system contributes to neuroprotective action of fibroblast growth factor 21. *Theranostics, 10*(18), 8430.

Szablewski, L. (2025). Human Glucose Transporters in Health and Selected Neurodegenerative Diseases. *International Journal of Molecular Sciences, 26*(15), 7392.

Tansey, M. G., Wallings, R. L., Houser, M. C., Herrick, M. K., Keating, C. E., & Joers, V. (2022). Inflammation and immune dysfunction in Parkinson disease. *Nature Reviews. Immunology, 22*(11), 657–673.

Tolleson, C. M., & Fang, J. Y. (2013). Advances in the mechanisms of Parkinson's disease. *Discovery Medicine, 15*(80), 61–66.

Tong, S., Wang, R., Li, H., Tong, Z., Geng, D., Zhang, X., et al. (2025). Executive dysfunction in Parkinson's disease: From neurochemistry to circuits, genetics and neuroimaging. *Progress in Neuro-Psychopharmacology and Biological Psychiatry*, 111272.

Toomey, C. E., Heywood, W. E., Evans, J. R., Lachica, J., Pressey, S. N., Foti, S. C., et al. (2022). Mitochondrial dysfunction is a key pathological driver of early stage Parkinson's. *Acta Neuropathol Commun, 10*(1), 134.

Trošt, M., Perovnik, M., & Pirtošek, Z. (2019). Correlations of neuropsychological and metabolic brain changes in Parkinson's disease and other α-synucleinopathies. *Frontiers in Neurology, 10*, 1204.

van der Flier, W. M., Pijnenburg, Y. A., Fox, N. C., & Scheltens, P. (2011). Early-onset versus late-onset Alzheimer's disease: the case of the missing APOE ε4 allele. *The Lancet Neurology, 10*(3), 280–288.

Veloz Castillo, M. F., Magistretti, P. J., & Cali, C. (2021). l-Lactate: food for thoughts, memory and behavior. *Metabolites, 11*(8), 548.

Verma, M., & Howard, R. J. (2012). Semantic memory and language dysfunction in early Alzheimer's disease: a review. *International Journal of Geriatric Psychiatry, 27*(12), 1209–1217.

Villain, N., Desgranges, B., Viader, F., de la Sayette, V., Mézenge, F., Landeau, B., et al. (2008). Relationships between hippocampal atrophy, white matter disruption, and gray matter hypometabolism in Alzheimer's disease. *The Journal of neuroscience: the official journal of the Society for Neuroscience, 28*(24), 6174–6181.

Villain, N., Fouquet, M., Baron, J. C., Mézenge, F., Landeau, B., de La Sayette, V., et al. (2010). Sequential relationships between grey matter and white matter atrophy and brain metabolic abnormalities in early Alzheimer's disease. *Brain: a journal of neurology, 133*(11), 3301–3314.

Vyhnálek, M., Marková, H., Laczó, J., De Beni, R., & Di Nuovo, S. (2019). Assessment of memory impairment in early diagnosis of Alzheimer's disease. *Current Alzheimer Research, 16*(11), 975–985.

Walker, L., Stefanis, L., & Attems, J. (2019). Clinical and neuropathological differences between Parkinson's disease, Parkinson's disease dementia and dementia with Lewy bodies–current issues and future directions. *Journal of Neurochemistry, 150*(5), 467–474.

Wang, H.-Y., Ren, L., Yan, Z., Zhou, T., & Liang, Z. (2025). Neural basis of dysexecutive and visuospatial impairments in Parkinson's disease with MCI: a task-based fNIRS study. *npj Parkinson's Disease, 11*(1), 163.

Watanabe, H., Shima, S., Kawabata, K., Mizutani, Y., Ueda, A., & Ito, M. (2025). Brain network and energy imbalance in Parkinson's disease: linking ATP reduction and α-synuclein pathology. *Frontiers in Molecular Neuroscience, 17*, 1507033.

Wenk, G. L. (2006). Neuropathologic changes in Alzheimer's disease: potential targets for treatment. *Journal of Clinical Psychiatry, 67*, 3.

Wong, K. Y., Roy, J., Fung, M. L., Heng, B. C., Zhang, C., & Lim, L. W. (2020). Relationships between mitochondrial dysfunction and neurotransmission failure in Alzheimer's disease. *Aging and disease, 11*(5), 1291.

Wu, T., Dejanovic, B., Gandham, V. D., Gogineni, A., Edmonds, R., Schauer, S., et al. (2019). Complement C3 is activated in human AD brain and is required for neurodegeneration in mouse models of amyloidosis and tauopathy. *Cell reports, 28*(8), 2111–2123.e6.

Xiang, X., Wind, K., Wiedemann, T., Blume, T., Shi, Y., Briel, N., et al. (2021). Microglial activation states drive glucose uptake and FDG-PET alterations in neurodegenerative diseases. *Science translational medicine, 13*(615), eabe5640.

Xiaopeng, Z., Jing, Y., Xia, L., Xingsheng, W., Juan, D., Yan, L., et al. (2025). Global Burden of Alzheimer's disease and other dementias in adults aged 65 years and older, 1991-2021: population-based study. *Front Public Health, 13*, 1585711.

Xie, A., Gao, J., Xu, L., & Meng, D. (2014). Shared mechanisms of neurodegeneration in Alzheimer's disease and Parkinson's disease. *Biomed Res Int, 2014*, 648740.

Yang, C., Pan, R.-Y., Guan, F., & Yuan, Z. (2024). Lactate metabolism in neurodegenerative diseases. *Neural Regeneration Research, 19*(1), 69–74.

You, W., Li, Y., Liu, K., Mi, X., Li, Y., Guo, X., et al. (2024). Latest assessment methods for mitochondrial homeostasis in cognitive diseases. *Neural Regeneration Research, 19*(4), 754–768.

Yuan, Y., Zhao, G., & Zhao, Y. (2024). Dysregulation of energy metabolism in Alzheimer's disease. *Journal of Neurology, 272*(1), 2.

Yuan, Y., Zhao, G., & Zhao, Y. (2025). Dysregulation of energy metabolism in Alzheimer's disease. *Journal of Neurology, 272*(1), 2.

Zhang, J., Zhang, Y., Wang, J., Xia, Y., Zhang, J., & Chen, L. (2024). Recent advances in Alzheimer's disease: mechanisms, clinical trials and new drug development strategies. *Signal Transduction and Targeted Therapy, 9*(1), 211.

Zhang, W., Xiao, D., Mao, Q., & Xia, H. (2023). Role of neuroinflammation in neurodegeneration development. *Signal transduction and targeted therapy, 8*(1), 267.

Zvěřová, M. (2019). Clinical aspects of Alzheimer's disease. *Clinical Biochemistry, 72*, 3–6.

CHAPTER THREE

Energetic crisis, mitochondrial vulnerability and disruption of lactate shuttle in Alzheimer's disease

Gowsika Baskar[a] and Mahesh Kandasamy[a,b,*]
[a]Laboratory of Stem Cells and Neuroregeneration, Department of Animal Science, School of Life Sciences, Bharathidasan University, Tiruchirappalli, Tamil Nadu, India
[b]University Grants Commission-Faculty Recharge Programme (UGC-FRP), New Delhi, India
*Corresponding author. e-mail address: pkmahesh5@gmail.com; mahesh.kandasamy@bdu.ac.in

Contents

Abstract

Alzheimer's disease (AD) has traditionally been characterized by amyloid-beta (Aβ) plaques and neurofibrillary tangles. Emerging evidence reveals that metabolic dysfunction represents a key pathological feature central to disease progression. Mitochondrial dysfunction in AD leads to impaired electron transport chain activity and reduced level of adenosine triphosphate (ATP) synthesis, preceding neurodegeneration and structural abnormalities in cognitive centres of the brain. Early glucose hypometabolism and lactate deprivation or interference with their utilization represent a primary bioenergetic failure driving mitochondrial dysfunction and neuroinflammation prior to the clinical manifestation of AD. While the brain relies on lactate as a prominent energy substrate, astrocytic metabolic defects lead to impaired neuronal energy homeostasis, thereby promoting neurodegeneration. In turn, this metabolic uncoupling could also be associated with defects in

International Review of Neurobiology, Volume 188
ISSN 0074-7742, https://doi.org/10.1016/bs.irn.2026.05.012

regenerative mechanisms by impairing adult neurogenesis in the hippocampus due to energy deprivation, accounting for memory deficits. This chapter discusses the evidence for the energetic crisis in AD, focusing on the disruption of the astrocyte-neuron lactate shuttle (ANLS), hypometabolism of glucose, and mitochondrial vulnerability, as interconnected pathogenic mechanisms. We emphasise cerebral bioenergetic metabolic failure as a central driver of cognitive decline, arising from irreversible reactive gliosis and neuroblastosis mechanisms and highlight various therapeutic options, including restoration of ANLS to mitigate the pathogenesis and memory loss in AD.

1. Introduction

Alzheimer's disease (AD) represents the most common cause of dementia, marked by progressive degeneration of cholinergic neurons mainly in the hippocampus (Kumar et al., 2025). Accumulation of amyloid beta (Aβ) plaques and hyperphosphorylated tau tangles has been linked to the neuropathogenic events, including microglial activation in AD (Rajmohan & Reddy, 2017). Besides, impaired regulation of hippocampal neurogenesis has also been recognized as an underlying mechanism of memory loss in AD (Kim et al., 2022). While several comorbid pathogenic mechanisms have been proposed for the onset and progression of AD, accumulating evidence indicates the occurrence of metabolic dysfunction at a very early stage, before detectable structural brain deformities, even long before the onset of cognitive impairment, could be highly relevant (Amidfar et al., 2024; Constantino et al., 2025; Maciejewska et al., 2021). AD has also been considered as type 3 diabetes due to impairments in brain insulin signaling and glucose metabolism, leading to disrupted neuronal energy homeostasis (Nguyen et al., 2020). Mitochondrial abnormalities leading to altered bioenergetics and excessive production of reactive oxygen species (ROS) have been strongly implicated in neurodegenerative events in AD (Guo et al., 2013). Accumulating evidence indicates that neurons rely predominantly on lactate, supplied via the astrocyte–neuron lactate shuttle (ANLS), as a major energy source (Y. Kim et al., 2025). Astrocytes are a major class of glial cells in the brain that play a crucial role in maintaining cerebral homeostasis by providing metabolic support and regulating energy supply and gas exchange (Beard et al., 2022). Under physiological conditions, astrocytes metabolise glucose into lactate and supply it to neurons through the ANLS to maintain ATP production (Beard et al., 2022).

Notably, microglia act as metabolic sentinels that help in maintaining energy coupling between astrocytes and neurons through the secretion of insulin-like growth factor (IGF) and transforming growth factor (TGF)-β (Vainchtein & Molofsky, 2020). Whereas in AD, ANLS dysfunction restricts astrocyte-derived lactate delivery to neurons, leading to impaired neuronal energy metabolism that underlies synaptic failure and neurodegeneration (Wu et al., 2023). This metabolic impairment results in excess accumulation of unutilized glucose in the brain. Activated microglia appear to uptake excessive glucose, engage the pentose phosphate pathway, and secrete pro-inflammatory cytokines, thereby creating a neuroinflammatory microenvironment and energy-deficient conditions (Miao et al., 2023). Eventually, this results in a neuroenergetic disconnection due to impairment in neurovascular coupling, exacerbating the pathogenesis of AD (Nelson et al., 2016). Considering these facts, the viewpoint on the etiopathology and underlying mechanisms of AD has been revisited, with metabolic and energy dysfunction emphasized as central contributors to Aβ and tau-related pathogenic cascades accounting for neurodegeneration (Ratan et al., 2023). Moreover, this metabolic disharmony may underlie the adverse microenvironment to neurogenic processes in AD, leading to regenerative failure and progressive memory loss. From a therapeutic standpoint, recent advances in understanding the underlying mechanisms of AD suggest that the energy crisis represents a promising therapeutic target to restore mitochondrial dynamics and cellular bioenergetic homeostasis (Constantino et al., 2025). This book chapter presents how mitochondrial regulation, glial to neuronal metabolic support, and in concert with Aβ and tau related mechanism, influence the course of AD. This chapter also discusses how re-energizing these interconnected networks may offer therapeutic strategies for neuroprotection and neuroregeneration, ultimately helping to mitigate cognitive decline in AD.

2. Brain energy metabolism and the role of mitochondria in the physiological state

The brain primarily relies on substrates and metabolites derived from dietary nutrients from the circulating blood as its major source of energy (Clarke & Sokoloff, 1999). Generally, glucose serves as the primary source of energy for the whole body under physiological conditions. It enters the brain through endothelial glucose transporter type (GLUT)-1 transporters and is taken up by neurons via GLUT3, which is subsequently oxidized in mitochondria through the tricarboxylic acid (TCA) cycle and oxidative phosphorylation (Koepsell, 2020). In addition to glucose, alternative substrates such as lactate, ketone bodies, and certain amino acids can also contribute to the energy metabolism of the brain, particularly during development, fasting, exercise, and pathological conditions (Jensen et al., 2020). Among these, lactate has long been regarded as a by-product of anaerobic glycolytic metabolism, recognized as a main energy source often preferred over glucose in the brain (Wu et al., 2023). Besides, lactate functions as an important signaling molecule that regulates synaptic plasticity and adult neurogenesis, thereby contributing to memory consolidation and retention (Lev-Vachnish et al., 2019). Considering the high metabolic demand effectuated by highly complex synaptic networks and continuous electrical impulses, mitochondria of the brain need to produce a great amount of ATP despite their active participation in the regulation of neural transmission, calcium homeostasis, lipid synthesis, redox signaling, apoptotic pathways and neuroimmune functions (Song et al., 2024). Among their diverse cellular functions, mitochondria play a pivotal role in regulating intrinsic apoptosis via the cytochrome-c–mediated caspase activation cascade (Wang & Youle, 2009). Controlled activation of this pathway is essential for maintaining neuronal survival. Therefore, mitochondrial health is increasingly recognized as an important determinant of neuronal viability and long-term brain function. Notably, mitochondria also generate endogenous antioxidants such as glutathione to balance cellular redox state, a crucial mechanism of neuroprotection (Marí et al., 2009). In neurons, biogenesis of mitochondria occurs in the cell body and is transported to the synapse by motor proteins through microtubule tracks. Studies using genetically encoded ATP reporter systems have revealed that nerve terminals contain exceptionally high densities of mitochondria strategically positioned near active zones, where neurotransmitter-containing synaptic vesicles fuse with the presynaptic membrane (Zaninello et al., 2024). Kinetic studies demonstrate that mitochondrial calcium uptake plays a critical buffering role by preventing excessive accumulation of Ca^{2+} in the cytoplasm (Walters & Usachev, 2023). Defects in this process impair intracellular Ca^{2+} homeostasis and can trigger excitotoxic cell death, as observed in neurodegenerative disorders (Nicholls, 2009). Thus, mitochondrial dysfunction has been attributed to synaptic impairment, leading to defective neurotransmission under neuropathological conditions, including AD. Pulsatile releases of synaptic vesicles and subsequent recycling demand large amounts of ATP, rendering presynaptic terminals among the most energy-intensive cellular compartments in the body (Pathak et al., 2015). In order to achieve the high-frequency stimulation and maintain periodic neurotransmission, mitochondrial ATP production at synapses must increase 2–3-fold at every cycle. Defects in mitochondrial physiology or metabolic deprivation can result in synaptic depression and transmission failure (Pathak et al., 2015).

Astrocytes are key metabolic supporters of neurons, as they uptake glucose via GLUT1 transporters localized at their perivascular end feet and preferentially engage in glycolysis (Ardanaz et al., 2024). Astrocytic glycolytic activity is stimulated by neuronal activity–associated signals, including glutamate release, elevations in extracellular K^+, and nitric oxide (Pellerin & Magistretti, 1994). This process is facilitated by the selective expression of the glycolytic regulator 6-phosphofructo-2-kinase/fructose-2,6-bisphosphatase-(PFKFB)-3 in astrocytes, which increases intracellular fructose-2,6-bisphosphate levels and thereby activates phosphofructokinase (PFK)-1, enhancing glycolytic flux (Bonvento & Bolaños, 2021). Alternatively, astrocytic expression of pyruvate dehydrogenase kinase (PDK)-4 inactivates pyruvate dehydrogenase (PDH), thereby preventing pyruvate from entering the TCA cycle and promoting its conversion into lactate (Liu et al., 2024). In the absence or low levels of glucose, lactate enters the brain across the blood–brain barrier (BBB) primarily via monocarboxylate transporter (MCT)-1 expressed on endothelial cells and astrocytic end feet (Ueno et al., 2023). Astrocytes can convert glucose into glycogen, which is stored predominantly within astrocytes in the brain (Dienel & Cruz, 2015). During intense mental stimulation and cognitive activity, astrocytes preferentially produce lactate through glycogenolysis, a process stimulated by neuromodulators such as norepinephrine, adenosine, and elevations in extracellular potassium (K^+) (Newman et al., 2011). The lactate produced is subsequently exported from astrocytes via MCT1 and MCT4. In the brain parenchyma, lactate is taken up by neurons predominantly through MCT2, where it is converted to pyruvate by lactate dehydrogenase and subsequently enters the TCA cycle to support oxidative phosphorylation and ATP production (Brooks, 2018).

Likewise, oligodendrocytes, the myelinating macroglia of the brain, depend largely on glycolytic metabolism to sustain their energy requirements and to supply metabolic substrates essential for axonal integrity and function (Funfschilling et al., 2012). Importantly, oligodendrocyte-derived lactate serves as a key energy substrate for axonal mitochondria, especially for the long, heavily myelinated axons, where direct access to extracellular glucose is limited (Li & Sheng, 2023). The resulting lactate sustains axonal energy metabolism, enabling ATP-dependent synaptic vesicle cycling and neurotransmitter release. In addition to glucose metabolism, oligodendrocytes are highly involved in the synthesis of lipids and cholesterol to support the structural and functional aspects of myelin sheaths responsible for the action potential of the neurons (Zhang et al., 2025). Under conditions of energy deprivation, oligodendrocytes can engage fatty acid β-oxidation, catabolizing intracellular lipids within peroxisomes and mitochondria, thereby utilizing alternative energy sources to sustain cellular and axonal metabolic demands (Liu et al., 2024).

With reference to microglial energy metabolism, it is not primarily oriented to provide metabolic support of neurons; rather, it is intrinsically utilized to sustain immune surveillance, facilitate phagocytic clearance of apoptotic cells and pathological protein aggregates, enable activity-dependent synaptic pruning and remodelling, and initiate as well as propagate neuroinflammatory responses (Hassan et al., 2025). In the healthy brain, the neuroprotective phenotype of microglia predominantly relies on glucose-derived pyruvate, which is directed into mitochondrial oxidative phosphorylation for ATP production in a PDH–dependent manner (Pisani & Blondeau, 2025). Microglia can also metabolize lipids derived from myelin debris, apoptotic cells, and lipid droplets. β-oxidation in mitochondria supports ATP generation during phagocytosis (Sun et al., 2025). Additionally, microglia can exploit astrocyte-derived lactate, which enters mitochondria via MCTs and is converted to pyruvate, linking glial metabolism to immune function. During learning and memory formation, microglia transiently enhance glycolytic flux to meet acute energy demands associated with phagocytic remodelling of synaptic elements, while rapidly restoring mitochondrial respiration to prevent disproportionate inflammatory activation (Li et al., 2025). This metabolic flexibility ensures that microglial engagement with synapses remains adaptive rather than pathological. Upon infection or during disease conditions, pathogenic activation of microglia is accompanied by adaptive metabolic reprogramming that meets the heightened bioenergetic and biosynthetic demands required to execute inflammatory, phagocytic, and immune effector functions (Li et al., 2025). Disruption of metabolic plasticity of microglia, characterized by progressive glycolytic deterioration, mitochondrial dysfunction, or impaired mitophagy, leads to aberrant synaptic pruning and chronic neuroinflammation, which are strongly associated with cognitive decline in aging and neurodegenerative disorders such as AD (Mani et al., 2025).

The cellular energy and mitochondrial dynamics have emerged as central regulators of neural regeneration of the brain, as it sustains maintenance of neural stem cells (NSCs), orchestrate lineage specification, and ensure the functional integration of newly generated neurons (Garone et al., 2024). NSCs exhibit a distinct metabolic program tightly linked to their proliferative and lineage-determining states. In the quiescent state, NSCs primarily rely on glycolysis with minimal ATP without involving mitochondria to avoid oxidative phosphorylation (Sánchez-Ramírez et al., 2024). During proliferation, NSCs predominantly utilize aerobic glycolysis, which supplies rapid ATP and essential biosynthetic precursors for macromolecular synthesis and biogenesis of subcellular organelles required for effective propagation (Petridi et al., 2022). Following fate commitment, NSCs undergoing neuronal differentiation exhibit a pronounced metabolic shift toward mitochondrial substrate oxidation and oxidative phosphorylation, accompanied by mitochondrial biogenesis (Iwata & Vanderhaeghen, 2021). During this transition, enhanced pyruvate transport into mitochondria, activation of PDH, and increased electron transport chain activity collectively meet the high energy demands of neurite outgrowth, synaptogenesis, and functional maturation. Notably, facilitation of these metabolic processes enhances neurogenic events and the functional integration of newly generated neurons, which are considered essential for active learning and memory processes, particularly in spatial and non-spatial contexts (Landry & Huang, 2021).

Evidence from genetic and experimental models demonstrates that prominent disruption of brain energy metabolism, including mitochondrial dysfunction, functional loss of GLUTs and MCTs, impairment of glycolytic pathways, and defective lactate transfer have been tightly linked to neurodegenerative processes as well as neuroregenerative failure. Disruption of astrocyte–vascular coupling and reactive astrogliosis under neuropathogenic conditions severely impairs the ANLS, limiting the metabolic support required for neuronal function and survival. ATP depletion causes collapse of the mitochondrial membrane potential, leading to dysregulated Ca^{2+} and other ion fluxes across the mitochondrial membrane. Such bioenergetic stress promotes mitochondrial dysfunction and suppresses mitochondrial biogenesis. This metabolic uncoupling represents a key pathogenic hallmark of neurodegenerative diseases, notably AD, wherein Aβ and tau-induced mitochondrial dysfunction is coupled with reduced lactate metabolism.

3. Metabolic energy crisis in Alzheimer's disease

The pathogenesis of AD has been extensively characterized by the accumulation of Aβ aggregates and hyperphosphorylated tau proteins (Zhang et al., 2021). Growing evidence indicates that Aβ and tau-driven neurodegeneration is closely associated with extreme instability in cerebral energy metabolism, leading to mitochondrial vulnerability and dysfunction. Notably, a progressive reduction in the metabolic rate of glucose in the hippocampus, cerebral cortex, and other susceptible brain regions has been linked to neurodegeneration resulting in cognitive deficits. Metabolic positron emission tomography (PET) imaging studies demonstrate a marked reduction in cerebral glucose uptake in the brains of AD subjects (Marcus et al., 2014). Unlike peripheral tissues, neuronal glucose uptake occurs predominantly through insulin-independent GLUTs. While insulin-dependent glucose transport mechanisms play a negligible role in basal cerebral glucose uptake, accumulating evidence indicates that brain insulin resistance arising from downregulation or desensitized insulin receptors, rather than defective glucose transport, contributes significantly to cerebral hypometabolism in AD (González et al., 2022). This insulin resistance and its downstream signaling dysfunction, including phosphoinositide-3-kinase–protein kinase B (PI3K–PKB) are conceptualized AD as a form of type 3 diabetes frequently accompanied by systemic metabolic comorbidities such as dyslipidaemia (Wang et al., 2025). Moreover, chronic insulin resistance, characterized by hyperinsulinemia, competitively interferes with Aβ clearance, as insulin and Aβ share common degradative mechanisms, particularly via the insulin-degrading enzyme (IDE) (Cerasuolo et al., 2025). This competitive inhibition reduces Aβ degradation and promotes its accumulation. In addition, impaired insulin clearance and IDE dysfunction may indirectly exacerbate cerebral glucose hypometabolism by disrupting insulin signaling and glucose transporter regulation, thereby further compromising neuronal glucose uptake in AD (Cerasuolo et al., 2025). Moreover, increased activity of glycogen synthase kinase (GSK)-3β, resulting from impaired insulin signaling associated with insulin resistance, has been strongly linked to enhanced tau hyperphosphorylation, thereby contributing to the progression of AD (Hobday & Parmar, 2021). Nevertheless, transcriptomic and proteomic studies also indicate that altered expression and disease-associated epigenetic modifications of glucose transporters, particularly GLUT1 and GLUT3, contribute to impaired cerebral glucose transport, at least in part accounting for the reduced glucose uptake observed in AD (Kyrtata et al., 2021). Aβ and hyperphosphorylation of tau appear to interfere with the expression, trafficking, and functional activity of GLUTs, thereby disrupting glucose uptake in AD (Kyrtata et al., 2021).

Unlike glucose transport, lactate uptake via MCTs is insulin-independent and occurs through proton-coupled diffusion driven by lactate concentration and pH gradients (Zhang et al., 2026). Surprisingly, increased lactate levels in the cerebrospinal fluid (CSF) have been reported in patients with AD, particularly at the stage of mild cognitive impairment (MCI) (Zebhauser et al., 2022). Moreover, microglia in transgenic models of AD exhibit elevated lactate production, consistent with a shift toward a glycolytic, pro-inflammatory metabolic phenotype (Sangineto et al., 2023). This may result from altered regulation of the lactate transport, buffering, or clearance systems in AD, as the accumulation of unused lactate is considered detrimental to the brain. Notably, acidification of the environment of neurons and astrocytes induced by Aβ and tau pathology leads to an impaired pH buffering system, which has been linked to altered expression and function of proton-regulating systems, including Na^+/H^+ exchangers, proton pumps, carbonic anhydrases and bicarbonate transporters (Prasad & Rao, 2015). Moreover, microglial activation leading to neuroinflammation is known to induce chronic tissue acidosis, thereby disturbing pH gradient–dependent mechanisms in association with disrupted lysosomal homeostasis in AD (Wang et al., 2025). These alterations likely reflect impaired lactate utilization and transport rather than a modulation in lactate generation. Astrocyte-mediated metabolic support becomes progressively impaired in AD, as multiple studies have demonstrated that MCTs responsible for lactate shuttling are markedly downregulated or dysfunctional. Consistent with this, in the 3 ×Tg-AD model, resting lactate production via aerobic glycolysis appears to be altered in astrocytes, indicating intrinsic astrocytic metabolic impairment (Le Douce et al., 2020). Furthermore, studies in AD mouse models and post-mortem human brain tissues reveal decreased expression of MCT1, MCT2, and MCT4, resulting in inefficient lactate transfer from astrocytes to neurons (Zhang et al., 2018). In the mutant amyloid precursor protein and presenilin 1 (APP/PS1) mouse model of AD, reduced MCT expression results in defective lactate shuttling from glial cells to neurons, culminating in neuronal energy deficits and heightened vulnerability to degeneration (Zhang et al., 2018). Collectively, these findings indicate that altered lactate dynamics in AD arise from a breakdown of glia–neuron metabolic coupling, linking impaired lactate transport with Aβ and tau pathology, progressive energy failure, and neurodegeneration.

4. Mitochondrial defects in Alzheimer's disease

When energy deficits occur, mitochondria become increasingly vulnerable due to insufficient ATP production, which in turn exacerbates the activation of oxidative stress and cell death signaling pathways. Mitochondria in the brains of subjects with AD have been reported to exhibit marked structural and functional deficits. Electron microscopy studies reveal that mitochondria in neurons from AD subjects display disrupted cristae architecture, reduced size, and abnormal morphology, reflecting severe mitochondrial dysfunction (Baloyannis, 2011). Importantly, this mitochondrial impairment is closely associated with synaptic pathology, including loss of dendritic spines and pronounced attenuation of dendritic arborization (Baloyannis, 2011).

The electron transport chain (ETC), located in the inner mitochondrial membrane, is central to cellular energy production, generating ATP through oxidative phosphorylation. It consists of Complexes I–IV and ATP synthase (Complex V), with each complex transferring electrons and pumping protons to establish a mitochondrial membrane potential. Proper ETC function is essential for neuronal energy demands, synaptic transmission, and maintenance of dendritic architecture (Jonckheere et al., 2012). Substantial experimental evidence indicates that mitochondrial ETC components are structurally and functionally impaired in AD, particularly in neurons of the hippocampus and cortex (Wang et al., 2020). The Q-cytochrome c oxidoreductase, known as Complex III catalyses the transfer of electrons from ubiquinol to cytochrome c, while simultaneously pumping protons across the inner mitochondrial membrane to help establish the proton gradient necessary for ATP synthesis. In AD brains, activity of Q-cytochrome c oxidoreductase has been reported to be decreased, leading to enhanced generation of reactive oxygen species (ROS), particularly superoxide, which damages mitochondrial DNA, lipids, and proteins, thereby exacerbating oxidative stress (Misrani et al., 2021). Cytochrome c oxidase (Complex IV) is a key enzyme in the mitochondrial ETC, and it plays a central role in cellular energy production. It catalyses the transfer of electrons from cytochrome c to molecular oxygen, reducing O_2 to water. This process is coupled to proton pumping across the inner mitochondrial membrane, contributing to the proton gradient used by ATP synthase to generate ATP. In AD, the function of the activity of cytochrome c oxidase is hindered, leading to impaired mitochondrial respiration and energy deficits (Rak et al., 2016). This dysfunction contributes directly to mitochondrial degeneration, characterized by disrupted cristae, reduced mitochondrial size, and abnormal morphology. The altered function of cytochrome c oxidase may also contribute to acidification of the intracellular environment by disrupting the H^+ gradient, thereby impairing proton-coupled lactate transport. In addition, metabolic uncoupling in AD is associated with altered expression of uncoupling proteins (UCPs), particularly UCP2 and UCP4, in the brain, which dissipates the mitochondrial proton gradient and reduces ATP synthesis (Kwok et al., 2010). These proteins allow protons to bypass ATP synthase, releasing energy as heat rather than storing it in ATP bonds. While mild uncoupling provides neuroprotection by limiting oxidative stress, excessive or dysregulated uncoupling in AD exacerbates energy deficits, leading to synaptic failure and neurodegeneration.

Moreover, early studies demonstrated that hippocampal neurons exposed to Aβ exhibit disrupted anterograde mitochondrial trafficking to axons. Aβ impairs transport by interacting with motor proteins, leading to stalled mitochondria (Calkins & Reddy, 2011). In parallel, hyperphosphorylated tau detaches from microtubules, destabilizing the cytoskeletal network (Reddy, 2011). Evidence from tau transgenic mice shows reduced mitochondrial density in axons and synaptic terminals (Calkins & Reddy, 2011). Taken together, Aβ and tau-mediated neuropathogenesis strongly correlates with energy deficits and mitochondrial dysfunction, driving neurodegeneration and implicating altered ANLS activity. However, the precise mechanisms linking impaired energy metabolism with disruptions in glucose and lactate metabolism, along with other pathogenic determinants such as cell cycle events, remain to be fully elucidated.

5. Mechanistic insights into energy deficits-driven mitochondrial dysfunction linking aberrant cell-cycle events and impaired neurogenesis in Alzheimer's disease

Pathogenic events initiated by Aβ and tau are strongly associated with neurodegeneration resulting from mitochondrial dysfunction and cerebral energy crisis, making them a major focus of research investigation. The involvement of immune function and adult neurogenesis in the brain represents another critical dimension of neuropathology in AD. In this context, reactive gliosis and neuroblastosis constitute two distinct cellular responses observed in the pathological state of the brain (De Sousa, 2022; Poornimai Abirami et al., 2020). Upon pathogenic triggers, on one hand, the first line neuroimmune defence becomes activated, while on the other, neurogenesis mechanisms attempt to compensate for neurodegeneration as an intrinsic defence response. In response to Aβ and tau pathology, astrocytes and microglia undergo activation and adopt neuroimmune phenotypes that are initially neuroprotective by facilitating anti-inflammatory signaling (Deng et al., 2024). As AD advances, persistent astrocytic gliosis and microglial activation become maladaptive, driving aberrant proliferation and releasing harmful proinflammatory signaling to neuronal survival (Cai et al., 2022). Notably, Aβ and tau-driven microglial activation induces a metabolic shift toward glycolysis, accompanied by impaired mitochondrial oxidative phosphorylation, leading to excessive ROS production and sensitizing neuronal mitochondria to cytochrome c release, thereby triggering apoptotic neuronal cell-death pathways (Li et al., 2025). Besides, astrocytic activation alters glycolytic flux and lactate production, while mitochondrial oxidative phosphorylation is diminished, leading to reduced metabolic support for neurons and NSCs (Liu et al., 2024). These abnormal cell-cycle events in glial cells can lead to bioenergetic failure, exacerbating neuronal mitochondrial stress and synaptic dysfunction, and ultimately contributing to impaired neurogenesis and progressive neurodegeneration. While animal models of AD and postmoderm brains of AD have been characterized by increased proliferation of neuronal precursor at an early stage, indicating a compensatory attempt via neurogenic mechanisms (Demars et al., 2013; Zhou et al., 2023). Studies in transgenic animal models of AD, including APP/PS1, depict an early increase in proliferation of neuroblasts in the hippocampus during pre-plaque stages (Unger et al., 2016). Although this transient surge of reactive neuroblastosis is thought to represent an endogenous compensatory response to early Aβ and tau-mediated neuronal loss, it progressively diminishes with disease advancement, ultimately resulting in impaired neurogenesis as reported in 3xTg-AD mice (Babcock et al., 2021; Liu et al., 2023).

Notably, induced expression of neuroinflammatory factors and cytokines, including TGF-β, has been associated with both quiescence of NSCs and fate commitment and differentiation of neuroblasts (Kandasamy et al., 2011, 2014). However, these mechanisms have not been adequately explored in the context of cerebral energy crisis and mitochondrial dysfunction. Specifically, reactive astrogliosis may lead to the retraction or loss of astrocytic end feet from cerebral blood vessels, resulting in metabolic disconnection at the neurovascular unit. This disruption compromises glucose transport and lactate delivery from the vasculature to neurons and NSCs. Notably, pathological activation and aberrant proliferative responses of astrocytes and microglia markedly increase their metabolic demand, leading these glial cells to preferentially consume most glucose and lactate. Consequently, the neurogenic niche would experience profound energy deprivation, exacerbating mitochondrial dysfunction. Under these circumstances, NSCs preferentially enter a quiescent state, whereas aberrantly cycling neuroblasts engage in metabolically demanding, reactive cell-cycle events that exacerbate energy depletion and ultimately lead to mitochondrial deformation and dysfunction. Eventually, reactive gliosis and aberrant neuroblastosis are likely to adopt a Warburg-like metabolic shift, particularly under conditions of insufficient or dysregulated neuronal energy supply. Such metabolic reprogramming may aberrantly activate cell-cycle–related pathways in post-mitotic neurons, triggering unscheduled cell-cycle re-entry. This maladaptive response can culminate in neuronal loss through apoptosis or promote pathological transdifferentiation, thereby exacerbating neuropathogenesis in AD (Manickam et al., 2020; Poteet et al., 2013). Consequently, this pathogenic state would culminate in impaired terminal neurogenesis, limiting neuronal replacement and synaptic plasticity, thereby contributing to memory loss in addition to progressive neurodegeneration in AD. Dysbiosis resulting from an impaired gut–brain axis (GBA) has been speculated as a potent pathogenic contributor in AD (Lu et al., 2024). Abnormal gut microbiota can disrupt intestinal physiology, leading to systemic inflammation and malabsorption of key nutrients and coenzymes, cofactors required for cerebral energy metabolism. Reduced availability of B vitamins, short-chain fatty acids, and essential lipids can impair mitochondrial oxidative phosphorylation and disrupt astrocyte–neuron metabolic coupling (Wan et al., 2022). Thus, disease-modifying therapies targeting Aβ and tau pathology may be required in combination with strategies that restore functional neurogenesis by mitigating metabolic demand and improving energy homeostasis, which should be actively pursued to effectively counteract neurodegeneration and dementia in AD.

6. Possible therapeutic strategies to mitigate energy crisis and mitochondrial defects in Alzheimer's disease

Numerous therapeutic interventions have been explored for the treatment of AD, with particular emphasis on mitigating oxidative stress and neuroinflammation, two central contributors to mitochondrial dysfunction and neuronal vulnerability. Among these, antioxidant supplementation aimed at counteracting excessive ROS has gained considerable attention. Compounds such as resveratrol, a polyphenolic molecule with potent antioxidant properties, have proven to provide significant neuroprotection by enhancing mitochondrial biogenesis, modulating sirtuin signaling pathways, and improving cellular redox homeostasis (Ren et al., 2025; Surya et al., 2025). In parallel, anti-inflammatory strategies targeting chronic microglial and astrocytic activation have been extensively investigated, given the strong link between sustained neuroinflammation and progressive synaptic and neuronal loss in AD (Kamila et al., 2025). Beta-site amyloid precursor protein cleaving enzyme (BACE)-1 is the rate-limiting enzyme that initiates amyloidogenic processing of amyloid precursor protein (APP). Cleavage of APP by BACE1 generates the C99 fragment, which is subsequently processed by γ-secretase to produce Aβ peptides. Thus, the identification and characterization of BACE1 inhibitors, such as Verubecestat (MK-8931) and Lanabecestat (AZD3293), aimed at reducing the generation of Aβ has also been considered in AD treatment regimens (Koelsch, 2017). In parallel, preclinical studies have explored small molecules such as methylthioninium chloride (LMTX), which are designed to prevent tau misfolding and aggregation, thereby mitigating neurofibrillary tangle pathology in AD (Congdon et al., 2023). Additionally, inhibition of key tau-related kinases, including glycogen synthase kinase-3 beta (GSK3β), cyclin-dependent kinase (CDK)-5, and microtubule affinity-regulating kinase (MARK), has been investigated to reduce tau hyperphosphorylation, stabilize microtubule dynamics, and protect neuronal and synaptic function (Annadurai et al., 2017; Liu et al., 2016; Sayas & Ávila, 2021). Acetylcholine (ACh) restorative therapies have been extensively explored to counteract the cholinergic deficit characteristic of AD (Chen et al., 2022). In particular, pharmacological inhibition of acetylcholinesterase (AChE) and agonistic choline acetyltransferase (ChAT)-associated pathways has also been experimentally and clinically exploited to restore ACh levels (Čolović et al., 2013; Ferreira-Vieira et al., 2016). Among these, AChE inhibitors such as donepezil, rivastigmine, and galantamine reduce enzymatic ACh degradation, thereby prolonging its availability within the synaptic cleft and partially rescuing cognitive function (Walczak-Nowicka & Herbet, 2021). However, strategies aimed at directly counteracting the cerebral energy crisis in AD are limited. Given that impaired glucose uptake, defective lactate shuttling, and mitochondrial dysfunction collectively minimize the availability of ATPs, pharmacological approaches that mimic or enhance cellular energy homeostasis have gained increasing attention. ATP-mimetic compounds and agents that stabilize intracellular bioenergetic signaling by donating phosphate groups, such as creatine and acetyl-L-carnitine, have been proposed as potential therapeutic strategies to restore neuronal energy homeostasis and preserve synaptic function in AD (Owen & Sunram-Lea, 2011). In parallel, pharmacological strategies that enhance the expression or functional activity of GLUTs, may improve cerebral glucose transport and utilization. Restoration of GLUT-mediated glucose transport could partially reverse the hypometabolic state observed in vulnerable brain regions such as the hippocampus and cortex. Antidiabetic drugs such as metformin, pioglitazone, liraglutide, and exenatide may help reverse insulin resistance and improve glucose transport into cells in AD (Corraliza-Gomez et al., 2025). Similarly, modulation of MCTs by drugs such as Rosiglitazone, a peroxisome proliferator-activated receptor (PPAR) agonist, represents a promising approach that could improve proper utilization of lactate, thereby ensuring metabolic coupling and sustaining neuronal energy metabolism (Nelson et al., 2023). A proper diet is beneficial in treating AD by supporting neuroprotection and neurogenesis (Bhuiyan et al., 2023; Han et al., 2023). Intermittent fasting, which induces a metabolic switch from glucose utilization to ketone body production, has been explored as a potential therapeutic intervention in AD (Ye et al., 2024). Besides, emerging evidence demonstrates that astrocytes naturally transfer healthy mitochondria to damaged neurons through tunneling nanotubes, representing an endogenous neuroprotective mechanism that becomes impaired in AD (Dilna et al., 2021). Thus, transplantation of tunnelling nanotube-loaded extracellular vesicles and/or NSCs could represent a promising therapeutic strategy for AD. Taken together, the development of combination therapies that simultaneously target Aβ and tau pathology, along with restoring metabolic support to optimize energy supply, mitochondrial health, and neurotrophic factors, has become essentials form AD.

7. Conclusion

AD has been recognized as a disorder of cerebral energy failure, in which mitochondrial vulnerability and disruption of ANLS lead to synaptic dysfunction and neuronal loss responsible for cognitive decline. Besides, metabolic dysregulation and energy deficiency profoundly impair hippocampal neurogenesis, thereby impeding neuroregenerative plasticity. Although many concepts have been established, ranging from impaired absorption of glucose and lactate to defects in the expression of molecular transporters, this book chapter emphasizes that reactive gliosis and aberrant neuroblastosis in the brain may act as major metabolic consumers. The abnormal cell cycle events of the brain upon pathogenic episodes can deplete local glucose and lactate availability, thereby depriving substrates essential for ATP synthesis associated with mitochondrial structural abnormalities and functional impairment, ultimately leading to synaptic dysfunction and progressive neurodegeneration. Therefore therapeutic strategies aimed at restoring mitochondrial resilience, enhancing lactate production and transport, and re-establishing astrocyte–neuron metabolic coupling may therefore offer disease-modifying potential. Collectively, restoring brain energy metabolism lost during the pathogenesis of AD represents a promising therapeutic strategy to safeguard neuronal integrity and enhance adult neurogenesis, thereby improving cognitive function and mitigating dementia.

Acknowledgement

M.K. has been supported by the Faculty Recharge Programme, University Grants Commission (UGC-FRP), New Delhi, India. 005266, Government of India, for the financial support. The author acknowledges the support of the University Grants Commission-Special Assistance Program (UGC-SAP) and Department of Science and Technology-Fund for Improvement of S&T Infrastructure (DST-FIST) for providing infrastructure to the Department of Animal Science, Bharathidasan University and Anusandhan National Research Foundation (ANRF)/Science Engineering Research Board (SERB) CRG/2023 for the financial support.

Disclosure

The authors would like to disclose that ChatGPT and grammar-checking tools were used solely for language editing and improvement of the manuscript.

References

Amidfar, M., Askari, G., & Kim, Y.-K. (2024). Association of metabolic dysfunction with cognitive decline and Alzheimer's disease: A review of metabolomic evidence. *Progress in Neuro-psychopharmacology & Biological Psychiatry, 128*, 110848. https://doi.org/10.1016/j.pnpbp.2023.110848.

Annadurai, N., Agrawal, K., Džubák, P., Hajdúch, M., & Das, V. (2017). Microtubule affinity-regulating kinases are potential druggable targets for Alzheimer's disease. *Cellular and Molecular Life Sciences: CMLS, 74*(22), 4159–4169. https://doi.org/10.1007/s00018-017-2574-1.

Ardanaz, C. G., de la Cruz, A., Minhas, P. S., Hernández-Martín, N., Pozo, M. Á., Valdecantos, M. P., ... Solas, M. (2024). Astrocytic GLUT1 reduction paradoxically improves central and peripheral glucose homeostasis. *Science Advances, 10*(42), eadp1115. https://doi.org/10.1126/sciadv.adp1115.

Babcock, K. R., Page, J. S., Fallon, J. R., & Webb, A. E. (2021). Adult hippocampal neurogenesis in aging and Alzheimer's disease. *Stem Cell Reports, 16*(4), 681–693. https://doi.org/10.1016/j.stemcr.2021.01.019.

Baloyannis, S. J. (2011). Mitochondria are related to synaptic pathology in Alzheimer's disease. *International Journal of Alzheimer's Disease, 2011*, 305395. https://doi.org/10.4061/2011/305395.

Beard, E., Lengacher, S., Dias, S., Magistretti, P. J., & Finsterwald, C. (2022). Astrocytes as key regulators of brain energy metabolism: New therapeutic perspectives. *Frontiers in Physiology, 12*, 825816. https://doi.org/10.3389/fphys.2021.825816.

Bhuiyan, N. Z., Hasan, M. K., Mahmud, Z., Hossain, M. S., & Rahman, A. (2023). Prevention of Alzheimer's disease through diet: An exploratory review. *Metabolism Open, 20*, 100257. https://doi.org/10.1016/j.metop.2023.100257.

Bonvento, G., & Bolaños, J. P. (2021). Astrocyte-neuron metabolic cooperation shapes brain activity. *Cell Metabolism, 33*(8), 1546–1564. https://doi.org/10.1016/j.cmet.2021.07.006.

Brooks, G. A. (2018). The science and translation of lactate shuttle theory. *Cell Metabolism, 27*(4), 757–785. https://doi.org/10.1016/j.cmet.2018.03.008.

Cai, Y., Liu, J., Wang, B., Sun, M., & Yang, H. (2022). Microglia in the neuroinflammatory pathogenesis of Alzheimer's disease and related therapeutic targets. *Frontiers in Immunology, 13*, 856376. https://doi.org/10.3389/fimmu.2022.856376.

Calkins, M. J., & Reddy, P. H. (2011). Amyloid beta impairs mitochondrial anterograde transport and degenerates synapses in Alzheimer's disease neurons. *Biochimica et Biophysica Acta, 1812*(4), 507–513. https://doi.org/10.1016/j.bbadis.2011.01.007.

Cerasuolo, M., Auriemma, M. C., Di Meo, I., Lenti, C., Papa, M., Paolisso, G., & Rizzo, M. R. (2025). Understanding the insulin-degrading enzyme: A new look at Alzheimer's disease and Aβ plaque management. *International Journal of Molecular Sciences, 26*(14), 6693. https://doi.org/10.3390/ijms26146693.

Chen, Z.-R., Huang, J.-B., Yang, S.-L., & Hong, F.-F. (2022). Role of cholinergic signaling in Alzheimer's disease. *Molecules (Basel, Switzerland), 27*(6), 1816. https://doi.org/10.3390/molecules27061816.

Clarke, D. D., & Sokoloff, L. (1999). *Substrates of cerebral metabolism. Basic neurochemistry: Molecular, cellular and medical aspects* (6th ed.). Lippincott-Raven. https://www.ncbi.nlm.nih.gov/books/NBK28048/.

Čolović, M. B., Krstić, D. Z., Lazarević-Pašti, T. D., Bondžić, A. M., & Vasić, V. M. (2013). Acetylcholinesterase inhibitors: Pharmacology and toxicology. *Current Neuropharmacology, 11*(3), 315–335. https://doi.org/10.2174/1570159×11311030006.

Congdon, E. E., Ji, C., Tetlow, A. M., Jiang, Y., & Sigurdsson, E. M. (2023). Tau-targeting therapies for Alzheimer disease: Current status and future directions. *Nature Reviews. Neurology, 19*(12), 715–736. https://doi.org/10.1038/s41582-023-00883-2.

Constantino, N. J., Ashley, C. C., & Macauley, S. L. (2025). The energetic collapse of the Alzheimer's brain: Metabolic inflexibility across cells and networks. *Journal of Neurochemistry, 169*(11), e70294. https://doi.org/10.1111/jnc.70294.

Corraliza-Gomez, M., Vargas-Soria, M., & Garcia-Alloza, M. (2025). Effect of antidiabetic drugs in Alzheimer's disease: A systematic review of preclinical and clinical studies. *Molecular Neurodegeneration, 20*, 112. https://doi.org/10.1186/s13024-025-00894-1.

De Sousa, R. A. L. (2022). Reactive gliosis in Alzheimer's disease: A crucial role for cognitive impairment and memory loss. *Metabolic Brain Disease, 37*(4), 851–857. https://doi.org/10.1007/s11011-022-00953-2.

Demars, M. P., Hollands, C., Zhao, K. D. (Tommy), & Lazarov, O. (2013). Soluble amyloid precursor protein-α rescues age-linked decline in neural progenitor cell proliferation. *Neurobiology of Aging, 34*(10), 2431–2440. https://doi.org/10.1016/j.neurobiolaging.2013.04.016.

Deng, Q., Wu, C., Parker, E., Liu, T. C.-Y., Duan, R., & Yang, L. (2024). Microglia and astrocytes in Alzheimer's disease: Significance and summary of recent advances. *Aging and Disease, 15*(4), 1537–1564. https://doi.org/10.14336/AD.2023.0907.

Dienel, G. A., & Cruz, N. F. (2015). Contributions of glycogen to astrocytic energetics during brain activation. *Metabolic Brain Disease, 30*(1), 281–298. https://doi.org/10.1007/s11011-014-9493-8.

Dilna, A., Deepak, K. V., Damodaran, N., Kielkopf, C. S., Kagedal, K., Ollinger, K., & Nath, S. (2021). Amyloid-β induced membrane damage instigates tunneling nanotube-like conduits by p21-activated kinase dependent actin remodulation. *Biochimica Et Biophysica Acta. Molecular Basis of Disease, 1867*(12), 166246. https://doi.org/10.1016/j.bbadis.2021.166246.

Ferreira-Vieira, T. H., Guimaraes, I. M., Silva, F. R., & Ribeiro, F. M. (2016). Alzheimer's disease: Targeting the cholinergic system. *Current Neuropharmacology, 14*(1), 101–115. https://doi.org/10.2174/1570159×13666150716165726.

Funfschilling, U., Supplie, L. M., Mahad, D., Boretius, S., Saab, A. S., Edgar, J., ... Nave, K.-A. (2012). Glycolytic oligodendrocytes maintain myelin and long-term axonal integrity. *Nature, 485*(7399), 517–521. https://doi.org/10.1038/nature11007.

Garone, C., De Giorgio, F., & Carli, S. (2024). Mitochondrial metabolism in neural stem cells and implications for neurodevelopmental and neurodegenerative diseases. *Journal of Translational Medicine, 22*, 238. https://doi.org/10.1186/s12967-024-05041-w.

González, A., Calfio, C., Churruca, M., & Maccioni, R. B. (2022). Glucose metabolism and AD: Evidence for a potential diabetes type 3. *Alzheimer's Research & Therapy, 14*, 56. https://doi.org/10.1186/s13195-022-00996-8.

Guo, C., Sun, L., Chen, X., & Zhang, D. (2013). Oxidative stress, mitochondrial damage and neurodegenerative diseases. *Neural Regeneration Research, 8*(21), 2003–2014. https://doi.org/10.3969/j.issn.1673-5374.2013.21.009.

Han, H., Zhao, Y., Du, J., Wang, S., Yang, X., Li, W., ... Chen, L. (2023). Exercise improves cognitive dysfunction and neuroinflammation in mice through Histone H3 lactylation in microglia. *Immunity & Ageing: I & A, 20*, 63. https://doi.org/10.1186/s12979-023-00390-4.

Hassan, H., Rawlinson, C., Lan, Y.-L., Jenkins, S., & Chen, R. (2025). Microglia-mediated phagocytosis in Alzheimer's disease: Mechanisms, heterogeneity, and therapeutic insights. *Biomolecules, 15*(11), 1629. https://doi.org/10.3390/biom15111629.

Hobday, A. L., & Parmar, M. S. (2021). The link between diabetes mellitus and tau hyperphosphorylation: Implications for risk of Alzheimer's disease. *Cureus, 13*(9), e18362. https://doi.org/10.7759/cureus.18362.

Iwata, R., & Vanderhaeghen, P. (2021). Regulatory roles of mitochondria and metabolism in neurogenesis. *Current Opinion in Neurobiology, 69*, 231–240. https://doi.org/10.1016/j.conb.2021.05.003.

Jensen, N. J., Wodschow, H. Z., Nilsson, M., & Rungby, J. (2020). Effects of ketone bodies on brain metabolism and function in neurodegenerative diseases. *International Journal of Molecular Sciences, 21*(22), 8767. https://doi.org/10.3390/ijms21228767.

Jonckheere, A. I., Smeitink, J. A. M., & Rodenburg, R. J. T. (2012). Mitochondrial ATP synthase: Architecture, function and pathology. *Journal of Inherited Metabolic Disease, 35*(2), 211–225. https://doi.org/10.1007/s10545-011-9382-9.

Kamila, P., Kar, K., Chowdhury, S., Chakraborty, P., Dutta, R., S, S., ... Prajapati, B. G. (2025). Effect of neuroinflammation on the progression of Alzheimer's disease and its significant ramifications for novel anti-inflammatory treatments. *IBRO Neuroscience Reports, 18*, 771–782. https://doi.org/10.1016/j.ibneur.2025.05.005.

Kandasamy, M., Lehner, B., Kraus, S., Sander, P. R., Marschallinger, J., Rivera, F. J., ... Aigner, L. (2014). TGF-beta signalling in the adult neurogenic niche promotes stem cell quiescence as well as generation of new neurons. *Journal of Cellular and Molecular Medicine, 18*(7), 1444–1459. https://doi.org/10.1111/jcmm.12298.

Kandasamy, M., Reilmann, R., Winkler, J., Bogdahn, U., & Aigner, L. (2011). Transforming growth factor-beta signaling in the neural stem cell niche: A therapeutic target for Huntington's disease. *Neurology Research International, 2011*(1), 124256. https://doi.org/10.1155/2011/124256.

Kim, T. A., Syty, M. D., Wu, K., & Ge, S. (2022). Adult hippocampal neurogenesis and its impairment in Alzheimer's disease. *Zoological Research, 43*(3), 481–496. https://doi.org/10.24272/j.issn.2095-8137.2021.479.

Kim, Y., Dube, S. E., & Park, C. B. (2025). Brain energy homeostasis: The evolution of the astrocyte-neuron lactate shuttle hypothesis. *The Korean Journal of Physiology & Pharmacology: Official Journal of the Korean Physiological Society and the Korean Society of Pharmacology, 29*(1), 1–8. https://doi.org/10.4196/kjpp.24.388.

Koelsch, G. (2017). BACE1 function and inhibition: Implications of intervention in the amyloid pathway of Alzheimer's disease pathology. *Molecules: A Journal of Synthetic Chemistry and Natural Product Chemistry, 22*(10), 1723. https://doi.org/10.3390/molecules22101723.

Koepsell, H. (2020). Glucose transporters in brain in health and disease. *Pflugers Archiv, 472*(9), 1299–1343. https://doi.org/10.1007/s00424-020-02441-x.

Kumar, A., Sidhu, J., Lui, F., & Tsao, J. W. (2025). *Alzheimer disease. StatPearls.* StatPearls Publishing, http://www.ncbi.nlm.nih.gov/books/NBK499922/.

Kwok, K. H.-H., Ho, P. W.-L., Chu, A. C.-Y., Ho, J. W.-M., Liu, H.-F., Yiu, D. C.-W., ... Ho, S.-L. (2010). Mitochondrial UCP5 is neuroprotective by preserving mitochondrial membrane potential, ATP levels, and reducing oxidative stress in MPP+ and dopamine toxicity. *Free Radical Biology & Medicine, 49*(6), 1023–1035. https://doi.org/10.1016/j.freeradbiomed.2010.06.017.

Kyrtata, N., Emsley, H. C. A., Sparasci, O., Parkes, L. M., & Dickie, B. R. (2021). A systematic review of glucose transport alterations in Alzheimer's disease. *Frontiers in Neuroscience, 15*, 626636. https://doi.org/10.3389/fnins.2021.626636.

Landry, T., & Huang, H. (2021). Mini review: The relationship between energy status and adult hippocampal neurogenesis. *Neuroscience Letters, 765*, 136261. https://doi.org/10.1016/j.neulet.2021.136261.

Le Douce, J., Maugard, M., Veran, J., Matos, M., Jégo, P., Vigneron, P.-A., ... Bonvento, G. (2020). Impairment of glycolysis-derived l-serine production in astrocytes contributes to cognitive deficits in Alzheimer's disease. *Cell Metabolism, 31*(3), 503–517.e8. https://doi.org/10.1016/j.cmet.2020.02.004.

Lev-Vachnish, Y., Cadury, S., Rotter-Maskowitz, A., Feldman, N., Roichman, A., Illouz, T., ... Okun, E. (2019). L-Lactate promotes adult hippocampal neurogenesis. *Frontiers in Neuroscience, 13*, 403. https://doi.org/10.3389/fnins.2019.00403.

Li, S., & Sheng, Z.-H. (2023). Oligodendrocyte-derived transcellular signaling regulates axonal energy metabolism. *Current Opinion in Neurobiology, 80*, 102722. https://doi.org/10.1016/j.conb.2023.102722.

Li, X., Fang, C., Li, Y., Xiong, X., Xu, X., & Gu, L. (2025). Glycolytic reprogramming during microglial polarization in neurological diseases. *Frontiers in Immunology, 16*, 1648887. https://doi.org/10.3389/fimmu.2025.1648887.

Liu, S.-L., Wang, C., Jiang, T., Tan, L., Xing, A., & Yu, J.-T. (2016). The role of Cdk5 in Alzheimer's disease. *Molecular Neurobiology, 53*(7), 4328–4342. https://doi.org/10.1007/s12035-015-9369-x.

Liu, Y., Bilen, M., McNicoll, M.-M., Harris, R. A., Fong, B. C., Iqbal, M. A., ... Slack, R. S. (2023). Early postnatal defects in neurogenesis in the 3xTg mouse model of Alzheimer's disease. *Cell Death & Disease, 14*(2), 138. https://doi.org/10.1038/s41419-023-05650-1.

Liu, Z., Guo, Y., Zhang, Y., Gao, Y., & Ning, B. (2024). Metabolic reprogramming of astrocytes: Emerging roles of lactate. *Neural Regeneration Research, 21*(2), 421–432. https://doi.org/10.4103/NRR.NRR-D-24-00776.

Lu, S., Zhao, Q., Guan, Y., Sun, Z., Li, W., Guo, S., & Zhang, A. (2024). The communication mechanism of the gut-brain axis and its effect on central nervous system diseases: A systematic review. *Biomedicine & Pharmacotherapy = Biomedecine & Pharmacotherapie, 178*, 117207. https://doi.org/10.1016/j.biopha.2024.117207.

Maciejewska, K., Czarnecka, K., & Szymański, P. (2021). A review of the mechanisms underlying selected comorbidities in Alzheimer's disease. *Pharmacological Reports, 73*(6), 1565–1581. https://doi.org/10.1007/s43440-021-00293-5.

Mani, S., Wasnik, S., Shandilya, C., Srivastava, V., Khan, S., & Singh, K. K. (2025). Pathogenic synergy: Dysfunctional mitochondria and neuroinflammation in neurodegenerative diseases associated with aging. *Frontiers in Aging, 6*, 1615764. https://doi.org/10.3389/fragi.2025.1615764.

Manickam, N., Radhakrishnan, R. K., Vergil Andrews, J. F., Selvaraj, D. B., & Kandasamy, M. (2020). Cell cycle re-entry of neurons and reactive neuroblastosis in Huntington's disease: Possibilities for neural-glial transition in the brain. *Life Sciences, 263*, 118569. https://doi.org/10.1016/j.lfs.2020.118569.

Marcus, C., Mena, E., & Subramaniam, R. M. (2014). Brain PET in the diagnosis of Alzheimer's disease. *Clinical Nuclear Medicine, 39*(10), e413–e426. https://doi.org/10.1097/RLU.0000000000000547.

Marí, M., Morales, A., Colell, A., García-Ruiz, C., & Fernández-Checa, J. C. (2009). Mitochondrial glutathione, a key survival antioxidant. *Antioxidants & Redox Signaling, 11*(11), 2685–2700. https://doi.org/10.1089/ars.2009.2695.

Miao, J., Chen, L., Pan, X., Li, L., Zhao, B., & Lan, J. (2023). Microglial metabolic reprogramming: Emerging insights and therapeutic strategies in neurodegenerative diseases. *Cellular and Molecular Neurobiology, 43*(7), 3191–3210. https://doi.org/10.1007/s10571-023-01376-y.

Misrani, A., Tabassum, S., & Yang, L. (2021). Mitochondrial dysfunction and oxidative stress in Alzheimer's disease. *Frontiers in Aging Neuroscience, 13*, 617588. https://doi.org/10.3389/fnagi.2021.617588.

Nelson, A. R., Sweeney, M. D., Sagare, A. P., & Zlokovic, B. V. (2016). Neurovascular dysfunction and neurodegeneration in dementia and Alzheimer's disease. *Biochimica et Biophysica Acta, 1862*(5), 887–900. https://doi.org/10.1016/j.bbadis.2015.12.016.

Nelson, M. L., Pfeifer, J. A., Hickey, J. P., Collins, A. E., & Kalisch, B. E. (2023). Exploring Rosiglitazone's potential to treat Alzheimer's disease through the modulation of brain-derived neurotrophic factor. *Biology, 12*(7), 1042. https://doi.org/10.3390/biology12071042.

Newman, L. A., Korol, D. L., & Gold, P. E. (2011). Lactate produced by glycogenolysis in astrocytes regulates memory processing. *PLoS One, 6*(12), e28427. https://doi.org/10.1371/journal.pone.0028427.

Nguyen, T. T., Ta, Q. T. H., Nguyen, T. K. O., Nguyen, T. T. D., & Van Giau, V. (2020). Type 3 diabetes and its role implications in Alzheimer's disease. *International Journal of Molecular Sciences, 21*(9), 3165. https://doi.org/10.3390/ijms21093165.

Nicholls, D. G. (2009). Mitochondrial calcium function and dysfunction in the central nervous system. *Biochimica et Biophysica Acta, 1787*(11), 1416–1424. https://doi.org/10.1016/j.bbabio.2009.03.010.

Owen, L., & Sunram-Lea, S. I. (2011). Metabolic agents that enhance ATP can improve cognitive functioning: A review of the evidence for glucose, oxygen, pyruvate, creatine, and L-carnitine. *Nutrients, 3*(8), 735–755. https://doi.org/10.3390/nu3080735.

Pathak, D., Shields, L. Y., Mendelsohn, B. A., Haddad, D., Lin, W., Gerencser, A. A., ... Nakamura, K. (2015). The role of mitochondrially derived ATP in synaptic vesicle recycling. *The Journal of Biological Chemistry, 290*(37), 22325–22336. https://doi.org/10.1074/jbc.M115.656405.

Pellerin, L., & Magistretti, P. J. (1994). Glutamate uptake into astrocytes stimulates aerobic glycolysis: A mechanism coupling neuronal activity to glucose utilization. *Proceedings of the National Academy of Sciences of the United States of America, 91*(22), 10625–10629. https://doi.org/10.1073/pnas.91.22.10625.

Petridi, S., Dubal, D., Rikhy, R., & van den Ameele, J. (2022). Mitochondrial respiration and dynamics of in vivo neural stem cells. *Development (Cambridge, England), 149*(23), dev200870. https://doi.org/10.1242/dev.200870.

Pisani, D. F., & Blondeau, N. (2025). Deciphering the brain glucose metabolism: A gateway to innovative stroke therapies. *Journal of Cerebral Blood Flow & Metabolism, 45*(9), 1635–1653.

Poornimai Abirami, G. P., Radhakrishnan, R. K., Johnson, E., Roshan, S. A., Yesudhas, A., Parveen, S., ... Kandasamy, M. (2020). The regulation of reactive neuroblastosis, neuroplasticity, and nutraceuticals for effective management of autism spectrum disorder. *Advances in Neurobiology, 24*, 207–222. https://doi.org/10.1007/978-3-030-30402-7_8.

Poteet, E., Choudhury, G. R., Winters, A., Li, W., Ryou, M.-G., Liu, R., ... Yang, S.-H. (2013). Reversing the warburg effect as a treatment for glioblastoma. *The Journal of Biological Chemistry, 288*(13), 9153–9164. https://doi.org/10.1074/jbc.M112.440354.

Prasad, H., & Rao, R. (2015). The Na+/H+ exchanger NHE6 modulates endosomal pH to control processing of amyloid precursor protein in a cell culture model of Alzheimer disease. *The Journal of Biological Chemistry, 290*(9), 5311–5327. https://doi.org/10.1074/jbc.M114.602219.

Rajmohan, R., & Reddy, P. H. (2017). Amyloid beta and phosphorylated tau accumulations cause abnormalities at synapses of Alzheimer's disease neurons. *Journal of Alzheimer's Disease: JAD, 57*(4), 975–999. https://doi.org/10.3233/JAD-160612.

Rak, M., Bénit, P., Chrétien, D., Bouchereau, J., Schiff, M., El-Khoury, R., ... Rustin, P. (2016). Mitochondrial cytochrome c oxidase deficiency. *Clinical Science (London, England: 1979), 130*(6), 393–407. https://doi.org/10.1042/CS20150707.

Ratan, Y., Rajput, A., Maleysm, S., Pareek, A., Jain, V., Pareek, A., ... Singh, G. (2023). An insight into cellular and molecular mechanisms underlying the pathogenesis of neurodegeneration in Alzheimer's disease. *Biomedicines, 11*(5), 1398. https://doi.org/10.3390/biomedicines11051398.

Reddy, P. H. (2011). Abnormal tau, mitochondrial dysfunction, impaired axonal transport of mitochondria, and synaptic deprivation in Alzheimer's disease. *Brain Research, 1415*, 136–148. https://doi.org/10.1016/j.brainres.2011.07.052.

Ren, Z., Zheng, S., Sun, Z., Luo, Y., Wang, Y., Yi, P., ... Xiao, W. (2025). Resveratrol: Molecular mechanisms, health benefits, and potential adverse effects. *MedComm, 6*(6), e70252. https://doi.org/10.1002/mco2.70252.

Sánchez-Ramírez, E., Ung, T. P. L., Stringari, C., & Aguilar-Arnal, L. (2024). Emerging functional connections between metabolism and epigenetic remodeling in neural differentiation. *Molecular Neurobiology, 61*(9), 6688–6707. https://doi.org/10.1007/s12035-024-04006-w.

Sangineto, M., Ciarnelli, M., Cassano, T., Radesco, A., Moola, A., Bukke, V. N., ... Serviddio, G. (2023). Metabolic reprogramming in inflammatory microglia indicates a potential way of targeting inflammation in Alzheimer's disease. *Redox Biology, 66*, 102846. https://doi.org/10.1016/j.redox.2023.102846.

Sayas, C. L., & Ávila, J. (2021). GSK-3 and Tau: A key duet in Alzheimer's disease. *Cells, 10*(4), 721. https://doi.org/10.3390/cells10040721.

Song, N., Mei, S., Wang, X., Hu, G., & Lu, M. (2024). Focusing on mitochondria in the brain: From biology to therapeutics. *Translational Neurodegeneration, 13*, 23. https://doi.org/10.1186/s40035-024-00409-w.

Sun, Y., Wei, K., Liao, X., Wang, J., Gao, L., & Pang, B. (2025). Lipid metabolism in microglia: Emerging mechanisms and therapeutic opportunities for neurodegenerative diseases (Review). *International Journal of Molecular Medicine, 56*(3), 139. https://doi.org/10.3892/ijmm.2025.5580.

Surya, K., Rathinam, A., Abubakkar, M. N., Jayachandran, K. S., Kandasamy, M., & Anusuyadevi, M. (2025). Resveratrol mitigates activated astrocytes and microglia preventing Alzheimer's disease (AD) progression and facilitates neuronal communication in Amyloid-β25-35 induced rat model for AD: A special emphasis on non-neuronal involvement in AD pathophysiology. *Psychopharmacology, 242*(11), 2529–2545. https://doi.org/10.1007/s00213-025-06814-x.

Ueno, M., Chiba, Y., Murakami, R., Miyai, Y., Matsumoto, K., Wakamatsu, K., ... Yanase, K. (2023). Distribution of Monocarboxylate transporters in brain and choroid plexus epithelium. *Pharmaceutics, 15*(8), 2062. https://doi.org/10.3390/pharmaceutics15082062.

Unger, M. S., Marschallinger, J., Kaindl, J., Höfling, C., Rossner, S., Heneka, M. T., ... Aigner, L. (2016). Early changes in hippocampal neurogenesis in transgenic mouse models for Alzheimer's disease. *Molecular Neurobiology, 53*(8), 5796–5806. https://doi.org/10.1007/s12035-016-0018-9.

Vainchtein, I. D., & Molofsky, A. V. (2020). Astrocytes and microglia: In sickness and in health. *Trends in Neurosciences, 43*(3), 144–154. https://doi.org/10.1016/j.tins.2020.01.003.

Walczak-Nowicka, J., & Herbet, M. (2021). Acetylcholinesterase Inhibitors in the treatment of neurodegenerative diseases and the role of acetylcholinesterase in their pathogenesis. *International Journal of Molecular Sciences, 22*(17), 9290. https://doi.org/10.3390/ijms22179290.

Walters, G. C., & Usachev, Y. M. (2023). Mitochondrial calcium cycling in neuronal function and neurodegeneration. *Frontiers in Cell and Developmental Biology, 11*, 1094356. https://doi.org/10.3389/fcell.2023.1094356.

Wan, Z., Zheng, J., Zhu, Z., Sang, L., Zhu, J., Luo, S., ... He, H. (2022). Intermediate role of gut microbiota in vitamin B nutrition and its influences on human health. *Frontiers in Nutrition, 9*, 1031502. https://doi.org/10.3389/fnut.2022.1031502.

Wang, C., & Youle, R. J. (2009). The role of mitochondria in apoptosis. *Annual Review of Genetics, 43*, 95–118. https://doi.org/10.1146/annurev-genet-102108-134850.

Wang, S., Shi, Y., Xin, R., Kang, H., Xiong, H., & Ren, J. (2025). Exploring the role of insulin resistance in bridging the metabolic syndrome and Alzheimer's disease-a review of mechanistic studies. *Frontiers in Endocrinology, 16*, 1614006. https://doi.org/10.3389/fendo.2025.1614006.

Wang, W., Zhao, F., Ma, X., Perry, G., & Zhu, X. (2020). Mitochondria dysfunction in the pathogenesis of Alzheimer's disease: Recent advances. *Molecular Neurodegeneration, 15*, 30. https://doi.org/10.1186/s13024-020-00376-6.

Wang, X., Shao, X., Yu, L., Sun, J., Yin, X.-S., Chen, Z., ... Ma, C. (2025). Changes in the pH value of the human brain in Alzheimer's disease pathology correlated with CD68-positive microglia: A community-based autopsy study in Beijing, China. *Molecular Brain, 18*(1), 10. https://doi.org/10.1186/s13041-025-01180-3.

Wu, A., Lee, D., & Xiong, W.-C. (2023). Lactate metabolism, signaling, and function in brain development, synaptic plasticity, angiogenesis, and neurodegenerative diseases. *International Journal of Molecular Sciences, 24*(17), 13398. https://doi.org/10.3390/ijms241713398.

Ye, Y.-C., Chai, S.-F., Li, X.-R., Wu, M.-N., Cai, H.-Y., & Wang, Z.-J. (2024). Intermittent fasting and Alzheimer's disease-targeting ketone bodies as a potential strategy for brain energy rescue. *Metabolic Brain Disease, 39*(1), 129–146. https://doi.org/10.1007/s11011-023-01288-2.

Zaninello, M., Baptista, P., & Duarte, F. V. (2024). Mitochondrial dynamics and mRNA translation: A local synaptic tale. *Biology, 13*(9), 746. https://doi.org/10.3390/biology13090746.

Zebhauser, P. T., Berthele, A., Goldhardt, O., Diehl-Schmid, J., Priller, J., Ortner, M., & Grimmer, T. (2022). Cerebrospinal fluid lactate levels along the Alzheimer's disease continuum and associations with blood-brain barrier integrity, age, cognition, and biomarkers. *Alzheimer's Research & Therapy, 14*, 61. https://doi.org/10.1186/s13195-022-01004-9.

Zhang, H., Wei, W., Zhao, M., Ma, L., Jiang, X., Pei, H., ... Li, H. (2021). Interaction between Aβ and Tau in the pathogenesis of Alzheimer's disease. *International Journal of Biological Sciences, 17*(9), 2181–2192. https://doi.org/10.7150/ijbs.57078.

Zhang, M., Cheng, X., Dang, R., Zhang, W., Zhang, J., & Yao, Z. (2018). Lactate deficit in an Alzheimer disease mouse model: The relationship with neuronal damage. *Journal of Neuropathology and Experimental Neurology*, 77(12), 1163–1176. https://doi.org/10.1093/jnen/nly102.

Zhang, N., Yi, R., Zhong, F., Lu, Y., Chen, W., Ke, Z., ... Li, W. (2025). Oligodendrocytes and myelination: Pioneering new frontiers in cognitive neuroscience. *Frontiers in Neuroscience, 19*, 1618468. https://doi.org/10.3389/fnins.2025.1618468.

Zhang, S., Wang, J., Xu, Z., Cheng, Z., Shao, B., & Yu, J. (2026). Lactate: Elucidating its indispensable role in human health. *Molecular Cancer, 25*, 2. https://doi.org/10.1186/s12943-025-02519-z.

Zhou, Y., Wang, X., Liu, Y., Gu, Y., Gu, R., Zhang, G., & Lin, Q. (2023). Mechanisms of abnormal adult hippocampal neurogenesis in Alzheimer's disease. *Frontiers in Neuroscience, 17*, 1125376. https://doi.org/10.3389/fnins.2023.1125376.

CHAPTER FOUR

Astroglial dysregulation and gliotransmitter imbalance in schizophrenia

Mercy Priyadharshini Babu Deva Irakkam[a] **and Mahesh Kandasamy**[a,b,*]

[a]Laboratory of Stem Cells and Neuroregeneration, Department of Animal Science, Bharathidasan University, Tiruchirappalli, Tamil Nadu, India
[b]University Grants Commission-Faculty Recharge Programme (UGC-FRP), New Delhi, India
*Corresponding author. e-mail address: pkmahesh5@gmail.com; mahesh.kandasamy@bdu.ac.in

Contents

Abstract

Schizophrenia is a neuropsychiatric illness characterized by progressive deterioration of thought processes and marked behavioral abnormalities arising from an unknown pathogenesis underlying the false perceptions, primarily hallucinations and delusions. The clinical symptoms of schizophrenia also include apathy, communication disorders and suicidal thoughts. Although the etiopathogenesis of schizophrenia remains not fully elucidated, the development of schizophrenia has been linked to adverse pregnancy, obstetric complications, neurodevelopmental disorders, neurotransmission imbalance and aberrant neurogenic events. Astrocytes are the most predominant glial cell type in the central nervous system (CNS), where they provide metabolic support, regulate neuroimmune mechanisms, facilitate neurotransmitter reuptake, and sustain synaptic homeostasis. Abnormal neural transmission is

International Review of Neurobiology, Volume 188
ISSN 0074-7742, https://doi.org/10.1016/bs.irn.2026.05.013

considered central to the pathogenesis of schizophrenia. Recent studies have highlighted that malformations in glial cells, particularly dysfunctional or reactive astrocytes, play a crucial role in the pathophysiology. Astrocytic dysregulation in schizophrenia is likely to result in synaptic dysfunction due to altered levels of key gliotransmitters such as glutamate, gamma-aminobutyric acid (GABA) and D-serine leading to altered neurobehavioral outcomes. Thus, insight into the scientific concepts that interrelate the pathophysiology of schizophrenia with astroglia dysregulation at the level of gliotransmitter imbalance could provide innovative hints for developing therapeutic strategies for the treatment of schizophrenia. This chapter describes the key roles of astrocytes and emphasizes imbalances in gliotransmitters as main contributors to the pathophysiology of schizophrenia.

1. Introduction

Schizophrenia is a multifaceted neuropsychiatric disorder that involves complex interplay between genetic abnormalities, epigenetic variables, developmental defects, and altered environmental factors (Föcking et al., 2019). Around 1 % of the global population is estimated to be affected by schizophrenia (Hany & Rizvi, 2026). Considering the clinical trajectory, schizophrenia has been characterized to exhibit (1) positive symptoms like delusions, disorganized thought, hallucinations, (2) negative symptoms like emotional disconnection, anhedonia, avolition and (3) cognitive symptoms, which include dysregulation in perception and memory loss (Orsolini et al., 2022). Among various mental deformities, hallucination and delusions are the most robust and consistently reported symptomatological manifestations of schizophrenia (Anwar et al., 2025). Besides, accumulating evidence indicates that cognitive impairment is increasingly recognized as a core pathological signature of the disorder, contributing to progressive deterioration of interpersonal skills and a considerable degree of memory impairment (Vita et al., 2024). However, the neurodegenerative events that are typically associated with memory loss are less evident in schizophrenia (Wilson et al., 2024). Thus, the pathophysiology of schizophrenia is speculated to arise from a complex interplay among metabolic disturbances, genetic variations, genomic alterations, abnormal neurodevelopmental processes, environmental influences and altered neurogenic processes (Luvsannyam et al., 2022) Most pathogenic alterations in schizophrenia converge on disruptions in the neural circuit arrangements and aberrant neurotransmission (Lewis & Sweet, 2009). While evidence from studies using dopamine receptor antagonists indicates that abnormal activation of the dopaminergic system in the mesolimbic pathway underlies positive symptoms, including hallucinations and delusions, mesocortical hypodopaminergia appears to contribute to cognitive dysfunction and memory loss (Brisch et al., 2014). Reports from studies using N-methyl-D-aspartate (NMDA) receptor antagonists such as ketamine and phencyclidine (PCP) show that reduced glutamatergic signaling leads to cortical and subcortical dysfunction, which in turn contributes to behavioral alterations in schizophrenia (Iasevoli et al., 2014). Besides, impaired cortical inhibitory control resulting from dysfunction of parvalbumin interneurons has been linked to clinical manifestations of schizophrenia (Lewis et al., 2012). Also, aberrant synaptic pruning and neural circuit disconnection, mediated by dysregulation of the complement cascade and neuroimmune signaling appear to be involved in the pathophysiology of schizophrenia (Eltokhi et al., 2020). Moreover, neurodevelopmental changes affecting dynamic of neural precursor cells have been recognized to contribute to schizophrenia (Fatemi & Folsom, 2009). Considering the above-mentioned facts, the etiology and pathophysiology of schizophrenia are multifactorial, making it difficult to pursue or target a single underlying mechanism. Therefore, identifying a central regulatory factor that serves as an axis in integrating the diverse pathophysiological mechanisms is of considerable importance.

Astrocytes are a predominant subtype of glial cells in the CNS (Gradisnik & Velnar, 2023). To note, radial glia have been extensively reported to exhibit abnormalities associated with aberrant neural development and pathological states (Viola et al., 2024). Astrocytes have the potential to regulate neuronal function through physical support and via the release of their secretome, including gliotransmitters and signaling molecules (Ota et al., 2013). Activation of astrocytes during disease states can contribute to pathogenesis by disrupting the neurotransmitter balance and physiology of gap-junction. Remarkably, neuroimaging studies and several post-mortem data strongly indicate that abnormal astroglial signatures appear to be a prominent pathogenic feature of schizophrenia (Tarasov et al., 2020). Marked alterations in astrocyte density and aberrant transcriptomic signatures of astrocytes across distinct brain regions have been reported in the brains of individuals with schizophrenia (Toker et al., 2018). Despite substantial evidence demonstrating abnormalities in neurotransmission and altered neurodevelopment in schizophrenia, a deeper understanding of astroglia dysfunction and impaired gliotransmission could provide significant novel insights into disease pathophysiology and facilitate the development of more precise therapeutic interventions.

2. The biological significance of astroglia in a healthy brain

Astrocytes are the most abundant and functionally diverse glial cells in the CNS (Westergard & Rothstein, 2020). Astrocytes play a key role in development, as they serve as multipotent NPCs, particularly during early neural development when they exhibit radial glial morphology (Lattke & Guillemot, 2022). Radial glial astrocytes have been considered the neural stem cells (NSCs) and contribute to the cytoarchitecture of the developing brain (Howard et al., 2008). Radial astroglia can generate neurons and macroglial cells and provide guidance for the structural framework of the brain by facilitating the orderly migration of neuroblasts from the ventricular zone to their designated cortical layers (Gengatharan et al., 2016). This astrocyte-guided neural migration is essential for the proper formation of the three-dimensional neuroarchitecture of the various brain regions, including the cortex (Miyajima et al., 2023). In the adult brain, astrocytes assume different roles to execute multiple tasks in maintaining both organizational and functional integrity essential for neuroplasticity (Sidoryk-wegrzynowicz et al., 2011). Astrocytes play a crucial role in the regulation of extracellular ion balance, water equilibrium and maintain neurotransmitter levels through molecular internalization, reutilization and recycling mechanisms (Weber & Barros, 2015). Astrocytes contribute structural and functional support to the blood–brain barrier, regulate cerebral blood flow via neurovascular coupling, and supply metabolic substrates to neurons (Schiera et al., 2024). While the expression of glial fibrillary acidic protein (GFAP), S100 calcium-binding protein B (S100β), and vimentin are considered the classical structural markers of astrocytes, their functional status is assessed by the presence of aldehyde dehydrogenase family 1, member L(ALDH1L)-1, a cytosolic enzyme involved in folate metabolism, glutamate-aspartate transporter (GLAST), key glutamate transporters involved in synaptic clearance and prevention of excitotoxicity, Kir4.1, an inwardly rectifying $K2^+$ channel, reflects spatial potassium buffering capacity, aquaporin (AQP)-4 in perivascular end feet, indicates water homeostasis and neurovascular coupling and connexin 43 as a principal gap junction protein mediating astrocytic network communication (Jurga et al., 2021). Following injury or pathological stimuli, astrocytes upregulate complement component 3 (C3) and serine protease inhibitor A3N (SerpinA3N), indicative of a phenotypic and functional shift toward reactive astrogliosis (Gomes et al., 2024). Given the potent gliocrine capacity, astrocytes actively participate in the regulation of neuronal survival, synaptic plasticity, vascular coupling, immune signaling, and tissue remodelling through the release of a wide range of signaling molecules (Yamamoto & Takano, 2025). Astrocytes release key neurotrophic elements, such as brain-derived neurotrophic factor (BDNF), glial cell line–derived neurotrophic factor (GDNF), nerve growth factor (NGF), and ciliary neurotrophic factor (CNTF), which collectively support neuronal survival, synaptic plasticity, dendritic remodelling and neurogenesis (Linnerbauer & Rothhammer, 2020). In addition to trophic support, astrocytes function as immune-competent cells by secreting cytokines such as interleukin (IL)-1β, IL-6, and tumor necrosis factor-alpha (TNF)-α, as well as chemokines like C-C motif ligand (CCL)-2, thereby modulating neuroinflammatory responses synchronized with microglia, endothelial cells and pericytes (Giovannoni & Quintana, 2020). The astrocytes mediated tissue remodelling and repair appear to be mediated through the release of growth factors such as vascular endothelial growth factor (VEGF), transforming growth factor (TGF)-β, and fibroblast growth factor (FGF), which contribute to angiogenesis and cell cycle events of NSCs (Puebla et al., 2022). The glial-derived factors are essential for maintaining stemness, promoting proliferation, driving neuronal differentiation, enhancing cell survival, and supporting regenerative and neuroprotective processes underlying the cognitive process in the CNS (Gallo & Deneen, 2014). Moreover, astrocytes secrete extracellular vesicle components such as thrombospondins, laminin, and tenascin-C, which influence synaptogenesis, neurite outgrowth, and structural plasticity of the brain (Baldwin & Eroglu, 2017). Metabolically, astrocytes sustain neuronal activity by supplying lactate, ketone bodies, cholesterol, and amino acids as alternative substrates to glucose for high-energy synaptic transmission (Beard et al., 2022). Through the release of extracellular vesicles containing proteins, lipids, and regulatory microRNAs, astrocytes facilitate intercellular communication and modulate various aspects of neuroplasticity (Upadhya et al., 2020). Collectively, this multifaceted gliocrine function designates astrocytes as dynamic cellular orchestrators of brain function at various aspects, whereas dysregulation of these pathways may contribute to the underlying pathophysiology in various neurodegenerative and psychiatric disorders.

An additional dimension of astrocyte biology lies in their involvement in dynamic modulation of neuronal signaling and network activity via the release of gliotransmitters, which tightly regulate neuronal excitability and synaptic transmission (Purushotham & Buskila, 2023). Prominent examples of gliotransmitters include glutamate, adenosine triphosphate (ATP), D-serine, and gamma-aminobutyric acid (GABA). These gliotransmitters play a critical role in shaping cognitive behaviour, including learning and memory (Harada et al., 2016). Abnormal release or dysregulation of pathways related to gliotransmitters has been increasingly implicated in the disease mechanisms in the development and progression of various neurological and psychiatric disorders, including schizophrenia (Won et al., 2025). While the role of trophic factors has been extensively studied in neurodegenerative disorders, deeper insight into astrocytic gliotransmission could provide a new perspective on disease pathogenesis and unveil novel therapeutic regimens to overcome and manage neuropsychiatric disorders.

3. Reactive astrogliosis in neurodegenerative disorders and schizophrenia

The most prevalent neurodegenerative disorders, such as Alzheimer's disease (AD), Parkinson's disease (PD), Huntington's disease (HD), Spinocerebellar Ataxia (SCA) and Amyotrophic lateral sclerosis (ALS) have been characterized by the progressive degeneration of specific neuronal populations in distinct brain regions (Ahmad et al., 2017). These neurodegenerative disorders are highly debilitating pathogenic conditions as they display progressively worsening motor, non-motor, and cognitive symptoms (Lamptey et al., 2022). While the etiology of many of these disorders remains unclear, it is hypothesized that they arise, at least in part, from non–cell-autonomous mechanisms associated with glial cell abnormalities in the brain (Lobsiger & Cleveland, 2007). Substantial experimental evidence indicates that aberrant proliferative and dystrophic signatures in astrocytes drive their transition into a reactive state, representing an early hallmark of disease progression across a broad spectrum of neurological, mood, affective, and psychiatric disorders (Verkhratsky et al., 2023). This reactive transition of astrocytes can disrupt calcium signaling, promote neuroinflammatory and apoptotic pathways, contributing to progressive synaptic dysfunction and destabilization of neuronal networks (Koizumi et al., 2021). In AD, activated astrocytes have been identified as a key source of amyloid-β and also serve as a reservoir for the release of neuroinflammatory mediators (Zhao et al., 2011). Moreover, astrocyte-derived apolipoprotein E has been shown to induce the formation of poorly clearable amyloid-β aggregates, accounting for neurodegeneration-mediated dementia (Strickland et al., 2024). Similarly, in PD, α-synuclein, a key pathogenic driver, is internalized by astrocytes, where its accumulation induces a reactive phenotype marked by abnormal elevations in intracellular calcium, along with increased secretion of C3, pro-inflammatory cytokines, and oxidative mediators, thereby promoting non–cell-autonomous neuronal cell death (Sorrentino et al., 2019). In spinocerebellar ataxia, Bergmann glia in the cerebellum exhibits a dystrophic phenotype driven by activation of the nuclear factor kappa B (NF-κB) signaling pathway (Edamakanti et al., 2023). This reactive transformation is accompanied by elevated intracellular Ca^{2+} signaling, which enhances NMDA receptor–mediated excitotoxicity and adversely affects Purkinje neurons, ultimately contributing to motor deficits (Armada-Moreira et al., 2020). Moreover, in HD, selective expression of mutant huntingtin (mHTT) in astrocytes has been reported to induce neurodegeneration, leading to motor and cognitive deficits (Molero et al., 2016). Although the primary pathogenic trigger intrinsically targets neurons, accumulating evidence suggests that astrocyte-mediated alterations may be more influential in initiating and propagating neurodegeneration (Oksanen et al., 2019). Given the active participation of astroglial abnormalities in various conditions, structural and functional variations of astrocytes could be key pathogenic features of many neuropsychiatric disorders, including schizophrenia.

4. Astroglial pathology in schizophrenia

Astroglial pathology is increasingly recognized as a key malicious component of schizophrenia pathobiology, reframing the disorder from a purely neuronal synaptopathy to a disturbance of integrated neuron–glia network homeostasis (Verkhratsky et al., 2014). In schizophrenia, converging data from post-mortem analyses, neuroimaging, peripheral biomarkers, and experimental models point to astroglial dysfunction as a central pathological feature contributing to circuit dysconnectivity and cognitive behavioral impairment (Ricci et al., 2026). Consistent reductions in glial cell density and disrupted maturation across several brain regions indicate a disturbance of neuron-glial interactions rather than a classical neurodegenerative process (Bernstein et al., 2025). Neurochemical investigations further support this concept by providing the supportive evidence for alterations in astrocyte-associated metabolites (Oksanen et al., 2019). Notably, medial prefrontal myo-inositol levels shows modest but significant reductions, suggesting impaired astrocytic metabolism within specific patient subgroups (Das et al., 2018). Peripheral studies extend these observations, revealing increased levels of astrocytic markers such as GFAP in circulating exosomes and plasma, which correlate with disease severity and treatment resistance (Forró et al., 2024). Notably, astrocytic abnormalities intersect with neurodevelopmental inflammatory pathways, altered gliotransmission, and deficient synaptic maintenance, thereby influencing dopaminergic and glutamatergic signaling central to schizophrenia pathophysiology (Vellucci, et al., 2025). Post-mortem and translational studies report reduced glial cell numbers across cortical and subcortical regions, with astrocytic deficits accompanying more pronounced oligodendroglial loss, indicating impaired circuit maintenance rather than classical neurodegeneration (Bogus et al., 2026). Experimental models demonstrate that astrocyte-restricted molecular alterations can drive schizophrenia-like phenotypes; developmental overexpression of heme oxygenase (HMOX)-1 induces mesolimbic hyperdopaminergia and behavioral abnormalities reversible by antipsychotic treatment, highlighting astrocytes as upstream modulators of dopaminergic dysregulation (Tavitian et al., 2020). Further pharmacological data show that clozapine, a widely used antipsychotic drug, directly modifies signaling cascades in astroglial cells by enhancing connexin-43–mediated release of glutamate and D-serine (Fukuyama et al., 2020). Transcriptomic and preclinical findings also implicate astrocyte immune crosstalk and complement-associated synaptic remodelling, with schizophrenia-like models exhibiting enrichment of astroglial and microglial immune genes reversible by cannabidiol. Importantly, these alterations span an orchestration from astrocytopathy to maladaptive inflammatory activation, rather than uniform reactive astrogliosis, indicating stage- and subtype-specific glial states (Vellucci, Mazza, Barone, Nasti, De Simone, et al., 2025). A 20–40 % decrease in astrocytic excitatory amino acid glutamate transporters (EAAT)1/2 transporter expression impairs glutamate clearance at the molecular level rendering to excitotoxicity (Todd & Hardingham, 2020). Cortical imbalance and dopaminergic hyperactivity are caused by dysregulated monoamine oxidase (MAO)-B activity, which also disrupts dopamine metabolism and increases oxidative stress (Graves et al., 2020). Collectively, current evidence positions astrocytes as active determinants of synaptic regulation, metabolic support, and network connectivity, and identifies astroglial dysfunction as a tractable target for biomarker development and glia-directed therapeutic strategies in schizophrenia.

5. Aberrant gliotransmission and schizophrenia

Recent recognition of astrocytes as active regulators of neuroenergetic balance and synaptic transmission has gradually changed the long-standing view on neuron-centric understanding of the pathophysiology of schizophrenia. Astrocytes regulate excitatory-inhibitory equilibrium, redox balance, and calcium signaling across cortical and subcortical circuits by releasing important gliotransmitters, including glutamate, D-serine, ATP, and GABA (Liu et al., 2021). Glutamatergic signaling at the tripartite synapse is now understood as a dynamic, astrocyte-regulated process rather than a purely neuronal event, with astrocytes actively shaping excitatory transmission through coordinated uptake, metabolism, and activity-dependent glutamate release (Cuellar-Santoyo et al., 2023).

D-serine is now recognized as a critical modulator of excitatory neurotransmission at the tripartite synapse, where coordinated interactions between astrocytes and neurons regulate NMDA receptor (NMDAR) mediated signaling pathways and synaptic plasticity (Wang et al., 2026). Although originally described as a Ca^{2+} dependent astrocytic gliotransmitter, subsequent evidence demonstrates that D-serine arises from a tightly coupled metabolic and signaling partnership between these two cell types (Martineau et al., 2014). Astrocytes sense synaptic activity through receptor-driven Ca^{2+} signaling and readily supply substrates, particularly L-serine, that sustain neuronal D-serine synthesis and determine availability of co-agonist for NMDAR (Kang et al., 2013). This astrocyte–neuron serine exchange functionally links gliotransmission to brain energy metabolism (Murat & García-Cáceres, 2021). Activity-dependent shifts in astrocytic metabolism, including glycolytic reprogramming and lactate signaling, regulate D-serine production and thereby influence synaptic and cognitive processes (Fernández-Moncada et al., 2024). Functionally, D-serine signaling contributes to long-term potentiation, circuit engagement, thereby determining the behavior across species, whereas disruption of astroglial release pathways can leads to stress-related and neuropsychiatric phenotypes (Wang et al., 2026).

GABA, the principal inhibitory neurotransmitter not only secreted by neurons but also hyperpolarizes neurons via Cl^- influx or K^+ efflux through GABA receptors; in contrast, activation of astrocytic GABA receptors raises intracellular Ca^{2+} and initiates the release of gliotransmitters, including glutamate, ATP and D-serine, thereby establishing bidirectional neuron–glia communication (Nawafleh et al., 2022). Astrocytic dysfunction is linked to the cognitive and behavioural symptoms of schizophrenia upon disruption of the signaling pathways, resulting in impaired glutamate-glutamine cycling, dysregulated NMDA receptor co-activation, and a failure in synaptic integration (Mei et al., 2018). Recent literature suggests that mitochondrial dysfunction, abnormal calcium signaling, and decreased expression of astrocytic transporters together contribute to oxidative stress and excitotoxicity. Maladaptive neuroinflammatory processes and gliotransmission are also facilitated by reactive astrocytes (Satarker et al., 2022). A promising treatment strategy to restore synaptic stability and energy balance is pharmacological modulation of astrocyte-derived signals, namely by L-β-aminoisobutyric acid (L-BAIBA) activation of GABA-B and group III metabotropic glutamate (mGlu) receptors (Fukuyama et al., 2023). Astrocytic mechanisms linked to schizophrenia are characterized by: (1) the glutamate-glutamine cycle and excitotoxicity, which emphasize energy deficits and metabolic impairment; (2) dysregulation of D-serine, ATP, and GABA release, indicating the disruption of gliotransmission; and (3) impaired calcium signaling and synaptic crosstalk, denoting altered Ca^{2+} dynamics and their effect on network synchronization (Mei et al., 2018; Vellucci, Mazza, Barone, Nasti, De Simone, et al., 2025).

5.1 Glutamate–glutamine cycle and excitotoxicity

By coordinating the metabolism of neurons and astrocytes, the glutamate-glutamine cycle preserves excitatory neurotransmission and shields neurons from glutamate-induced damage (Sidoryk-Węgrzynowicz et al., 2024). Glutamate synthetase (GS) transforms extracellular glutamate into glutamine, which astrocytes recycle back to neurons after clearing more than 90 % of it using the EAAT1 and EAAT2- Glutamate Transporter-(GLT)-1 transporters (Zhang et al., 2024). These transporter expression decreases 20–40 % in schizophrenia, which impairs glutamate clearance, resulting in overstimulation of NMDA and α-amino-3-hydroxy-5-methyl-4-isoxazolepropionic acid (AMPA) receptors and increased extracellular glutamate (Rubio et al., 2012). This overactivation weakens cortical connections by causing oxidative stress, dendritic damage, and Ca^{2+} overload. Neuronal energy supply is decreased by redox imbalance and mitochondrial dysfunction, which also impair ATP-dependent glutamate uptake and the astrocyte-neuron lactate shuttle (Trubalski et al., 2025). In the meantime, excitatory overdrive is maintained by hemichannel and pannexon overactivation, which increases glutamate and ATP release (Orellana et al., 2016). When dysregulated, D-serine, an astrocyte-derived NMDA co-agonist, contributes to excitotoxicity but typically supports neuroplasticity (Coyle et al., 2020). Persistent hypofunction of the NMDA receptor is linked to anomalies in D-serine metabolism is the key hallmark of schizophrenia (Balu, 2016). According to Fukuyama et al., 2023, clozapine reduces excitotoxic stress and thalamocortical glutamatergic hyperactivity via improving L-BAIBA signaling via GABA-B and group III mGlu receptors (Fukuyama et al., 2023). Thus, disruption of the glutamate-glutamine cycle appears to be an important mechanism connecting astrocytic metabolic failure to excitotoxic damage. Therefore, addressing receptor cross-talk and astrocyte glutamate metabolism may aid in reestablishing excitatory-inhibitory balance and synaptic resilience.

5.2 Dysregulation of D-Serine, ATP, and GABA release

In schizophrenia, excitatory-inhibitory signaling is disrupted by an imbalance in gliotransmission (Lányi et al., 2024). NMDA receptor activity is regulated by D-serine, which is produced by serine racemase; wherein its disrupted level leads to dysregulation in plasticity and excitotoxicity (Billard, 2018). Adenosine produced from ATP is released by hemichannels and pannexons, modifying purinergic signaling and energy coupling (Dahl, 2015). Activation in these channels by inflammation leads to ATP efflux and oxidative stress (Dosch et al., 2018). In addition, reactive astrocytes release tonic GABA, which typically regulates excitability but, in excess, impairs cognitive ability (Jo et al., 2014). Adenosine 2A–Dopamine 2 (A2A–D2) receptor heteromers disrupt the glutamate–GABA balance that is essential to psychosis in the striatum by changing gliotransmitter output and astroglial Ca^{2+} dynamics (Pelassa et al., 2019). A convergent astrocytic axis that combines neuroinflammation, oxidative stress, and synaptic disarray is defined by dysregulated D-serine, ATP, and GABA release (Illes et al., 2025).

5.3 Impaired calcium signaling and synaptic crosstalk

Astrocytic Ca^{2+} signaling uses microdomain oscillations that are transmitted through Cx43 gap junctions to coordinate neuron–glia communication (Wang et al., 2025). Ca^{2+} dynamics are essential for synaptic coordination. Synaptic remodelling in cortical areas and astrocyte loss in schizophrenia decrease Ca^{2+} wave propagation and impair network synchronization. Through inflammatory hemichannel opening, aberrant Ca^{2+} signaling raises cytosolic Ca^{2+} and oxidative stress and modifies gliotransmitter release (Chen et al., 2025). In striatal astrocytes, receptor complexes like A2A–D2 heteromers further disrupt the excitatory–inhibitory balance (Pelassa et al., 2019). While the exact mechanisms behind Ca^{2+} dependent exocytosis are still up for debate, accumulating data suggest that synaptic regulation and plasticity are hampered by disturbed Ca^{2+} signaling (Pang & Südhof, 2010). Cortical dysconnectivity is reinforced by these anomalies, which exacerbate excitotoxic and metabolic stress in conjunction with decreased glutamate (McCutcheon et al., 2020).

Considering these facts, abnormal gliotransmission is now recognized as a key non-neuronal regulatory mechanism in schizophrenia, with astrocytes actively shaping synaptic and extrasynaptic communication through Ca^{2+}-dependent release of gliotransmitters such as glutamate, D-serine, ATP, adenosine, and GABA (Yamamoto & Takano, 2025). These signals regulate excitation–inhibition balance, synaptic plasticity, and network oscillations that underlie cognition and behavior (Murat & García-Cáceres, 2021). Under physiological conditions, astrocytic signaling maintains proper neuromodulation and synaptic homeostasis; however, in schizophrenia, this system becomes dysregulated, leading to impaired astrocyte–neuron coupling, altered NMDA receptor co-agonism, and reduced plasticity associated with cognitive and negative symptoms (de Oliveira Figueiredo et al., 2022). Biochemical studies demonstrate altered central and peripheral levels of D-amino acids, particularly D-serine and D-aspartate, which are critical astrocyte-derived modulators of NMDA receptor activity, supporting a model of glial-driven hypoglutamatergia (Rampino et al., 2024). Concurrently, glial dysfunction is closely associated with neuroinflammation, oxidative stress, and excitotoxicity, pathological processes that both result from and exacerbate disrupted gliotransmission (Quincozes-Santos et al., 2021). A tripartite synapse regulation involving neuronal, astrocytic, and metabolic interactions, in which disturbed gliotransmitter dynamics impair NMDA-dependent plasticity, cortical synchrony, and neurochemical stability, thereby contributing to psychosis (Noriega-Prieto & Araque, 2021). Ketamine and other NMDA antagonists consistently cause hallucinations and interfere with astrocytic calcium signaling, indicating that poor neuron–astrocyte coupling plays a role in perceptual distortions (Illes et al., 2025). Furthermore, abnormal sensory firing patterns and cortical hyperexcitability seen in psychotic states can be exacerbated by extrasynaptic glutamate elevation caused by malfunctioning of astrocytic EAAT1/2 (Bera et al., 2025). Further regulating thalamocortical oscillations is ATP–adenosine signaling from astrocytes; decreased adenosine tone has been linked to sensory gating abnormalities, a defining feature of diseases that are prone to hallucinations (Illes et al., 2025).

6. Translational insights and biomarker potential of gliotransmission in schizophrenia

The combination of cellular, computational, and molecular methods has revolutionized our comprehension of glial dysfunction as a systems-level cause of neuropsychiatric disorders (Braff & Braff, 2013). Converging evidence now links astrocytic and microglial dysregulation as a key factor in synaptic disconnection, oxidative stress, and neuroinflammation in schizophrenia and related disorders, whereas traditional neuropharmacology focused on neuronal signaling (Comer et al., 2020). These establish glial dysfuntion as a major factor in cognitive dysfunction and network instability (Vishnumukkala et al., 2025). Glial pathology is a quantifiable basis for target validation, patient categorization, and treatment monitoring from a translational standpoint (Garden & Campbell, 2016). Cortical dysconnectivity and cognitive decline are closely associated with abnormal astrocyte density, as determined by postmortem and imaging data (García-San-Martín et al., 2025). Astroglial phenotypes unique to each patient can now be established by induced pluripotent stem cells (iPSCs)-derived brain organoids, allowing for pharmacodynamic monitoring in line with clinical results (Zheng et al., 2026). Using prefrontal-based social cognition tasks in conjunction with behavioural z-scoring converts molecular dysregulation into quantifiable functional deficits (Uliana & Grace, 2026). N-acetylcysteine improves glutathione and EAAT2 function with ~30–40 % symptom reduction, and vitamin D deficiency is observed in 30–50 % of patients, supporting its inclusion in endocrine immune biomarker panels (Bradlow et al., 2022). Multimodal evaluation of therapy response is in coherence with mechanistic biomarkers that span molecular and imaging modalities, including positron emission tomography (PET) and functional magnetic resonance imaging (fMRI), such as astrocytic MAO-B activity, 5-HT_2A network connection, and Cx43 coupling (Sun et al., 2026). In order to enhance real-world translational validity, implementation science frameworks also recommend integrating these biomarkers into affordable, adherence-optimized regimens, such as long-acting injectables (Guastaferro et al., 2025). Precision pharmacology and biomarker-driven clinical trials are strongly supported by glia-centered, multi-omic biomarkers that connect neuroinflammation, mitochondrial metabolism, and, considering the identification of astroglia-like cells in peripheral tissues, abnormalities in astroglia and astroglial transmitters can be reflected in peripheral tissues and blood (García-Domínguez, 2026). Therefore, the altered glial transmission can be observed in a non-invasive manner in schizophrenia through blood biomarkers. Moreover, the treatment-related changes reflected in the gliotransmission can also be monitored.

7. Conclusion

Neuropsychiatric illnesses have been reframed as multicellular system pathologies by recent developments in glial biology, which have positioned oligodendrocytes, microglia, and astrocytes as active modulators of neurotransmission, metabolism, and neuroimmune homeostasis. Astrocyte abnormalities in schizophrenia lead to impairment of glutamate clearance, lactate shuttling, and gap-junctional signaling, resulting in mitochondrial dysfunction, oxidative stress, and cortical dysconnectivity. A self-sustaining cycle of redox imbalance, neuroinflammation, and gliotransmission failure is interlinked. Overactive MAO-B raises hydrogen peroxide and interferes with dopamine metabolism. Through the absorption, metabolism, and release of gliotransmitters such as glutamate, D-serine, ATP, and GABA, astrocytes actively control synaptic transmission in addition to their conventional supporting functions. The excitatory–inhibitory balance necessary for neural communication is disrupted when these pathways are dysregulated. NMDAR hypofunction, synaptic instability, and abnormal brain circuitry underlying cognitive, positive, and negative symptoms in schizophrenia are caused by deprived astrocytic glutamate clearance, decreased D-serine availability, and altered gliotransmitter release. This book chapter emphasizes that astroglia dysfunction and gliotransmission are quantifiable causes of circuit instability and progressive pathogenesis in schizophrenia and the novel link paves a potential target for precision pharmacology, which aims to restore metabolic integrity, glial neuronal coupling, and redox balance in schizophrenia.

Acknowledgement

Mahesh Kandasamy has been supported by University Grants Commission–Faculty Recharge Programme (UGC-FRP), New Delhi, India and also acknowledges Anusandhan National Research Foundation/Science Engineering Research Board (CRG/2023/005266), Department of Science and Technology, Government of India, for financial support. Mercy Priyadharshini Babu Deva Irakkam was supported as a project associate from the ANRF-SERB project grant (CRG/2023/005266).

Declaration

The authors would like to disclose that ChatGPT and grammar-checking tools were used solely for language editing and improvement of the manuscript.

References

Ahmad, K., Baig, M. H., Mushtaq, G., Kamal, M. A., Greig, N. H., & Choi, I. (2017). Commonalities in biological pathways, genetics, and cellular mechanism between alzheimer disease and other neurodegenerative diseases: An in silico-updated overview. *Current Alzheimer Research, 14*(11), 1190–1197. https://doi.org/10.2174/1567205014666170203141151.

Anwar, A., Mustafa, A. M., Abdou, K., A. Rabie, M., El-Shiekh, R. A., & El-Dessouki, A. M. (2025). A comprehensive review on schizophrenia: Epidemiology, pathogenesis, diagnosis, conventional treatments, and proposed natural compounds used for management. *Naunyn-Schmiedeberg's Archives of Pharmacology, 398*(11), 15231–15255. https://doi.org/10.1007/s00210-025-04351-0.

Armada-Moreira, A., Gomes, J. I., Pina, C. C., Savchak, O. K., Gonçalves-Ribeiro, J., Rei, N., ... Vaz, S. H. (2020). Going the extra (synaptic) mile: Excitotoxicity as the road toward neurodegenerative diseases. *Frontiers in Cellular Neuroscience, 14*, 90. https://doi.org/10.3389/fncel.2020.00090.

Baldwin, K. T., & Eroglu, C. (2017). Molecular mechanisms of astrocyte-induced synaptogenesis. *Current Opinion in Neurobiology, 45*, 113–120. https://doi.org/10.1016/j.conb.2017.05.006.

Balu, D. T. (2016). The NMDA receptor and schizophrenia: From pathophysiology to treatment. *Advances in Pharmacology (San Diego, Calif.), 76*, 351–382. https://doi.org/10.1016/bs.apha.2016.01.006.

Beard, E., Lengacher, S., Dias, S., Magistretti, P. J., & Finsterwald, C. (2022). Astrocytes as key regulators of brain energy metabolism: New therapeutic perspectives. *Frontiers in Physiology, 12*, 825816. https://doi.org/10.3389/fphys.2021.825816.

Bera, K., Looger, L. L., Proekt, A., & Cichon, J. (2025). Cortical mechanisms contributing to ketamine-induced dissociation. *The Neuroscientist, 32*(1), 72–88. https://doi.org/10.1177/10738584251403946.

Bernstein, H.-G., Nussbaumer, M., Vasilevska, V., Dobrowolny, H., Nickl-Jockschat, T., Guest, P. C., & Steiner, J. (2025). Glial cell deficits are a key feature of schizophrenia: Implications for neuronal circuit maintenance and histological differentiation from classical neurodegeneration. *Molecular Psychiatry, 30*(3), 1102–1116. https://doi.org/10.1038/s41380-024-02861-6.

Billard, J.-M. (2018). Changes in serine racemase-dependent modulation of NMDA receptor: Impact on physiological and pathological brain aging. *Frontiers in Molecular Biosciences, 5*, 106. https://doi.org/10.3389/fmolb.2018.00106.

Bogus, K., Marchesi, N., Campagnoli, L. I. M., Pascale, A., & Pałasz, A. (2026). Glial cells as key mediators in the pathophysiology of neurodegenerative diseases. *International Journal of Molecular Sciences, 27*(2), 884. https://doi.org/10.3390/ijms27020884.

Bradlow, R. C. J., Berk, M., Kalivas, P. W., Back, S. E., & Kanaan, R. A. (2022). The potential of N-acetyl-l-cysteine (NAC) in the treatment of psychiatric disorders. *CNS Drugs, 36*(5), 451–482. https://doi.org/10.1007/s40263-022-00907-3.

Braff, L., & Braff, D. L. (2013). The neuropsychiatric translational revolution still very early and still very challenging. *JAMA Psychiatry, 70*(8), 777–779. https://doi.org/10.1001/jamapsychiatry.2013.2184.

Brisch, R., Saniotis, A., Wolf, R., Bielau, H., Bernstein, H.-G., Steiner, J., ... Gos, T. (2014). The role of dopamine in schizophrenia from a neurobiological and evolutionary perspective: Old fashioned, but still in vogue. *Frontiers in Psychiatry, 5*, 47. https://doi.org/10.3389/fpsyt.2014.00047.

Chen, Y., Ye, Y., Jia, J., Long, B., Dou, T., & Yan, X. (2025). Advances in astrocytic calcium signaling research. *Frontiers in Cellular Neuroscience, 19*, 1719532. https://doi.org/10.3389/fncel.2025.1719532.

Comer, A. L., Carrier, M., Tremblay, M.-È., & Cruz-Martín, A. (2020). The inflamed brain in schizophrenia: The convergence of genetic and environmental risk factors that lead to uncontrolled neuroinflammation. *Frontiers in Cellular Neuroscience, 14*. https://doi.org/10.3389/fncel.2020.00274.

Coyle, J. T., Balu, D., & Wolosker, H. (2020). D-Serine, the shape-shifting NMDA receptor Co-agonist. *Neurochemical Research, 45*(6), 1344–1353. https://doi.org/10.1007/s11064-020-03014-1.

Cuellar-Santoyo, A. O., Ruiz-Rodríguez, V. M., Mares-Barbosa, T. B., Patrón-Soberano, A., Howe, A. G., Portales-Pérez, D. P., ... Estrada-Sánchez, A. M. (2023). Revealing the contribution of astrocytes to glutamatergic neuronal transmission. *Frontiers in Cellular Neuroscience, 16*, 1037641. https://doi.org/10.3389/fncel.2022.1037641.

Dahl, G. (2015). ATP release through pannexon channels. *Philosophical Transactions of the Royal Society B: Biological Sciences, 370*(1672), 20140191. https://doi.org/10.1098/rstb.2014.0191.

Das, T. K., Dey, A., Sabesan, P., Javadzadeh, A., Théberge, J., Radua, J., & Palaniyappan, L. (2018). Putative astroglial dysfunction in schizophrenia: A meta-analysis of 1H-MRS studies of medial prefrontal myo-inositol. *Frontiers in Psychiatry, 9*, 438. https://doi.org/10.3389/fpsyt.2018.00438.

de Oliveira Figueiredo, E. C., Calì, C., Petrelli, F., & Bezzi, P. (2022). Emerging evidence for astrocyte dysfunction in schizophrenia. *Glia, 70*(9), 1585–1604. https://doi.org/10.1002/glia.24221.

Dosch, M., Gerber, J., Jebbawi, F., & Beldi, G. (2018). Mechanisms of ATP release by inflammatory cells. *International Journal of Molecular Sciences, 19*(4), 1222. https://doi.org/10.3390/ijms19041222.

Edamakanti, C. R., Mohan, V., & Opal, P. (2023). Reactive Bergmann glia play a central role in spinocerebellar ataxia inflammation via the JNK pathway. *Journal of Neuroinflammation, 20*, 126. https://doi.org/10.1186/s12974-023-02801-1.

Eltokhi, A., Janmaat, I. E., Genedi, M., Haarman, B. C. M., & Sommer, I. E. C. (2020). Dysregulation of synaptic pruning as a possible link between intestinal microbiota dysbiosis and neuropsychiatric disorders. *Journal of Neuroscience Research, 98*(7), https://doi.org/10.1002/jnr.24616.

Fatemi, S. H., & Folsom, T. D. (2009). The neurodevelopmental hypothesis of schizophrenia, revisited. *Schizophrenia Bulletin, 35*(3), 528–548. https://doi.org/10.1093/schbul/sbn187.

Fernández-Moncada, I., Lavanco, G., Fundazuri, U. B., Bollmohr, N., Mountadem, S., Dalla Tor, T., ... Marsicano, G. (2024). A lactate-dependent shift of glycolysis mediates synaptic and cognitive processes in male mice. *Nature Communications, 15*, 6842. https://doi.org/10.1038/s41467-024-51008-2.

Föcking, M., Doyle, B., Munawar, N., Dillon, E. T., Cotter, D., & Cagney, G. (2019). Epigenetic factors in schizophrenia: Mechanisms and experimental approaches. *Molecular Neuropsychiatry, 5*(1), 6–12. https://doi.org/10.1159/000495063.

Forró, T., Manu, D. R., Băjenaru, O.-L., & Bălaşa, R. (2024). GFAP as astrocyte-derived extracellular vesicle cargo in acute ischemic stroke patients—A pilot study. *International Journal of Molecular Sciences, 25*(11), 5726. https://doi.org/10.3390/ijms25115726.

Fukuyama, K., Motomura, E., & Okada, M. (2023). Enhanced L-β-Aminoisobutyric Acid Is Involved in the Pathophysiology of Effectiveness for Treatment-Resistant Schizophrenia and Adverse Reactions of Clozapine. *Biomolecules, 13*(5), 862. https://doi.org/10.3390/biom13050862.

Fukuyama, K., Okubo, R., Murata, M., Shiroyama, T., & Okada, M. (2020). Activation of astroglial connexin is involved in concentration-dependent double-edged sword clinical action of clozapine. *Cells, 9*(2), 414. https://doi.org/10.3390/cells9020414.

Gallo, V., & Deneen, B. (2014). Glial development: The crossroads of regeneration and repair in the CNS. *Neuron, 83*(2), 283–308. https://doi.org/10.1016/j.neuron.2014.06.010.

García-Domínguez, M. (2026). Glial cell dynamics in neuroinflammation: Mechanisms, interactions, and therapeutic implications. *Biomedicines, 14*(1), 115. https://doi.org/10.3390/biomedicines14010115.

García-San-Martín, N., Bethlehem, R. A., Segura, P., Mihalik, A., Seidlitz, J., Sebenius, I., ... Romero-García, R. (2025). Reduced brain structural similarity is associated with maturation, neurobiological features, and clinical status in schizophrenia. *Nature Communications, 16*(1), 8745. https://doi.org/10.1038/s41467-025-63792-6.

Garden, G. A., & Campbell, B. M. (2016). Glial biomarkers in human central nervous system disease. *Glia, 64*(10), 1755–1771. https://doi.org/10.1002/glia.22998.

Gengatharan, A., Bammann, R. R., & Saghatelyan, A. (2016). The role of astrocytes in the generation, migration, and integration of new neurons in the adult olfactory bulb. *Frontiers in Neuroscience, 10*. https://doi.org/10.3389/fnins.2016.00149.

Giovannoni, F., & Quintana, F. J. (2020). The role of astrocytes in CNS inflammation. *Trends in Immunology, 41*(9), 805–819. https://doi.org/10.1016/j.it.2020.07.007.

Gomes, C., Huang, K.-C., Harkin, J., Baker, A., Hughes, J. M., Pan, Y., ... Meyer, J. S. (2024). Induction of astrocyte reactivity promotes neurodegeneration in human pluripotent stem cell models. *Stem Cell Reports, 19*(8), 1122–1136. https://doi.org/10.1016/j.stemcr.2024.07.002.

Gradisnik, L., & Velnar, T. (2023). Astrocytes in the central nervous system and their functions in health and disease: A review. *World Journal of Clinical Cases, 11*(15), 3385–3394. https://doi.org/10.12998/wjcc.v11.i15.3385.

Graves, S. M., Xie, Z., Stout, K. A., Zampese, E., Burbulla, L. F., Shih, J. C., ... Surmeier, D. J. (2020). Dopamine metabolism by a monoamine oxidase mitochondrial shuttle activates the electron transport chain. *Nature Neuroscience, 23*(1), 15–20. https://doi.org/10.1038/s41593-019-0556-3.

Guastaferro, K., Moucheraud, C., Purtle, J., Collins, L. M., & Shelley, D. (2025). Integrating implementation science and intervention optimization. *Implementation Science: IS, 20*, 41. https://doi.org/10.1186/s13012-025-01457-0.

Hany, M., & Rizvi, A. (2026). *Schizophrenia. StatPearls*. StatPearls Publishing, http://www.ncbi.nlm.nih.gov/books/NBK539864/.

Harada, K., Kamiya, T., & Tsuboi, T. (2016). Gliotransmitter release from astrocytes: Functional, developmental, and pathological implications in the brain. *Frontiers in Neuroscience, 9*, 499. https://doi.org/10.3389/fnins.2015.00499.

Howard, B. M., Mo, Z., Filipovic, R., Moore, A. R., Antic, S. D., & Zecevic, N. (2008). Radial glia cells in the developing human brain. *The Neuroscientist: A Review Journal Bringing Neurobiology, Neurology and Psychiatry, 14*(5), 459–473. https://doi.org/10.1177/1073858407313512.

Iasevoli, F., Tomasetti, C., Buonaguro, E. F., & Bartolomeis, A. de (2014). The glutamatergic aspects of schizophrenia molecular pathophysiology: Role of the postsynaptic density, and implications for treatment. *Current Neuropharmacology, 12*(3), 219. https://doi.org/10.2174/1570159X12666140324183406.

Illes, P., Rubini, P., Ulrich, H., Yin, H., & Tang, Y. (2025). Dysregulation of astrocytic ATP/adenosine release in the hippocampus cause cognitive and affective disorders: Molecular mechanisms, diagnosis, and therapy. *MedComm, 6*(5), e70177. https://doi.org/10.1002/mco2.70177.

Jo, S., Yarishkin, O., Jin Hwang, Y., Eun Chun, Y., Park, M., Ho Woo, D., ... Lee, C. J. (2014). GABA from reactive astrocytes impairs memory in mouse models of Alzheimer's disease. *Nature Medicine, 20*(8), 886–896. https://doi.org/10.1038/nm.3639.

Jurga, A. M., Paleczna, M., Kadluczka, J., & Kuter, K. Z. (2021). Beyond the GFAP-astrocyte protein markers in the brain. *Biomolecules, 11*(9), 1361. https://doi.org/10.3390/biom11091361.

Kang, N., Peng, H., Yu, Y., Stanton, P. K., Guilarte, T. R., & Kang, J. (2013). Astrocytes release D-serine by a large vesicle. *Neuroscience, 240*, 243–257. https://doi.org/10.1016/j.neuroscience.2013.02.029.

Koizumi, S., Shigetomi, E., Sano, F., Saito, K., Kim, S. K., & Nabekura, J. (2021). Abnormal Ca2+ signals in reactive astrocytes as a common cause of brain diseases. *International Journal of Molecular Sciences, 23*(1), 149. https://doi.org/10.3390/ijms23010149.

Lamptey, R. N. L., Chaulagain, B., Trivedi, R., Gothwal, A., Layek, B., & Singh, J. (2022). A review of the common neurodegenerative disorders: Current therapeutic approaches and the potential role of nanotherapeutics. *International Journal of Molecular Sciences, 23*(3), 1851. https://doi.org/10.3390/ijms23031851.

Lányi, O., Koleszár, B., Schulze Wenning, A., Balogh, D., Engh, M. A., Horváth, A. A., ... Csukly, G. (2024). Excitation/inhibition imbalance in schizophrenia: A meta-analysis of inhibitory and excitatory TMS-EMG paradigms. *Schizophrenia, 10*(1), 56. https://doi.org/10.1038/s41537-024-00476-y.

Lattke, M., & Guillemot, F. (2022). Understanding astrocyte differentiation: Clinical relevance, technical challenges, and new opportunities in the omics era. *Wires Mechanisms of Disease, 14*(5), e1557. https://doi.org/10.1002/wsbm.1557.

Lewis, D. A., Curley, A. A., Glausier, J., & Volk, D. W. (2012). Cortical Parvalbumin Interneurons and Cognitive Dysfunction in Schizophrenia. *Trends in Neurosciences, 35*(1), 57–67. https://doi.org/10.1016/j.tins.2011.10.004.

Lewis, D. A., & Sweet, R. A. (2009). Schizophrenia from a neural circuitry perspective: Advancing toward rational pharmacological therapies. *The Journal of Clinical Investigation, 119*(4), 706–716. https://doi.org/10.1172/JCI37335.

Linnerbauer, M., & Rothhammer, V. (2020). Protective functions of reactive astrocytes following central nervous system insult. *Frontiers in Immunology, 11*. https://doi.org/10.3389/fimmu.2020.573256.

Liu, X., Ying, J., Wang, X., Zheng, Q., Zhao, T., Yoon, S., ... Hua, F. (2021). Astrocytes in neural circuits: Key factors in synaptic regulation and potential targets for neurodevelopmental disorders. *Frontiers in Molecular Neuroscience, 14*. https://doi.org/10.3389/fnmol.2021.729273.

Lobsiger, C. S., & Cleveland, D. W. (2007). Glial cells as intrinsic components of non-cell autonomous neurodegenerative disease. *Nature Neuroscience, 10*(11), 1355–1360. https://doi.org/10.1038/nn1988.

Luvsannyam, E., Jain, M. S., Pormento, M. K. L., Siddiqui, H., Balagtas, A. R. A., Emuze, B. O., & Poprawski, T. (2022). Neurobiology of Schizophrenia: A Comprehensive Review. *Cureus, 14*(4), e23959.

Martineau, M., Parpura, V., & Mothet, J.-P. (2014). Cell-type specific mechanisms of D-serine uptake and release in the brain. *Frontiers in Synaptic Neuroscience, 6*, 12. https://doi.org/10.3389/fnsyn.2014.00012.

McCutcheon, R., Pillinger, T., Jauhar, S., Pepper, F., Rogdaki, M., & Howes, O. (2020). S172. Glutamate related connectivity disturbances of the salience and default mode networks in psychosis. *Schizophrenia Bulletin, 46*(1), S102–S103. https://doi.org/10.1093/schbul/sbaa031.238.

Mei, Y.-Y., Wu, D. C., & Zhou, N. (2018). Astrocytic regulation of glutamate transmission in schizophrenia. *Frontiers in Psychiatry, 9*, 544. https://doi.org/10.3389/fpsyt.2018.00544.

Miyajima, M., Tabata, H., & Nakajima, K. (2023). Migratory mode transition of astrocyte progenitors in the cerebral cortex: An intrinsic or extrinsic cell process? *Neural Regeneration Research, 19*(3), 471–472. https://doi.org/10.4103/1673-5374.380886.

Molero, A. E., Arteaga-Bracho, E. E., Chen, C. H., Gulinello, M., Winchester, M. L., Pichamoorthy, N., ... Mehler, M. F. (2016). Selective expression of mutant huntingtin during development recapitulates characteristic features of Huntington's disease. *Proceedings of the National Academy of Sciences of the United States of America, 113*(20), 5736–5741. https://doi.org/10.1073/pnas.1603871113.

Murat, C. D. B., & García-Cáceres, C. (2021). Astrocyte gliotransmission in the regulation of systemic metabolism. *Metabolites, 11*(11), 732. https://doi.org/10.3390/metabo11110732.

Nawafleh, S., Qaswal, A. B., Suleiman, A., Alali, O., Zayed, F. M., Al-Adwan, M. A. O., & Bani Ali, M. (2022). GABA receptors can depolarize the neuronal membrane potential via quantum tunneling of chloride ions: A quantum mathematical study. *Cells, 11*(7), 1145. https://doi.org/10.3390/cells11071145.

Noriega-Prieto, J. A., & Araque, A. (2021). Sensing and regulating synaptic activity by astrocytes at tripartite synapse. *Neurochemical Research, 46*(10), 2580–2585. https://doi.org/10.1007/s11064-021-03317-x.

Oksanen, M., Lehtonen, S., Jaronen, M., Goldsteins, G., Hämäläinen, R. H., & Koistinaho, J. (2019). Astrocyte alterations in neurodegenerative pathologies and their modeling in human induced pluripotent stem cell platforms. *Cellular and Molecular Life Sciences: CMLS, 76*(14), 2739–2760. https://doi.org/10.1007/s00018-019-03111-7.

Orellana, J. A., Retamal, M. A., Moraga-Amaro, R., & Stehberg, J. (2016). Role of astroglial hemichannels and pannexons in memory and neurodegenerative diseases. *Frontiers in Integrative Neuroscience, 10*, 26. https://doi.org/10.3389/fnint.2016.00026.

Orsolini, L., Pompili, S., & Volpe, U. (2022). Schizophrenia: A narrative review of etiopathogenetic, diagnostic and treatment aspects. *Journal of Clinical Medicine, 11*(17), 5040. https://doi.org/10.3390/jcm11175040.

Ota, Y., Zanetti, A. T., & Hallock, R. M. (2013). The role of astrocytes in the regulation of synaptic plasticity and memory formation. *Neural Plasticity, 2013*, 185463. https://doi.org/10.1155/2013/185463.

Pang, Z. P., & Südhof, T. C. (2010). Cell biology of Ca2+-triggered exocytosis. *Current Opinion in Cell Biology, 22*(4), 496–505. https://doi.org/10.1016/j.ceb.2010.05.001.

Pelassa, S., Guidolin, D., Venturini, A., Averna, M., Frumento, G., Campanini, L., ... Marcoli, M. (2019). A2A-D2 heteromers on striatal astrocytes: Biochemical and biophysical evidence. *International Journal of Molecular Sciences, 20*(10), 2457. https://doi.org/10.3390/ijms20102457.

Puebla, M., Tapia, P. J., & Espinoza, H. (2022). Key role of astrocytes in postnatal brain and retinal angiogenesis. *International Journal of Molecular Sciences, 23*(5), 2646. https://doi.org/10.3390/ijms23052646.

Purushotham, S. S., & Buskila, Y. (2023). Astrocytic modulation of neuronal signalling. *Frontiers in Network Physiology, 3*, 1205544. https://doi.org/10.3389/fnetp.2023.1205544.

Quincozes-Santos, A., Santos, C. L., de Souza Almeida, R. R., da Silva, A., Thomaz, N. K., Costa, N. L. F., ... Bobermin, L. D. (2021). Gliotoxicity and glioprotection: The dual role of glial cells. *Molecular Neurobiology, 58*(12), 6577–6592. https://doi.org/10.1007/s12035-021-02574-9.

Rampino, A., Garofalo, M., Nuzzo, T., Favia, M., Saltarelli, S., Masellis, R., ... Usiello, A. (2024). Variations of blood D-serine and D-aspartate homeostasis track psychosis stages. *Schizophrenia, 10*(1), 115. https://doi.org/10.1038/s41537-024-00537-2.

Ricci, V., Martinotti, G., Mosca, A., & Maina, G. (2026). Bioenergetic impairment in schizophrenia: Role of mitochondrial signaling in synaptic dysfunction - a systematic review. *Frontiers in Cell and Developmental Biology, 14*, 1740079. https://doi.org/10.3389/fcell.2026.1740079.

Rubio, M. D., Drummond, J. B., & Meador-Woodruff, J. H. (2012). Glutamate receptor abnormalities in schizophrenia: Implications for innovative treatments. *Biomolecules & Therapeutics, 20*(1), 1–18. https://doi.org/10.4062/biomolther.2012.20.1.001.

Satarker, S., Bojja, S. L., Gurram, P. C., Mudgal, J., Arora, D., & Nampoothiri, M. (2022). Astrocytic glutamatergic transmission and its implications in neurodegenerative disorders. *Cells, 11*(7), 1139. https://doi.org/10.3390/cells11071139.

Schiera, G., Di Liegro, C. M., Schirò, G., Sorbello, G., & Di Liegro, I. (2024). Involvement of astrocytes in the formation, maintenance, and function of the Blood–Brain barrier. *Cells, 13*(2), 150. https://doi.org/10.3390/cells13020150.

Sidoryk-wegrzynowicz, M., Wegrzynowicz, M., Lee, E., Bowman, A. B., & Aschner, M. (2011). Role of astrocytes in brain function and disease. *Toxicologic Pathology, 39*(1), 115–123. https://doi.org/10.1177/0192623310385254.

Sidoryk-Węgrzynowicz, M., Adamiak, K., & Strużyńska, L. (2024). Astrocyte–Neuron interaction via the glutamate–Glutamine cycle and its dysfunction in tau-dependent neurodegeneration. *International Journal of Molecular Sciences, 25*(5), 3050. https://doi.org/10.3390/ijms25053050.

Sorrentino, Z. A., Giasson, B. I., & Chakrabarty, P. (2019). α-Synuclein and astrocytes: Tracing the pathways from homeostasis to neurodegeneration in Lewy body disease. *Acta Neuropathologica, 138*(1), 1–21. https://doi.org/10.1007/s00401-019-01977-2.

Strickland, M. R., Rau, M., Summers, B., Basore, K., Wulf, J., Jiang, H., ... Holtzman, D. M. (2024). Apolipoprotein E secreted by astrocytes forms antiparallel dimers in discoidal lipoproteins. *Neuron, 112*(7), 1100–1109.e5. https://doi.org/10.1016/j.neuron.2023.12.018.

Sun, N., Liu, Y., Lin, H., Xiong, J., He, Y., Yang, H., ... He, J. (2026). Multimodal PET/MR imaging of prolonged disorders of consciousness: A pilot feasibility study. *Frontiers in Neuroscience, 20*. https://doi.org/10.3389/fnins.2026.1761097.

Tarasov, V. V., Svistunov, A. A., Chubarev, V. N., Sologova, S. S., Mukhortova, P., Levushkin, D., ... Aliev, G. (2020). Alterations of astrocytes in the context of schizophrenic dementia. *Frontiers in Pharmacology, 10*. https://doi.org/10.3389/fphar.2019.01612.

Tavitian, A., Cressatti, M., Song, W., Turk, A. Z., Galindez, C., Smart, A., ... Schipper, H. M. (2020). Strategic timing of glial HMOX1 expression results in either schizophrenia-like or Parkinsonian behavior in mice. *Antioxidants & Redox Signaling, 32*(17), 1259–1272. https://doi.org/10.1089/ars.2019.7937.

Todd, A. C., & Hardingham, G. E. (2020). The regulation of astrocytic glutamate transporters in health and neurodegenerative diseases. *International Journal of Molecular Sciences, 21*(24), 9607. https://doi.org/10.3390/ijms21249607.

Toker, L., Mancarci, B. O., Tripathy, S., & Pavlidis, P. (2018). Transcriptomic evidence for alterations in astrocytes and parvalbumin interneurons in subjects with bipolar disorder and schizophrenia. *Biological Psychiatry, 84*(11), 787–796. https://doi.org/10.1016/j.biopsych.2018.07.010.

Trubalski, M., Markiewicz-Gospodarek, A., Żerebiec, M., Poleszak, J., Szczotka, M., Markiewicz, R., ... Szymańczyk, S. (2025). Oxidative stress-mediated neuroinflammation in the pathophysiology of schizophrenia. *International Journal of Molecular Sciences, 26*(22), 11139. https://doi.org/10.3390/ijms262211139.

Uliana, D. L., & Grace, A. A. (2026). Prefrontal cortex dysfunction as a precipitating factor for schizophrenia and depression. *Journal of Neurochemistry, 170*(3), e70409. https://doi.org/10.1111/jnc.70409.

Upadhya, R., Zingg, W., Shetty, S., & Shetty, A. K. (2020). Astrocyte-derived extracellular vesicles: Neuroreparative properties and role in the pathogenesis of neurodegenerative disorders. *Journal of Controlled Release: Official Journal of the Controlled Release Society, 323*, 225–239. https://doi.org/10.1016/j.jconrel.2020.04.017.

Vellucci, L., Mazza, B., Barone, A., Nasti, A., De Simone, G., Iasevoli, F., & de Bartolomeis, A. (2025). The role of astrocytes in the molecular pathophysiology of schizophrenia: Between neurodevelopment and neurodegeneration. *Biomolecules, 15*(5), 615. https://doi.org/10.3390/biom15050615.

Verkhratsky, A., Butt, A., Li, B., Illes, P., Zorec, R., Semyanov, A., ... Sofroniew, M. V. (2023). Astrocytes in human central nervous system diseases: A frontier for new therapies. *Signal Transduction and Targeted Therapy, 8*(1), 396. https://doi.org/10.1038/s41392-023-01628-9.

Verkhratsky, A., Rodríguez, J. J., & Steardo, L. (2014). Astrogliopathology: A central element of neuropsychiatric diseases? *The Neuroscientist: A Review Journal Bringing Neurobiology, Neurology and Psychiatry, 20*(6), 576–588. https://doi.org/10.1177/1073858413510208.

Viola, V., Chinnappa, K., & Francis, F. (2024). Radial glia progenitor polarity in health and disease. *Frontiers in Cell and Developmental Biology, 12*, 1478283. https://doi.org/10.3389/fcell.2024.1478283.

Vishnumukkala, T., Che Mohd Nassir, C. M. N., Hein, Z. M., Kalerammana Gopalakrishna, P., Karikalan, B., Alkatiri, A., ... Kamaruzzaman, M. A. (2025). Glial cells as emerging therapeutic targets in neurodegenerative diseases: Mechanistic insights and translational perspectives. *Cells, 14*(19), 1497. https://doi.org/10.3390/cells14191497.

Vita, A., Barlati, S., Cavallaro, R., Mucci, A., Riva, M. A., Rocca, P., ... Galderisi, S. (2024). Definition, assessment and treatment of cognitive impairment associated with schizophrenia: Expert opinion and practical recommendations. *Frontiers in Psychiatry, 15*. https://doi.org/10.3389/fpsyt.2024.1451832.

Wang, H., Chen, C., Lin, Y., Tian, Z., Yan, Z., Zeng, X., ... Chen, N. (2025). Connexin43 and its regulation of astrocyte Gap junction function: Influencing depression progression by mediating electrical and chemical signals. *CNS Neuroscience & Therapeutics, 31*(9), e70600. https://doi.org/10.1111/cns.70600.

Wang, J., Guo, Y., Han, W., Zhang, H., & Jiang, P. (2026). D-Serine: A multitalented neuromodulator in brain function, systemic homeostasis, and disease. *Cellular and Molecular Neurobiology, 46*, 65. https://doi.org/10.1007/s10571-026-01696-9.

Weber, B., & Barros, L. F. (2015). The astrocyte: Powerhouse and recycling center. *Cold Spring Harbor Perspectives in Biology,* 7(12), a020396. https://doi.org/10.1101/cshperspect.a020396.

Westergard, T., & Rothstein, J. D. (2020). Astrocyte diversity: Current insights and future directions. *Neurochemical Research, 45*(6), 1298–1305. https://doi.org/10.1007/s11064-020-02959-7.

Wilson, J. C., Liu, K. Y., Jones, K., Mahmood, J., Arya, U., & Howard, R. (2024). Biomarkers of neurodegeneration in schizophrenia: Systematic review and meta-analysis. *BMJ Mental Health, 27*(1), e301017. https://doi.org/10.1136/bmjment-2024-301017.

Won, W., Bhalla, M., Lee, J.-H., & Lee, C. J. (2025). Astrocytes as key regulators of neural signaling in health and disease. *Annual Review of Neuroscience, 48*(1), 251–276. https://doi.org/10.1146/annurev-neuro-112723-035356.

Yamamoto, M., & Takano, T. (2025). Astrocyte-mediated plasticity: Multi-scale mechanisms linking synaptic dynamics to learning and memory. *Cells, 14*(24), 1936. https://doi.org/10.3390/cells14241936.

Zhang, D., Hua, Z., & Li, Z. (2024). The role of glutamate and glutamine metabolism and related transporters in nerve cells. *CNS Neuroscience & Therapeutics, 30*(2), e14617. https://doi.org/10.1111/cns.14617.

Zhao, J., O'Connor, T., & Vassar, R. (2011). The contribution of activated astrocytes to Aβ production: Implications for Alzheimer's disease pathogenesis. *Journal of Neuroinflammation, 8*, 150. https://doi.org/10.1186/1742-2094-8-150.

Zheng, Y., Zhou, W., Chang, H., & Zheng, K. (2026). Brain organoids as precision models for neurodegenerative diseases: From disease modeling to drug discovery. *Frontiers in Neuroscience, 20*, 1764964. https://doi.org/10.3389/fnins.2026.1764964.

CHAPTER FIVE

Molecular insights of peroxisome proliferator-activated receptor-γ signalling in amyotrophic lateral sclerosis and Huntington's disease

Gopika Venu, Vignesh Pandi A, and Antony Justin*
Department of Pharmacology, JSS College of Pharmacy, JSS Academy of Higher Education & Research, Ooty, The Nilgiris, Tamil Nadu, India
*Corresponding author. e-mail address: justin@jssuni.edu.in

Contents

Abstract

Progressive neuronal loss is a hallmark of neurodegenerative diseases like Huntingtons disease (HD) and Amyotrophic lateral sclerosis (ALS) which are caused by convergent mechanisms such as oxidative stress, mitochondrial dysfunction, neuroinflammation, impaired autophagy and dysregulated cell death pathways. Both conditions share significant disruptions in metabolic and inflammatory signalling despite having different genetic origins and clinical manifestations; underscoring the necessity of pathway-oriented treatment approaches. In the central nervous system, peroxisome proliferator-activated receptor-γ (PPAR-γ), a ligand-activated nuclear receptor has become an

International Review of Neurobiology, Volume 188
ISSN 0074-7742, https://doi.org/10.1016/bs.irn.2026.05.014

important regulator of inflammation, redox homeostasis, mitochondrial biogenesis and cellular stress responses. After giving a thorough overview of PPAR-γ structure activation and transcriptional regulation and the PGC-1α–mediated mitochondrial biogenesis axis, this chapter delves deeply into its interactions with major signalling pathways such as NF-κB, Wnt/β-catenin Nrf2/ARE and the autophagy–apoptosis networks. With a focus on experimental data showing PPAR-γ signaling's neuroprotective, anti-inflammatory, antioxidant and metabolic regulatory roles the pathophysiology of ALS and HD is critically investigated. Lastly the need for improved biomarkers, tailored multi-target strategies and selective modulators is highlighted in the discussion of current therapeutic limitations and translational difficulties.

Abbreviations

AF1	Activation function 1.
AF2	Activation function 2.
ALS	Amyotrophic Lateral Sclerosis.
AMBRA1	Activating Molecule in Beclin-1-Regulated Autophagy.
AMPK	AMP-activated protein kinase.
Apaf-1	Apoptotic protease-activating factor-1.
APC	Adenomatous polyposis coli.
ARE	Antioxidant response element.
Atg13	Autophagy-related Protein 13.
ATP	Adenosine triphosphate.
BAT	Brown adipose tissue.
CBD	Catenin binding domain.
CBP	CREB-binding protein.
CK1	Casein kinase 1.
CNS	Central Nervous System.
COX-2	Cyclooxygenase 2.
DBD	DNA binding domain.
DISC	Death-inducing signaling complex.
DNA	Deoxyribonucleic acid.
Drp1	Dynamin-related protein.
Dsh	Dishevelled.
ERRα	Estrogen-related receptor alpha.
FIP200	FAK-family Interacting Protein of 200 kD.
Fz	Frizzled.
GPx3	Glutathione peroxidase 3.
GSK3β	Glycogen Synthase Kinase-3β.
HD	Huntington's disease.
HO-1	Heme oxygenase-1.
IKK	IκB kinase.
iNOS	Inducible Nitric Oxide Synthase.
KEAP1	Kelch-like ECH-associated protein 1.
LBD	Ligand-binding domain.
LC3	Light chain 3.
LRP5/6	Low-density lipoproteinrelated protein 5/6.

Mfn	Mitofusin.
mHtt	Mutant huntingtin.
MnSOD	Manganese superoxide dismutase.
mTOR	Mammalian target of rapamycin.
NF-κB	Nuclear Factor kappa-light-chain-enhancer of activated B cells.
NRF-1	Nuclear Respiratory Factor 1.
NrF2	Nuclear factor erythroid 2-related factor.
Opa1	Optic atrophy 1.
PGC-1α	Peroxisome proliferator-activated receptor γ coactivator 1α.
PI3K	Phosphatidylinositol-3-kinase.
PPAR	Peroxisome Proliferator-Activated Receptors.
PPRE	Peroxisome proliferator response elements.
RNA	Ribonucleic acid.
ROS	Reactive oxygen species.
RXR	Retinoid X receptor.
SIRT1	Sirtuin 1.
SOD	Superoxide dismutase.
SPPARMs	Selective PPAR-γ modulators.
TFAM	Mitochondrial transcription factor A.
TZDs	Thiazolidinediones.
ULK	Unc-51-Like Kinase.
VPS15	Vacuolar Protein Sorting 15.
WAT	White adipose tissue.
β-TrCP	β-transducin repeat-containing protein.

1. Introduction

Neurodegenerative disorders encompass a diverse set of progressive conditions in which neurons progressively lose their functional capacity and viability, eventually resulting in widespread neuronal degeneration (Gadhave et al., 2024). Amyotrophic Lateral Sclerosis (ALS) and Huntington's disease (HD) are two severe neurodegenerative diseases that share overlapping cellular vulnerabilities and mechanistic disturbances despite their distinct clinical presentations (Tada et al., 2012). In ALS, the degeneration predominantly affects upper and lower motor neurons, producing progressive muscle weakness and atrophy that ultimately lead to paralysis and shortened survival, often within a few years of diagnosis. Most ALS cases arise sporadically, whereas familial cases frequently involve mutations in genes such as C9orf72, SOD1, TARDBP, or FUS. Despite their genetic differences, both forms show similar clinical and pathological features. Currently, riluzole (a glutamate modulator) and edaravone (antioxidant) is approved for ALS treatment in some countries, but it only modestly extends survival (Feldman et al., 2022). On the other hand, Huntington's Disease (HD) is a hereditary condition caused by CAG repeat expansion in the huntingtin (*HTT*) gene and the resultant Mutant huntingtin (mHtt) protein with expanded polyglutamine tracts forms intracellular aggregates (in cortex and striatum) and induces the degeneration of striatal and cortical neurons. Clinically, HD manifests with chorea, cognitive decline, psychiatric symptoms and progressive weight loss; patients usually succumb within 15–25 years after onset. Like ALS, no disease-modifying therapies exist for HD (Tabrizi et al., 2022).

In recent years, increasing attention has been directed toward the role of nuclear receptors, a class of ligand-activated transcription factors, in modulating cellular responses in neurodegenerative conditions (Moutinho et al., 2019). Among these, the Peroxisome Proliferator-Activated Receptors (PPARs) have gained interest for their involvement in regulating inflammation, mitochondrial function, oxidative stress, and lipid metabolism—pathways crucial in the progression of ALS and HD. Recent studies suggest that PPAR-γ signaling is dysregulated in both ALS and HD, with potential implications for pathogenesis and therapy (Kiaei, 2008).

The potential of PPAR-γ as an integrative molecular hub and a promising therapeutic target in ALS and Huntington's disease is highlighted collectively in this chapter. The role of peroxisome proliferator-activated receptor-γ (PPAR-γ) signaling in the pathophysiology of Huntington's disease and amyotrophic lateral sclerosis is the main topic of this book chapter. It summarizes the molecular structure and transcriptional roles of PPAR-γ alongside PGC-1α-mediated energy regulation and emphasizes how it interacts with important neurodegenerative pathways such as oxidative stress, neuroinflammation, mitochondrial dysfunction, autophagy-apoptosis balance. The chapter also addresses current translational challenges and future perspectives critically assesses preclinical evidence supporting PPAR-γ as a potential therapeutic target and discusses disease-specific and shared pathogenic mechanisms in ALS and Huntington's disease.

2. Peroxisome proliferator-activated receptor-γ (PPAR-γ): A molecular overview

The PPARG gene encodes the PPAR-γ receptor, which is one of the three PPAR isoforms (α, β/δ and γ) that are members of the nuclear receptor superfamily (Małodobra-Mazur et al., 2024). The most studied receptor is PPAR-γ which has two isoforms due to alternative splicing and alternative translation initiation. The ligand-independent activation domain (Activation function 1 (AF1) region and A/B-domain), the DNA binding domain (DBD) (C-domain), the hinge region (D-domain) and the ligand-dependent ligand-binding domain (LBD) (E/F-domain) are the five domains that PPARγ like other nuclear receptors has from N to C terminal (Fig. 1). AF1 is linked to ligand-independent coregulator binding; it is not well conserved and varies greatly among nuclear receptors. The highly conserved C domain contains two zinc finger proteins that recognise particular DNA half-sites called peroxisome proliferator response elements (PPRE) (Zhang, Tang, & Shang, 2024). However, the D domain which has a nuclear localization signal and functions as a flexible hinge that allows rotation between the DNA binding domain and the ligand binding domain is a poorly conserved region (Ramanan et al., 2008). After the DNA binding domain, the ligand binding domain (E domain), the largest domain in PPARγ is the second most conserved domain in nuclear receptors. A second dimerization interface, the ligand binding pocket, a coregulator binding surface and activation function 2 (AF2) are the four primary roles of the ligand binding domain. After ligand binding stabilizes, the LBDs structure AF2 serves as a binding site for coregulator proteins which facilitates their interaction with chromatin to recruit transcriptional machinery and remodel chromatin, ultimately leading to the expression of genes (Kroker & Bruning, 2015; Ramanan et al., 2008).

Small lipophilic substances activate PPARs, which functions as heterodimer with the retinoid X receptor-a (RXR) in the nucleus. The PPAR/RXR heterodimer modulates transcriptional activity by binding to the particular PPRE on the promoter region of PPAR target genes following activation. Hundreds of gene cascades involved in various biological processes are activated when PPAR binds specifically to DNA

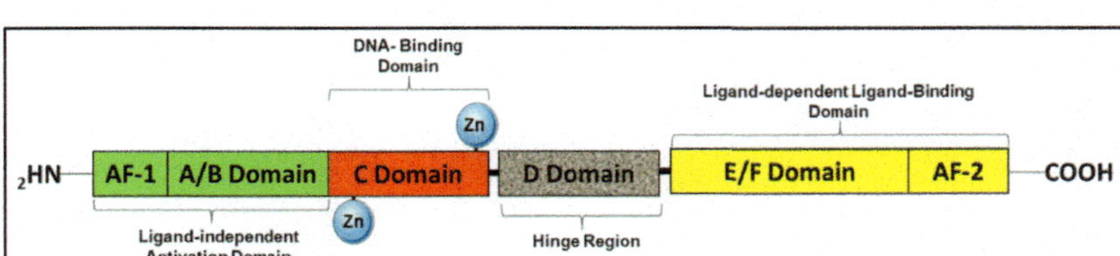

Fig. 1 PPAR-γ receptor.

sequences (Fig. 2). PPAR and RXR heterodimers are bound to co-repressor complexes and inhibit gene transcription when ligands are not present. Transcriptional activation of PPAR-γ target genes is made possible by a conformational shift brought about by ligand binding which displaces corepressors and attracts coactivators (Kim et al., 2013). Additionally, PPAR-γ also has transrepressive effects. It can prevent the transcription of inflammatory genes by interfering with transcription factors like AP-1 and NF-κB through mechanisms other than direct DNA binding (Ricote & Glass, 2007).

Oxidized lipid mediators, prostaglandin derivatives and polyunsaturated fatty acids are examples of natural PPAR-γ ligands. Artificial agonists i. e., Thiazolidinediones (TZDs) including troglitazone, rosiglitazone and pioglitazone have a strong affinity for the ligand binding domain. Human reproductive organs, primary insulin target tissues, liver, white and brown adipose tissue (WAT and BAT respectively), skeletal muscle, cardiac tissue, immune cells and cells of the central nervous system; all express PPAR-γ genes. Lipid and glucose homeostasis, insulin sensitization, control of cell proliferation and differentiation, tissue damage and repair, inflammation and immunology are just a few of the biological processes in which they have been implicated (Costa et al., 2010).

PPARγ controls inflammation in the central nervous system and is expressed in astrocytes and microglia (Bright et al., 2008). Genes linked to

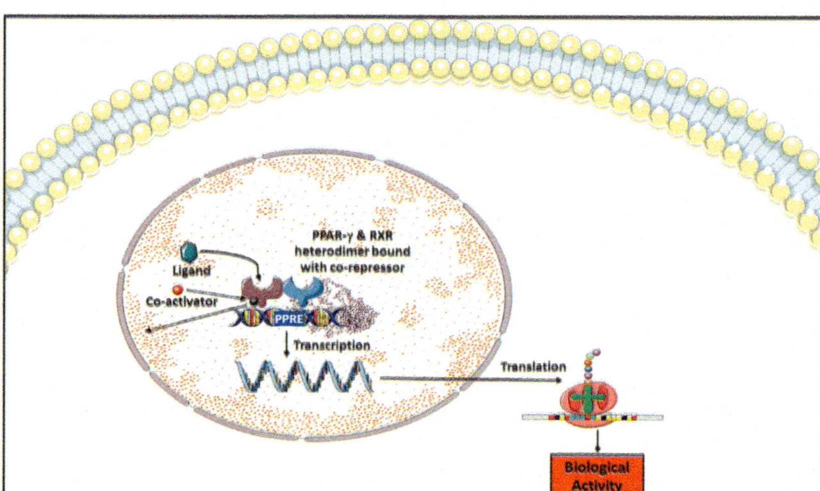

Fig. 2 PPAR gamma activation.

inflammation, antioxidant defense and the brain's metabolism of fats and carbohydrates are regulated by PPAR-γ. In brain cells, while blocking NF-κB signalling which in turn suppresses pro-inflammatory mediators (such as cytokines, iNOS and COX-2), PPAR-γ activation increases anti-apoptotic proteins (like Bcl-2) and antioxidant enzymes (like SOD and catalase). PPAR-γ is therefore frequently regarded as a transcription factor with anti-inflammatory and neuroprotective qualities in the central nervous system. Nonetheless, PPARγ and the canonical WNT/β-catenin pathway typically operate in opposition to each other. As seen in many CNS disorders, if one of them is upregulated the other is downregulated and vice versa (Mannan et al., 2021).

3. PPAR-γ and PGC-1α in mitochondrial biogenesis and cellular energy regulation

One important regulator of neuronal energy metabolism and cellular stress resilience in the central nervous system is the PPAR-γ/peroxisome proliferator-activated receptor γ coactivator 1α (PGC-1α) signalling axis (Guo et al., 2025). Cellular metabolic homeostasis, oxidative phosphorylation and mitochondrial biogenesis are all regulated by the transcriptional complex that PPAR-γ and its coactivator PGC-1α form. PGC-1α improves PPAR-γs capacity to activate genes related to antioxidant responses energy production and adaptive metabolic programs. This axis is crucial for preserving mitochondrial integrity, sustaining synaptic activity and promoting long-term neuronal survival in neurons where energy demand is extraordinarily high (Jamwal, Blackburn, & Elsworth, 2021). PPAR-γ/PGC-1α signalling plays a role in glial cells' metabolic control and neuroinflammatory response modulation (Tang et al., 2025). Increased PGC-1α activity in microglia encourages a change toward an anti-inflammatory phenotype and lowers oxidative burst activity, whereas it supports lactate shuttling and metabolic coupling with neurons in astrocytes (Liang & Ward, 2006). When this pathways dysregulated in the central nervous system; mitochondrial function is compromised, ATP availability is decreased, oxidative stress is elevated and neuroinflammation is more likely to occur. Therefore, restoring PPAR-γ/PGC-1α signalling is a crucial therapeutic approach to support neuronal energy homeostasis and shield the central nervous system from degenerative processes.

PGC-1α is a member of the PGC family of transcriptional regulators (Liang & Ward, 2006). 798 amino acids make up the human PGC-1α protein which has a molecular weight of 91 kDa. The activation domain, inactivation domain, short serine/arginine-rich stretches (RS) domain and RNA recognition motif are just a few of its numerous functional domains (Abu Shelbayeh et al., 2023). PGC-1α coactivates several nuclear transcription factors including Nuclear Respiratory Factor 1 (NRF-1), NRF-2, estrogen-related receptor alpha (ERRα) and PPARγ upon activation by upstream sensors like AMP-activated protein kinase (AMPK) and Sirtuin 1 (SIRT1) via deacetylation. This causes the expression of genes necessary for transcription, protein import and mitochondrial DNA (mtDNA) replication (Abu Shelbayeh et al., 2023). PGC-1α promotes efficient oxidative phosphorylation by upregulating mitochondrial transcription factor A (TFAM) which drives mtDNA transcription and replication through NRF-1/2. It also regulates components of complexes I–IV and cytochrome c (Abu Shelbayeh et al., 2023). Beyond biogenesis, PGC-1α promotes selective mitophagy to eliminate damaged organelles and modulates fusion–fission dynamics by controlling mitofusin (Mfn) 1/2, optic atrophy 1 (Opa1) and dynamin-related protein (Drp1) (Li et al., 2025).

When PPAR-γ is activated, PGC-1α expression is increased and its coactivator activity is stimulated. This increases the transcription of genes related to oxidative phosphorylation, mtDNA replication factors and parts of the mitochondrial import machinery (Uguccioni, D'souza, & Hood, 2010). Through this interaction, PPAR-γ increases the assembly of electron transport chain complexes, boosts ATP production and mitochondrial membrane potential and supports TFAM-mediated mtDNA maintenance (Abu Shelbayeh et al., 2023). By inhibiting excessive Drp1-mediated fission and promoting PGC-1α-dependent expression of fusion proteins like Mfn1 and Mfn2 which collectively maintain mitochondrial structure and stop fragmentation-induced energy loss, PPAR-γ also affects mitochondrial dynamics (Jamwal, Blackburn, & Elsworth, 2021). PPAR-γ keeps PGC-1α stable and stops cytokine-induced suppression of its antioxidant gene targets by lowering oxidative stress and inflammatory signalling especially by inhibiting NF-κB (Jamwal, Blackburn, & Elsworth, 2021). Therefore, PPAR-γ maintains energy metabolism mitochondrial health and resistance to inflammatory and metabolic stress by acting both upstream and in concert with PGC-1α.

4. PPAR-γ crosstalk with other molecular pathways

4.1 PPAR-γ and NF-κB in neuroinflammation

By opposing NF-κB and other pathways PPAR-γ activation reduces the expression of pro-inflammatory genes in microglia and astrocytes (Liu et al., 2017). NF-κB is an inducible transcription factor that is important in inflammatory responses and controls immunological processes. NF-κB helps regulate inflammasomes and influences a number of pro-inflammatory genes such as chemokines and cytokines. NF-κB controls the development, activation and survival of inflammatory T cells and innate immune cells (Dana, Vaseghi, & Haghjooy-Javanmard, 2019). As a result, the NF-B pathway is regarded as a key target in inflammation and needs to be blocked in order to reduce or stop inflammation. Numerous studies show that PPAR-agonists can reduce inflammation by blocking the NF-B pathway (Kannan et al., 2025).

Endotoxins, oxidative stress and cytokines are among the many inflammatory triggers that activate the NF-κB pathway. NF-κB is typically found in the cytosol where it is complexed with the inhibitory protein IκBα, rendering it inactive (Hou et al., 2012). IκB kinase (IKK) is activated when stimuli attach to cell surface receptors and phosphorylate IκB. When the IκBα protein is phosphorylated, it is ubiquitinated, separated from NF-κB and destroyed by the proteasome. This allows NF-κB (primarily the p65/p50 heterodimer) to translocate into the nucleus where it triggers the transcription of genes that promote inflammation. However, a variety of interconnected biochemical pathways cause these inflammatory genes to be transcriptionally repressed upon PPAR-γ activation (Ricote & Glass, 2007).

When PPAR-γ is activated, it directly interacts with the p65 subunit of NF-κB to form an inactive complex that prevents NF-κB from binding to DNA and starting the transcription of genes that promote inflammation (Carvalho et al., 2021). By competing with NF-κB, it also restricts the availability of essential coactivators for NF-κB-mediated transcription such as p300 and CREB-binding protein (CBP) (Ju et al., 2019). Additionally, PPAR-γ maintains NF-κB in the cytoplasm by increasing IκBα production. By further suppressing IκB kinase (IKK) activity PPAR-γ agonists stabilize IκBα and sustain NF-κB suppression (Lecarpentier & Vallée, 2016). SUMOylation of ligand-bound PPAR-γ is another regulatory mechanism that maintains chromatin in a repressed state by permitting it to bind to transcriptional repressor complexes on inflammatory gene promoters is

(Ju et al., 2019). When taken as a whole these molecular processes result in anti-inflammatory activity and reduce the expression of pro-inflammatory cytokines.

4.2 PPAR-γ and Wnt/β-catenin in neurogenesis and synaptic plasticity

Numerous neurological disorders are influenced by the interaction between the nuclear receptor PPAR-γ and the Wnt/β-catenin pathway in the central nervous system (CNS). One is typically suppressed when the other is activated (He, Liao, & Pan, 2018). By attracting presynaptic and postsynaptic components, Wnt signaling is crucial for axon pathfinding, dendritic development and synapse assembly in the central nervous system. Excessive Wnt signaling can result in aberrant neurogenesis and inflammation despite the fact that this pathway is essential for brain development, adult neurogenesis, learning and memory (Cai et al., 2018). Conversely PPAR-γ activation promotes metabolic homeostasis, anti-inflammatory reactions and neuronal maturation over proliferation (Yang & Zhang, 2020). Therefore, through its antagonistic action on Wnt/β-catenin signaling, balanced PPAR-γ action can guarantee appropriate synaptic stability and neuroprotection (Liu et al., 2022). Thus, neurogenesis and synaptic plasticity are finely tuned by the dynamic interaction between PPAR-γ and Wnt/β-catenin signaling with implications for neurodegenerative and psychiatric disorders where this balance is upset.

The Wnt signaling pathway can be categorized as either the canonical or noncanonical pathway based on the different roles that β-catenin plays in the downstream processes. Because the canonical Wnt pathway depends on β-catenin, it is also called the Wnt/β-catenin pathway (Anand, Khan, & Kar, 2023). When Wnt is in the Wnt-ON state, secreted Wnt binds to the cysteine-rich domain of the Frizzled (Fz) receptor and its coreceptor, low-density lipoproteinrelated protein 5/6 (LRP5/6). This activates Dishevelled (Dsh) which breaks up the destruction complex made up of Axin, adenomatous polyposis coli (APC), casein kinase 1 (CK1) and Ser/Thr kinase, Glycogen Synthase Kinase (GSK) -3β. When GSK-3β is inhibited, a complicated chain of events is triggered that stabilizes β-catenin by stopping its breakdown. This enables it to build up in the cytoplasm and go through nuclear translocation where it binds with TCF/LEF to trigger target genes like metalloproteinases, c-Myc, cyclin D1 and PPAR-γ. These genes then control inflammation, neurogenesis, cell survival and proliferation (Colozza, & Koo, 2021). On the other hand in the Wnt-OFF state, β-catenin is phosphorylated by CK1 and GSK-3β leading to its ubiquitination by the E3 ligase, β-transducin repeat-containing protein (β-TrCP) and subsequent degradation by proteasomes. As a result, target gene expression is suppressed and it is prevented from entering the nucleus (Lecarpentier et al., 2017).

The TCF/LEF binding domain of beta-catenin and a catenin binding domain (CBD) within PPAR gamma are involved in the functional interaction between PPAR gamma and the canonical Wnt/beta-catenin signaling (Liu, Wang, Zuo, & Farmer, 2006). By blocking the Wnt/β-catenin signaling pathway, PPAR-γ suppresses cell proliferation and promotes differentiation through a variety of interconnected molecular mechanisms. One of the primary mechanisms is the breakdown of β-catenin whereby active ligand-bound PPAR-γ stimulates GSK3β-dependent phosphorylation and subsequent proteasomal degradation of β-catenin, decreasing its availability for nuclear signalling (Lecarpentier & Vallée, 2016). Furthermore, by competing with β-catenin for binding to TCF/LEF transcription factors PPAR-γ successfully inhibits the activation of Wnt target genes (Vallée, & Lecarpentier, 2018). Furthermore, it can inhibit the transcriptional activity of DNA by directly interacting with its β-catenin–TCF/LEF complex. By doing these things, PPAR-γ inhibits the expression of downstream Wnt target genes such as c-myc and Sox934 that support survival and proliferation. Furthermore PPAR-γ activation can downregulate the transcriptional and translational expression of Wnt pathway components including β-catenin itself, strengthening the pathways' overall inhibition and promoting cellular differentiation and cell cycle arrest (Gupta et al., 2020).

4.3 Interactions with Nrf2/ARE signaling in antioxidant defense

Reactive oxygen species (ROS) production is influenced by PPAR-γ agonists at various cellular levels (Villapol, 2018). Naturally generated during metabolism, reactive oxygen species (ROS) are a class of unstable reactive compounds with an unpaired electron that can harm DNA and cells. Antioxidants neutralize free radicals and reactive oxygen species (ROS) by chelating metal ions that catalyze oxidation or by donating electrons or hydrogen atoms. Oxidative stress occurs when the body's ability to fight off reactive oxygen species (ROS) with antioxidants is out of balance (Chandimali et al., 2025). Oxidative stress in the CNS overwhelms the CNSs antioxidant defenses and results in neuronal malfunction and death and is a major cause of neurodegenerative diseases (Dash et al., 2025). By increasing the synthesis of enzymes like superoxide dismutase (SOD), catalase, manganese superoxide dismutase (MnSOD), glutathione peroxidase 3 (GPx3) and heme oxygenase-1 (HO-1), PPARγ activation strengthens antioxidant defences (Korbecki et al., 2019).

By catalysing the breakdown of the extremely reactive superoxide anion into the less dangerous hydrogen peroxide and molecular oxygen, SOD acts as the first line of defense against oxygen radicals. After that, catalase works in tandem with SOD to quickly break down the hydrogen peroxide generated, turning it into oxygen and water to stop it from building up and creating hazardous hydroxyl radicals through the Fenton reaction (Ighodaro, & Akinloye, 2018). By scavenging superoxide radicals produced as byproducts of the electron transport chain, MnSOD (also known as SOD2) a particular mitochondrial isoform of SOD plays an essential role inside the mitochondrial matrix and supports the function of energy-demanding organs like the heart and brain (Grujicic, & Allen, 2025). Using glutathione as a cofactor, GPx3 which is primarily present in extracellular fluids and plasma and is highly expressed in the kidneys further converts hydrogen peroxide and lipid hydroperoxides to water or corresponding alcohols (Pei et al., 2023). As a component of the later lines of antioxidant defense, HO-1 is an inducible enzyme that breaks down heme into biliverdin, free iron and carbon monoxide molecules that have significant antioxidant and cellular signalling roles, helping to maintain redox balance and provide adaptive defense during oxidative stress and inflammation (Otterbein et al., 2016).

PPAR-γ triggers the Kelch-like ECH-associated protein 1/Nuclear factor erythroid 2-related factor (KEAP1/NrF2) pathway which are involved in the production of antioxidants (Polvani, Tarocchi, & Galli, 2012). Keap1 binds to Nrf2 during rest and promotes its ubiquitination and degradation which keeps intracellular Nrf2 levels low. The cysteine residues in Keap1 undergo changes that prevent Keap1 from binding and degrading Nrf2 when cellular stress levels rise. Consequently, stabilized Nrf2 builds up and moves into the nucleus where it binds to the antioxidant response element (ARE) in DNA and forms a complex with small Maf proteins. Numerous cytoprotective genes including those for glutathione synthesis and other enzymes that boost antioxidant components like MnSOD, GPx3 and HO-1 (Lee, 2017) are activated by this interaction. Multiple studies provide compelling evidence for the mutual regulation of the Nrf2 and PPAR-γ pathways showing that they enhance each other's expression through a positive feedback loop that concurrently regulates transcription factors and the antioxidant target genes downstream (Polvani, Tarocchi, & Galli, 2012). Additionally, PPAR-γ regulates catalase through PPREs that have the canonical direct repeat-1 (DR 1) domain (Girnun et al., 2002).

4.4 Modulation of autophagy and apoptosis-related pathways

The primary function of autophagy is cytoprotection, allowing neurons and glial cells to use lysosomal pathways to break down misfolded proteins, damaged organelles and toxic aggregates (Valencia et al., 2021). Dysregulated autophagy plays a major role in Alzheimers, Parkinsons, ALS, HD and other neurodegenerative disorders and contributes to the accumulation of pathogenic proteins. Efficient autophagy is essential for preventing neurodegeneration, maintaining synaptic function and supporting adaptation to metabolic and oxidative stress (Valencia et al., 2021). On the other hand, apoptosis is a strictly controlled type of programmed cell death that is necessary for the proper development of the CNS including the pruning of neurons and the removal of damaged or undesirable cells. However, after injury or in disease states, excessive or inappropriate activation of apoptosis causes a progressive loss of glia and neurons (Hollville et al., 2019). Significantly autophagy and apoptosis are related: autophagic flux is regulated by some apoptotic regulators while insufficient or excessive autophagy can push cells toward apoptosis (Gump, & Thorburn, 2011). Whether or not CNS cells survive stress-induced programmed death depends on how these processes are balanced. Therefore, maintaining neural homeostasis and halting the progression of disease depend on autophagy and apoptosis being properly coordinated. By controlling important molecular pathways that control cellular survival and stress responses PPAR-γ activation plays a crucial role in regulating the balance between autophagy and apoptosis.

The Unc-51-Like Kinase (ULK) complex which is made up of ULK, Autophagy-related Protein 13 (Atg13) and FAK-family Interacting Protein of (FIP200) is activated when the mammalian target of rapamycin (mTOR) inhibits autophagy and promotes cell growth under stress or nutrient starvation. Following the phosphorylation of Activating Molecule in Beclin-1-Regulated Autophagy (AMBRA1) by the ULK complex the phosphatidylinositol-3-kinase (PI3K) complex containing Vacuolar Protein Sorting 15 (VPS15), VPS34 Beclin-1 and AMBRA1 is activated (Panwar et al., 2023). This results in the phosphorylation of Beclin-1 and its release from dynein motor complexes to start the formation of autophagosomes.

Phagophore assembly starts with membrane nucleation after autophagy is triggered. Membrane sources come from the mitochondria endoplasmic reticulum trans-Golgi network late endosomes and plasma membrane (Zhen, & Stenmark, 2023). The Atg12-Atg5-Atg16 system in which Atg12 is conjugated to Atg5 via Atg7 and Atg10 and the Microtubule-associated protein 1A/1B-light chain 3 (LC3)/Atg8 system in which LC3-I is processed by Atg4 Atg7 and Atg3 to produce LC3-II, a characteristic autophagosome marker control phagophore elongation and expansion (Zhen, & Stenmark, 2023). Cytoskeletal elements and lysosomal membrane proteins like LAMP1/2 regulate the process of autophagy as the autophagosome unites with the lysosome to form an autolysosome. Amino acids and other catabolic products are released into the cytoplasm for recycling and reuse after lysosomal hydrolases inside the autolysosome break down the inner cargo (Khandia et al., 2019).

Conversely, apoptosis is a controlled process that is marked by nuclear shrinkage, chromatin condensation and the formation of apoptotic bodies. It is triggered by either receptor-mediated signaling or mitochondrial stress and it proceeds through two main pathways: intrinsic and extrinsic (Elmore, 2007). Stimuli including hypoxia, DNA damage, oxidative stress and growth factor deprivation triggers the intrinsic pathway and cause mitochondrial outer membrane permeabilization; which is regulated by the Bcl-2 family proteins that share a Bcl-2 Homology (BH3) domain such as anti-apoptotic members (Bcl-2 Bcl-XL Mcl-1 Bcl-W A1/Bfl-1) and pro-apoptotic members (Bak Bax Bid Bad Noxa PUMA) (Singh, Letai, & Sarosiek, 2019). Anti-apoptotic Bcl-2 proteins bind and neutralize Bax and Bak which undergo conformational changes and polymerize on the mitochondrial membrane to promote permeabilization in response to apoptotic stimulation. Cytochrome-c which is released during mitochondrial permeabilization binds to apoptotic protease-activating factor-1 (Apaf-1) and promotes the formation of the apoptosome, a platform that attracts and activates caspase-9 which in turn activates executioner caspases-3 and -7 (Biswas, Roy, Ghosh, & Chakrabarti, 2024). The cell experiences-controlled disintegration after caspase activation which is characterized by chromatin condensation, nuclear disintegration and CAD-mediated DNA fragmentation. Phagocytes use exposed phosphatidylserine to identify and eliminate the apoptotic bodies that result from the bleeding of the plasma membrane (Jan & Chaudhry, 2019). On the other hand, the extrinsic pathway starts when trimeric TNF family ligands activate death receptors such as Fas/CD95, TNF-R1, TRAIL-R1, TRAIL-R2, DR-3 and DR-6. This causes the receptors to trimerize and attract adaptor proteins that contain death domains like FADD and TRADD. In order to cause irreversible cellular degradation these adaptors recruit and activate caspase-8 and caspase-10 within the death-inducing signaling complex (DISC) which then activates executioner caspases-3 and -7 (Nair, Lu, Petersen, & Ashkenazi, 2014).

PPAR-γ regulates autophagy and apoptosis in a context-dependent manner which affects cellular survival in both healthy and diseased states (Elrod, & Sun, 2008; Mahmood et al., 2011). PPAR-γ activation suppresses autophagy in nutrient-rich environments by increasing insulin sensitivity and boosting mTOR signalling which inhibits the ULK1 complex and stops autophagosome initiation (Villapol, 2018). However, PPAR-γ switches to a cytoprotective function during metabolic, oxidative or inflammatory stress by inhibiting mTOR activity, upregulating genes linked to autophagy and aiding in the removal of damaged proteins and organelles (Hao et al., 2023). Concurrently PPAR-γ tightly controls apoptosis in a manner that is likewise context-specific. Its activation suppresses NF-κB-mediated inflammation under normal or protective states, lowers ROS by inducing antioxidant enzymes like HO-1 and GPx3 and stabilizes mitochondrial integrity by blocking Bax translocation, cytochrome-c release and downstream caspase-3/9 activation. PPAR-γ promotes neuronal and glial survival by downregulating pro-apoptotic mediators and upregulating survival genes like Bcl-2 (Villapol, 2018). Thus PPAR-γ functions as an essential integrator of cellular stress responses that preserves CNS homeostasis and guards against the advancement of neurodegenerative diseases by concurrently promoting adaptive autophagy under stress and limiting inappropriate apoptotic signalling.

5. Amyotrophic lateral sclerosis (ALS): Pathophysiology and molecular targets

The deadly adult-onset neurodegenerative disease, amyotrophic lateral sclerosis (ALS) is characterized by the progressive degeneration of both upper and lower motor neurons which causes muscle weakness, atrophy, dysarthria, dysphagia and eventually respiratory failure. The majority of patients pass away within two to five years of the onset of symptoms; though each person's disease course is unique (Verma, 2021). With limb-onset and bulbar-onset presentations as well as overlap with the ALS–frontotemporal dementia spectrum, clinical heterogeneity is common and reflects the various biological processes that underlie the progression of the disease (Achi, & Rudnicki, 2012).

Both familial (10–15 %) and sporadic (85–90 %) forms of ALS present with similar downstream molecular pathologies. More than 30 pathogenic variants have been found to cause familial ALS including SOD1, C9orf72, TARDBP (encoding TDP-43) and FUS. The most prevalent genetic cause is C9orf72 hexanucleotide repeat expansions (González-Sánchez et al., 2025). Age-related cellular vulnerability, environmental exposures and genetic susceptibility interact intricately to cause sporadic ALS which does not follow a Mendelian inheritance pattern. Most cases have similar mechanistic disruptions despite these differences in etiology (Gros-Louis, Gaspar, & Rouleau, 2006).

A multifactorial network of neuronal dysfunction is formed by the multiple intersecting pathways that drive the pathogenesis of ALS. Mutant SOD1 proteins cause oxidative stress, interfere with axonal transport, misfold and aggregate and disturb mitochondrial function (Rubino et al., 2023). Nearly all ALS cases have TDP-43 pathology which causes nuclear depletion, cytoplasmic aggregation, defects in RNA metabolism and prion-like propagation (Suk, & Rousseaux, 2020). Through glutathione depletion, iron dysregulation, mitochondrial impairment and aberrant reactive species production, oxidative stress worsens neuronal damage. In line with these mechanisms, motor neuron loss is accelerated by neuroinflammation caused by activated microglia, malfunctioning astrocytes and immune cell infiltration (Obrador et al., 2020) Furthermore, calcium overload, mitochondrial collapse and the activation of cell-death pathways are encouraged by glutamate excitotoxicity resulting from compromised EAAT2-mediated clearance and calcium-permeable AMPA receptor expression (Arnold et al., 2024).

Despite a great deal of research, there are still few treatments for ALS and the benefits of the disease-modifying medications currently available are minimal (Arnold et al., 2024). Riluzole, a glutamate inhibitor slows the progression of the disease by modifying glutamatergic transmission whereas edaravone provides antioxidant protection mainly in cases that are early and progress quickly. Although more recent developments such as the SOD1-targeted antisense oligonucleotide, tofersen show promise for multi-pathway and precision-based interventions their effectiveness is still limited by genetic specificity inconsistent clinical responses and inconsistent results from confirmatory trials (Guan et al., 2025). AMX0035, a fixed-dose combination of taurursodiol (ursodoxicoltaurine) and sodium phenylbutyrate was used to treat ALS by reducing neuronal death by addressing endoplasmic reticulum stress and mitochondrial dysfunction (Roberts, 2023). However, after a Phase 3 trial failed to demonstrate significant efficacy, it was discontinued.

Improving patient quality of life still depends on symptomatic management which includes multidisciplinary care, respiratory and nutritional support and communication tools. However, ALS heterogeneity, lack of biomarkers, lack of understanding of the underlying mechanisms and frequent translational failures in clinical trials continue to impede therapeutic advancement (Brent, & Ajroud-Driss, 2019).

The convergence of various pathogenic mechanisms in ALS highlights the critical need for multitarget therapeutic approaches. Since oxidative stress, metabolic failure, inflammation, protein misfolding and glial dysfunction are among the interrelated pathways that cause ALS, medications that target a single node of pathology have not demonstrated much therapeutic benefit. As a result the focus of current research has shifted to multi-target and pathway-modulating interventions putting metabolic and anti-inflammatory regulators like PPAR-γ agonists at the forefront of new approaches (Bonafede, & Mariotti, 2017).

6. Role of PPAR-γ in ALS

PPAR-γ has a complex role in the pathophysiology of ALS, mainly due to its involvement in antioxidant and anti-inflammatory mechanisms, regulation of mitochondrial function and modulation of glial cell activation (Kiaei, 2008). PPAR-γ signalling is changed and generally downregulated in ALS as shown by a number of experimental models. Reduced PPAR-γ expression and decreased transcriptional activity including decreased activation of PPRE-driven reporters have been reported in studies conducted in human ALS tissue and SOD1-G93A mice (Qi et al., 2015). According to the opposite interplay theory, ALS is classified as a condition marked by concurrent suppression of PPAR-γ activity and increased Wnt/β-catenin signalling (Vallée, & Lecarpentier, 2018). Endogenous PPAR-γ is necessary for cellular stress resilience because PPAR-γ knockout models in vivo exhibit increased neuronal vulnerability: injury-induced neuronal loss is exacerbated by neuron-specific deletion of PPAR-γ (Quintanilla et al., 2014). PPAR-γ function is also compromised by oxidative stress (Park, & Yang, 2021) excitotoxicity and kinase activation linked to ALS pathology according to in vitro research (Kiaei, 2008).

PPAR-γ is a major transcriptional regulator that suppresses CNS inflammatory pathways (Liu et al., 2017). It inhibits the expression of cytokines and mediators such as TNF-α, IL-1β, iNOS, and COX-2, and by blocking NF-κB and other pro-inflammatory transcriptional programs, it also promotes a neuroprotective state in glial cells (Huang, Zhu, & Zhu, 2025). The repression of NF-κB in microglia and astrocytes through activation of PPAR-γ prevents their shift to a pro-inflammatory phenotype (Pearson et al., 2024). In ALS, PPAR-γ agonists exert strong suppression of microgliosis and glial inflammatory production; pioglitazone potentiates IκB accumulation in glial cells and decreases NF-κB signalling (Kiaei et al., 2005). Simultaneously, PPAR-γ activation promotes antioxidant defenses, in part via the coactivation of PGC-1α, to upregulate genes facilitating redox homeostasis (Jamwal, Blackburn, & Elsworth, 2021). Such anti-inflammatory and antioxidant activities together attenuate neuronal apoptosis induced by stress. PPAR-γ has a major impact on mitochondrial biology by promoting biogenesis, enhancement of oxidative phosphorylation efficiency, and improved resistance towards ROS. In ALS, where excitotoxicity and mutant SOD1 induce mitochondrial dysfunction and activate p38 MAPK and related stress kinases, PPAR-γ agonists counteract these pathological cascades (Corona, & Duchen, 2016).

Thiazolidinedione (TZD), PPAR-γ agonists such as troglitazone, rosiglitazone, and pioglitazone, are demonstrated to hold a strong neuroprotective action in preclinical models of ALS (Quintanilla et al., 2014; Park & Yang, 2021; Kiaei, 2008). Pioglitazone (10–50 mg/kg) delays disease onset in SOD1-G93A mice, maintains motor neuron integrity, improves muscle fiber morphology, and increases survival odds (Schütz et al., 2005). All of these are associated with upregulation of PPAR-γ target genes and restoration of metabolic signaling pathways. Mechanistically, protective effects are consistently linked to reduced glial activation, enhanced mitochondrial stability, and reduction of oxidative and inflammatory stress factors that contribute to motor neuron death (Qi et al., 2015; Vallée, & Lecarpentier, 2018; Quintanilla et al., 2014; Park & Yang, 2021; Kiaei, 2008). Pioglitazone decreases p38 phosphorylation and lowers neuronal ROS content in SOD1-G93A mice (Schütz et al., 2005) and TZDs stabilize mitochondrial membranes and inhibit ROS accumulation in cultured neural cells (Ghosh et al., 2007). Pioglitazone has been shown to substantially suppress microgliosis and astrocytic inflammatory responses in ALS models, promoting local toxicity reduction across the motor neuron microenvironment (Schütz et al., 2005). Such glial modulation plays a key role in understanding the therapeutic benefit seen in PPAR-γ agonist-treated animals. PPAR-γ activation promotes mitochondrial resilience and mitigates oxidative-related cellular damage as the basis for ALS (Fig. 3).

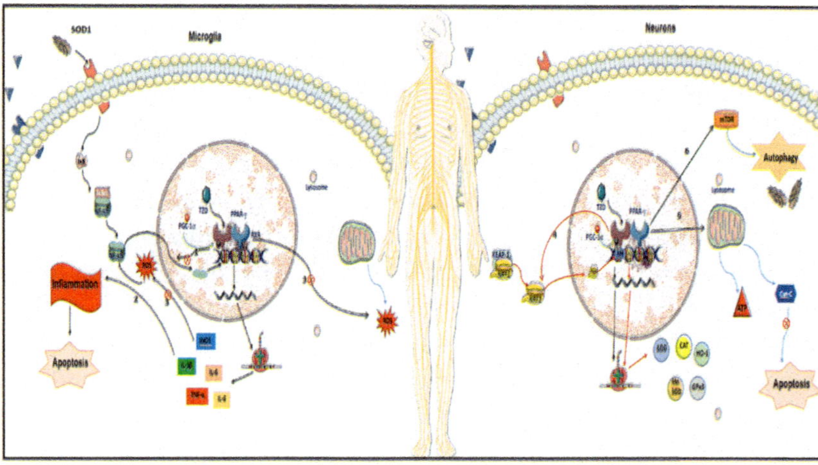

Fig. 3 In microglia, PPAR-γ activation suppresses NF-κB signaling, reducing ROS production and the expression of pro-inflammatory mediators, including TNF-α, IL-1β, IL-6, IL-8, and iNOS, thereby limiting non–cell-autonomous motor neuron toxicity. In neurons, PPAR-γ promotes NRF2–ARE–dependent antioxidant gene expression, enhances mitochondrial biogenesis and oxidative phosphorylation via PGC-1α, increases ATP production, and induces autophagy through mTOR modulation, collectively protecting neurons from oxidative stress and apoptosis.

Although preclinical data are encouraging, translation has been ineffective. In a phase II clinical trial of pioglitazone for ALS, no therapeutic benefit was confirmed. The non-efficacy may be due either to the poor CNS penetration of the drug, its inadequate target engagement, off-target metabolic or side effects, or potentially harmful drug interactions. Moreover, safety concerns such as edema, weight gain, heart failure risk, and bladder cancer associations restrict long-term clinical utility (Dupuis et al., 2012). Currently, targeted strategies focus on designing selective PPAR-γ modulators (SPPARMs) and dual PPAR-γ/α agonists, which yield superior safety data and improved bioavailability to the CNS (Pizcueta et al., 2023).

7. Huntington's disease (HD): Pathophysiology and molecular targets

A combination of motor abnormalities cognitive decline and psychiatric disturbances characterize Huntingtons disease (HD), an inherited neurodegenerative disease that progresses over time. It results in an elongated polyglutamine tract in huntingtin and a toxic gain-of-function of mutant huntingtin (mHTT) due to an expanded CAG trinucleotide repeat in exon 1 of the HTT gene (Ajitkumar, Lui, & de Jesus, 2025). Huntingtin is expressed throughout the body, but the hallmark; motor, cognitive and behavioural symptoms of HD neuropathology are caused by the early and selective degeneration of striatal medium spiny neurons followed by progressive loss of cortical neurons. This region-specific vulnerability is a reflection of transcriptional changes, metabolic load, loss of trophic support, metabolic burden, and stress sensitivity in striatal neurons. (Ehrlich, 2012).

The unstable expansion of CAG repeats within HTT is the defining genetic cause of HD. Longer repeats are linked to earlier onset and faster disease progression. Mutant huntingtin compromises neuronal survival by interfering with a number of cellular functions such as transcriptional regulation, synaptic signalling, axonal transport and mitochondrial homeostasis (Jimenez-Sanchez et al., 2017). Because of their high metabolic demand, reliance on cortical BDNF input and specific transcriptional programs that are dysregulated by mHTT; striatal medium spiny neurons show increased susceptibility. Cortical neurons also deteriorate as the illness worsens which exacerbates behavioural abnormalities and cognitive impairment. The fundamental pathogenic framework of HD is composed of both region-specific cellular vulnerability and genetic determinism (Ehrlich, 2012).

Numerous interrelated cellular disruptions that reinforce one another are the driving forces behind HD pathogenesis. Increased production of reactive oxygen species, weakened antioxidant defenses and respiratory abnormalities in the mitochondria cause oxidative stress which damages lipids, proteins and nucleic acids over time (Irfan et al., 2022). Because mHTT interferes with mitochondrial biogenesis, modifies dynamics, hinders calcium handling and reduces ATP production; it causes a chronic energy deficit in susceptible neurons, making mitochondrial dysfunction a key pathological characteristic (Joshi et al., 2025). Toxicity is made worse by autophagy impairment; mHTT builds up in autophagic vesicles and obstructs lysosomal degradation which lowers the removal of misfolded proteins and causes a gradual collapse of proteostasis (Ciechanover & Kwon, 2015). These anomalies work in concert with transcriptional dysregulation, excitotoxicity and glial-mediated inflammation to cause striatal and cortical neurons to degenerate selectively.

Even though our understanding of HD pathophysiology has advanced significantly, there is currently no disease-modifying treatment and the majority of treatment is still symptomatic. The intricacy of HD in which toxic mHTT causes early extensive and multifactorial cellular dysfunction presents significant obstacles to the development of treatments (Tong et al., 2024). Small molecules, antisense oligonucleotides and RNA-targeting techniques are examples of HTT-lowering strategies that have shown promise in preclinical models by enhancing mitochondrial function, BDNF expression and behavioural outcomes. However, these strategies necessitate careful balancing of mutant and wild-type HTT reduction (Tabrizi, Ghosh, & Leavitt, 2019). Autophagy enhancers (Maiuri, & Kroemer, 2019), mitochondrial biogenesis and metabolic modulators and transcriptional regulators targeted at reestablishing disrupted neuronal programs are other tactics being studied (Uittenbogaard, & Chiaramello, 2014). For significant disease modification, multi-target and pathway-integrated therapeutic approaches are necessary; but obtaining safe long-lasting and CNS-wide delivery continues to be a significant translational barrier.

The limitations of traditional single-target therapeutic approaches are highlighted by the convergent mechanisms of oxidative stress, impaired autophagy, mitochondrial dysfunction and transcriptional dysregulation in Huntingtons disease (Tabrizi et al., 2022). Interventions targeted at specific molecular nodes are unlikely to reverse the pathological environment created by the widespread metabolic instability, persistent inflammatory changes and progressive collapse of proteostatic machinery caused by mutant huntingtin. These difficulties underscore the necessity of approaches that can regulate energy metabolism, redox balance, inflammatory reactions and transcriptional resilience in a coordinated manner. Nuclear receptor pathways, especially those controlled by PPAR-γ have drawn more attention in this regard because of their extensive regulatory functions that closely correspond with the central metabolic and inflammatory abnormalities typical of Huntingtons disease.

8. Role of PPAR-γ in Huntington's disease

PPAR-γ is a crucial regulator of antioxidant gene expression, oxidative phosphorylation and mitochondrial biogenesis. These functions are primarily mediated by its coactivator PGC-1α. Mutant huntingtin (mHtt) suppresses the PPAR-γ/PGC-1α axis in Huntingtons disease (HD) which results in downregulation of mitochondrial enzymes, impaired calcium homeostasis, reduced ATP production and compromised antioxidant defences (Johri, Chandra, & Flint Beal, 2013). Reactive oxygen species (ROS) are increased, mitochondrial depolarization occurs and neuronal energy homeostasis is generally compromised as a result of this repression (Johri, Chandra, & Flint Beal, 2013). Evidence from R6/2 HD mice demonstrates a significant decrease in PPAR-γ transcripts whereas thiazolidinediones (TZDs) treatment improves mitochondrial membrane potential and ATP levels, normalizes mitochondrial Ca^{2+} handling, restores PPAR-γ activity and normalizes expression of PGC-1α and downstream mitochondrial genes (Chiang, Chern, & Juo, 2011). PPAR-γ agonists reverse ROS accumulation and stabilize mitochondrial function in cells derived from HD patients exhibiting similar improvements (Quintanilla et al., 2008). Together these results demonstrate how PPAR-γ and PGC-1α are essential for preserving mitochondrial integrity and emphasize the therapeutic potential of focusing on this pathway to lessen bioenergetic failure in HD.

In HD models, neuroinflammatory markers like IL-1β, TNF-α and MMP-9 are elevated which increases neuronal vulnerability (Rauf et al., 2022). By inhibiting NF-κB activation and decreasing cytokine production in mHtt-expressing cells, PPAR-γ agonists lessen this inflammatory phenotype (Rauf et al., 2022). PPAR-γ serves as a vital endogenous brake on inflammation through these mechanisms and its suppression in HD may worsen glial activation and neuronal damage.

Impaired PPAR-γ signalling in HD is strongly supported by in vivo evidence. Reduced cortical PPAR-γ mRNA levels are seen in R6/2 mice and mHtt can aberrantly sequester PPAR-γ protein into aggregates reducing its functional availability (Chiang, Chern, & Huang, 2012). By reducing PGC-1α and oxidative enzyme expression, PPAR-γ deficiency speeds up metabolic and mitochondrial dysfunction. Rosiglitazone corrects defects in mitochondrial Ca^{2+} handling and increased depolarization in mutant striatal cells (Quintanilla et al., 2008). Chronic TZD treatment increases neuronal stress resistance, decreases mHtt aggregation, decreases weight loss and improves motor function in HD mice (Chiang et al., 2010). Targeting this signalling axis is supported by the metabolic abnormalities reported in human HD despite the lack of clinical trials testing PPAR-γ agonists in HD patients.

Across HD models, PPAR-γ agonists show widespread neuroprotective effects. In mHtt-expressing cells, rosiglitazone treatment restores mitochondrial ATP levels, enhances membrane potential, decreases ROS accumulation and inhibits inflammatory signalling (Quintanilla et al., 2008; Jin et al., 2013). In vivo long-term TZD treatment reduces brain mHtt aggregation in R6/2 mice, restores mitochondrial biochemistry, improves

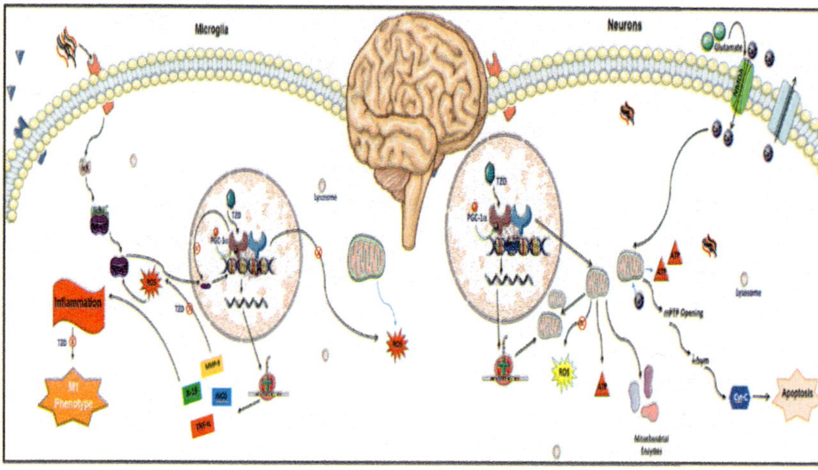

Fig. 4 In microglia, PPAR-γ activation suppresses NF-κB–driven inflammatory signaling, reducing ROS production and the expression of pro-inflammatory mediators such as TNF-α, IL-1β, iNOS, and MMP-9, thereby limiting neurotoxic microglial activation. In neurons, restoration of the PPAR-γ/PGC-1α axis enhances mitochondrial biogenesis and oxidative phosphorylation, increases ATP production, and activates NRF2–ARE-dependent antioxidant defenses, preventing mitochondrial dysfunction, cytochrome-c release, and apoptosis.

motor coordination and lessens weight loss (Chiang, Chern, & Huang, 2012; Quintanilla et al., 2008; Chiang et al., 2010). Additionally, pioglitazone shields striatal neurons from excitotoxic damage (e. g. A g. quinolinic acid) (Kalonia, Kumar, & Kumar, 2010). Together these results demonstrate that pharmacological activation of PPAR-γ improves several pathological domains including cell death inflammation and metabolic insufficiency indicating its potential as a therapeutic target for HD (Fig. 4).

9. Therapeutic implications and future prospects

PPAR-γ agonists such as thiazolidinediones (TZDs), including pioglitazone and rosiglitazone, have drawn attention to their potential for repurposing in neurodegenerative disease due to their anti-inflammatory, antioxidant, and mitochondrial-supportive properties. Systemic long-term usage is limited by adverse effects including edema, risk of heart failure, bone fractures, and potential oncogenic concerns. A similar emphasis on selective PPAR-γ modulators (SPPARMs) and partial agonists, which provide neuroprotective and anti-inflammatory effects while limiting metabolic and cardiovascular liabilities, has followed. Brain-penetrant, low-dose, or biased agonists may offer safer therapeutic windows for chronic neurological conditions.

Nevertheless, given the multifactorial nature of neurodegeneration, PPAR-γ activation is not likely to be sufficient by itself. Rational combination therapies are another avenue forward in this regard, including by combining PPAR-γ agonists with antioxidants, mitochondrial enhancers, autophagy modulators, or other anti-inflammatory agents. In ALS and HD, combinations should be specifically designed to avoid pathway antagonism—for instance, risk of counter-regulation of PPAR-γ signalling and Wnt/β-catenin signaling in patients on riluzole. The potential synergistic efficacy of targeting inflammation, oxidative stress, mitochondrial dysfunction, and protein aggregation on disease-modifying drugs is high.

While PPAR-γ agonists have been effective in preclinical studies, many barriers restrict clinical translation: uncertain clinical therapeutic window, limited information about cell-specific PPAR-γ activity in CNS, and insufficient biomarkers to assess target engagement in humans. By the time clinical symptoms develop, irreversible neuronal loss might have already occurred, and late-stage intervention may therefore be less effective. Moreover, systemic activation of PPAR-γ could lead to off-target effects, emphasizing requirements of CNS-selective or cell-type-targeted delivery systems (nano-based systems or gene therapy).

Overcoming these challenges requires a shift towards personalized therapeutic modalities. Inter-individual variation in PPAR-γ pathway function, induced by gene polymorphisms, differential neuronal versus glial expression, metabolic state, and disease stage, probably drives heterogeneous responses to PPAR-γ therapeutics. Identification of patients early—using genetic risk profiling, metabolic signatures, or biomarkers based on PPAR-γ dysfunction—may allow for stratification of the individuals more likely to benefit from PPAR-γ–targeted interventions. These precision-based approaches may also guide optimal timing, dosing, and therapeutic duration to augment therapeutic effect.

Given that small and largely inconclusive clinical trials have examined PPAR-γ modulation in neurodegenerative diseases, the therapeutic potential of this strategy is unknown. Future studies must also incorporate clinically relevant biomarkers from blood, cerebrospinal fluid, imaging, or transcriptomic analysis to ensure pathway engagement and biological activity. Further clinically appropriate strategies include careful patient selection, intervention in early stage, and rational combination therapies targeting convergent disease mechanisms. At a final point, longitudinal, robust studies will be key to identify if personalized modulation of PPAR-γ may have the capacity to affect disease progression and not just the transient or symptomatic effects.

10. Conclusion

Huntingtons disease and amyotrophic lateral sclerosis are examples of complex neurodegenerative disorders where a number of pathogenic pathways come together to cause progressive loss and dysfunction of neurons. Both diseases share fundamental mechanisms involving oxidative stress, mitochondrial impairment, chronic neuroinflammation, disrupted autophagy and dysregulated apoptosis despite differences in their genetic and clinical characteristics. The limitations of single-target therapeutic approaches are highlighted by these overlapping features which also highlight the significance of modulating pathways.

PPAR-γ is a central transcriptional regulator that can integrate stress-response, inflammatory and metabolic signaling in the central nervous system. PPAR-γ has broad neuroprotective effects that are highly relevant to both ALS and HD pathology through its interactions with NF-κB, Wnt/β-catenin, Nrf2/ARE, autophagy–apoptosis networks and the PGC-1α–dependent mitochondrial biogenesis machinery. According to experimental data from cellular and animal models, restoring or activating PPAR-γ signaling reduces neuroinflammation, strengthens antioxidant defences, maintains mitochondrial integrity and increases neuronal survival.

PPAR-γ-based therapies have had limited clinical translation despite strong preclinical support. This is due to issues with safety, CNS bioavailability, cell-type specificity and insufficient biomarkers of target engagement. These drawbacks underscore the necessity of next-generation approaches such as CNS-targeted delivery systems, rational combination therapies, selective PPAR-γ modulators and customized treatment paradigms informed by metabolic and genetic profiling.

To sum up, PPAR-γ is a promising molecular hub that can modulate several disease-relevant pathways in Huntingtons disease and ALS. Future research concentrating on accurate context-dependent PPAR-γ signaling modulation may offer a promising path toward disease-modifying therapies that go beyond treating symptoms to address the fundamental processes causing neurodegeneration.

References

Abu Shelbayeh, O., Arroum, T., Morris, S., & Busch, K. B. (2023). PGC-1α Is a master regulator of mitochondrial lifecycle and ROS stress response. *Antioxidants, 12*(5), 1075.

Achi, E. Y., & Rudnicki, S. A. (2012). ALS and frontotemporal dysfunction: A review. *Neurology Research International, 2012*, 1–9.

Ajitkumar, A., Lui, F., & De Jesus, O. (2025, April 6). *Huntington disease*. In StatPearls [Internet]. StatPearls Publishing. https://www.ncbi.nlm.nih.gov/books/NBK559166/.

Anand, A. A., Khan, M. V. M., & Kar, D. (2023). The molecular basis of Wnt/β-catenin signaling pathways in neurodegenerative diseases. *International Journal of Cell Biology, 2023*, 1–13.

Arnold, F. J., Putka, A. F., Raychaudhuri, U., Hsu, S., Bedlack, R. S., Bennett, C. L., et al. (2024). Revisiting glutamate excitotoxicity in amyotrophic lateral sclerosis and age-related neurodegeneration. *International Journal of Molecular Sciences, 25*(11), 5587.

Biswas, U., Roy, R., Ghosh, S., & Chakrabarti, G. (2024). The interplay between autophagy and apoptosis: Its implication in lung cancer and therapeutics. *Cancer Letters, 585*, 216662.

Bonafede, R., & Mariotti, R. (2017). ALS pathogenesis and therapeutic approaches: The role of mesenchymal stem cells and extracellular vesicles. *Frontiers in Cellular Neuroscience, 11*, 80.

Brent, J. R., & Ajroud-Driss, S. (2019). Amyotrophic lateral sclerosis: Integrated care provides the best quality of life. *Practical neurology (Fort Washington, Pa), 18*, 46–49.

Bright, J. J., Kanakasabai, S., Chearwae, W., & Chakraborty, S. (2008). PPAR regulation of inflammatory signaling in CNS diseases. *PPAR Research, 2008*(1).

Cai, W., Yang, T., Liu, H., Han, L., Zhang, K., Hu, X., et al. (2018). Peroxisome proliferator-activated receptor γ (PPARγ): A master gatekeeper in CNS injury and repair. *Progress in Neurobiology, 163–164*, 27–58.

Carvalho, M. V. de, Gonçalves-de-Albuquerque, C. F., & Silva, A. R. (2021). PPAR gamma: From definition to molecular targets and therapy of lung diseases. *International Journal of Molecular Sciences, 22*(2), 805.

Chandimali, N., Bak, S. G., Park, E. H., Lim, H. J., Won, Y. S., Kim, E. K., et al. (2025). Free radicals and their impact on health and antioxidant defenses: A review. *Cell Death Discovery, 11*(1), 19.

Chiang, M. C., Chen, C. M., Lee, M. R., Chen, H. W., Chen, H. M., Wu, Y. S., et al. (2010). Modulation of energy deficiency in Huntington's disease via activation of the peroxisome proliferator-activated receptor gamma. *Human Molecular Genetics, 19*(20), 4043–4058.

Chiang, M. C., Chern, Y., & Huang, R. N. (2012). PPARgamma rescue of the mitochondrial dysfunction in Huntington's disease. *Neurobiology of Disease, 45*(1), 322–328.

Chiang, M. C., Chern, Y., & Juo, C. G. (2011). The dysfunction of hepatic transcriptional factors in mice with Huntington's Disease. *Biochimica et Biophysica Acta (BBA) - Molecular Basis of Disease, 1812*(9), 1111–1120.

Ciechanover, A., & Kwon, Y. T. (2015). Degradation of misfolded proteins in neurodegenerative diseases: Therapeutic targets and strategies. *Experimental & Molecular Medicine, 47*(3), e147 e147.

Colozza, G., & Koo, B. (2021). Wnt/β-catenin signaling: Structure, assembly and endocytosis of the signalosome. *Development, Growth & Differentiation, 63*(3), 199–218.

Corona, J. C., & Duchen, M. R. (2016). PPARγ as a therapeutic target to rescue mitochondrial function in neurological disease. *Free Radical Biology & Medicine, 100*, 153–163.

Costa, V., Gallo, M. A., Letizia, F., Aprile, M., Casamassimi, A., & Ciccodicola, A. (2010). PPARG: Gene expression regulation and next-generation sequencing for unsolved issues. *PPAR Research, 2010*, 1–17.

Dana, N., Vaseghi, G., & Haghjooy-Javanmard, S. (2019). Crosstalk between peroxisome proliferator-activated receptors andtoll-like receptors: A systematic review. *Advanced Pharmaceutical Bulletin, 9*(1), 12–21.

Dash, U. C., Bhol, N. K., Swain, S. K., Samal, R. R., Nayak, P. K., Raina, V., et al. (2025). Oxidative stress and inflammation in the pathogenesis of neurological disorders: Mechanisms and implications. *Acta Pharmaceutica Sinica B, 15*(1), 15–34.

Dupuis, L., Dengler, R., Heneka, M. T., Meyer, T., Zierz, S., Kassubek, J., et al. (2012). A randomized, double blind, placebo-controlled trial of pioglitazone in combination with riluzole in amyotrophic lateral sclerosis. *PLoS One,* 7(6), e37885.

Ehrlich, M. E. (2012). Huntington's disease and the striatal medium spiny neuron: Cell-autonomous and non-cell-autonomous mechanisms of disease. *Neurotherapeutics: the journal of the American Society for Experimental NeuroTherapeutics, 9*(2), 270–284.

Elmore, S. (2007). Apoptosis: A review of programmed cell death. *Toxicologic Pathology, 35*(4), 495–516.

Elrod, H. A., & Sun, S. Y. (2008). PPAR γ and apoptosis in cancer. *PPAR Research, 2008*(1).

Feldman, E. L., Goutman, S. A., Petri, S., Mazzini, L., Savelieff, M. G., Shaw, P. J., et al. (2022). Amyotrophic lateral sclerosis. *The Lancet, 400*(10360), 1363–1380.

Gadhave, D. G., Sugandhi, V. V., Jha, S. K., Nangare, S. N., Gupta, G., Singh, S. K., et al. (2024). Neurodegenerative disorders: Mechanisms of degeneration and therapeutic approaches with their clinical relevance. *Ageing Research Reviews, 99,* 102357.

Ghosh, S., Patel, N., Rahn, D., McAllister, J., Sadeghi, S., Horwitz, G., et al. (2007). The thiazolidinedione pioglitazone alters mitochondrial function in human neuron-like cells. *Molecular Pharmacology, 71*(6), 1695–1702.

Girnun, G. D., Domann, F. E., Moore, S. A., & Robbins, M. E. C. (2002). Identification of a functional peroxisome proliferator-activated receptor response element in the rat catalase promoter. *Molecular Endocrinology, 16*(12), 2793–2801.

González-Sánchez, M., Ramírez-Expósito, M. J., & Martínez-Martos, J. M. (2025). Pathophysiology, clinical heterogeneity, and therapeutic advances in amyotrophic lateral sclerosis: A comprehensive review of molecular mechanisms, diagnostic challenges, and multidisciplinary management strategies. *Life (Chicago, Ill.: 1978), 15*(4), 647.

Gros-Louis, F., Gaspar, C., & Rouleau, G. A. (2006). Genetics of familial and sporadic amyotrophic lateral sclerosis. *Biochimica et Biophysica Acta (BBA) - Molecular Basis of Disease, 1762*(11–12), 956–972.

Grujicic, J., & Allen, A. R. (2025). Manganese superoxide dismutase: Structure, function, and implications in human disease. *Antioxidants, 14*(7), 848.

Guan, S., Wang, S., Shi, Y., Leng, Y., Ming, Y., Hou, Z., et al. (2025). Comparative safety analysis of riluzole, edaravone and tofersen in ALS management: Insights from FAERS database. *Frontiers in Pharmacology, 16.*

Gump, J. M., & Thorburn, A. (2011). Autophagy and apoptosis: What is the connection? *Trends in Cell Biology, 21*(7), 387–392.

Guo, P., Lu, Q., Ye, X., Martínez, M. A., Lopez-Torres, B., Martínez, M., et al. (2025). Regulating peroxisome proliferator activated receptor gamma coactivator-1alpha (PGC-1α): A potential role in neurological disorder and treatment. *European Journal of Pharmacology, 1001,* 177750.

Gupta, V., Dhull, D. K., Joshi, J., Kaur, S., & Kumar, A. (2020). Neuroprotective potential of azilsartan against cerebral ischemic injury: Possible involvement of mitochondrial mechanisms. *Neurochemistry International, 132,* 104604.

Hao, K., Wang, J., Yu, H., Chen, L., Zeng, W., Wang, Z., et al. (2023). Peroxisome proliferator-activated receptor γ regulates lipid metabolism in sheep trophoblast cells through mTOR pathway-mediated autophagy. *PPAR Research, 2023,* 1–13.

He, C. W., Liao, C. P., & Pan, C. L. (2018). Wnt signalling in the development of axon, dendrites and synapses. *Open Biology, 8*(10).

Hollville, E., Romero, S. E., & Deshmukh, M. (2019). Apoptotic cell death regulation in neurons. *The FEBS Journal, 286*(17), 3276–3298.

Hou, Y., Moreau, F., & Chadee, K. (2012). PPARγ is an E3 ligase that induces the degradation of NFκB/p65. *Nature Communications, 3*(1), 1300.

Huang, Y., Zhu, Y., & Zhu, Z. (2025). Recent advancements in the research of the mechanism by which PPAR -γ Is involved in postoperative neurocognitive disorder. *The FASEB Journal, 39*(22).

Ighodaro, O. M., & Akinloye, O. A. (2018). First line defence antioxidants-superoxide dismutase (SOD), catalase (CAT) and glutathione peroxidase (GPX): Their fundamental role in the entire antioxidant defence grid. *Alexandria Journal of Medicine, 54*(4), 287–293.

Irfan, Z., Khanam, S., Karmakar, V., Firdous, S. M., el Khier, B. S. I. A., Khan, I., et al. (2022). Pathogenesis of Huntington's disease: An emphasis on molecular pathways and prevention by natural remedies. *Brain Sciences, 12*(10).

Jamwal, S., Blackburn, J. K., & Elsworth, J. D. (2021). PPARγ/PGC1α signaling as a potential therapeutic target for mitochondrial biogenesis in neurodegenerative disorders. *Pharmacology & Therapeutics, 219*, 107705.

Jan, R., & Chaudhry, G. E. S. (2019). Understanding apoptosis and apoptotic pathways targeted cancer therapeutics. *Advanced Pharmaceutical Bulletin, 9*(2), 205–218.

Jimenez-Sanchez, M., Licitra, F., Underwood, B. R., & Rubinsztein, D. C. (2017). Huntington's disease: Mechanisms of pathogenesis and therapeutic strategies. *Cold Spring Harbor Perspectives in Medicine, 7*(7).

Jin, J., Albertz, J., Guo, Z., Peng, Q., Rudow, G., Troncoso, J. C., et al. (2013). Neuroprotective effects of PPAR-γ agonist rosiglitazone in N171-82Q mouse model of Huntington's disease. *Journal of Neurochemistry, 125*(3), 410–419.

Johri, A., Chandra, A., & Flint Beal, M. (2013). PGC-1α, mitochondrial dysfunction, and Huntington's disease. *Free Radical Biology & Medicine, 62*, 37–46.

Joshi, D. C., Chavan, M. B., Gurow, K., Gupta, M., Dhaliwal, J. S., & Ming, L. C. (2025). The role of mitochondrial dysfunction in Huntington's disease: Implications for therapeutic targeting. *Biomedicine & Pharmacotherapy, 183*, 117827.

Ju, Z., Su, M., Li, D., Hong, J., Im, D. S., Kim, S., et al. (2019). An Algal metabolite-based PPAR-γ agonist displayed anti-inflammatory effect via inhibition of the NF-κB pathway. *Marine Drugs, 17*(6), 321.

Kalonia, H., Kumar, P., & Kumar, A. (2010). Pioglitazone ameliorates behavioral, biochemical and cellular alterations in quinolinic acid induced neurotoxicity: Possible role of peroxisome proliferator activated receptor-ϒ (PPARϒ) in Huntington's disease. *Pharmacology, Biochemistry, and Behavior, 96*(2), 115–124.

Kannan, G., Paul, B. M., & Thangaraj, P. (2025). Stimulation, regulation, and inflammaging interventions of natural compounds on nuclear factor kappa B (NF-kB) pathway: A comprehensive review. *Inflammopharmacology, 33*(1), 145–162.

Khandia, R., Dadar, M., Munjal, A., Dhama, K., Karthik, K., Tiwari, R., et al. (2019). A comprehensive review of autophagy and its various roles in infectious, non-infectious, and lifestyle diseases: Current knowledge and prospects for disease prevention, novel drug design, and therapy. *Cells, 8*(7), 674.

Kiaei, M. (2008). Peroxisome proliferator-activated receptor-γ in amyotrophic lateral sclerosis and Huntington's disease. *PPAR Research, 2008*(1).

Kiaei, M., Kipiani, K., Chen, J., Calingasan, N., & Beal, M. (2005). Peroxisome proliferator-activated receptor-gamma agonist extends survival in transgenic mouse model of amyotrophic lateral sclerosis. *Experimental Neurology, 191*(2), 331–336.

Kim, T. H., Kim, M. Y., Jo, S. H., Park, J. M., & Ahn, Y. H. (2013). Modulation of the transcriptional activity of peroxisome proliferator-activated receptor gamma by protein-protein interactions and post-translational modifications. *Yonsei Medical Journal, 54*(3), 545.

Korbecki, J., Bobiński, R., & Dutka, M. (2019). Self-regulation of the inflammatory response by peroxisome proliferator-activated receptors. *Inflammation Research, 68*(6), 443–458.

Kroker, A. J., & Bruning, J. B. (2015). Review of the structural and dynamic mechanisms of PPAR *γ* partial agonism. *PPAR Research, 2015*, 1–15.

Lecarpentier, Y., Claes, V., Vallée, A., & Hébert, J. L. (2017). Interactions between PPAR gamma and the canonical Wnt/beta-catenin pathway in type 2 diabetes and colon cancer. *PPAR Research, 2017*, 1–9.

Lecarpentier, Y., & Vallée, A. (2016). Opposite Interplay between PPAR gamma and canonical Wnt/Beta-catenin pathway in amyotrophic lateral sclerosis. *Frontiers in Neurology, 7*.

Lee, C. (2017). Collaborative power of Nrf2 and PPAR *γ* activators against metabolic and drug-induced oxidative injury. *Oxidative Medicine and Cellular Longevity, 2017*(1).

Li, W., Gui, Y., Guo, C., Huang, Y., Liu, Y., Yu, X., et al. (2025). Molecular mechanisms of mitochondrial quality control. *Translational Neurodegeneration, 14*(1), 45.

Liang, H., & Ward, W. F. (2006). PGC-1α: A key regulator of energy metabolism. *Advances in Physiology Education, 30*(4), 145–151.

Liu, J., Wang, H., Zuo, Y., & Farmer, S. R. (2006). Functional interaction between peroxisome proliferator-activated receptor γ and β-catenin. *Molecular and Cellular Biology, 26*(15), 5827–5837.

Liu, J., Xiao, Q., Xiao, J., Niu, C., Li, Y., Zhang, X., et al. (2022). Wnt/β-catenin signalling: Function, biological mechanisms, and therapeutic opportunities. *Signal Transduction and Targeted Therapy, 7*(1), 3 3.

Liu, T., Zhang, L., Joo, D., & Sun, S. C. (2017). NF-κB signaling in inflammation. *Signal Transduction and Targeted Therapy, 2*(1), 17023.

Mahmood, D. F. D., Jguirim-Souissi, I., Khadija, E. H., Blondeau, N., Diderot, V., Amrani, S., et al. (2011). Peroxisome proliferator-activated receptor γ induces apoptosis and inhibits autophagy of human monocyte-derived macrophages via induction of cathepsin L. *Journal of Biological Chemistry, 286*(33), 28858–28866.

Maiuri, M. C., & Kroemer, G. (2019). Therapeutic modulation of autophagy: Which disease comes first? *Cell Death & Differentiation, 26*(4), 680–689.

Małodobra-Mazur, M., Ołdakowska, M., & Dobosz, T. (2024). Exploring PPAR gamma and PPAR Alpha's regulation role in metabolism via epigenetics mechanism. *Biomolecules, 14*(11), 1445.

Mannan, A., Garg, N., Singh, T. G., & Kang, H. K. (2021). Peroxisome proliferator-activated receptor-gamma (PPAR-γ): Molecular effects and its importance as a novel therapeutic target for cerebral ischemic injury. *Neurochemical Research, 46*(11), 2800–2831.

Moutinho, M., Codocedo, J. F., Puntambekar, S. S., & Landreth, G. E. (2019). Nuclear receptors as therapeutic targets for neurodegenerative diseases: Lost in translation. *Annual Review of Pharmacology and Toxicology, 59*(1), 237–261.

Nair, P., Lu, M., Petersen, S., & Ashkenazi, A. (2014). Apoptosis initiation through the cell-extrinsic pathway. *Methods in Enzymology,* 99–128.

Obrador, E., Salvador, R., López-Blanch, R., Jihad-Jebbar, A., Vallés, S. L., & Estrela, J. M. (2020). Oxidative stress, neuroinflammation and mitochondria in the pathophysiology of amyotrophic lateral sclerosis. *Antioxidants, 9*(9), 901.

Otterbein, L. E., Foresti, R., & Motterlini, R. (2016). Heme oxygenase-1 and carbon monoxide in the heart. *Circulation Research, 118*(12), 1940–1959.

Panwar, V., Singh, A., Bhatt, M., Tonk, R. K., Azizov, S., Raza, A. S., et al. (2023). Multifaceted role of mTOR (mammalian target of rapamycin) signaling pathway in human health and disease. *Signal Transduction and Targeted Therapy, 8*(1), 375.

Park, H. R., & Yang, E. J. (2021). Oxidative stress as a therapeutic target in amyotrophic lateral sclerosis: Opportunities and limitations. *Diagnostics, 11*(9), 1546.

Pearson, A., Koprivica, M., Eisenbaum, M., Ortiz, C., Browning, M., Vincennie, T., et al. (2024). PPARγ activation ameliorates cognitive impairment and chronic microglial activation in the aftermath of r-mTBI. *Journal of Neuroinflammation, 21*(1), 194.

Pei, J., Pan, X., Wei, G., & Hua, Y. (2023). Research progress of glutathione peroxidase family (GPX) in redoxidation. *Frontiers in Pharmacology, 14*.

Pizcueta, P., Vergara, C., Emanuele, M., Vilalta, A., Rodríguez-Pascau, L., & Martinell, M. (2023). Development of PPARγ agonists for the treatment of neuroinflammatory and neurodegenerative diseases: Leriglitazone as a promising candidate. *International Journal of Molecular Sciences, 24*(4), 3201.

Polvani, S., Tarocchi, M., & Galli, A. (2012). PPAR and oxidative stress: Con catenating NRF2 and FOXO. *PPAR Research, 2012*, 1–15.

Qi, Y., Yin, X., Wang, S., Wang, J., Jiang, H., Wang, X., et al. (2015). Differential peroxisome proliferator activated receptors activity in a rodent model of amyotrophic lateral sclerosis. *International Journal of Clinical and Experimental Medicine, 8*(3), 3743–3751.

Quintanilla, R. A., Jin, Y. N., Fuenzalida, K., Bronfman, M., & Johnson, G. V. W. (2008). Rosiglitazone treatment prevents mitochondrial dysfunction in mutant huntingtin-expressing cells. *Journal of Biological Chemistry, 283*(37), 25628–25637.

Quintanilla, R. A., Utreras, E., & Cabezas-Opazo, F. A. (2014). Role of PPAR γ in the differentiation and function of neurons. *PPAR Research, 2014*, 1–9.

Ramanan, S., Kooshki, M., Zhao, W., Hsu, F. C., & Robbins, M. E. (2008). PPARα ligands inhibit radiation-induced microglial inflammatory responses by negatively regulating NF-κB and AP-1 pathways. *Free Radical Biology and Medicine, 45*(12), 1695–1704.

Rauf, A., Badoni, H., Abu-Izneid, T., Olatunde, A., Rahman, Md. M., Painuli, S., et al. (2022). Neuroinflammatory markers: Key indicators in the pathology of neurodegenerative diseases. *Molecules (Basel, Switzerland), 27*(10), 3194.

Ricote, M., & Glass, C. K. (2007). PPARs and molecular mechanisms of transrepression. *Biochimica et Biophysica Acta (BBA) - Molecular and Cell Biology of Lipids, 1771*(8), 926–935.

Roberts, E. (2023). A review of oral fixed-dose sodium phenylbutyrate and ursodoxicoltaurine in people with amyotrophic lateral sclerosis. *EMJ Neurology, 2*(12).

Rubino, V., la Rosa, G., Pipicelli, L., Carriero, F., Damiano, S., Santillo, M., et al. (2023). Insights on the multifaceted roles of wild-type and mutated superoxide dismutase 1 in amyotrophic lateral sclerosis pathogenesis. *Antioxidants, 12*(9), 1747.

Schütz, B., Reimann, J., Dumitrescu-Ozimek, L., Kappes-Horn, K., Landreth, G. E., Schürmann, B., et al. (2005). The oral antidiabetic pioglitazone protects from neurodegeneration and amyotrophic lateral sclerosis-like symptoms in superoxide dismutase-G93A transgenic mice. *The Journal of Neuroscience, 25*(34), 7805–7812.

Singh, R., Letai, A., & Sarosiek, K. (2019). Regulation of apoptosis in health and disease: The balancing act of BCL-2 family proteins. *Nature Reviews. Molecular Cell Biology, 20*(3), 175–193.

Suk, T. R., & Rousseaux, M. W. C. (2020). The role of TDP-43 mislocalization in amyotrophic lateral sclerosis. *Molecular Neurodegeneration, 15*(1), 45.

Tabrizi, S. J., Estevez-Fraga, C., van Roon-Mom, W. M. C., Flower, M. D., Scahill, R. I., Wild, E. J., et al. (2022). Potential disease-modifying therapies for Huntington's disease: Lessons learned and future opportunities. *The Lancet Neurology, 21*(7), 645–658.

Tabrizi, S. J., Ghosh, R., & Leavitt, B. R. (2019). Huntingtin lowering strategies for disease modification in Huntington's disease. *Neuron, 101*(5), 801–819.

Tada, M., Coon, E. A., Osmand, A. P., Kirby, P. A., Martin, W., Wieler, M., et al. (2012). Coexistence of Huntington's disease and amyotrophic lateral sclerosis: A clinicopathologic study. *Acta Neuropathologica, 124*(5), 749–760.

Tang, M. B. O., Liu, Y. X., Hu, Z. W., Luo, H. Y., Zhang, S., Shi, C. H. E., et al. (2025). Study insights in the role of PGC-1α in neurological diseases: Mechanisms and therapeutic potential. *Frontiers in Aging Neuroscience, 16.*

Tong, H., Yang, T., Xu, S., Li, X., Liu, L., Zhou, G., et al. (2024). Huntington's disease: Complex pathogenesis and therapeutic strategies. *International Journal of Molecular Sciences, 25*(7), 3845.

Uguccioni, G., D'souza, D., & Hood, D. A. (2010). Regulation of PPAR γ coactivator-1 α function and expression in muscle: Effect of exercise. *PPAR Research, 2010,* 1–7.

Uittenbogaard, M., & Chiaramello, A. (2014). Mitochondrial biogenesis: A therapeutic target for neurodevelopmental disorders and neurodegenerative diseases. *Current Pharmaceutical Design, 20*(35), 5574–5593.

Valencia, M., Kim, S. R., Jang, Y., & Lee, S. H. (2021). Neuronal autophagy: Characteristic features and roles in neuronal pathophysiology. *Biomolecules & Therapeutics, 29*(6), 605–614.

Vallée, A., & Lecarpentier, Y. (2018). Crosstalk between peroxisome proliferator-activated receptor gamma and the canonical WNT/β-catenin pathway in chronic inflammation and oxidative stress during carcinogenesis. *Frontiers in Immunology, 9.*

Verma, A. (2021). Clinical manifestation and management of amyotrophic lateral sclerosis. *Amyotrophic Lateral Sclerosis.* Exon Publications, 1–14.

Villapol, S. (2018). Roles of peroxisome proliferator-activated receptor gamma on brain and peripheral inflammation. *Cellular and Molecular Neurobiology, 38*(1), 121–132.

Yang, Y., & Zhang, Z. (2020). Microglia and Wnt pathways: Prospects for inflammation in Alzheimer's disease. *Frontiers in Aging Neuroscience, 12.*

Zhang, J., Tang, M., & Shang, J. (2024). PPARγ modulators in lung cancer: Molecular mechanisms, clinical prospects, and challenges. *Biomolecules, 14*(2), 190.

Zhen, Y., & Stenmark, H. (2023). Autophagosome biogenesis. *Cells, 12*(4), 668.

CHAPTER SIX

A comprehensive review on therapeutics and diagnostic agents targeting acetyl and butyryl cholinesterases for AD

Vishal Chaurasia[a,1], Sanjib Das Modak[a,1], Gireesh Kumar Singh[b], and Gyan Modi[a,*]

[a]Department of Pharmaceutical Engineering & Technology, Indian Institute of Technology (BHU), Varanasi, Uttar Pradesh, India

[b]Department of Pharmacy, School of Health Science, Central University of South Bihar, Gaya, Bihar, India

*Corresponding author. e-mail address: gpmodi.phe@itbhu.ac.in

Contents

Abstract

Alzheimer's disease (AD) is a progressive, irreversible, and multifaceted neurodegenerative disorder characterized by cognitive decline, memory loss, and behavioral impairment, posing a major global health challenge. Its multifactorial pathology includes cholinergic dysfunction, amyloid-β deposition, tau hyperphosphorylation, oxidative stress, and neuroinflammation. Among these, impairment of the cholinergic system, characterized by reduced acetylcholine levels, plays a crucial role in cognitive deficits. The enzymes acetylcholinesterase (AChE) and butyrylcholinesterase (BChE),

[1] These authors contributed equally.

International Review of Neurobiology, Volume 188
ISSN 0074-7742, https://doi.org/10.1016/bs.irn.2026.05.017

which hydrolyze acetylcholine, are closely involved in disease progression and serve as important therapeutic and diagnostic targets in AD. This book chapter provides a comprehensive overview of therapeutic and diagnostic agents targeting AChE and BChE in AD, and discusses small-molecule inhibitors, multifunctional ligands, and emerging strategies to modulate cholinesterase activity and restore cholinergic neurotransmission, alleviating disease symptoms. In addition, the chapter highlights advances in diagnostic approaches using fluorescent probes, particularly near-infrared (NIR) probes, for selective detection and imaging of AChE and BChE, including their molecular design, photophysical properties, enzyme selectivity, and mechanisms of action, all of which are critically examined. Targeting AChE and BChE offers a dual advantage in AD by enabling both symptomatic treatment and early-stage diagnosis. This chapter aims to present a clear and comprehensive overview of recent advances in therapeutic and diagnostic approaches, offering meaningful insights for researchers in developing effective strategies for the treatment and monitoring of AD.

1. Introduction

Alzheimer's disease (AD) is progressive, irreversible, and one of the most prevalent neurodegenerative diseases, marked by a progressive decline in cognitive function and memory, primarily affecting the elderly population (>65 years of age), and represents the leading cause of dementia worldwide (Li et al., 2025a; Shankar et al., 2025). AD accounts for approximately 60–70 % of all dementia cases, and the global burden is expected to rise substantially by 2050, affecting more than 100 million people worldwide (Ramachandran et al., 2025; Sharma et al., 2025). Some common symptoms of mild AD include memory and thinking decline, cognitive deficits, personality changes, cognitive impairment, poor judgment, language problems, and difficulty in performing normal daily tasks (Singh et al., 2021). The exact pathophysiology of AD remains uncertain, although several factors, including the deposition of extracellular Aβ (amyloid-β) plaques in the form of senile plaques, formed through amyloidogenic APP cleavage, as well as the formation of intracellular NFTs (neurofibrillary tangles) arising from the hyperphosphorylated tau proteins, increased oxidative stress, neuroinflammation, neuronal and mitochondrial dysfunction, metal ion toxicity, and cholinergic neurodegeneration (Jevtić et al., 2024; Kumar & Modi, 2025). Due to the complex, multifactorial pathophysiology of AD, numerous hypotheses have been proposed; among them, the cholinergic hypothesis remains one of the most promising targets for therapeutic intervention in AD, particularly for the development of potent anti-AD drugs (Li et al. 2021a, 2025a). According to the cholinergic hypothesis, the level of acetylcholine (ACh), an essential neurotransmitter for memory and learning, is markedly reduced in cognitive brain regions of AD patients due to the action of the cholinesterase (ChE) enzyme (Asadipour et al., 2024). In a healthy brain, there are mainly two types of ChE enzymes: acetylcholinesterase (AChE) and butyrylcholinesterase (BChE). The AChE enzyme is predominantly found in the brain, post-synaptic neuromuscular junction, cholinergic neurons, and muscle tissues, whereas the BChE enzyme is generally present in the brain, blood, neurons, peripheral circulation system, glial cells, and liver (Makhaeva et al., 2020; Noori et al., 2024; Shankar et al., 2024). These enzymes are serine hydrolases that hydrolyze the neurotransmitter acetylcholine (ACh), producing choline and acetic acid, and share approximately 65 % structural similarity (Singh et al., 2020; Verma et al., 2024). However, in a healthy brain, ACh is predominantly (80 %) hydrolyzed by AChE, while BChE acts as a secondary enzyme, playing a supplementary role as a co-regulator of AChE activity (Makhaeva et al., 2020). As AD progresses, AChE activity decreases, while BChE activity gradually increases in the brain, reaching approximately 165 % of normal levels in the late stages. These changes are particularly evident in the hippocampus, a brain region essential for memory and cognitive function (Giacobini, 2003; Li et al., 2021a; Panek et al., 2023). Nevertheless, both enzymes remain key targets within the cholinergic hypothesis for the development of a potent therapeutic anti-Alzheimer's drug for early- and late-stage AD intervention. As of now, there is no therapy to stop or even slow the AD. The available treatments primarily focus on alleviating symptoms rather than on halting or slowing disease progression (Khan et al., 2025; Singh et al., 2024a). Currently, five medications have been approved by the U.S. Food and Drug Administration (FDA) for the management of moderate to severe AD. These treatments include four currently approved cholinesterase inhibitors (ChEIs), like donepezil (Aricept) **(1)**, rivastigmine (Exelon) **(2)**, galantamine (Reminyl) **(3)**, and tacrine (Cognex) **(5)** (was retracted from the market in 2013 due to its hepatotoxicity side effects), and one N-methyl-D-aspartate (NMDA) receptor antagonist, memantine (Namenda) **(4)**, is the latest drug approved for the management of the disease (He & Tam, 2024; Singh et al., 2024b). Additionally, two monoclonal antibodies, aducanumab and donanemab, have recently been approved in the USA for the elimination of amyloid plaques. Sodium oligomannate, a drug used to regulate neuroinflammation by modulating the gut microbiome, has also been approved in China (He & Tam, 2024; Verma et al., 2024). However, these drugs only provide temporary relief or alleviate the symptoms of illness, and none of these drugs can address the underlying causes of AD due to its complex pathophysiology and multifactorial characteristics (Li et al. 2021a, 2025a).

Although numerous therapeutic strategies are being investigated by researchers worldwide to alter the progression of AD, the most promising approach may lie in early detection and timely intervention before extensive brain damage. Nevertheless, clinical diagnosis of AD remains challenging, often difficult, and unreliable. To address this challenge, several diagnostic and imaging techniques, including positron emission tomography (PET), single-photon emission computed tomography (SPECT), and magnetic resonance imaging (MRI), have been utilized (Goud et al., 2021; Mandloi et al., 2026). A wide range of PET and SPECT imaging probes targeting amyloid-β (Aβ) aggregates, AChE, and BChE enzymes have been developed and investigated (DeBay et al., 2017; Oukoloff et al., 2015; Roy et al., 2016; Tiepolt et al., 2022). Amyloid beta is also a major hallmark of AD. For the detection of Aβ, various PET imaging agents have been developed, such as [^{11}C]PIB (also called [^{11}C]6-OH-BTA-1) **(6)**, [^{18}F]Flutemetamol **(7)**, and [^{11}C]AZD2184 **(8)**, [^{18}F]AV-45 **(9)**, [^{18}F]AV-1**(10)**, [^{11}C]SB-13 **(11)**, [^{11}C] BF-227 **(12)**, [^{18}F]BF-227 **(13)**, and [^{18}F]AZD4694 **(14)**. Additionally, SPECT imaging agents such as [^{123}I]-IMPY and ^{99m}Tc-ECD have been developed, as shown in Fig. 1C (Goud et al., 2021; Rai et al., 2022). Similarly, acetylcholine analogues have been developed as potential radiotracer agents and are clinically used to quantify regional cerebral ChE activity in AD patients using PET. AMP or MP4A (1-[^{11}C]methyl-4-piperidinyl acetate) **(15)** and PMP or MP4P (1-[^{11}C]methyl-4-piperidinyl propionate) **(16)** are used to measure AChE activity, whereas MP4B (*N*-[11 C] methylpiperidin-4-yl butyrate) **(17)** is used to measure BChE activity (Pappata et al., 1996; Roivainen et al., 2004; Shao et al., 2003; Snyder et al., 2001). In addition, many other radiotracer imaging agents have been developed for the detection of AChE in vitro and in vivo in the rat brain, including [^{11}C]CP-126,998 **(18)**, [^{11}C]Donepezil **(19)**, [^{11}C]Physostigmine **(20)**, and [^{11}C]Methyltacrine **(21)** (Gil-Rivas et al., 2023; Musachio et al., 2002). Despite their imaging potential, these techniques have several limitations, including high cost, unavoidable radiation exposure, and limited availability of narrow and short-lived radioisotopes, time-consuming data acquisition, and the requirement for highly skilled personnel (Rai et al., 2022, 2024). To overcome these issues, near-infrared fluorescence (NIRF) probes have emerged as a promising alternative, making them suitable for

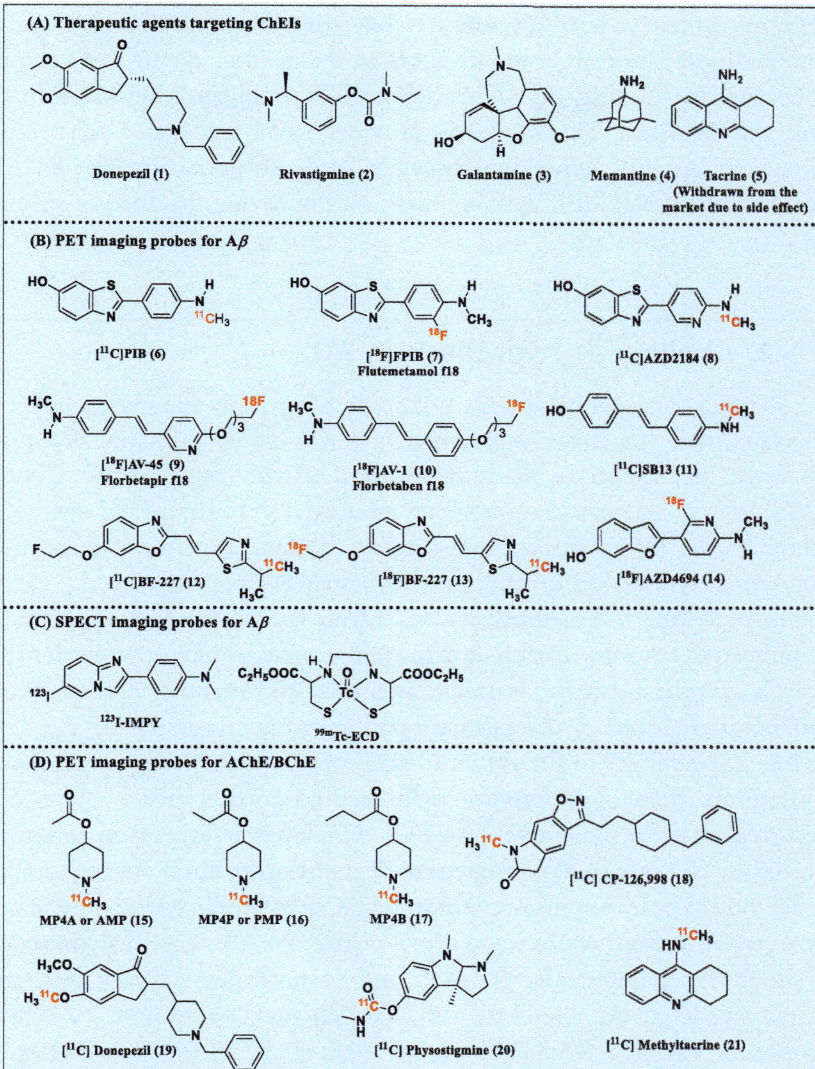

Fig. 1 Chemical structures of (A) Cholinesterase inhibitors (ChEIs) approved by the FDA for the treatment of AD, (B) PET imaging agents for Aβ, (C) SPECT imaging agents for Aβ, (D) PET imaging agents for AChE/BChE.

detecting ChE enzymes in the brain. Consequently, NIRF imaging holds considerable potential for early AD diagnosis, enabling timely therapeutic intervention and improved disease management.

This book chapter focuses on therapeutic and diagnostic agents targeting AChE and BChE enzymes in AD, emphasizing molecular design, inhibitory potency, and selectivity as key parameters for effective cholinesterase modulation. Recent advances in therapeutic agents and imaging probes are discussed to highlight their roles in cholinesterase-based treatment and diagnosis. This overview provides a comprehensive perspective on current progress, emerging strategies, and prospects for targeted AD management using ChEs, with emphasis on improving therapeutic efficacy and diagnostic accuracy.

2. Cholinergic hypothesis in AD

The cholinergic hypothesis is considered an early theoretical concept proposing that cognitive impairment in AD arises primarily from an imbalance in cholinergic neurotransmission (Bartus et al., 1982; Cuello, 2025). It suggests that brain regions essential for learning and memory, including the hippocampus and cerebral cortex, experience substantial cholinergic dysfunction in affected individuals (Whitehouse et al., 1982). Cognitive decline and memory loss result from the progressive degeneration of cholinergic neurons (Žužek, 2024). Ach is a crucial neurotransmitter that controls synaptic plasticity, learning, attention, and memory consolidation. Cholinergic neurons in the normal brain are primarily found in the basal forebrain, specifically in the nucleus basalis of Meynert (Schliebs & Arendt, 2011). From there, they project to limbic and cortical areas, where they affect higher-order cognitive processes. Cholinergic tone is significantly reduced in AD due to the progressive degradation of these neural circuits. Under normal physiological conditions, ACh is produced in presynaptic nerve terminals from choline and acetyl-coenzyme A by the catalytic action of choline acetyltransferase (ChAT) and is subsequently packaged into synaptic vesicles via the vesicular acetylcholine transporter (VAChT) (Amenta & Tayebati, 2008; H Ferreira-Vieira et al., 2016). ACh is released into the synaptic cleft after neuronal stimulation, where it communicates with postsynaptic membrane muscarinic and nicotinic receptors (H Ferreira-Vieira et al., 2016; Parsons et al., 1993).

Cholinesterase hydrolyzes ACh to choline and acetate, thereby terminating the signal quickly (Ferreira-Vieira et al., 2016) as shown in Fig. 2. These enzymes can also use butyrylcholine, a structurally similar ester, as a substrate, especially in pathological conditions (Prešern et al., 2025). The significance of cholinergic impairment in disease pathology is further supported by post-mortem analysis, which consistently shows a loss of

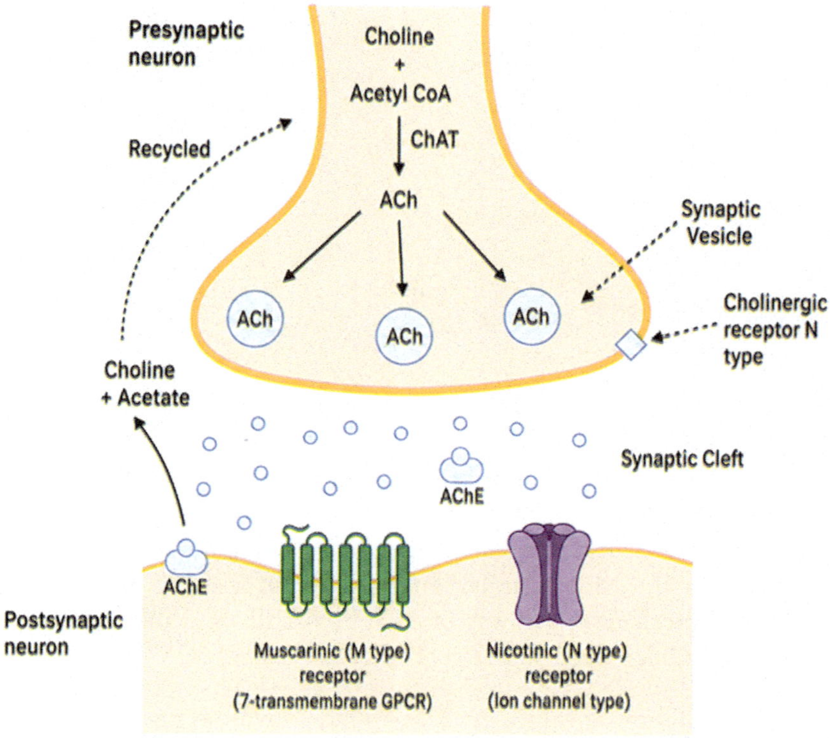

Fig. 2 Scematic illustration of the cholinergic hypothesis.

choline acetyltransferase activity, thereby decreasing Ach synthesis (Davies & Maloney, 1976). Two enzymes, AChE and BChE, play central roles in regulating synaptic ACh concentrations. In healthy brain tissue, AChE predominates in controlling ACh hydrolysis, while BChE has a comparatively limited function (Perry et al., 1978). However, in the AD brain, this enzymatic balance is significantly altered. AChE activity declines due to cholinergic neuronal loss, whereas BChE activity is preserved or even increased, particularly in the advanced stages of the disease (Kamal et al., 2009; Prešern et al., 2025). This shift leads to increased acetylcholine hydrolysis, further aggravating cholinergic deficiency and cognitive decline (Zhang et al., 2024). Beyond their role in neurotransmitter metabolism, ChEs, especially AChE, have been implicated in additional pathological processes associated with AD. In addition to hydrolyzing Ach, AChE is involved in AD pathology through its interaction with Aβ peptides, a process that accelerates fibril formation and increases neurotoxicity (Inestrosa et al., 2008). The resulting AChE-Aβ complexes are believed to

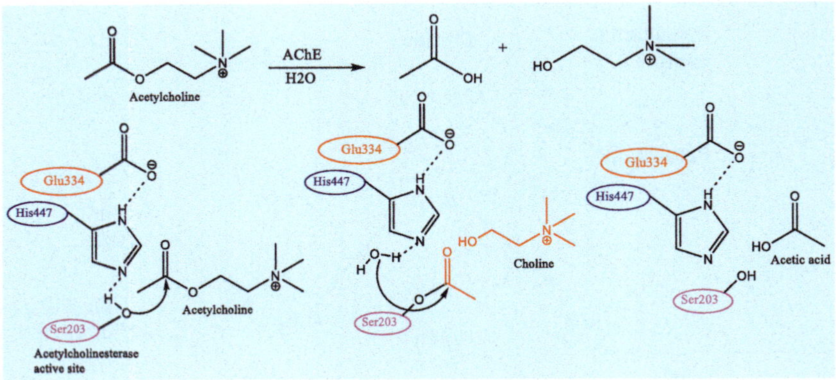

Fig. 3 Degradation mechanism pathway of ACh by AChE.

expedite amyloid plaque formation and intensify neurodegeneration. BChE has also been detected within amyloid plaques, suggesting that both enzymes contribute not only to cholinergic dysfunction but also to disease progression at the Molecular level (Darvesh, 2016; Perry et al., 1978). Degradation mechanism pathway of ACh by AChE are depicted in Fig. 3. Dais HE byAChhas been shown in Fig. 3.

3. Overview of therapeutic agents targeting cholinesterases

3.1 Clinically approved cholinesterase inhibitors

Cholinesterase inhibitors (ChEIs) are the most important class of drugs used for symptomatic treatment of mild to severe AD (Hansen et al., 2008). Their mechanism of action involves inhibition of BChE and/or AChE, thereby improving cholinergic neurotransmission and increasing ACh levels in the postsynaptic region. Among them, donepezil **(1)** is the most commonly prescribed drug. It interacts with both the catalytic active site (CAS) and the peripheral anionic site (PAS) of the enzyme, acting as a reversible and highly selective AChE inhibitor. This dual-site binding facilitates long-term enzyme inhibition and effective CNS penetration (Grabowska et al., 2025; Saxena & Dubey, 2019). Rivastigmine **(2)** is a pseudo-irreversible carbamate that targets both AChE and BChE, a mechanism that distinguishes it from other ChEIs (Wu et al., 2020a). It is believed to be beneficial in the later stages of AD, when BChE activity becomes more important. However, its plasma half-life is short, and it exhibits dose-dependent side effects, which may limit its therapeutic use (Mustazza et al., 2002). Galantamine **(3)**, originally isolated from natural alkaloid sources, is a reversible AChE inhibitor. In addition to enzyme inhibition, it exerts a secondary effect by functioning as an allosteric potentiating ligand of nicotinic acetylcholine receptors (Dajas-Bailador et al., 2003; Samochocki et al., 2003), thereby enhancing cholinergic neurotransmission through a complementary mechanism. Tacrine **(5)** exhibited potent AChE inhibition and was the first cholinesterase inhibitor to be effective in AD therapy (Knapp et al., 1994). However, due to severe liver toxicity, its clinical use was ultimately discontinued (Watkins et al., 1994). Tacrine is still essential in medicinal chemistry, nevertheless, because its fundamental structure makes it a useful pharmacophore for creating hybrid and multipurpose ChEIs. Overall, although currently available ChEIs (Fig. 1) play a vital role in symptomatic treatment, they do not halt disease progression, are less effective in advanced stages, and are often associated with peripheral cholinergic adverse effects. These limitations underscore the urgent need to develop novel synthetic, multifunctional ChEIs with improved efficacy and potential disease-modifying properties.

3.2 Naturally-inspired cholinesterase inhibitors

Several recent studies have explored natural product-inspired scaffolds for cholinesterase inhibition. Ferulic acid (FA) is a naturally occurring phenolic compound that has attracted interest in Alzheimer's research for its neuroprotective effects and promise as a starting point for developing anticholinesterase agents (Di Giacomo et al., 2022). In 2018, Sang and coworkers synthesized several FA-derived compounds as potential therapeutic agents for AD (Fig. 4). Among them, TM-10 **(22)** emerged as the most potent and selective BChE inhibitor, exhibiting an IC_{50} of 0.0089 μM against *equine* BChE, while showing comparatively lower inhibitory activity toward AChE than the parent compound, ferulic acid. Notably, TM-10 **(22)** demonstrated good blood-brain barrier permeability, an essential characteristic for CNS-targeted drugs. In a zebrafish model of AD, it significantly improved behavioral responsiveness and recovery, and in SH-SY5Y neural cells, it provided substantial neuroprotection against $A\beta_{1-42}$ induced toxicity (Sang et al., 2018). In a subsequent study, the same research group designed O-carbamoyl ferulamide derivatives as multitarget-directed ligands as anticholinesterase agents (Fig. 4). Among the synthesized compounds, 4f **(24)**, which incorporates a 1,2,3,4-tetrahydroisoquinoline fragment, demonstrated the highest

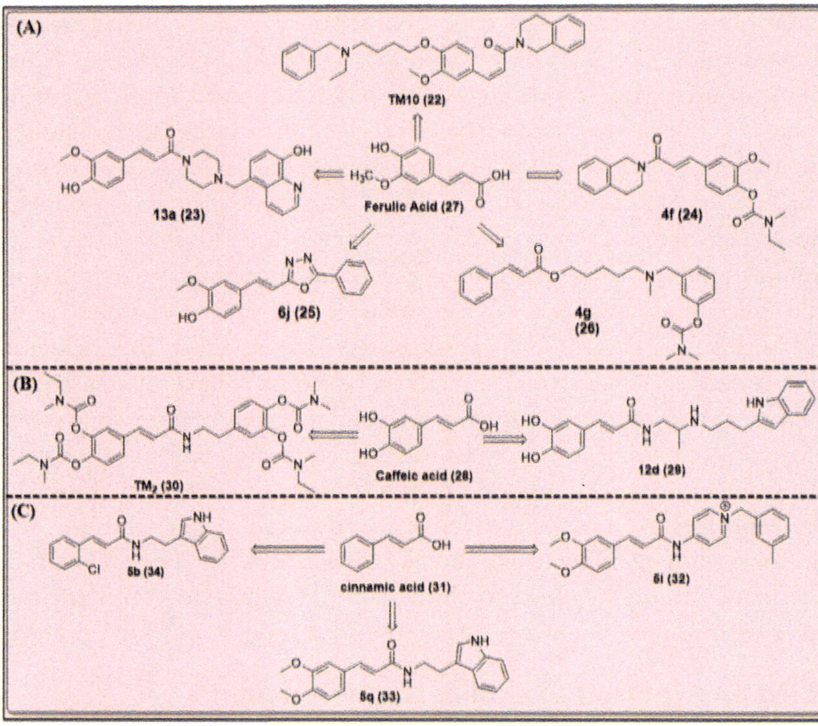

Fig. 4 Chemical structures of ferulic acid derivatives (A), Caffeic acid derivatives (B), and Cinnamic acid derivatives (C).

inhibitory potency against human BChE, with an IC_{50} of 0.97 μM. Despite its limited activity against AChE, compound 4f **(24)** was also permeable to the blood-brain barrier and improved scopolamine-induced cognitive deficits (Sang et al., 2020). In another approach, Tripathi and co-workers developed FA-based hybrids containing a 1,3,4-oxadiazole linker, aiming to enhance enzyme affinity and overall pharmacological performance by modifying the phenylpropanoid framework (Fig. 4). Biological assessment using Ellman's method revealed that several derivatives inhibited both *human* AChE and BChE (Tripathi et al., 2020). Among them, compound 6j **(25)** exhibited significant inhibitory activity against AChE (IC_{50} = 0.068 μM) along with considerable activity against BChE (IC_{50} = 0.218 μM). Structure-activity relationship (SAR) study suggested that electron-withdrawing substituents, such as trifluoromethyl groups on the terminal phenyl ring, enhanced cholinesterase inhibition, likely by strengthening interactions within the CAS and PAS. Enzyme kinetic studies further revealed a mixed-type inhibition mechanism for 6j **(25)**. Notably, this lead derivative demonstrated good blood-brain barrier permeability, reduced amyloid-β aggregation, and exerted neuroprotective effects in SH-SY5Y neuronal cells. Moreover, improvements in cognitive performance observed in in vivo models further highlight the potential of ferulic acid-derived hybrids as multifunctional, natural product-inspired therapeutic agents for AD (Tripathi et al., 2020). More recently, FA has remained a versatile scaffold for the development of multifunctional anti-Alzheimer agents. In 2024, Lan and co-workers synthesized a series of ferulic acid derivatives and evaluated their multi-target potential (Fig. 4). Among them, compound 4 g **(26)** showed significant inhibitory activity against human AChE (IC_{50} = 19.7 nM) and BChE (IC_{50} = 0.59 μM), exhibiting a mixed-type inhibition profile toward AChE. Molecular docking studies indicated that 4 g **(26)** interacts with both the PAS and the CAS domains of AChE. Furthermore, the compound improved cellular survival under H_2O_2-induced oxidative stress in PC12 cells and demonstrated notable neuroprotective effects at micromolar concentrations (Lan et al., 2020). In a related investigation during the same year, Singh et al. synthesized ferulic acid-piperazine hybrids. Within this group, the 8-hydroxyquinoline-containing derivative 13a **(23)** exhibited the most promising cholinesterase inhibition, with an IC_{50} of 0.59 μM against AChE and 5.02 μM against BChE.

Enzyme kinetic studies showed that it is a mixed-type inhibitor for both enzymes (Singh et al., 2024a). Beyond ferulic acid, other naturally occurring phenolic scaffolds have also drawn attention. Caffeic acid, for instance, has been explored as a promising template for the development of ChEIs. In 2025, Hu and colleagues reported caffeic acid-based derivatives, as depicted in Fig. 4, among which TM-2 **(30)**, bearing a carbamate moiety, showed potent inhibitory activity against BChE (IC_{50} = 0.36 μM). This compound also reduced $A\beta_{1\text{-}42}$ aggregation, crossed the blood-brain barrier in vitro, and significantly treated the memory impairment in a scopolamine-induced mouse model. However, its inhibitory effect on AChE was not so significant (Hu et al., 2025). In 2025, Kumar et al. synthesized several caffeic acid derivatives and identified compound 12d **(29)**, a tryptamine-containing analog, as the most active against both AChE and BChE. Enzyme kinetic analysis indicated a mixed-type inhibition of *human* AChE and *equine* BChE. A propidium iodide displacement assay further suggested interaction with the peripheral anionic site, with 12d **(29)** producing 23.58 % displacement compared with 26.15 % for donepezil at the same concentration (50 μM). Permeability evaluation using the PAMPA method demonstrated favorable blood-brain barrier penetration (Pe > 2.0) (Kumar & Modi, 2025). Being structurally related to caffeic acid, cinnamic acid has also attracted attention as a useful phenylpropanoid scaffold for the design of anticholinesterase compounds. In 2017, Lan et al. synthesized derivatives containing a cinnamic acid core with an N-benzyl pyridinium moiety. Compound 5I **(32)** bearing a fluoro atom at the 4th position on the benzyl group showed the most promising cholinesterase inhibition towards human AChE and BChE (IC_{50} 8.6 nM and 4.4 μM, respectively). Enzyme kinetic studies suggest that compound 5I **(32)** behaves as a mixed-type inhibitor of AChE, meaning it can bind to the enzyme at multiple binding sites. The compound showed no noticeable cytotoxicity under the tested conditions in PC12 neuroblastoma cells. Additional experiments also showed that 5I **(32)** reduced the self-aggregation of $A\beta_{1-42}$ peptides, an essential factor in Alzheimer's pathology. Results from the PAMPA assay further suggested that the molecule can cross the blood-brain barrier, highlighting its potential suitability for central nervous system activity (Lan et al., 2017). In 2020, Ghafary et. al. had synthesized a series of novel cinnamic acid derivatives, among which compound 5q **(33)** showed the most potent AChE inhibitory activity (IC_{50} = 11.51 μM), and 5b **(34)** showed promising BChE inhibitory activity (IC_{50} = 1.95 μM). 5q **(33)** exhibited mild neuroprotective activity at 10 μM but showed no significant neuroprotective activity at 1 and 100 μM against H_2O_2-induced damage in PC12 cells. Molecular docking studies revealed that compounds **5q** and **5b** both interacted with the CAS and PAS of AChE and BChE, respectively (Ghafary et al., 2020). Zhang et al. developed and synthesized some cinnamic acid-based derivatives by structural modification of the natural phenylpropanoid scaffold to enhance cholinesterase inhibition for AD therapy. The compounds were evaluated against AChE and BChE using Ellman's assay, and several derivatives exhibited dual inhibitory activity, the most potent analogue showing IC_{50} values of approximately 0.12 μM for AChE and 0.19 μM for BChE, respectively. SAR analysis indicated that extended π-conjugation and optimal aromatic substitution significantly contributed to enhanced enzyme inhibition (Zhang et al., 2018). Overall, this study establishes cinnamic acid as a promising natural product-derived scaffold for the development of potent ChEIs. Alkaloid-derived frameworks have also been extensively studied as sources of anticholinesterase drugs, along with phenolic natural compounds. Because of their proven therapeutic utility and structural flexibility, galantamine-based derivatives remain the most appealing of these. In 2025, Li and co-workers formed a series of galantamine analogues by introducing a benzylpyridine fragment at the hydroxyl position (Fig. 5). Among the synthesized compounds, compound C2 **(35)** was the most active derivative, displaying submicromolar inhibitory potency toward both human AChE (IC_{50} = 513.90 nM) and human BuChE (IC_{50} = 357.77 nM). Biological studies further indicated minimal cytotoxicity in PC12 neuronal cells and demonstrated neuroprotective effects under H_2O_2-induced oxidative stress conditions. Molecular modeling supported favorable interactions of C2 within the catalytic regions of both ChEs, consistent with its inhibitory activity (Li et al., 2025b).

Similarly, in a related study, Wang and co-workers explored C1-aryl-modified galantamine derivatives as multifunctional molecules inspired by natural scaffolds. Among the compounds prepared, **4k (36)**, which contains a meta-thiomethyl-substituted benzene ring, showed notable

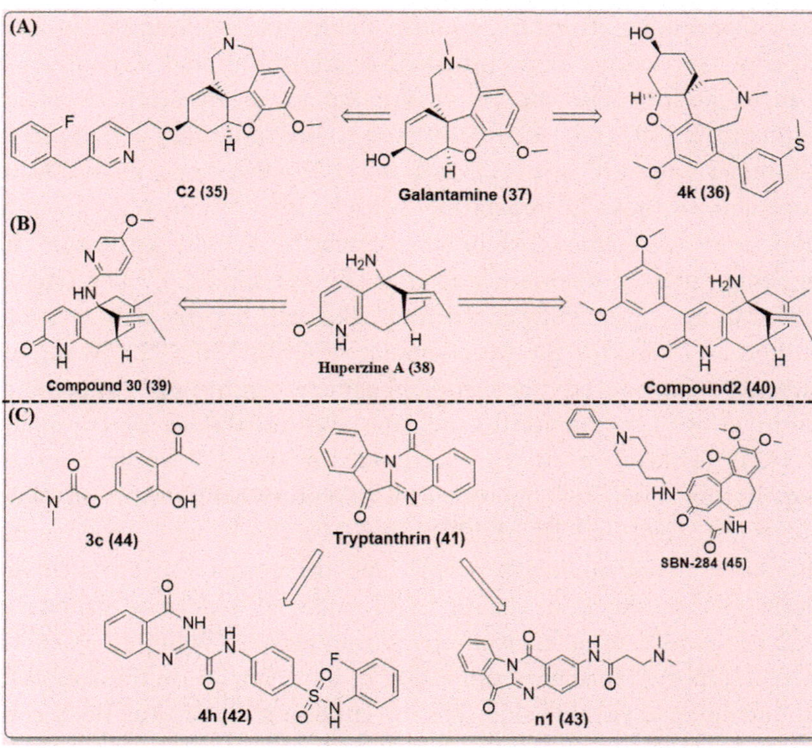

Fig. 5 Chemical structures of galantamine analogs (A), Huperazine analogs (B), Tryptanthrin derivatives and rivastigmine analogs (C).

inhibitory activity toward both *ee*AChE (IC_{50} = 0.47 μM) and *ee*BChE (IC_{50} = 2.98 μM). Beyond cholinesterase inhibition, this compound was also evaluated for its biological effects at the cellular level. It showed protective activity against H_2O_2-induced oxidative stress in SH-SY5Y neuronal cells, indicating that its role may extend beyond enzyme inhibition to include neuroprotective properties (Wang et al., 2025a). Huperazine A (HPA), a naturally derived anti-Alzheimer agent, has also served as a valuable template for structural modification. In 2021, Miao and colleagues prepared a series of Pd-catalyzed N-aryl derivatives of HPA to enhance its cholinesterase inhibitory properties (Fig. 5). Within this set, compound 30 **(39)**, incorporating a 5-methoxypyridine substituent, demonstrated the most vigorous activity toward AChE, showing approximately 7.6-fold greater potency (IC_{50} = 1.5 μM) compared with the parent drug HPA (IC_{50} = 11 μM). In addition to enzyme inhibition, this derivative exhibited notable neuroprotective effects by reducing H_2O_2-induced cytotoxicity in SH-SY5Y neuronal cells (Miao et al., 2021). Continuing efforts to optimize the huperazine scaffold, in 2024, Zhu and co-workers investigated C-3 aryl-substituted derivatives to enhance cholinesterase inhibition. Among the synthesized molecules, compounds 2 **(40)** and 15 demonstrated AChE inhibitory activity in the micromolar range (IC_{50} ≈ 1.205 μM and 0.225 μM, respectively), values comparable to those of huperazine A itself. In addition to AChE inhibition, these derivatives exhibited noteworthy activity against BChE, suggesting potential dual-target activity (Zhu et al., 2024). The tryptanthrin scaffold, a naturally occurring alkaloid, has recently attracted attention as a basis for the development of ChEIs. In 2023, Wang and colleagues introduced benzenesulfonamide-substituted tryptanthrin derivatives as multifunctional candidates. Within this series, compound 4 h **(42)**, featuring a fluoro substituent at the C2 position of the aromatic ring, displayed vigorous dual inhibitory activity toward AChE (IC_{50} = 0.13 μM) and BChE (IC_{50} = 6.11 μM). Kinetic investigations indicated a mixed inhibition profile, and the compound provided significant neuroprotection against H_2O_2-induced oxidative damage in PC12 neuronal cells at micromolar concentrations (Wang et al., 2023). Xia et al. reported further optimization of the tryptanthrin framework by introducing an acylamide side chain to modify its structure. With compound n1**(43)** emerging as a highly effective inhibitor (IC_{50} = 12.17 nM) and showing a strong preference for BChE, this technique yielded derivatives with greater selectivity for AChE. Analysis of the structure-activity connection revealed that linker length and substitution at the tryptanthrin core's 2-position were essential factors influencing activity, most likely as a result of better accommodation inside the AChE active gorge. The enzyme kinetic and docking results confirmed a mixed-type, reversible inhibitory mechanism (Xia et al., 2024). All of these results support tryptanthrin as a viable template for the design of selective ChEIs derived from natural products. Hybridization strategies have also been explored using other pharmacophores. In 2022, Deng and co-workers developed 2-acetylphenol–rivastigmine hybrids, identifying compound 3c **(44)** as a selective *h*AChE inhibitor (IC_{50} = 8.7 μM). The compound showed no detectable cytotoxicity in MTT assays, provided protection against H_2O_2-induced PC12 cell injury, and demonstrated efficient blood-brain barrier permeability in a PAMPA assay (Deng et al., 2022). Reddy et al. followed a two-pharmacophore hybrid approach to design and synthesize a series of colchicine derivatives as noncytotoxic cholinesterase inhibitors (Fig. 5). Compound SBN-284 **(45)** was found to be the lead, showing an IC_{50} of 0.313 μM against *h*AChE and 0.29 μM against *eq*BChE. The lead compound displayed non-competitive inhibition in the enzyme kinetic assay. It is permeable across the BBB and nontoxic to cells (Reddy et al., 2024).

3.3 Synthetic and hybrid cholinesterase inhibitors

The discovery of natural alkaloids such as physostigmine and galantamine first demonstrated the therapeutic usefulness of cholinesterase inhibition. Still, their use has been constrained by issues including poor selectivity, short duration of action, and cholinergic side effects. These restrictions have enabled the development of synthetic and hybrid inhibitors with improved pharmacological properties. Modern medicinal chemistry increasingly focuses on rationally designed compounds that can interact with both the CAS and PAS of AChE, thereby improving inhibitory potency and affording the potential for additional disease-modifying effects.

3.3.1 Coumarin-based inhibitors

Because of their versatile benzopyrone core, which can engage the deep active gorge of AChE through π–π stacking and hydrophobic interactions while also providing chemical handles for functionalization, coumarin and its derivatives have been a topic of interest in AD drug discovery. In recent years, medicinal chemists have used this framework to produce both simple synthetic analogues and hybrid compounds (Fig. 6) that combine cholinesterase inhibition with

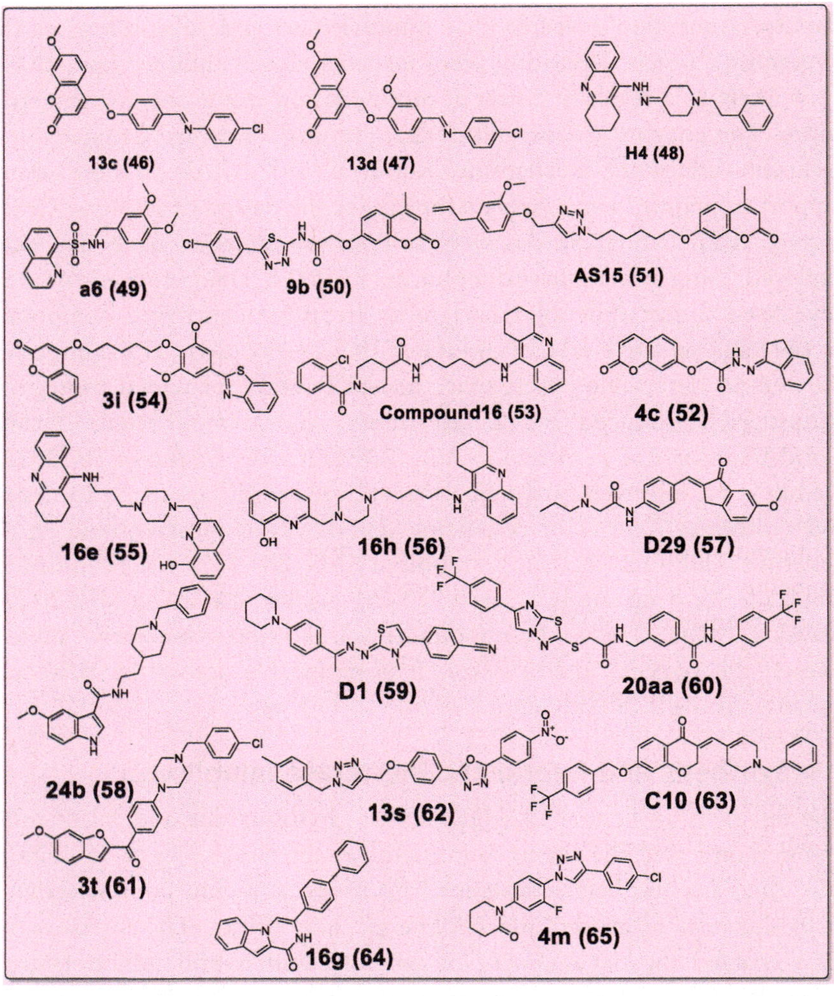

Fig. 6 Chemical structures of coumarin, tacrine, donepezil-based, and various heterocyclic synthetic cholinesterase inhibitors.

multi-target characteristics. Earlier in 2023, Hasan et al. synthesized a group of compounds, called Schiff base-linked coumarin hybrids, to assess their inhibitory activity against ChEs. Overall, the compounds showed promising inhibition of AChE. Notably, derivatives bearing a chloro substituent, which decreases electron density on the aromatic ring, demonstrated enhanced potency. In particular, compounds 13c **(46)** and 13d **(47)** exhibited vigorous activity with IC_{50} values of 0.232 μM and 0.190 μM, respectively. Among the series, compounds **13c** and **13d** showed the most favorable binding profiles, consistent with their higher inhibitory activity (Hasan et al., 2023). Coumarin-based structures have been widely investigated for their potential as ChEIs. For example, Kamel and co-workers prepared a series of 2-oxo-chromene-7-oxymethylene acetohydrazide derivatives, among which the indene-containing analogue 4c **(52)** showed the strongest AChE inhibition (IC_{50} = 0.802 μM) compared with donepezil as a reference. In behavioral evaluation using a T-maze model, AD animals treated with compound **4c** required less time to reach food, suggesting a possible improvement in cognitive performance (Kamel et al., 2023). Coumarin scaffolds have also been modified through hybridization approaches. In 2024, Jaber et al. designed coumarin-oxathiadiazole hybrids using a structure-guided strategy, and compound 9b **(50)** emerged as the most active molecule (IC_{50} = 0.75 μM). The docking studies suggested that these hybrids may interact with both the CAS and PAS of AChE, likely due to the combination of the coumarin core with heterocyclic fragments such as thiazole and thiadiazole (Jaber et al., 2024). Further structural variation was explored by Asadipour *et al.*, who synthesized amino-7,8-dihydro-4H-chromene derivatives that inhibit both AChE and BChE (Asadipour et al., 2024). Along similar lines, Mirjali and colleagues developed 4-(benzo[*d*]thiazole-2-yl) phenol derivatives derived from 4-hydroxycoumarin, among which compound 3i **(54)** demonstrated the highest AChE inhibitory activity (IC_{50} = 2.7 μM) (Mirjalili et al., 2024). More recent studies have focused on converting coumarin derivatives into multi-target-directed ligands (MTDLs). In 2025, Singh Atamjit et al. reported triazole-containing coumarin-eugenol hybrids designed to target both ChEs and amyloid-β peptides. In this series, compound AS15 **(51)**, which includes a 4-methyl-7-hydroxycoumarin unit linked to a triazole ring through a six-carbon spacer, showed selective inhibition of AChE (IC_{50} = 0.047 μM) over BChE. Kinetic studies suggested a mixed inhibition mechanism, while toxicity evaluation suggested that the compound is relatively safe, and behavioral testing in an AD mouse model showed noticeable improvements in memory and learning performance (Singh et al., 2025). Menna et al. explored the coumarin scaffold as a versatile pharmacophore for the development of ChEIs targeting AD. A series of coumarin-based derivatives was synthesized and evaluated, showing notable in vitro acetylcholinesterase inhibitory activity, with several compounds exhibiting activity in the low micromolar to sub-micromolar range. SAR analysis highlighted the importance of appropriate substitution on the coumarin core for enhancing enzyme inhibition, supporting coumarin hybrids as promising scaffolds in cholinesterase-focused drug design (Abd El-Mageed et al., 2025).

3.3.2 Tacrine/quinoline-based hybrid cholinesterase inhibitors

Tacrine represents the earliest clinically introduced acetylcholinesterase inhibitor, and its planar acridine framework allows strong interaction within the CAS of AChE. However, its therapeutic application became limited because of hepatotoxicity and related adverse effects (Bubley et al., 2023). Despite these safety concerns, tacrine continues to serve as an essential pharmacophoric unit in drug design. Researchers frequently incorporate it into the hybrid molecules to preserve cholinesterase affinity while simultaneously introducing additional biological properties such as neuroprotection or antioxidant effects. With the growing interest in MTDLs, tacrine-based hybrids have received considerable attention. One strategy involved combining tacrine with quinolinetrione fragments. In this context, in 2023, Uliassi and co-workers reported several hybrids with nanomolar AChE inhibitory activity. Beyond enzyme inhibition, these molecules reduced intracellular reactive oxygen species in neuroblastoma cells subjected to oxidative stress, suggesting an additional protective role. Another hybridization approach linked tacrine to the benzylpiperidine portion of donepezil through a hydrazone bridge (Uliassi et al., 2023). In 2024, Bayraktar et al. demonstrated that many of these conjugates inhibited both AChE and BChE at submicromolar concentrations. The lead compound, H4 **(48)**, showed IC_{50} values of 0.501 μM toward *h*AChE and 0.045 μM toward *h*BChE. Importantly, these derivatives did not display hepatotoxicity in HepG2 cells or cytotoxicity in human neuroblastoma cells. In addition to enzyme inhibition, the lead analogue exhibited notable neuroprotective potential and favourable blood-brain barrier permeability. Molecular modelling indicated dual-site engagement, where the tacrine portion participated in π-π stacking with PAS residues such as Trp286 and Tyr72, while the piperidine and phenylethyl fragments interacted with CAS residues, including Tyr341, Tyr337, and Trp86 (Bayraktar et al., 2024). Structural modification of tacrine has also been explored by incorporating N-aryl-piperidine-4-carboxamide units. In 2024, Jevtić et al. developed a series of such derivatives using a pentamethylene linker. Among them, compound 16 **(53)** bearing a 2-chlorophenyl substituent emerged as the most active, with low-nanomolar inhibition against *ee*AChE and *eq*BChE. Kinetic studies revealed a mixed inhibition pattern, and the compound also reduced intracellular ROS levels under H_2O_2-induced oxidative conditions. Notably, effective cholinesterase inhibition occurred at concentrations below those associated with cytotoxic effects (Jevtić et al., 2024). Beyond tacrine itself, quinoline-based frameworks have also been investigated. Quinoline-sulfonamide hybrids described by Jalil et al. in 2024, displayed dual inhibitory activity against ChEs and monoamine oxidases (MAO), highlighting their multitarget potential (Jalil et al., 2024). More recently, in 2025, Wang and colleagues introduced tacrine-quinoline hybrids containing 8-hydroxyquinoline 9b **(50)** and AS15 **(51)** as metal-chelating motifs. These compounds not only inhibited AChE and BChE in the submicromolar range but also showed the ability to reduce $A\beta_{1-42}$ aggregation and modulate metal ion imbalance through Cu^{2+}/Zn^{2+} chelation. Within this series, compounds 16e **(55)** and 16 h **(56)** were identified as the most potent representatives (Wang et al., 2025b).

3.3.3 Donepezil/indanone-based hybrid cholinesterase inhibitors

Among second-generation acetylcholinesterase inhibitors, donepezil remains one of the most clinically successful drugs for AD, and its structural design has strongly influenced modern hybrid drug strategies (Cacabelos, 2007). The molecule is fascinating because of its dual binding behavior: the indanone ring system is oriented toward the PAS, whereas the benzylpiperidine fragment penetrates deep into the CAS of AChE (Kryger et al., 1999). This simultaneous interaction with both regions of the enzyme gorge explains the high potency and selectivity of donepezil. Owing to this modular arrangement, donepezil is widely regarded as a pharmacophoric template that can be extended or fused with other bioactive motifs to obtain MTDLs (Fig. 6). Building on this concept, in 2022, Saglik et al. explored the indanone moiety of donepezil in greater depth and designed a series of indan-1-one derivatives as multifunctional anti-AD candidates. Several compounds, particularly D29 **(57)** exhibited notable inhibition of AChE and BChE. Docking studies revealed that these molecules preserved key interactions observed with donepezil, including π–π stacking between the indanone ring and Trp286 and hydrogen bonding between the carbonyl oxygen and active-site residues, underscoring the significance of this scaffold. Following a similar design strategy, Banoo et al. (2024) developed indole–piperidine amide derivatives incorporating a donepezil-like pharmacophore. Among them, compound 24b **(58)** demonstrated balanced dual inhibition of AChE and BChE ($IC_{50} \approx 2\ \mu M$ for both enzymes) and showed good blood-brain barrier permeability. Enzyme kinetic analysis indicated a mixed inhibition mechanism, while docking studies revealed that the indole ring engaged the PAS through π–π interaction with Trp286, and the N-benzylpiperidine moiety interacted with CAS residues such as Trp86 and Tyr337 via π–π and cation–π interactions, closely mimicking the binding orientation of donepezil itself (Banoo et al., 2024). More recently, in 2025, Ali et al. reported dihydrothiazolyl hydrazone hybrids derived from the donepezil scaffold and incorporating phenylpiperidine units. Among these, compound D1 **(59)** emerged as particularly potent against AChE (IC_{50} = 0.039 μM) and was found to be non-toxic toward NIH3T3 fibroblast cells at active concentrations. Docking analysis indicated favorable interactions at both PAS and CAS, suggesting that combining the donepezil pharmacophore with additional heterocyclic elements can further enhance enzyme affinity while maintaining safety (Ali et al., 2025). Overall, these studies highlight the versatility of the donepezil structural framework as a platform for hybrid design. By retaining its dual-site binding characteristics and introducing new functional fragments, researchers can improve potency, broaden target profiles, and move toward more effective multifunctional anti-Alzheimer agents.

3.4 Heterocyclic synthetic cholinesterase inhibitors

Exploration of heterocyclic scaffolds has become an important direction in the search for new anticholinesterase agents, as these rings offer structural versatility and enable multiple interactions within the active sites of AChE and BChE. Their ability to combine enzyme inhibition with antioxidant or neuroprotective properties makes them particularly attractive for the design of AD drugs. For instance, in 2025, Rong Du and co-workers reported imidazo [2,1-b] (Li et al., 2025a; Ramachandran et al., 2025; Sharma et al., 2025) thiadiazole derivatives as dual ChEIs. Their results suggested that the imidazothiadiazole nucleus plays a central role in AChE inhibition, whereas benzamide substitution contributes more to BChE activity. The most active compound 20aa **(60)**, showed IC_{50} values of 0.75 μM against *ee*AChE and 4.11 μM against eqBChE. Besides enzyme inhibition, these molecules also demonstrated antioxidant potential, and docking analysis supported their ability to interact with both CAS and PAS regions of the enzymes (Du et al., 2025). Benzoxazole chemistry has also shown promising outcomes. Kuzu et al. developed 2-aryl-6-carboxamide benzoxazole derivatives that bind both CAS and PAS. Their lead compound displayed potent inhibition of AChE and BChE, even surpassing donepezil in potency, highlighting how rigid heteroaromatic frameworks can enhance binding efficiency within the active-site gorge (Kuzu et al., 2025). Similarly, in 2025, Iraji et al. introduced 1,2,3–triazole–1,3,4–oxadiazole hybrids. In which compound 13 s **(62)**, containing a nitro substituent, was the most effective in that series against BChE and behaved as a competitive inhibitor. Detailed docking studies revealed multiple π–π, π- cation, and hydrogen-bond interactions involving residues in the catalytic triad and the oxyanion hole, explaining its activity. Notably, the compound showed no toxicity toward SH-SY5Y neuroblastoma cells at the tested concentrations. Other heteroaromatic systems have also been explored (Iraji et al., 2025). Pourtaher et al. investigated polyaromatic pyrrole-based molecules, in which increased π-conjugation enhanced AChE inhibition (Pourtaher et al., 2022). Li et al. designed chromanone-tetrahydropyridine hybrids, and their lead compound, C10 **(63)**, not only inhibited AChE at submicromolar concentrations but also showed MAO-B inhibition and improved cognitive performance in animal models (Li et al., 2022). Benzofuran analogues of donepezil developed by Sahin et al. demonstrated that scaffold hybridization can maintain strong AChE binding while modifying the interaction profile inside the enzyme gorge (Sahin et al., 2022). Additional contributions include pyrazolopyridine derivatives (Waly et al., 2023), 1,2,4-oxadiazole analogues (Ayoup et al., 2024), benzothiazole-oxadiazole acetamides (Nasab et al., 2024), and pyrazinoindole derivatives (Kuzu & Demir, 2025), all of which showed notable cholinesterase inhibition with favorable docking interactions. In many of these studies, structural features such as extended aromaticity, electron-withdrawing substituents, and flexible linkers were repeatedly associated with stronger enzyme binding and, in some cases, blood-brain barrier permeability. More recent efforts also include heterostilbene and triazole oxime systems, as well as lactam-triazole hybrids synthesized via green click-chemistry approaches (Shiri, 2021). Some of these compounds displayed balanced dual inhibition and mixed-type kinetic behavior, supporting their potential as multitarget agents. Overall, these findings collectively demonstrate how heterocyclic scaffold diversity-from thiadiazoles and benzoxazoles to triazoles, indoles, and oxadiazoles-offers a rich platform for developing potent and multifunctional cholinesterase inhibitors. Rather than acting solely as enzyme blockers, many of these molecules also address oxidative stress, metal imbalance, or amyloid aggregation, aligning well with the multifactorial nature of AD.

4. Overview of diagnostic agents for targeting ChEs enzymes

Cholinesterases (ChEs) exist mainly as two isoforms, AChE and BChE, which catalyze the hydrolysis of Ach to produce choline and acetic acid at cholinergic synapses (Singh et al., 2020). Abnormal expression or excessive production of ChEs disrupts ACh metabolism, leading to impaired cholinergic signaling in the brain and contributing to the development of neurological conditions, such as dementia and AD (Ma et al., 2020a; Mineur et al., 2013). Dysregulation of ChEs is closely associated with both synaptic dysfunction and cognitive decline during AD progression. Nevertheless, the exact molecular roles of ChEs within the highly complex brain environment of AD remain incompletely understood (Kumar & Modi, 2025). Therefore, the development of efficient imaging strategies for ChEs in the brain is crucial for elucidating their physiological and pathological roles in AD. Currently, various methods are available to detect ChEs activity, including calorimetric methods, chemiluminescent methods, spectrophotometry, electrochemistry, and nanotechnology (Lin et al., 2024). However, these methods have some disadvantages, including their inability to detect ChEs activity in living organisms, the need for expensive equipment, and relatively low detection sensitivity and selectivity. Additionally, the inability to perform real-time monitoring limits the broader applicability of these techniques (Andreani et al., 2008; Li et al., 2014; Lin et al., 2024). Recently, fluorescence imaging technology has emerged as an attractive and promising alternative for diagnosing neurodegenerative disease progression due to its simplicity, rapidity, non-invasiveness, deep tissue penetration, low cost, high biosafety, high spatiotemporal resolution, high compatibility, high sensitivity, selectivity, and real-time monitoring both in vivo and in vitro (Lin et al., 2024). We will now discuss various classes of developed fluorescent probes for the detection and imaging of ChEs, both in vitro and in vivo.

4.1 Developed fluorescent probes targeting AChE

Huprine is a powerful AChE inhibitor obtained by hybridizing huperzine A and tacrine (Camps et al., 2000; Ronco et al., 2011). Huprine has been used in recent years as a functionalized backbone for the synthesis of MTDLs, promising new AD therapies (Sola et al., 2015; Viayna et al., 2014). To avoid tissue autofluorescence, in 2016, Chao et al. reported the design of two near-infrared (NIR) fluorescent probes, HupNIR1 (66) and HupNIR2 (67), based on the huprine moiety and the cyanine 5.0 fluorophore. These probes were evaluated in situ for AChE detection across various tissues, and exhibited strong affinity and selectivity for AChE over BChE, with IC_{50} values of 80 ± 8.5 and 31 ± 1.5 nM for HupNIR1 and HupNIR2, respectively. However, HupNIR2 appears to be the best probe for AChE detection in the striatum, a brain region with low AChE levels, because of its high selectivity and sensitivity (Chao et al., 2016). Recently, several fluorescent probes have been developed for in vivo detection of AChE activity. For example, in 2019, Wang et al. developed, for the first time, a two-photon (TP) fluorescent probe, MCYN **(68)**, for the real-time detection of AChE activity in vitro and in vivo in a mouse model (Fig. 7A). The elevation of AChE activity in PC12 cells under oxidative stress was confirmed by in situ visualization. Moreover, MYCN was also used to assess AChE levels in the brains of mice exhibiting depression phenotypes (Wang et al., 2019). Nevertheless, this probe exhibited short-wavelength emission (560 nm), resulting in limited sensitivity and insufficient tissue penetration, which may limit its practical application in in vivo imaging.

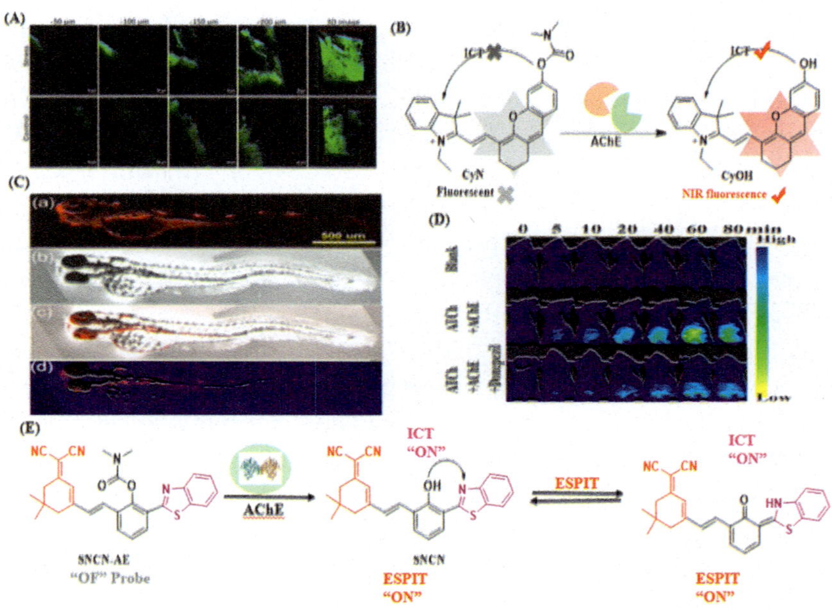

Fig. 7 (A) In situ TP fluorescence imaging in the brains of mice; **Stress:** mice exposed to 14 consecutive days of chronic-restraint stress; **Control:** mice not subjected to chronic-restraint stress. (B) Recognition Mechanism of CyN for sensing AChE. (C) (a) Confocal images of CyN response to endogenic AChE in 3-day-old zebrafish; (b) bright-field of zebrafish; (c) merged pattern of (a,b); (d) blue background modified zebrafish to highlight the red fluorescence. (D) Fluorescence images of mice injected with NFL-SF, after intraperitoneal injection of PBS, ATCh + AChE, or ATCh + AChE + Donepezil. (E) Sensing mechanism of probe SNCN-AE for detecting AChE. *(A) Reprinted with permission from ref. (Wang et al., 2019). Copyright 2019 American Chemical Society. (C) Reprinted with permission from ref. (Ma et al., 2019). Copyright 2020 American Chemical Society. (D) Reprinted with permission from ref. (Zhao et al., 2022). Copyright 2022 Elsevier. (E) Reproduced from ref. (Wei et al., 2023).*

To overcome this limitation, in 2020, Ma et al. reported a hemicyanine-based off-on probe, CyN **(69)**, with an extended emission wavelength for detecting endogenous AChE activity in PC12 cells under hypoxic stimulation, as shown in Fig. 7B. CyN demonstrated outstanding sensing performance, characterized by high selectivity toward AChE, minimal cytotoxicity, good cellular and tissue penetration, and lower background interference. Moreover, CyN has been used to image the distribution of AChE in living zebrafish, as shown in Fig. 7C (Ma et al., 2019). In 2020, Fortibui et al. developed a fluorescent probe, EW3 **(70)**, that exhibits NIR emission to detect AChE, with high sensitivity and selectivity, and a fast reaction time, and was used to image AChE in living cells and in the zebrafish model (Fortibui et al., 2022). In 2022, Zhao et al. developed an NIRF probe NFL-SF **(71)** for monitoring both exogenous and endogenous AChE activity associated with chemotherapeutic drug-induced apoptosis in living cells. Notably, this probe was used to image AChE activity in vivo in nude mice (Fig. 7D) (Zhao et al., 2022). In 2023, Wei et al. developed two fluorescent probes (SNCN-AE **(72)** and SNC-AE) that exhibit significant spectral shifts via ESIPT and ICT mechanisms for AChE detection (Fig. 7E). These probes were developed by linking the electron acceptor dicyanoisophorone (CN) to the ESIPT fluorophore 2-(2′-Hydroxyphenyl) benzothiazole (SN-1), resulting in large Stokes shifts (285 nm for SNCN-AE and 270 nm for SNC-AE) and extended emission wavelengths. Among them, SNCN-AE demonstrated superior detection, rapid response, good photostability, and minimal cytotoxicity, enabling in situ imaging of endogenous AChE in living systems, and was further applied to monitor AChE variations in PC12 cells and zebrafish models under oxidative stress (Wei et al., 2023). Currently, developing specific substrates for selective AChE imaging is challenging due to interference from BChE and carboxylesterase (CE).

To address this issue, in 2020, Wu et al. designed a highly specific substrate, P10 **(73)** by integrating dimethylcarbamate choline with a self-immolative scaffold to effectively eliminate the BChE and CE interference, and it was used to enable highly sensitive and selective imaging of AChE activity in different cells and was further successfully applied for in vivo and ex vivo visualization of AChE activity in the brain of living mice (Wu et al., 2020b). In 2023, Sidhu et al. investigated the structural characteristics of AChE, which contains two key subsites: an anionic region that engages in electrostatic interactions with the quaternary ammonium moiety of acetylcholine, and an esteratic region that functions as the catalytic center for ester bond hydrolysis.

Based on these features, a naphthalimide-based fluorescent probe was rationally designed using an acetylcholine-mimicking strategy. The probe incorporates a quaternary ammonium moiety to enhance binding to the anionic subsite and an ester group positioned 2–3 carbons away to maintain a compact structure. Both recognition elements were directly attached to the fluorophore, enabling efficient detection of AChE activity. Three probes (Naph-1, Naph-2, and Naph-3) were developed; Naph-3 **(74)** exhibited high selectivity for endogenous AChE, demonstrated effective AChE tracing in live Neuro-2a cells due to its high cell membrane permeability, and can be used for screening AChE inhibitors (Fig. 8A and B) (Sidhu et al., 2023). Currently, most fluorescent probes used to track AChE activity rely on traditional polycyclic aromatic fluorophores, which suffer from aggregation-caused quenching (ACQ) in aqueous media due to strong π–π interactions, leading to diminished fluorescence intensity and elevated background signals, particularly in vivo applications (Gao & Tang, 2020; Li et al., 2021b; Mei et al., 2015). In contrast, probes that employ the aggregation-induced emission (AIE) mechanism to overcome these drawbacks by exhibiting intensified fluorescence upon aggregation due to restricted intramolecular motions, while offering high sensitivity, excellent

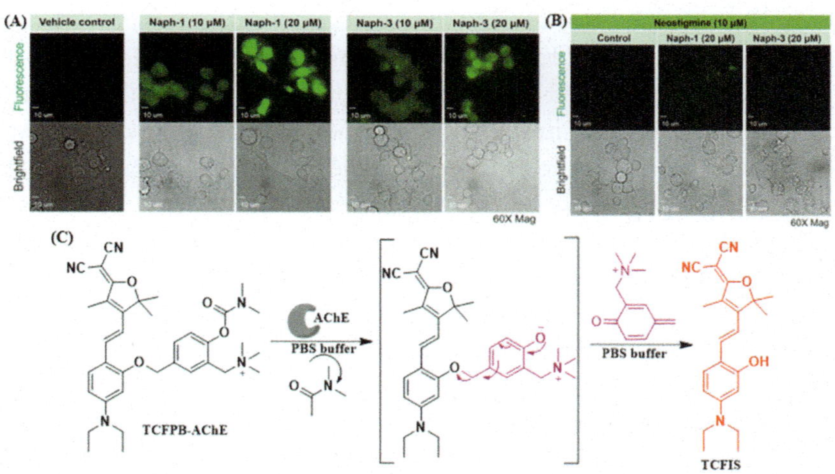

Fig. 8 (A) Fluorescence and brightfield images of Neuro-2a cells treated with vehicle control (DMSO), Naph-1, and Naph-3 at different concentrations. (B) Fluorescence and brightfield images of Neuro-2a cells that were pre-treated with neostigmine before the probe treatment. (C) Structure of TCFPB-AChE and its detection mechanism for AChE. *(B) Reprinted with permission from ref. (Sidhu et al., 2023). Copyright 2023 American Chemical Society. (C) Reproduced from ref. (Xiang et al., 2022a).*

photostability, good biocompatibility, and low background interference in living systems (Li et al., 2021b; Liu et al., 2021; Luo et al., 2001). Despite these advantages, AChE-responsive AIEgen had not yet been reported.

In 2022, Xiang et al. introduced an AIE-based fluorescent probe, TCFPB-AChE **(75)**, whose mechanism is shown in Fig. 8C, and demonstrated its utility for visualizing AChE activity in vitro, ex vivo, and in vivo. Notably, this probe exhibited a highly selective fluorescence response toward AChE with an ultralow detection limit (0.009355 U/mL). Additionally, this probe was used effectively to detect endogenous AChE activity in U87 cell lines and AD mouse brain tissues, and to visualize AChE at different concentrations in living normal mouse brains in vivo, with a high signal-to-noise ratio (Xiang et al., 2022a). Several fluorescent probes incorporating a dimethyl carbamate fragment as an AChE recognition element have been used effectively in living biological systems (He et al., 2021; Ma et al., 2019; Meng et al., 2021; Wang et al., 2019; Wu et al., 2020b; Zhao et al., 2022). But most of these probes exhibit small Stokes shifts (<100 nm), and probes with large Stokes shifts can significantly reduce self-absorption and prevent interference from incident light. Although a number of fluorescent probes based on AIE have been developed for AChE detection (Chang et al., 2016; Xiang et al., 2022a, 2025). In 2025, Xiang et al. developed a red-emissive ratiometric fluorescent probe, DPHP-AChE **(76)**, that integrates AIE with an excited-state intramolecular proton transfer (ESIPT) mechanism and exhibits a large Stokes shift of 188 nm. This probe was successfully applied to visualize AChE activity in U87MG cell lines and AD mouse brain tissue (Fig. 9A and B). The results suggested that the probe performs excellently for visualizing AChE activity in both cell cultures and in vivo brain tissues (Xiang et al., 2025). However, most AChE fluorescent probes exhibit one-photon (OP) excitation and high background fluorescence. To address this issue, Tang et al. developed a fluorescent probe suitable for two-photon (TP) excitation (Wang et al., 2019). Additionally, emission wavelength critically influences tissue imaging capability (Sidhu et al., 2023). Nevertheless, currently available TP fluorescent probes for AChE detection generally emit at wavelengths below 600 nm, limiting their penetration depth and preventing fluorescence imaging of deep neural tissue (Wang et al., 2019; Wu et al., 2020b). In 2024, An et al. developed an isophorone-based red-emitting TP fluorescent probe, WZ-AChE **(77)**, for highly selective and sensitive detection of AChE both in vitro and in vivo, as shown in Fig. 9C. This probe was further employed to visualize alterations in AChE activity in models of Parkinson's disease (PD) and depression, highlighting its potential

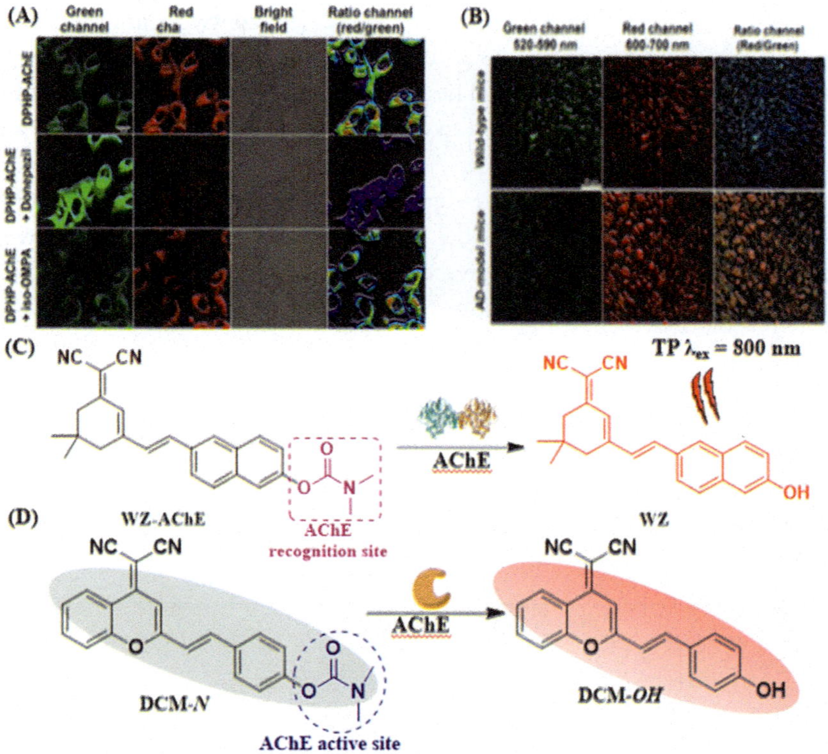

Fig. 9 (A) DPHP-AChE: U87MG cells treated with DPHP-AChE (1 μM) for 3 h; Donepezil: cells pretreated with 50 μM donepezil (AChE inhibitor) and iso-OMPA (BChE inhibitor) for 5 h, and then incubated with DPHP-AChE (1 μM) for 3 h. (B) Ratiometric images of AD mice and wild-type mice brain tissues obtained by confocal imaging DPHP-AChE = 1 μM. (C) Schematic representation of the probe WZ-AChE for the detection of AChE. (D) Detection mechanism of DCM-N for AChE. *(B) Reprinted with permission from ref. (Xiang et al., 2025). Copyright 2025 American Chemical Society. (C) Reproduced from ref. (An et al., 2024). (D) Reproduced from ref. (Zeng et al., 2025).*

utility for probing AChE involvement in complex physiological and pathological processes (An et al., 2024). Curcumin, a low-cost natural compound extracted from turmeric, is well recognized for its ability to trigger apoptosis in various cancer cell types, including lung cancer cells (Laali et al., 2016; Zhang et al., 2017). Curcumin structure stabilization can be achieved through coordination with BF2 at the 3- and 5-positions to form a six-membered ring. Notably, this cyclized form of curcumin exhibits remarkable fluorescence properties and pronounced anti-proliferative activity against lung, colorectal, and leukemia cells (Abonia et al., 2020; Laali et al., 2018; Laali et al., 2019). To achieve probes with suitable oil-water partition coefficients, in 2024, Lin et al. developed two curcumin-based "on-off" fluorescent probes, BF2-cur-But (**78**) and BF2-cur-Ben (**79**), for the selective AChE detection. These probes readily penetrated cell membranes and were applied to detect exogenous and intracellular AChE, as well as to monitor its overexpression during PMA-induced apoptosis in lung cancer cells. In addition, their in vivo applicability was validated by imaging AChE activity in a zebrafish model (Lin et al., 2024). In 2025, Zeng et al. reported a novel turn-on fluorescent probe, DCM-N **(80)**, that offers superior performance in emission, Stokes shifts, LOD, and response time compared with previously developed probes, with great prospects for real-time tracking of AChE activity both in vitro and in vivo, detection mechanism shown in Fig. 9D (Zeng et al., 2025). The chemical structures of all fluorescent probes targeting AChE are illustrated in Fig. 10, while their photophysical properties, detection type, and corresponding mechanisms are summarized in Table 1.

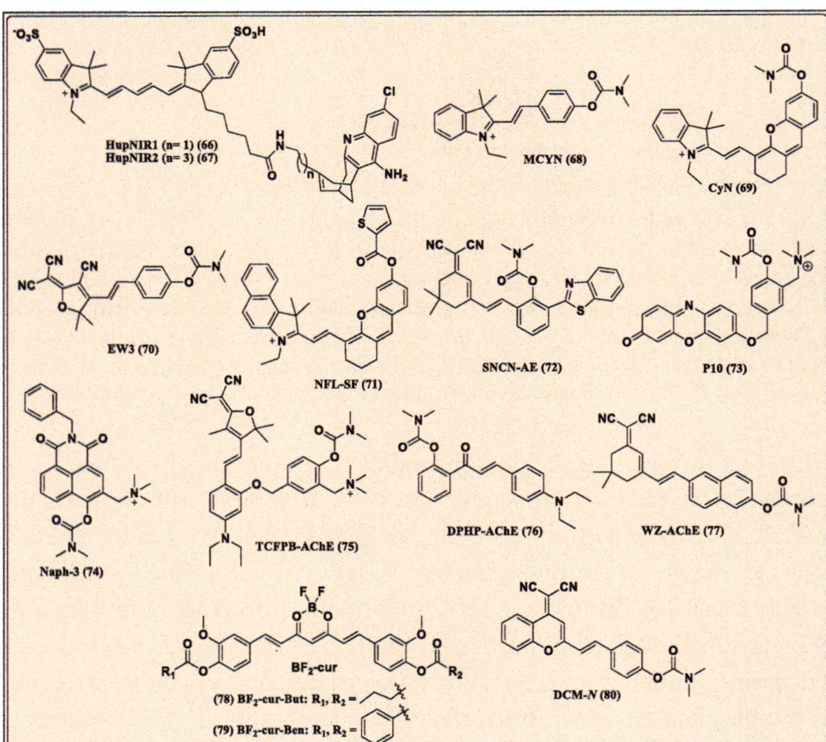

Fig. 10 Molecular structures of carboxylic ester-based probes for AChE.

Table 1 Representative probes for AChE, BChE, and for both AChE & BChE, and their fluorescent properties.

Sr. no.	Probe	Exi (λex)/Emi (λem)	Detection type/mechanism	Limit of detection	References
1.1 Representative probes for AChE and their fluorescent properties					
1.	HupNIR2 **(67)**	650/665	NR	NR	Chao et al. (2016)
2.	MCYN **(68)**	520/560	Turn-on/Esterase reaction (Dimethyl carbamate hydrolysis)	0.36 U/mL	Wang et al. (2019)
3.	CyN **(69)**	670/685	Off-on/Esterase reaction (Dimethyl carbamate hydrolysis)	0.1173 U/mL	Ma et al. (2019)
4.	EW3 **(70)**	560/620 & 690	Turn on/Esterase reaction (Dimethyl carbamate hydrolysis)	0.17 U/mL	Fortibui et al. (2022)
5.	NFL-SF **(71)**	680/745	Torn-on/PET (Photo-induced electron transfer)	0.2 mU/mL	Zhao et al. (2022)
6.	SNCN-AE **(72)**	425/710	Turn-on/Esterase reaction (Dimethyl carbamate hydrolysis)	0.13 U/mL	Wei et al. (2023)
7.	P10 **(73)**	478/582	-/Esterase reaction (Dimethyl carbamate hydrolysis)	0.017 U/mL	Wu et al. (2020b)
8.	Naph-3 **(74)**	460/550	Turn-on/ICT ON	0.1 U/mL	Sidhu et al. (2023)
9.	TCFPB-AChE **(75)**	575/634	Turn on/Ether bond cleavage	0.009355 U/mL	Xiang et al. (2022a)

(continued)

Table 1 Representative probes for AChE, BChE, and for both AChE BChE, and their fluorescent properties. *(cont'd)*

Sr. no.	Probe	Exi (λex)/Emi (λem)	Detection type/mechanism	Limit of detection	References
10.	DPHP-AChE **(76)**	450/638	Turn-on/Esterase reaction (Dimethyl carbamate hydrolysis)	0.11 U/mL	Xiang et al. (2025)
11.	WZ-AChE **(77)**	445/620	Turn-on (Two-photon)/ Esterase reaction (Dimethyl carbamate hydrolysis)	0.46 U/mL	An et al. (2024)
12.	BF_2-cur-Ben **(79)**	476/533	On-off/NR	0.031 U/mL	Lin et al. (2024)
13.	DCM-N **(80)**	460/700	Turn-on/Esterase reaction (Dimethyl carbamate hydrolysis)	0.06 U/mL	Zeng et al. (2025)
1.2 Representative probes for BChE and their fluorescent properties					
14.	1 or N-CTDA **(81)**	355/520	"Turn-off"/Intramolecular PET process	NR	Kang et al. (2016)
15.	DSC **(82)**	340/485	Turn-on/Ester bond hydrolysis	0.0012 U/mL	(Yoo & Han, 2019)
16.	BChE-FP **(83)**	455/515	"off-on"/Ester bond hydrolysis	NR	Yang et al. (2017)
17.	BChE-NIRFP **(84)**	665/705	Turn-on/Esterase reaction (Cyclopropyl acetate hydrolysis)	NR	Liu et al. (2018)
18.	CyClCP **(85)**	670/708	Turn-on/Esterase reaction (Cyclopropyl acetate hydrolysis)	3.75 U/L	Ma et al. (2020b)

19.	BChE-NBD **(86)**	580/628	Turn-on/Esterase reaction (Cyclopropyl acetate hydrolysis)	29 ng/mL	Cao et al. (2021)
20.	DCD-P1 **(87)**	464/562	Turn-on/Esterase reaction (Cyclopropyl acetate hydrolysis)	0.12 μg/mL	Yang et al. (2022a)
21.	DX-2 **(89)**	555/665	"off-on"/Pull-push effect	0.08 U/mL	Zhang et al. (2022a)
22.	FDCM-BChE **(90)**	585/820	Turn-on/Esterase reaction (Cyclopropyl acetate hydrolysis)	0.05 U/mL	Jiang et al. (2024)
23.	Chy-1 **(91)**	670/715	Turn-on/Esterase reaction (Cyclopropyl acetate hydrolysis)	0.12 ng/mL	Yang et al. (2022b)
24.	W1 **(93)**	415/556	Ratiometric detection/Esterase reaction (Cyclopropyl acetate hydrolysis)	0.077 μg/mL	Wan et al. (2022)
25.	TB-BChE **(94)**	575/626	Ratiometric detection/ aggregation-induced emission (AIE)	39.24 ng/mL	Xiang et al. (2022b)
26.	HBT-BChE **(95)**	365/457	Turn-on/ AIEE, ESIPT	7.540×10^{-4} U/mL	Pei et al. (2023)
27.	S1 **(96)**	450/500	Two-photon imaging/Esterase reaction (Cyclopropyl acetate hydrolysis)	8.28 U/L (for BChE), 3.09 μM (for HClO)	Guo et al. (2023)

(continued)

Table 1 Representative probes for AChE, BChE, and for both AChE BChE, and their fluorescent properties. (*cont'd*)

Sr. no.	Probe	Exi (λex)/Emi (λem)	Detection type/mechanism	Limit of detection	References
28.	DTNP **(97)**	665/710	Turn-on/Esterase reaction (Cyclopropyl acetate hydrolysis)	0.173 U/mL	Wang et al. (2024b)
29.	HCy-Me **(98)** HCy-Et **(99)**	4759(abs)/647 (HCy-Me), 480(abs)/622 (HCy-Et)	Turn-on (Dual response)/ Esterase reaction (Cyclopropyl acetate hydrolysis)	0.0086 U/mL (for HCy-Me), 0.010 U/mL 622 (for HCy-Et)	Ji et al. (2025)
30.	BZ-BChE **(100)**	400/520	Ratiometric detection/Esterase reaction (Cyclopropyl acetate hydrolysis)	22.85 mU/mL	Liu et al. (2026)
31.	Probe-BChE **(101)**	600/676	Turn-on/Esterase reaction (Dimethyl carbamate hydrolysis)	0.14 U/mL	Zhang et al. (2025)
1.3 Representative probes for Both AChE and BChE and their fluorescent properties					
32.	IPAN **(102)**	545/654	Turn-on/Esterase reaction (Dimethyl carbamate hydrolysis)	0.127 U/mL (For AChE) and 0.0117 U/mL (For BChE)	Ma et al. (2020a)
33.	OHPD **(103)**	550/586	Turn-on/Esterase reaction (Dimethyl carbamate hydrolysis)	0.24 U/mL (For AChE) and 0.01 U/mL (For BChE)	Tang et al. (2024)

*Abbreviation word is NR, Not reported.

4.2 Developed fluorescent probes targeting BChE

In 2016, Kang et al. designed a "turn-off" fluorescent probe, 1 or N-CTDA **(81)**, firstly for the direct measurement of BChE activity. Notably, N-CTDA was selectively catalyzed by BChE even in the presence of abundant AChE, suggesting its suitability for a specific BChE assay. The probe's fluorescence signal was quenched via intramolecular photoinduced electron transfer (PET), as shown in Fig. 11A. Owing to its operational simplicity, high selectivity, and rapid response, N-CTDA was further applied to high-throughput screening for BChE inhibitors, underscoring its promise for drug discovery (Kang et al., 2016). In 2019, Yoo et al. developed a turn-on fluorescent probe, DSC **(82)**, derived from a Dansyl-L-sarcosine-choline (DSC) fluorophore, for highly sensitive and selective detection of BChE activity in human serum, demonstrating clear discrimination from AChE.

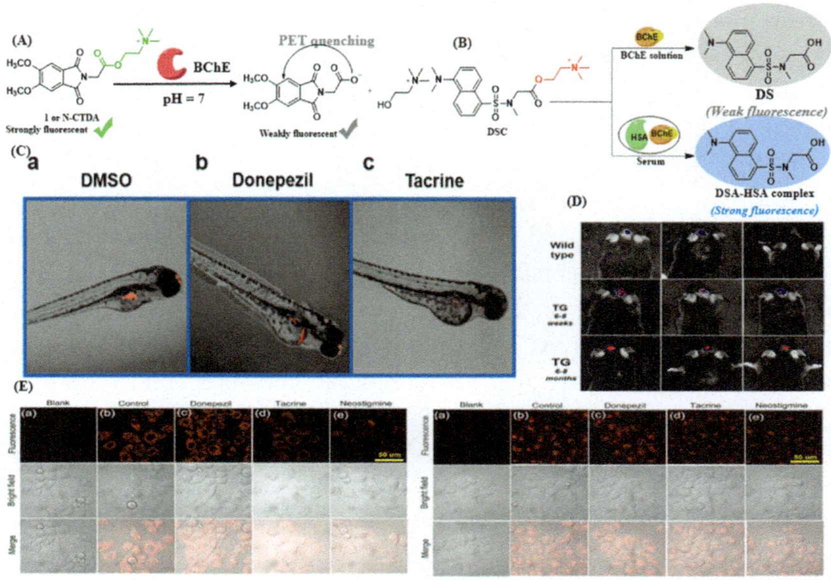

Fig. 11 (A) Representation of a direct BChE assay using 1 or N-CTDA. (B) Representation of the BChE assay based on a probe comprising a fluorophore with a specific binding affinity for albumin. (C) Confocal imaging of BChE in zebrafish embryos after a 2 h incubation with BChE-NIRFP (5 μM). (D) Confocal imaging of stained brain sections from the three groups of mice 30 min after the administration of 100 μL of the BChE-NIRFP probe (100 μM). (E) Fluorescence images of HepG2 cells and LO2 cells using 5.0 μM CyClCP. *(A) Reproduced from ref. (Kang et al., 2016). (B) Reproduced from ref. (Yoo & Han, 2019). (D) Reprinted from ref. (Liu et al., 2018). (E) Reprinted with permission from ref. (Ma et al., 2020b). Copyright 2020 Elsevier.*

The probe was also used for the high-throughput screening for BChE inhibitors. The fluorescence of DSC was significantly increased when incubated with BChE in the presence of human serum albumin (HAS), as DSC attached to BChE, released free DS, and then bound to HSA to emit intense fluorescence, as shown in Fig. 11B (Yoo & Han, 2019). In 2017, Yang et al. designed an "off-on" fluorescence probe, BChE-FP **(83)**, for targeting BChE and first achieved effective distinction between BChE and AChE by simulating the natural substrate butyrocholine and screening recognition groups, focusing on the structural differences between the active cavities of AChE and BChE. BChE-FP exhibited excellent specificity toward BChE (275-fold) over AChE. Furthermore, the probe was successfully used to monitor endogenous BChE in living cells and to perform high-throughput screening of BChE inhibitors. However, this probe has limited biological applications due to its short excitation and emission wavelengths (Yang et al., 2017). Probes with the NIR region significantly increase the tissue penetration depth and spatial resolution of fluorescence imaging due to their small tissue absorption, low autofluorescence background, and negligible biological toxicity in living tissue (Chen et al., 2023; Li et al., 2020; Wang et al., 2024a).

In 2018, Yang et al. reported a NIRF probe, BChE-NIRFP **(84)**, for the monitoring of BChE over AChE (more than 30-fold). The probe demonstrated effective sensing and imaging performance for tracking endogenous BChE in human cells, zebrafish, and AD mouse models in vivo, as shown in Fig. 11C and D (Liu et al., 2018). Unfortunately, this probe also showed a slight response towards AChE; therefore, a new fluorescent probe that specifically targets BChE and has significant anti-interference properties against AChE still needs to be developed. In 2020, Ma et al. developed a novel NIRF probe, CyClCP **(85)**, designed by incorporating a chlorine substituent at the ortho-position of the phenolic hydroxyl group of BChE-NIRFP on the basis of structural characteristics of AChE and BChE. Compared with previously reported BChE-NIRFP, CyClCP showed greater selectivity for BChE. In addition, the probe demonstrated good membrane permeability and was effective for imaging endogenous BChE levels in HepG2 and LO2 cells (Fig. 11E) (Ma et al., 2020b).

However, despite their ability to detect BChE activity, none of the aforementioned probes has been investigated for BChE distribution or in vivo imaging. In 2018, Yang et al. used the probe BChE-NIRFP **(84)** to investigate BChE expression and imaging in the brain tissue of transgenic AD mice for the first time (Yang et al., 2017). In 2021, Cao et al. developed a "turn-on" fluorescent probe, BChE-NBD **(86)**, for targeting BChE; mitochondrial localization and in vivo detection and imaging were unaffected by other analytes. For the first time, this probe specifically tracked and imaged endogenous BChE in organelles, such as mitochondria, across different cancer cells. More significantly, BChE-NBD was used to confirm the variations in BChE expression in various tissues and organs in vivo, as well as its extensive presence in liver tissues, as shown in Fig. 12A and B (Cao et al., 2021). In 2022, Yang et al. developed two novel DCD-based fluorescent probes, DCD-P1 **(87)** and DCD-P2 **(88)**, for tracking the BChE activity in living cells with excellent selectivity and specificity. Among them, DCD-P1 has good membrane permeability and biocompatibility, as demonstrated by

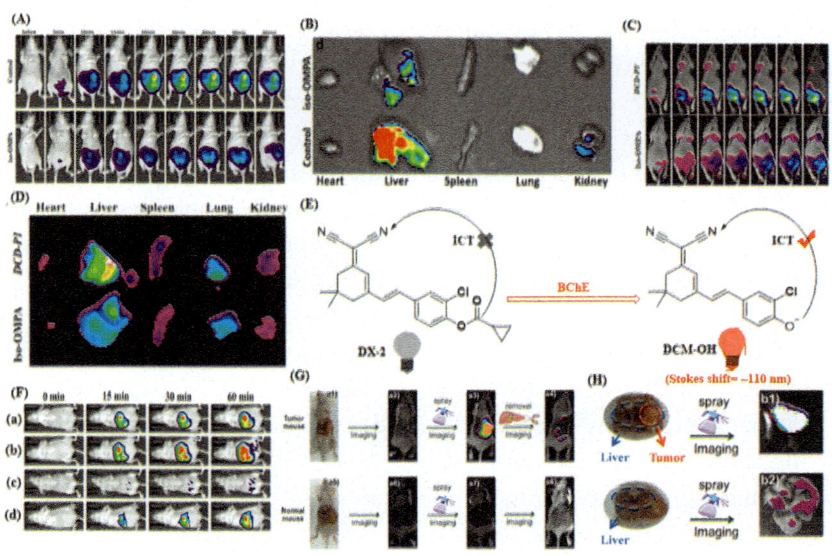

Fig. 12 (A) Time-dependent images. (B) The imaging of the main organs of control and iso-OMPA groups. (C) The mice groups were treated with and without ISO-OMPA, and then DCD-P1 (1 mM, 100 μL) was injected. (D) Fluorescence imaging of organs with and without ISO-OMPA. (E) Detection principle of probe DX-2 for BChE. (F) Fluorescence images of mice after intravenous injection with FDCM-BChE (200 μM, 100 μL): (a) control group, (b) diabetes group, (c) diabetic with iso- OMPA group, and (d) diabetes-treatment group. (G) Fluorescence imaging of a tumor-bearing mouse and a healthy mouse after Chy−1 was sprayed on the tumor-bearing mouse's bare belly. (H) Fluorescence imaging of the tumor, liver, and healthy liver in vitro. *(B) Reprinted with permission from ref. (Cao et al., 2021). Copyright 2021 Elsevier. (D) Reprinted with permission from ref. (Yang et al., 2022a). Copyright 2022 Elsevier. (E) Reproduced from ref. (Zhang et al., 2022a). (F) Reprinted with permission from ref. (Jiang et al., 2024). Copyright 2024 Elsevier. (H) Reprinted with permission from ref. (Yang et al., 2022b). Copyright 2022 American Chemical Society.*

its effective use for real-time imaging of BChE activity in brain tissues and in a mouse model (Fig. 12C and D), suggesting that DCD-P1 could qualitatively track BChE activity in hepatic metabolic and neurodegenerative diseases (Yang et al., 2022a). Previously existing NIRF probes for BChE detection typically exhibit Stokes shifts below 60 nm, which can restrict their effectiveness in BChE detection and biological imaging due to interference with the emission wavelength from scattered excitation light. In 2022, Zhang et al. designed an "off-on" far-red/NIRF probe, DX-2 **(89)**, for the detection of BChE, having a large Stokes shift (110 nm), and a mechanism shown in Fig. 12E. This probe has minimal toxicity, good membrane permeability, enabling highly sensitive, selective, and rapid in vivo monitoring of BChE activity, and it shows excellent optical performance in both normal and tacrine-pre-treated cells. Additionally, DX-2 has been used effectively to visualize BChE activity in real time in HepG2 tumor-bearing mice (Zhang et al., 2022a). Nevertheless, most existing studies focus on basic BChE imaging rather than on assessing its expression levels across different disease models. In 2024, Yu et al. reported an ultra-sensitive NIRF probe, FDCM-BChE **(90)**, for selective BChE detection, featuring rapid response, high stability, good selectivity, an ultralow detection limit, and a large Stokes shift of 235 nm, effectively minimizing excitation-emission spectral overlap and thereby reducing self-quenching and autofluorescence. This probe was validated for BChE measurement in human serum, and, due to its low cytotoxicity, was further used to detect endogenous BChE in different cells and in diabetic mouse models, depicted in Fig. 12F, revealing increased BChE activity in diabetes and reduced levels in liver cancer and hepatic injury, highlighting the probe's potential utility for investigating BChE-associated pathological conditions (Jiang et al., 2024). However, previously reported NIR fluorescent probes for BChE do not clearly differentiate between tumor and normal cells. In 2022, Yang et al. designed two novel "turn-on" NIRF probes, Chy-1 **(91)** and Bhy-1 **(92)**, by converting one of the six-membered rings of probe BChE-FP into a five-membered ring for the rapid and accurate detection of BChE activity. As shown in Fig. 12G, the tumor tissue region was clearly highlighted in red fluorescence compared to normal tissue by in situ Chy-1 spraying in tumor-bearing mouse models, suggesting that Chy-1 serves as an effective tool for the diagnosis of clinical BChE-related tumors, surgical navigation, and tumor resection. The fluorescence intensity in the liver was higher than in other organs in the primary ex vivo organ imaging, as seen in Fig. 12H, suggesting that Chy-1 can be used to image BChE activity in the liver region.

Furthermore, Chy-1 was effectively used to detect endogenous BChE activity in AD mouse models (Yang et al., 2022b). Compared with single-wavelength fluorescent probes, ratiometric probes provide enhanced sensitivity, stability, and quantitative accuracy by utilizing the fluorescence intensity ratio of two emission bands. In addition, this ratiometric detection strategy enables self-calibration by simultaneously monitoring emission changes at two wavelengths, thereby effectively minimizing systematic measurement errors (Guo et al., 2012; Wang et al., 2020). In 2022, Wan et al. reported a roatiometric fluorescent probe, W1 **(93)**, for the determination of BChE activity. This probe displayed excellent selectivity toward BChE, with no interference from other biological species, including AChE, and can be used to effectively image BChE activity in living cells and animal tissues (Wan et al., 2022). Fluorescent probes with aggregation-induced emission (AIE) properties have been widely explored for chemical sensing and bioimaging applications, due to their excellent photostability, low biotoxicity, switchable fluorescence, and long-term imaging capabilities (Mei et al., 2015; Ni et al., 2018; Peng et al., 2017; Shi et al., 2017). Nevertheless, AIE-type fluorescent probes specifically designed for BChE detection have not yet been reported. In 2022, Xiang et al. developed a novel AIE-active fluorescence probe, TB-BChE **(94)**, for the ratiometric detection of endogenous BChE both in vitro and in vivo, detection mechanism shown in Fig. 13A. The probe exhibited high sensitivity and selectivity toward BChE and was successfully applied to image BChE activity in a nonalcoholic fatty liver disease (NAFLD) model (Xiang et al., 2022b). Recent studies indicate that elevated BChE activity is closely associated with NAFLD progression, underscoring the urgent need for simple, effective in vivo imaging strategies for this disease (Gu et al., 2023). In addition, fluorescent probes that combine aggregation-induced enhanced emission (AIEE) and excited-state intramolecular proton transfer (ESIPT) for BChE detection are rarely reported (Chen et al., 2017a, 2017b; Zhang et al., 2022b). In 2023, Pei et al. developed a turn-on fluorescent probe, HBT-BChE **(95)**, for the detection and imaging of BChE on the basis of a benzothiazole scaffold with AIEE and ESIPT properties (Fig. 13B) that help to prevent self-absorption of the fluorescent dye, which typically results in a large Stokes shift (>100 nm), thereby markedly reducing background interference. Additionally, HBT-BChE was effectively applied to detect and track BChE in HeLa cells and in normal and NAFLD zebrafish liver tissues, offering a potent tool for diagnosing NAFLD and screening for BChE inhibitors, as shown in

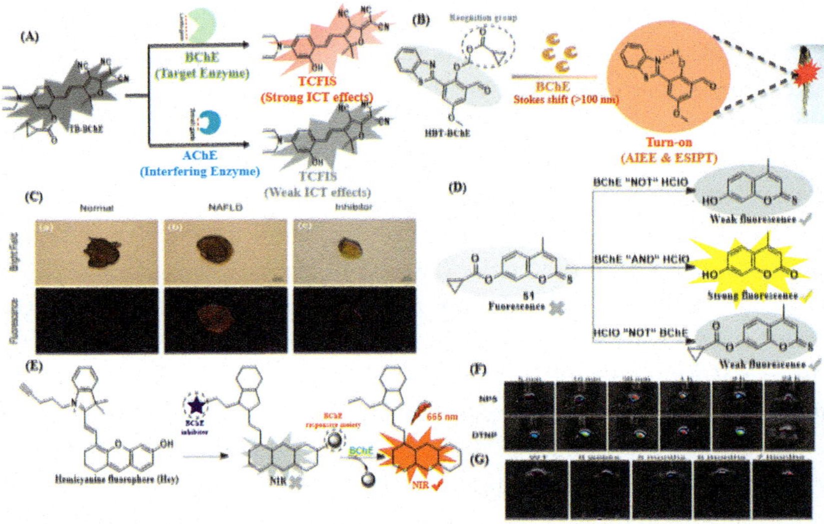

Fig. 13 (A) The schematic illustration shows that the ratiometric probe has excellent selectivity for BChE. (B) Response mechanisms of probe HBT-BChE for BChE. (C) Part of the liver of zebrafish confocal imaging by HBT-BChE (10 μM) with normal zebrafish (a) and NAFLD zebrafish (b) for 2 h. (c) Part of the liver of zebrafish, confocal imaging, showing that zebrafish were injected with tacrine (20 μM) for 2 h and then injected with HBT-BChE (10 μM) for 2 h. (D) Detection mechanism of probe (S1) for BChE. (E) Detection mechanism of BChE-activated near-infrared probes. (F) In vivo imaging of 6-month-old AD mouse brain areas at different times after intravenous injection of probes (100 μM). (G) Brain imaging of WT and AD-model mice in various stages, 1 h after intravenous injection of DTNP. *(A) Reproduced from ref (Xiang et al., 2022b). (C) Reprinted with permission from ref. (Chen et al., 2017a). Copyright 2022 Elsevier. (D) Reproduced from ref. (Samanta & Govindaraju, 2019). (G) Reprinted with permission from ref. (Cui et al., 2021). Copyright 2024 American Chemical Society.*

Fig. 13C (Pei et al., 2023). In AD, hypochlorous acid (HClO), involved in the removal of invading bacteria or pathogens, and BChE, involved in the hydrolysis of the acetylcholine neurotransmitter, are both relatively significantly altered in AD (Samanta & Govindaraju, 2019). However, there are few probes for dual detection of both HClO and BChE. Additionally, single-response probes exhibit significant off-target effects, while NIRF probes face difficulties crossing the blood-brain barrier due to their high molecular weight. Due to its advantages of deeper penetration, nontissue autofluorescence, and reduced light scattering, two-photon fluorescence imaging technology has been used in brain imaging in recent years (Cui et al., 2021; Zhang et al., 2021). Therefore, it is crucial to develop an AND logic-gate probe based on two-photon absorption to diagnose brain-related diseases. In 2023, Li et al. reported a TP fluorescent probe, S1 **(96)** that recognizes HClO and BChE with high specificity, high stability, and low background fluorescence, using a dual-lock strategy (Fig. 13D). This probe produced excellent imaging in PC12 cells, and deep two-photon imaging was demonstrated in AD mice brains following tail vein injection, suggesting that the probe may provide a useful tool for more accurate diagnosis of AD (Guo et al., 2023). Nevertheless, the aforementioned fluorescent probes for BChE focus on imaging of BChE activity, and no theranostic fluorescent probes have been reported. Thus, a probe that can track AD development in vivo is extremely important, as it could improve clinical drug administration strategies by enabling real-time visualization of BChE levels in the brain.

In 2024, Sun et al. first developed a BChE-activated NIR probe, DTNP **(97)**, for detecting BChE activity in AD. This probe has a cyclopropylic carboxylic ester moiety and an NIR fluorophore hemicyanine moiety, serving as the recognition unit and therapeutic unit, respectively. The inhibitor moiety in DTNP exhibits strong "turn-on" NIR fluorescence, binds BChE with high affinity, and hydrolyzes the recognition group, as shown in Fig. 13E. With an IC_{50} of 29.7 nM, DTNP can significantly increase acetylcholine release, reduce BChE activity, lower Aβ1–42 levels, reduce oxidative stress, and restore learning deficits and cognitive impairment. In vivo imaging (Fig. 13F and G) demonstrates that DTNP is a viable technique for initial AD diagnosis, owing to its high blood-brain barrier permeability and long-term tracking capabilities. The primary organs, as assessed by hematoxylin and eosin (H&E) staining, showed no obvious damage following DTNP treatment (Wang et al., 2024b). In recent years, dual-mode sensing techniques have attracted considerable attention for their ability to reduce false-positive signals and provide integrated self-calibration (Dong et al., 2023, 2025). However, most advanced sensing systems remain focused on single targets. In AD, whose pathogenesis involves interactions among multiple pathological factors, relying on a single biomarker may be insufficient due to disease heterogeneity. Based on this rationale, in 2026, Ji et al. reported two D-π-A probes, HCy-Me **(98)** and HCy-Et **(99)**, for the specific detection of BChE and Aβ. Both probes effectively visualized the spatial distribution of BChE in ex vivo brain tissue from APP/PS1 transgenic mice, while *Aβ* plaques in AD mouse brain were selectively labeled by the hydrolysis products HMy-OH and HEy-OH, as shown in Fig. 14A. Moreover, an in vivo imaging study demonstrated that HCys possessed excellent imaging capability for BChE, despite the limited BBB permeability

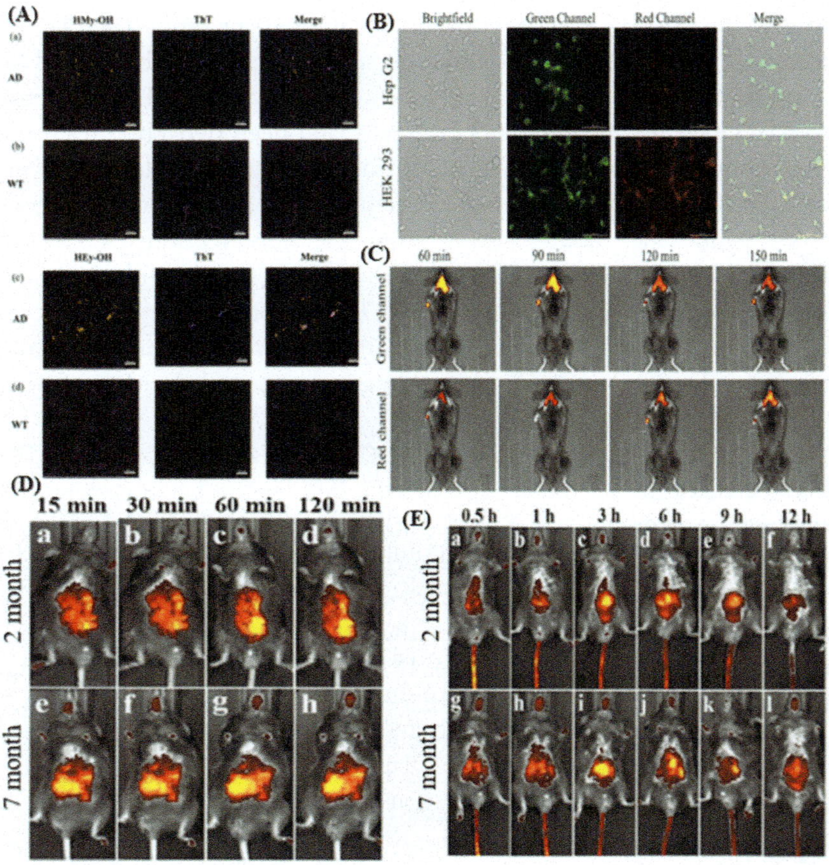

Fig. 14 (A) (a, b) APP/PS1 transgenic mice (12 months old, male) brain tissue was stained histologically using probes HMy-OH and HEy-OH with commercial dye ThT, (c, d) WT mice (12 months old, male) brain tissue was stained histologically using probes HMy-OH and HEy-OH with commercial dye ThT. (B) Confocal fluorescence images of Hep G2 cells and HEK 293 cells incubated with 10 μM BZ-BChE. (C) Whole body fluorescence images of normal mice after injection with BZ-BChE from 60 to 150 min (D) Time-dependent in vivo fluorescence imaging of APP/PS1 mice after intraperitoneal injection of Probe-BChE (200 μL, 100 μM). (E) Time-dependent in vivo fluorescence imaging of APP/PS1 mice after tail vein injection of Probe-BChE (200 μL, 100 μM). *(C) Reprinted with permission from ref. (Wang et al., 2024b). Copyright 2026 Elsevier and ref. (Dong et al., 2023). Copyright 2026 Elsevier. (E) Reprinted with permission from ref. (Cheng et al., 2024). Copyright 2025 Elsevier.*

of the hydrolytic probes (HMy-OH and HEy-OH) (Ji et al., 2025). In 2026, Liu et al. reported a ratiometric fluorescent probe, BZ-BChE **(100)**, for sensitive and selective detection of BChE activity. Employing benzothiazole as the fluorophore core provided notable advantages over conventional ratiometric probes, including a large Stokes shift, an excited-state intramolecular proton transfer (ESIPT) process, and straightforward synthesis. The probe was successfully applied to visualize BChE in living cell lines and in the brains of AD mouse models (Fig. 14B and C), highlighting its potential utility for early-stage AD diagnosis (Liu et al., 2026). NIRF probes are better suited for in vivo optical visualization of BChE activity due to their advantageous characteristics, including deep tissue penetration, minimal biological toxicity, and resistance to environmental interference (Cheng et al., 2024; Wang et al., 2022; Yang et al., 2020). Among the fluorescence probes developed for monitoring BChE activity, only three fluorescence probes with NIR fluorescence properties were reported (Liu et al., 2018; Ma et al., 2020b; Zhang et al., 2022a). Although these probes enable sensitive detection of BChE, however, they have certain disadvantages, including a long reaction time and interference from AChE. Therefore, βit is crucial to develop NIR fluorescent probes featuring a novel recognition group to detect and image BChE with high sensitivity and selectivity in complex biological systems. In 2025, Zhang et al. developed a NIRF probe, Probe-BChE **(101)**, for BChE activity detection by exploiting structural differences in the active sites and binding pockets of AChE and BChE. The probe demonstrated excellent selectivity and sensitivity for BChE over AChE, with a low detection limit of 0.14 U/mL and a rapid response time (~3 s). In addition, Probe-BChE was applied to visualize endogenous BChE in APP/PS1 AD mouse models, revealing higher BChE expression in the frontal cortex and major organs of 7-month-old mice compared with 2-month-old mice, as illustrated in Fig. 14D and E (Zhang et al., 2025). The chemical structures of all fluorescent probes targeting BChE are illustrated in Fig. 15, while their photophysical properties, detection type, and corresponding mechanisms are summarized in Table 1.

4.3 Developed fluorescent probes targeting cholinesterase both AChE and BChE

ChEs primarily exist as two isoenzymes, AChE and BChE, which catalyze the breakdown of acetylcholine into choline and acetic acid at cholinergic synapses. Excessive production of ChEs directly disrupts ACh metabolism, leading to disruption of brain neurotransmission and contributing to neurological disorders such as depression and AD. Additionally, ChEs' abnormal expression may promote cell division and contribute to tumor development (Fig. 16).

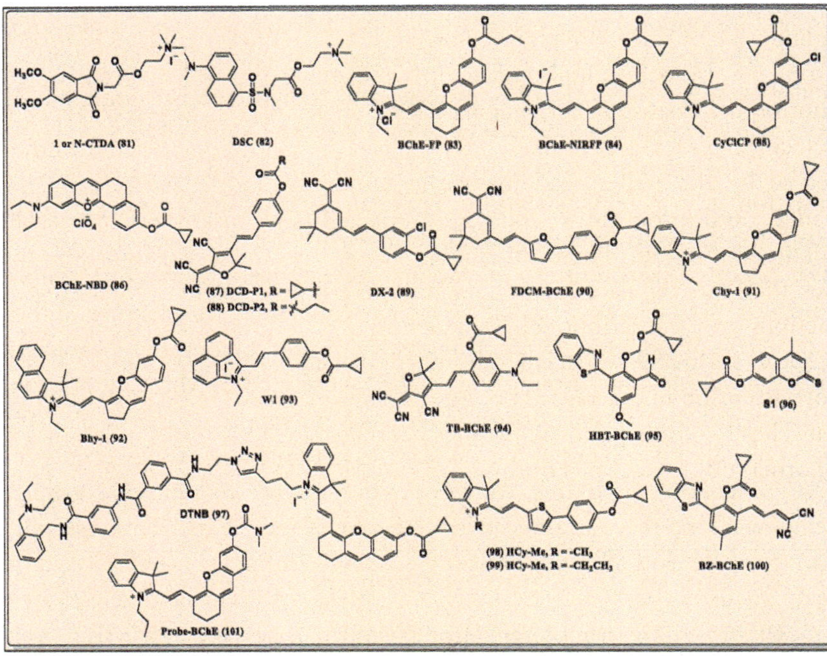

Fig. 15 Molecular structures of carboxylic ester-based probes for BChE.

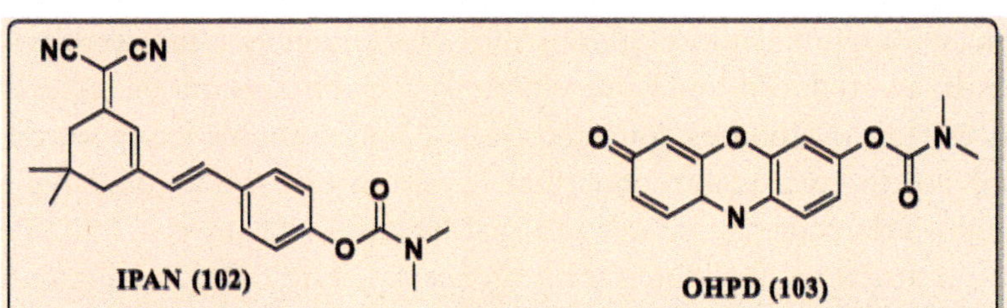

Fig. 16 Molecular structures of carboxylic ester-based probes for cholinesterases (both AChE and BChE).

Despite these correlations, the molecular mechanisms governing ChEs function in cancer, particularly within the complex brain environment, remain poorly understood (Ma et al., 2020a). Consequently, the development of effective in situ imaging strategies for ChEs in the brain is crucial for elucidating their physiological and pathological roles.

Although several fluorescent probes have recently been introduced for detecting ChE activity, further improvements are needed to establish an in vivo link between ChE and disease (Deng et al., 2019; Liu et al., 2018;

Xu et al., 2019). For example, probes with large Stokes shifts and near-infrared emission can facilitate deep tissue imaging while minimizing self-absorption of fluorescence, and rapid response kinetics are essential for detecting real-time fluctuations in ChE activity. Notably, fluorescence-based detection of ChE activity in brain gliomas has not yet been developed. In 2020, Ma et al. reported an NIRF probe, IPAN **(102)**, for ChEs detection, exhibiting high sensitivity, an ultralow detection limit, fast response, and a large Stokes shift, as shown in Fig. 17A. Additionally, the probe was further successfully applied to monitor variations in ChE activity in living glioma and zebrafish models

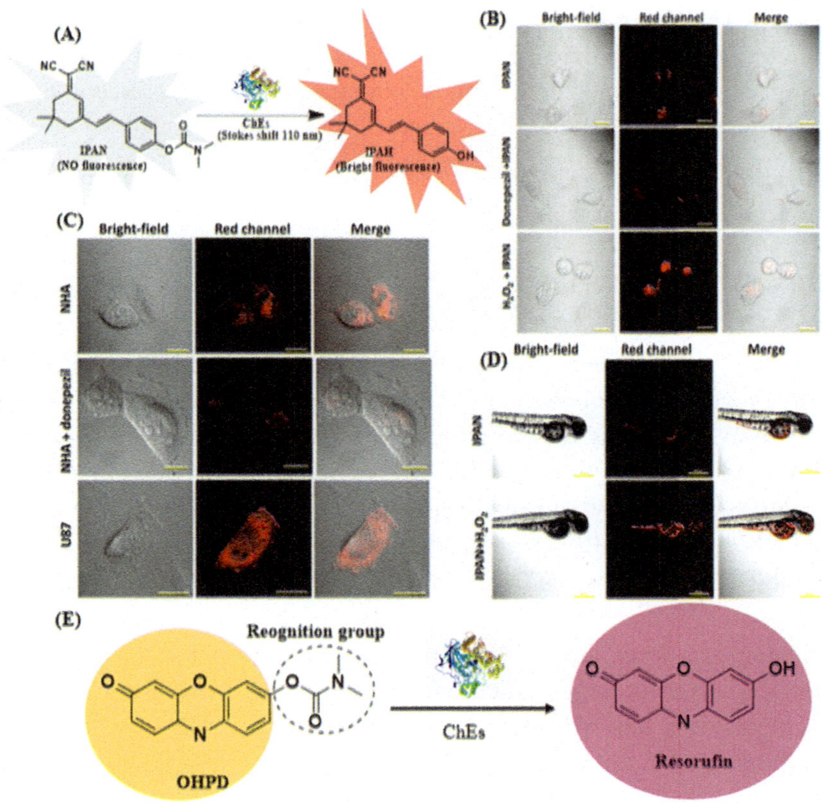

Fig. 17 (A) Sensing Mechanism of the Probe IPAN for the Detection of ChEs. (B) & (C) Fluorescence imaging of ChEs in living cells by sing IPAN (10 μM). (D) Fluorescence imaging of ChE activity in the zebrafish model using IPAN (10 μM). (E) Proposed activation mechanism of probe OHPD with AChE or BChE. *(D) Reprinted with permission from ref. (Ma et al., 2020a). Copyright 2021 American Chemical Society. (E) Reproduced from ref. (Zhang et al., 2025)*

during apoptosis, as depicted in Fig. 17B, C, and D (Ma et al., 2020a). In 2019, Wang et al. reported a TP fluorescent probe for the selective detection of AChE (Wang et al., 2019). Despite these developments, fluorescent probes with outstanding characteristics, such as a long analytical wavelength (>550 nm), good water solubility, high selectivity, and sensitivity for AChE/BChE detection and real-time imaging, remain scarce. In 2024, Tang et al. developed a turn-on fluorescent probe, OHPD **(103)**, for real-time monitoring of cholinesterases in vitro and in vivo, as shown in Fig. 17E. This probe exhibited low background interference, extended wavelength emission, excellent photostability, deep tissue penetration, and high selectivity for AChE/BChE. Additionally, OHPD was applied for imaging in zebrafish, suggesting that endogenous AChE/BChE may be elevated in the yolk sac of zebrafish (Tang et al., 2024). The chemical structures of all fluorescent probes targeting both AChE and BChE are illustrated in Fig. 16, while their photophysical properties, detection type, and corresponding mechanisms are summarized in Table 1.

5. Conclusion and outlook

Cholinesterase enzymes (ChEs), mainly AChE and BChE, play a key role in regulating cholinergic neurotransmission by hydrolyzing ACh, and their dysregulation is closely associated with the pathogenesis and progression of AD. In AD, reduced acetylcholine levels due to excessive cholinesterase activity contribute to cognitive impairment, while changes in AChE and BChE expression across disease stages underscore their diagnostic and therapeutic significance. Consequently, both enzymes have emerged as important molecular targets for therapeutic and diagnostic development. This book chapter presents a comprehensive overview of therapeutic agents targeting ChEs, with particular emphasis on rational design strategies and structure-activity relationships (SAR). Various classes of inhibitors, including carbamate derivatives, heterocyclic scaffolds, dual-binding site inhibitors, and multi-target-directed ligands (MTDLs), have been discussed. The incorporation of key pharmacophores capable of interacting with both the catalytic active site (CAS) and peripheral anionic site (PAS) of AChE, or the broader active gorge of BChE, has enabled the development of highly potent inhibitors. In this chapter, therapeutic molecules with promising, low IC_{50} values against AChE and BChE are systematically highlighted to demonstrate their potency and effectiveness in targeting ChEs in AD. Additionally, this chapter has also discussed diagnostic agents for the detection and imaging of AChE and BChE by using fluorescent probes, along with their design strategies, photophysical properties, and their detection mechanisms, such as turn-on or turn-off fluorescence responses, providing valuable tools for real-time monitoring of ChE activity in vitro and in vivo. These probes not only enable sensitive and selective detection but also contribute to understanding enzyme distribution and disease progression in biological models. The ability to measure ChE activity in cellular and animal models significantly enhances their potential for early diagnosis, drug screening, and therapeutic monitoring. Collectively, the integration of therapeutic inhibitors and diagnostic probes targeting ChEs represents a synergistic, comprehensive approach to managing AD, highlighting the crucial role of cholinesterase-based approaches in improving both treatment and early diagnosis of the disease.

Acknowledgment

GPM is thankful to the Anushandhan National Research Foundation, earlier Science and Engineering Research Board under the Core Research Grant (ANRF-CRG/2022/007126).

References

Abd El-Mageed, M. M., et al. (2025). Rational design, synthesis and computational studies of multi-targeted anti-Alzheimer's agents integrating coumarin scaffold. *Bioorganic Chemistry, 154*, 108024.

Abonia, R., et al. (2020). A flexible strategy for modular synthesis of curcuminoid-BF2/curcuminoid pairs and their comparative antiproliferative activity in human cancer cell lines. *ChemMedChem, 15*(4), 354–362.

Ali, S. H., Osmaniye, D., & Kaplancıklı, Z. A. (2025). Synthesis and biological evaluation of novel hydrazone derivatives for the treatment of Alzheimer's disease. *RSC Advances, 15*(53), 45729–45743.

Amenta, F., & Tayebati, S. K. (2008). Pathways of acetylcholine synthesis, transport and release as targets for treatment of adult-onset cognitive dysfunction. *Current Medicinal Chemistry, 15*(5), 488–498.

An, W., et al. (2024). A highly specific two-photon fluorescent probe for real-time monitoring of acetylcholinesterase in neurogenic disorders in vivo. *Analytica Chimica Acta, 1331*, 343309.

Andreani, A., et al. (2008). Chemiluminescent high-throughput microassay applied to imidazo [2, 1-b] thiazole derivatives as potential acetylcholinesterase and butyrylcholinesterase inhibitors. *European Journal of Medicinal Chemistry, 43*(3), 657–661.

Asadipour, A., et al. (2024). Amino-7, 8-dihydro-4H-chromenone derivatives as potential inhibitors of acetylcholinesterase and butyrylcholinesterase for Alzheimer's disease management; in vitro and in silico study. *BMC Chemistry, 18*(1), 70.

Ayoup, M. S., et al. (2024). Design, synthesis, and biological evaluation of 1, 2, 4-oxadiazole-based derivatives as multitarget anti-Alzheimer agents. *RSC Medicinal Chemistry, 15*(6), 2080–2097.

Banoo, R., et al. (2024). Design, synthesis, and pharmacological evaluation of indole-piperidine amides as Blood− brain barrier permeable dual cholinesterase and β-secretase inhibitors. *European Journal of Medicinal Chemistry, 266*, 116131.

Bartus, R. T., et al. (1982). The cholinergic hypothesis of geriatric memory dysfunction. *Science (New York, N. Y.), 217*(4558), 408–414.

Bayraktar, G., et al. (2024). Novel multifunctional tacrine–donepezil hybrids against Alzheimer's disease: Design synthesis and bioactivity studies. *Archiv der Pharmazie, 357*(7), 2300575.

Bubley, A., et al. (2023). Tacrine-based hybrids: Past, present, and future. *International Journal of Molecular Sciences, 24*(2), 1717.

Cacabelos, R. (2007). Donepezil in Alzheimer's disease: From conventional trials to pharmacogenetics. *Neuropsychiatric Disease and Treatment, 3*(3), 303–333.

Camps, P., et al. (2000). New tacrine− huperzine A hybrids (huprines): Highly potent tight-binding acetylcholinesterase inhibitors of interest for the treatment of Alzheimer's disease. *Journal of Medicinal Chemistry, 43*(24), 4657–4666.

Cao, T., et al. (2021). A highly butyrylcholinesterase selective red-emissive mitochondria-targeted fluorescent indicator imaging in liver tissue of mice. *Sensors and Actuators B: Chemical, 330*, 129348.

Chang, J., et al. (2016). based fluorescent sensor for rapid naked-eye detection of acetylcholinesterase activity and organophosphorus pesticides with high sensitivity and selectivity. *Biosensors and Bioelectronics, 86*, 971–977.

Chao, S., et al. (2016). A selective and sensitive near-infrared fluorescent probe for acetylcholinesterase imaging. *Chemical Communications, 52*(77), 11599–11602.

Chen, Y., et al. (2017a). A benzothiazole-based fluorescent probe for ratiometric detection of Al3+ in aqueous medium and living cells. *Industrial & Engineering Chemistry Research, 56*(43), 12267–12275.

Chen, Y., et al. (2017b). A mitochondria-targeted fluorescent probe for ratiometric detection of hypochlorite in living cells. *Chinese Chemical Letters, 28*(10), 1957–1960.

Chen, Y., Wang, S., & Zhang, F. (2023). Near-infrared luminescence high-contrast in vivo biomedical imaging. *Nature Reviews Bioengineering, 1*(1), 60–78.

Cheng, Z., et al. (2024). Enzyme-activatable near-infrared hemicyanines as modular scaffolds for in vivo photodynamic therapy. *Angewandte Chemie International Edition, 63*(30), e202404587.

Cuello, A. C. (2025). The involvement of the cholinergic system in Alzheimer disease. *Handbook of Clinical Neurology, 211*, 63–79.

Cui, Y., et al. (2021). Highly selective two-photon fluorescent off–on probes for imaging tyrosinase activity in living cells and tissues. *Chemical Communications, 57*(56), 6911–6914.

Dajas-Bailador, F. A., Heimala, K., & Wonnacott, S. (2003). The allosteric potentiation of nicotinic acetylcholine receptors by galantamine is transduced into cellular responses in neurons: Ca2+ signals and neurotransmitter release. *Molecular Pharmacology, 64*(5), 1217–1226.

Darvesh, S. (2016). Butyrylcholinesterase as a diagnostic and therapeutic target for Alzheimer's disease. *Current Alzheimer Research, 13*(10), 1173–1177.

Davies, P., & Maloney, A. (1976). Selective loss of central cholinergic neurons in Alzheimer's disease. *The Lancet, 308*(8000), 1403.

DeBay, D. R., et al. (2017). Targeting butyrylcholinesterase for preclinical single photon emission computed tomography (SPECT) imaging of Alzheimer's disease. *Alzheimer's & Dementia: Translational Research & Clinical Interventions, 3*(2), 166–176.

Deng, C., et al. (2022). Design, synthesis, and biological evaluation of novel 2-acetylphenol-rivastigmine hybrids as potential multifunctional agents for the treatment of Alzheimer's disease. *Medicinal Chemistry Research, 31*(6), 1035–1048.

Deng, W., et al. (2019). Enzymatic reaction modulated synthesis of quantum dots for visual detection of cholinesterase activity and inhibitor. *Sensors and Actuators B: Chemical, 292*, 180–186.

Dong, H., et al. (2023). Dual-mode ratiometric electrochemical and turn-on fluorescent detection of butyrylcholinesterase utilizing a single probe for the diagnosis of Alzheimer's disease. *Analytical Chemistry, 95*(21), 8340–8347.

Dong, H., et al. (2025). High selectivity fluorescence and electrochemical dual-mode detection of glutathione in the serum of Parkinson's disease model mice and humans. *Analytical Chemistry, 97*(2), 1318–1328.

Du, W.-R., et al. (2025). Design, synthesis, biological evaluation, and in silico studies of imidazo [2, 1-b][1, 3, 4] thiadiazole derivatives as cholinesterase inhibitors. *Medicinal Chemistry Research, 34*(4), 910–928.

Fortibui, M. M., et al. (2022). Near-infrared fluorescence probe for specific detection of acetylcholinesterase and imaging in live cells and zebrafish. *ACS Applied Bio Materials, 5*(5), 2232–2239.

Gao, M., & Tang, B. Z. (2020). AIE-based cancer theranostics. *Coordination Chemistry Reviews, 402*, 213076.

Ghafary, S., et al. (2020). Design, synthesis, and evaluation of novel cinnamic acid-tryptamine hybrid for inhibition of acetylcholinesterase and butyrylcholinesterase. *DARU Journal of Pharmaceutical Sciences, 28*(2), 463–477.

Giacobini, E. (2003). Cholinesterases: New roles in brain function and in Alzheimer's disease. *Neurochemical Research, 28*(3), 515–522.

Di Giacomo, S., et al. (2022). Recent advances in the neuroprotective properties of ferulic acid in Alzheimer's disease: A narrative review. *Nutrients, 14*(18), 3709.

Gil-Rivas, A., et al. (2023). New advances in the exploration of esterases with PET and fluorescent probes. *Molecules (Basel, Switzerland), 28*(17), 6265.

Goud, N. S., et al. (2021). Carbon-11: Radiochemistry and target-based PET molecular imaging applications in oncology, cardiology, and neurology. *Journal of Medicinal Chemistry, 64*(3), 1223–1259.

Grabowska, W., et al. (2025). Acetylcholinesterase as a multifunctional target in amyloid-driven neurodegeneration: From dual-site inhibitors to anti-agregation strategies. *International Journal of Molecular Sciences, 26*(17), 8726.

Gu, H., et al. (2023). Excited-state intramolecular proton transfer (ESIPT)-based fluorescent probes for biomarker detection: Design, mechanism, and application. *Chemical Communications, 59*(15), 2056–2071.

Guo, J., et al. (2023). Developing a two-photon "AND" logic probe and its application in Alzheimer's disease differentiation. *Analytical Chemistry, 95*(46), 16868–16876.

Guo, Z., et al. (2012). A highly selective ratiometric near-infrared fluorescent cyanine sensor for cysteine with remarkable shift and its application in bioimaging. *Chemical Science, 3*(9), 2760–2765.

H Ferreira-Vieira, T., et al. (2016). Alzheimer's disease: Targeting the cholinergic system. *Current Neuropharmacology, 14*(1), 101–115.

Hansen, R. A., et al. (2008). Efficacy and safety of donepezil, galantamine, and rivastigmine for the treatment of Alzheimer's disease: A systematic review and meta-analysis. *Clinical Interventions in Aging, 3*(2), 211–225.

Hasan, A. H., et al. (2023). Discovery of novel coumarin-schiff base hybrids as potential acetylcholinesterase inhibitors: Design, synthesis, enzyme inhibition, and computational studies. *Pharmaceuticals, 16*(7), 971.

He, J., & Tam, K. Y. (2024). Dual-target inhibitors of cholinesterase and GSK-3β to modulate Alzheimer's disease. *Drug Discovery Today, 29*(4), 103914.

He, N., et al. (2021). Near-infrared fluorescent probe for evaluating the acetylcholinesterase effect in the aging process and dietary restriction via fluorescence imaging. *Journal of Materials Chemistry B, 9*(11), 2623–2630.

Hu, K., et al. (2025). Development of novel caffeic acid derivatives as multifunctional agents for the treatment of AD. *Medicinal Chemistry Research, 34*(6), 1364–1376.

Inestrosa, N. C., Dinamarca, M. C., & Alvarez, A. (2008). Amyloid–cholinesterase interactions: Implications for Alzheimer's disease. *The FEBS Journal, 275*(4), 625–632.

Iraji, A., et al. (2025). Design and synthesis of new 1, 2, 3-triazole-methoxyphenyl-1, 3, 4-oxadiazole derivatives: Selective butyrylcholinesterase inhibitors against Alzheimer's disease. *BMC Chemistry, 19*(1), 97.

Jaber, T., et al. (2024). Design, synthesis and evaluation of novel coumarin-oxa/thiadiazole hybrids as AChE inhibitors for the treatment of alzheimer's disease. *Indian Journal of Chemical Technology (IJCT), 31*(3), 483–493.

Jalil, S., et al. (2024). Quinoline–sulfonamides as a multi-targeting neurotherapeutic for cognitive decline: In vitro, in silico studies and ADME evaluation of monoamine oxidases and cholinesterases inhibitors. *RSC Advances, 14*(13), 8905–8920.

Jevtić, I. I., et al. (2024). Multi-target potential of newly designed tacrine-derived cholinesterase inhibitors: Synthesis, computational and pharmacological study. *Bioorganic & Medicinal Chemistry, 101*, 117649.

Ji, S., et al. (2025). A novel enzyme-activated tandem fluorescent probe for dual detection of BChE and A plaques in Alzheimer's disease. *Talanta*, 129169.

Jiang, Y., Cui, H., & Yu, Q. (2024). A novel near-infrared fluorescent probe for high-sensitivity detection of butyrylcholinesterase in various pathological states. *Spectrochimica Acta Part A: Molecular and Biomolecular Spectroscopy, 308*, 123801.

Kamal, M., Greig, N., & Reale, M. (2009). Anti-Inflammatory properties of acetylcholinesterase inhibitors administred in Alzheimer's disease. *Anti-Inflammatory & Anti-Allergy Agents in Medicinal Chemistry (Formerly Current Medicinal Chemistry-Anti-Inflammatory and Anti-Allergy Agents, 8*(1), 85–100.

Kamel, N. N., et al. (2023). Anti-Alzheimer activity of new coumarin-based derivatives targeting acetylcholinesterase inhibition. *RSC Advances, 13*(27), 18496–18510.

Kang, S., et al. (2016). A direct assay of butyrylcholinesterase activity using a fluorescent substrate. *Organic & Biomolecular Chemistry, 14*(37), 8815–8820.

Khan, M. S., et al. (2025). Synthesis and neurobehavioral evaluation of a potent multi-targeted inhibitor for the treatment of Alzheimer's disease. *Molecular Neurobiology, 62*(2), 1558–1576.

Knapp, M. J., et al. (1994). A 30-week randomized controlled trial of high-dose tacrine in patients with Alzheimer's disease. *JAMA: the Journal of the American Medical Association, 271*(13), 985–991.

Kryger, G., Silman, I., & Sussman, J. L. (1999). Structure of acetylcholinesterase complexed with E2020 (Aricept®): Implications for the design of new anti-Alzheimer drugs. *Structure (London, England: 1993),* 7(3), 297–307.

Kumar, S., & Modi, G. (2025). Design, synthesis, and biological evaluation of caffeic acid-based novel multifunctional molecules for the management of Alzheimer's disease. *European Journal of Medicinal Chemistry, 296*, 117831.

Kuzu, B., et al. (2025). Structure-based inhibition of acetylcholinesterase and butyrylcholinesterase with 2-Aryl-6-carboxamide benzoxazole derivatives: Synthesis, enzymatic assay, and in silico studies. *Molecular Diversity, 29*(1), 671–693.

Kuzu, B., & Demir, Y. (2025). Synthesis of novel pyrazino [1, 2-a] indol-1 (2H)-one derivatives as potent cholinesterase inhibitors and their in vitro and in silico evaluations. *Archives of Biochemistry and Biophysics, 771*, 110504.

Laali, K. K., et al. (2016). Fluoro-curcuminoids and curcuminoid-BF2 adducts: Synthesis, X-ray structures, bioassay, and computational/docking study. *Journal of Fluorine Chemistry, 191*, 29–41.

Laali, K. K., et al. (2018). Novel fluorinated curcuminoids and their pyrazole and isoxazole derivatives: Synthesis, structural studies, Computational/Docking and in-vitro bioassay. *Journal of Fluorine Chemistry, 206*, 82–98.

Laali, K. K., et al. (2019). Deuterated curcuminoids: Synthesis, structures, computational/docking and comparative cell viability assays against colorectal cancer. *ChemMedChem, 14*(12), 1173–1184.

Lan, J.-S., et al. (2017). Design, synthesis and evaluation of novel cinnamic acid derivatives bearing N-benzyl pyridinium moiety as multifunctional cholinesterase inhibitors for Alzheimer's disease. *Journal of Enzyme Inhibition and Medicinal Chemistry, 32*(1), 776–788.

Lan, J.-S., et al. (2020). Design, synthesis and evaluation of novel ferulic acid derivatives as multi-target-directed ligands for the treatment of Alzheimer's disease. *Bioorganic Chemistry, 94*, 103413.

Li, C., et al. (2020). Advanced fluorescence imaging technology in the near-infrared-II window for biomedical applications. *Journal of the American Chemical Society, 142*(35), 14789–14804.

Li, H., et al. (2021b). Activity-based smart AIEgens for detection, bioimaging, and therapeutics: recent progress and outlook. *Aggregate, 2*(4), e51.

Li, M., et al. (2025b). Design, synthesis and biological evaluation of galantamine analogues for cognitive improvement in Alzheimer's disease. *European Journal of Medicinal Chemistry, 284*, 117198.

Li, Q., et al. (2021a). Highly potent and selective butyrylcholinesterase inhibitors for cognitive improvement and neuroprotection. *Journal of Medicinal Chemistry, 64*(10), 6856–6876.

Li, X., et al. (2022). Design, synthesis, and biological evaluation of novel chromanone derivatives as multifunctional agents for the treatment of Alzheimer's disease. *ACS Chemical Neuroscience, 13*(23), 3488–3501.

Li, Z., et al. (2014). Unmodified silver nanoparticles for rapid analysis of the organophosphorus pesticide, dipterex, often found in different waters. *Sensors and Actuators B: Chemical, 193*, 205–211.

Li, Z. H., et al. (2025a). Design, synthesis and evaluation of vanillin derivatives as dual-target inhibitors for the treatment of Alzheimer's disease. *Bioorganic & Medicinal Chemistry, 129*, 118296.

Lin, X., et al. (2024). Two fluorescent probes for recognition of acetylcholinesterase: Design, synthesis, and comparative evaluation. *Molecules (Basel, Switzerland), 29*(9), 1961.

Liu, S., et al. (2021). Recent advances of AIE light-up probes for photodynamic therapy. *Chemical Science, 12*(19), 6488–6506.

Liu, S.-Y., et al. (2018). Discovery of butyrylcholinesterase-activated near-infrared fluorogenic probe for live-cell and in vivo imaging. *ACS Sensors, 3*(10), 2118–2128.

Liu, Z., et al. (2026). 2-(2-Hydroxyphenyl) benzoazole-based ratiometric fluorescence probe for the detection and imaging of butyrylcholinesterase activity in Alzheimer's disease mice brain. *Dyes and Pigments, 246*, 113354.

Luo, J., et al. (2001). Aggregation-induced emission of 1-methyl-1, 2, 3, 4, 5-pentaphenylsilole. *Chemical Communications*, (18), 1740–1741.

Ma, J., et al. (2019). Near-infrared fluorescence probe for evaluating acetylcholinesterase activity in PC12 cells and in situ tracing AChE distribution in zebrafish. *ACS Sensors, 5*(1), 83–92.

Ma, J., et al. (2020b). Rational design of a near-infrared fluorescence probe for highly selective sensing butyrylcholinesterase (BChE) and its bioimaging applications in living cell. *Talanta, 219*, 121278.

Ma, Y., et al. (2020a). Observation of the elevation of cholinesterase activity in brain glioma by a near-infrared emission chemsensor. *Analytical Chemistry, 92*(19), 13405–13410.

Makhaeva, G. F., et al. (2020). New hybrids of 4-amino-2, 3-polymethylene-quinoline and p-tolylsulfonamide as dual inhibitors of acetyl-and butyrylcholinesterase and potential multifunctional agents for Alzheimer's disease treatment. *Molecules (Basel, Switzerland), 25*(17), 3915.

Mandloi, U., et al. (2026). Theranostic advances in Alzheimer's disease: Structure-guided design of near-infrared fluorescent probes targeting amyloid-β and cholinergic dysfunction. *RSC Medicinal Chemistry*.

Mei, J., et al. (2015). Aggregation-induced emission: Together we shine, united we soar!. *Chemical Reviews, 115*(21), 11718–11940.

Meng, W., et al. (2021). Two birds with one stone: The detection of nerve agents and AChE activity with an ICT-ESIPT-based fluorescence sensor. *Journal of Hazardous Materials, 410*, 124811.

Miao, S.-X., et al. (2021). Pd-catalyzed direct diversification of natural anti-alzheimer's disease drug: Synthesis and biological evaluation of N-aryl huperzine A analogues. *Journal of Natural Products, 84*(8), 2374–2379.

Mineur, Y. S., et al. (2013). Cholinergic signaling in the hippocampus regulates social stress resilience and anxiety-and depression-like behavior. *Proceedings of the National Academy of Sciences, 110*(9), 3573–3578.

Mirjalili, B. B. F., Fazeli Attar, S. A., & Shiri, F. (2024). Synthesis, biological evaluation and in silico study of 4-(benzo [d] thiazole-2-yl) phenols based on 4-hydroxy coumarin as acetylcholinesterase inhibitors. *Scientific Reports, 14*(1), 26459.

Musachio, J. L., et al. (2002). Radiosynthesis and mouse brain distribution studies of [11C] CP-126,998: a PET ligand for in vivo study of acetylcholinesterase. *Nuclear Medicine and Biology, 29*(5), 547–552.

Mustazza, C., et al. (2002). Synthesis and cholinesterase activity of phenylcarbamates related to Rivastigmine, a therapeutic agent for Alzheimer's disease. *European Journal of Medicinal Chemistry, 37*(2), 91–109.

Nasab, N. H., et al. (2024). Exploring the anti-Alzheimer potential: Design, synthesis, biological activity, and molecular docking study of benzothiazol-1, 3, 4-oxadiazole-acetamide compounds. *Journal of Molecular Structure, 1318*, 139307.

Ni, X., et al. (2018). Near-infrared afterglow luminescent aggregation-induced emission dots with ultrahigh tumor-to-liver signal ratio for promoted image-guided cancer surgery. *Nano Letters, 19*(1), 318–330.

Noori, M., et al. (2024). Isoindolinedione-benzamide pyridinium derivatives for targeting Alzheimer's disease. *ACS omega, 9*(49), 48032–48043.

Oukoloff, K., et al. (2015). PET and SPECT radiotracers for Alzheimer's disease. *Current Medicinal Chemistry, 22*(28), 3278–3304.

Panek, D., et al. (2023). Discovery of new, highly potent and selective inhibitors of BuChE-design, synthesis, in vitro and in vivo evaluation and crystallography studies. *European Journal of Medicinal Chemistry, 249*, 115135.

Pappata, S., et al. (1996). vivo imaging of human cerebral acetylcholinesterase. *Journal of Neurochemistry, 67*(2), 876–879.

Parsons, S. M., Prior, C., & Marshall, I. G. (1993). Acetylcholine transport, storage, and release. *International Review of Neurobiology, 35*, 279–390.

Pei, X., et al. (2023). A turn-on fluorescent probe based on ESIPT and AIEE mechanisms for the detection of butyrylcholinesterase activity in living cells and in non-alcoholic fatty liver of zebrafish. *Spectrochimica Acta Part A: Molecular and Biomolecular Spectroscopy, 287*, 122044.

Peng, L., et al. (2017). Rational design of a red-emissive fluorophore with AIE and ESIPT characteristics and its application in light-up sensing of esterase. *Analytical Chemistry, 89*(5), 3162–3168.

Perry, E. K., et al. (1978). Correlation of cholinergic abnormalities with senile plaques and mental test scores in senile dementia. *British Medical Journal, 2*(6150), 1457–1459.

Pourtaher, H., Hasaninejad, A., & Iraji, A. (2022). Design, synthesis, in silico and biological evaluations of novel polysubstituted pyrroles as selective acetylcholinesterase inhibitors against Alzheimer's disease. *Scientific Reports, 12*(1), 15236.

Prešern, U., Goličnik, M., & Bavec, A. (2025). Genetic and functional dynamics of Butyrylcholinesterase in Alzheimer's disease: From mechanisms to clinical relevance. *Chemico-Biological Interactions*, 111809.

Rai, H., et al. (2022). Near-infrared fluorescent probes as imaging and theranostic modalities for amyloid-beta and tau aggregates in Alzheimer's disease. *Journal of Medicinal Chemistry, 65*(13), 8550–8595.

Rai, H., et al. (2024). Rhodanine composite fluorescence probes to detect pathological hallmarks in Alzheimer's disease models. *Sensors and Actuators B: Chemical, 407*, 135364.

Ramachandran, A. K., et al. (2025). Design, synthesis, and preclinical evaluation of novel 1-benzylpiperidine derivatives as multitarget agents against Alzheimer's disease. *European Journal of Pharmacology*, 178192.

Reddy, C. N., et al. (2024). Synthesis and biological evaluation of colchicine— Aryl/alkyl amine hybrids as potential noncytotoxic cholinesterase inhibitors: Identification of SBN-284 as a dual inhibitor of cholinesterases and NLRP3 inflammasome. *ACS Chemical Neuroscience, 15*(15), 2779–2794.

Roivainen, A., et al. (2004). Biodistribution and blood metabolism of 1-11C-methyl-4-piperidinyl n-butyrate in humans: an imaging agent for in vivo assessment of butyrylcholinesterase activity with PET. *Journal of Nuclear Medicine, 45*(12), 2032–2039.

Ronco, C., et al. (2011). New huprine derivatives functionalized at position 9 as highly potent acetylcholinesterase inhibitors. *ChemMedChem, 6*(5), 876–888.

Roy, R., et al. (2016). Cholinergic imaging in dementia spectrum disorders. *European Journal of Nuclear Medicine and Molecular Imaging, 43*(7), 1376–1386.

Sahin, Z., et al. (2022). Novel benzofurane carbonyl analogs of donepezil as acetylcholinesterase inhibitors. *Journal of Molecular Structure, 1264*, 133193.

Samanta, S., & Govindaraju, T. (2019). Unambiguous detection of elevated levels of hypochlorous acid in double transgenic AD mouse brain. *ACS Chemical Neuroscience, 10*(12), 4847–4853.

Samochocki, M., et al. (2003). Galantamine is an allosterically potentiating ligand of neuronal nicotinic but not of muscarinic acetylcholine receptors. *The Journal of Pharmacology and Experimental Therapeutics, 305*(3), 1024–1036.

Sang, Z., et al. (2018). Design, synthesis, and evaluation of novel ferulic acid derivatives as multi-target-directed ligands for the treatment of Alzheimer's disease. *ACS Chemical Neuroscience, 10*(2), 1008–1024.

Sang, Z., et al. (2020). Design, synthesis and biological evaluation of novel O-carbamoyl ferulamide derivatives as multi-target-directed ligands for the treatment of Alzheimer's disease. *European Journal of Medicinal Chemistry, 194*, 112265.

Saxena, M., & Dubey, R. (2019). Target enzyme in Alzheimer's disease: Acetylcholinesterase inhibitors. *Current Topics in Medicinal Chemistry, 19*(4), 264–275.

Schliebs, R., & Arendt, T. (2011). The cholinergic system in aging and neuronal degeneration. *Behavioural Brain Research, 221*(2), 555–563.

Shankar, G., et al. (2024). Discovery of novel substituted (*Z*)-N′-hydroxy-3-(3-phenylureido) benzimidamide derivatives as multifunctional molecules targeting pathological hallmarks of Alzheimer's disease. *European Journal of Medicinal Chemistry, 280*, 116959.

Shankar, G., et al. (2025). Discovery of novel hybrid tryptamine-rivastigmine molecules as potent AChE and BChE inhibitors exhibiting multifunctional properties for the management of Alzheimer's disease. *European Journal of Medicinal Chemistry, 283*, 117066.

Shao, X., et al. (2003). N-methylpiperidinemethyl, N-methylpyrrolidyl and N-methylpyrrolidinemethyl esters as PET radiotracers for acetylcholinesterase activity. *Nuclear Medicine and Biology, 30*(3), 293–302.

Sharma, A., et al. (2025). Recent advancements in the therapeutic approaches for Alzheimer's disease treatment: Current and future perspective. *RSC Medicinal Chemistry, 16*(2), 652–693.

Shi, J., et al. (2017). Enzyme-responsive bioprobes based on the mechanism of aggregation-induced emission. *ACS Applied Materials & Interfaces, 10*(15), 12278–12294.

Shiri, P. (2021). Novel hybrid molecules based on triazole-β-lactam as potential biological agents. *Mini Reviews in Medicinal Chemistry, 21*(5), 536–553.

Sidhu, J. S., et al. (2023). Acetylcholine structure-based small activatable fluorogenic probe for specific detection of acetylcholinesterase. *Analytical Chemistry, 95*(19), 7594–7602.

Singh, A., et al. (2025). Development of coumarin-inspired bifunctional hybrids as a new class of anti-Alzheimer's agents with potent in vivo efficacy. *RSC Medicinal Chemistry, 16*(5), 1986–2003.

Singh, G., et al. (2024a). Design, synthesis, and biological evaluation of ferulic acid-piperazine derivatives targeting pathological hallmarks of Alzheimer's disease. *ACS Chemical Neuroscience, 15*(15), 2756–2778.

Singh, G., et al. (2024b). Design, synthesis, and biological evaluation of ferulic acid template-based novel multifunctional ligands targeting nlrp3 inflammasome for the management of alzheimer's disease. *ACS Chemical Neuroscience, 15*(7), 1388–1414.

Singh, Y. P., et al. (2020). Design, synthesis and biological evaluation of novel naturally-inspired multifunctional molecules for the management of Alzheimer's disease. *European Journal of Medicinal Chemistry, 198*, 112257.

Singh, Y. P., et al. (2021). Further SAR studies on natural template based neuroprotective molecules for the treatment of Alzheimer's disease. *Bioorganic & Medicinal Chemistry, 46*, 116385.

Snyder, S. E., et al. (2001). Radiolabeled cholinesterase substrates: In vitro methods for determining structure-activity relationships and identification of a positron emission tomography radiopharmaceutical for in vivo measurement of butyrylcholinesterase activity. *Journal of Cerebral Blood Flow & Metabolism, 21*(2), 132–143.

Sola, I., et al. (2015). Novel levetiracetam derivatives that are effective against the Alzheimer-like phenotype in mice: Synthesis, in vitro, ex vivo, and in vivo efficacy studies. *Journal of Medicinal Chemistry, 58*(15), 6018–6032.

Tang, X., et al. (2024). Detection of acetylcholinesterase and butyrylcholinesterase in vitro and in vivo using a new fluorescent probe. *Chemical Communications, 60*(15), 2082–2085.

Tiepolt, S., et al. (2022). PET imaging of cholinergic neurotransmission in neurodegenerative disorders. *Journal of Nuclear Medicine, 63*(ement 1), 33S–44S.

Tripathi, A., et al. (2020). Design, synthesis, and biological evaluation of ferulic acid based 1, 3, 4-oxadiazole hybrids as multifunctional therapeutics for the treatment of Alzheimer's disease. *Bioorganic Chemistry, 95*, 103506.

Uliassi, E., et al. (2023). Quinolinetrione-tacrine hybrids as multi-target-directed ligands against Alzheimer's disease. *Bioorganic & Medicinal Chemistry, 91*, 117419.

Verma, A., et al. (2024). Design, synthesis, and biological investigation of quinazoline derivatives as multitargeting therapeutics in Alzheimer's disease therapy. *ACS Chemical Neuroscience, 15*(4), 745–771.

Viayna, E., et al. (2014). Synthesis and multitarget biological profiling of a novel family of rhein derivatives as disease-modifying anti-Alzheimer agents. *Journal of Medicinal Chemistry, 57*(6), 2549–2567.

Waly, O. M., et al. (2023). Multi-targeted anti-Alzheimer's agents: Synthesis, biological evaluation, and molecular modeling study of some pyrazolopyridine hybrids. *European Journal of Medicinal Chemistry, 262*, 115880.

Wan, C., et al. (2022). Ratiometric fluorescence assay for butyrylcholinesterase activity based on a hemicyanine and its application in biological imaging. *Dyes and Pigments, 197*, 109874.

Wang, G., et al. (2023). Discovery of novel tryptanthrin derivatives with benzenesulfonamide substituents as multi-target-directed ligands for the treatment of Alzheimer's disease. *Pharmaceuticals, 16*(10), 1468.

Wang, L., et al. (2020). A FRET-based near-infrared ratiometric fluorescent probe for detection of mitochondria biothiol. *Talanta, 219*, 121296.

Wang, L., et al. (2024a). Benzobisthiadiazole-based small molecular near-infrared-II fluorophores: From molecular engineering to nanophototheranostics. *ACS nano, 18*(6), 4683–4703.

Wang, L., et al. (2024b). Butyrylcholinesterase-activated near-infrared fluorogenic probe for in vivo theranostics of Alzheimer's disease. *Journal of Medicinal Chemistry, 67*(8), 6793–6809.

Wang, Q., et al. (2022). Near-infrared fluorescence probe with a new recognition moiety for specific detection and imaging of aldehyde dehydrogenase expecting the identification and isolation of cancer stem cells. *Analytical Chemistry, 94*(49), 17328–17333.

Wang, S.-M., et al. (2025a). Design, synthesis, and bio-evaluation of C1-aryl galantamine derivatives. *Fitoterapia, 183*, 106535.

Wang, X., et al. (2019). Observation of acetylcholinesterase in stress-induced depression phenotypes by two-photon fluorescence imaging in the mouse brain. *Journal of the American Chemical Society, 141*(5), 2061–2068.

Wang, X., et al. (2025b). Multifunctional Tacrine–Quinoline hybrids as cholinesterase inhibitors, Aβ aggregation blockers, and metal chelators for Alzheimer's therapy. *Molecules (Basel, Switzerland), 30*(17), 3489.

Watkins, P. B., et al. (1994). Hepatotoxic effects of tacrine administration in patients with Alzheimer's disease. *JAMA: the Journal of the American Medical Association, 271*(13), 992–998.

Wei, X., et al. (2023). Monitoring acetylcholinesterase level changes under oxidative stress through ESIPT-ICT-based near-infrared fluorescent probe. *Sensors and Actuators B: Chemical, 380*, 133392.

Whitehouse, P. J., et al. (1982). Alzheimer's disease and senile dementia: Loss of neurons in the basal forebrain. *Science (New York, N. Y.), 215*(4537), 1237–1239.

Wu, J., et al. (2020a). Design, synthesis and biological evaluation of novel carbamates as potential inhibitors of acetylcholinesterase and butyrylcholinesterase. *Bioorganic & Medicinal Chemistry, 28*(5), 115324.

Wu, X., et al. (2020b). A molecular approach to rationally constructing specific fluorogenic substrates for the detection of acetylcholinesterase activity in live cells, mice brains and tissues. *Chemical Science, 11*(41), 11285–11292.

Xia, J., et al. (2024). Design, synthesis, and biological evaluation of novel tryptanthrin derivatives as selective acetylcholinesterase inhibitors for the treatment of Alzheimer's disease. *Bioorganic Chemistry, 143*, 106980.

Xiang, C., et al. (2022a). A responsive AIE-active fluorescent probe for visualization of acetylcholinesterase activity in vitro and in vivo. *Materials Chemistry Frontiers, 6*(11), 1515–1521.

Xiang, C., et al. (2022b). Ratiometric imaging of butyrylcholinesterase activity in mice with nonalcoholic fatty liver using an AIE-based fluorescent probe. *Journal of Materials Chemistry B, 10*(22), 4254–4260.

Xiang, C., et al. (2025). Ratiometric red-emitting probe with large stokes shift and AIE/ESIPT characteristics for acetylcholinesterase imaging. *Analytical Chemistry, 97*(19), 10139–10144.

Xu, X., et al. (2019). A ratiometric fluorescence probe based on carbon dots for discriminative and highly sensitive detection of acetylcholinesterase and butyrylcholinesterase in human whole blood. *Biosensors and Bioelectronics, 131*, 232–236.

Yang, S.-H., et al. (2017). Discovery of a butyrylcholinesterase-specific probe via a structure-based design strategy. *Chemical Communications, 53*(28), 3952–3955.

Yang, Y., et al. (2020). A near-infrared fluorescence probe for imaging of pantetheinase in cells and mice in vivo. *Chemical Science, 11*(47), 12802–12806.

Yang, Y., et al. (2022a). Real-time fluorescent determination and biological imaging in living models via a butyrylcholinesterase-activated fluorescent probe. *Dyes and Pigments, 206*, 110596.

Yang, Y., et al. (2022b). Diagnosis of Alzheimer's disease and in situ biological imaging via an activatable near-infrared fluorescence probe. *Analytical Chemistry, 94*(39), 13498–13506.

Yoo, S., & Han, M. S. (2019). A fluorescent probe for butyrylcholinesterase activity in human serum based on a fluorophore with specific binding affinity for human serum albumin. *Chemical Communications, 55*(97), 14574–14577.

Zeng, K., et al. (2025). Real-time monitoring and in vivo visualization of acetylcholinesterase activity with a near-infrared fluorescent probe. *Journal of Pharmaceutical Analysis*101204.

Zhang, F., et al. (2022b). A combined "AIE+ ESIPT" fluorescent probe for detection of lipase activity. *Bioorganic Chemistry, 128*, 106026.

Zhang, J., et al. (2024). Recent advances in Alzheimer's disease: Mechanisms, clinical trials and new drug development strategies. *Signal Transduction and Targeted Therapy, 9*(1), 211.

Zhang, J., et al. (2025). A new near-infrared fluorescence probe for highly selective and sensitive detection and imaging of Butyrylcholinesterase in Alzheimer's disease mice. *Talanta, 285*, 127377.

Zhang, P., et al. (2021). AND-logic strategy for accurate analysis of Alzheimer's disease via fluorescent probe lighted up by two specific biomarkers. *Analytical Chemistry, 93*(32), 11337–11345.

Zhang, S., et al. (2017). Synergistic inhibitory effect of traditional Chinese medicine astragaloside IV and curcumin on tumor growth and angiogenesis in an orthotopic nude-mouse model of human hepatocellular carcinoma. *Anticancer Research, 37*(2), 465–473.

Zhang, W.-d, et al. (2022a). A far-red/near-infrared fluorescence probe with large Stokes shift for monitoring butyrylcholinesterase (BChE) in living cells and in vivo. *Analytica Chimica Acta, 1235*, 340540.

Zhang, X., et al. (2018). A review on the hybrids of hydroxycinnamic acid as multi-target-directed ligands against Alzheimer's disease. *Bioorganic & Medicinal Chemistry, 26*(3), 543–550.

Zhao, C., et al. (2022). Near-infrared fluorescent probe for in vivo monitoring acetylcholinesterase activity. *Sensors and Actuators B: Chemical, 360*, 131647.

Zhu, X.-X., Gao, F., & Wan, L.-X. (2024). Design, synthesis and anti-Alzheimer's disease activity evaluation of C-3 arylated huperzine A derivatives. *Fitoterapia, 177*, 106141.

Žužek, M. C. (2024). Advances in cholinesterase inhibitor research—An overview of preclinical studies of selected organoruthenium (II) complexes. *International Journal of Molecular Sciences, 25*(16), 9049.

CHAPTER SEVEN

Intestinal microbiota in neurodegeneration and ageing: Mechanisms, pathways, and therapeutic interventions

Gopal Khodve, Shaswat Raval, and Sugato Banerjee*
Department of Pharmacology and Toxicology, National Institute of Pharmaceutical Education and Research, Kolkata, India
*Corresponding author. e-mail address: sbanerjee@niperkolkata.ac.in

Contents

Abstract

The human gut microbiota represents a complex ecosystem of trillions of microorganisms with profound implications for neurological health. Emerging evidence demonstrates that dysbiosis, an imbalance in microbial composition and function, plays a crucial role in the pathogenesis of neurodegenerative diseases and age-related cognitive decline. This chapter summarizes current knowledge of the microbiota-gut-brain axis (MGBA) and elucidates how intestinal microbes and their metabolites communicate with the central nervous system via neural, immune, endocrine,

International Review of Neurobiology, Volume 188
ISSN 0074-7742, https://doi.org/10.1016/bs.irn.2026.05.015

and metabolic pathways. We examine the mechanistic links between gut dysbiosis and specific neurodegenerative conditions, including Alzheimer's disease (AD), Parkinson's disease (PD), Amyotrophic lateral sclerosis (ALS), Huntington's disease (HD), and Multiple sclerosis (MS). Furthermore, we explore age-related changes in the microbiota and their contributions to neuroinflammation, immunosenescence, and cognitive decline. Finally, we evaluate therapeutic interventions targeting the microbiota, including probiotics, prebiotics, synbiotics, and dietary modulation as promising strategies to prevent and ameliorate neurodegenerative pathology. The chapter provides a comprehensive summary of how microbiota-targeted approaches may delay ageing and neurodegeneration.

1. Introduction

The gut microbiota, comprising approximately 37.2 trillion microbial cells representing over 1000 bacterial species, constitutes a "metabolic organ" integral to host physiology. For decades, neuroscience and microbiology remained largely separate disciplines; however, contemporary research has unveiled intimate connections between intestinal microbes and brain function (Sastre et al., 2024). The discovery of the microbiota-gut-brain axis has revolutionized our understanding of neurodegenerative disease aetiology, shifting focus from purely cell-autonomous mechanisms to encompass the dynamic interactions between the microbiota and host nervous system (Chen, Wang, Li, & Zhang, 2025).

Neurodegenerative diseases represent a growing public health burden, with the World Health Organization projecting a doubling of dementia prevalence by 2050. Similarly, Parkinson's disease incidence increases exponentially with age, affecting approximately 1 % of individuals over 60 years (Wang, Jiang, Yang, Meng, & Zhang, 2024). Conventional therapeutic approaches targeting amyloid-β, tau pathology, or dopaminergic deficits have yielded limited clinical success, highlighting the need for novel intervention strategies. The emerging recognition of dysbiosis as a modifiable risk factor offers unprecedented opportunities for early intervention and disease prevention (Pini, Imbimbo, Griffa, Allali, & Pomara, 2026).

This chapter summarizes current evidence on how intestinal microbiota composition and function influence neurodegeneration through multiple interconnected pathways. We examine the mechanisms by which dysbiosis compromises intestinal barrier integrity, promotes neuroinflammation, disrupts neurotransmitter synthesis, alters metabolite production, and facilitates prion-like propagation of pathogenic proteins from the periphery

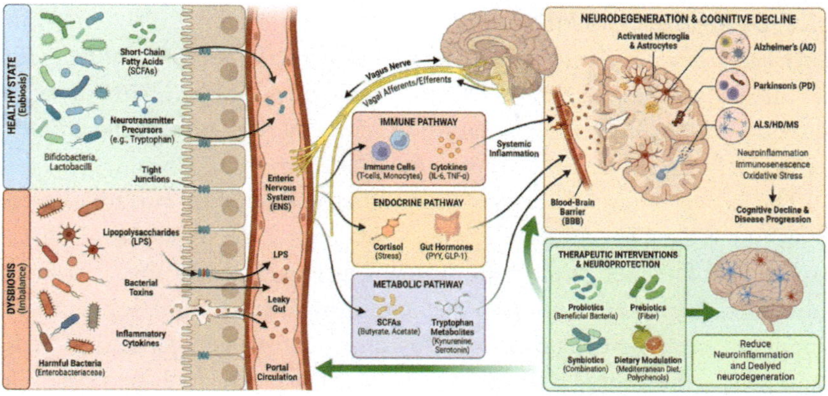

Fig. 1 Connecting intestinal dysbiosis with neuroinflammation, blood–brain barrier dysfunction, and cognitive decline and microbiota-targeted therapies against neurodegenerative disorders.

to the brain. Subsequently, we review dysbiotic patterns specific to individual neurodegenerative diseases and ageing-related cognitive decline, followed by an evaluation of microbiota-targeted therapeutic interventions supported by preclinical and clinical evidence (Fig. 1).

2. The gut-brain axis: Structure and function

2.1 Role of gut microbes

The gut microbiota functions as a sophisticated biological interface between the environment and the host CNS. The microbiota influences neural development, synaptic plasticity, and adult cognitive function through multiple mechanisms. Commensal bacteria produce neurotrophic factors, modulate immune tolerance at mucosal sites, and generate metabolites with neuroactive properties (Loh et al., 2024). The composition of the bacterial community directly determines metabolic output. Key microbial taxa include:

2.1.1 Beneficial bacteria

Bifidobacterium, *Lactobacillus*, *Akkermansia muciniphila*, *Faecalibacterium prausnitzii*, and *Roseburia* species producers of short-chain fatty acids (SCFAs), tryptophan metabolites, and immunoregulatory compounds. These taxa maintain intestinal barrier integrity, produce anti-inflammatory cytokines, and support mucus layer formation (Kumar et al., 2025).

2.1.2 Dysbiosis-associated bacteria

Enterobacteriaceae, Escherichia coli, Clostridioides difficile, and pro-inflammatory *Clostridium* species generate lipopolysaccharides (LPS), trimethylamine N-oxide (TMAO), and secondary bile acids, thereby promoting intestinal permeability and systemic inflammation (Shen et al., 2025).

The composition of the gut microbiota is shaped by multiple factors, including diet, age, antibiotic exposure, stress, sleep patterns, physical activity, and genetic predisposition. During ageing, the microbiota undergoes characteristic shifts: a decline in diversity, a reduction in SCFA-producing taxa, an expansion of pro-inflammatory genera, and increased intestinal permeability, a phenomenon termed age-related dysbiosis (Hou et al., 2022).

2.2 Role of gut metabolites

Microbial metabolites represent the principal mechanism through which the microbiota influences distant organ function. These postbiotics include:

Short-chain fatty acids (SCFAs): Butyrate, propionate, and acetate are produced via bacterial fermentation of dietary fiber, reaching concentrations of 100–200 mM in the colonic lumen. Butyrate serves as the primary energy substrate for colonocytes, while all SCFAs are absorbed into systemic circulation, where they exert neuroactive and immunomodulatory effects (O'Riordan et al., 2022).

Butyrate acts as a histone deacetylase (HDAC) inhibitor, promoting chromatin remodeling and altering gene expression to enhance anti-inflammatory responses. At the blood-brain barrier (BBB), butyrate preserves tight junction protein expression (claudin-5, occludin, ZO-1), maintaining barrier integrity. In the brain, butyrate enhances mitochondrial function, reduces reactive oxygen species (ROS) production, and promotes the differentiation of regulatory T cells (Rasouli-Saravani et al., 2023).

Tryptophan metabolites: Dietary tryptophan undergoes bacterial metabolism via multiple pathways:

- The aryl hydrocarbon receptor (AhR) pathway: bacterial conversion to indole derivatives (indole-3-propionic acid, indole-3-aldehyde) activates AhR, promoting IL-22 production and intestinal barrier maintenance (Ghiboub et al., 2020).
- The kynurenine pathway: bacterial tryptophanase produces kynurenine, an AhR agonist that modulates immune tolerance (Seo & Kwon, 2023).

- Neurotransmitter precursor: tryptophan serves as a substrate for serotonin synthesis, with microbial tryptophanase activity influencing serotonin bioavailability (Tang, Krushelnycky, Shaqo, & Cho, 2025).

Neurotransmitters and neuromodulators: Lactobacillus and Bifidobacterium species synthesize GABA, the principal inhibitory neurotransmitter; Bacillus and Serratia produce dopamine; *Escherichia*, *Bacillus*, and Saccharomyces generate serotonin. These microbial neurotransmitters modulate intestinal motility and, through vagal signaling, influence CNS function and mood (Taj et al., 2025).

Bile acid metabolites: Bacterial β-glucuronidase deconjugates primary bile acids into secondary bile acids, which activate farnesoid X receptor (FXR) and Takeda G-protein receptor 1 (TGR5), modulating immune responses and intestinal permeability (Cai, Rimal, Jiang, Chiang, & Patterson, 2022).

Lipopolysaccharide (LPS): Outer membrane component of Gram-negative bacteria that, when translocated across a compromised intestinal barrier, acts as a pathogen-associated molecular pattern (PAMP), activating toll-like receptor 4 (TLR4) and triggering systemic and CNS inflammation (Page, Kell, & Pretorius, 2022).

2.3 Pathways involved in the gut-brain axis

The MGBA encompasses four principal communication systems:

2.3.1 Neural pathways

The Vagus Nerve: The primary neural conduit of the gut-brain axis, the subdiaphragmatic vagus nerve (cranial nerve X) forms bidirectional connections between the enteric nervous system (ENS) and brainstem. Vagal afferent fibers, innervating the intestinal mucosa, transmit chemical signals (bacterial metabolites, immune mediators) to the nucleus of the solitary tract in the medulla. Efferent vagal fibers carry CNS signals regulating gastrointestinal motility and secretion (Aljeradat et al., 2024).

Emerging evidence indicates that pathogenic signals can be undergo retrograde transported via vagal projections. In Parkinson's disease, misfolded α-synuclein originating in enteric neurons can be transported to the substantia nigra, contributing to dopaminergic neuron loss. Preclinical studies demonstrate that vagotomized animals exhibit reduced PD incidence and severity, supporting the critical role of vagal signaling in α-synuclein propagation (Kim et al., 2019).

The Enteric Nervous System (ENS): Comprising 500 million neurons organized into myenteric and submucosal plexuses, the ENS regulates gastrointestinal motility, secretion, and immune function. ENS neurons directly interact with the microbiota, responding to bacterial antigens and metabolites through pattern recognition receptors (PRRs), including TLRs and nucleotide-binding oligomerisation domain (NOD)-like receptors (NLRs) (Wang, Cheng, & Yang, 2025).

2.3.2 Immune pathways

Intestinal Barrier Integrity: The intestinal epithelium, comprising a single layer of columnar enterocytes connected by tight junctions, provides the primary physical barrier against microbial translocation. Dysbiosis compromises barrier function through multiple mechanisms:

- Reduced SCFA production decreases energy availability for colonocyte metabolism and histone acetylation of barrier-maintaining genes (Silva, Bernardi, & Frozza, 2020)
- Reduced mucus-layer thickness (controlled by Akkermansia *muciniphila*-dependent glycosylation) permits direct bacterial contact with epithelium (Paone et al., 2026)
- Decreased IL-22 production (dependent on AhR-activating bacterial metabolites) impairs antimicrobial peptide synthesis (Mar et al., 2023)
- Increased zonula occludens-1 (ZO-1) degradation via inflammatory protease production (Yu, He, & Xie, 2023)

Mucosal Immune System: Dysbiosis skews mucosal immune responses toward proinflammatory Th17 cell differentiation. In healthy individuals, commensal bacteria promote the differentiation of regulatory T cells (Tregs) by producing SCFAs and presenting microbial polysaccharide A (PSA). A dysbiotic microbiota enriched in pathobionts and depleted of SCFA producers shifts the balance toward pathogenic Th17 cells and pathogenic innate lymphoid cells, promoting mucosal inflammation (Shim, Ryu, Jo, & Hong, 2023).

Systemic Immunity: LPS translocate across a compromised intestinal barrier and enters systemic circulation, activating TLR4 on circulating monocytes, dendritic cells, and B cells, triggering the release of pro-inflammatory cytokines (TNF-α, IL-6, IL-1β). This "metabolic endo toxemia" promotes activation of peripheral immune cells and their infiltration into the CNS, contributing to neuroinflammation (Ghosh, Wang, Yannie, & Ghosh, 2020).

2.3.3 Endocrine pathways

Intestinal hormone secretion: Enteroendocrine cells (comprising ~1 % of the intestinal epithelium) produce hormones, including glucagon-like peptide-1 (GLP-1), peptide YY (PYY), and cholecystokinin (CCK), in response to microbial metabolites, particularly SCFAs. These hormones regulate glucose homeostasis, appetite, and stress responses. GLP-1, for instance, crosses the BBB and enhances neurogenesis, synaptic plasticity, and neuroprotection effects that are reduced in dysbiotic states characterized by decreased GLP-1 secretion (Chao, Coleman, Keating, & Martin, 2025).

Hypothalamic-pituitary-adrenal (HPA) axis: Dysbiosis impairs HPA axis regulation, leading to elevated basal cortisol and blunted stress responsiveness. Dysbiotic microbiota-derived LPS hyperactivates TLR4 on CNS dendritic cells and microglia, promoting corticotrophin-releasing hormone (CRH) production and downstream HPA activation. Chronic HPA axis dysregulation contributes to neuroinflammation and neurodegeneration (Sheng et al., 2021).

2.3.4 Metabolic pathways

Microbial metabolite circulation: SCFA and tryptophan metabolites, absorbed across the intestinal epithelium, enter portal circulation and distribute systemically. Acetate crosses the BBB at the highest efficiency, while propionate and butyrate rely on monocarboxylate transporters. Within the brain parenchyma, SCFAs modulate glial cell function, HDAC activity, and neuronal gene expression (Lee, Lee, Song, & Choi, 2024).

3. Role of intestinal microbiota in neurodegenerative diseases

3.1 Alzheimer's disease

Alzheimer's disease (AD) is a progressive neurodegenerative disorder, clinically characterized by progressive cognitive decline and diminishing cognitive functions such as memory and language (Alzheimer's disease facts and figures, 2016; Yadav et al., 2023). Pathological hallmarks of AD are chronic neuroinflammation, extracellular deposition of amyloid-β (Aβ) plaques, and intracellular accumulation of neurofibrillary tangles composed of hyperphosphorylated tau. Amyloid beta (Aβ), mainly Aβ40 and Aβ42, that are generated through sequential proteolytic cleavage of amyloid precursor protein (APP), readily undergo self-aggregation to form oligomers and fibrillary assemblies within the central nervous system, which further initiates cascades of synaptic dysfunction, inflammatory activation, and neuronal toxicity (Khatoon, Kalam, Rashid, & Bano, 2023). Growing evidence implicates microbial dysbiosis and infection as contributing factors in AD pathogenesis. Disruptions in the gut microbiome and microbial-derived products have been shown to influence neuroinflammatory signaling and promote amyloid genic processes, suggesting that host microbe interactions may play a major role in the initiation and progression of AD.

A large-scale meta-analysis integrating 378 healthy controls and 427 individuals with Alzheimer's disease (AD) across 11 independent studies has discovered significant taxonomic alterations in the gut microbiota of AD patients. Precisely, abundance of *Proteobacteria*, *Bifidobacterium*, and *Phascolarctobacterium* were found to have increased, and amounts of the genera *Firmicutes, Clostridiaceae, Lachnospiraceae,* and *Rikenellaceae* were found to be decreased (Hung, Chang, Huang, Nouchi, & Cheng, 2022). It is also discovered that gut microbiome composition is altered even at the preclinical stages of AD, with specific microbial taxa correlating with amyloid-β (Aβ) and tau biomarkers in the brain, as well as with plasma Aβ42/Aβ40 ratios. These findings propose that gut microbiome perturbations may lead to neurodegeneration and arise early in the disease continuum (Ferreiro et al., 2023; Sheng et al., 2022). Mechanistic studies provide further insights into how central Aβ pathology affects gut homeostasis. Experimental administration of Aβ into the lateral ventricle induces cognitive impairment as well as suppresses the vagus nerve-mediated cholinergic anti-inflammatory pathway by downregulating M1 acetylcholine receptors in the brain (Qian, Liu, Chen, Chen, & Tang, 2022). This central inhibition further reduces acetylcholine secretion and decreases α7 nicotinic acetylcholine receptor expression in intestinal macrophages (Qian et al., 2022). These alterations shift macrophages toward a pro-inflammatory phenotype, causing enteric neuronal damage, disruption of the gut mucosal barrier, increased amyloidogenic processes, and gut dysbiosis (Qian et al., 2022). Collectively, these observations support a direct, bidirectional influence of brain Aβ pathology on the integrity of the peripheral gut ecosystem.

Perturbations in the gut microbiota contribute to disease progression and exacerbate AD pathology. Studies have found that Fecal microbiota transplantation (FMT) from AD patients or transgenic AD mouse models, such as 5 ×FAD, into healthy recipient mice intensifies Aβ production and deposition as well as elevated neuroinflammatory responses, resulting in cognitive deficits; when no such disabilities were found on FMT from healthy controls

(Chen et al., 2022; Kim et al., 2021; Zhang et al., 2023). In reverse, reductions in gut microbiota achieved either by germ-free rearing or antibiotic treatment ameliorate Aβ and tau accumulation as well as reduce cognitive impairments in certain AD mouse models such as APP/PS1, Tau/APOE4 (Dodiya et al., 2022; Harach et al., 2017; Seo et al., 2023). These crucial discoveries underscore the fundamental role of the gut microbiome and its metabolites in modulating microglial phenotypes and function, thereby defining the burden of Aβ pathology in the AD brain.

Growing evidence has also demonstrated the interplay between gut microbiota and microglia in AD. In a study using the triple transgenic AD (3xTg-AD) mouse model, the gut microbiome was found to influence the development of AD pathologies, including Aβ plaque formation, hyperphosphorylation of tau, synaptic dysfunction, and microglial activation. These results were further supported by the fact that Specific Pathogen Free (SPF) 3xTg-AD mice displayed more severe AD pathologies than Germ Free (GF) 3xTg-AD mice. Notably, FMT from AD patients to GF 3xTg-AD mice restored the main AD pathologies and microglial activation (Loh et al., 2024).

In the triple-transgenic (3xTg-AD) mouse model, which displays all the key AD pathologies, including Aβ deposition, hyperphosphorylation of tau, synaptic dysfunction, and microglial activation, the gut microbiome configuration significantly impacts disease severity. Specific-pathogen-free (SPF) variants of these triple transgenic mouse models exhibited more severe pathological features than their germ-free (GF) counterparts, indicating that microbial signals exacerbate neurodegeneration in AD. Notably, faecal microbiota transplantation (FMT) from AD patients into GF 3xTg-AD mice restores major AD pathologies and reactivates microglial responses, further demonstrating that gut microbial communities can modulate neuroinflammatory pathways and accelerate AD progression (Chen et al., 2022). Similar results have been concluded in various studies. Studies using the APP/PS1–21 mouse model have shown that germ-free (GF) conditions provide notable protection against amyloid-β accumulation and microglial activation. Interestingly, this protective effect is significantly reduced when faecal microbiota from 12-month-old conventionally raised APP/PS1–21 mice are transplanted into 4-month-old GF APP/PS1–21 recipients (Harach et al., 2017). These findings highlight the gut microbiota's capacity to modulate neuroinflammatory processes and to stimulate AD-related pathology.

To determine the temporal effects of the gut microbiome on AD progression, Dodiya et al. administered a short 7-day antibiotic treatment from postnatal day 14 to day 21 in APP/PS1–21 mice and evaluated them at 9 weeks of age. Consistent with earlier findings, short-term microbiota depletion significantly attenuated Aβ pathology, decreased the population of disease-associated microglia (DAM), and downregulated microglial sensome gene expression in male APP/PS1–21 mice. These protective effects were fully reversed following fecal microbiota transplantation from age-matched mice, reaffirming the causal role of gut microbial communities in early amyloidogenesis (Dodiya et al., 2022). Interestingly, depletion of microglia using the colony-stimulating factor 1 receptor (CSF1R) inhibitor PLX5622 eliminated the previously observed positive effects of antibiotic treatment, suggesting that microglia are vital intermediaries in mediating microbiota–gut–brain interactions. These findings, hence, prove that microglia are the key effector cells through which gut microbiota interacts and influences AD pathologies (Dodiya et al., 2022).

Gut microbiome's role in promoting microglial activation is also observed in models of tau-associated neurodegeneration. A current study has revealed a functional link between microbial composition, tau pathology, and APOE genotype in Alzheimer's disease. In this study, Seo and colleagues manipulated P301S tau transgenic mice to express either the human APOE3 or APOE4 gene and maintained the manipulated animals under standard specific-pathogen-free (SPF) conditions and in germ-free (GF) conditions. Germ-free P301S mice carrying human APOE4 (TE4 mice) displayed markedly reduced neurodegenerative outcomes, including diminished brain volume loss and showed lower levels of microglial and astrocytic activation compared with conventionally raised TE4 mice. Importantly, transferring faecal microbiota from sex-matched conventionally housed TE4 donors into GF TE4 recipients abolished these protective effects (Seo et al., 2023). Together, these data indicate that gut microbial signals are essential contributors to the onset and amplification of tau-driven neurodegenerative processes.

Reports indicate that chronic *Helicobacter pylori* infection may promote the production of pro-inflammatory cytokines and amyloidogenic peptides in individuals with Alzheimer's disease (Roubaud-Baudron et al., 2019). In vitro studies have shown that exposure to *H. pylori* filtrates induces abnormal tau hyperphosphorylation (Wang et al., 2015). Several other microorganisms also generate extracellular amyloids. For example, *Escherichia coli* produces curli fibres composed largely of the CsgA subunit, while other microbial species synthesize structurally analogous amyloids, including phenol-soluble modulins from *Staphylococcus aureus*, TasA from *Bacillus subtilis*, chaplins from *Streptomyces coelicolor*, FapC from *Pseudomonas fluorescens*, MccE492 from *Klebsiella pneumoniae*, and CsgA from *Salmonella spp* (Schwartz, Syed, Stephenson, Rickard, & Boles, 2012). Harach and colleagues, in an experiment on APP-transgenic mouse models, discovered that the gut microbiome of these transgenic mice was significantly altered and, based on these findings, concluded that these microbial alterations contributed to the pathogenesis of AD neurodegeneration. On comparing traditionally raised APP mice and germ-free reared APP mice, a significant reduction in cerebral amyloid deposition in germ-free reared mice was observed. (Harach et al., 2017). Also, recent studies have revealed that *Chlamydia pneumoniae* infection of astrocytes stimulates Aβ production and thereby potentiates neurotoxicity (Al-Atrache, Lopez, Hingley, & Appelt, 2019).

Based on this dynamic interplay between gut microbiome and brain, probiotic interventions have been proposed as a potential therapy for AD. administered *Lactobacillus* and *Bifidobacterium* strains in an Aβ-induced mouse model. They discovered that deficits in learning and memory were significantly ameliorated. Correspondingly, Wang, Liang, and colleagues studied that administration of *Lactobacillus fermentum* NS9 and *Lactobacillus helveticus* NS8 improved spatial memory performance under chronic restraint stress and reduced ampicillin-induced cognitive impairment (Liang et al., 2015). Akbari et al. (2016) reported that mini-mental state examination (MMSE) scores improved significantly following a 12-week intervention involving fermented milk enriched with *Lactobacillus* and *Bifidobacterium* species in a randomized clinical trial comprising 60 patients with Alzheimer's disease (Akbari et al., 2016). Clinical studies have also shown that eradicating *H. pylori* infection with standard triple-therapy regimens improves cognitive performance in AD patients (Kountouras et al., 2009). Altogether, these observations suggest that manipulating the gut microbiota, either with probiotics or antibiotics, may lead to more effective therapeutic strategies for the management of AD (Abraham et al., 2019). (Fig. 2).

3.2 Parkinson's disease

Parkinson's disease (PD) is a neurodegenerative disorder characterised by the intracellular accumulation of α-synuclein (α-syn) aggregates (also called Lewy bodies) in the central nervous system (CNS) and the enteric nervous system (ENS) (Vascellari et al., 2020). It is increasingly acknowledged that

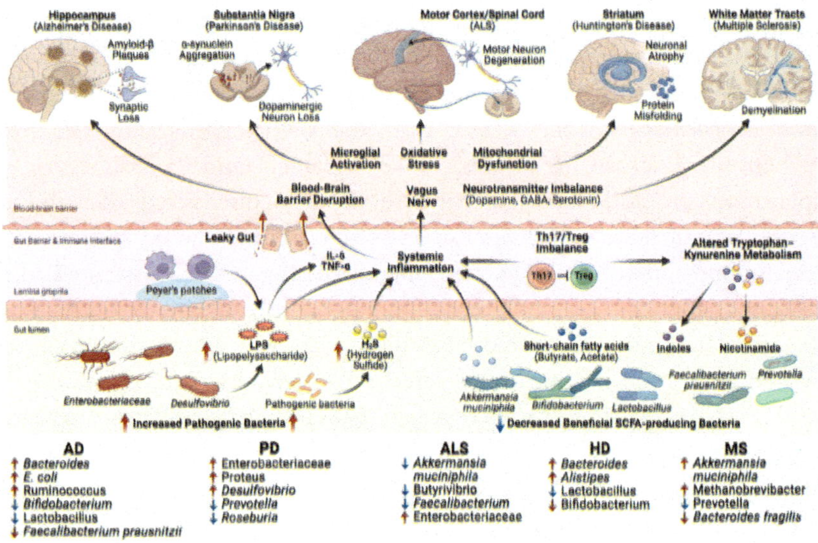

Fig. 2 Molecular mechanisms involved in gut dysbiosis associated neurodegenerative disorders and role of intestinal microbes.

the neurodegeneration in PD co-occurs with serious gastrointestinal disturbances. However, it isn't clear whether these alterations occur before or after the emergence of motor deficits (Abbott et al., 2001). Constipation is the most prevalent premotor manifestation of PD, affecting over 70 % of patients, and may contribute to disease pathogenesis more than a decade prior to the appearance of classical motor symptoms (Fasano, Visanji, Liu, Lang, & Pfeiffer, 2015). Notably, the accumulation and aggregation of misfolded α-synuclein (α-syn) within the gut, accompanied by enteric neurodegeneration, appear to begin as early as 20 years before detectable central nervous system (CNS) pathology, lending strong support to the "gut-to-brain" hypothesis (Cersosimo & Benarroch, 2012). Bidirectional communication between the intestine and the brain is mediated by the intestinal microbiota through immunological, neuroendocrine, and direct neurochemical pathways. In this context, gut microorganisms and their metabolic products may influence PD pathogenesis by fostering a proinflammatory milieu within the gut. Moreover, dysbiosis-associated intestinal inflammation is increasingly implicated in facilitating α-syn misfolding (Houser & Tansey, 2017).

There is substantial evidence that the gut microbiome in PD patients is disturbed (Boertien, Pereira, Aho, & Scheperjans, 2019), accompanied by alterations in SCFAs (Unger et al., 2016). Lewy body pathology in PD is found in structures of the central nervous system (CNS) as well as in the peripheral autonomic and enteric nervous systems (PANES). PD patients frequently exhibit non-motor symptoms, including signs and symptoms of gastrointestinal dysmotility (e.g. delayed gastric emptying, constipation). Lewy body pathology in the ENS might represent the pathohistological correlate of gastrointestinal symptoms in PD (Unger et al., 2016). LP (α-synuclein aggregates) exists in the olfactory bulb, spinal cord, peripheral autonomic ganglia, submandibular gland, cardiac nerves, and ENS before it appears in the substantia nigra, pars compacta, and before neuronal loss (Warnecke, Schäfer, Claus, Del Tredici, & Jost, 2022). The study by Sampson et al. demonstrated that the development of α-synuclein pathology, microglial activation, and motor deficits in α-synuclein-overexpressing (ASO) mice is influenced by the gut microbiome. This is evident in SPF ASO mice, which exhibit greater PD pathology than their GF and antibiotic-treated counterparts. Interestingly, the main PD symptoms, including α-synuclein-mediated motor dysfunction, were restored in germ-free ASO mice after transplantation of faecal matter from PD patients (Sampson et al., 2016). In another study involving transgenic rat models overexpressing α-synuclein, age-dependent progression of gut dysbiosis, and brief exposure to antibiotics, further diminished α-synuclein expression in the forebrain (Singh et al., 2023). In parallel, FMT derived from 1-methyl-4-phenyl-1,2,3,6-tetrahydropyridine (MPTP)-treated mice caused PD-like symptoms in healthy recipients. On the other hand, FMT from healthy donors to MPTP-induced PD models resulted in alleviated gut dysbiosis and attenuated multiple PD phenotypes (Sun et al., 2018). In another study, normally raised mice developed gastrointestinal dysfunction and motor abnormalities on chronic exposure to rotenone. Whereas no such deficits were observed in germ-free models (Bhattarai et al., 2021). Altogether, these studies emphasize the pivotal role of gut microbiota in modulating PD pathology and related outcomes.

In the ASO mice model of PD, increasing SCFA production by giving prebiotic supplements has resulted in the development of neuroprotective phenotypes of microglia along with decreased motor impairments and diminished α-synuclein accumulation within the substantia nigra (Abdel-Haq et al., 2022). In a similar pattern, administering the probiotic strain Bifidobacterium breve CCFM1067 to MPTP-induced PD mice decreased microglial activation and oxidative stress and improved motor performance (Li et al., 2022). Also, many studies have reported neuroprotective effects of sodium butyrate in MPTP-induced PD models, mainly by suppressing microglial hyperactivation through inhibition of TLR4/MyD88/NF-κB and

MAPK pathways. (Guo et al., 2023; Hou et al., 2021; Srivastav et al., 2019). SCFA supplementation in germ-free ASO mice resulted in increased microglial activation, α-synuclein deposition, and the onset of motor impairments. On the other hand, SCFA depletion in MPTP-induced PD mice improved motor performance and reduced microglial and astrocyte activation. (Sun et al., 2018; Zhong et al., 2021). This bidirectional influence of SCFAs on PD is of growing interest to researchers and is being studied to elucidate its mechanisms.

3.3 Amyotrophic lateral sclerosis

Amyotrophic lateral sclerosis (ALS) is a neuronal disorder defined by the progressive degeneration of motor neurons in both the spinal cord and the brain. Clinical symptoms commonly include muscle fasciculations, cramps, rigidity, impaired coordination, and progressive muscle weakness. In severe cases, difficulty in breathing, speech and swallowing is also seen. ALS may be sporadic or familial. Sporadic ALS accounts for the majority of cases and occurs due to unknown causes, while familial ALS is due to inherited genetic mutations (Toepfer et al., 1997).

The G93A SOD1 (superoxide dismutase 1) transgenic mouse model is most commonly used to study the effects of gut microbiota on ALS pathophysiology. In this ALS mouse model, disease progression can be outlined in three phases, viz, (i) Pre-onset phase that occurs between day 60–70, (ii) Onset phase observed between days 90–100, and (iii) Progressive phase spanning from day 120–130 (Zhou et al., 2015). These mice display increased intestinal permeability and compromised blood-brain barrier (BBB) integrity during the pre-onset phase. Increased intestinal permeability results from impaired expression of tight junction proteins, such as E-cadherin and zonula occludens-1 (ZO-1). Faecal analyses from SOD1^G93A mice showed diminished levels of butyrate-producing bacteria *Peptostreptococcus* and *Butyrivibrio fibrosolvens*, along with increased concentrations of pro-inflammatory cytokines- interleukin-17 (IL-17), in both serum and intestinal tissues (Fang et al., 2016; Wu, Yi, Zhang, Zhou, & Sun, 2015). In a study, Zhang and his team discovered that adding 2% butyrate to the drinking water of SOD1 transgenic mice can delay the progression of ALS. They thought this was a consequence of the restoration of *B. fibrosolvens* abundance and further improvement of intestinal barrier integrity (Zhang et al., 2017). Interestingly, faecal samples from ALS patients showed reduced levels of butyrate-producing *Anaerostipes*, *Oscillibacter*, and *Lachnospira,* and increased levels of ethanol-producing bacteria, *Dorea* (Fang et al., 2016).

Studies have shown that ALS patients have altered plasma levels of lipopolysaccharide (LPS) and β-methylamino-L-alanine (BMAA). These freely circulating LPA and BMAA can penetrate the BBB, leading to neuroinflammation and impaired BBB integrity. (Bengmark, 2013; Catanzaro et al., 2015). Impaired BBB appears to be a persistent feature across human and animal ALS models, underscoring its pivotal role in disease onset and progression (Garbuzova-Davis & Sanberg, 2014).

3.4 Huntington's disease

Huntington's disease is an inherited neurodegenerative disorder that results in progressive motor decline, cognitive dysfunction, and neuropsychiatric symptoms (Walker, 2007). The cause of HD is the expansion (often age-dependent) of GAC sequence repeats in exon 1 of the Huntington (HTT) gene, which is ubiquitously expressed. This expanded GAC repeat leads to the production of an abnormal polyglutamine (polyQ) tract, which further leads to an increase in the amyloidogenic properties of the exon 1 of the HTT gene that disrupts the normal functioning of numerous cellular pathways in the nervous system by forming oligomeric assemblies (Bates et al., 2015; A novel gene containing a trinucleotide repeat that is expanded and unstable on Huntington's disease chromosomes. The Huntington's Disease Collaborative Research Group, 1993). Because the HTT gene is ubiquitously expressed, it is postulated that mHTT (mutant HTT) gene toxicity in peripheral organs may contribute to the neurodegeneration in HD patients (Chuang & Demontis, 2021).

Gut microbiome disruptions have also been observed in mouse models expressing the mutant Huntington gene (mHTTex1) (Kong et al., 2020). For example, R6/1 transgenic HD mice at 12 weeks of age exhibit greater gut microbial diversity, with a higher proportion of *Bacteroidetes* (a Gram-negative bacterium) and a lower proportion of *Firmicutes* (a Gram-positive bacterium). Interestingly, male R6/1 mice exhibit gut dysbiosis accompanied by motor problems despite consuming relatively more food (Kong et al., 2020; Kong et al., 2021). Metabolic studies in these mice have shown higher levels of butanoate (a substance produced by certain gut bacteria) before the emergence of defined motor and cognitive symptoms (Kong et al., 2021). R6/2 mouse model, which has an even longer mutant gene, also shows a higher *Bacteroidetes* to *Firmicutes* ratio, which is linked to weight loss, shorter colons, and a leakier gut. Additionally, R6/1 HD mice exhibit alterations in gut fungi, which may interact with bacterial imbalances (Stan et al., 2020).

Studies on *Drosophila* models engineered to express the mutant huntingtin protein (mHTTex1) only in the nervous system show that these flies develop gut dysbiosis, suggesting that nervous system disruptions may lead to gut dysbiosis (Chongtham et al., 2022). In these flies, interestingly, altering or eliminating certain gut microbes reduces the buildup and harmful effects of mHTTex1, suggesting a two-way relationship between the gut and the brain. Gut dysbiosis may occur due to an altered nervous system, as evidenced by the gradual accumulation of mHTTex1 in neurons of people with Huntington's disease (HD). Inversely, there have been cases in which people have shown signs of gut dysbiosis long before HD symptoms appear, indicating that these changes precede the onset of brain and mental symptoms, just as early immune changes do (Tai et al., 2007; Wasser et al., 2020). *Drosophila* models that ubiquitously express the full-length human huntingtin protein also exhibit gut imbalances, such as an increase in gram-negative bacteria relative to gram-positive species, that occur before any motor problems arise (Chongtham et al., 2022). Gut microbiome assembly begins with the vertical acquisition of bacterial species from the birth canal and matures by adulthood, resulting in the most diverse microbial population. Since mHTT (mutant HTT gene) is active from the embryo stage and causes nervous system developmental problems, it's most likely that it could also affect how a baby's gut bacteria develop over time (Barnat et al., 2020; Ratsika, Cruz Pereira, Lynch, Clarke, & Cryan, 2023). In mouse models of HD, the most noticeable changes in gut bacteria don't appear until around 12 weeks of age, when movement problems usually begin (Kong et al., 2021). In another study using transgenic *Drosophila* models of HD, researchers discovered that colonizing these flies with E. coli accelerated the aggregation of an amyloidogenic HTT 586 (N-terminal 586 amino acids with 120Qs) fragment, which induced motor impairments and shortened the flies' lifespan. Acceleration in the progression of HD pathology was observed due to elevated levels of *Acetobacter,* in addition to E. coli (Chongtham et al., 2022). Additionally, the brains of transgenic *Drosophila* HD models (with the mHTT gene) exhibited substantial reductions in the expression of genes linked to neurodevelopment, neurotransmission, and synaptogenesis when infected with *E. coli*, suggesting that gut disruptions may accelerate mHTT-induced toxicity (Khoshnan, 2024).

Bacteria in the gut have been shown to affect the brain directly and indirectly. Studies have shown that microbes are translocated between the gut and the brain via the vagus nerve in rodent models of neurodegenerative diseases such as AD and PD (Khoshnan, 2024). In HD patients' brain tissue, fungal antigens and bacterial genome sequences have been discovered in postmortem studies. Specifically, DNA sequences from gram-negative bacteria, including *Pseudomonas, Acinetobacter, and Burkholderia,* were detected in HD brains (Alonso, Pisa, & Carrasco, 2019). Bacteria affect brain chemistry indirectly by modulating neuroimmune pathways. For instance, it has been suggested that hyperactivation of toll-like receptor 4 (TLR4) in the gut influences neurodegeneration in HD and PD (Perez-Pardo et al., 2019).

Curli amyloids (functional bacterial amyloids) are the amyloid-like fibril proteins produced by members of *Enterobacteriaceae*, like *E. coli*, *Salmonella*, and *Shigella*. Curil amyloids trigger inflammatory pathways linked to gut, urinary, and autoimmune inflammation. Also, these curil amyloids have been implicated in the pathogenesis of several neurological disorders, such as PD (Matheoud et al., 2019; Wang, Lau, Ma, & Zheng, 2021). Interestingly, inflammation and the induction of autoimmunity by curli amyloids activate the IKK/NF-B pathway, which further exacerbates the binding of mHTT in immune cells (Träger et al., 2014; Tursi et al., 2017). These curil amyloids aggravate the aggregation of the mHTTex1 gene in mammalian cells and lead to significantly reduced survival in *Drosophila* HDD models (Chongtham et al., 2022). Interestingly, yeast prions are found to induce mHTTex1 and influence the formation of toxic oligomers (Gropp, Klaips, & Hartl, 2022).

3.5 Multiple sclerosis

Multiple sclerosis is an autoimmune neuroinflammatory disease characterized by chronic inflammation and demyelination. The presence of inflammatory plaques and focal areas of demyelination in both the grey and white matter of the brain is a key characteristic of multiple sclerosis. The experimental autoimmune encephalomyelitis (EAE) model developed for MS revealed the involvement of CD4+ T lymphocytes in the pathogenesis of the disease, particularly CD4+ T helper 1 (Th1) and T helper 17 (Th17) lymphocytes (Fletcher, Lalor, Sweeney, Tubridy, & Mills, 2010). These inflammatory cells lead to chronic inflammation and related tissue damage through factors such as TNFα. BBB selectivity is also altered, leading to infiltration of certain immune cells that can recognize myelin antigens (CD4+ & CD8+ cells) and to inflammatory demyelination (Frohman, Racke, & Raine, 2006).

The gut microbiota is intimately linked to physiological health, and alterations in its composition or function can trigger immune responses. This immune activation by the gut microbiome is implicated in playing a vital role in neuroinflammatory disorders, such as MS. The pioneering studies using transgenic mice model of MS (expressing myelin-specific T cell receptor) concluded the role of gut bacteria in CNS inflammatory disease when these mice models developed spontaneous EAE in non-sterile environment but not in specific-pathogen free environment (Correale, Hohlfeld, & Baranzini, 2022). In a study using antibiotic treatment to reduce natural gut flora, it was observed that the treatment reduced levels of mesenteric TH17 cells, resulting in reduced EAE severity (Wang & Kasper, 2014). In another study, antibiotic treatment downregulated pro-inflammatory mechanisms and upregulated anti-inflammatory mechanisms, as observed when antibiotic treatment protected against induced EAE (Ochoa-Repáraz et al., 2009). These pioneering studies established the definitive roles of gut commensal bacteria in the progression and severity of inflammatory CNS diseases like MS.

Cosorich et al. compared RR-MS patients gut microbiota with that of healthy individuals using 16S ribosomal sequencing (Cosorich et al., 2017). This study showed that RNA sequencing of microbiota isolated from the gut of RR-MS patients revealed an increase in the phylum Firmicutes and a decrease in the phylum. Also, *Prevotella,* which produces propionate, an anti-inflammatory metabolite, was found to be decreased, while *Streptococcus mitis (S. mitis)* and *Streptococcus oralis* were found to have increased (Schwiertz et al., 2010). Decreased *Prevotella* levels were also linked to the expansion of Th17 cells and disease progression in RR-MS (Miyake et al., 2015). In another study conducted by Miyake et al., a reduction in the genus *Clostridium* (Clostridia clusters XIV and IV) was observed in patients with RR-MS compared with healthy subjects (Miyake et al., 2015). *Clostridium* bacteria produce short-chain fatty acids, which in turn promote the development of T regulatory (Treg) cells. Hence, this regulatory T cell production is compromised, leading to an unregulated immune system in MS due to decreased levels of Clostridium-produced SCFAs (Arpaia et al., 2013).

Very early studies on human subjects, comparing MS patients with normal controls, revealed that certain bacterial genera, such as *Akkermansia, Prevotella,* and *Methanobrevibacter,* are altered in MS patients, providing the first evidence that components of the human gut microbiota may contribute to CNS-specific autoimmunity (Pröbstel & Baranzini, 2018). Another interesting study involved 34 pairs of monozygotic twins, with one being diagnosed with MS in each pair. Analysis of the composition of gut microbiome of the subjects revealed no significant differences among the diseased and healthy ones except some bacterial genera, such as

Akkermansia, which were increased in untreated individuals with MS. Further, on FMT from MS patients and controls to germ free transgenic spontaneous EAE mice models, the incidence of spontaneous EAE was significantly increased among mice that received bacteria from the MS patients than among those that received bacteria from the healthy twins (Berer et al., 2017). Another large-scale study involving 71 untreated MS patients and 71 healthy ones was performed. Consistent with a previous study, no major differences in microbiome composition were observed between the groups, except for some taxonomic groups. These were MS-associated bacteria *Akkermansia muciniphila* and *Acinetobacter calcoaceticus*. These bacteria were found to induce pro-inflammatory responses in human peripheral blood mononuclear cells and in monocolonized mice. In contrast, *Parabacteroides distasonis*, which induces an anti-inflammatory T cell response, was reduced in the MS population. Also, FMT from MS-diagnosed patients into germ-free mice induced spontaneous EAE and decreased Treg cell numbers and IL-10 levels. These responses were not observed when faecal transplantation was done from healthy individuals (Cekanaviciute et al., 2017). The bacterial genus *Akkermansia* is of particular interest in MS patients. *Akkermansia* bacteria are found to be in abundance at the peak of disease. Several studies, both in vivo and in vitro, have discovered that these bacteria can induce proinflammatory responses (Cekanaviciute et al., 2017). In contrast, an EAE model study found that *Akkermansia* bacteria induced Treg differentiation by activating dendritic cells, thereby ameliorating disease severity (Liu et al., 2019).

A SCFAs-producing *Butyricoccus* was found to be reduced in primary progressive MS, resulting in impaired anti-inflammatory effects due to the bacteria's induction of Treg cells (Reynders et al., 2020). Primary progressive MS is also associated with increased levels of *Enterobacteriaceae* and *Clostridium*, and decreased levels of *Blautea* and *Agathobaculum* (Cox et al., 2021). Across all these studies, the *Akkermansia* bacterial species were found to be increased in all forms of MS (Cox et al., 2021).

4. Role of intestinal microbiota in ageing

Ageing represents the progressive decline in organ function and increasing vulnerability to disease. The gut microbiota undergoes significant age-related alterations that contribute to immunosenescence, inflammation, and cognitive decline in healthy ageing.

4.1 Age-related dysbiosis

In healthy adults, the gut microbiota remains relatively stable over long periods; however, this homeostasis progressively deteriorates with ageing, giving rise to an age-associated dysbiotic state that is now considered a hallmark of advanced age (Tseng & Wu, 2025). This dysbiosis is characterized by a significant reduction in microbial diversity, with alpha diversity declining notably after 65–70 years, reflecting a loss of species richness within individuals (Bradley & Haran, 2024). Concurrently, there is a reversal of the typical adult *Firmicutes/Bacteroidetes* ratio, marked by an expansion of Firmicutes and a contraction of Bacteroidetes (Magne et al., 2020). Ageing is also associated with a pronounced depletion of short-chain fatty acid-producing taxa, *including Faecalibacterium prausnitzii, Roseburia*, and butyrate-producing Clostridium clusters, compromising metabolic and anti-inflammatory functions. Beneficial commensals such as *Bifidobacterium* and *Lactobacillus* decline dramatically, often by 10−100 fold, further weakening colonization resistance and host-microbe mutualism (Loh et al., 2024). In parallel, there is an expansion of pathobionts, particularly members of *Enterobacteriaceae, Proteobacteria*, and potentially pathogenic Clostridia, contributing to low-grade inflammation. Additionally, reduced abundance of *Akkermansia muciniphila* leads to impaired mucus-layer glycosylation and thinning of the intestinal mucus barrier, thereby increasing epithelial vulnerability and susceptibility to inflammation in the ageing gut (Oliveira & Pamer, 2023).

4.2 Age-related dysbiosis and cognitive decline

Emerging evidence demonstrates that aged microbiota causally contributes to age-related cognitive decline:

Faecal microbiota transplantation studies: Transfer of microbiota from naturally aged mice to young recipients reproduces age-related cognitive impairment and hippocampal synaptic loss. Importantly, transfer of microbiota from young animals to aged mice reverses age-associated cognitive deficits, demonstrating that dysbiotic microbiota, not chronological age alone, drives cognitive decline (Li et al., 2025).

***Bifidobacterium pseudolongum* deficiency and microglia-mediated synapse loss:** Detailed mechanistic studies identify B. *pseudolongum deficiency* in aged microbiota as a critical contributor to cognitive decline.*B. pseudolongum produces* indole-3-aldehyde (IAA), an AhR agonist that prevents microglial engulfment of synapses through AhR-dependent mechanisms. Supplementation with B. *pseudolongum* restores synaptic density and reverses cognitive impairment in aged animals (Zhu et al., 2022).

Reduced SCFA production and neuroinflammation: Age-related dysbiosis markedly reduces faecal and colonic SCFA levels. Reduced butyrate impairs BBB integrity, allows systemic endotoxemia, and permits microglial overactivation. In aged animals, SCFA supplementation reduces neuroinflammatory markers (microglial activation, pro-inflammatory cytokines) and improves cognitive function (H. Guo et al., 2025).

Dysregulated tryptophan metabolism: Aged dysbiosis reduces AhR-activating tryptophan metabolites, impairing IL-22 production, intestinal barrier maintenance, and microglia anti-inflammatory polarisation (Gabela, Mthembu, & Hadebe, 2026).

Altered microbial amyloid and amyloid-β interactions: Dysbiotic microbiota enriched in E. *coli and* other Gram-negative bacteria produce elevated curli amyloids. In aged animals, these bacterial amyloids promote Aβ aggregation, cross-seeding with host amyloid pathology (Miller, Bessho, Grando, & Tükel, 2021).

4.3 Sex-specific ageing effects

Recent evidence reveals sex-specific differences in age-related dysbiosis. Aged females exhibit more pronounced dysbiosis and cognitive decline compared to aged males of the same chronological age (Han et al., 2024). These differences reflect estrogen's role in maintaining gut barrier integrity, promoting SCFA-producing bacteria, and regulating immune responses (Kverka & Stepan, 2024). Dysbiosis-driven neuroinflammation in females may partially explain why women comprise 65 % of Alzheimer's disease cases (Korf, Ganesh, & McCullough, 2022).

5. Therapeutic interventions and translational insights

Targeting the microbiota-gut-brain axis has emerged as a promising therapeutic strategy for preventing and mitigating neurodegeneration, with accumulating evidence from preclinical models and early clinical studies supporting the efficacy of probiotics, prebiotics, symbiotics, and dietary modulation (Khan, Khodve, Yadav, Mallick, & Banerjee, 2025; O'Riordan, Moloney, Keane, Clarke, & Cryan, 2025). Prebiotics such as inulin, fructo-oligosaccharides, arabinoxylan oligosaccharides, and galacto-oligosaccharides selectively enhance beneficial taxa, particularly SCFA-producing bacteria, leading to measurable improvements in executive function, memory performance, and physical frailty in ageing populations after sustained supplementation (Dalile et al., 2025; Mallick, Khodve, Ruwatia, & Banerjee, 2025). These cognitive benefits are mechanistically linked to increased SCFA production, underscoring their role in neuroprotection, although inter-individual variability, gastrointestinal intolerance in dysbiotic states, and the need for prolonged intervention remain key limitations. Symbiotics, by combining probiotics with complementary prebiotics, appear to exert synergistic effects on microbial engraftment, metabolite production, and systemic inflammation, with early trials demonstrating superior cognitive and anti-inflammatory outcomes compared to single interventions (Wang, Tian, Fabi, & de Vos, 2025). Among all approaches, long-term dietary modulation, particularly adherence to a Mediterranean-style diet rich in dietary fibre, polyphenols, and fermented foods, exerts the most robust and sustained impact on microbiota composition and function, positioning diet as a central, scalable, and translational strategy for microbiota-mediated neuroprotection (Jiménez-Arroyo, Molinero, & Moreno-Arribas, 2025).

6. Conclusion

The microbiota-gut-brain axis has emerged as a central nexus for neurodegeneration and cognitive ageing. Dysbiosis, characterized by reduced diversity, diminished SCFA production, elevated pro-inflammatory taxa, and compromised intestinal barrier integrity, contributes to neuroinflammation, amyloid/tau pathology, α-synuclein aggregation, and ultimately neuronal loss. Evidence demonstrates that dysbiosis is not merely a consequence of neurodegeneration but rather a contributing causal factor with therapeutic implications. The recognition that the microbiota is a modifiable factor offers unprecedented opportunities for early intervention and disease prevention. Individuals in the early stages of neurodegeneration, such as those with isolated REM sleep behavior disorder (pre-PD), mild cognitive impairment (pre-AD), or MS remission on therapy, may benefit from microbiota-targeted preventive interventions. Probiotics, prebiotics, postbiotic and dietary modulation toward Mediterranean-style patterns, and potentially FMT represent evidence-based approaches that are safe, accessible, and cost-effective.

However, substantial knowledge gaps remain. Mechanistic understanding of how specific microbial taxa influence individual neurodegenerative disease phenotypes requires further investigation. Biomarkers predicting responders to microbiota interventions would enable precision microbiota therapy. Large prospective randomized controlled trials in human populations, particularly employing multi-omics approaches integrating metagenomics, metabolomics, and neuroimaging, are necessary to establish clinical utility. Gender-specific effects, timing of intervention initiation in disease progression, and optimal intervention combinations require further research. Furthermore, harnessing the microbiota-gut-brain axis will likely require multidisciplinary approaches integrating gastroenterology, neurology, gerontology, and immunology. Future therapies may combine microbiota-targeted approaches with conventional neuroprotective pharmacology, physical rehabilitation, cognitive training, and sleep optimization, collectively targeting the multifactorial basis of neurodegeneration.

References

Alzheimer's disease facts and figures. (2016). *Alzheimer's Dement, 12*(4), 459–509. https://doi.org/10.1016/j.jalz.2016.03.001.

Abbott, R. D., Petrovitch, H., White, L. R., Masaki, K. H., Tanner, C. M., Curb, J. D., ... Ross, G. W. (2001). Frequency of bowel movements and the future risk of Parkinson's disease. *57*(3), 456–462. https://doi.org/10.1212/WNL.57.3.456.

Abdel-Haq, R., Schlachetzki, J. C. M., Boktor, J. C., Cantu-Jungles, T. M., Thron, T., Zhang, M., & Mazmanian, S. K. (2022). A prebiotic diet modulates microglial states and motor deficits in α-synuclein overexpressing mice. *Elife, 11*. https://doi.org/10.7554/eLife.81453.

Abraham, D., Feher, J., Scuderi, G. L., Szabo, D., Dobolyi, A., Cservenak, M., ... Radak, Z. (2019). Exercise and probiotics attenuate the development of Alzheimer's disease in transgenic mice: Role of microbiome. *Experimental Gerontology, 115*, 122–131. https://doi.org/10.1016/j.exger.2018.12.005.

Akbari, E., Asemi, Z., Daneshvar Kakhaki, R., Bahmani, F., Kouchaki, E., Tamtaji, O. R., ... Salami, M. (2016). Effect of probiotic supplementation on cognitive function and metabolic status in Alzheimer's disease: A randomized, double-blind and controlled trial. *Frontiers in Aging Neuroscience, 8*, 256. https://doi.org/10.3389/fnagi.2016.00256.

Al-Atrache, Z., Lopez, D. B., Hingley, S. T., & Appelt, D. M. (2019). Astrocytes infected with Chlamydia pneumoniae demonstrate altered expression and activity of secretases involved in the generation of β-amyloid found in Alzheimer disease. *BMC Neuroscience, 20*(1), 6. https://doi.org/10.1186/s12868-019-0489-5.

Aljeradat, B., Kumar, D., Abdulmuizz, S., Kundu, M., Almealawy, Y. F., Batarseh, D. R., ... Weinand, M. (2024). Neuromodulation and the gut-brain axis: Therapeutic mechanisms and implications for gastrointestinal and neurological disorders. *Pathophysiology: the Official Journal of the International Society for Pathophysiology/ISP, 31*(2), 244–268. https://doi.org/10.3390/pathophysiology31020019.

Alonso, R., Pisa, D., & Carrasco, L. (2019). Brain microbiota in Huntington's disease patients. *Frontiers in Microbiology, 10*, 2622. https://doi.org/10.3389/fmicb.2019.02622.

Arpaia, N., Campbell, C., Fan, X., Dikiy, S., van der Veeken, J., deRoos, P., ... Rudensky, A. Y. (2013). Metabolites produced by commensal bacteria promote peripheral regulatory T-cell generation. *Nature, 504*(7480), 451–455. https://doi.org/10.1038/nature12726.

Barnat, M., Capizzi, M., Aparicio, E., Boluda, S., Wennagel, D., Kacher, R., ... Humbert, S. (2020). Huntington's disease alters human neurodevelopment. *Science (New York, N. Y.), 369*(6505), 787–793. https://doi.org/10.1126/science.aax3338.

Bates, G. P., Dorsey, R., Gusella, J. F., Hayden, M. R., Kay, C., Leavitt, B. R., ... Tabrizi, S. J. (2015). Huntington disease. *Nature Reviews Disease Primers, 1*, 15005. https://doi.org/10.1038/nrdp.2015.5.

Bengmark, S. (2013). Gut microbiota, immune development and function. *Pharmacological Research: the Official Journal of the Italian Pharmacological Society, 69*(1), 87–113. https://doi.org/10.1016/j.phrs.2012.09.002.

Berer, K., Gerdes, L. A., Cekanaviciute, E., Jia, X., Xiao, L., Xia, Z., ... Wekerle, H. (2017). Gut microbiota from multiple sclerosis patients enables spontaneous autoimmune encephalomyelitis in mice. *Proceedings of the National Academy of Sciences of the United States of America, 114*(40), 10719–10724. https://doi.org/10.1073/pnas.1711233114.

Bhattarai, Y., Si, J., Pu, M., Ross, O. A., McLean, P. J., Till, L., ... Kashyap, P. C. (2021). Role of gut microbiota in regulating gastrointestinal dysfunction and motor symptoms in a mouse model of Parkinson's disease. *Gut Microbes, 13*(1), 1866974. https://doi.org/10.1080/19490976.2020.1866974.

Boertien, J. M., Pereira, P. A. B., Aho, V. T. E., & Scheperjans, F. (2019). Increasing comparability and utility of gut microbiome studies in Parkinson's disease: A systematic review. *Journal of Parkinson's Disease, 9*(s2), S297–s312. https://doi.org/10.3233/jpd-191711.

Bradley, E., & Haran, J. (2024). The human gut microbiome and aging. *Gut Microbes, 16*(1), 2359677. https://doi.org/10.1080/19490976.2024.2359677.

Cai, J., Rimal, B., Jiang, C., Chiang, J. Y. L., & Patterson, A. D. (2022). Bile acid metabolism and signaling, the microbiota, and metabolic disease. *Pharmacology & Therapeutics, 237*, 108238. https://doi.org/10.1016/j.pharmthera.2022.108238.

Catanzaro, R., Anzalone, M., Calabrese, F., Milazzo, M., Capuana, M., Italia, A., ... Marotta, F. (2015). The gut microbiota and its correlations with the central nervous system disorders. *Panminerva Medica, 57*(3), 127–143.

Cekanaviciute, E., Yoo, B. B., Runia, T. F., Debelius, J. W., Singh, S., Nelson, C. A., ... Baranzini, S. E. (2017). Gut bacteria from multiple sclerosis patients modulate human T cells and exacerbate symptoms in mouse models. *Proceedings of the National Academy of Sciences of the United States of America, 114*(40), 10713–10718. https://doi.org/10.1073/pnas.1711235114.

Cersosimo, M. G., & Benarroch, E. E. (2012). Pathological correlates of gastrointestinal dysfunction in Parkinson's disease. *Neurobiology of Disease, 46*(3), 559–564. https://doi.org/10.1016/j.nbd.2011.10.014.

Chao, J., Coleman, R. A., Keating, D. J., & Martin, A. M. (2025). Gut microbiome regulation of gut hormone secretion. *Endocrinology, 166*(4), https://doi.org/10.1210/endocr/bqaf004.

Chen, C., Liao, J., Xia, Y., Liu, X., Jones, R., Haran, J., ... Ye, K. (2022). Gut microbiota regulate Alzheimer's disease pathologies and cognitive disorders via PUFA-associated neuroinflammation. *Gut, 71*(11), 2233–2252. https://doi.org/10.1136/gutjnl-2021-326269.

Chen, C., Wang, G. Q., Li, D. D., & Zhang, F. (2025). Microbiota-gut-brain axis in neurodegenerative diseases: Molecular mechanisms and therapeutic targets. *Molecular Biomedicine, 6*(1), 64. https://doi.org/10.1186/s43556-025-00307-1.

Chongtham, A., Yoo, J. H., Chin, T. M., Akingbesote, N. D., Huda, A., Marsh, J. L., & Khoshnan, A. (2022). Gut bacteria regulate the pathogenesis of Huntington's disease in drosophila model. *Frontiers in Neuroscience, 16*, 902205. https://doi.org/10.3389/fnins.2022.902205.

Chuang, C. L., & Demontis, F. (2021). Systemic manifestation and contribution of peripheral tissues to Huntington's disease pathogenesis. *Ageing Research Reviews, 69*, 101358. https://doi.org/10.1016/j.arr.2021.101358.

Correale, J., Hohlfeld, R., & Baranzini, S. E. (2022). The role of the gut microbiota in multiple sclerosis. *Nature Reviews Neurology, 18*(9), 544–558. https://doi.org/10.1038/s41582-022-00697-8.

Cosorich, I., Dalla-Costa, G., Sorini, C., Ferrarese, R., Messina, M. J., Dolpady, J., ... Falcone, M. (2017). High frequency of intestinal T(H)17 cells correlates with microbiota alterations and disease activity in multiple sclerosis. *Science Advances, 3*(7), e1700492. https://doi.org/10.1126/sciadv.1700492.

Cox, L. M., Maghzi, A. H., Liu, S., Tankou, S. K., Dhang, F. H., Willocq, V., ... Weiner, H. L. (2021). Gut microbiome in progressive multiple sclerosis. *Annals of Neurology, 89*(6), 1195–1211. https://doi.org/10.1002/ana.26084.

Dalile, B., Boyle, N. B., Ruiz, F. T., Chakrabarti, A., Respondek, F., Dodd, G. F., ... Schellekens, H. (2025). Targeting cognitive resilience through prebiotics: A focused perspective. *Advances in Nutrition, 16*(1), 100343. https://doi.org/10.1016/j.advnut.2024.100343.

Dodiya, H. B., Lutz, H. L., Weigle, I. Q., Patel, P., Michalkiewicz, J., Roman-Santiago, C. J., ... Sisodia, S. S. (2022). Gut microbiota-driven brain Aβ amyloidosis in mice requires microglia. *The Journal of Experimental Medicine, 219*(1), https://doi.org/10.1084/jem.20200895.

Fang, X., Wang, X., Yang, S., Meng, F., Wang, X., Wei, H., & Chen, T. (2016). Evaluation of the microbial diversity in amyotrophic lateral sclerosis using high-throughput sequencing. *Frontiers in Microbiology, 7*, 1479. https://doi.org/10.3389/fmicb.2016.01479.

Fasano, A., Visanji, N. P., Liu, L. W. C., Lang, A. E., & Pfeiffer, R. F. (2015). Gastrointestinal dysfunction in Parkinson's disease. *The Lancet Neurology, 14*(6), 625–639. https://doi.org/10.1016/S1474-4422(15)00007-1.

Ferreiro, A. L., Choi, J., Ryou, J., Newcomer, E. P., Thompson, R., Bollinger, R. M., ... Dantas, G. (2023). Gut microbiome composition may be an indicator of preclinical Alzheimer's disease. *Science Translational Medicine, 15*(700), eabo2984. https://doi.org/10.1126/scitranslmed.abo2984.

Fletcher, J. M., Lalor, S. J., Sweeney, C. M., Tubridy, N., & Mills, K. H. (2010). T cells in multiple sclerosis and experimental autoimmune encephalomyelitis. *Clinical and Experimental Immunology, 162*(1), 1–11. https://doi.org/10.1111/j.1365-2249.2010.04143.x.

Frohman, E. M., Racke, M. K., & Raine, C. S. (2006). Multiple sclerosis–the plaque and its pathogenesis. *The New England Journal of Medicine, 354*(9), 942–955. https://doi.org/10.1056/NEJMra052130.

Gabela, A. M., Mthembu, N., & Hadebe, S. (2026). Tryptophan metabolism in health and disease- implications for non-communicable diseases. *Immunology Letters, 277*, 107093. https://doi.org/10.1016/j.imlet.2025.107093.

Garbuzova-Davis, S., & Sanberg, P. R. (2014). Blood-CNS barrier impairment in ALS patients versus an animal model. *Frontiers in Cellular Neuroscience, 8*, 21. https://doi.org/10.3389/fncel.2014.00021.

Ghiboub, M., Verburgt, C. M., Sovran, B., Benninga, M. A., de Jonge, W. J., & Van Limbergen, J. E. (2020). Nutritional therapy to modulate tryptophan metabolism and aryl hydrocarbon-receptor signaling activation in human diseases. *Nutrients, 12*(9), https://doi.org/10.3390/nu12092846.

Ghosh, S. S., Wang, J., Yannie, P. J., & Ghosh, S. (2020). Intestinal barrier dysfunction, LPS translocation, and disease development. *Journal of the Endocrine Society, 4*(2), bvz039. https://doi.org/10.1210/jendso/bvz039.

Gropp, M. H. M., Klaips, C. L., & Hartl, F. U. (2022). Formation of toxic oligomers of polyQ-expanded Huntingtin by prion-mediated cross-seeding. *Molecular Cell, 82*(22), 4290–4306.e4211. https://doi.org/10.1016/j.molcel.2022.09.031.

Guo, H., Tang, X., He, X., Weng, Y., Zhang, Q., Fang, Q., & Zhang, L. (2025). A comprehensive review of the role of the microbiota-gut-brain axis via neuroinflammation: Advances and therapeutic implications for ischemic stroke. *Biomolecules, 15*(7), https://doi.org/10.3390/biom15070920.

Guo, T. T., Zhang, Z., Sun, Y., Zhu, R. Y., Wang, F. X., Ma, L. J., ... Liu, H. D. (2023). Neuroprotective effects of sodium butyrate by restoring gut microbiota and inhibiting TLR4 signaling in mice with MPTP-induced Parkinson's disease. *Nutrients, 15*(4), https://doi.org/10.3390/nu15040930.

Han, Y. Y., Li, K., Hu, J. Y., Wu, J. C., Li, X., Liu, D. X., & Li, C. H. (2024). Gender differences in dendritic damage, gut microbiota dysbiosis, and cognitive impairment during aging processes. *CNS Neuroscience & Therapeutics, 30*(12), e70164. https://doi.org/10.1111/cns.70164.

Harach, T., Marungruang, N., Duthilleul, N., Cheatham, V., Mc Coy, K. D., Frisoni, G., ... Bolmont, T. (2017). Reduction of Abeta amyloid pathology in APPPS1 transgenic mice in the absence of gut microbiota. *Scientific Reports,* 7, 41802. https://doi.org/10.1038/srep41802.

Hou, K., Wu, Z.-X., Chen, X.-Y., Wang, J.-Q., Zhang, D., Xiao, C., ... Chen, Z.-S. (2022). Microbiota in health and diseases. *Signal Transduction and Targeted Therapy,* 7(1), 135. https://doi.org/10.1038/s41392-022-00974-4.

Hou, Y., Li, X., Liu, C., Zhang, M., Zhang, X., Ge, S., & Zhao, L. (2021). Neuroprotective effects of short-chain fatty acids in MPTP induced mice model of Parkinson's disease. *Experimental Gerontology, 150*, 111376. https://doi.org/10.1016/j.exger.2021.111376.

Houser, M. C., & Tansey, M. G. (2017). The gut-brain axis: is intestinal inflammation a silent driver of Parkinson's disease pathogenesis? *NPJ Parkinson's Disease, 3*(1), 3. https://doi.org/10.1038/s41531-016-0002-0.

Hung, C. C., Chang, C. C., Huang, C. W., Nouchi, R., & Cheng, C. H. (2022). Gut microbiota in patients with Alzheimer's disease spectrum: A systematic review and meta-analysis. *Aging (Albany NY), 14*(1), 477–496. https://doi.org/10.18632/aging.203826.

Jiménez-Arroyo, C., Molinero, N., & Moreno-Arribas, M. V. (2025). Gut microbiota and nutrition in nursing homes: Challenges and translational approaches for healthy aging. *Advances in Nutrition, 16*(11), 100520. https://doi.org/10.1016/j.advnut.2025.100520.

Khan, M. F., Khodve, G., Yadav, S., Mallick, K., & Banerjee, S. (2025). Probiotic treatment improves post-traumatic stress disorder outcomes in mice. *Behavioural Brain Research, 476*, 115246. https://doi.org/10.1016/j.bbr.2024.115246.

Khatoon, S., Kalam, N., Rashid, S., & Bano, G. (2023). Effects of gut microbiota on neurodegenerative diseases. *Frontiers in Aging Neuroscience, 15*, 1145241. https://doi.org/10.3389/fnagi.2023.1145241.

Khoshnan, A. (2024). Gut microbiota as a modifier of Huntington's disease pathogenesis. *Journal of Huntington's Disease, 13*(2), 133–147. https://doi.org/10.3233/jhd-240012.

Khoshnan, A. (2024). Gut microbiota as a modifier of Huntington's disease pathogenesis. *Journal of Huntington's Disease, 13*. https://doi.org/10.3233/JHD-240012.

Kim, N., Jeon, S. H., Ju, I. G., Gee, M. S., Do, J., Oh, M. S., & Lee, J. K. (2021). Transplantation of gut microbiota derived from Alzheimer's disease mouse model impairs memory function and neurogenesis in C57BL/6 mice. *Brain, Behavior, and Immunity, 98*, 357–365. https://doi.org/10.1016/j.bbi.2021.09.002.

Kim, S., Kwon, S.-H., Kam, T.-I., Panicker, N., Karuppagounder, S. S., Lee, S., ... Ko, H. S. (2019). Transneuronal propagation of pathologic α-synuclein from the gut to the brain models Parkinson's disease. *Neuron, 103*(4), 627–641.e627. https://doi.org/10.1016/j.neuron.2019.05.035.

Kong, G., Cao, K. L., Judd, L. M., Li, S., Renoir, T., & Hannan, A. J. (2020). Microbiome profiling reveals gut dysbiosis in a transgenic mouse model of Huntington's disease. *Neurobiology of Disease, 135*, 104268. https://doi.org/10.1016/j.nbd.2018.09.001.

Kong, G., Ellul, S., Narayana, V. K., Kanojia, K., Ha, H. T. T., Li, S., ... Hannan, A. J. (2021). An integrated metagenomics and metabolomics approach implicates the microbiota-gut-brain axis in the pathogenesis of Huntington's disease. *Neurobiology of Disease, 148*, 105199. https://doi.org/10.1016/j.nbd.2020.105199.

Korf, J. M., Ganesh, B. P., & McCullough, L. D. (2022). Gut dysbiosis and age-related neurological diseases in females. *Neurobiology of Disease, 168*, 105695. https://doi.org/10.1016/j.nbd.2022.105695.

Kountouras, J., Boziki, M., Gavalas, E., Zavos, C., Grigoriadis, N., Deretzi, G., ... Venizelos, I. (2009). Eradication of Helicobacter pylori may be beneficial in the management of Alzheimer's disease. *Journal of Neurology, 256*(5), 758–767. https://doi.org/10.1007/s00415-009-5011-z.

Kumar, S., Mukherjee, R., Gaur, P., Leal, É., Lyu, X., Ahmad, S., ... Pandey, R. P. (2025). Unveiling roles of beneficial gut bacteria and optimal diets for health. *Frontiers in Microbiology, 16*. https://doi.org/10.3389/fmicb.2025.1527755.

Kverka, M., & Stepan, J. J. (2024). Associations among estrogens, the gut microbiome and osteoporosis. *Current Osteoporosis Reports, 23*(1), 2. https://doi.org/10.1007/s11914-024-00896-w.

Lee, B., Lee, S. M., Song, J. W., & Choi, J. W. (2024). Gut microbiota metabolite messengers in brain function and pathology at a view of cell type-based receptor and enzyme reaction. *Biomol Ther (Seoul), 32*(4), 403–423. https://doi.org/10.4062/biomolther.2024.009.

Li, M., Ren, J., Bao, Y., Wei, W., Yu, X., He, X., ... Li, H. (2025). Aged gut microbiota contributes to cognitive impairment and hippocampal synapse loss in mice. *Aging Cell, 24*(7), e70064. https://doi.org/10.1111/acel.70064.

Li, T., Chu, C., Yu, L., Zhai, Q., Wang, S., Zhao, J., ... Tian, F. (2022). Neuroprotective effects of bifidobacterium breve CCFM1067 in MPTP-induced mouse models of Parkinson's disease. *Nutrients, 14*(21), https://doi.org/10.3390/nu14214678.

Liang, S., Wang, T., Hu, X., Luo, J., Li, W., Wu, X., ... Jin, F. (2015). Administration of Lactobacillus helveticus NS8 improves behavioral, cognitive, and biochemical aberrations caused by chronic restraint stress. *Neuroscience, 310*, 561–577. https://doi.org/10.1016/j.neuroscience.2015.09.033.

Liu, S., Rezende, R. M., Moreira, T. G., Tankou, S. K., Cox, L. M., Wu, M., ... Weiner, H. L. (2019). Oral administration of miR-30d from Feces of MS patients suppresses MS-like symptoms in mice by expanding akkermansia muciniphila. *Cell Host & Microbe, 26*(6), 779–794.e778. https://doi.org/10.1016/j.chom.2019.10.008.

Loh, J. S., Mak, W. Q., Tan, L. K. S., Ng, C. X., Chan, H. H., Yeow, S. H., ... Khaw, K. Y. (2024). Microbiota–gut–brain axis and its therapeutic applications in neurodegenerative diseases. *Signal Transduction and Targeted Therapy, 9*(1), 37. https://doi.org/10.1038/s41392-024-01743-1.

Magne, F., Gotteland, M., Gauthier, L., Zazueta, A., Pesoa, S., Navarrete, P., & Balamurugan, R. (2020). The firmicutes/bacteroidetes ratio: A relevant marker of gut dysbiosis in obese patients? *Nutrients, 12*(5), https://doi.org/10.3390/nu12051474.

Mallick, K., Khodve, G., Ruwatia, R., & Banerjee, S. (2025). Gut microbes: Therapeutic Target for neuropsychiatric disorders. *Journal of Psychiatric Research, 184*, 27–38. https://doi.org/10.1016/j.jpsychires.2025.02.031.

Mar, J. S., Ota, N., Pokorzynski, N. D., Peng, Y., Jaochico, A., Sangaraju, D., ... Keir, M. E. (2023). IL-22 Alters gut microbiota composition and function to increase aryl hydrocarbon receptor activity in mice and humans. *Microbiome, 11*(1), 47. https://doi.org/10.1186/s40168-023-01486-1.

Matheoud, D., Cannon, T., Voisin, A., Penttinen, A. M., Ramet, L., Fahmy, A. M., ... Desjardins, M. (2019). Intestinal infection triggers Parkinson's disease-like symptoms in Pink1(-/-) mice. *Nature, 571*(7766), 565–569. https://doi.org/10.1038/s41586-019-1405-y.

Miller, A. L., Bessho, S., Grando, K., & Tükel, Ç. (2021). Microbiome or infections: Amyloid-containing biofilms as a trigger for complex human diseases. *Frontiers in immunology, 12*. https://doi.org/10.3389/fimmu.2021.638867.

Miyake, S., Kim, S., Suda, W., Oshima, K., Nakamura, M., Matsuoka, T., ... Yamamura, T. (2015). Dysbiosis in the gut microbiota of patients with multiple sclerosis, with a striking depletion of species belonging to clostridia XIVa and IV clusters. *PLoS One, 10*(9), e0137429. https://doi.org/10.1371/journal.pone.0137429.

A novel gene containing a trinucleotide repeat that is expanded and unstable on Huntington's disease chromosomes. The Huntington's Disease Collaborative Research Group. (1993). *Cell, 72*(6), 971–983. https://doi.org/10.1016/0092-8674(93)90585-e.

O'Riordan, K. J., Collins, M. K., Moloney, G. M., Knox, E. G., Aburto, M. R., Fülling, C., ... Cryan, J. F. (2022). Short chain fatty acids: Microbial metabolites for gut-brain axis signalling. *Molecular and Cellular Endocrinology, 546*, 111572. https://doi.org/10.1016/j.mce.2022.111572.

O'Riordan, K. J., Moloney, G. M., Keane, L., Clarke, G., & Cryan, J. F. (2025). The gut microbiota-immune-brain axis: Therapeutic implications. *Cell Reports Medicine, 6*(3), 101982. https://doi.org/10.1016/j.xcrm.2025.101982.

Ochoa-Repáraz, J., Mielcarz, D. W., Ditrio, L. E., Burroughs, A. R., Foureau, D. M., Haque-Begum, S., & Kasper, L. H. (2009). Role of gut commensal microflora in the development of experimental autoimmune encephalomyelitis. *Journal of Immunology, 183*(10), 6041–6050. https://doi.org/10.4049/jimmunol.0900747.

Oliveira, R. A., & Pamer, E. G. (2023). Assembling symbiotic bacterial species into live therapeutic consortia that reconstitute microbiome functions. *Cell Host & Microbe, 31*(4), 472–484. https://doi.org/10.1016/j.chom.2023.03.002.

Page, M. J., Kell, D. B., & Pretorius, E. (2022). The role of lipopolysaccharide-induced cell signalling in chronic inflammation. *Chronic Stress, 6*, 24705470221076390. https://doi.org/10.1177/24705470221076390.

Paone, P., Petitfils, C., Puel, A., Latousakis, D., de Vos, W. M., Delzenne, N. M., ... Cani, P. D. (2026). Akkermansia muciniphila modulates intestinal mucus composition to counteract high-fat diet-induced obesity in mice. *Gut Microbes, 18*(1), 2612580. https://doi.org/10.1080/19490976.2025.2612580.

Perez-Pardo, P., Dodiya, H. B., Engen, P. A., Forsyth, C. B., Huschens, A. M., Shaikh, M., ... Keshavarzian, A. (2019). Role of TLR4 in the gut-brain axis in Parkinson's disease: a translational study from men to mice. *Gut, 68*(5), 829–843. https://doi.org/10.1136/gutjnl-2018-316844.

Pini, L., Imbimbo, B. P., Griffa, A., Allali, G., & Pomara, N. (2026). Targeting brain connectivity in Alzheimer's disease with repurposed drugs. *Ageing Research Reviews, 114*, 103001. https://doi.org/10.1016/j.arr.2025.103001.

Pröbstel, A. K., & Baranzini, S. E. (2018). The role of the gut microbiome in multiple sclerosis risk and progression: Towards characterization of the "MS Microbiome". *Neurotherapeutics: the Journal of the American Society for Experimental NeuroTherapeutics, 15*(1), 126–134. https://doi.org/10.1007/s13311-017-0587-y.

Qian, X. H., Liu, X. L., Chen, G., Chen, S. D., & Tang, H. D. (2022). Injection of amyloid-β to lateral ventricle induces gut microbiota dysbiosis in association with inhibition of cholinergic anti-inflammatory pathways in Alzheimer's disease. *Journal of Neuroinflammation, 19*(1), 236. https://doi.org/10.1186/s12974-022-02599-4.

Rasouli-Saravani, A., Jahankhani, K., Moradi, S., Gorgani, M., Shafaghat, Z., Mirsanei, Z., ... Mirzaei, R. (2023). Role of microbiota short-chain fatty acids in the pathogenesis of autoimmune diseases. *Biomedicine & Pharmacotherapy, 162*, 114620. https://doi.org/10.1016/j.biopha.2023.114620.

Ratsika, A., Cruz Pereira, J. S., Lynch, C. M. K., Clarke, G., & Cryan, J. F. (2023). Microbiota-immune-brain interactions: A lifespan perspective. *Current Opinion in Neurobiology, 78*, 102652. https://doi.org/10.1016/j.conb.2022.102652.

Reynders, T., Devolder, L., Valles-Colomer, M., Van Remoortel, A., Joossens, M., De Keyser, J., ... Raes, J. (2020). Gut microbiome variation is associated to Multiple Sclerosis phenotypic subtypes. *Annals of Clinical and Translational Neurology, 7*(4), 406–419. https://doi.org/10.1002/acn3.51004.

Roubaud-Baudron, C., Ruiz, V. E., Swan, A. M., Jr., Vallance, B. A., Ozkul, C., Pei, Z., ... Blaser, M. J. (2019). Long-term effects of early-life antibiotic exposure on resistance to subsequent bacterial infection. *mBio, 10*(6), https://doi.org/10.1128/mBio.02820-19.

Sampson, T. R., Debelius, J. W., Thron, T., Janssen, S., Shastri, G. G., Ilhan, Z. E., ... Mazmanian, S. K. (2016). Gut microbiota regulate motor deficits and neuroinflammation in a model of Parkinson's disease. *Cell, 167*(6), 1469–1480.e1412. https://doi.org/10.1016/j.cell.2016.11.018.

Sastre, D. E., Sultana, N., M, V. A. S. N., Huliciak, M., Du, J., Cifuente, J. O., ... Sundberg, E. J. (2024). Human gut microbes express functionally distinct endoglycosidases to metabolize the same N-glycan substrate. *Nature Communications, 15*(1), 5123. https://doi.org/10.1038/s41467-024-48802-3.

Schwartz, K., Syed, A. K., Stephenson, R. E., Rickard, A. H., & Boles, B. R. (2012). Functional amyloids composed of phenol soluble modulins stabilize Staphylococcus aureus biofilms. *PLoS Pathogens, 8*(6), e1002744. https://doi.org/10.1371/journal.ppat.1002744.

Schwiertz, A., Taras, D., Schäfer, K., Beijer, S., Bos, N. A., Donus, C., & Hardt, P. D. (2010). Microbiota and SCFA in lean and overweight healthy subjects. *Obesity (Silver Spring), 18*(1), 190–195. https://doi.org/10.1038/oby.2009.167.

Seo, D. O., O'Donnell, D., Jain, N., Ulrich, J. D., Herz, J., Li, Y., ... Holtzman, D. M. (2023). ApoE isoform- and microbiota-dependent progression of neurodegeneration in a mouse model of tauopathy. *Science (New York, N. Y.), 379*(6628), eadd1236. https://doi.org/10.1126/science.add1236.

Seo, S.-K., & Kwon, B. (2023). Immune regulation through tryptophan metabolism. *Experimental & Molecular Medicine, 55*(7), 1371–1379. https://doi.org/10.1038/s12276-023-01028-7.

Shen, Y., Fan, N., Ma, S. X., Cheng, X., Yang, X., & Wang, G. (2025). Gut microbiota dysbiosis: Pathogenesis, diseases, prevention, and therapy. *MedComm (2020), 6*(5), e70168. https://doi.org/10.1002/mco2.70168.

Sheng, C., Yang, K., He, B., Du, W., Cai, Y., & Han, Y. (2022). Combination of gut microbiota and plasma amyloid-β as a potential index for identifying preclinical Alzheimer's disease: A cross-sectional analysis from the SILCODE study. *Alzheimer's Research & Therapy, 14*(1), 35. https://doi.org/10.1186/s13195-022-00977-x.

Sheng, J. A., Bales, N. J., Myers, S. A., Bautista, A. I., Roueinfar, M., Hale, T. M., & Handa, R. J. (2021). The hypothalamic-pituitary-adrenal axis: Development, programming actions of hormones, and maternal-fetal interactions. *Frontiers in Behavioral Neuroscience, 14*. https://doi.org/10.3389/fnbeh.2020.601939.

Shim, J. A., Ryu, J. H., Jo, Y., & Hong, C. (2023). The role of gut microbiota in T cell immunity and immune mediated disorders. *International Journal of Biological Sciences, 19*(4), 1178–1191. https://doi.org/10.7150/ijbs.79430.

Silva, Y. P., Bernardi, A., & Frozza, R. L. (2020). The role of short-chain fatty acids from gut microbiota in gut-brain communication. *Frontiers in Endocrinology, 11*.

Singh, Y., Trautwein, C., Romani, J., Salker, M. S., Neckel, P. H., Fraccaroli, I., ... Casadei, N. (2023). Overexpression of human alpha-Synuclein leads to dysregulated microbiome/metabolites with ageing in a rat model of Parkinson disease. *Molecular Neurodegeneration, 18*(1), 44. https://doi.org/10.1186/s13024-023-00628-1.

Srivastav, S., Neupane, S., Bhurtel, S., Katila, N., Maharjan, S., Choi, H., ... Choi, D. Y. (2019). Probiotics mixture increases butyrate, and subsequently rescues the nigral dopaminergic neurons from MPTP and rotenone-induced neurotoxicity. *The Journal of Nutritional Biochemistry, 69*, 73–86. https://doi.org/10.1016/j.jnutbio.2019.03.021.

Stan, T. L., Soylu-Kucharz, R., Burleigh, S., Prykhodko, O., Cao, L., Franke, N., ... Björkqvist, M. (2020). Increased intestinal permeability and gut dysbiosis in the R6/2 mouse model of Huntington's disease. *Scientific Reports, 10*(1), 18270. https://doi.org/10.1038/s41598-020-75229-9.

Sun, M. F., Zhu, Y. L., Zhou, Z. L., Jia, X. B., Xu, Y. D., Yang, Q., ... Shen, Y. Q. (2018). Neuroprotective effects of fecal microbiota transplantation on MPTP-induced Parkinson's disease mice: Gut microbiota, glial reaction and TLR4/TNF-α signaling pathway. *Brain, Behavior, and Immunity, 70*, 48–60. https://doi.org/10.1016/j.bbi.2018.02.005.

Tai, Y. F., Pavese, N., Gerhard, A., Tabrizi, S. J., Barker, R. A., Brooks, D. J., & Piccini, P. (2007). Microglial activation in presymptomatic Huntington's disease gene carriers. *Brain, 130*(Pt 7), 1759–1766. https://doi.org/10.1093/brain/awm044.

Taj, T., Kaushik, M., Islam, A., Das, J., Kumar, B., Hussain, M. S., ... Ansari, M. Y. (2025). Microbiota-brain interaction: The role of gut-derived proteins in addressing various neurological disorders including Parkinson's (PD) and Alzheimer's diseases (AD). *Biomedicine & Pharmacotherapy, 193*, 118861. https://doi.org/10.1016/j.biopha.2025.118861.

Tang, J., Krushelnycky, L., Shaqo, A., & Cho, C. E. (2025). A comprehensive review of nutritional influences on the serotonergic system. *Advances in Nutrition, 16*(11), 100524. https://doi.org/10.1016/j.advnut.2025.100524.

Toepfer, M., Schroeder, M., Klauser, A., Lochmüller, H., Hirschmann, M., Riepl, R. L., ... Müller-Felber, W. (1997). Delayed colonic transit times in amyotrophic lateral sclerosis assessed with radio-opaque markers. *European Journal of Medical Research, 2*(11), 473–476.

Träger, U., Andre, R., Lahiri, N., Magnusson-Lind, A., Weiss, A., Grueninger, S., ... Tabrizi, S. J. (2014). HTT-lowering reverses Huntington's disease immune dysfunction caused by NFκB pathway dysregulation. *Brain, 137*(Pt 3), 819–833. https://doi.org/10.1093/brain/awt355.

Tseng, C. H., & Wu, C. Y. (2025). From dysbiosis to longevity: A narrative review into the gut microbiome's impact on aging. *Journal of Biomedical Science, 32*(1), 93. https://doi.org/10.1186/s12929-025-01179-x.

Tursi, S. A., Lee, E. Y., Medeiros, N. J., Lee, M. H., Nicastro, L. K., Buttaro, B., ... Tükel, Ç. (2017). Bacterial amyloid curli acts as a carrier for DNA to elicit an autoimmune response via TLR2 and TLR9. *PLoS Pathogens, 13*(4), e1006315. https://doi.org/10.1371/journal.ppat.1006315.

Unger, M. M., Spiegel, J., Dillmann, K. U., Grundmann, D., Philippeit, H., Bürmann, J., ... Schäfer, K. H. (2016). Short chain fatty acids and gut microbiota differ between patients with Parkinson's disease and age-matched controls. *Parkinsonism & Related Disorders, 32*, 66–72. https://doi.org/10.1016/j.parkreldis.2016.08.019.

Vascellari, S., Palmas, V., Melis, M., Pisanu, S., Cusano, R., Uva, P., ... Manzin, A. (2020). Gut microbiota and metabolome alterations associated with Parkinson's disease. *mSystems, 5*(5), 00561-00520. https://doi.org/10.1128/msystems.00561-20.

Walker, F. O. (2007). Huntington's disease. *Lancet, 369*(9557), 218–228. https://doi.org/10.1016/s0140-6736(07)60111-1.

Wang, C., Lau, C. Y., Ma, F., & Zheng, C. (2021). Genome-wide screen identifies curli amyloid fibril as a bacterial component promoting host neurodegeneration. *Proceedings of the National Academy of Sciences of the United States of America, 118*(34), https://doi.org/10.1073/pnas.2106504118.

Wang, J., Cheng, W., & Yang, R. (2025). Nervous system–gut microbiota–immune system axis: Future directions for preventikng tumor. *Frontiers in immunology, 16*. https://doi.org/10.3389/fimmu.2025.1535955.

Wang, S., Jiang, Y., Yang, A., Meng, F., & Zhang, J. (2024). The expanding burden of neurodegenerative diseases: An unmet medical and social need. *Aging and Disease, 16*(5), 2937–2952. https://doi.org/10.14336/ad.2024.1071.

Wang, W., Gao, Y., Iribarren, P., Lei, Y., Xiang, Y., Zhang, G., & Lu, A. (2015). Wang et al. 2015.

Wang, Y., & Kasper, L. H. (2014). The role of microbiome in central nervous system disorders. *Brain, Behavior, and Immunity, 38*, 1–12. https://doi.org/10.1016/j.bbi.2013.12.015.

Wang, Y., Tian, L., Fabi, J. P., & de Vos, P. (2025). The potential of prebiotics, probiotics, and synbiotics for ameliorating intestinal barrier dysfunction and modulating inflammatory responses as dietary supplements in diabetes mellitus management. *Food Bioscience, 72*, 107539. https://doi.org/10.1016/j.fbio.2025.107539.

Warnecke, T., Schäfer, K. H., Claus, I., Del Tredici, K., & Jost, W. H. (2022). Gastrointestinal involvement in Parkinson's disease: Pathophysiology, diagnosis, and management. *NPJ Parkinson's Disease, 8*(1), 31. https://doi.org/10.1038/s41531-022-00295-x.

Wasser, C. I., Mercieca, E. C., Kong, G., Hannan, A. J., McKeown, S. J., Glikmann-Johnston, Y., & Stout, J. C. (2020). Gut dysbiosis in Huntington's disease: Associations among gut microbiota, cognitive performance and clinical outcomes. *Brain Communications, 2*(2), fcaa110. https://doi.org/10.1093/braincomms/fcaa110.

Wu, S., Yi, J., Zhang, Y. G., Zhou, J., & Sun, J. (2015). Leaky intestine and impaired microbiome in an amyotrophic lateral sclerosis mouse model. *Physiological Reports, 3*(4), https://doi.org/10.14814/phy2.12356.

Yadav, S., Aggarwal, P., Khan, F., Khodve, G., Padhy, D. S., Yadav, P., & Banerjee, S. (2023). β-sitosterol protects against aluminium chloride-mediated neurotoxicity. *Current Alzheimer Research, 20*(1), 29–37. https://doi.org/10.2174/1567205020666230308151443.

Yu, S., He, J., & Xie, K. (2023). Zonula occludens proteins signaling in inflammation and tumorigenesis. *International Journal of Biological Sciences, 19*(12), 3804–3815. https://doi.org/10.7150/ijbs.85765.

Zhang, Y., Shen, Y., Liufu, N., Liu, L., Li, W., Shi, Z., ... Xie, Z. (2023). Transmission of Alzheimer's disease-associated microbiota dysbiosis and its impact on cognitive function: Evidence from mouse models and human patients. *Research Square*. https://doi.org/10.21203/rs.3.rs-2790988/v1.

Zhang, Y. G., Wu, S., Yi, J., Xia, Y., Jin, D., Zhou, J., & Sun, J. (2017). Target intestinal microbiota to alleviate disease progression in amyotrophic lateral sclerosis. *Clinical Therapeutics, 39*(2), 322–336. https://doi.org/10.1016/j.clinthera.2016.12.014.

Zhong, Z., Chen, W., Gao, H., Che, N., Xu, M., Yang, L., ... Ye, M. (2021). Fecal microbiota transplantation exerts a protective role in MPTP-induced Parkinson's disease via the TLR4/PI3K/AKT/NF-κB pathway stimulated by α-synuclein. *Neurochemical Research, 46*(11), 3050–3058. https://doi.org/10.1007/s11064-021-03411-0.

Zhou, Y., Lu, Y., Fang, X., Zhang, J., Li, J., Li, S., ... Xu, R. (2015). An astrocyte regenerative response from vimentin-containing cells in the spinal cord of amyotrophic lateral sclerosis's disease-like transgenic (G93A SOD1) mice. *Neuro-Degenerative Diseases, 15*(1), 1–12. https://doi.org/10.1159/000369466.

Zhu, G., Guo, M., Zhao, J., Zhang, H., Wang, G., & Chen, W. (2022). Bifidobacterium breve intervention combined with environmental enrichment alleviates cognitive impairment by regulating the gut microbiota and microbial metabolites in Alzheimer's disease mice. *Frontiers in Immunology, 13*, 1013664. https://doi.org/10.3389/fimmu.2022.1013664.

CHAPTER EIGHT

Brain energy crisis in Alzheimer's and Parkinson's disease: Nanotechnology as a therapeutic strategy

Mohini Milmile, Shweta Singh, Ajay Pandey, Gajanan Pawar, Pradnya Petkar, Yogesh Chander, Ranjan Mishra, and Ruchi Chawla*

Department of Pharmaceutical Engineering and Technology, IIT(BHU), Varanasi, India

*Corresponding author. e-mail address: rchawla.phe@iitbhu.ac.in

Contents

International Review of Neurobiology, Volume 188

ISSN 0074-7742, https://doi.org/10.1016/bs.irn.2026.05.016

Abstract

Alzheimer's disease and Parkinson's disease are increasingly recognized as disorders marked not only by protein aggregation but by a sustained failure of brain energy metabolism. Years before overt cognitive or motor symptoms emerge, neurons begin to experience impaired glucose utilization, mitochondrial dysfunction, and declining ATP production. Because the brain is highly energy-dependent, even subtle metabolic disturbances can disrupt synaptic function, impair neuronal signaling, and trigger oxidative stress. As mitochondrial efficiency declines, reactive oxygen species accumulate, inflammatory pathways become chronically activated, and damaged cellular components are insufficiently cleared, creating a vicious cycle that accelerates neurodegeneration. Conventional metabolic therapies, including antioxidants and mitochondrial cofactors, have produced limited clinical success, largely due to poor penetration across the blood–brain barrier and lack of specificity for vulnerable neuronal populations and intracellular targets such as mitochondria. Nanotechnology introduces a more precise therapeutic strategy by enabling targeted delivery of metabolic modulators directly to the brain. Engineered nanocarriers can be designed to cross biological barriers, enhance drug stability, and release therapeutic agents in a controlled or stimuli-responsive manner within diseased regions. Advanced systems including polymeric nanoparticles, lipid-based carriers, intranasal nano formulations, biomimetic vesicles, and catalytic nanozymes offer the ability to simultaneously modulate oxidative stress, restore mitochondrial function, and regulate neuroinflammation. By integrating delivery precision with metabolic intervention, neuro-nanomedicine shifts the therapeutic focus from symptomatic management toward addressing the underlying bioenergetic crisis. Although challenges remain in safety validation, scalability, and clinical translation, targeted nanotherapeutic strategies hold significant promise for transforming the treatment landscape of these debilitating neurodegenerative disorders.

1. Introduction: Brain energy crisis and delivery challenges

1.1 Overview of metabolic failure in AD and PD

Metabolic dysfunction is now recognized as a foundational pathological mechanism in both Alzheimer's disease (AD) and Parkinson's disease (PD). In Alzheimer's disease, cerebral glucose hypometabolism is one of the earliest detectable abnormalities, appearing years before overt cognitive decline (Raut et al., 2023). Fluorodeoxyglucose positron emission tomography (FDG-PET) studies consistently demonstrate reduced glucose uptake in the posterior cingulate cortex, precuneus, temporoparietal cortex, and hippocampus. This reduction reflects impaired neuronal energy metabolism rather than simply neuronal loss (Coutinho et al., 2015). The human brain consumes approximately 20 % of total body glucose despite representing only about 2 % of body weight, underscoring its extreme dependence on uninterrupted ATP production (Mergenthaler et al., 2013). Reduced expression of glucose transporters such as GLUT1 and GLUT3, insulin resistance within the brain, mitochondrial respiratory chain inefficiency, and altered astrocyte-neuron metabolic coupling together diminish ATP availability (Szablewski, 2021). In Parkinson's disease, dopaminergic neurons of the substantia nigra exhibit marked mitochondrial complex I deficiency, resulting in impaired oxidative phosphorylation and increased reactive oxygen species generation. These neurons are especially vulnerable because their autonomous pacemaking activity imposes sustained calcium influx and high metabolic demand (Trist et al., 2019). Mitochondrial DNA deletions accumulate with aging in these neurons, further compromising respiratory capacity (Krishnan et al., 2007). Across both disorders, impaired mitochondrial dynamics, defective mitophagy, oxidative stress amplification, and reduced metabolic flexibility create a vicious cycle in which ATP depletion promotes protein aggregation and inflammatory activation, which in turn further damages mitochondrial function. Thus, AD and PD can be conceptualized not only as proteinopathies but as chronic disorders of cerebral energy insufficiency (Bender et al., 2006).

1.2 Limitations of conventional metabolic therapeutics

Therapeutic efforts aimed at correcting metabolic dysfunction in Alzheimer's disease (AD) and Parkinson's disease (PD) have included antioxidant supplementation, mitochondrial cofactors, insulin sensitizers, and ketogenic interventions (Zilberter & Zilberter, 2023); however, despite strong mechanistic rationale derived from experimental models, clinical outcomes have been inconsistent and often modest. A major limitation is restricted blood-brain barrier (BBB) permeability, as the tightly regulated endothelial interface prevents most systemically administered agents from achieving therapeutically relevant concentrations within the central nervous system (Zhang et al., 2022). Lipophilic antioxidants such as vitamin E and coenzyme-Q10, although biologically plausible for reducing mitochondrial reactive oxygen species (ROS), demonstrate limited CNS accumulation relative to peripheral exposure, resulting in insufficient mitochondrial redox modulation at tolerable doses (Bender et al., 2006). Furthermore, conventional agents lack cellular and subcellular specificity; neurodegeneration in AD and PD is regionally selective and mitochondria-centered, yet most pharmacologic compounds distribute non-selectively and do not preferentially accumulate within mitochondria, where the mitochondrial membrane potential and structural barriers require specialized targeting strategies. Brain insulin resistance, characterized by impaired insulin receptor signaling and downstream PI3K/Akt pathway dysfunction, further complicates systemic metabolic correction, as peripheral insulin sensitizers do not reliably restore neuronal insulin signaling due to differential transport and regulatory mechanisms across the BBB (Talbot et al., 2012). Ketogenic approaches attempt to bypass glycolytic impairment by providing ketone bodies as alternative substrates for oxidative phosphorylation; although cerebral ketone uptake remains relatively preserved in AD, sustained ketosis is difficult to maintain, interindividual metabolic variability influences brain ketone availability, and such strategies do not directly repair mitochondrial structural damage or defective mitophagy (Cunnane et al., 2016). Additionally, most clinical trials are conducted during moderate or advanced stages of disease, when significant neuronal and synaptic loss has already occurred, thereby limiting the potential for metabolic rescue to restore network integrity. Finally, conventional metabolic therapies often target isolated components of a complex, interdependent system in which oxidative stress, mitochondrial dysfunction, neuroinflammation, and glucose hypometabolism amplify one another in a self-perpetuating cycle. Collectively, these translational and biological constraints demonstrate that effective bioenergetic therapy in AD and PD requires precise BBB penetration, selective neuronal and mitochondrial targeting, and coordinated modulation of interconnected metabolic pathways-capabilities not inherently provided by traditional pharmacological formulations (Beal, 1998; Murphy & Smith, 2007).

1.3 Rationale for nanotechnology-based intervention

Nanotechnology offers a mechanistically precise strategy to restore cerebral bioenergetics in Alzheimer's disease and Parkinson's disease by overcoming the delivery limitations of conventional metabolic therapies. Engineered nanoparticles (1–200 nm) can be functionalized with ligands targeting transferrin or insulin receptors to enable receptor-mediated transcytosis across the blood–brain barrier, thereby improving central nervous system drug accumulation (Pardridge, 2012). Encapsulation protects labile metabolic cofactors such as NAD^+ precursors and coenzyme Q10 from rapid degradation and enhances pharmacokinetic stability (Covarrubias et al., 2021).

Surface modification with lipophilic cations like triphenylphosphonium drives selective mitochondrial accumulation by exploiting the organelle's negative membrane potential, directly targeting the primary site of ATP depletion and reactive oxygen species generation (Murphy & Smith, 2007). Nanocarriers also enable combinatorial delivery of antioxidants, metabolic substrates, anti-inflammatory agents, and gene modulators within a single platform, addressing the interconnected nature of mitochondrial dysfunction, oxidative stress, and neuroinflammation (Pegoraro et al., 2024). Additionally, stimuli-responsive systems incorporating redox-sensitive linkers allow selective drug release in oxidatively stressed neurons (Li et al., 2020), while exosome-based carriers provide enhanced biocompatibility and intrinsic BBB permeability (El-Andaloussi et al., 2012). By enabling BBB penetration, neuronal specificity, mitochondrial targeting, and controlled multi-pathway intervention, nanotechnology provides a biologically coherent framework for correcting brain energy failure in neurodegenerative disorders.

2. Therapeutic targets in brain bioenergetics

2.1 Mitochondrial dysfunction and ATP depletion

Mitochondria are central regulators of neuronal survival, integrating oxidative phosphorylation, calcium homeostasis, apoptosis signaling, and redox balance. In both AD and PD, mitochondrial dysfunction represents a convergent pathological mechanism. Reduced activity of electron transport chain (ETC) complexes has been documented in postmortem brain tissue complex IV deficiency in AD (Nicholson et al., 2010) and complex I deficiency in PD (Schapira et al., 1990). Impaired electron transport reduces proton motive force across the inner mitochondrial membrane, directly limiting ATP synthase activity and cellular energy availability. ATP depletion activates AMPK, a cellular energy sensor responsive to increased AMP/ATP ratios (Hardie, 2008). Under physiological stress, AMPK promotes catabolic pathways, enhances mitochondrial biogenesis via PGC-1α activation, and suppresses anabolic signaling through inhibition of mTOR complex 1 (mTORC1). However, chronic mitochondrial dysfunction in neurodegeneration leads to maladaptive AMPK activation. Sustained AMPK signaling can impair synaptic plasticity and exacerbate tau phosphorylation through downstream kinase modulation (Caccamo et al., 2011). Conversely, impaired AMPK signaling may reduce mitochondrial biogenesis, limiting compensatory capacity. mTOR signaling plays a pivotal role in coordinating cellular growth and autophagy. Hyperactivation of mTORC1, observed in AD brains, suppresses autophagic clearance of damaged mitochondria and misfolded proteins (Caccamo et al., 2011). In PD, mutations affecting PINK1 and Parkin disrupt mitophagy, leading to accumulation of dysfunctional mitochondria and further ATP decline (Pickrell & Youle, 2015). Defective mitophagy amplifies oxidative damage and propagates mitochondrial DNA instability, particularly in dopaminergic neurons vulnerable to metabolic stress (Xu et al., 2025a). SIRT1, a NAD^+-dependent deacetylase, links cellular redox state to mitochondrial function. Reduced NAD^+ availability in aging and neurodegeneration impairs SIRT1 activity, limiting deacetylation of PGC-1α and decreasing mitochondrial biogenesis (Houtkooper et al., 2012). Declining NAD^+ pools therefore compromise mitochondrial renewal and ATP generation capacity. Restoration of NAD^+ levels has been shown experimentally to improve mitochondrial respiration and neuronal resilience. Thus, mitochondrial dysfunction in AD and PD is not merely structural but embedded within AMPK–mTOR–SIRT1 regulatory networks (Amjad et al., 2021). Therapeutic strategies targeting ETC stabilization, NAD^+ restoration, mitophagy enhancement, and balanced AMPK modulation represent central approaches to restoring ATP homeostasis.

2.2 Oxidative stress and redox imbalance

Oxidative stress is a central and self-propagating mechanism in the pathogenesis of Alzheimer's disease (AD) and Parkinson's disease (PD), tightly coupled to mitochondrial dysfunction and impaired cellular metabolism. Reactive oxygen species (ROS) are generated physiologically during oxidative phosphorylation, primarily at complexes I and III of the electron transport chain (ETC), where small amounts of electron leakage produce superoxide radicals. Under normal conditions, these species are neutralized by endogenous antioxidant systems, including superoxide dismutase (SOD), catalase, and the glutathione (GSH) pathway, thereby maintaining redox homeostasis. In AD and PD, however, ETC inefficiency increases electron leakage and amplifies ROS production beyond buffering capacity (Reddy & Beal, 2008). In AD, amyloid-β peptides interact with mitochondrial membranes and respiratory chain components, promoting oxidative injury and lipid peroxidation, while tau-mediated disruption of axonal mitochondrial transport exacerbates localized synaptic oxidative stress. In PD, dopaminergic neurons are particularly vulnerable because dopamine auto-oxidation generates hydrogen peroxide and quinones, and iron accumulation within the substantia nigra catalyzes Fenton reactions that produce highly reactive hydroxyl radicals (Dexter et al., 1989a, 1989b). Excess ROS damage mitochondrial DNA, oxidize respiratory enzymes, and destabilize membrane phospholipids, thereby impairing ATP production and establishing a feed-forward cycle in which mitochondrial dysfunction further increases ROS generation. Redox imbalance also intersects with major metabolic signaling pathways: ROS can activate AMP-activated protein kinase (AMPK) as a stress response, while chronic activation of mTORC1-reported in AD-suppresses autophagic clearance of damaged mitochondria, permitting accumulation of oxidized organelles (Caccamo et al., 2011). Furthermore, declining NAD^+ levels in aging and neurodegeneration reduce SIRT1 activity, limiting deacetylation of transcription factors such as PGC-1α and FOXO that regulate antioxidant gene expression (Houtkooper et al., 2012). Oxidative stress additionally activates redox-sensitive transcription factors such as NF-κB, enhancing pro-inflammatory cytokine production and reinforcing a bidirectional relationship between mitochondrial dysfunction and neuroinflammation.

2.3 Impaired glucose and ketone metabolism

Impaired glucose and ketone metabolism in Alzheimer's disease (AD) and Parkinson's disease (PD) reflects a complex disruption of substrate transport, intracellular metabolic flux, and energy-sensing signaling networks that converge on mitochondrial bioenergetic failure. In AD, cerebral glucose hypometabolism is detectable decades before dementia onset and correlates with synaptic dysfunction rather than neuronal loss alone (Mosconi et al., 2008). At the molecular level, reduced expression and functional impairment of glucose transporter 1 (GLUT1) at the blood–brain barrier and GLUT3 in neurons limit intracellular glucose availability (Veys et al., 2020). This restriction diminishes glycolytic throughput, reducing pyruvate production and consequently acetyl-CoA entry into the tricarboxylic acid (TCA) cycle. As a result, generation of NADH and $FADH_2$ declines, impairing electron transport chain (ETC) activity and ATP synthesis (Arnold & Finley, 2023). Concurrently, brain insulin resistance-characterized by serine phosphorylation of insulin receptor substrate-1 (IRS-1) and impaired PI3K/Akt signaling-disrupts glucose uptake and suppresses downstream survival pathways (Talbot et al., 2012). Akt normally inhibits glycogen synthase kinase-3β (GSK-3β); therefore, impaired signaling enhances tau phosphorylation and contributes to cytoskeletal instability. Insulin resistance also dysregulates mTOR complex

1 (mTORC1), altering protein synthesis and autophagic balance, thereby coupling metabolic stress to proteostatic dysfunction. Energy depletion increases the AMP/ATP ratio, activating AMP-activated protein kinase (AMPK), which attempts to restore metabolic equilibrium by promoting catabolic pathways and mitochondrial biogenesis through peroxisome proliferator activated receptor gamma coactivator-1α (PGC-1α) activation (Hardie et al., 2012). However, chronic metabolic stress can produce maladaptive AMPK signaling, impairing synaptic maintenance and exacerbating tau pathology (Galasso et al., 2017). Moreover, aberrant mTORC1 activation observed in AD-suppresses autophagy, limiting clearance of damaged mitochondria and reinforcing bioenergetic decline (Caccamo et al., 2011). In PD, mitochondrial complex I deficiency further constrains the capacity to oxidize glucose-derived substrates, rendering dopaminergic neurons-already burdened by high pacemaking related calcium influx-particularly sensitive to fluctuations in metabolic supply. Ketone metabolism provides a partially preserved alternative pathway. Ketone bodies-β-hydroxybutyrate and acetoacetate-are transported via monocarboxylate transporters (MCT1 and MCT2), which remain relatively intact even when GLUT-mediated transport declines (Cunnane et al., 2016). Within mitochondria, ketones are converted to acetyl-CoA independent of glycolysis, bypassing insulin-dependent regulatory steps. Beyond serving as fuel, β-hydroxybutyrate functions as a signaling metabolite that inhibits class I histone deacetylases, modulates FOXO and PGC-1α transcriptional activity, and enhances oxidative stress resistance (Kolb et al., 2021). However, effective ketone utilization still requires functional mitochondrial oxidative capacity; thus, substrate substitution alone cannot fully compensate for ETC impairment. Additionally, declining nicotinamide adenine dinucleotide (NAD^+) pools in aging and neurodegeneration reduce SIRT1 activity, impairing deacetylation of PGC-1α and diminishing mitochondrial biogenesis (Houtkooper et al., 2012). Reduced NAD^+ availability also limits redox buffering and compromises glycolytic enzyme function.

2.4 Immunometabolic dysregulation and neuroinflammation

Immunometabolic dysregulation constitutes a critical mechanistic bridge between chronic neuroinflammation and progressive bioenergetic failure in Alzheimer's disease (AD) and Parkinson's disease (PD). Microglia, the principal innate immune cells of the central nervous system, undergo profound metabolic reprogramming upon activation by pathological stimuli such as amyloid-β in AD or misfolded α-synuclein in PD. Under physiological conditions, microglia rely predominantly on mitochondrial oxidative phosphorylation for ATP generation; however, inflammatory activation induces a shift toward aerobic glycolysis, a process analogous to the Warburg effect observed in peripheral immune cells (Orihuela et al., 2016). This glycolytic phenotype supports rapid ATP production and biosynthetic demands required for cytokine secretion but is metabolically inefficient and associated with increased mitochondrial reactive oxygen species (ROS) production. Mitochondrial ROS serve as signaling intermediates that activate redox-sensitive transcription factors such as nuclear factor kappa B (NF-κB), promoting transcription of pro-inflammatory cytokines including tumor necrosis factor-α (TNF-α) and interleukin-1β (IL-1β). In AD, mitochondrial dysfunction and ROS accumulation contribute to activation of the NLRP3 inflammasome, leading to caspase-1–mediated maturation of IL-1β and IL-18, thereby amplifying neuroinflammatory cascades (Heneka et al., 2013). Similarly, α-synuclein aggregates in PD stimulate microglial pattern recognition receptors and inflammasome activation, reinforcing oxidative injury within dopaminergic regions. Energy sensing pathways tightly regulate this immunometabolic shift. AMP-activated protein kinase (AMPK) normally promotes oxidative metabolism and suppresses inflammatory signaling by inhibiting NF-κB activation, whereas hyperactivation of mechanistic target of rapamycin complex 1 (mTORC1) enhances glycolytic flux and inflammatory gene expression while suppressing autophagy (Saxton & Sabatini, 2017). Dysregulation of the AMPK–mTOR axis therefore sustains a pro-inflammatory metabolic phenotype and impairs clearance of damaged mitochondria. Furthermore, declining nicotinamide adenine dinucleotide (NAD^+) levels in aging and neurodegeneration reduce activity of the NAD^+-dependent deacetylase SIRT1, which normally suppresses inflammation by deacetylating the p65 subunit of NF-κB and promotes mitochondrial biogenesis via PGC-1α activation (Xie et al., 2025). Reduced SIRT1 activity thereby links metabolic insufficiency directly to heightened inflammatory signaling. Importantly, neuronal mitochondrial damage releases damage-associated molecular patterns (DAMPs), including oxidized mitochondrial DNA, which further activate microglia and perpetuate inflammatory cycles (Deus et al., 2022). Activated microglia generate nitric oxide and ROS that inhibit neuronal electron transport chain complexes, deepening ATP depletion and establishing a bidirectional amplification loop between metabolic stress and immune activation (Simpson & Oliver, 2020). Thus, immunometabolic dysregulation in AD and PD represents an integrated disturbance of glycolytic reprogramming, mitochondrial ROS signaling, AMPK–mTOR imbalance, and NAD^+-SIRT1 decline, making restoration of coordinated immune–metabolic homeostasis a central therapeutic objective.

3. Nanotechnology design principles for brain energy therapy

3.1 Blood brain barrier (BBB) crossing strategies

The BBB is the dominant pharmacokinetic barrier limiting bioenergetic therapeutics for Alzheimer's disease and Parkinson's disease, formed by tight junction-sealed endothelial cells supported by pericytes and astrocytic end-feet, with abundant efflux transporters (e.g., P-gp, BCRP) that restrict CNS accumulation (Kim et al., 2025). Nanocarriers are therefore designed to exploit *active* transport routes rather than passive diffusion. Receptor-mediated transcytosis is the most widely engineered mechanism, using ligands or antibodies directed toward transferrin receptor (TfR), insulin receptor, or LRP1 to shuttle nanoparticles across brain endothelium and improve parenchymal delivery (Pardridge, 2012; Ulbrich et al., 2011). Adsorptive-mediated transcytosis can be induced by cationic or amphipathic surfaces that interact with the negatively charged glycocalyx, but it requires careful optimization to avoid toxicity and nonspecific uptake. Additional BBB strategies include biomimetic coatings (e.g., apolipoprotein E–mimetic peptides), intranasal nanocarriers that leverage olfactory/trigeminal pathways, and extracellular vesicles (exosomes) that exhibit intrinsic BBB interaction and favorable biocompatibility (Ahmed et al., 2025; De et al., 2026; Ding et al., 2025; Saraiva et al., 2016; Ulbrich et al., 2011). Across these approaches, particle size (often ~50–150 nm), stealth properties (PEGylation or zwitterionic coatings), and controlled ligand density are tuned to balance circulation time, endothelial binding, transcytosis efficiency, and avoidance of rapid reticuloendothelial clearance (Ouyang et al., 2024).

3.2 Cellular and mitochondrial targeting approaches

After BBB transit, the therapeutic requirement shifts from tissue access to cell- and organelle-scale precision, because bioenergetic failure is heterogeneous across cell types and is mitochondria-centered. Neuronal targeting can be achieved using ligands to receptors enriched on neurons or diseased endothelium (e.g., LRP1-binding peptides), while activated microglia can be targeted via scavenger receptors or other inflammation-associated markers, enabling immunometabolic modulation (Ruan et al., 2018). Intracellular delivery is then enhanced by endosomal escape motifs (pH-buffering "proton sponge" polymers, fusogenic lipids, or membrane-active peptides) to prevent lysosomal degradation (Shete et al., 2014). For mitochondrial targeting, a dominant design principle is exploiting the negative mitochondrial membrane potential using lipophilic cations such as triphenylphosphonium (TPP^+), which drives electrophoretic accumulation of nanoparticle-associated cargo within mitochondria and increases local concentration where ATP depletion and ROS overproduction originate (Murphy & Smith, 2007). Complementary strategies include mitochondria-penetrating peptides, cardiolipin-binding motifs for inner-membrane association, and mitochondria-targeted antioxidants or redox modulators formulated into lipid/polymeric carriers to couple substrate delivery with ETC stabilization (Alia et al., 2026; Apostolova & Victor, 2015; Szeto & Liu, 2018; Torchilin, 2009). Effective designs integrate targeting with safety constraints, because excessive cationic charge can disrupt membranes and exacerbate mitochondrial stress.

3.3 Controlled and stimuli-responsive release systems

Brain energy therapy benefits from nanocarriers that release cargo selectively within diseased microenvironments, thereby increasing therapeutic index and reducing off-target systemic exposure. Stimuli-responsive systems are engineered to respond to biochemical hallmarks of neurodegeneration-elevated reactive oxygen species (ROS), altered redox state (high glutathione gradients), inflammatory enzymes, and local pH changes in endosomes/lysosomes (Liu et al., 2023; Rajes et al., 2020). ROS-responsive polymers and linkers (e.g., thioketal, boronic ester, or peroxalate-based motifs) enable drug liberation preferentially in oxidatively stressed neurons and glia, matching release kinetics to pathological ROS burden (Wan et al., 2023). Redox-responsive disulfide crosslinks can be cleaved intracellularly, supporting cytosolic delivery of NAD^+ precursors, mitochondrial peptides, or anti-inflammatory agents. pH-responsive liposomes and polymers facilitate endosomal escape and timed release, while enzyme-responsive coatings (e.g., matrix metalloproteinase-sensitive peptides in inflamed tissue) can trigger local unmasking of targeting ligands (Lee & Thompson, 2017). These programmable systems are particularly valuable in AD/PD where oxidative stress and neuroinflammation fluctuate spatially and temporally, and where sustained, low-level mitochondrial support may be preferable to bolus dosing (Mura et al., 2013).

3.4 Multi-cargo and combinatorial nanoformulations

Because AD and PD bioenergetic collapse arises from intertwined mechanisms-ETC impairment, NAD^+ decline, oxidative stress, impaired glucose/ketone utilization, and immunometabolic activation-single-agent metabolic therapies often fail to deliver durable benefit. Nanotechnology enables *multi-cargo formulations* that co-encapsulate complementary agents (e.g., NAD^+ boosters + mitochondria-targeted antioxidants + anti-inflammatory drugs) with synchronized pharmacokinetics and coordinated delivery to the same target cell/compartment (Xu et al., 2025b). Layered or compartmentalized architectures (core-shell nanoparticles, liposome-in-polymer systems, or lipid–polymer hybrids) can separate cargo classes and program sequential release-such as early anti-inflammatory payload release followed by sustained mitochondrial substrate delivery-supporting systems-level correction of energy failure (Li et al., 2024). Nucleic-acid cargos (siRNA/ASO/mRNA) can be combined with small molecules to modulate pathways such as AMPK–mTOR–SIRT1/PGC-1α while simultaneously providing metabolic substrates, enabling mechanistic synergy rather than additive effects (Mitchell et al., 2021). Exosome-based or biomimetic carriers further facilitate multi-modal payload delivery with reduced immunogenicity and improved cellular communication fidelity (El-Andaloussi et al., 2012; Lin et al., 2013). Rational combinatorial design in neuroenergetics therefore prioritizes (i) mechanistic complementarity, (ii) matched intracellular destinations (e.g., mitochondria + cytosol), (iii) controlled ratios of cargos, and (iv) safety-by-design to avoid cumulative oxidative or inflammatory stress from the carrier itself.

4. Nanocarrier platforms for metabolic intervention

4.1 Polymeric nanoparticles

Polymeric nanoparticles are among the most extensively investigated nanocarriers for metabolic intervention in Alzheimer's and Parkinson's disease. As summarized in Table 1, various polymeric systems formulated from biodegradable materials such as PLGA, PEG, chitosan, and PBCA have been developed to enhance brain delivery of therapeutic agents. These nanoparticles protect metabolically active compounds from degradation, improve pharmacokinetics, and enable controlled release. Surface functionalization with targeting ligands further enhances blood–brain barrier penetration and neuronal uptake. By delivering antioxidants,

Table 1 >Polymeric nanoparticles used in CNS disorders.

S.No	Type of polymeric nanoparticle	Polymer used	Target disease	References
1	Tet-1 peptide-conjugated curcumin NP	PLGA-PEG-Tet-1	AD	Mathew et al. (2012)
2	Curcumin-loaded NP	PLGA	AD	Doggui et al. (2012)
3	Curcumin PEG-PLGA-B6 peptide NP	PLGA-PEG-B6	AD	Fan et al. (2018)
4	Tacrine NP	Chitosan	AD	Wilson et al. (2010)
5	Rivastigmine brain-targeted NP	PBCA + polysorbate 80	AD	Wilson et al. (2008)
6	Donepezil sustained-release NP	PLGA	AD	Md et al. (2014)
7	Galantamine NP	PLGA	AD	Fornaguera et al. (2015)
8	Curcumin osmotic NP	PLGA	AD	Tiwari et al. (2014)
9	Levodopa intranasal NP	PLGA + chitosan	PD	Ahmad et al. (2022)
10	Levodopa NP (optimized)	PLGA	PD	Ahmad et al. (2022)
11	Levodopa redox-responsive NP	PLGA	PD	Ahmad et al. (2022)
12	Levodopa surface-modified NP	PLGA	PD	Ahmad et al. (2022)
13	Galantamine NP	PLGA	AD	Hanumanthappa et al. (2025)

mitochondrial cofactors, anti-inflammatory agents, or gene modulators, polymeric nanoparticles support mitochondrial function, reduce oxidative stress, and help restore neuronal energy balance.

4.2 Lipid-based systems

Lipid-based nanocarriers represent a clinically advanced and highly adaptable platform for metabolic intervention in Alzheimer's and Parkinson's disease, systems such as liposomes, solid lipid nanoparticles (SLNs), nanostructured lipid carriers (NLCs), transferosomes are been extensively investigated for brain-targeted delivery. Composed of physiologically compatible lipids, these carriers offer excellent biocompatibility, low toxicity, and enhanced ability to encapsulate both lipophilic and hydrophilic therapeutic agents. Lipid-based systems can improve drug stability, facilitate blood-brain barrier penetration, and enable sustained or controlled release profiles. Surface modification with targeting ligands or mucoadhesive polymers further enhances brain accumulation, particularly via intranasal routes. By delivering antioxidants, cholinesterase inhibitors, dopaminergic agents, and mitochondrial modulators, lipid-based nanocarriers contribute to reducing oxidative stress, supporting mitochondrial bioenergetics, and modulating neuroinflammation in neurodegenerative disorders Tables 2 and 3.

4.3 Exosomes and biomimetic nanocarriers

4.3.1 Emerging nanozyme and hybrid systems

Emerging nanozyme and hybrid nanocarrier systems introduce a catalytic dimension to metabolic intervention in Alzheimer's and Parkinson's disease. Nanozymes such as cerium oxide, manganese dioxide, and Prussian blue nanoparticlesexhibit intrinsic enzyme-mimetic activities, including superoxide dismutase- and catalase-like functions. Previously reported nanozymes are mentioned in Table 4. Unlike conventional antioxidants, these catalytic systems continuously scavenge reactive oxygen species, thereby addressing oxidative stress more sustainably. Hybrid platforms that combine inorganic nanozymes with polymers, lipids, or biomimetic coatings enhance stability, targeting efficiency, and biocompatibility. Some systems integrate drug delivery with catalytic antioxidation, enabling simultaneous mitochondrial protection and metabolic support. By reducing oxidative damage, modulating neuroinflammation, and improving mitochondrial function, nanozyme-based and hybrid nanoplatforms represent a promising next-generation strategy for restoring neuronal energy homeostasis in neurodegenerative disorders.

Table 2 Lipid nanoparticles used in CNS disorders.

S.No	Type of lipid-based system	Lipids used	Target disease	Therapeutic moiety	References
1.	Liposomes (intranasal, brain-targeted)	Phospholipid/cholesterol liposomes (formulation per study)	Alzheimer's disease (AD)	Rivastigmine; nose-to-brain delivery	Wavikar et al. (2017)
2.	Liposomes in situ nasal gel (intranasal)	Donepezil-loaded liposomes + in situ gel matrix	Alzheimer's disease (AD)	Donepezil HCl; PK/biodistribution after intranasal dosing	Rajput and Butani (2022)
3.	Nanostructured lipid carriers (NLCs), chitosan-coated (intranasal)	Compritol + Capryol 90; Poloxamer 188; chitosan coating	Alzheimer's disease (AD)	Donepezil; nose-to-brain delivery	Yasir et al. (2022)
4.	NLCs co-loaded (intranasal)	Glyceryl palmitostearate (solid lipid) + oleic acid (liquid lipid)	Alzheimer's disease (AD)	Donepezil + astaxanthin; AD-like rat model (AlCl3)	Shehata et al. (2023)
5.	NLC-loaded in situ gel (intranasal)	DPZ-NLC embedded in gellan gum/Lutrol F127 in situ gel	Alzheimer's disease (AD)	Donepezil; trimethyl tin (TMT) neurodegeneration model	Sonawane and Pokharkar (2024)
6.	Cubosomes in in situ nasal gel	Cubosomes embedded in in situ gel (per study)	Alzheimer's disease (AD)	Donepezil HCl; controlled nose-to-brain delivery	Patil et al. (2019)

(continued)

Table 2 Lipid nanoparticles used in CNS disorders. (*cont'd*)

S.No	Type of lipid-based system	Lipids used	Target disease	Therapeutic moiety	References
7.	Solid lipid nanoparticles (SLNs) /Nanostructured lipid carriers (NLCs)	SLN/NLC lipid matrices (per study)	Alzheimer's disease (AD)/ CNS delivery	Curcumin; pharmacokinetic evaluation of brain accumulation	Sadegh Malvajerd et al. (2019)
8.	Solid lipid nanoparticles (SLNs)	Resveratrol-containing SLNs (optimized by design)	Alzheimer's disease (AD)	Resveratrol; in vitro + in vivo neurobehavioral/ biochemical readouts	Khishvand et al. (2024)
9.	Liposomal + siRNA/ drug combination (dual/triple cargo)	Liposomes (formulation per study)	Alzheimer's disease (AD)	Donepezil + memantine + BACE1 siRNA (reported combination approach)	Lee et al. (2024)
10.	Solid lipid nanoparticles (SLN) and Nanostructured lipid carriers (NLC)	Ropinirole lipid nanoparticles; hydrogel using Carbopol 934	Parkinson's disease (PD)	Ropinirole; haloperidol-induced PD model; PK/ PD evaluation	Dudhipala and Gorre (2020)
11.	Solid lipid nanoparticles (SLN) / Nanostructured lipid carriers (NLC)	Tristearin/tricaprin (NLC) and SLN variants (per study)	Parkinson's disease (PD)	Bromocriptine; 6-OHDA hemilesioned rat model	Dudhipala and Gorre (2020)

12.	Bromocriptine-loaded SLN and NLC (brain targeting)	Lipid nanoparticles prepared by high-pressure homogenization	Parkinson's disease (PD)	Bromocriptine; plasma PK + brain distribution	Esposito et al. (2008)
13.	Transferosomes in in situ nasal gel (intranasal)	Phosphatidylcholine + edge activator (e.g., sodium deoxycholate) in gel	Parkinson's disease (PD)	Rasagiline mesylate; brain targeting & PK after intranasal dosing	K M et al. (2024)
14.	Solid lipid nanoparticles (SLN) – nanoscale solid lipid particles	SLN lipid matrix (microemulsion method per study)	Parkinson's disease (PD)	Rasagiline mesylate; formulation & in vitro release	ElShagea et al. (2023)
15.	Liposomal/vesicular system + in situ gel (intranasal)	Vesicular transferosomes embedded into thermoresponsive gel	Parkinson's disease (PD)	Rasagiline mesylate; enhanced brain exposure	ElShagea et al. (2023)
16.	SLN (ropinirole)	Ropinirole-loaded solid lipid nanoparticles (RP-SLN)	Parkinson's disease (PD)	Ropinirole; PK/PD in haloperidol-induced PD model	Dudhipala and Gorre (2020)
17.	NLC (ropinirole)	Ropinirole-loaded nanostructured lipid carriers (RP-NLC)	Parkinson's disease (PD)	Ropinirole; PK/PD in haloperidol-induced PD model	Dudhipala and Gorre (2020)

(continued)

Table 2 Lipid nanoparticles used in CNS disorders. *(cont'd)*

S.No	Type of lipid-based system	Lipids used	Target disease	Therapeutic moiety	References
18.	SLN hydrogel (ropinirole)	RP-SLN embedded in Carbopol 934 hydrogel (RP-SLN-C)	Parkinson's disease (PD)	Ropinirole; improved bioavailability in vivo	Dudhipala and Gorre (2020)
19.	NLC hydrogel (ropinirole)	RP-NLC embedded in Carbopol 934 hydrogel (RP-NLC-C)	Parkinson's disease (PD)	Ropinirole; improved bioavailability in vivo	Dudhipala and Gorre (2020)
20.	SLN (bromocriptine; lipidic component variant)	Solid lipid nanoparticles (various lipidic components)	Parkinson's disease (PD)	Bromocriptine; controlled release (dialysis)	Esposito et al. (2008)
21.	NLC (bromocriptine; tristearin/tricaprin)	Tristearin/tricaprin nanostructured lipid carriers	Parkinson's disease (PD)	Bromocriptine; improved onset and prolonged effect (bar test)	K M et al. (2024)
22.	Transferosomes (rasagiline)	Transferosomes (phosphatidylcholine + edge activator)	Parkinson's disease (PD)	Rasagiline mesylate; ex vivo permeation + PK	ElShagea et al. (2023)
23.	Donepezil-NLCs (optimized)	Transcutol/Precirol NLC system (per study)	Alzheimer's disease (AD)	Donepezil NLCs; oral vs intranasal comparisons (as described)	Sonawane and Pokharkar (2024)

24.	Rivastigmine liposomes (alternate formulation reported within same primary paper)	Liposomes with formulation variables (per paper)	Alzheimer's disease (AD)	Rivastigmine; intranasal delivery	Yang et al. (2013)
25.	Resveratrol SLNs (scopolamine-induced cognitive deficit model)	Resveratrol-loaded SLNs	Alzheimer's disease (AD)	Resveratrol; mitigation of neurodegenerative symptoms	Khishvand et al. (2024)
26.	Curcumin SLNs (brain delivery)	Solid lipid nanoparticles for curcumin	Alzheimer's disease (AD) / CNS delivery	Curcumin; brain accumulation	Sadegh Malvajerd et al. (2019)

Table 3 Exosomes and biomimetic nanocarriers in CNS disorders.

Sr. No.	Type of exosome/ biomimetic nanocarrier	Material	Therapeutic moiety	Target disease	References
1.	Targeted exosomes (RVG-Lamp2b engineered)	Dendritic-cell-derived exosomes displaying RVG peptide	siRNA against BACE1 (gene-silencing)	Alzheimer's disease (AD)	Alvarez-Erviti et al. (2011)
2.	Curcumin-primed exosomes (Exo-cur)	Exosomes derived from curcumin-treated donor cells	Intrinsic curcumin-associated exosomal cargo (Tau phosphorylation modulation)	Alzheimer's disease (AD)	Wang et al. (2019)
3.	Dual-drug loaded exosomes	Engineered exosomes	Curcumin + methylene blue (EXO-Cur+MB)	Alzheimer's disease (AD)	Yang et al. (2025)
4.	Drug-loaded exosomes (macrophage/ monocyte exosomes)	Exosomes secreted by monocytes/ macrophages	Catalase (antioxidant enzyme)	Parkinson's disease (PD)	Haney et al. (2015)
5.	Blood exosomes for brain targeting	Blood-derived exosomes (Tf-TfR-mediated targeting)	Dopamine	Parkinson's disease (PD)	Qu et al. (2018)

6.	Targeted exosomes (RVG-exosomes)	RVG-functionalized exosomes	siRNA against α-synuclein (SNCA)	Parkinson's disease (PD)	Izco et al. (2019)
7.	Designer exosomes (EXOtic)	Implanted engineered producer cells generating designer exosomes	Catalase mRNA (therapeutic mRNA delivery)	Parkinson's disease (PD)	Kojima et al. (2018)
8.	Small extracellular vesicles (sEVs)	Umbilical cord blood mononuclear cell-derived sEVs	miR-124-3p (neuroprotective miRNA delivery)	Parkinson's disease (PD)	Esteves et al. (2022)
9.	Cell-mediated nanozyme delivery (biomimetic, macrophage carrier)	Bone-marrow-derived macrophages carrying catalase 'nanozymes'	Catalase packaged in PEI-PEG block ionomer complex (nanozyme)	Parkinson's disease (PD)	Batrakova et al. (2007)
10.	Exosome-mimetic nanovesicles (biomimetic platform)	Cell-derived nanovesicles generated by extrusion (exosome-mimetic)	Platform study (demonstrates exosome-like vesicles for delivery)	Biomimetic platform (widely cited; adaptable to CNS)	Jang et al. (2013)

(continued)

Table 3 Exosomes and biomimetic nanocarriers in CNS disorders. (*cont'd*)

Sr. No.	Type of exosome/biomimetic nanocarrier	Material	Therapeutic moiety	Target disease	References
11.	Microglia membrane-coated biomimetic nanoparticles	Microglia membrane coating over drug-loaded nanoparticles	Anti-neuroinflammatory/neuromodulatory payload (per study)	Parkinson's disease (PD) /neuroinflammation-targeting	Li et al. (2025)
12.	Dopamine-loaded blood exosomes (IV administration)	Blood exosomes (Tf–TfR targeting mechanism described)	Dopamine (IV delivery across BBB)	Parkinson's disease (PD)	Qu et al. (2018)
13.	Catalase-loaded exosomes (intranasal/alternative loading methods reported)	Macrophage/monocyte exosomes	Catalase (various loading approaches reported)	Parkinson's disease (PD)	Qu et al. (2018)
14.	RVG-exosome siRNA (systemic administration)	RVG-modified exosomes	α-Synuclein siRNA (systemic delivery)	Parkinson's disease (PD)	Cooper et al. (2014)
15.	RVG-exosome shRNA minicircles (systemic administration)	RVG-modified exosomes	Anti-α-synuclein shRNA minicircle (long-term knockdown)	Parkinson's disease (PD)	Izco et al. (2019)

Table 4 Nanozymes and hybrid systems in CNS disorders.

S No	Nanozyme / hybrid system	Catalytic (enzyme-mimetic) function (s) / key mechanism	Therapeutic moiety	Target disease	References
1.	Mn-doped CeO_2 hollow nanoparticles loaded with resveratrol (LMC-RES)	ROS-scavenging (SOD/CAT-like) via Ce^{3+}/Ce^{4+} redox cycling; Nrf2/HO-1 activation; inhibits Aβ aggregation	Drug-loaded nanozyme (resveratrol) enabling BBB crossing and neuronal enrichment	AD	Hu et al. (2023)
2.	Brain-penetrating MnO_2–polymer–lipid hybrid nanoparticles (Ab-TP-MDNPs)	Catalase-like O_2 generation + ROS quenching; pH normalization; remodels hypoxia/oxidative stress loop; reduces Aβ	Polymer–lipid hybrid; brain-penetrating + disease-site targeting (antibody/ligand)	AD	Park et al. (2023)
3.	Albumin–MnO_2 nanocomposites (HMn NCs)	SOD/CAT-mimetic ROS scavenging; suppresses Aβ fibrillization; anti-neuroinflammation	Protein–inorganic biomineralized hybrid (HSA template)	AD	Gao et al. (2023)
4.	Au nanorods with spatially separated CeO_2 deposition (plasmonic-enhanced nanozyme/PTT)	CeO_2 antioxidant catalysis + plasmonic/photothermal enhancement; improves BBB permeability; anti-Aβ/anti-oxidative stress	Hybrid plasmonic–ceria nanozyme enabling NIR photothermal therapy + catalytic antioxidation	AD	Ge et al. (2022)

(continued)

Table 4 Nanozymes and hybrid systems in CNS disorders. *(cont'd)*

S No	Nanozyme / hybrid system	Catalytic (enzyme-mimetic) function (s) / key mechanism	Therapeutic moiety	Target disease	References
5.	US-responsive nanozyme neuromodulation strategy (reviewed platform class)	Stimulus-triggered modulation of oxidative stress/inflammation using ultrasound as external trigger	Hybrid/triggered nanozyme concept (US activation)	AD	Karthika et al. (2024)
6.	Prussian blue nanozyme (PBzyme)	Multienzyme-like ROS scavenging; inhibits NLRP3 inflammasome/ caspase-1; blocks microglial pyroptosis; protects dopaminergic neurons	Intra-cerebroventricular delivery used in MPTP PD model (study design feature)	PD	Ma et al. (2022)
7.	Yb^{3+}/Er^{3+} co-doped CeO_{2-x} upconversion nanozyme (UCNPs)	Enhanced enzyme-like catalytic antioxidation (ROS scavenging) with upconversion-enabled functionality	Rare-earth doped nanozyme (theranostic-capable)	PD	Li et al. (2021)

8.	Ceria nanoparticles mitigating Parkinson-like toxicity (nanoceria)	Antioxidant nanozyme activity supporting dopamine metabolism and cellular survival	Nanoceria as stand-alone catalytic antioxidant	PD (Parkinson-like)	Pinna et al. (2015)
9.	CeO_2 nanoparticles suppressing α-synuclein-induced toxicity (yeast model)	Antioxidant nanozyme activity; mitigates α-syn toxicity at cellular level	Stand-alone CeO_2 nanozyme	PD	Ruotolo et al. (2020)
10.	Microglia membrane-coated biomimetic nanoparticles	Neuroinflammation regulation and neuroprotection (platform; payload-specific)	Cell-membrane-coated hybrid biomimetic system for inflammatory lesion targeting	PD	Li et al. (2025)

5. Route of administration and translational considerations

5.1 Intranasal nose-to-brain delivery

Intranasal administration represents a transformative strategy for bypassing the stringent limitations of the BBB, which traditionally restricts over 98 % of small-molecule drugs and nearly 100 % of large-molecule therapeutics from entering the brain (Wu et al., 2025). The nose-to-brain pathway utilizes two distinct routes: the olfactory nerve pathway and the trigeminal nerve pathway (Drath et al., 2025). The olfactory bulb, which is located in the anterior fossa of the skull, is in direct contact with the cerebrospinal fluid (CSF) and the olfactory receptor neurons. Nanocarriers deposited in the olfactory epithelium can enter the brain via either intracellular transport (endocytosis into the olfactory neurons and subsequent axonal transport) or paracellular transport (diffusion through the tight junctions between the epithelial cells into the subarachnoid space). The trigeminal nerve pathway, which innervates the nasal cavity, provides an additional route to the brainstem and other regions (Bahadur et al., 2020). This route is particularly efficacious for delivering therapeutic agents directly to deep brain structures involved in PD, such as the substantia nigra. To maximize the efficiency of this delivery, nanocarriers must be formulated with specific rheological properties-typically a low-viscosity liquid or a thermo-responsive gel-that allows for rapid coverage of the nasal mucosa upon administration. Furthermore, the inclusion of permeation enhancers and mucoadhesive polymers is critical to ensure that the carriers remain in the nasal cavity long enough for transport to occur, rather than being cleared by mucociliary action. This approach is highly promising for the treatment of AD and PD because it provides rapid, high-concentration delivery of metabolic enhancers, such as insulin mimetics or antioxidants, directly into the CNS, effectively avoiding first-pass metabolism and systemic side effects. However, the dose-volume is limited by the capacity of the nasal cavity, necessitating the development of concentrated, high-efficiency nanoformulations (Khatri et al., 2023).

5.2 Intravenous and systemic targeting

Systemic administration, typically via intravenous infusion, remains the conventional route for delivering therapeutics that require a broad distribution throughout the brain. However, achieving effective concentrations in the parenchyma necessitates overcoming the tightly regulated endothelial interface of the BBB (Hersh et al., 2016). This is achieved through the design of functionalized nanocarriers capable of receptor mediated transcytosis (RMT). By targeting endogenous transport receptors such as the transferrin receptor (TfR), insulin receptor, or the low-density lipoprotein receptor related protein 1 (LRP1) nanocarriers can be "shuttled" across the brain endothelium (Gampa et al., 2026). The efficiency of this process is highly dependent on ligand density and affinity. For instance, high affinity ligands may lead to "trapping" within the endothelial cells, whereas optimal affinity ligands ensure successful transcytosis into the brain interstitium. Once across, targeting can be further refined to ensure the nanocarriers selectively reach neurons or glia (Sierri et al., 2024). This involves surface engineering with ligands that recognize receptors upregulated on the diseased cell surface, such as the sigma-1 receptor or specific integrins (Soriani & Kourrich, 2019). Systemic targeting is particularly beneficial for diseases like AD, where widespread pathology across multiple cortical regions necessitates a global distribution of the therapeutic agen (Desai, 2015). Furthermore, the use of "smart" nanocarriers that only release their cargo in response to specific triggers, such as the acidic pH of the brain tumor microenvironment or the ROS-rich environment of inflamed neurons, significantly enhances safety by minimizing off-target effects. Despite the challenges, systemic targeting provides the most reliable route for delivering combinatorial therapies that require precise control over the pharmacokinetics of multiple, potentially interacting agents (Hossen et al., 2019).

5.3 Comparative translational feasibility

The transition of nanotherapeutics from bench to bedside is dictated by a trade-off between targeting efficacy, manufacturing complexity, and regulatory feasibility. Intranasal delivery offers a lower barrier to entry in terms of manufacturing, as it requires less stringent sterility requirements compared to intravenous formulations and avoids the complex pharmacokinetics associated with systemic circulation (Nunziata et al., 2026). However, the intranasal route faces challenges regarding the reproducibility of dose delivery, which is influenced by anatomical variability and the presence of nasal congestion or inflammation (Koo et al., 2024). Conversely, intravenous systems offer more predictable dosing and are better suited for large-scale clinical application, but they necessitate significant investment in sophisticated surface engineering and rigorous safety testing to prevent immune system activation (Mitragotri et al., 2014). From a translational perspective, simple, stable, and scalable platforms like lipid-based systems or polymeric nanoparticles currently lead the field. These systems benefit from established regulatory pathways for existing liposomal and polymer-based drugs. In contrast, complex, multi-component "theranostic" platforms-which combine diagnosis and therapy-face significant hurdles in the "Chemistry, Manufacturing, and Controls" (CMC) arena. Ensuring the consistency of multi-functional nanoparticles, particularly those incorporating complex targeting ligands or multiple drug types, is an immense challenge for large-scale production. Ultimately, the successful translation of these therapies will likely depend on a stepwise approach: starting with well-characterized, single-cargo carriers that address fundamental metabolic failures, and subsequently advancing to more complex, multi-modal platforms as manufacturing and regulatory protocols mature (Naser et al., 2024).

6. Challenges and future directions in neuro-nanomedicine

6.1 Standardization of bioenergetic endpoints

Understanding and comparing bioenergetic outcomes across studies is crucial to advancing nanomedicine for neurodegenerative diseases like Alzheimer's disease (AD) and Parkinson's disease (PD). Neurons are extraordinarily reliant on continuous ATP production via mitochondrial oxidative phosphorylation; even modest imbalances between energy supply and demand can accelerate synaptic failure and neurodegeneration. Indeed, decades of work have established mitochondrial dysfunction and bioenergetic failure as key features of multiple neurodegenerative disorders, with evidence of impaired respiratory chain activity, mitochondrial DNA damage, and elevated oxidative stress in both AD and PD patient tissues and model systems. For example, early research demonstrated substantial complex I defects and increased oxidative damage in PD brains, as well as reduced complex IV activity and ATP generation in AD samples, implicating a feedback loop between impaired energy metabolism and oxidative stress in disease progression. Despite the centrality of bioenergetic failure to neuropathology, there is no universally accepted set of bioenergetic outcome measures for therapeutic studies, particularly those evaluating nanocarrier-mediated interventions. Researchers use a diverse array of assays-mitochondrial membrane potential ($\Delta\Psi m$), ATP quantification, oxygen consumption rate (OCR), extracellular acidification rate (ECAR), reactive oxygen species

(ROS) levels, and metabolic flux analyses often without standardized protocols or reference controls (Strope et al., 2022). This heterogeneity makes it difficult to compare effectiveness between nanocarrier designs or to aggregate data across laboratories. Standardization is further complicated by differences in model systems (e.g., primary neurons, iPSC-derived neurons, animal models) that inherently vary in metabolic baseline and adaptability. Recent work underscores that even within the same model type, methodological differences in how OCR or ATP assays are conducted can produce divergent results, complicating interpretation. Stronger consensus is needed on quantitative benchmarks for "metabolic rescue," such as minimum fold-changes in ATP production relative to untreated controls, standardized ROS metrics corrected for cell density, and reproducible measures of mitochondrial network integrity. Validated bioenergetic endpoints aligned with clinical biomarkers (e.g., PET-based glucose metabolism) would bridge preclinical and translational work, accelerating meaningful comparisons and reducing the risk of false positives or non-translatable findings. Ultimately, internationally agreed standards for bioenergetic measurements in neuro-nanomedicine will be foundational to regulatory confidence and clinical translation (Sharma et al., 2025).

6.2 Stage-specific and personalized nanotherapies

Neurodegenerative diseases such as AD and PD are not homogeneous static entities; they unfold dynamically over decades with evolving molecular phenotypes, variable rates of progression, and significant inter-individual variability. AD typically begins with subtle metabolic changes marked by cerebral glucose hypometabolism in temporoparietal regions years before overt cognitive decline, while PD often features early dopaminergic neuronal dysfunction and oxidative stress long before motor symptoms manifest. This heterogeneity has critical implications for nanotherapy design, highlighting the necessity for stage-specific and personalized approaches rather than one-size-fits-all solutions (Xie et al., 2014). Nanocarriers can be engineered to deliver distinct therapeutic cargoes tailored to disease stage: early AD may benefit from metabolic enhancers that bolster mitochondrial biogenesis and glucose utilization, whereas later stages might require combined anti-inflammatory and proteinopathy-modulating payloads. Similarly, early PD interventions may prioritize mitochondrial complex I support and antioxidant delivery, whereas advanced disease might necessitate dopaminergic replacement or synaptic support (Yadav et al., 2025). Personalized nanotherapies also demand integration with patient-specific biomarkers that reflect metabolic status, such as FDG-PET measures of glucose uptake, peripheral blood metabolites, or emerging mitochondrial function assays. Biomarker-guided stratification could ensure that patients receive nano-formulations suited to their underlying pathology and metabolic requirements, improving therapeutic efficacy and reducing unnecessary exposure to ineffective interventions (Kim et al., 2013). Although personalized nanomedicine has progressed significantly in oncology, its application to neurodegeneration is nascent. Integrating nanocarrier therapeutics with personalized diagnostic metrics, including genomic risk profiles (e.g., APOE genotype in AD, LRRK2 mutations in PD) and dynamic metabolic biomarkers, could transform treatment paradigms. However, this approach requires standardized biomarker validation and robust clinical studies that correlate specific bioenergetic profiles with nanocarrier response, enabling precision neuro-nanotherapy that is adaptive to disease staging and individual physiology (Alghamdi et al., 2022).

6.3 Theranostic and multi-functional nanoplatforms

Nanomedicine uniquely enables theranostic platforms that combine therapeutic and diagnostic functionalities within a single nanoscale system, a paradigm particularly attractive for complex disorders like AD and PD where early detection and ongoing monitoring are crucial. Theranostic nanocarriers can integrate drug delivery with imaging agents (e.g., MRI contrast materials, PET tracers, fluorescent markers) or biosensors that report on local metabolic status (e.g., ROS levels, pH changes), facilitating real-time assessment of treatment response (Toader et al., 2024). Such platforms offer major advantages over conventional therapies by enabling targeted delivery of therapeutics while simultaneously tracking biodistribution, target engagement, and therapeutic efficacy. In neurodegeneration, these multi-functional nanoplatforms can be designed to cross the blood–brain barrier and localize to regions of interest (e.g., amyloid plaques in AD or dopamine-depleted striatum in PD), allowing clinicians to visualize and quantify pathological hallmarks and metabolic restoration over time (Chen et al., 2025). Beyond imaging, theranostic systems can integrate sensors responsive to oxidative stress or mitochondrial dysfunction, enabling dynamic feedback that could guide dosage or trigger environment-responsive drug release. Integration with artificial intelligence and computational analytics further augments the capacity for personalized therapeutic decisions, as real-time data streams from these platforms could inform adaptive treatment adjustments. While much of the work in neurotheranostics remains at the preclinical stage, the conceptual promise is supported by rapid advances in multifunctional nanocarrier engineering that enable simultaneous targeting, controlled release, and signal reporting. However, realizing clinical impact will necessitate addressing challenges in multi-component safety, manufacturing scalability, and regulatory approval pathways tailored for integrated diagnostic-therapeutic agents (Chen et al., 2024).

6.4 Path toward clinical translation

Translating neuro-nanomedicine from bench to bedside has proven more challenging in AD and PD than in areas like oncology or infectious disease, largely due to the complexity of the central nervous system and the stringent requirements for safety, efficacy, and reproducibility. Clinical translation of nanomedicines for brain disorders must overcome the blood–brain barrier through validated delivery strategies, demonstrate meaningful improvements in patient-relevant outcomes, and satisfy regulatory expectations for long-term safety and manufacturability. More than 50 nanomedicines are approved globally for various indications, and over a hundred are under clinical investigation, including formulations aimed at brain disorders; however, none yet fully address metabolic rescue in neurodegeneration. A successful clinical path will likely require early engagement with regulatory agencies to define acceptable bioenergetic endpoints and surrogate markers linked to clinical benefit. Rigorous phase I and II trials need to demonstrate safety, biodistribution, target engagement (e.g., modulation of mitochondrial function in cerebrospinal fluid or PET imaging), and preliminary efficacy, ideally guided by robust biomarkers that reflect metabolic and pathological changes. Manufacturing challenges include ensuring batch-to-batch consistency of complex nanostructures, scalability of production, and stability of multifunctional cargoes. Immune and off-target effects must be thoroughly evaluated in long-term toxicology studies, especially for inorganic nanozymes or biomimetic systems that may accumulate over repeated dosing. Moreover, clinical translation will benefit from interdisciplinary collaborations spanning nanotechnology, neurology, imaging sciences, and regulatory science, ensuring that nanocarrier design aligns with real clinical needs and regulatory pathways. Ultimately, advancing neuro-nanomedicine toward clinical impact will depend on bridging robust preclinical evidence with well-designed clinical studies, standardized biomarkers, and scalable manufacturing, thereby fulfilling the promise of targeted metabolic rescue in AD and PD (Gong et al., 2023; Saraiva et al., 2016).

7. Conclusion

The bioenergetic crisis in Alzheimer's and Parkinson's disease represents a fundamental pathological pillar that traditional, systemic pharmacological agents have failed to address. The limitations of the blood-brain barrier, coupled with the organelle-specific nature of mitochondrial failure, have historically rendered conventional metabolic support ineffective. Nanotechnology provides a precise, multi-dimensional framework to overcome these challenges. By enabling site-specific targeting, stimuli-responsive release, and the simultaneous modulation of interconnected pathways (e.g., ETC stabilization, NAD^+ restoration, and immunometabolic balance), nanomedicine transforms the prospect of metabolic rescue from a theoretical possibility into a scientifically grounded therapeutic objective. While the field faces ongoing hurdles in standardization, long-term neurotoxicity assessment, and clinical translation, the integration of biomimetic systems and theranostic platforms offers a promising future for personalized, precision neuro-nanomedicine.

References

Ahmad, M. M. Z., Sabri, A. H. B., Anjani, Q. K., Domínguez-Robles, J., Abdul Latip, N., & Hamid, K. A. (2022). Design and development of levodopa loaded polymeric nanoparticles for intranasal delivery. *Pharmaceuticals, 15*, 370. https://doi.org/10.3390/ph15030370.

Ahmed, S. S., Pande, A. H., & Sharma, S. S. (2025). *Therapeutic use of ApoE-mimetic peptides. Apolipoprotein E.* Cham: Springer Nature Switzerland. 1–20. https://doi.org/10.1007/978-3-031-52641-1_43-1.

Alghamdi, M. M. A., Fallica, A. N., Virzì, N., Kesharwani, P., Pittalà, V., & Greish, K. (2022). The promise of nanotechnology in personalized medicine. *Journal of Personalized Medicine, 12*, 673. https://doi.org/10.3390/jpm12050673.

Alia, S. S., Pedriali, G., Compagnucci, P., Valeri, Y., Membrino, V., Di Crescenzo, T., ... Casella, M. (2026). Mitochondria at the crossroads of cardiovascular disease: Mechanistic drivers and emerging therapeutic strategies. *Cells, 15*, 372. https://doi.org/10.3390/cells15040372.

Alvarez-Erviti, L. L., Seow, Y., Yin, H., Betts, C., Lakhal, S., & Wood, M. J. A. (2011). Delivery of siRNA to the mouse brain by systemic injection of targeted exosomes. *Nature Biotechnology, 29*, 341–345. https://doi.org/10.1038/nbt.1807.

Amjad, S. S., Nisar, S., Bhat, A. A., Shah, A. R., Frenneaux, M. P., Fakhro, K., ... Bagga, P. (2021). Role of NAD+ in regulating cellular and metabolic signaling pathways. *Molecular Metabolism, 49*, 101195. https://doi.org/10.1016/j.molmet.2021.101195.

Apostolova, N. N., & Victor, V. M. (2015). Molecular strategies for targeting antioxidants to mitochondria: Therapeutic implications. *Antioxidants & Redox Signaling, 22*, 686–729. https://doi.org/10.1089/ars.2014.5952.

Arnold, P. P. K., & Finley, L. W. S. (2023). Regulation and function of the mammalian tricarboxylic acid cycle. *Journal of Biological Chemistry, 299*, 102838. https://doi.org/10.1016/j.jbc.2022.102838.

Bahadur, S. S., Pardhi, D. M., Rautio, J., Rosenholm, J. M., & Pathak, K. (2020). Intranasal nanoemulsions for direct nose-to-brain delivery of actives for CNS Disorders. *Pharmaceutics, 12*, 1230. https://doi.org/10.3390/pharmaceutics12121230.

Batrakova, E. E. V., Li, S., Reynolds, A. D., Mosley, R. L., Bronich, T. K., Kabanov, A. V., & Gendelman, H. E. (2007). A macrophage−nanozyme delivery system for Parkinson's disease. *Bioconjugate Chemistry, 18*, 1498–1506. https://doi.org/10.1021/bc700184b.

Beal, M. M. F. (1998). Mitochondrial dysfunction in neurodegenerative diseases. *Biochimica et Biophysica Acta (BBA) - Bioenergetics, 1366*, 211–223. https://doi.org/10.1016/S0005-2728(98)00114-5.

Bender, A. A., Krishnan, K. J., Morris, C. M., Taylor, G. A., Reeve, A. K., Perry, R. H., ... Turnbull, D. M. (2006). High levels of mitochondrial DNA deletions in substantia nigra neurons in aging and Parkinson disease. *Nature Genetics, 38*, 515–517. https://doi.org/10.1038/ng1769.

Caccamo, A. A., Maldonado, M. A., Majumder, S., Medina, D. X., Holbein, W., Magrí, A., & Oddo, S. (2011). Naturally secreted amyloid-β increases mammalian target of rapamycin (mTOR) activity via a PRAS40-mediated mechanism. *Journal of Biological Chemistry, 286*, 8924–8932. https://doi.org/10.1074/jbc.M110.180638.

Chen, L., Guan, Y., Wang, S., Han, X., Guo, F., & Wang, Y. (2025). Engineered nanoplatforms for brain-targeted co-delivery of phytochemicals in Alzheimer's disease: Rational design, blood-brain barrier penetration, and multi-target therapeutic synergy. *Neurotherapeutics: the Journal of the American Society for Experimental NeuroTherapeutics, 22*, e00722. https://doi.org/10.1016/j.neurot.2025.e00722.

Chen, S., Li, Q., Shi, H., Li, F., Duan, Y., & Guo, Q. (2024). New insights into the role of mitochondrial dynamics in oxidative stress-induced diseases. *Biomedicine & Pharmacotherapy, 178*, 117084. https://doi.org/10.1016/j.biopha.2024.117084.

Cooper, J. J. M., Wiklander, P. B. O., Nordin, J. Z., Al-Shawi, R., Wood, M. J., Vithlani, M., ... Alvarez-Erviti, L. (2014). Systemic exosomal siRNA delivery reduced alpha-synuclein aggregates in brains of transgenic mice. *Movement Disorders, 29*, 1476–1485. https://doi.org/10.1002/mds.25978.

Coutinho, A. M. N., Porto, F. H. G., Zampieri, P. F., Otaduy, M. C., Perroco, T. R., Oliveira, M. O., ... Buchpiguel, C. A. (2015). Analysis of the posterior cingulate cortex with [18 F]FDG-PET and Naa/mI in mild cognitive impairment and Alzheimer's disease: Correlations and differences between the two methods. *Dementia & Neuropsychologia, 9*, 385–393. https://doi.org/10.1590/1980-57642015DN94000385.

Covarrubias, A. A. J., Perrone, R., Grozio, A., & Verdin, E. (2021). NAD+ metabolism and its roles in cellular processes during ageing. *Nature Reviews. Molecular Cell Biology, 22*, 119–141. https://doi.org/10.1038/s41580-020-00313-x.

Cunnane, S. S. C., Courchesne-Loyer, A., St-Pierre, V., Vandenberghe, C., Pierotti, T., Fortier, M., ... Castellano, C. (2016). Can ketones compensate for deteriorating brain glucose uptake during aging? Implications for the risk and treatment of Alzheimer's disease. *Annals of the New York Academy of Sciences, 1367*, 12–20. https://doi.org/10.1111/nyas.12999.

De, R. R., Shi, S., & Kim, K. (2026). Crossing the blood–brain barrier with molecularly imprinted polymeric nanocarriers: An emerging frontier in brain disease therapy. *Advanced Science, 13*. https://doi.org/10.1002/advs.202517004.

Desai, P. P. (2015). Therapeutic targets and delivery challenges for Alzheimer's disease. *World Journal of Pharmacology, 4*, 236. https://doi.org/10.5497/wjp.v4.i3.236.

Deus, C. C. M., Tavares, H., Beatriz, M., Mota, S., & Lopes, C. (2022). Mitochondrial damage-associated molecular patterns content in extracellular vesicles promotes early inflammation in neurodegenerative disorders. *Cells, 11*, 2364. https://doi.org/10.3390/cells11152364.

Dexter, D. D. T., Carter, C. J., Wells, F. R., Javoy-Agid, F., Agid, Y., Lees, A., ... Marsden, C. D. (1989b). Basal lipid peroxidation in substantia nigra is increased in Parkinson's disease. *Journal of Neurochemistry, 52*, 381–389. https://doi.org/10.1111/j.1471-4159.1989.tb09133.x.

Dexter, D. D. T., Wells, F. R., Lee, A. J., Agid, F., Agid, Y., Jenner, P., & Marsden, C. D. (1989a). Increased nigral iron content and alterations in other metal ions occurring in brain in Parkinson's disease. *Journal of Neurochemistry, 52*, 1830–1836. https://doi.org/10.1111/j.1471-4159.1989.tb07264.x.

Ding, F. F., Hou, R., Han, B., & Fang, X. (2025). Cell membrane- and vesicle-based bionic nanodrugs: Applications in central nervous system diseases and exploration of nasal–Cerebral delivery. *Gels, 11*, 846. https://doi.org/10.3390/gels11110846.

Doggui, S. S., Sahni, J. K., Arseneault, M., Dao, L., & Ramassamy, C. (2012). Neuronal uptake and neuroprotective effect of curcumin-loaded PLGA nanoparticles on the human SK-N-SH cell line. *Journal of Alzheimer's Disease, 30*, 377–392. https://doi.org/10.3233/JAD-2012-112141.

Drath, I. I., Richter, F., & Feja, M. (2025). Nose-to-brain drug delivery: From bench to bedside. *Translational Neurodegeneration, 14*, 23. https://doi.org/10.1186/s40035-025-00481-w.

Dudhipala, N. N., & Gorre, T. (2020). Neuroprotective effect of ropinirole lipid nanoparticles enriched hydrogel for Parkinson's disease: In vitro, ex vivo, pharmacokinetic and pharmacodynamic evaluation. *Pharmaceutics, 12*, 448. https://doi.org/10.3390/pharmaceutics12050448.

El-Andaloussi, S. S., Lee, Y., Lakhal-Littleton, S., Li, J., Seow, Y., Gardiner, C., ... Wood, M. J. A. (2012). Exosome-mediated delivery of siRNA in vitro and in vivo. *Nature Protocols, 7*, 2112–2126. https://doi.org/10.1038/nprot.2012.131.

ElShagea, H. H. N., Makar, R. R., Salama, A. H., Elkasabgy, N. A., & Basalious, E. B. (2023). Investigating the targeting power to brain tissues of intranasal rasagiline mesylate-loaded transferosomal in situ gel for efficient treatment of Parkinson's disease. *Pharmaceutics, 15*, 533. https://doi.org/10.3390/pharmaceutics15020533.

Esposito, E. E., Fantin, M., Marti, M., Drechsler, M., Paccamiccio, L., Mariani, P., ... Cortesi, R. (2008). Solid lipid nanoparticles as delivery systems for bromocriptine. *Pharmaceutical Research, 25*, 1521–1530. https://doi.org/10.1007/s11095-007-9514-y.

Esteves, M. M., Abreu, R., Fernandes, H., Serra-Almeida, C., Martins, P. A. T., Barão, M., ... Bernardino, L. (2022). MicroRNA-124-3p-enriched small extracellular vesicles as a therapeutic approach for Parkinson's disease. *Molecular Therapy, 30*, 3176–3192. https://doi.org/10.1016/j.ymthe.2022.06.003.

Fan, S. S., Zheng, Y., Liu, X., Fang, W., Chen, X., Liao, W., ... Liu, J. (2018). Curcumin-loaded PLGA-PEG nanoparticles conjugated with B6 peptide for potential use in Alzheimer's disease. *Drug Delivery, 25*, 1091–1102. https://doi.org/10.1080/10717544.2018.1461955.

Fornaguera, C. C., Feiner-Gracia, N., Calderó, G., García-Celma, M. J., & Solans, C. (2015). Galantamine-loaded PLGA nanoparticles, from nano-emulsion templating, as novel advanced drug delivery systems to treat neurodegenerative diseases. *Nanoscale, 7*, 12076–12084. https://doi.org/10.1039/C5NR03474D.

Galasso, A. A., Cameron, C. S., Frenguelli, B. G., & Moffat, K. G. (2017). An AMPK-dependent regulatory pathway in tau-mediated toxicity. *Biology Open*. https://doi.org/10.1242/bio.022863.

Gampa, G. G., Vadlakonda, R., Stefanich, E., Kamath, A. V., Sadekar, S., & Shivva, V. (2026). Bridging the blood-brain barrier: Strategies to improve delivery of biologics to tumors in the brain. *Fluids Barriers CNS, 23*, 25. https://doi.org/10.1186/s12987-026-00760-2.

Gao, W. W., Liu, W., Dong, X., & Sun, Y. (2023). Albumin–manganese dioxide nanocomposites: A potent inhibitor and ROS scavenger against Alzheimer's β-amyloid fibrillogenesis and neuroinflammation. *Journal of Materials Chemistry B, 11*, 10482–10496. https://doi.org/10.1039/D3TB01763J.

Ge, K. K., Mu, Y., Liu, M., Bai, Z., Liu, Z., Geng, D., & Gao, F. (2022). Gold nanorods with spatial separation of CeO $_2$ deposition for plasmonic-enhanced antioxidant stress and photothermal therapy of Alzheimer's disease. *ACS Applied Materials & Interfaces, 14*, 3662–3674. https://doi.org/10.1021/acsami.1c17861.

Gong, B., Zhuang, J., Ji, W., Chen, X., Li, P., Cheng, W., ... Yin, Y. (2023). The long and the short of current nanomedicines for treating Alzheimer's disease. *Journal of Translational Internal Medicine, 10*, 294–296. https://doi.org/10.2478/jtim-2021-0054.

Haney, M. M. J., Klyachko, N. L., Zhao, Y., Gupta, R., Plotnikova, E. G., He, Z., ... Batrakova, E. V. (2015). Exosomes as drug delivery vehicles for Parkinson's disease therapy. *Journal of Controlled Release, 207*, 18–30. https://doi.org/10.1016/j.jconrel.2015.03.033.

Hanumanthappa, R. R., Heggannavar, G. B., Banakar, A., Achari, D. D., Ramesh Karoshi, V., Krushna, B. R. R., ... Devaraju, K. S. (2025). A levodopa-encapsulated poly-ε-caprolactone nanocomposite improves the motor symptoms and neurochemical changes in a rotenone-induced mouse model of Parkinson's disease. *ACS Omega, 10*, 19682–19696. https://doi.org/10.1021/acsomega.5c00691.

Hardie, D. D. G. (2008). AMPK: A key regulator of energy balance in the single cell and the whole organism. *International Journal of Obesity, 32*, S7–S12. https://doi.org/10.1038/ijo.2008.116.

Hardie, D. D. G., Ross, F. A., & Hawley, S. A. (2012). AMPK: A nutrient and energy sensor that maintains energy homeostasis. *Nature Reviews. Molecular Cell Biology, 13*, 251–262. https://doi.org/10.1038/nrm3311.

Heneka, M. M. T., Kummer, M. P., Stutz, A., Delekate, A., Schwartz, S., Vieira-Saecker, A., ... Golenbock, D. T. (2013). NLRP3 is activated in Alzheimer's disease and contributes to pathology in APP/PS1 mice. *Nature, 493*, 674–678. https://doi.org/10.1038/nature11729.

Hersh, D. D. S., Wadajkar, A. S., Roberts, N. B., Perez, J. G., Connolly, N. P., Frenkel, V., ... Kim, A. J. (2016). Evolving drug delivery strategies to overcome the blood brain barrier. *Current Pharmaceutical Design, 22*, 1177–1193. https://doi.org/10.2174/1381612822666151221150733.

Hossen, S. S., Hossain, M. K., Basher, M. K., Mia, M. N. H., Rahman, M. T., & Uddin, M. J. (2019). Smart nanocarrier-based drug delivery systems for cancer therapy and toxicity studies: A review. *Journal of Advanced Research, 15*, 1–18. https://doi.org/10.1016/j.jare.2018.06.005.

Houtkooper, R. R. H., Pirinen, E., & Auwerx, J. (2012). Sirtuins as regulators of metabolism and healthspan. *Nature Reviews. Molecular Cell Biology, 13*, 225–238. https://doi.org/10.1038/nrm3293.

Hu, Y. Y., Guo, H., Cheng, S., Sun, J., Du, J., Liu, X., ... Tian, H. (2023). Functionalized cerium dioxide nanoparticles with antioxidative neuroprotection for Alzheimer's disease. *International Journal of Nanomedicine, 18*, 6797–6812. https://doi.org/10.2147/IJN.S434873.

Izco, M. M., Blesa, J., Schleef, M., Schmeer, M., Porcari, R., Al-Shawi, R., ... Alvarez-Erviti, L. (2019). Systemic exosomal delivery of shRNA minicircles prevents parkinsonian pathology. *Molecular Therapy, 27*, 2111–2122. https://doi.org/10.1016/j.ymthe.2019.08.010.

Jang, S. S. C., Kim, O. Y., Yoon, C. M., Choi, D.-S., Roh, T.-Y., Park, J., ... Gho, Y. S. (2013). Bioinspired exosome-mimetic nanovesicles for targeted delivery of chemotherapeutics to malignant tumors. *ACS Nano, 7*, 7698–7710. https://doi.org/10.1021/nn402232g.

K M, A. A. S., Angolkar, M., Rahamathulla, M., Thajudeen, K. Y., Ahmed, M. M., Farhana, S. A., ... Natarajan, J. (2024). Box-Behnken design-based optimization and evaluation of lipid-based nano drug delivery system for brain targeting of bromocriptine. *Pharmaceuticals, 17*, 720. https://doi.org/10.3390/ph17060720.

Karthika, V. V., Sridharan, B., Nam, J. W., Kim, D., & Gyun Lim, H. (2024). Neuromodulation by nanozymes and ultrasound during Alzheimer's disease management. *Journal of Nanobiotechnology, 22*, 139. https://doi.org/10.1186/s12951-024-02406-7.

Khatri, D. D. K., Preeti, K., Tonape, S., Bhattacharjee, S., Patel, M., Shah, S., ... Singh, S. B. (2023). Nanotechnological advances for nose to brain delivery of therapeutics to improve the Parkinson therapy. *Current Neuropharmacology, 21*, 493–516. https://doi.org/10.2174/1570159X20666220507022701.

Khishvand, M. M. A., Yeganeh, E. M., Zarei, M., Soleimani, M., Mohammadi, M., & Mahjub, R. (2024). Development, statistical optimization, and characterization of resveratrol-containing solid lipid nanoparticles (SLNs) and determination of the efficacy in reducing neurodegenerative symptoms related to Alzheimer's disease: In vitro and in vivo study. *BioMed Research International, 2024*. https://doi.org/10.1155/2024/7877265.

Kim, S. S., Jung, U. J., & Kim, S. R. (2025). The crucial role of the blood–brain barrier in neurodegenerative diseases: Mechanisms of disruption and therapeutic implications. *Journal of Clinical Medicine, 14*, 386. https://doi.org/10.3390/jcm14020386.

Kim, T. T. H., Lee, S., & Chen, X. (2013). Nanotheranostics for personalized medicine. *Expert Review of Molecular Diagnostics, 13*, 257–269. https://doi.org/10.1586/erm.13.15.

Kojima, R. R., Bojar, D., Rizzi, G., Hamri, G. C.-E., El-Baba, M. D., Saxena, P., ... Fussenegger, M. (2018). Designer exosomes produced by implanted cells intracerebrally deliver therapeutic cargo for Parkinson's disease treatment. *Nature Communications, 9*, 1305. https://doi.org/10.1038/s41467-018-03733-8.

Kolb, H. H., Kempf, K., Röhling, M., Lenzen-Schulte, M., Schloot, N. C., & Martin, S. (2021). Ketone bodies: From enemy to friend and guardian angel. *BMC Medicine, 19*, 313. https://doi.org/10.1186/s12916-021-02185-0.

Koo, J. J., Lim, C., & Oh, K. T. (2024). Recent advances in intranasal administration for brain-targeting delivery: A comprehensive review of lipid-based nanoparticles and stimuli-responsive gel formulations. *International Journal of Nanomedicine, 19*, 1767–1807. https://doi.org/10.2147/IJN.S439181.

Krishnan, K. J., Greaves, L. C., Reeve, A. K., & Turnbull, D. (2007). The ageing mitochondrial genome. *Nucleic Acids Research, 35*, 7399–7405. https://doi.org/10.1093/nar/gkm635.

Lee, D. D., Shen, A. M., Shah, M., Garbuzenko, O. B., & Minko, T. (2024). Vivo evaluation of nose-to-brain delivery of liposomal donepezil, memantine, and BACE-1 siRNA for Alzheimer's disease therapy. *International Journal of Molecular Sciences, 25*, 10357. https://doi.org/10.3390/ijms251910357.

Lee, Y. Y., & Thompson, D. H. (2017). Stimuli-responsive liposomes for drug delivery. *WIREs Nanomedicine and Nanobiotechnology, 9*. https://doi.org/10.1002/wnan.1450.

Li, L. L., Sun, C., Cai, S., Hang, Z., Gao, X., Hou, L., ... Du, H. (2025). Biomimetic nanoparticles enhance recovery of movement disorders in Parkinson's disease by improving microglial mitochondrial homeostasis and suppressing neuroinflammation. *ACS Applied Materials & Interfaces, 17*, 23536–23552. https://doi.org/10.1021/acsami.4c22181.

Li, R. R., Peng, F., Cai, J., Yang, D., & Zhang, P. (2020). Redox dual-stimuli responsive drug delivery systems for improving tumor-targeting ability and reducing adverse side effects. *Asian Journal of Pharmaceutical Sciences, 15*, 311–325. https://doi.org/10.1016/j.ajps.2019.06.003.

Li, Y. Y., Li, X., Wei, L., & Ye, J. (2024). Advancements in mitochondrial-targeted nanotherapeutics: Overcoming biological obstacles and optimizing drug delivery. *Frontiers in Immunology, 15*. https://doi.org/10.3389/fimmu.2024.1451989.

Li, Y. Y., Li, Y., Wang, H., & Liu, R. (2021). Yb $^{3+}$, Er $^{3+}$ codoped cerium oxide upconversion nanoparticles enhanced the enzymelike catalytic activity and antioxidative activity for Parkinson's disease treatment. *ACS Applied Materials & Interfaces, 13*, 13968–13977. https://doi.org/10.1021/acsami.1c00157.

Lin, Y. Y.-L., Yuksel Durmaz, Y., Nör, J. E., & ElSayed, M. E. H. (2013). Synergistic combination of small molecule inhibitor and RNA interference against antiapoptotic Bcl-2 protein in head and neck cancer cells. *Molecular Pharmaceutics, 10*, 2730–2738. https://doi.org/10.1021/mp4001662.

Liu, J. J., Jia, B., Li, Z., & Li, W. (2023). Reactive oxygen species-responsive polymer drug delivery systems. *Frontiers in Bioengineering and Biotechnology, 11*. https://doi.org/10.3389/fbioe.2023.1115603.

Ma, X. X., Hao, J., Wu, J., Li, Y., Cai, X., & Zheng, Y. (2022). Prussian blue nanozyme as a pyroptosis inhibitor alleviates neurodegeneration. *Advanced Materials, 34*. https://doi.org/10.1002/adma.202106723.

Mathew, A. A., Fukuda, T., Nagaoka, Y., Hasumura, T., Morimoto, H., Yoshida, Y., ... Kumar, D. S. (2012). Curcumin loaded-PLGA nanoparticles conjugated with Tet-1 peptide for potential use in Alzheimer's disease. *PLoS One, 7*, e32616. https://doi.org/10.1371/journal.pone.0032616.

Md, S. S., Ali, M., Baboota, S., Sahni, J. K., Bhatnagar, A., & Ali, J. (2014). Preparation, characterization, in vivo biodistribution and pharmacokinetic studies of donepezil-loaded PLGA nanoparticles for brain targeting. *Drug Development and Industrial Pharmacy, 40*, 278–287. https://doi.org/10.3109/03639045.2012.758130.

Mergenthaler, P., Lindauer, U., Dienel, G. A., & Meisel, A. (2013). Sugar for the brain: The role of glucose in physiological and pathological brain function. *Trends in Neurosciences, 36*, 587–597. https://doi.org/10.1016/j.tins.2013.07.001.

Mitchell, M. M. J., Billingsley, M. M., Haley, R. M., Wechsler, M. E., Peppas, N. A., & Langer, R. (2021). Engineering precision nanoparticles for drug delivery. *Nature Reviews. Drug Discovery, 20*, 101–124. https://doi.org/10.1038/s41573-020-0090-8.

Mitragotri, S. S., Burke, P. A., & Langer, R. (2014). Overcoming the challenges in administering biopharmaceuticals: Formulation and delivery strategies. *Nature Reviews. Drug Discovery, 13*, 655–672. https://doi.org/10.1038/nrd4363.

Mosconi, L. L., Tsui, W. H., Herholz, K., Pupi, A., Drzezga, A., Lucignani, G., ... de Leon, M. J. (2008). Multicenter standardized ^{18} F-FDG PET diagnosis of mild cognitive impairment, Alzheimer's disease, and other dementias. *Journal of Nuclear Medicine, 49*, 390–398. https://doi.org/10.2967/jnumed.107.045385.

Mura, S. S., Nicolas, J., & Couvreur, P. (2013). Stimuli-responsive nanocarriers for drug delivery. *Nature Materials, 12*, 991–1003. https://doi.org/10.1038/nmat3776.

Murphy, M. M. P., & Smith, R. A. J. (2007). Targeting antioxidants to mitochondria by conjugation to lipophilic cations. *Annual Review of Pharmacology and Toxicology, 47*, 629–656. https://doi.org/10.1146/annurev.pharmtox.47.120505.105110.

Naser, S. S. S., Gupta, A., Choudhury, A., Yadav, A., Sinha, A., Kirti, A., ... Verma, S. K. (2024). Biophysical translational paradigm of polymeric nanoparticle: Embarked advancement to brain tumor therapy. *Biomedicine & Pharmacotherapy, 179*, 117372. https://doi.org/10.1016/j.biopha.2024.117372.

Nicholson, R. R. M., Kusne, Y., Nowak, L. A., LaFerla, F. M., Reiman, E. M., & Valla, J. (2010). Regional cerebral glucose uptake in the 3xTG model of Alzheimer's disease highlights common regional vulnerability across AD mouse models. *Brain Research, 1347*, 179–185. https://doi.org/10.1016/j.brainres.2010.05.084.

Nunziata, G. G., Pizzetti, F., Veglianese, P., Forloni, G., Balducci, C., & Rossi, F. (2026). Can nanoparticle-based intranasal delivery systems revolutionize treatment of central nervous system diseases? *Expert Opinion on Drug Delivery, 23*, 7–11. https://doi.org/10.1080/17425247.2025.2569644.

Orihuela, R. R., McPherson, C. A., & Harry, G. J. (2016). Microglial M1/M2 polarization and metabolic states. *British Journal of Pharmacology, 173*, 649–665. https://doi.org/10.1111/bph.13139.

Ouyang, X. X., Liu, Y., Zheng, K., Pang, Z., & Peng, S. (2024). Recent advances in zwitterionic nanoscale drug delivery systems to overcome biological barriers. *Asian Journal of Pharmaceutical Sciences, 19*, 100883. https://doi.org/10.1016/j.ajps.2023.100883.

Pardridge, W. W. M. (2012). Drug transport across the blood–brain barrier. *Journal of Cerebral Blood Flow & Metabolism, 32*, 1959–1972. https://doi.org/10.1038/jcbfm.2012.126.

Park, E. E., Li, L. Y., He, C., Abbasi, A. Z., Ahmed, T., Foltz, W. D., ... Wu, X. Y. (2023). Brain-penetrating and disease site-targeting manganese dioxide-polymer-lipid hybrid nanoparticles remodel microenvironment of Alzheimer's disease by regulating multiple pathological pathways. *Advanced Science, 10*. https://doi.org/10.1002/advs.202207238.

Patil, R. R. P., Pawara, D. D., Gudewar, C. S., & Tekade, A. R. (2019). Nanostructured cubosomes in an in situ nasal gel system: An alternative approach for the controlled delivery of donepezil HCl to brain. *Journal of Liposome Research, 29*, 264–273. https://doi.org/10.1080/08982104.2018.1552703.

Pegoraro, C. C., Domingo-Ortí, I., Conejos-Sánchez, I., & Vicent, M. J. (2024). Unlocking the mitochondria for nanomedicine-based treatments: Overcoming biological barriers, improving designs, and selecting verification techniques. *Advanced Drug Delivery Reviews, 207*, 115195. https://doi.org/10.1016/j.addr.2024.115195.

Pickrell, A. A. M., & Youle, R. J. (2015). The roles of PINK1, Parkin, and mitochondrial fidelity in Parkinson's disease. *Neuron, 85*, 257–273. https://doi.org/10.1016/j.neuron.2014.12.007.

Pinna, A. A., Malfatti, L., Galleri, G., Manetti, R., Cossu, S., Rocchitta, G., ... Innocenzi, P. (2015). Ceria nanoparticles for the treatment of Parkinson-like diseases induced by chronic manganese intoxication. *RSC Advances, 5*, 20432–20439. https://doi.org/10.1039/C4RA16265J.

Qu, M. M., Lin, Q., Huang, L., Fu, Y., Wang, L., He, S., ... Sun, X. (2018). Dopamine-loaded blood exosomes targeted to brain for better treatment of Parkinson's disease. *Journal of Controlled Release, 287*, 156–166. https://doi.org/10.1016/j.jconrel.2018.08.035.

Rajes, K. K., Walker, K., Hadam, S., Zabihi, F., Rancan, F., Vogt, A., & Haag, R. (2020). Redox-responsive nanocarrier for controlled release of drugs in inflammatory skin diseases. *Pharmaceutics, 13*, 37. https://doi.org/10.3390/pharmaceutics13010037.

Rajput, A. A., & Butani, S. (2022). Donepezil HCl liposomes: Development, characterization, cytotoxicity, and pharmacokinetic study. *AAPS PharmSciTech, 23*, 74. https://doi.org/10.1208/s12249-022-02209-9.

Raut, S., Bhalerao, A., Powers, M., Gonzalez, M., Mancuso, S., & Cucullo, L. (2023). Hypometabolism, Alzheimer's disease, and possible therapeutic targets: An overview. *Cells, 12*, 2019. https://doi.org/10.3390/cells12162019.

Reddy, P. P. H., & Beal, M. F. (2008). Amyloid beta, mitochondrial dysfunction and synaptic damage: Implications for cognitive decline in aging and Alzheimer's disease. *Trends in Molecular Medicine, 14*, 45–53. https://doi.org/10.1016/j.molmed.2007.12.002.

Ruan, H. H., Chai, Z., Shen, Q., Chen, X., Su, B., Xie, C., ... Lu, W. (2018). A novel peptide ligand RAP12 of LRP1 for glioma targeted drug delivery. *Journal of Controlled Release, 279*, 306–315. https://doi.org/10.1016/j.jconrel.2018.04.035.

Ruotolo, R. R., De Giorgio, G., Minato, I., Bianchi, M., Bussolati, O., & Marmiroli, N. (2020). Cerium oxide nanoparticles rescue α-synuclein-induced toxicity in a yeast model of Parkinson's disease. *Nanomaterials, 10*, 235. https://doi.org/10.3390/nano10020235.

Sadegh Malvajerd, S. S., Azadi, A., Izadi, Z., Kurd, M., Dara, T., Dibaei, M., ... Hamidi, M. (2019). Brain delivery of curcumin using solid lipid nanoparticles and nanostructured lipid carriers: Preparation, optimization, and pharmacokinetic evaluation. *ACS Chemical Neuroscience, 10*, 728–739. https://doi.org/10.1021/acschemneuro.8b00510.

Saraiva, C. C., Praça, C., Ferreira, R., Santos, T., Ferreira, L., & Bernardino, L. (2016). Nanoparticle-mediated brain drug delivery: Overcoming blood–brain barrier to treat neurodegenerative diseases. *Journal of Controlled Release, 235*, 34–47. https://doi.org/10.1016/j.jconrel.2016.05.044.

Saxton, R. R. A., & Sabatini, D. M. (2017). mTOR signaling in growth, metabolism, and disease. *Cell, 168*, 960–976. https://doi.org/10.1016/j.cell.2017.02.004.

Schapira, A. A. H. V., Cooper, J. M., Dexter, D., Clark, J. B., Jenner, P., & Marsden, C. D. (1990). Mitochondrial complex I deficiency in Parkinson's disease. *Journal of Neurochemistry, 54*, 823–827. https://doi.org/10.1111/j.1471-4159.1990.tb02325.x.

Sharma, E. E., Fotooh Abadi, L., Kombe Kombe, J. A., Kandala, M., Parker, J., Winicki, N., & Kelesidis, T. (2025). Overview of methods that determine mitochondrial function in human disease. *Metabolism: Clinical and Experimental, 170*, 156300. https://doi.org/10.1016/j.metabol.2025.156300.

Shehata, M. M. K., Ismail, A. A., & Kamel, M. A. (2023). Combined donepezil with astaxanthin via nanostructured lipid carriers effective delivery to brain for Alzheimer's disease in rat model. *International Journal of Nanomedicine, 18*, 4193–4227. https://doi.org/10.2147/IJN.S417928.

Shete, H. H. K., Prabhu, R. H., & Patravale, V. B. (2014). Endosomal escape: A bottleneck in intracellular delivery. *Journal of Nanoscience and Nanotechnology, 14*, 460–474. https://doi.org/10.1166/jnn.2014.9082.

Sierri, G. G., Patrucco, M., Ferrario, D., Renda, A., Comi, S., Ciprandi, M., ... Re, F. (2024). Targeting specific brain districts for advanced nanotherapies: A review from the perspective of precision nanomedicine. *WIREs Nanomedicine and Nanobiotechnology, 16*. https://doi.org/10.1002/wnan.1991.

Simpson, D. D. S. A., & Oliver, P. L. (2020). ROS generation in microglia: Understanding oxidative stress and inflammation in neurodegenerative disease. *Antioxidants, 9*, 743. https://doi.org/10.3390/antiox9080743.

Sonawane, D. D., & Pokharkar, V. (2024). Nose to brain targeting of the donepezil nanostructured lipid carrier in situ gel: Formulation, in vitro, *ex vivo*, in vivo pharmacokinetic and pharmacodynamic characterization. *RSC Pharmaceutics, 1*, 820–840. https://doi.org/10.1039/D4PM00174E.

Soriani, O. O., & Kourrich, S. (2019). The sigma-1 receptor: When adaptive regulation of cell electrical activity contributes to stimulant addiction and cancer. *Frontiers in Neuroscience, 13*. https://doi.org/10.3389/fnins.2019.01186.

Strope, T. T. A., Birky, C. J., & Wilkins, H. M. (2022). The role of bioenergetics in neurodegeneration. *International Journal of Molecular Sciences, 23*, 9212. https://doi.org/10.3390/ijms23169212.

Szablewski, L. (2021). Brain glucose transporters: Role in pathogenesis and potential targets for the treatment of Alzheimer's disease. *International Journal of Molecular Sciences, 22*, 8142. https://doi.org/10.3390/ijms22158142.

Szeto, H. H. H., & Liu, S. (2018). Cardiolipin-targeted peptides rejuvenate mitochondrial function, remodel mitochondria, and promote tissue regeneration during aging. *Archives of Biochemistry and Biophysics, 660*, 137–148. https://doi.org/10.1016/j.abb.2018.10.013.

Talbot, K. K., Wang, H.-Y., Kazi, H., Han, L.-Y., Bakshi, K. P., Stucky, A., ... Arnold, S. E. (2012). Demonstrated brain insulin resistance in Alzheimer's disease patients is associated with IGF-1 resistance, IRS-1 dysregulation, and cognitive decline. *Journal of Clinical Investigation, 122*, 1316–1338. https://doi.org/10.1172/JCI59903.

Tiwari, S. S. K., Agarwal, S., Seth, B., Yadav, A., Nair, S., Bhatnagar, P., ... Gupta, K. C. (2014). Curcumin-loaded nanoparticles potently induce adult neurogenesis and reverse cognitive deficits in Alzheimer's disease model via canonical Wnt/β-catenin pathway. *ACS Nano, 8*, 76–103. https://doi.org/10.1021/nn405077y.

Toader, C. C., Dumitru, A. V., Eva, L., Serban, M., Covache-Busuioc, R.-A., & Ciurea, A. V. (2024). Nanoparticle strategies for treating CNS disorders: A comprehensive review of drug delivery and theranostic applications. *International Journal of Molecular Sciences, 25*, 13302. https://doi.org/10.3390/ijms252413302.

Torchilin, V. V. (2009). Multifunctional and stimuli-sensitive pharmaceutical nanocarriers. *European Journal of Pharmaceutics and Biopharmaceutics, 71*, 431–444. https://doi.org/10.1016/j.ejpb.2008.09.026.

Trist, B. G., Hare, D. J., & Double, K. L. (2019). Oxidative stress in the aging substantia nigra and the etiology of Parkinson's disease. *Aging Cell, 18*. https://doi.org/10.1111/acel.13031.

Ulbrich, K. K., Michaelis, M., Rothweiler, F., Knobloch, T., Sithisarn, P., Cinatl, J., & Kreuter, J. (2011). Interaction of folate-conjugated human serum albumin (HSA) nanoparticles with tumour cells. *International Journal of Pharmaceutics, 406*, 128–134. https://doi.org/10.1016/j.ijpharm.2010.12.023.

Veys, K. K., Fan, Z., Ghobrial, M., Bouché, A., García-Caballero, M., Vriens, K., ... De Bock, K. (2020). Role of the GLUT1 glucose transporter in postnatal CNS angiogenesis and blood-brain barrier integrity. *Circulation Research, 127*, 466–482. https://doi.org/10.1161/CIRCRESAHA.119.316463.

Wan, Y. Y., Wang, W., Lai, Q., Wu, M., & Feng, S. (2023). Advances in cell-penetrating poly(disulfide)s for intracellular delivery of therapeutics. *Drug Discovery Today, 28*, 103668. https://doi.org/10.1016/j.drudis.2023.103668.

Wang, H. H., Sui, H., Zheng, Y., Jiang, Y., Shi, Y., Liang, J., & Zhao, L. (2019). Curcumin-primed exosomes potently ameliorate cognitive function in AD mice by inhibiting hyperphosphorylation of the Tau protein through the AKT/GSK-3β pathway. *Nanoscale, 11*, 7481–7496. https://doi.org/10.1039/C9NR01255A.

Wavikar, P. P., Pai, R., & Vavia, P. (2017). Nose to brain delivery of rivastigmine by in situ gelling cationic nanostructured lipid carriers: Enhanced brain distribution and pharmacodynamics. *Journal of Pharmaceutical Sciences, 106*, 3613–3622. https://doi.org/10.1016/j.xphs.2017.08.024.

Wilson, B. B., Samanta, M. K., Santhi, K., Kumar, K. P. S., Paramakrishnan, N., & Suresh, B. (2008). Poly(n-butylcyanoacrylate) nanoparticles coated with polysorbate 80 for the targeted delivery of rivastigmine into the brain to treat Alzheimer's disease. *Brain Research, 1200*, 159–168. https://doi.org/10.1016/j.brainres.2008.01.039.

Wilson, B. B., Samanta, M. K., Santhi, K., Kumar, K. P. S., Ramasamy, M., & Suresh, B. (2010). Chitosan nanoparticles as a new delivery system for the anti-Alzheimer drug tacrine. *Nanomedicine: Nanotechnology, Biology, and Medicine, 6*, 144–152. https://doi.org/10.1016/j.nano.2009.04.001.

Wu, H. H., Li, C., Yuan, H., Zhao, J., & Li, S. (2025). Brain delivery strategies for biomacromolecular drugs: Intranasal administration. *International Journal of Nanomedicine, 20*, 6463–6487. https://doi.org/10.2147/IJN.S520768.

Xie, A. A., Gao, J., Xu, L., & Meng, D. (2014). Shared mechanisms of neurodegeneration in Alzheimer's disease and Parkinson's disease. *BioMed Research International, 2014*, 1–8. https://doi.org/10.1155/2014/648740.

Xie, Z. Z., Zhang, X., Li, Y., & Zhu, R. (2025). Mitochondrial dysfunction drives cellular senescence: Molecular mechanisms of inter-organelle communication. *Experimental Gerontology, 211*, 112913. https://doi.org/10.1016/j.exger.2025.112913.

Xu, K. K., Wang, L., Lv, T., Jin, C., Wu, Q., Zhou, Y., ... Yang, J. (2025b). Nanomedicines against mitochondrial dysfunction-induced metabolic diseases. *Advanced Science, 12*. https://doi.org/10.1002/advs.202514522.

Xu, X. X., Pang, Y., & Fan, X. (2025a). Mitochondria in oxidative stress, inflammation and aging: From mechanisms to therapeutic advances. *Signal Transduction and Targeted Therapy, 10*, 190. https://doi.org/10.1038/s41392-025-02253-4.

Yadav, V. V. K., Dhanasekaran, S., Choudhary, N., Nathiya, D., Thakur, V., Gupta, R., ... Patel, A. (2025). Recent advances in nanotechnology for Parkinson's disease: Diagnosis, treatment, and future perspectives. *Frontiers in Medicine (Lausanne), 12*. https://doi.org/10.3389/fmed.2025.1535682.

Yang, D. D., Deng, Z., Zhou, H., Zhang, Q., Zhang, X., & Gong, J. (2025). Exosome-mediated dual drug delivery of curcumin and methylene blue for enhanced cognitive function and mechanistic elucidation in Alzheimer's disease therapy. *Frontiers in Cell and Developmental Biology, 13*. https://doi.org/10.3389/fcell.2025.1562565.

Yang, Z. Z.-Z., Zhang, Y.-Q., Wang, Z.-Z., Wu, K., Lou, J.-N., & Qi, X.-R. (2013). Enhanced brain distribution and pharmacodynamics of rivastigmine by liposomes following intranasal administration. *International Journal of Pharmaceutics, 452*, 344–354. https://doi.org/10.1016/j.ijpharm.2013.05.009.

Yasir, M. M., Zafar, A., Noorulla, K. M., Tura, A. J., Sara, U. V. S., Panjwani, D., ... Dalecha, D. D. (2022). Nose to brain delivery of donepezil through surface modified NLCs: Formulation development, optimization, and brain targeting study. *Journal of Drug Delivery Science and Technology, 75*, 103631. https://doi.org/10.1016/j.jddst.2022.103631.

Zhang, S. S., Gan, L., Cao, F., Wang, H., Gong, P., Ma, C., ... Lin, X. (2022). The barrier and interface mechanisms of the brain barrier, and brain drug delivery. *Brain Research Bulletin, 190*, 69–83. https://doi.org/10.1016/j.brainresbull.2022.09.017.

Zilberter, Y. Y., & Zilberter, T. (2023). Metabolic correction of neurodegenerative pathologies: The role of macronutrients and timing. *Exploration of Neuroscience*, 67–81. https://doi.org/10.37349/en.2023.00013.

CHAPTER NINE

The potential of HDAC inhibitors for Alzheimer's disease

Vedika Jain and Sharda Bharti*
Department of Biotechnology, National Institute of Technology (NIT) Raipur, Raipur, Chhattisgarh, India
*Corresponding author. e-mail address: sbharti.bt@nitrr.ac.in

Contents

Abstract

Alzheimer's disease is a progressive neurodegenerative condition characterized by cognitive deterioration, memory loss, and persistent neuroinflammation. Notwithstanding considerable scientific advancements, current therapy strategies predominantly address symptoms and are ineffective in arresting illness progression. Recent studies have demonstrated the crucial role of epigenetic changes, especially histone modifications, in the pathophysiology of Alzheimer's disease. Removal of the acetyl group from histones and non-histone proteins by histone deacetylases (HDACs) plays a pivotal role in the regulation of gene expression, synaptic plasticity, and neuronal survival. Such changes lead to dysregulated HDAC activity, which is further associated with significant clinical characteristics of Alzheimer's disease, including amyloid-beta accumulation, tau hyperphosphorylation, oxidative stress, and neuroinflammation. In Alzheimer's disease and other neurodegenerative diseases, the histone acetylation equilibrium is markedly disrupted, resulting in a shift towards hypoacetylation, which further inhibits the production of neuroprotective genes. Pharmacological inhibition of HDACs can reinstate hyperacetylation, therefore facilitating neuroprotective effects. This chapter explores the therapeutic potential of HDAC inhibitors in relation to Alzheimer's disease. This chapter also

International Review of Neurobiology, Volume 188
ISSN 0074-7742, https://doi.org/10.1016/bs.irn.2026.01.005

focuses on various HDAC isoforms associated with disease progression and explores the detailed mechanism by which HDAC inhibitors affect the epigenetic regulation and neuronal function. Preclinical investigations focusing on the role of HDAC inhibitors in mitigating neuroinflammation and Alzheimer's diseases, with a special focus on HDAC inhibitors in clinical trials, present intriguing opportunities for therapeutic advancement. The chapter further explores various challenges such as off-target effects, restricted isoform specificity, and inadequate blood–brain barrier permeability. To address these constraints, various strategies such as isoform-selective inhibitors, targeted delivery methods, and combination treatments are also explored. Thus, the chapter provides in-depth information on the role of HDAC inhibitors, which hold significant potential as disease-modifying agents in the treatment of Alzheimer's disease.

1. Introduction

Alzheimer's disease (AD), a dementia that was discovered in the year 1906 by Alois Alzheimer, is a cognitive disturbance associated with increasing age and can lead to death (Soria Lopez et al., 2019). An estimated 5.8 million Americans have Alzheimer's dementia, a number projected to increase to 13.8 million by mid-century due to the aging baby boom generation. In 2017, AD was the sixth leading cause of death in the U.S., with 121,404 recorded deaths. Between 2000 and 2017, deaths from AD increased by 145 %. In 2018, over 16 million unpaid caregivers provided approximately 18.5 billion hours of care, valued at nearly $234 billion, highlighting the emotional and physical toll on these caregivers. Medicare payments for beneficiaries aged 65 and older with dementia are more than three times higher than for those without. By 2050, the total number of individuals with AD dementia is expected to reach 13.8 million, including 7.0 million who will be 85 years old or older. Overall, healthcare costs, including long-term and hospice services, for this demographic totaled approximately $290 billion in 2019. Early detection and efficient treatment of Alzheimer's can benefit individuals, families, and society as a whole ("2015 Alzheimer's Disease Facts and Figures," 2015; "2019 Alzheimer's Disease Facts and Figures," 2019; Hebert et al., 2013).

Formerly, the diagnosis of AD was limited to the dementia stage, which is a clinical condition defined by significant progressive cognitive decline that impacts various areas, or neurobehavioral symptoms that are severe enough to affect daily life noticeably. A person experiencing dementia cannot maintain full independence, and this decrease in independence is the main characteristic that distinguishes dementia from mild cognitive impairment. In light of advancements in the biomarker field and the goal of implementing them in diagnostic practices, researchers classified the biomarkers into three categories: A (amyloid), T (phosphorylated tau), and N (neurodegeneration, assessed by total tau when applicable), known as the ATN framework (Lane et al., 2018; Scheltens et al., 2021). Therapeutic approaches to AD can be categorized into symptomatic, disease-modifying, and regenerative therapies. Current approved treatments, like cholinesterase inhibitors and glutamate antagonists, are symptomatic and mainly improve cognitive function. There is a significant need for more effective symptomatic treatments that address cognitive decline and related issues. Most current research is focused on disease modification, aiming to delay onset or slow progression to help individuals live longer without developing AD (Fish et al., 2019). Over the past few years, the primary focus of research on AD treatment has been on novel pharmacotherapies. The aim of these treatments is to decrease the production of beta-amyloid (Aβ) by inhibiting the enzymes β-secretase and γ-secretase, as well as to encourage the breakdown of existing Aβ plaques in the brain. Aβ plaques are believed to play a role in the progression of Alzheimer's disease (AD). However, this approach has shown only modest effectiveness (Folch et al., 2016).

Histone deacetylases (HDACs) constitute a superfamily of enzymes responsible for removing acetyl functional groups from lysine residues on both histone and non-histone proteins (Bertrand, 2010). DNA methylation, covalent modifications of histones, non-covalent modifications of histones, and non-coding RNAs represent the four main categories of epigenetic changes. While DNA methylation results in the silencing of transcription, histone modifications can cause either activation or repression of transcription; additionally, these modifications may occur alongside genetic abnormalities. The possibility of reversing these epigenetic changes has sparked significant interest in creating agents that specifically target these variations (Ramakrishnan & Pili, 2013). Mammals have a total of 18 histone deacetylases (HDACs), which are divided into four classes based on their primary similarity to yeast HDACs. Notably, Class I and Class II HDACs are essential for the deacetylation of lysine residues in the N-terminal tails of histones. In mammals, HDACs play a critical role in regulating gene transcription and also influence cell growth, survival, and proliferation. Additionally, HDACs play a significant role in modulating the expression of inflammatory genes (Kulthinee et al., 2022). HDACs are crucial for cognition and are linked to various psychiatric and neurologic disorders. Inhibiting specific isoforms of HDACs has shown benefits for conditions like Alzheimer's, Huntington's disease, traumatic brain injury, PTSD, stress, depression, and addiction (Volmar & Wahlestedt, 2015).

Neuronal function and differentiation are regulated by the genome and epigenome, with environmental factors influencing epigenetic changes. Neurodegeneration arises from dysregulation of these systems. Modifications such as chromatin changes, DNA methylation, and regulatory RNAs significantly impact neurodegenerative diseases and memory, contributing to conditions like Alzheimer's, Parkinson's, and Huntington's diseases (Ghosh & Saadat, 2023). Epigenetic regulation plays a crucial role in human health, as changes in chromatin are associated with various diseases, particularly cancer. Drugs targeting DNA methylation and histone deacetylation have been approved by the FDA for treatment. In the brain, mutations in chromatin-related factors are associated with neurological conditions such as autism spectrum disorder, mental disability, intellectual disability, and epilepsy, emphasizing the significance of epigenetic processes in brain development and function (Berson et al., 2018). HDAC activity dysregulation is present in AD. Therapeutic HDAC inhibition has shown potential in preclinical models, namely lower phosphorylation, reduced amyloid beta plaque formation, reduced neuroinflammation, reduced tau protein aggregation, greater microtubule stability, and improved metabolic homeostasis (Pereira et al., 2024). The development of HDAC inhibitors (HDACi) for therapy is of great interest to researchers and biotech entrepreneurs. Ongoing studies focus on their specificities and mechanisms of action. These inhibitors achieve transcriptional reprogramming through complex mechanisms, including histone and transcription factor deacetylation, chromatin remodelling, and effects on transcription initiation (Delcuve et al., 2012).

In this chapter, the role of HDACs in the context of AD progression is discussed, followed by the mechanistic insight of HDACi in the treatment of Alzheimer's disease. Furthermore, the therapeutic agent targeting HDAC and its status in clinical applications were discussed. Lastly, the challenges with HDACi, strategies to overcome the challenges, and future outlooks were highlighted. Overall, this chapter will help to address the potential of HDACi for dealing with Alzheimer's disease.

2. Role of HDACs in Alzheimer's disease pathogenesis

During the late onset of AD development, the histone modification plays a major role and has gathered attention (Kumar et al., 2022; Zhang et al., 2024). HDACs contribute to transcriptional repression through the process of deacetylation, which subsequently results in mitochondrial dysfunction, neuroinflammation, and apoptosis. The pathogenesis of AD progression is shown in Fig. 1. These effects ultimately lead to disruptions in brain homeostasis and neuronal death. Additionally, elevated levels of HDACs are associated with cognitive decline, which can promote the onset of AD by hindering the removal of β-amyloid, damaging synaptic structures, suppressing the expression of memory-related genes in neurons and synapses, and interfering with calcium ion pathways in organelles. Currently, research on the HDAC protein family in AD primarily focuses on subtypes HDAC 2, 3, 4, 5, and 6, with HDAC2 being the most extensively studied (Cao et al., 2018; Chen et al., 2024; Zhang et al., 2024).

HDACs, having an epigenic role, are classified into four classes. Class 1 HDACs (HDACs 1, 2, 3, and 8) are primarily nuclear and often found in multiprotein complexes, with HDAC8 being unique for its smooth muscle specificity. These enzymes serve as major nuclear deacetylases with varying efficiencies (Bondarev et al., 2021). Class 2 HDACs exhibit tissue-specific expression: HDAC4 and HDAC5 are predominantly expressed in the brain, heart, and skeletal muscles; HDAC7 is primarily found in the heart, lungs, and pancreas; and HDAC9 is mainly expressed in the brain and skeletal muscle. HDAC6 is cytoplasmic, affecting non-histone proteins, while HDAC10 functions as a transcriptional repressor in both the nucleus and cytoplasm and can relocate based on cellular signals. HDAC11, the sole

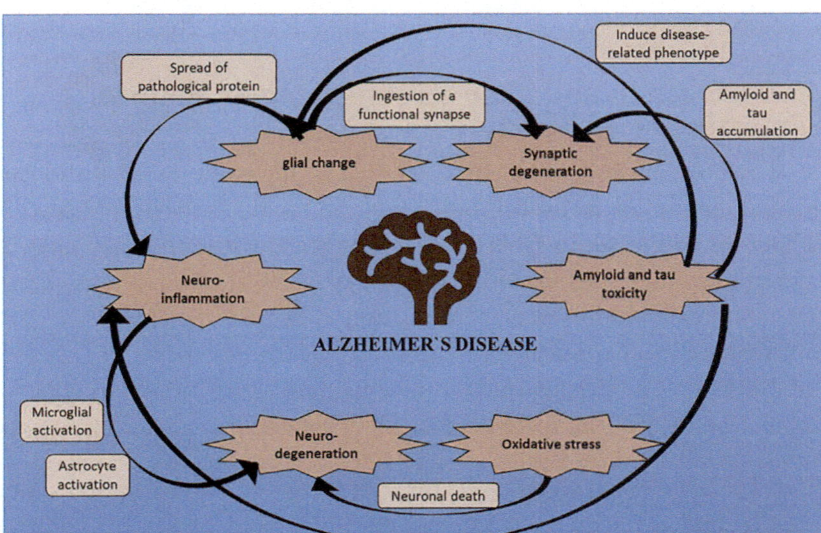

Fig. 1 Schematics showing the pathogenesis of AD development.

Class 4 enzyme, is nuclear, expressed in various tissues, and noted for its high fatty-acid acylase efficiency, showing similarity to HDAC8. Class 3 is also known as sirtuins (SIRT) and includes SIRT1, 2, 3, 4, 5, 6, and 7 (Bondarev et al., 2021). The classification of HDACs is shown in Fig. 2. Class 1, Class 2, and Class 4 HDACs are zinc-dependent, while Class 3 HDACs are nicotinamide adenine dinucleotide (NAD)-dependent or SIRTs (Li et al., 2022).

HDAC1, a class I HDAC, plays dual roles—being both neurotoxic and neuroprotective. Activating HDAC1 is considered a promising therapeutic approach for AD and other neurodegenerative diseases. Among the HDAC isoforms, HDAC3 is predominantly expressed in neurons and has harmful effects on synaptic plasticity and cognitive functions. Additionally, HDAC3 promotes neuroinflammation and neurotoxicity, contributing to the development of AD (Li et al., 2025; Watson et al., 2024). Recent experiments

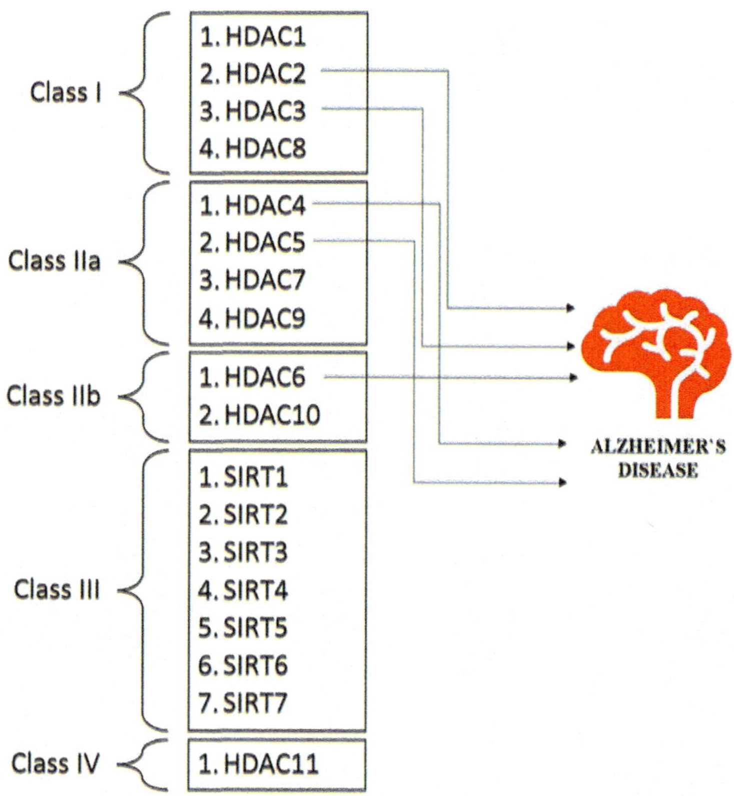

Fig. 2 Classification of HDAC and its role in AD.

with isoform-selective inhibitors and molecular genetic techniques have demonstrated that HDAC3 plays a significant role in multiple stages of neurodevelopment. This includes processes such as neurogenesis, gliogenesis, glial cell fate decisions, and myelination in oligodendrocytes and Schwann cells. However, specific post-translational modifications and alterations in its binding partners can transform HDAC3 from a beneficial protein into a neurotoxic one. HDAC3 plays a role in promoting neurodegeneration and hindering recovery after nerve injury, as well as regulating memory in the adult and aging brain (D'Mello, 2020). Among the various HDACs, HDAC3 has been identified as a potential therapeutic target for Alzheimer's disease regulation. HDAC3 is highly expressed in neurons and has several detrimental effects on brain function, as previously mentioned (Li et al., 2025).

Dysregulated HDAC2 activity plays a role in initiating and advancing various complex processes that lead to synaptic dysfunction, the significant accumulation of amyloid-beta (Aβ) plaques, tauopathies, the development of neurofibrillary tau tangles, neuroinflammation, and oxidative stress. (Asadi et al., 2025). The reduction of HDAC2 levels enhances mitochondrial respiration and lowers the concentrations of neurotoxic amyloid beta peptides. Overall, research reveals HDAC2's involvement in differentiating hiPSC-derived neurons, identifies neuron-specific isoforms of Endophilin-B1 as a marker for this differentiation, and shows that HDAC2 plays a regulatory role in essential neuronal and mitochondrial pathways within hiPSC-derived neurons (Frankowski et al., 2021). Mahady et al. (2019) reported that nuclear HDAC2 is associated with AD progression (Mahady et al., 2019). Thus, the HDAC2 inhibitors may promote brain health.

HDAC4, a class IIa HDAC member, is a potential therapeutic target for inflammatory and metabolic conditions, as it deacetylates non-histone proteins, including transcription factors, influencing various biological pathways. HDAC4 activity is controlled at multiple levels: transcriptional, post-transcriptional, and post-translational (Lin et al., 2025). Its functions vary depending on tissue type and are influenced by endogenous and exogenous factors and their specific substrates. HDAC4 interacts with non-histone targets, including transcription factors like myocyte enhancer factor 2, hypoxia-inducible factor, STAT1, and forkhead box proteins, which play vital roles in regulating inflammation and metabolism. The HDAC4 has a role in inflammation, insulin signaling, glucose metabolism, and cardiac muscle function, which are key areas of ongoing research and development (Kang et al., 2024). It has been reported that in AD treatment, a potential therapeutic effect of an inhibitor of the HDAC4 on neuronal cells was demonstrated by the regulation of synaptic plasticity and memory-related genes (Chen et al., 2022).

Formisano et al. (2020) identified a key epigenetic mechanism involved in stroke-induced neuronal damage: the transcriptional repressor DREAM. By recruiting HDAC4 and HDAC5, DREAM downregulates NCX3 in the brain, which is a critical neuroprotective factor in stroke (Formisano et al., 2020). Agis-Balboa et al. (2012) examined HDAC5's role in memory and AD development. They found that losing HDAC5 impairs memory but has little effect on AD pathology in a mouse model with amyloid pathology. This study reveals a role of HDAC5 in memory consolidation, and future AD treatments should avoid targeting HDAC5 with more selective HDACi (Agis-Balboa et al., 2012). Muñoz-Cobo et al. (2018) support the validity of the SH-SY5Y cell model for studying epigenetic changes linked to synaptic plasticity induced by antidepressants and the effects of selective HDACi. Specifically, antidepressant-mediated nucleo-cytoplasmic export of HDAC5 and SIRT2 downregulation may enhance markers of synaptic plasticity, contributing to antidepressant effects (Muñoz-Cobo et al., 2018).

HDAC6 is involved in the transport of polyubiquitinated protein aggregates and has a domain that interacts with proteins such as tau, IIp45 (invasion inhibitory protein 45), and EGFR (epidermal growth factor receptor). Its deacetylase activity targets several proteins, including tubulin, cortactin, heat shock proteins 83/90, and Bruchpilot proteins. In studies related to neurodegenerative diseases, inhibiting HDAC6 pharmacologically helps restore the acetylation of alpha-tubulin and improves mitochondrial transport (LoPresti, 2020). Regarding neurodegenerative diseases, the function of HDAC6 varies depending on the specific disease model. For instance, it can play a protective role, as seen in a Parkinson's disease model, where it aids in the formation of alpha-synuclein inclusions. On the other hand, the absence of HDAC6 may hinder tau-related axon degeneration and improve cognitive deficits in models of Alzheimer's disease. These seemingly opposing effects likely stem from HDAC6's involvement in various cellular pathways, making its loss of function have different impacts based on the disease context (Perry et al., 2017). Various clinical and preclinical research demonstrates that HDAC6 inhibition is a potential therapeutic approach for the treatment of autoimmune diseases, cancer, and neurological disorders (Kalinski et al., 2019; LoPresti, 2020; Shen & Kozikowski, 2020). In the present section, the different types of

HDACs and their role in brain metabolism and AD were discussed. Followed by the section, which will focus on HDACi and their mechanism of action in the treatment of AD progression.

3. Mechanistic action of HDAC inhibitors for Alzheimer's disease

HDACi can often lead to growth arrest, differentiation, and death in cancer cells. Several HDAC inhibitors have been approved to treat cutaneous or peripheral T-cell lymphoma and multiple myeloma. There is more and more proof that HDAC plays a big part in keeping neurons in shape and the brain in balance. HDAC may be related to several neurodegenerative illnesses, such as AD, Parkinson's disease (PD), Amyotrophic lateral sclerosis (ALS), and Huntington's disease (HD). As a result, HDAC inhibitors or degraders may become therapeutic medications for Alzheimer's disease (Li et al., 2022). HDAC superfamily inhibitors are classified into four categories based on structural and functional characteristics: hydroxamate inhibitors (e.g., SAHA), cyclic tetrapeptides (e.g., apicidin), short-chain fatty acids (e.g., butyrate), and benzamides (e.g., MS-275). At present, the histone deacetylase inhibitors (HDACi) approved by the United States Food and Drug Administration (FDA) specifically vorinostat, romidepsin, chidamide, panobinostat, and belinostat are predominantly employed in the management of cutaneous T-cell lymphoma and multiple myeloma (Zhang et al., 2024). Various HDACi were discussed in the subsequent section.

In recent days, it has been frequently seen that HDACi have certain properties, such as neurotrophic, neuroprotective, and anti-inflammatory, which can reduce neurodegeneration and improve neurological performance and memory (Chuang et al., 2009). HDACi are effective for treating neurodegenerative disorders by regulating brain metabolism, such as neural regeneration, memory formation, ischemic stroke, and fear conditioning (Ahmad Ganai et al., 2016; Pal et al., 2023). HDAC isoform`s function in neurological disorders and mechanism of HDACi as a therapeutic agent is shown in Fig. 3. Histone modifications play a crucial role in regulating gene expression, which helps cells adapt to environmental changes, particularly in neuronal development, plasticity, and memory. These modifications, known as "histone codes," influence chromatin structure and gene expression, promoting synaptic formation and long-term memory in the hippocampus. The key forms of synaptic plasticity, long-term potentiation (LTP) and long-term

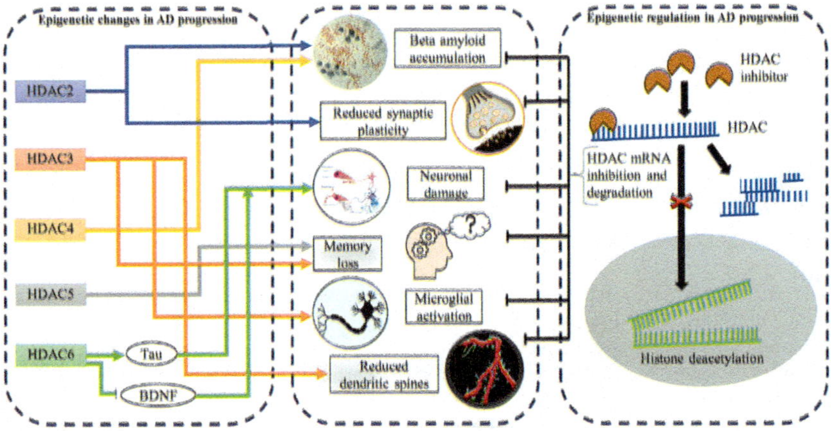

Fig. 3 HDACi mechanism of action in brain metabolism and for AD treatment.

depression (LTD), are essential for learning and memory, which rely on synaptic connections between neurons. Histone acetylation and deacetylation affect chromatin and gene expression related to synaptic plasticity. HDAC2 are particularly enriched at neuroplasticity-associated genes like BDNF IV, Arc, and others, highlighting the importance of histone modifications in neuronal plasticity and memory consolidation (Geng et al., 2021; Xhemalce et al., 2011).

HDAC inhibition reduced overall HDAC activity in the motor cortex, leading to increased acetylation levels of histone H4 and H3. Exercise also elevated acetylation of these histones in the same region. Furthermore, HDAC inhibition boosted the transcription of key genes involved in neuroplasticity, creating a more favourable environment for neurorehabilitation (Maejima et al., 2021). The study demonstrated that repeated HDACi treatments improved cognitive function, and combining HDAC inhibition with exercise had a synergistic effect on learning and memory. This combination was associated with enhanced transcription of genes important for neuronal plasticity in the hippocampus. (Kitahara et al., 2021). HDAC2 inhibition was found to be effective as an AD treatment therapy by ameliorating neuronal impairments induced by Amyloid β. HDAC2 inhibition also increased the dendrite length and potentiated the memory (Nakatsuka et al., 2021). Similarly, JRM-28, an HDAC2 inhibitor, has been shown to improve post-synaptic neurotransmission in hippocampal neurons, augment spine density, and dendritic strength (Rahman et al., 2024).

Several researchers have found that the HDACi play a significant role in enhancing cognition and improving behavioural changes in people with neurological disorders. RG2833 is an HDACi, which was studied for AD therapeutics, showing that HDACi are a potential drug in improving cognition (Ndukwe et al., 2025). Smies et al. (2024) studied that blocking HDAC3 enhanced subthreshold updates without affecting memory retention of the original information, allowing a balanced expression of both. This study implies that HDAC3 may play a role in age-related deficits in memory updating and could influence the strength of a memory update in young mice, impacting the balance between original and new information during testing. (Smies et al., 2024). Class I HDACi mediated activation of the brain-derived neurotrophic factor (BDNF)-tropomycin receptor kinase B (TrkB) pathway (BDNF/TrkB pathway), which was studied to improve memory (object recognition) and found that HDAC inhibition potentiates long-term memory consolidation (Ramirez-Mejia et al., 2021). Moreover, PT3 (benzamide) is an HDAC3 inhibitor and was found to be effective in improving long-term memory (Pulya et al., 2021). Thus, based on the existing literature, it is proven that HDAC inhibition, especially HDAC3, enhances learning and memory behaviours.

The level of histone acetylation in chromatin is primarily regulated by HDACs; therefore, histone acetylation increases with HDAC inhibition. The HDACi play a crucial role in altering the function of proteins involved in AD progression, such as amyloid β, Glycogen synthase kinase-3β (GSK-3β), and tau protein. GSK-3β, which primarily acts as a tau phosphokinase, gets activated by amyloid β and hyperphosphorylates tau protein (Xu et al., 2011). GSK-3 regulates the metabolism of amyloid precursor protein (APP) and Aβ production, while also promoting neuronal death induced by Aβ. In various mouse models of AD, various compounds inhibiting GSK-3 have been reported to decrease Amyloid β deposition and the formation of neuritic plaques, reduce tau phosphorylation, and enhance cognitive performance as assessed by behavioural tests. In contrast, mice that conditionally overexpress GSK-3β in their forebrain (Tet/GSK-3β mice) exhibit hyperphosphorylation of tau at sites relevant to Alzheimer's disease, which correlates with the accumulation of tau in the somatodendritic region of hippocampal neurons, neurodegeneration, and memory deficit (Avila et al., 2012; Sayas & Ávila, 2021). Therefore, blocking HDAC will be beneficial in reducing the hyperphosphorylation of tau and the function of a gene that contributes to memory deficit. The mechanistic action of HDACi in improving memory and cognition is depicted in Fig. 4. Dysregulation of certain genes may lead to memory

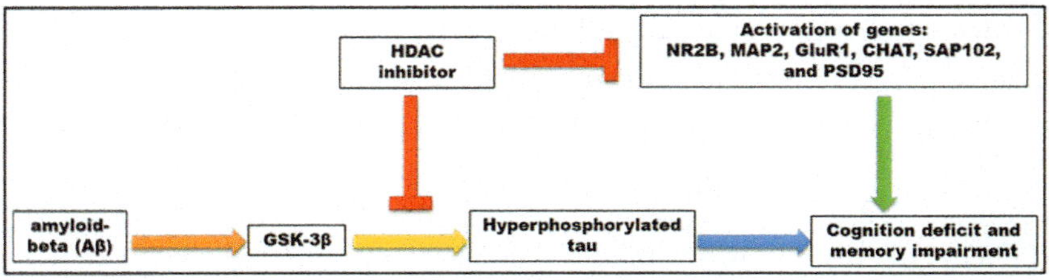

Fig. 4 Mechanism of HDACi in AD treatment: HDACi inhibit the hyperphosphorylation of tau protein and important genes involved in learning and memory weakening.

deficits by disrupting synaptic plasticity, neuronal damage, neurotransmission destruction, etc. This important genes are N-methyl-D-aspartate (NMDA) receptor 2B (NR2B), microtubule-associated protein 2 (MAP2), α-amino-3-hydroxy-5-methyl-4-isoxazolepropionic acid (AMPA) receptor subunit (GluR1), choline acetyltransferase (CHAT), synapse-associated protein 102 (SAP102), and postsynaptic density protein 95 (PSD95) (Coley & Gao, 2018; Kay et al., 2022; Sultana et al., 2010; Yu et al., 2017). Mostly, HDACi chelate the zinc ion present in the active site of HDACs; such a strong interaction blocks the activity of HDAC (Cheng et al., 2025; Liu & Liou, 2023).

4. Therapeutic applications and clinical development of HDAC inhibitors

The HDACi as therapeutics are largely classified into benzamides, hydroxamic acid derivatives, carboxylic acid derivatives, cyclic peptides, including cyclic depsipeptides, and electrophilic ketones (Bondarev et al., 2021). Benzamide has been identified as a possible zinc-binding group. For over ten years, scientists have shown interest in developing benzamide-derived HDAC3 inhibitors (Amin et al., 2019). Hydroximate derivatives for HDAC inhibition exhibit a good therapeutic efficacy for the treatment of cancer (Cheng et al., 2025). Varinostat is a well-known HDACi, which is a derivative of hydroxamic acid used as an anticancer drug to treat T-cell lymphoma and as a neuroprotective agent (Neganova et al., 2021). Cyclic peptides are a complex type of HDACi, which generally include dipeptides, cyclic hydroxamic acid peptides, and apicidin. Among all cyclic peptides, dipeptides are of great importance as it is approved by the Food and Drug Administration (FDA) (Kim & Bae, 2011). They are prodrugs, they are reduced in the intracellular environment, and their reduced functional form contains a sulfydryl group, which interacts with zinc present in the active site of HDACs (Furumai et al., 2002). The role of carboxylic acid derivatives in HDAC inhibition is less explored. Although it is thought that it possesses a carboxylic acid function group which binds with the metal group of HDAC. There are only a few carboxylic acid-based HDACi, such as PBA (phenylbutyric acid), BA (butyric acid), and VPA (valproic acid), which are in their phase trials for the treatment of neurodegenerative disorders and cancer (Bora-Tatar et al., 2009; Ibrahim et al., 2020). Electrophilic ketone-based HDACi, such as trifluoromethyl ketone, are efficient in inhibiting several proteases and zinc-dependent enzymes such as HDAC (Mottamal et al., 2015). Moreover, short-chain fatty acids, particularly sodium butyrate, generated during microbial fermentation under anaerobic conditions in the gastrointestinal system, are known to block several classes of HDAC. Compounds such as butyrate and propionate have been shown to induce apoptosis in neutrophils through HDAC inhibition. Both forms of these chemicals are going through clinical investigations for various kinds of disorders (Aoyama et al., 2010; Mottamal et al., 2015; Zou et al., 2021).

At present, the FDA has approved only four small-molecule drugs for the treatment of AD: Donepezil, Galantamine, Rivastigmine, and Memantine. However, all of these are single-target medications that address cognitive and memory issues rather than functioning as disease-modifying treatments. Monoamine oxidase (MAO), which includes the two isoenzymes MAO-A and MAO-B, is a flavin-dependent metabolic enzyme found in the outer membrane of mitochondria. In which MAO-B has garnered significant interest as a potential target for AD treatment. The four small-molecule MAO-B inhibitors currently in clinical use are Selegiline, Rasagiline, Safinamide, and Pargyline. Vorinostat (SAHA), Romidepsin, Belinostat, Panobinostat, and Chidamide are FDA-approved HDACi utilised in various cancer treatments. Additionally, over twenty specific HDACi are undergoing clinical trials, such as Entinostat, AR-42, and ITF-2357 (Yao et al., 2022).

PT3 has high blood-brain barrier permeability and can improve long-term memory in a mouse model of novel object recognition (NOR). It is more selective for HDAC3 than for HDAC1, HDAC6, and HDAC8. In vivo, PT3 administration led to elevated expressions of H3K9 acetylation, CREB 1, BDNF, TRKB, Nr4a2, c-fos, PKA, GAP 43, and MMP9 in mouse neuronal (N2A) cells. Their preclinical results indicated that PT3 is a promising selective inhibitor of HDAC3 that can traverse the blood–brain barrier and enhance long-term memory in C57/BL6 mice. We regard PT3 as a prospective treatment agent for age-related cognitive

Table 1 Listing HDACi in clinical trials dealing with neurological disorders, including Alzheimer's disease.

HDACi	Related disorders	Status	Identification number/reference
MBF-015	Brain Diseases Central Nervous System Diseases Nervous System Diseases Dementia Heredodegenerative Disorders, Neurodegenerative Diseases, Genetic Diseases, Inborn Congenital, Hereditary, and Neonatal Diseases and Abnormalities Cognition Disorders Neurocognitive Disorders Mental Disorders, Huntington's Disease, Alzheimer's disease,	Phase 2	NCT06469853
CKD-504		Phase1	NCT03713892
sodium valproate		Phase 2	NCT01619644
sodium phenylbutyrate		Phase2	NCT00212316
nicotinamide, or vitamin B3		Phase 1 Phase 2	NCT00580931
Panobinostat		Phase 1	NCT01324635
AR-42		Early Phase 1	NCT02282917
Atezolizuma, Carboplatin, Entinostat, Etoposide		Phase 1	NCT04631029
Suberoylanilide hydroxamic acid (SAHA, Vorinostat), tamoxifen citrate (Tamoxifen)		Phase 2	NCT00365599
MS-275		Phase 1	NCT00020579
FRM-0334		Phase 2	NCT02149160
Pembrolizumab, Vorinostat, Temozolomide		Phase 1	NCT03426891
sodium phenylbutyrate		Phase 1 Phase 2	NCT00107770
MBF-015 16 mg oral capsules		Phase 2	NCT06469853
RGFP966		Preclinical	(L. Xu et al., 2024)

decline and other conditions associated with memory impairment, including Alzheimer's disease (Pulya et al., 2021). Tubastatin A and ricolinostat (ACY-1215), both of which are HDAC6 inhibitors, have been shown to enhance behavioural deficits in mouse models of Alzheimer's disease. Valproic acid, which inhibits class I and II HDACs, has demonstrated neuroprotective properties in Alzheimer's disease. For example, it can reduce the activity of GSK-3β, a mechanism linked to various neuropathological processes (Pereira et al., 2024). Mondal et al. investigated a novel HDAC6 inhibitor, PB118, in microglia BV2 cells and three-dimensional human neural culture models associated with Alzheimer's. This HDACi facilitated the stability of microtubules, inhibited NFT formation, reduced the generation of Aβ, and lowered levels of neuroinflammatory cytokines and chemokines (Mondal et al., 2024). Moreover, several HDACi are in clinical trials, some of which are listed in Table 1, and may be effective for AD treatment and other related disorders.

5. Challenges with HDAC inhibitors and strategies to reduce those challenges

HDACi as a therapeutic for AD has certain challenges, they are developmental challenges, side effects and drug efficacy (Yang et al., 2017). Developmental challenges arise due to low permeability of the blood-brain barrier, cytotoxicity of the inhibitor, and strong HDAC inhibition homology (Zhang et al., 2024). Dehydration, thrombocytopenia, diarrhoea, thrombosis, fatigue, anorexia, nausea/vomiting, hypocalcemia, anaemia, cardiac arrhythmia, fever, leukopenia, neutropenia, sick sinus syndrome, elevated creatinine, elevated uric acid, decreased potassium, status epilepticus, paresthesia, vasculitis, renal failure, pyrexia, loss of appetite, and dyspnea are some of the side effects of HDACi (Talom et al., 2025). Drug efficacy means the effectiveness of the drug; enhanced drug efficacy will be able to reduce AD pathology, enhance neuroprotection, improve cognition, high brain permeability of the drug, and so on (Yang et al., 2017). Challenges and strategies to overcome those changes associated with HDACi for AD are schematically shown in Fig. 5. The strategies combating challenges associated with HDACi are discussed in detail in the subsequent section:

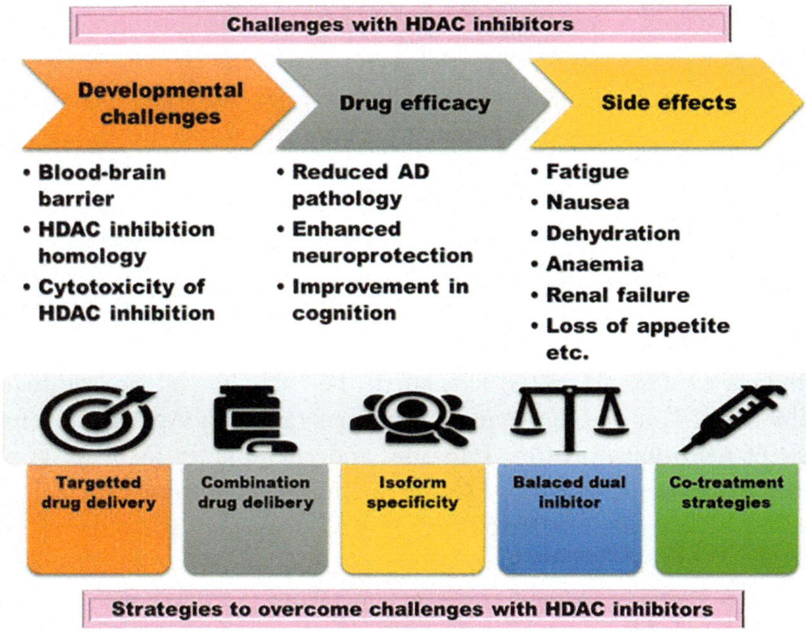

Fig. 5 Challenges and strategies to overcome those changes associated with HDACi for AD.

5.1 Selective specificity of HDAC-isoforms

Isoform specificity for HDACi is crucial in reducing side effects and enhancing drug efficacy. As a result, it is important to understand the role of all the HDAC isoforms involved in AD. For instance, HDAC3 represents a fascinating target for AD therapy. It is significantly present in neurons and acts as a crucial negative regulator of memory formation and synaptic plasticity. In Alzheimer's disease, HDAC3 levels are increased, contributing to the disease's progression by not only worsening synaptic damage but also by triggering neuroinflammation in glial cells. Since the selective inhibition or depletion of HDAC3 reduces damage to the blood-brain barrier, lowers amyloid production, decreases tau phosphorylation, and improves cognitive function, targeting HDAC3 pharmacologically offers a promising approach for Alzheimer's treatment.(Li et al., 2025). Researching neuro-immune interactions and HDAC6 pathophysiology aims to develop selective HDACi with therapeutic potential in animal and cellular models, which may expedite clinical trials (Zhang et al., 2025).

Similarly, HDAC5 have a role in memory consolidation; therefore, HDACi should be specific and should avoid targeting HDAC5 for treatment of AD (Agis-Balboa et al., 2012).

5.2 Targeted delivery of HDAC inhibitors

To increase the drug efficacy and reduce off-target effects, the targeted delivery of HDAC becomes important. At present, the targeted delivery of HDACi has been much explored for cancer treatment rather than AD, but it can be considered for AD. Nanocarriers (K C et al., 2021), epigenetic therapy (Hussain et al., 2025), antibody drug conjugates (Cianferotti et al., 2021), etc., are means for the targeted delivery of HDACi. LAQ824-Doxorubicin nanoparticles (LD NPs) demonstrated a response to the tumor microenvironment, and the tumor-targeted therapy facilitated by LD NPs for the delivery of HDACi could greatly enhance the tumor microenvironment (Cianferotti et al., 2021). Exosome-mediated resolved drug delivery system showed a high loading efficiency of SAHA and enhanced the epigenetic inhibition for cancer treatment (Cianferotti et al., 2021). Additionally, a drug delivery system has been designed to efficiently transport HDACi to latent HIV-infected cells. Since HDACi are less effective in hypoxic environments, oxygen-containing nanosomes are used as carriers. These nanosomes can enhance chemotherapy effectiveness by providing vital oxygen to the cells (Cianferotti et al., 2021). Furthermore, Cianferotti et al. (2021) communicated that HDACi hydroxamates SAHA and Dacinostat, as antitumor agents, were linked to an antibody named cetuximab and trastuzumab. These antibody drug conjugates were effective in inhibiting HDAC in numerous tumour cell lines (Cianferotti et al., 2021).

Nanoparticles offer several benefits, such as improved drug effectiveness, better biological compatibility, enhanced availability, and targeted delivery to areas affected by AD. Functionalizing ligands on nanoparticles can aid in penetrating the blood-brain barrier and enable controlled drug release, increasing drug concentrations at AD sites. The use of nanotechnology for targeted therapy in AD holds significant potential and could revolutionise treatment strategies (Pawar et al., 2025). Therefore, nano-formulation-mediated delivery of HDACi as a targeted therapy in AD might be advantageous, as it has been in considerations for treating cancer and tumors. The advanced drug delivery systems for AD include nanoparticles, hydrogel-based systems, network systems and dendrimers, vesicular systems, and biologics (Greeny et al., 2024).

5.3 Combinatorial treatment method

The current treatment strategies in modulating AD only slow down the progression of the disease; thus, the combination drug therapy was thought to be more effective than mono-drug therapy (Badrikoohi et al., 2022). There are several such studies that demonstrate that the combination therapy for AD treatment is comparatively more effective; some of these studies are discussed later.

Synergistic treatment efficiency of radiotherapy with a low dose of HDACi in the treatment of AD was studied. It has been reported that this synergistic approach effectively reduced the neuroinflammation, increased neurotrophic factor gene expression, and processed amyloid β precursor protein (Ricciardi et al., 2023). It has been studied that the combinatorial treatment of an epigenetic modulator, vorinostat, in combination with rosiglitazone can enhance neurotrophic factors activity and reduce biochemical and cellular alterations linked with AD, mainly by encapsulating these drugs in a surface-modified nanocarrier system for improved bioavailability and to enhance therapeutic efficacy (K C et al., 2021). Moreover, simultaneous administration of JQ1 (bromodomain protein, BRD4 inhibitor) and MS275 (HDACi) increased the expression of cAMP-response element-binding protein (CREB) and downregulated the expression of tumor necrosis factor-α (TNF-α) in the rat model (Badrikoohi et al., 2022).

5.4 Other strategies

Treatment of AD using a balanced dual inhibitor was supposed to be effective. *In vitro* study of butyrylcholinesterase (BChE) inhibitor and HDAC6 inhibitor, a dual inhibitor, has shown a potential neuroprotective effect (Wang et al., 2023). Co-treatment strategies such as diabetes liked neurodegenerative disorder are also effective. An antidiabetic drug named Repaglinide (REP) has shown some challenges, such short half-life and poor permeability to the blood-brain barrier, but this drug has the property of reducing the neurodegeneration associated with AD and Huntington's disease. To evade these challenges, researchers developed and examined PEGylated nanocarriers for oral delivery of REP in the treatment of diabetes mellitus-associated neurodegenerative disorders (Wadhwa et al., 2023). For the current HDACi, the blood–brain barrier permeability and off-target effects are major translational barriers for these promising epigenetic therapies, which restrict long-term use of HDACi in CNS diseases (Asadi et al., 2025).

6. Research gap and future outlook

Understanding the role of HDACs in AD pathogenesis and HDACi action in controlling AD progression, the following are the major research gaps in the field of HDACi for the treatment of AD:

- Targeting HDAC inhibition to a specific and potent HDAC isoform to reduce the off-target effects and to increase drug efficacy (Pulya et al., 2023; Yang et al., 2017).
- Designing an HDACi that can cross the blood-brain barrier, reducing the pharmacokinetic challenges (W. Zhang et al., 2025).
- There are very less number studies under clinical trials, that are in advanced and completed stages in this area of research, which is really a great research gap (Joshi et al., 2025; Pulya et al., 2023).
- Understanding the mechanistic insight of HDAC inhibition, particularly for the treatment of AD, as there only less research conducted concerning AD. Mostly, the HDACi are studied for the treatment of cancer only.

Based on the literature review, the key areas of focus for the potential of HDACi for AD include examining dynamic HDAC changes in AD progression. Investigating the promising potent HDACi, therapeutic mechanisms for AD pathology. Developing blood-brain barrier (BBB)-penetrant HDACi, potentially using carrier-mediated prodrug strategies or brain-targeted nanoparticles. Furthermore, utilizing computational and AI technologies for improved drug design of HDAC6 and other potent HDACi for AD, including high-resolution crystal structure analysis and machine learning for assessing ADMET characteristics and key features in drug datasets. Thus, overcoming challenges addressed with HDACi for AD treatment is the research gap in research related to HDACi for AD and must be considered by researcher in their future research.

7. Conclusion

This chapter highlights the therapeutic potential of HDACi for Alzheimer's disease. The chapter focuses on various HDAC isoforms associated with disease progression and explores the detailed mechanism by which HDACi affect the disease progression. The chapter further explores Preclinical investigations and clinical trials focusing on the potential of HDACi in mitigating neuroinflammation and Alzheimer's diseases several challenges, such as off-target effects, restricted isoform specificity, and inadequate blood–brain barrier permeability, and several strategies, such as isoform-selective inhibitors, targeted delivery methods, and combination treatments, are also explored. Thus, the chapter provides in-depth information on the role of HDACi, which holds noteworthy potential as disease-modifying agents in the treatment of Alzheimer's disease.

Targeting specific HDAC isoforms, designing inhibitors that are permeable for the blood-brain barrier, to encourage advanced clinical trials focused on HDACi for AD treatment, are some of the research gaps that must be addressed. Limited research on the mechanisms of HDAC inhibition specifically for AD should be considered as an urgent research need as AD is increasing day by day. Key focus areas for potential HDACi in AD include examining dynamic HDAC changes in AD progression, investigating potent HDACi and understanding their therapeutic mechanisms. Addressing these challenges is crucial for future research on HDACi in AD treatment.

Acknowledgments

The authors would like to acknowledge the National Institute of Technology (NIT) Raipur, an Institute of National Importance, for the support in writing the chapter.

Declaration of Interests

The authors declare that they have no known competing financial interests.

References

2015 Alzheimer's disease facts and figures. (2015). *Alzheimer's & Dementia, 11*(3), 332–384. https://doi.org/10.1016/j.jalz.2015.02.003.

2019 Alzheimer's disease facts and figures. (2019). *Alzheimer's & Dementia, 15*(3), 321–387. https://doi.org/10.1016/j.jalz.2019.01.010.

Agis-Balboa, R. C., Pavelka, Z., Kerimoglu, C., & Fischer, A. (2012). Loss of HDAC5 impairs memory function: Implications for Alzheimer's disease. *Journal of Alzheimer's Disease, 33*(1), 35–44. https://doi.org/10.3233/JAD-2012-121009.

Ahmad Ganai, S., Ramadoss, M., & Mahadevan, V. (2016). Histone Deacetylase (HDAC) Inhibitors - emerging roles in neuronal memory, learning, synaptic plasticity and neural regeneration. *Current Neuropharmacology, 14*(1), 55–71. https://doi.org/10.2174/1570159×13666151021111609.

Amin, S. A., Adhikari, N., Kotagiri, S., Jha, T., & Ghosh, B. (2019). Histone deacetylase 3 inhibitors in learning and memory processes with special emphasis on benzamides. *European Journal of Medicinal Chemistry, 166*, 369–380. https://doi.org/10.1016/j.ejmech.2019.01.077.

Aoyama, M., Kotani, J., & Usami, M. (2010). Butyrate and propionate induced activated or non-activated neutrophil apoptosis via HDAC inhibitor activity but without activating GPR-41/GPR-43 pathways. *Nutrition, 26*(6), 653–661. https://doi.org/10.1016/j.nut.2009.07.006.

Asadi, K. K. R., Satarker, S., & Gurram, P. C. (2025). Histone deacetylase 2 in Alzheimer's disease: A comprehensive molecular blueprint for therapeutic targeting. *Molecular Neurobiology*. https://doi.org/10.1007/s12035-025-05264-y.

Avila, J., León-Espinosa, G., García, E., García-Escudero, V., Hernández, F., & DeFelipe, J. (2012). Tau phosphorylation by GSK3 in different conditions. *International Journal of Alzheimer's Disease, 2012*, 1–7. https://doi.org/10.1155/2012/578373.

Badrikoohi, M., Esmaeili-bandboni, A., & Babaei, P. (2022). Simultaneous administration of bromodomain and histone deacetylase I inhibitors alleviates cognition deficit in Alzheimer's model of rats. *Brain Research Bulletin, 179*, 49–56. https://doi.org/10.1016/j.brainresbull.2021.12.004.

Berson, A., Nativio, R., Berger, S. L., & Bonini, N. M. (2018). Epigenetic regulation in neurodegenerative diseases. *Trends in Neurosciences, 41*(9), 587–598. https://doi.org/10.1016/j.tins.2018.05.005.

Bertrand, P. (2010). Inside HDAC with HDAC inhibitors. *European Journal of Medicinal Chemistry, 45*(6), 2095–2116. https://doi.org/10.1016/j.ejmech.2010.02.030.

Bondarev, A. D., Attwood, M. M., Jonsson, J., Chubarev, V. N., Tarasov, V. V., & Schiöth, H. B. (2021). Recent developments of HDAC inhibitors: Emerging indications and novel molecules. *British Journal of Clinical Pharmacology, 87*(12), 4577–4597. https://doi.org/10.1111/bcp.14889.

Bora-Tatar, G., Dayangaç-Erden, D., Demir, A. S., Dalkara, S., Yelekçi, K., & Erdem-Yurter, H. (2009). Molecular modifications on carboxylic acid derivatives as potent histone deacetylase inhibitors: Activity and docking studies. *Bioorganic & Medicinal Chemistry, 17*(14), 5219–5228. https://doi.org/10.1016/j.bmc.2009.05.042.

Cao, T., Zhou, X., Zheng, X., Cui, Y., Tsien, J. Z., Li, C., & Wang, H. (2018). Histone deacetylase inhibitor alleviates the neurodegenerative phenotypes and histone dysregulation in presenilins-deficient mice. *Frontiers in Aging Neuroscience, 10*. https://doi.org/10.3389/fnagi.2018.00137.

Chen, T.-J., Hung, H.-S., Cheng, T.-L., & Wang, D.-C. (2024). Histone deacetylase inhibitor attenuates the effects of 27-hydroxycholesterol on the rat brain. *Neuroscience Letters, 818*, 137533. https://doi.org/10.1016/j.neulet.2023.137533.

Chen, Y.-A., Lu, C.-H., Ke, C.-C., Chang, C.-W., Yang, B.-H., Gelovani, J. G., & Liu, R.-S. (2022). *Monitoring HDAC4 Expression in Alzheimer's Disease Using [18F]TFAHA-PET*, 61–70. https://doi.org/10.1007/978-3-030-92786-8_8.

Cheng, Y., Zhuang, Y., Zhu, N., Yu, Y., & Hu, M. (2025). Hydroxamic acid hybrids as histone deacetylase inhibitors for cancer therapy: An update (2022 to date). *Bioorganic Chemistry, 163*, 108700. https://doi.org/10.1016/j.bioorg.2025.108700.

Chuang, D.-M., Leng, Y., Marinova, Z., Kim, H.-J., & Chiu, C.-T. (2009). Multiple roles of HDAC inhibition in neurodegenerative conditions. *Trends in Neurosciences, 32*(11), 591–601. https://doi.org/10.1016/j.tins.2009.06.002.

Cianferotti, C., Faltoni, V., Cini, E., Ermini, E., Migliorini, F., Petricci, E., Taddei, M., Salvini, L., Battistuzzi, G., Milazzo, F. M., Anastasi, A. M., Chiapparino, C., De Santis, R., & Giannini, G. (2021). Antibody drug conjugates with hydroxamic acid cargos for histone deacetylase (HDAC) inhibition. *Chemical Communications, 57*(7), 867–870. https://doi.org/10.1039/D0CC06131J.

Coley, A. A., & Gao, W.-J. (2018). PSD95: A synaptic protein implicated in schizophrenia or autism? *Progress in Neuro-Psychopharmacology and Biological Psychiatry, 82*, 187–194. https://doi.org/10.1016/j.pnpbp.2017.11.016.

D'Mello, S. R. (2020). Histone deacetylase-3: Friend and foe of the brain. *Experimental Biology and Medicine, 245*(13), 1130–1141. https://doi.org/10.1177/1535370220928278.

Delcuve, G. P., Khan, D. H., & Davie, J. R. (2012). Roles of histone deacetylases in epigenetic regulation: Emerging paradigms from studies with inhibitors. *Clinical Epigenetics, 4*(1), 5. https://doi.org/10.1186/1868-7083-4-5.

Fish, P. V., Steadman, D., Bayle, E. D., & Whiting, P. (2019). New approaches for the treatment of Alzheimer's disease. *Bioorganic & Medicinal Chemistry Letters, 29*(2), 125–133. https://doi.org/10.1016/j.bmcl.2018.11.034.

Folch, J., Petrov, D., Ettcheto, M., Abad, S., Sánchez-López, E., García, M. L., Olloquequi, J., Beas-Zarate, C., Auladell, C., & Camins, A. (2016). Current research therapeutic strategies for Alzheimer's disease treatment. *Neural Plasticity, 2016*, 1–15. https://doi.org/10.1155/2016/8501693.

Formisano, L., Laudati, G., Guida, N., Mascolo, L., Serani, A., Cuomo, O., Cantile, M., Boscia, F., Molinaro, P., Anzilotti, S., Pizzorusso, V., Di Renzo, G., Pignataro, G., & Annunziato, L. (2020). HDAC4 and HDAC5 form a complex with DREAM that epigenetically down-regulates NCX3 gene and its pharmacological inhibition reduces neuronal stroke damage. *Journal of Cerebral Blood Flow & Metabolism, 40*(10), 2081–2097. https://doi.org/10.1177/0271678×19884742.

Frankowski, H., Yeboah, F., Berry, B. J., Kinoshita, C., Lee, M., Evitts, K., Davis, J., Kinoshita, Y., Morrison, R. S., & Young, J. E. (2021). Knock-Down of HDAC2 in human induced pluripotent stem cell derived neurons improves neuronal mitochondrial dynamics, neuronal maturation and reduces amyloid beta peptides. *International Journal of Molecular Sciences, 22*(5), 2526. https://doi.org/10.3390/ijms22052526.

Furumai, R., Matsuyama, A., Kobashi, N., Lee, K.-H., Nishiyama, M., Nakajima, H., Tanaka, A., Komatsu, Y., Nishino, N., Yoshida, M., & Horinouchi, S. (2002). FK228 (depsipeptide) as a natural prodrug that inhibits class I histone deacetylases. *Cancer Research, 62*(17), 4916–4921. PMID: 12208741.

Geng, H., Chen, H., Wang, H., & Wang, L. (2021). The histone modifications of neuronal plasticity. *Neural Plasticity, 2021*, 1–7. https://doi.org/10.1155/2021/6690523.

Ghosh, P., & Saadat, A. (2023). Neurodegeneration and epigenetics: A review. *Neurología (English Edition), 38*(6), e62–e68. https://doi.org/10.1016/j.nrleng.2023.05.001.

Greeny, A., Nair, A., Sadanandan, P., Satarker, S., Famurewa, A. C., & Nampoothiri, M. (2024). Epigenetic alterations in Alzheimer's disease: Impact on insulin signaling and advanced drug delivery systems. *Biology, 13*(3), 157. https://doi.org/10.3390/biology13030157.

Hebert, L. E., Weuve, J., Scherr, P. A., & Evans, D. A. (2013). Alzheimer disease in the United States (2010–2050) estimated using the 2010 census. *Neurology, 80*(19), 1778–1783. https://doi.org/10.1212/WNL.0b013e31828726f5.

Hussain, M. S., Zhang, L., Rana, A. J., Maqbool, M., Ashique, S., Khan, Y., Jakhmola, V., Hanbashi, A., Mawkili, W., & Khan, G. (2025). Epigenetic therapy meets targeted protein degradation: HDAC-PROTACs in cancer treatment. *Future Medicinal Chemistry, 17*(14), 1725–1737. https://doi.org/10.1080/17568919.2025.2533113.

Ibrahim, T. S., Sheha, T. A., Abo-Dya, N. E., AlAwadh, M. A., Alhakamy, N. A., Abdel-Samii, Z. K., Panda, S. S., Abuo-Rahma, G. E.-D. A., & Mohamed, M. F. A. (2020). Design, synthesis and anticancer activity of novel valproic acid conjugates with improved histone deacetylase (HDAC) inhibitory activity. *Bioorganic Chemistry, 99*, 103797. https://doi.org/10.1016/j.bioorg.2020.103797.

Joshi, N., Khadse, P., Jadhav, S., & Alavala, R. R. (2025). Histone deacetylase inhibition in Alzheimer's disease: Molecular mechanisms, therapeutic potential, and future perspectives. *Current Topics in Medicinal Chemistry, 25*(28), 3215–3233. https://doi.org/10.2174/0115680266360664250606110220.

K C, S., Kakoty, V., Krishna, K. V., Dubey, S. K., Chitkara, D., & Taliyan, R. (2021). Neuroprotective efficacy of co-encapsulated rosiglitazone and vorinostat nanoparticle on streptozotocin induced mice model of Alzheimer disease. *ACS Chemical Neuroscience, 12*(9), 1528–1541. https://doi.org/10.1021/acschemneuro.1c00022.

Kalinski, A. L., Kar, A. N., Craver, J., Tosolini, A. P., Sleigh, J. N., Lee, S. J., Hawthorne, A., Brito-Vargas, P., Miller-Randolph, S., Passino, R., Shi, L., Wong, V. S. C., Picci, C., Smith, D. S., Willis, D. E., Havton, L. A., Schiavo, G., Giger, R. J., Langley, B., &

Twiss, J. L. (2019). Deacetylation of Miro1 by HDAC6 blocks mitochondrial transport and mediates axon growth inhibition. *Journal of Cell Biology, 218*(6), 1871–1890. https://doi.org/10.1083/jcb.201702187.

Kang, H., Park, Y.-K., Lee, J.-Y., & Bae, M. (2024). Roles of histone deacetylase 4 in the inflammatory and metabolic processes. *Diabetes & Metabolism Journal, 48*(3), 340–353. https://doi.org/10.4093/dmj.2023.0174.

Kay, Y., Tsan, L., Davis, E. A., Tian, C., Décarie-Spain, L., Sadybekov, A., Pushkin, A. N., Katritch, V., Kanoski, S. E., & Herring, B. E. (2022). Schizophrenia-associated SAP97 mutations increase glutamatergic synapse strength in the dentate gyrus and impair contextual episodic memory in rats. *Nature Communications, 13*(1), 798. https://doi.org/10.1038/s41467-022-28430-5.

Kim, H.-J., & Bae, S.-C. (2011). Histone deacetylase inhibitors: Molecular mechanisms of action and clinical trials as anti-cancer drugs. *American Journal of Translational Research, 3*(2), 166–179. http://www.ncbi.nlm.nih.gov/pubmed/21416059.

Kitahara, M., Inoue, T., Mani, H., Takamatsu, Y., Ikegami, R., Tohyama, H., & Maejima, H. (2021). Exercise and pharmacological inhibition of histone deacetylase improves cognitive function accompanied by an increase of gene expressions crucial for neuronal plasticity in the hippocampus. *Neuroscience Letters, 749*, 135749. https://doi.org/10.1016/j.neulet.2021.135749.

Kulthinee, S., Yano, N., Zhuang, S., Wang, L., & Zhao, T. C. (2022). Critical functions of Histone Deacetylases (HDACs) in modulating inflammation associated with cardiovascular diseases. *Pathophysiology, 29*(3), 471–485. https://doi.org/10.3390/pathophysiology29030038.

Kumar, V., Kundu, S., Singh, A., & Singh, S. (2022). Understanding the Role of histone deacetylase and their inhibitors in neurodegenerative disorders: Current targets and future perspective. *Current Neuropharmacology, 20*(1), 158–178. https://doi.org/10.2174/1570159×19666210609160017.

Lane, C. A., Hardy, J., & Schott, J. M. (2018). Alzheimer's disease. *European Journal of Neurology, 25*(1), 59–70. https://doi.org/10.1111/ene.13439.

Li, Y., Izhar, T., & Kanekiyo, T. (2025). HDAC3 as an emerging therapeutic target for Alzheimer's disease and other neurological disorders. *Molecular Neurobiology, 62*(8), 9573–9585. https://doi.org/10.1007/s12035-025-04866-w.

Li, Y., Lin, S., Gu, Z., Chen, L., & He, B. (2022). Zinc-dependent deacetylases (HDACs) as potential targets for treating Alzheimer's disease. *Bioorganic & Medicinal Chemistry Letters, 76*, 129015. https://doi.org/10.1016/j.bmcl.2022.129015.

Lin, Y., Li, Y., Ke, C., Jin, Y., Lao, W., Wu, Y., Liu, Y., Kong, X., Qiao, J., Zhai, A., & Bi, C. (2025). HDAC4: An emerging target in diabetes mellitus and diabetic complications. *European Journal of Medical Research, 30*(1), 429. https://doi.org/10.1186/s40001-025-02697-y.

Liu, Y.-M., & Liou, J.-P. (2023). An updated patent review of histone deacetylase (HDAC) inhibitors in cancer (2020 – present). *Expert Opinion on Therapeutic Patents, 33*(5), 349–369. https://doi.org/10.1080/13543776.2023.2219393.

LoPresti, P. (2020). HDAC6 in diseases of cognition and of neurons. *Cells, 10*(1), 12. https://doi.org/10.3390/cells10010012.

Maejima, H., Kitahara, M., Takamatsu, Y., Mani, H., & Inoue, T. (2021). Effects of exercise and pharmacological inhibition of histone deacetylases (HDACs) on epigenetic regulations and gene expressions crucial for neuronal plasticity in the motor cortex. *Brain Research, 1751*, 147191. https://doi.org/10.1016/j.brainres.2020.147191.

Mahady, L., Nadeem, M., Malek-Ahmadi, M., Chen, K., Perez, S. E., & Mufson, E. J. (2019). HDAC 2 dysregulation in the nucleus basalis of Meynert during the progression of Alzheimer's disease. *Neuropathology and Applied Neurobiology, 45*(4), 380–397. https://doi.org/10.1111/nan.12518.

Mondal, P., Bai, P., Gomm, A., Bakiasi, G., Lin, C. J., Wang, Y., Choi, S. H., Tanzi, R. E., Wang, C., & Zhang, C. (2024). Structure-based discovery of a small molecule inhibitor of Histone Deacetylase 6 (HDAC6) that significantly reduces Alzheimer's disease neuropathology. *Advanced Science, 11*(1), https://doi.org/10.1002/advs.202304545.

Mottamal, M., Zheng, S., Huang, T., & Wang, G. (2015). Histone deacetylase inhibitors in clinical studies as templates for new anticancer agents. *Molecules, 20*(3), 3898–3941. https://doi.org/10.3390/molecules20033898.

Muñoz-Cobo, I., Erburu, M. M., Zwergel, C., Cirilli, R., Mai, A., Valente, S., Puerta, E., & Tordera, R. M. (2018). Nucleocytoplasmic export of HDAC5 and SIRT2 downregulation: Two epigenetic mechanisms by which antidepressants enhance synaptic plasticity markers. *Psychopharmacology, 235*(10), 2831–2846. https://doi.org/10.1007/s00213-018-4975-8.

Nakatsuka, D., Izumi, T., Tsukamoto, T., Oyama, M., Nishitomi, K., Deguchi, Y., Niidome, K., Yamakawa, H., Ito, H., & Ogawa, K. (2021). Histone deacetylase 2 knockdown ameliorates morphological abnormalities of dendritic branches and spines to improve synaptic plasticity in an APP/PS1 transgenic mouse model. *Frontiers in Molecular Neuroscience, 14*. https://doi.org/10.3389/fnmol.2021.782375.

Ndukwe, K., Serrano, P. A., Rockwell, P., Xie, L., & Figueiredo-Pereira, M. E. (2025). Brain-penetrant histone deacetylase inhibitor RG2833 improves spatial memory in females of an Alzheimer's disease rat model. *Journal of Alzheimer's Disease, 104*(1), 173–190. https://doi.org/10.1177/13872877251314777.

Neganova, M. E., Klochkov, S. G., Aleksandrova, Y. R., & Aliev, G. (2021). The hydroxamic acids as potential anticancer and neuroprotective agents. *Current Medicinal Chemistry, 28*(39), 8139–8162. https://doi.org/10.2174/0929867328666201218123154.

Pal, D., Sahu, P., Mishra, A. K., Hagelgans, A., & Sukocheva, O. (2023). Histone deacetylase inhibitors as cognitive enhancers and modifiers of mood and behavior. *Current Drug Targets, 24*(9), 728–750. https://doi.org/10.2174/1389450124666221207090108.

Pawar, S., Rauf, M. A., Abdelhady, H., & Iyer, A. K. (2025). Tau-targeting nanoparticles for treatment of Alzheimer's disease. *Exploration, 5*(2), https://doi.org/10.1002/EXP.20230137.

Pereira, M., Cruz, M. T., Fortuna, A., & Bicker, J. (2024). Restoring the epigenome in Alzheimer's disease: Advancing HDAC inhibitors as therapeutic agents. *Drug Discovery Today, 29*(7), 104052. https://doi.org/10.1016/j.drudis.2024.104052.

Perry, S., Kiragasi, B., Dickman, D., & Ray, A. (2017). The role of histone deacetylase 6 in synaptic plasticity and memory. *Cell Reports, 18*(6), 1337–1345. https://doi.org/10.1016/j.celrep.2017.01.028.

Pulya, S., Ganorkar, A., Biswas, S., & Ghosh, B. (2023). Histone deacetylase inhibitors: Design, synthesis, and biological evaluation. *In Natural Product-based Synthetic Drug Molecules in Alzheimer's Disease*. Springer Nature Singapore, 427–447. https://doi.org/10.1007/978-981-99-6038-5_16.

Pulya, S., Mahale, A., Bobde, Y., Routholla, G., Patel, T., Swati, Biswas, S., Sharma, V., Kulkarni, O. P., & Ghosh, B. (2021). PT3: A novel benzamide class histone deacetylase 3 inhibitor improves learning and memory in novel object recognition mouse model. *ACS Chemical Neuroscience, 12*(5), 883–892. https://doi.org/10.1021/acschemneuro.0c00721.

Rahman, A. F. M. T., Bulbule, S., Belayet, J. Bin, Benko, A., Gottschalk, C. G., Frick, D. N., Arnold, L. A., Hossain, M. M., & Roy, A. (2024). JRM-28, a novel HDAC2 inhibitor, upregulates plasticity-associated proteins in hippocampal neurons and enhances morphological plasticity via activation of CREB: Implications for Alzheimer's disease. *Cells, 13*(23), 1964. https://doi.org/10.3390/cells13231964.

Ramakrishnan, S., & Pili, R. (2013). Histone deacetylase inhibitors and epigenetic modifications as a novel strategy in renal cell carcinoma. *The Cancer Journal, 19*(4), 333–340. https://doi.org/10.1097/PPO.0b013e3182a09e07.

Ramirez-Mejia, G., Gil-Lievana, E., Urrego-Morales, O., Soto-Reyes, E., & Bermúdez-Rattoni, F. (2021). Class I HDAC inhibition improves object recognition memory consolidation through BDNF/TrkB pathway in a time-dependent manner. *Neuropharmacology, 187*, 108493. https://doi.org/10.1016/j.neuropharm.2021.108493.

Ricciardi, N. R., Modarresi, F., Lohse, I., Andrade, N. S., Newman, I. R., Brown, J. M., Borja, C., Marples, B., Wahlestedt, C. R., & Volmar, C.-H. (2023). Investigating the synergistic potential of low-dose HDAC3 Inhibition and radiotherapy in Alzheimer's disease models. *Molecular Neurobiology, 60*(8), 4811–4827. https://doi.org/10.1007/s12035-023-03373-0.

Sayas, C. L., & Ávila, J. (2021). GSK-3 and Tau: A key duet in Alzheimer's Disease. *Cells, 10*(4), 721. https://doi.org/10.3390/cells10040721.

Scheltens, P., De Strooper, B., Kivipelto, M., Holstege, H., Chételat, G., Teunissen, C. E., Cummings, J., & van der Flier, W. M. (2021). Alzheimer's disease. *The Lancet, 397*(10284), 1577–1590. https://doi.org/10.1016/S0140-6736(20)32205-4.

Shen, S., & Kozikowski, A. P. (2020). A patent review of histone deacetylase 6 inhibitors in neurodegenerative diseases (2014-2019). *Expert Opinion on Therapeutic Patents, 30*(2), 121–136. https://doi.org/10.1080/13543776.2019.1708901.

Smies, C. W., Bellfy, L., Wright, D. S., Bennetts, S. G., Urban, M. W., Brunswick, C. A., Shu, G., & Kwapis, J. L. (2024). Pharmacological HDAC3 inhibition alters memory updating in young and old male mice. *Frontiers in Molecular Neuroscience, 17*. https://doi.org/10.3389/fnmol.2024.1429880.

Soria Lopez, J. A., González, H. M., & Léger, G. C. (2019). *Alzheimer's disease*, 231–255. https://doi.org/10.1016/B978-0-12-804766-8.00013-3.

Sultana, R., Banks, W. A., & Butterfield, D. A. (2010). Decreased levels of PSD95 and two associated proteins and increased levels of BCl 2 and caspase 3 in hippocampus from subjects with amnestic mild cognitive impairment: Insights into their potential roles for loss of synapses and memory, accumulation of Aβ. *Journal of Neuroscience Research, 88*(3), 469–477. https://doi.org/10.1002/jnr.22227.

Talom, A., Barhoi, A., Jirpu, T., Dawn, B., & Ghosh, A. (2025). Clinical progress and functional modalities of HDAC inhibitor-based combination therapies in cancer treatment. *Clinical and Translational Oncology*. https://doi.org/10.1007/s12094-025-03995-x.

Volmar, C.-H., & Wahlestedt, C. (2015). Histone deacetylases (HDACs) and brain function. *Neuroepigenetics, 1*, 20–27. https://doi.org/10.1016/j.nepig.2014.10.002.

Wadhwa, G., Krishna, K. V., Dubey, S. K., & Taliyan, R. (2023). PEGylated polymer–lipid hybrid nanoparticles to enhance in vivo exposure and uptake of repaglinide in brain cells to treat diabetes-linked neurodegenerative disorders. *ACS Applied Nano Materials, 6*(5), 3497–3512. https://doi.org/10.1021/acsanm.2c05272.

Wang, L., Sun, T., Wang, Z., Liu, H., Qiu, W., Tang, X., Guo, H., Yang, P., Chen, Y., & Sun, H. (2023). Design, synthesis, and proof of concept of balanced dual inhibitors of Butyrylcholinesterase (BChE) and histone deacetylase 6 (HDAC6) for the treatment of Alzheimer's disease. *ACS Chemical Neuroscience, 14*(17), 3226–3248. https://doi.org/10.1021/acschemneuro.3c00358.

Watson, N., Kuppuswamy, S., Ledford, W. L., & Sukumari-Ramesh, S. (2024). The role of HDAC3 in inflammation: Mechanisms and therapeutic implications. *Frontiers in Immunology, 15*. https://doi.org/10.3389/fimmu.2024.1419685.

Xhemalce, B., Dawson, M. A., & Bannister, A. J. (2011). Histone modifications. *In Encyclopedia of Molecular Cell Biology and Molecular Medicine*. Wiley-VCH Verlag GmbH & Co. KGaA, https://doi.org/10.1002/3527600906.mcb.201100004.

Xu, K., Dai, X.-L., Huang, H.-C., & Jiang, Z.-F. (2011). Targeting HDACs: A promising therapy for Alzheimer's disease. *Oxidative Medicine and Cellular Longevity, 2011*, 1–5. https://doi.org/10.1155/2011/143269.

Xu, L., An, T., Jia, B., Wu, Q., Shen, J., Jin, J., Liu, J., & Li, C. (2024). Histone deacetylase 3-specific inhibitor RGFP966 attenuates oxidative stress and inflammation after traumatic brain injury by activating the Nrf2 pathway. *Burns & Trauma, 12*. https://doi.org/10.1093/burnst/tkad062.

Yang, S., Zhang, R., Wang, G., & Zhang, Y. (2017). The development prospection of HDAC inhibitors as a potential therapeutic direction in Alzheimer's disease. *Translational Neurodegeneration, 6*(1), 19. https://doi.org/10.1186/s40035-017-0089-1.

Yao, C., Jiang, X., Zhao, R., Zhong, Z., Ge, J., Zhu, J., Ye, X.-Y., Xie, Y., Liu, Z., Xie, T., & Bai, R. (2022). HDAC1/MAO-B dual inhibitors against Alzheimer's disease: Design, synthesis and biological evaluation of N-propargylamine-hydroxamic acid/o-aminobenzamide hybrids. *Bioorganic Chemistry, 122*, 105724. https://doi.org/10.1016/j.bioorg.2022.105724.

Yu, Q., Du, F., Douglas, J. T., Yu, H., Yan, S. S., & Yan, S. F. (2017). Mitochondrial dysfunction triggers synaptic deficits via activation of p38 MAP kinase signaling in differentiated Alzheimer's disease trans-mitochondrial cybrid cells. *Journal of Alzheimer's Disease, 59*(1), 223–239. https://doi.org/10.3233/JAD-170283.

Zhang, L.-Y., Zhang, S.-Y., Wen, R., Zhang, T.-N., & Yang, N. (2024). Role of histone deacetylases and their inhibitors in neurological diseases. *Pharmacological Research, 208*, 107410. https://doi.org/10.1016/j.phrs.2024.107410.

Zhang, W., Zhang, Y., Wang, Y., Wang, C., & Zhang, C. (2025). Advancing histone deacetylase 6 (HDAC6) as a promising therapeutic target for Alzheimer's disease: from molecular insights to clinical prospects. *Frontiers in Drug Discovery, 5*. https://doi.org/10.3389/fddsv.2025.1662691.

Zou, F., Qiu, Y., Huang, Y., Zou, H., Cheng, X., Niu, Q., Luo, A., & Sun, J. (2021). Effects of short-chain fatty acids in inhibiting HDAC and activating p38 MAPK are critical for promoting B10 cell generation and function. *Cell Death & Disease, 12*(6), 582. https://doi.org/10.1038/s41419-021-03880-9.

CHAPTER TEN

The Tau-mitochondria connection and its impact on cellular energy metabolism in Alzheimer's disease

Abhay Kumar Singh[a], Supraja M.R.[a], Amal P.S.[a], Senthilkumar Krishnamoorthi[b], Ashok Iyaswamy[b,c], and Siva Sundara Kumar Durairajan[a,d,*]

[a]Molecular Mycology and Neurodegenerative Disease Research Laboratory, Department of Microbiology, Central University of Tamil Nadu, Tiruvarur, Tamil Nadu, India

[b]Mr. & Mrs Ko Chi-Ming Centre for Parkinson's Disease Research, School of Chinese Medicine, Hong Kong Baptist University, Kowloon, Hong Kong, P.R. China

[c]Department of Biochemistry, Karpagam Academy of Higher Education, Coimbatore, Tamil Nadu, India

[d]School of Biomedical Sciences, Li Ka Shing Faculty of Medicine, The University of Hong Kong, Hong Kong, P.R. China

*Corresponding author. e-mail address: d.sivasundarakumar@cutn.ac.in; dsskumar75@gmail.com

Contents

International Review of Neurobiology, Volume 188
ISSN 0074-7742, https://doi.org/10.1016/bs.irn.2026.06.002

Abstract

Alzheimer's disease (AD) is a complex multifactorial neurodegenerative disease process resulting in progressive cognitive deterioration and synaptic dysfunction. The primary research approach in AD has traditionally focused on amyloid- pathology however an increasingly evidence suggests that tau protein is a key mediator of neuronal damage via a direct action on mitochondrial bioenergetics. In this chapter we look at the nature of the tau-mitochondrial interface, and propose a paradigm of tau-induced energy failure in AD. Physiologically tau provides stability to the microtubules and is involved in transport mechanisms within cells. In AD, tau is excessively post-translationally modified hyperphosphorylated and truncated tau species form toxic oligomers that incorrectly translocate to mitochondria, interacting pathologically with critical proteins such as voltage-dependent anion channel 1 (VDAC1) and adenine nucleotide translocase (ANT), impeding the mitochondrial ATP/ADP exchange and reducing oxidative phosphorylation efficiency. Tau also further damages mitochondria by excessive fission, inhibition of axonal transport and Inhibition of mitophagy by interrupting PINK1-Parkin signaling. In turn, the build-up of dysfunctional mitochondria leads to ROS production, mtDNA damage and calcium imbalance creating a vicious cycle toward oxidative stress and tau pathology. At the cellular level they cause an energy depletion of the synapse and at the systems level cause glucose hypometabolism and activation of neuroinflammation. The chapter additionally discusses novel therapeutic approaches that target both tau and mitochondrial abnormalities, namely antisense oligonucleotides (ASO), mitochondria targeted compounds and mitophagy modifiers, stressing that it would be more effective to utilize a cocktail of these inhibitors. As a whole, in the context of decreased bioenergetics, the tau–mitochondria axis is an important factor to consider in the successful treatment of AD.

1. Introduction

1.1 Alzheimer's disease: Epidemiology and pathological hallmarks

The leading cause of neurodegenerative dementia globally is Alzheimer's disease (AD). At present there are approximately 56.9 million people worldwide living with the condition (Hampel et al., 2021), and with our global populations ageing, this is set to increase to 152.8 million people by 2050 (Qin et al., 2025). This is a major social and public health problem globally. The incident burden is consistently higher in females than males, particularly over the age of 65 (Mielke et al., 2022). This is not solely due to female longevity and is thought to involve both neurobiological differences between the sexes that influence disease etiology, and possibly include lower mitochondrial protection afforded by the decreased activity of oestrogen-related proteins after menopause; regional differences in tau spread; increased inflammatory responses to neurodegeneration (Buckley et al., 2019). The financial implications of AD are enormous and are escalating. Costs for care in 2025 are projected to reach $384 billion globally and will approach $1 trillion in 2050 (Alzheimer's Association, 2023), not including informal care costs of over $415 billion per year.

AD is characterised by two defining brain lesions observed at the neuropathological level.The first, involves the extracellular deposition of amyloid- β (A β) plaques, associated with abnormal amyloid precursor protein (APP) processing (Hampel et al., 2021). Such plaques are also associated with the accumulation of hyperphosphorylated tau protein (Hampel et al., 2021). The second, involves the intracellular accumulation of neurofibrillary tangles (NFTs), which are associated with axonal microtubule disruption (Sperling et al., 2011). Soluble A β, oligomers are considered to be the major neurotoxic drivers of AD pathology. In the natural history of AD, a pre-clinical presymptomatic period of 15–20 years occurs, from a predictable sequence of amyloid deposition, tau aggregation and neurodegeneration. Such biomarkers are now accessible through blood tests, such as plasma p-tau217, which now allow pre-symptomatic AD diagnosis with sensitivity and specificity >85 % in numerous cohorts in studies (Palmqvist et al., 2023). Plasma p-tau217 can be definitively identified 10–15 years prior to disease presentation, and has an area under the receiver operating characteristic curve (AUC) between 0.89 and 0.96 relative to amyloid PET, and correlates significantly pcognitive severity, with MMSE r values of−0.72 to −0.82. The plasma p-tau217 biomarker thus appears to be a key endpoint in potential tau targeted therapies (Barthélemy et al., 2020). A successful antisense oligonucleotide (ASO) approach would reduce p-tau217 levels by 40–50 %. The glial fibrillary acidic protein (GFAP) signal complements p-tau217 as a useful screening biomarker. GFAP is detectable up to 10 years before symptoms and achieves an AUC of 0.86 for amyloid positivity. Two amyloid therapies, lecanemab and donanemab, are currently available and clear plaques effectively. However, clinical slowing remains modest following their use (van Dyck et al., 2023). This suggests that tau-related neurodegeneration is a more direct driver of synaptic loss (Sims et al., 2023).

1.2 Neuronal energy crisis in Alzheimer's disease

Neurons exhibit an unusually high energy requirement relative to their size and mass, consuming nearly 20 % of the basal metabolic rate while composing only 2 % of body weight (Mink et al., 1981). This energy requirement is necessary for action potential generation, synaptic vesicle cycle and transport, and axonal transport (Rossi & Pekkurnaz, 2019). Energy requirement is supplied by mitochondrial oxidative phosphorylation.

Oxidation of glucose via the TCA cycle produces electrons that donate to respiratory chain complexes I-IV. The proton gradient is used to drive ATP synthase to form ATP. Neuronal dysfunction is also found at late stages of AD. Compromised bioenergetics, involving dysfunction in respiratory complexes I-V and enhanced ROS, has been found at very early stages, even prior to amyloid plaque accumulation in transgenic AD mouse models (Yao et al., 2009). Glucose hypometabolism in the brain can be detected prior to the onset of symptoms, years before the appearance of clinical diagnosis, among apolipoprotein E epsilon 4 (APOE4) carriers, even in cognitively normal individuals (Reiman et al., 2004).

1.3 Rationale and chapter scope

Recent evidence is now directing tau to a key, and causal, role in mitochondrial dysfunction. It is no longer considered to be solely a downstream marker of neurodegeneration (Eckert et al., 2014). Tau forms aggregates not just in microtubules but within mitochondria where it localises to a variety of compartments including the outer mitochondrial membrane, inner mitochondrial membrane, the matrix (Cieri et al., 2018) and binds to protein import machinery and ETC complexes, the subunits of ATP synthase, calcium transporting channels and the proteins that regulate mitochondrial dynamics (Torres et al., 2021). The link between tau and mitochondria is bi-directional as mitochondrial ROS and bioenergetic failure leads to the activation of kinase cascades that promote tau phosphorylation (Cheng & Bai, 2018) and subsequently toxic tau leads to exacerbated mitochondrial dysfunction, driving the development of this self-perpetuating pathology (Castillo-Casaña et al., 2025). In this chapter the molecular mechanisms by which tau causes dysfunction in bioenergetics, redox status, calcium transport, networks of organelles and the quality control processes of the cell will be discussed to provide the evidence for the neuronal energy crisis theory of AD.

2. Tau protein: From physiological function to pathological species

2.1 Normal Tau biology

Tau is one of six microtubule associated proteins (MAPs) expressed in the nervous system. The principle physiological role of tau is thought to be to perform stabilisation of the axonal cytoskeleton and control of intracellular transport (Liu et al., 2005). In the human nervous system, there are six Tau isoforms that can be generated through alternative splicing. These Isoforms have lengths between 352–441 amino acids and molecular weights between 36.7 and 45.8 kDa respectively (Buchholz & Zempel, 2024). They differ in two principal structural features. First, they contain either zero, one, or two N-terminal inserts. Second, they contain either three or four microtubule-binding repeats, referred to as 3R or 4R MTBRs. This distinction is determined by inclusion of exon 10 in the microtubule-binding domain. The 3R to 4R tau ratio also changes across development (Goedert et al., 1989). In the foetal brain, 3R tau predominates, whereas the adult human brain maintains an approximately 1:1 ratio.

These isoform distinctions have profound cell-biological and pathological consequences for tau. The predominant tau isoforms in the adult brain are 0N4R, 1N4R and 2N4R, comprising 85–95 % of adult tau, and have significantly higher tubulin-binding affinity than the 3R isoforms [KD≈50 nM for 4R isoforms vs. KD≈150 nM for 3R tau (Bruce et al., 2000)]. More specifically, 4R isoforms also have significantly higher binding affinity for voltage dependent anion channel 1 (VDAC1) present on the outer mitochondrial membrane. Particularly high-affinity (KD≈45 nM versus≈180 nM for 3R tau) is hyperphosphorylated 2N4R, and 0N4R (Castillo-Casaña et al., 2025). Such ontological differences help rationalize the differences we see between tauopathies. Prominent 4R (presence in oligodendrocytes) is seen in the radically different tauopathies progressive supranuclear palsy (PSP) and corticobasal degeneration (CBD), while the relatively weak VDAC1-mediated hypothesized metabolic gating dysfunction of 3 R-predominant Pick's disease is not associated with the same degree of oligodendrocyte mitochondrial involvement despite similar aggregate loads (Ferrer et al., 2019; Aliu Opeyemi et al., 2026).

Under physiological conditions, tau carries approximately 2–3 phosphates per molecule. In vivo phosphorylation occurs at only a limited set of serine and threonine residues (Liu et al., 2009). O-GlcNAcylation is a glucose-dependent post-translational modification that directly antagonises tau phosphorylation. It is tightly linked to neuronal glucose availability (Alhadidy & Kanaan, 2024) This reduction in cerebral glucose metabolism leads to increased O-GlcNAc levels which in turn inhibits tau O-GlcNAcylation and thus increases pathological hyperphosphorylation. O-GlcNAc is transferred by OGT at several sites including Thr231, Ser400 and Ser409, the substrate for this reaction being UDP-GlcNAc obtained from hexosamine biosynthesis. OGA removes O-GlcNAc, levels of which are 2.1 times increased in AD hippocampus (Gong et al., 2005). This means that in disease O-GlcNAc modification of tau becomes significantly reduced. Overall tau O-GlcNAc stoichiometry falls from between 1.8 and 2.2 mol O-GlcNAc/mol tau in normal subjects to 0.3 and 0.5 mol/mol in AD tangles, which equates to a 70–80 % decrease in modification and mirrors the percentage of reduction seen in cerebral glucose metabolism (Liu et al., 2004; Yuzwa et al., 2012).

Apart from O-GlcNAcylation, additional post-translational modifications regulate physiological tau. Acetylation at K174 is induced by CREB-binding protein (CBP) and p300 acetyltransferases (Cohen et al., 2011). K48-linked ubiquitination is induced by CHIP and CHIP-dependent, HSP70-dependent E3 ligase activity (Petrucelli et al., 2004). All of these together normally regulate tau half-life at approximately 12 hr and also prevent inappropriate tau transport into mitochondria.

2.2 Pathological Tau transformations and toxic species

The transition from physiological to pathological tau is driven by kinase dysregulation and phosphatase deficits. Glycogen synthase kinase 3 (GSK3) and cyclin-dependent kinase 5 (CDK5) are among the best-characterised tau kinases in AD. They activate a cascade of priming to cause long-term hyperphosphorylation of over 30 sites (Li & Paudel, 2006). Pathological tau contains a phosphate loading of about 9–10 mol phosphate/mol tau (Köpke et al., 1993). It is a characteristic biochemical characteristic of tau pathology. Oxidative damage also activates stress kinases such as JNK and p38 MAPK, they increase phosphorylation site and therefore enhances the iterative release of tau from microtubules (Lee et al., 2004; Lovell et al., 2004).

After being released, hyperphosphorylated tau can be proteolytically cleaved by caspases to shorter isoforms. Casepase-3-cleaved Asp421 tau is particularly prominent and it filaments 2-fold faster than full-length tau (Gamblin et al., 2003). This isoform is accumulated near the mitochondria and causes increased bioenergetic toxicity (Rissman et al., 2004). The major neurotoxic species is soluble tau oligomers rather than PHF ends. They enriched decades before insoluble PHF formation (Lasagna-Reeves et al., 2012) and it is more correlated with cognitive deterioration and synaptic loss than tangle load (Gerson et al., 2016).

Phosphorylation staging across disease progression follows a predictable and well-documented hierarchy. Each phosphorylation site carries distinct mitochondrial consequences for neuronal function. Thr231 phosphorylation is the earliest detectable plasma biomarker, emerging 8–12 years before symptom onset. Occupancy increases from 8 plus or minus 3 % in controls to

68 plus or minus 6 % in AD tangles. Mechanistically, phosphorylated Thr231 enhances TOMM40 binding and potentiates mitochondrial import blockade (Barthélemy et al., 2020). Ser262 phosphorylation is mediated by microtubule affinity-regulating kinases 1 through 4 (MARK1–4) at KXGS motifs in the microtubule-binding domain. It produces an immediate fourfold reduction in tubulin affinity. It also exposes a cryptic mitochondrial-targeting sequence-like region at residues 150–230. This increases outer mitochondrial membrane (OMM) localisation 2.8-fold and facilitates adenine nucleotide translocator (ANT) inhibition. The Ser396 and Ser404 doublet, known as the PHF-1 epitope, shows approximately 75 % occupancy in AD tangles. It underlies preferential synaptic mitochondrial targeting of phospho-tau. It reduces ATP and ADP flux through VDAC1 by 60–75 % and lowers the calcium threshold for mitochondrial permeability transition pore (mPTP) sensitisation. Thr217 phosphorylation represents the strongest plasma biomarker with an AUC of 0.89–0.96 (Milà-Alomà et al., 2022). It is especially informative because it occurs at a primary O-GlcNAcylation competition site. Its increase therefore closely mirrors the metabolic failure-driven loss of protective O-GlcNAcylation (Barthélemy et al., 2020).

Post-translational modification (PTM) pattern in disease context goes beyond just phosphorylation. It 's known that acetylation, particularly on microtubule-binding region and specifically on K280 is one of the key sites targeted by CBP and p300 in AD relevant 4R tau. It weakens electrostatic interactions with negatively charged tubulin C-terminal tails, modestly lowering binding affinity. It also markedly accelerates 4R tau oligomerisation by up to 2.3-fold and fibril formation by up to 1.8-fold. SIRT1 activity, which normally deacetylates these sites, is reduced by approximately 40 % in AD brain tissue. K63-linked ubiquitination at K311 and K317 is elevated 3.2-fold in soluble AD tau. This creates polyubiquitin scaffolds that promote tau oligomerisation and preferentially route tau to autophagosomes. However, this pathway is overwhelmed by parallel failure of K48-linked ubiquitin clearance. This failure is secondary to CHIP downregulation, which is reduced 40–55 % in the AD hippocampus. Nitrosylation and oxidative modifications at Cys291 and Cys322 are driven by ROS from tau-impaired respiratory chains (Quintanilla et al., 2009). These modifications lock tau into beta-sheet-rich conformations and generate intermolecular disulfide crosslinks that stabilise PHF structures. Oxidised tau also shows a 2.1-fold higher affinity for cardiolipin than reduced tau (Melov et al., 2007). This further strengthens pathogenic interactions with the inner mitochondrial membrane.

3. Mitochondrial biology in the neuronal context

3.1 Ultrastructural architecture and its functional significance

Neuronal mitochondria have a highly regulated architecture with important functional consequences. Their three-dimensional structure is tightly coupled to bioenergetics, calcium buffering, and vulnerability to pathogenic stress. The organelle is defined by two distinct membrane systems. The external mitochondrial membrane (OMM) is a semi-permeable barrier to the cytoplasm. The inner mitochondrial membrane (IMM) is very convoluted and functionally impermeable. This selectivity is necessary to sustain the electrochemical gradient that sustains ATP production (Frey & Mannella, 2000; Paradies et al., 2019). The IMM is again subdivided into two topologically different subdomains. The inner boundary membrane (IBM) is parallel to the OMM. It is a docking site of mitochondrial carrier proteins and nuclear-encoded translocases (Gilkerson et al., 2003). Cristae are membrane infolds in the form of lamellae or tubes that offer the major surface area with which oxidative phosphorylation occurs (Puff et al., 2021).

The MICOS (Mitochondrial Contact Site and Cristae Organising System) complex at cristae junctions and OPA1, a dynamin-related GTPase, with long and short versions regulated by stress-responsive proteases (YME1L, OMA1) regulate cristae organisation (Blum et al., 2019). ATP synthase dimers align along cristae ridges at an angle of about 86°, to create membrane curvature that assists in providing tubular cristae architecture; thus, structure and function come together within the same structure (Blum et al., 2019; Strauss et al., 2008). In AD, this coupling is disrupted by direct pathological tau interactions with ATP synthase subunits.

3.2 Cardiolipin: The mitochondria-specific lipid that integrates structure and function

The cardiolipin (CL) is a special dimeric phospholipid located almost exclusively in the IMM. It promotes a wide spectrum of structural and catalytic activities of the mitochondria. CL is composable of up to 20 % of IMM phospholipids. It has a unique geometry with two phosphate head groups as well as four acyl chains. This geometry endows biophysical properties not available to single phospholipids. Structural studies with high-resolution determine discrete CL-binding sites in respiratory chain complexes. They are located at lipid-protein interfaces in Complex I, Complex III, Complex IV, and ATP synthase (Schlame & Ren, 2009). At these locations, CL is a critical cofactor by stabilising transmembrane helices. It facilitates also the proton-transfer pathways between subunits and enhances quaternary assembly of functional enzyme complexes (Mileykovskaya & Dowhan, 2014). CL is also a local proton buffer at the membrane-water interface. This buffering assists to direct respiratory-chain to proton flux toward ATP synthase and enhance thermodynamic coupling efficiency. The interactions between tau and cytochrome c increase CL peroxidation in AD neurons by mitochondrially produced hydroxyl radicals (Paradies et al., 2019). This is a mechanistic explanation of early drops in synthetic efficiency of ATP in preclinical models (Szeto, 2014).

3.3 Respiratory supercomplexes and electron transfer efficiency

Respiratory-chain complexes embedded in the IMM self-assemble into stable quaternary structures. These structures are known as respirasomes or respiratory supercomplexes. In mammalian mitochondria, the dominant assembly typically includes one copy of Complex I at approximately 1 MDa. It also includes a Complex III dimer at approximately 500 kDa and one to four copies of Complex IV at approximately 200 kDa each (Acín-Pérez et al., 2008). Together, these form a supramolecular structure of approximately 1.7–2.0 MDa (Lapuente-Brun et al., 2013). This organisation offers major functional advantages for mitochondrial performance. It increases inter-complex electron-transfer efficiency by an estimated 5–10-fold through substrate channelling. It also reduces electron leak from mobile carriers such as ubiquinol and cytochrome c, thereby limiting superoxide generation. Supercomplex integrity is strongly dependent on cardiolipin availability and integrity. Tau-driven ROS accumulation in AD promotes peroxidation of the inner-membrane lipid environment, including bound cardiolipin (Petrosillo et al., 2008). This destabilises supercomplex architecture and favours disaggregation into isolated complexes. As a result, mitochondria shift from a channelling-enabled, low-ROS state to a diffusion-limited, high-ROS state (Acín-Pérez et al., 2008; Lapuente-Brun et al., 2013).

3.4 Bioenergetic requirements of the neuron and activity-coupled ATP synthesis

A single cortical neuron is estimated to consume approximately 4.7 billion ATP molecules per second at rest. This demand can more than double during prolonged electrical activity (Attwell & Laughlin, 2001). The energetic cost of restoring ion gradients after action potential firing drives this elevated demand (Zhu et al., 2012). The key coupling signal between synaptic activity and mitochondrial ATP production is calcium. Within seconds of entering presynaptic terminals and postsynaptic spines, calcium activates three major TCA-cycle dehydrogenases. They are pyruvate dehydrogenase, NAD-dependent isocitrate dehydrogenase, and oxoglutarate dehydrogenase (Denton, 2009). They enhance the production of ATP to match the synaptic energy requirements (Glancy & Balaban, 2012). ATP synthase has a membrane-bounded F0 sector that has the c-ring through which protons pass. It also has a catalytic F1 sector whose rotating stalk catalyzes the formation of ATP. When the force of proton motions fall below the ATP-synthesis threshold, the enzyme may turn the other way. The enzyme degrades ATP in the reverse mode instead of synthesising it. This reverse mode activity is inhibited by inhibitory factor 1 (IF1) under physiological conditions. However, this protective mechanism seems to be impaired in neurons with tau impairments (Rieger et al., 2021).

3.5 Mitochondrial dynamics: Fission, fusion, and positional control

The fission and fusion processes of neuronal mitochondria are regulated to continually remodel the local composition and organisation of the mitochondria. These processes are balanced to determine the maintenance of mitochondrial DNA, bioenergetic efficiency and predisposition to autophagic clearance (Chen & Chan, 2006).

Elongated mitochondria formed by fusion will typically exhibit increased oxidative phosphorylation, increased calcium-buffering, decreased ROS production and enhanced inter-mitochondrial complementation by sharing matrix content. The cytosolic dynamin-related protein 1 (Drp1) mediates fission by binding itself to the OMM through adaptor proteins (MFF, mitochondrial fission 1 protein [FIS1]) and MiD49 and MiD51. Reversible post-translational regulation of Drp1 activity, Ser616 phosphorylation activates the protein, Ser637 phosphorylation by PKA inhibits the protein, and S-nitrosylation at Cys644 (preferred in high nitric-oxidestates, as found in brain of AD patients) causes excessive fragmentation (Barsoum et al., 2006).

Mitochondria interact trans-membrane at the outer membrane through Mitofusin 1 and 2 (MFN1/2), and inner-membrane fusion is regulated by OPA1, a process that is physically related to cristae remodelling, tightening of junctions, and the elevation of cristae membrane potential (Ban et al., 2017; Mishra et al., 2014) The Miro-TRAK motor-adaptor system is a spatially organized system of attachment to kinesin-1 (anterograde) and dynein (retrograde) motors that involves the outer-membrane Miro acting as a calcium sensor in this process. Local calcium increase during synaptic activity binds Miro EF-hand domains, induces motor uncoupling and holds mitochondria in proximity to active synapses (MacAskill et al., 2009; Wang & Schwarz, 2009). Hyperphosphorylated tau disrupts this positioning network by targeting kinesin-1 by impairing the PAD-initiated signalling. This causes a progressive breakdown of mitochondria in the more remote synapses and in the soma.

3.6 Mitochondrial quality control: The PINK1-Parkin pathway and the mitochondrial unfolded protein response

PINK1-Parkin pathway is the canonical pathway of selective mitophagic removal of damaged mitochondria. Under physiological conditions, PINK1, a serine/threonine kinase, is imported through TIM23 and rapidly cleaved by presenilin-associated rhomboid-like (PARL) protease (Jin et al., 2010). This cleavage keeps mitochondrial PINK1 levels low under healthy conditions. When membrane potential dissipates, a universal marker of mitochondrial injury, import is halted. PINK1 then accumulates on the OMM, where it phosphorylates ubiquitin at Ser65 and promotes Parkin recruitment (Kazlauskaite et al., 2014; Lee et al., 2017). Parkin is an E3 ligase that is normally held in an autoinhibited state. Phospho-ubiquitin binding together with PINK1-dependent activation relieves this inhibition. This drives polyubiquitination of multiple OMM substrates (Kane et al., 2014). A feed-forward ubiquitin signal is thereby generated that commits damaged mitochondria to mitophagy (Seirafi et al., 2015). Under normal conditions, mitophagy proceeds more rapidly in axons than in the soma. The estimated half-life is approximately 4–6 h in axons versus 12–18 h in the soma. This enables efficient renewal of the synaptic mitochondrial pool, with a clearance rate of approximately 88 % in control neurons.

The mitochondrial unfolded protein response (UPRmt) complements this quality-control axis through an ATF5-driven transcriptional programme. Under import stress, this programme induces matrix chaperones including mtHSP70 and HSPA9 at 8.2-fold, and HSP60 and HSP10 at 6.5-fold (Fiorese et al., 2016). It also induces the protease LONP1 at 12.3-fold above baseline levels (Melber & Haynes, 2018). Together, these responses restore proteostasis without requiring whole-organelle destruction (Narendra & Youle, 2024).

4. Molecular mechanisms of Tau-mitochondria interactions

4.1 Routes of Tau entry into mitochondrial compartments

Tau is a cytosolic protein and its major physiological binding location is on the outside of microtubule protofilaments. Its location in guarded mitochondrial bodies thus is a significant mechanistic puzzle. Tau is detected in mitochondrial fractions in aged brains of animals and in the postmortem AD tissue in studies of subcellular fractionation. The increased level of tau is seen exclusively in synaptic mitochondria (Torres et al., 2021). Proteinase K protection experiments also indicate that a sub-population of this tau is resistant to digestion (Castillo-Casaña et al., 2025). This suggests that localisation is in the intermembrane space or matrix and not on the mitochondrial surface only. These studies have found both soluble and aggregated tau in mitochondria (Manczak & Reddy, 2012a).

Tau entry into the mitochondrial matrix may occur through TOM and TIM23 import machineries. Although tau lacks a canonical mitochondrial targeting sequence, chaperone-mediated import has been proposed as one mechanism. In this mechanism, cytosolic chaperone complexes partially unfold tau and dock it at TOM40 in an import-competent conformation (Tracy et al., 2022). Binding of tau to both TOMM40 and TIMM23 has been demonstrated by co-immunoprecipitation and proximity ligation assays (Castillo-Casaña et al., 2025). This supports the idea that tau hijacks mitochondrial translocases to gain pathogenic entry into the organelle (Fig. 1). The higher abundance of pathologically phosphorylated tau in synaptic versus non-synaptic mitochondria may reflect increased synaptic mitochondrial workload. Transient calcium spikes at active presynaptic terminals may also contribute to this preferential accumulation (Torres et al., 2021).

4.2 Tau-VDAC1 interaction: Metabolic gate disruption

The most common OMM protein is voltage-dependent anion channel 1 (VDAC1), which is the main route of small metabolite exchanges. It aids the exchange of the ATP, ADP, pyruvate, malate, succinate, and calcium across the OMM (Shoshan-Barmatz et al., 2018; Yang et al., 2024). The phosphorylated tau physically interacts with VDAC1 and alters channel gating to partially or fully closed states (Fig. 2). This decreases open probability which was about 0.82 to about 0.35. This forms a metabolic bottleneck at the OMM that disables the ATP and ADP transport between the cytosol and the

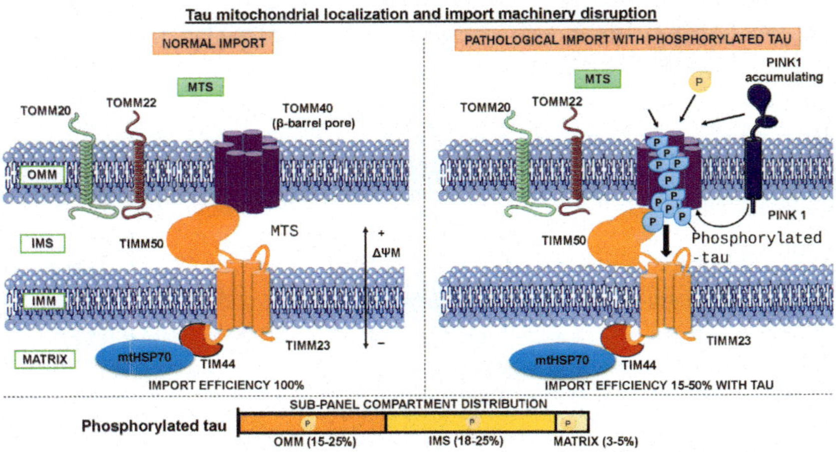

Fig. 1 Tau mitochondrial localisation and import disruption. Normal import of proteins into mitochondria via TOM and TIM complexes is shown on the left; the right panel illustrates how phosphorylated tau alters this machinery, reduces import efficiency, and changes the distribution of tau within mitochondrial compartments.

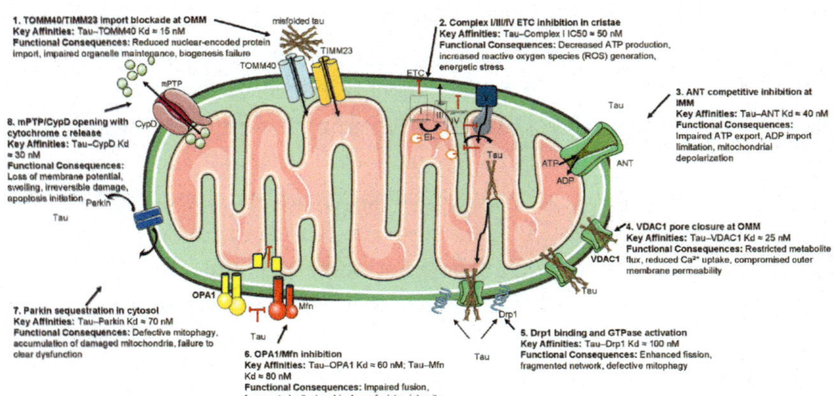

Fig. 2 Pathological tau interactions with mitochondrial components. Schematic of a neuron mitochondrion highlighting multiple tau-related sites of dysfunction, including TOM/TIM import block, respiratory chain inhibition in cristae, ANT-mediated ATP/ADP transport impairment, VDAC1 pore closure, Drp1-driven excessive fission, OPA1-dependent fusion defects, and Parkin sequestration leading to defective mitophagy.

intermembrane space (Manczak & Reddy, 2012b). Powerful in vivo data confirms VDAC1 as a critical centre of tau toxicity in neuronal functioning. In tau-transgenic models, partial VDAC1 haploinsufficiency improves mitophagy, mitochondrial dynamics, synaptic connectivity, and cognition (Vijayan et al., 2022).

4.3 Tau-ANT interaction: Impairment of ATP/ADP exchange across the inner membrane

Tau also inhibits the production of ATP in the IMM by direct inhibition of adenine nucleotide translocase (ANT). The most common neuronal isoform is ANT1 which controls the exchange of ADP and ATP across inner membrane. It uses one ADP in the cytosol and one transport cycle in the matrix in exchange, powered by the membrane electrochemical gradient. Atlante et al. (2008) indicated that an allosteric ANT1 site was high-affinity bound by a 19-residue N-terminal tau peptide (residues 26–44). Its inhibitory constant (Ki) was determined as 180 nM and the peptide is a strong competitive inhibitor of ADP and ATP exchange. This interaction, functionally, reduces the cytosolic ATP to ADP ratio and increases the matrix ATP levels. This acts as feedback on ATP synthase and drives the respiratory chain towards an idling condition. During this quiescence, there is a consumption of substrates but restricted net cytosolic gain in ATP (Atlante et al., 2008).

4.4 Tau-Drp1 interaction: Driving pathological fission

Pathogenic tau has a well-defined pathologically significant interaction with the mitochondrial fission GTPase Drp1. The co-immunoprecipitation experiments in postmortem AD cortex and transgenic AD mouse models show that the hyperphosphorylated tau and Drp1 are age-dependently correlated. Such models are APP/PS1 and 3xTg-AD mice, and this correlation could not be found in control tissue (Manczak & Reddy, 2012b). It has been reported that there is an increased Drp1 GTPase activity in AD cortical tissue as compared to healthy controls. This implies that pathological tau can increase Drp1 catalytic cycling and stabilise oligomerised Drp1 on the OMM, thus eliciting excessive fission. This interaction could be pharmacologically inhibited by P110 or mdivi-1, which reduce mitochondrial abnormalities and enhance cognitive performance in AD models (Joshi et al., 2018). Tau knockdown genetic methods also have the same effect on mitochondrial structure and functioning (Manczak et al., 2010).

4.5 Tau interactions with ATP synthase and respiratory complex subunits

Pathological tau also makes direct contacts with the mitochondrial ATP-synthesising machinery and its regulators. ATP5B, the oligomycin-sensitivity conferring protein (OSCP/ATP5O), is one of the most frequently validated mitochondrial tau interactors (Drummond et al., 2020; Torres et al., 2022). ATP5B forms one of three catalytic sites in the F1 head that phosphoryases ADP. OSCP not only links the head to the stalk, but maintains the paste-shaped Arrenius' machine optimum structure (Isei et al., 2025). In pluripotent neuronal bioenergetic systems that combine ATP-demand stress with Tau phosphomimicry, ATP and respiratory control ration are decreased. Cell biological studies establish the presence of tau-antagonism-induced IMM proton leak (Isei et al., 2025). Tau and OSCP association lowers the calcium threshold for irreversible calcium-induced mPTP opening. This is in stark contrast to healthy neurons, which undergo even uchronical calcium stress reversibly (Pérez et al., 2024).

4.6 Tau-mediated sequestration of Parkin and mitophagy blockade

Perhaps the most important feature of tau mitochondrial toxicity is two parallel and destructive events. Firstly, tau acts to destabilise mitochondria whilst simultaneously blocking their quality-control removal. This results in a trapped pool of phosphorylated mitochondria that generate the ROS, calcium dyshomeostasis and inflammatory signals. The second event is tau's inhibition of the ability of autophagic clearance machinery to access and remove this defective pool. Two mechanistically distinct PINK1-dependent pathways that sequester Parkin have been described in the literature. In the first pathway, tau-induced increase in the mitochondrial MMP (with a previously defined tau binding site in the AMLOGIC domain) prevents PINK1 accumulation on the OMM and hence PARK2 sequestration (Hu et al., 2016). The second pathway is a direct route whereby tau relays the binding of the cytosolic Parkin through N-terminal projection domain (KD = 110 nM) stabilising a tau–Parkin complex visible in proximity ligation assays in tau-transgenic neurons and binding affinity appears slightly enhanced in P301L tau (KD = 44.4 nM) (Cummins et al., 2019). However, despite stable PINK1 accumulation, in these neurons Parkin mitochondrial recruitment was abolished after 48 hr depolarisation time (Cummins et al., 2019).

5. Consequences for cellular energy metabolism

5.1 Electron transport chain dysfunction and the ROS-Tau amplification cycle

Tau-driven mitochondrial dysfunction does more than simply reduce ATP production. Tau-induced mitochondrial failure sets in motion a vicious positive feedback loop of oxidative stress and morbid tau aggregation (Tamagno et al., 2021) that actively worsens neurodegeneration over time (Alavi Naini & Soussi-Yanicostas, 2015). A key biochemical insult is the failure of progressive electron transfer through the ETC. While normally leak into superoxide is low (approx 0.1–0.2 %), tau-related dysfunction of ETC leads to both increased leak and generally higher oxidative burden (Eckert et al., 2014; Kozlov et al., 2024).

This tau-ROS cascade also produces oxidations which can further increase the rate of tau aggregation. Cys322 in tau repeat region is oxidized forming intermolecular disulfide bridges, which linkstaumonomers together and form larger oligomers of increasing seeding ability (Schweers et al., 1995). These larger oligomers were shown to have prionic-like propagating potential. By products of lipid peroxidation (such as 4-hydroxynonenal, 4-HNE and malondialdehyde) form covalent links with tau which are stable and are formed at lysine and histidine residue of tau and reduce proteasomal degradation of pathogenic tau (Sultana et al., 2013; Wojsiat et al., 2018). Several kinases are concurrently activated to close the cycle in vivo. GSK3 is activated 2.3-fold, JNK 3.8-fold, and p38 3.2-fold (Melov et al., 2007). This occurs alongside marked PP2A suppression via oxidation of catalytic-subunit methionine residues. Genetic support for this cycle comes from SOD2 knockout mice. These mice show marked increases in phospho-tau burden across multiple AD-relevant epitopes relative to littermate controls (Melov et al., 2007).

5.2 mtDNA damage and respiratory subunit deficiency

Mitochondrial DNA (mtDNA) is a circular DNA molecule which is short, contains thirteen important respiratory-chain and ATP-synthetase subunits and is particularly sensitive to oxidative damage, more so than nuclear DNA. This is due to the absence of protection by nucleosomes and its closeness to IMM ROS generating sites (Santos et al., 2012) as well as a somewhat poor base-excision repair capability. AD brain mtDNA shows increased oxidation by between 3 and 10 fold in relation to age matched controls (in the form of 8-hydroxydeoxyguanosine- 8-OHdG). Elevated 8-oxo-deoxyguanosine (8-oxo-dG), shed from damaged mitochondria, was also shown to be higher in AD brain and CSF by 2.5 fold respectively (Wang et al., 2005), correlates with tau and AD brain pathology and was suggested to act as a biomarker of efficacy in mitochondrial protective strategies. A 30 % or greater reduction within 12 weeks was projected to demonstrate efficacy of elamipretide (Wilkins & Morris, 2017; Wu et al., 2017; Karaa et al., 2023).

Simultaneously, amounts of the CSF ATP synthase components ATP5B and ATP5O are reduced 30–40 % in AD compared to control further evidence of both mitochondrial and cristae loss. It should thus be observed that this cristae-stabilizing therapy will up-regulate amounts to greater than 20% relative to control, and that beyond approx. 60–80 % heteroplasmy for a respiratory subunit gene, activity will be largely abrogated. This creates a progressive respiratory deficit that is further perpetuated by tau-impaired mitophagy (Santos et al., 2012).

5.3 Calcium homeostasis disruption and the ER-mitochondria interface

Multiple converging pathways by which tau impairs mitochondrial Ca 2+ buffering exist. These pathways are focused on Ca 2+ transferring machinery at mitochondria-associated membranes (MAMs). At the MAM, smooth ER and mitochondria can be found tightly apposed through proteins including the IP3R3, sigma-1 receptor, VDAC1 and Grp75 (Navarro & Montesinos, 2025) by the use of linking proteins. There are several ways which tau pathology destabilises this Ca 2+ signaling relay (Fernandes et al., 2023). Tau binds to both AHCYL1 and IRBIT, leading to reduced ER calcium-release dependence and decreased coordination between ER release and mitochondrial uptake. Short tau species found within the inter-membrane space are another route to MAM architecture disruption. They lead to a larger ER-OMM distance and a decreased efficiency of calcium microdomain formation (Cieri et al., 2018).

This disruption has two major effects. First it results in defective activity-dependent metabolic amplification (when influx of calcium at the synapse normally activate TCA dehydrogenases). Second it results in deficient buffering of mitochondrial calcium, leading to increased peaks of cytosolic calcium, calpain activation and subsequent cleavage of tau to more detrimental fragments. Pathological tau can then hyperactivate TPC2 in lysosomes, leading to excessive release of calcium from lysosomes. This will result in increase in cytosolic calcium and increase in intralysosomal alkalization and it also leads to defective degradation of damaged mitochondria by autophagy which further deteriorate quality control failure (Tong et al., 2022).

5.4 mPTP opening and the commitment to neuronal death

The mPTP is a calcium and oxidative stress-sensitive megachannel on IMM activated in conditions of long term mitochondrial stress which transports non-selective flux of molecules up to about 1.5 kDa through the normally ion-impermeable IMM (Quintanilla et al., 2009). At the moment the formation of pore is considered to be formed by F1FO ATP synthase, c-ring and OSCP being key components and CypD making the channel sensitive to calcium-induced activation (Pérez et al., 2024). Pathologicaltau forms have been shown to strongly sensitize mPTP opening: particularly caspase-3-cleaved tau (TauC3) has been characterized as having a potent effect on mPTP opening, transforming the regulated mitochondrial response to stress into uncontrolled permeability. Once transported to mitochondrial matrix these forms of tau bind directly to CypD and ATP synthase OSCP (Pérez et al., 2024).

The subsequent cellular consequences of mPTP opening are detrimental; they consist of rapidly followed mitochondrial fragmentation and depolarisation, ultimately loss of ATP synthesis. Matrix swelling, outer mitochondrial membrane rupture, release of cytochrome c and ultimately cell death occur. Morphological, functional and survival end-points can be significantly rescued by blockade of CypD with cyclosporin A, and similar protection observed following genetic ablation of CypD (Pérez et al., 2024).

5.5 Glucose metabolism, insulin resistance, and the metabolic-epigenetic coupling of Tau pathology

AD shows widespread cerebral glucose hypometabolism as measured by 18F-FDG PET. This hypometabolism can be seen up to 15 years prior to clinical presentation. This hypometabolism is most increased in the posterior cingulate, precuneus, and temporoparietal lobe regions. In these regions, the region shows around 18–30 % reduction in glucose uptake which shows a strong correlation with cognitive decline. MMSE correlations range from r = 0.78–0.85, establishing FDG-PET imaging as a vital biomarker and therapeutic outcome measure for metabolic rescue.

Glucose hypometabolism in AD results from additional molecular and cellular defects. First, decreased neuronal levels of the glucose transporters GLUT1 and GLUT3 impair glucose entry and correlate with tau hyperphosphorylation. Second, brain insulin resistance arises partly due to JNK and IKK kinase activation by A-beta oligomers, resulting in inhibitory serine phosphorylation of IRS-1, and thus impaired insulin downstream signalling. Third, lack of efficiency in the activity of PDH prevents pyruvate from glycolysis pathways being converted into mitochondrial acetyl-CoA, effectively uncoupling glycolysis from the TCA cycle. In the face of four separate metabolic deficiencies, cellular metabolism effectively bypasses oxidative phosphorylation which generates approximately 32 ATP per glucose molecule, and instead uses glycolysis which generates about 2 ATP per glucose. This shift exacerbates neuronal energy deficits considerably.

Glucose hypometabolism is also mechanismistically related to tau pathology via a metabolic-epigenetic regulation. The impaired glucose flux leads to reduced intracellular UDP-GlcNAc, thus, the decreased activity of OGT, resulting in reduced O-GlcNAcylation of tau protein. Normally, the latter antagonizes tau phosphorylation in microtubule-binding domain of tau protein. The decrease of the O-GlcNAcylation of tau leads to hyperphosphorylation of tau, causing dissociation and aggregation of tau protein and its abnormal binding to mitochondria, in which positive feedback between metabolism and tau pathology is built. This positive feedback further deranges both metabolic and mitochondrial function.

Although glucose is poorly utilized in the AD brain, the ability of the AD brain to utilize ketone bodies remains relatively spared and presents an avenue to potentially provide energy to neurons. The two main ketone bodies, beta-hydroxybutyrate and acetoacetate, enter the brain and cross the BBB via monocarboxylate transporters where they are converted to acetyl-CoA via the mitochondrial enzyme's beta-hydroxybutyrate dehydrogenase and OXCT1, circumventing the need for GLUT-mediated uptake and PDH entry into the TCA cycle. Indeed, the rates of ketone uptake remain unaffected even when the rates of glucose utilization are depressed by 20–40%. The implications for therapy are evident in the demonstration that randomized, placebo-controlled administration of medium-chain triglycerides result in the modest benefit in the cognition of mild-to-moderate AD (Taylor et al., 2018). Greater effects are observed in APOE4-negative individuals. Complementary 11C-acetoacetate PET data further support ketone utilisation as a viable metabolic rescue pathway. Mitochondrial Quality Control Failure in Tauopathies.

5.6 The PINK1-parkin pathway: Architecture and Tau-mediated disruption

Selective mitophagic removal of dysfunctional mitochondria is the final cellular defence against accumulation of defective organelles. In tauopathy, tau imposes two independent failures in the PINK1-Parkin pathway. In the first failure, tau induces paradoxical mitochondrial hyperpolarisation through ATP synthase inhibition and slowed F1FO rotation. This increases back-pressure and keeps PINK1 import and degradation near baseline levels (Hu et al., 2016). A population of hyperpolarised yet dysfunctional mitochondria is thereby created that evades mitophagy quality control entirely. In the second failure, even when mitochondria are depolarised and PINK1 accumulates on the OMM, tau-mediated sequestration of cytosolic Parkin blocks downstream pathway propagation. This paradox means that neurons with the highest tau burden can show the strongest PINK1 initiation signal but insufficient Parkin response.

In C. elegans neuronal mitophagy reporter models, axonal mitochondrial density rises 3.8-fold above control values. Mean mitochondrial length falls to 0.9 micrometres versus 2.4 micrometres in controls. This confirms accumulation of fragmented, dysfunctional organelles rather than their clearance. Orphaned PINK1 then aberrantly activates pro-death JNK signalling at 3.8-fold and p38 MAPK signalling at 3.2-fold above baseline. It also increases caspase-3 cleavage and promotes TauC3 generation, creating a self-amplifying positive feedback loop.

5.7 Consequences of mitophagy failure: Accumulation of dysfunctional mitochondria

Uncleared damaged mitochondria continue to oxidise respiratory substrates but with severely uncoupled function. They minimise their contribution to cellular metabolism and instead function as metabolic sinks. These dysfunctional organelles consume glucose-derived substrates and oxygen while providing very little ATP back to the cytosol (Castillo-Casaña et al., 2025). This increasingly parasitic metabolic burden deepens the neuronal bioenergetic crisis over time. Additionally, uncleared mitochondria continue to produce superoxide via disassembled respirasomes at proportionately elevated rates. This sustains a chronically elevated ROS environment within affected neurons (Melov et al., 2007). This ongoing ROS production represents another facet of the vicious amplification cycle described in Section 5.1 and drives yet more tau phosphorylation (Alquezar et al., 2022).

Perhaps the most surprising consequence of accumulated non-functional mitochondria is a paradoxical depletion of key mitophagy-initiating proteins. Rather than accumulation, these proteins are reduced in post-mortem human AD brain samples. In AD cortical tissue, the content of PINK1, Bcl2L13, BNIP3L/NIX, phospho-TBK1, and phospho-ULK1 are all suppressed in AD neurons (Fang et al., 2019). This is theorised to result from the cumulative resource expenditure of these scarce pathway proteins. They are consumed through repeated, failed engagement with the mounting number of damaged mitochondria.

5.8 Lysosomal dysfunction and the autophagy-lysosomal pathway collapse

Mitophagy ultimately targets damaged mitochondria for delivery to the lysosome. The lysosome is a membrane-bounded organelle that relies on the vacuolar H-ATPase (v-ATPase) proton pump to maintain intralysosomal pH in the range of 4.3–5.0 (Nixon & Rubinsztein, 2024). In AD, lysosomal acidification is profoundly affected by changes that converge on v-ATPase function. A-beta oligomers and pathological tau inhibit v-ATPase assembly and proton pump efficiency. This raises intralysosomal pH to a nearer-neutral range of 5.5–6.5, at which cathepsins are substantially less active. The magnitude of this lysosomal dysfunction is measurable and significant. Lysosomal pH rises from 4.6 to 5.4, reducing cathepsin D activity by approximately 75%. Lysosomal motility falls from 0.45 to 0.13 micrometres per second, representing a 70 % reduction. The mean autophagosome-lysosome distance increases from 2.8 to 8.5 micrometres, preventing effective cargo fusion.

Tau oligomers additionally overactivate Rab7, leading to endolysosomal membrane permeabilisation. This causes partially degraded tau seeds to escape to the cytosol. In doing so, the clearance mechanism is perversely converted into a propagation mechanism for tau pathology. The resulting lysosomal membrane permeabilisation (LMP) allows cathepsins to leak into the cytosol. There, they cleave cytoskeletal proteins including tau, producing further toxic fragments and closing the pathogenic loop (Nixon & Rubinsztein, 2024).

5.9 UPRmt activation and its limits in tauopathy

Uncleared dysfunctional mitochondria activate the mitochondrial unfolded protein response (UPRmt). This response begins as a neuroprotective compensatory mechanism to restore mitochondrial proteostasis and repair organelle biogenesis (Fiorese et al., 2016). The UPRmt transcriptional programme induces matrix chaperones such as HSP60, mtHSP70/GRP75, and HSP10. It also induces matrix proteases such as LONP1 and ClpP, as well as regulators of mitophagy (Liu et al., 2022). However, the compensatory capacity of the UPRmt becomes progressively insufficient as chronic tau-mediated mitochondrial damage intensifies. Accurate recognition of import stress during ATF5-mediated UPRmt transcriptional response, and recognition through the mitochondrial import pathway, is crucial. It is inhibited by the physical interaction of toxic tau with TOMM40 and TIMM23 (Tracy et al., 2022). Increased tau pathology is correlated with worsening lysosomal dysfunction, a subsequent block in proteostasis, and inhibited degradation of mitophagy substrates that are appropriately translocated. This causes further cycles of futile UPRmt signaling to futilely consume cellular energy.

6. Synaptic dysfunction and axonal transport deficits

6.1 The extraordinary metabolic demands of the synapse

Synapses place extreme metabolic demands on presynaptic terminals and postsynaptic spines and are arguably among the most energy-intensive cellular structures per unit volume in mammals. Quantitative estimates indicate that a single synaptic vesicle cycle requires roughly 164,000 ATP molecules (Rangaraju et al., 2014; Underwood et al., 2020). At high-frequency activity (e.g., 40 Hz), with approximately 10–40 vesicle cycles per second, the ATP requirement per synapse rises dramatically during peak network activity (Fig. 3). These calculations indicate that, beyond

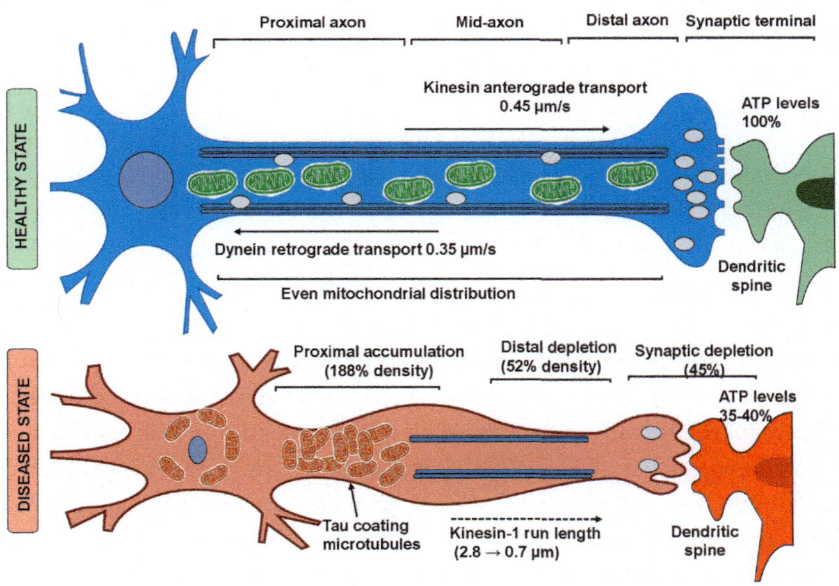

Fig. 3 Disrupted axonal mitochondrial transport and synaptic energy loss in tauopathy. Comparison of healthy neurons, with balanced kinesin and dynein transport and even mitochondrial distribution, to diseased neurons where tau-coated microtubules cause proximal mitochondrial accumulation, distal and synaptic depletion, reduced local ATP levels, and impaired dendritic spine function.

ATP supplied from somatic mitochondria, local presynaptic mitochondria must generate ATP directly at boutons via oxidative phosphorylation, on the order of 10 trillion molecules per second. Optical measurements using synaptically targeted ATeam ATP biosensors show that electrical stimulation rapidly consumes presynaptic ATP, followed by recovery within about one second through coordinated mitochondrial oxidative phosphorylation and glycolysis (Rangaraju et al., 2014).

Synapses are extremely demanding of both presynaptic terminals and postsynaptic spines metabolically and they are probably among the most energy demanding structures per unit volume in mammalian cells. Calculations suggest that 1 synaptic vesicle cycle demands about 164,000 ATPs (Rangaraju et al., 2014). The demand for ATP rises significantly at high-frequency activity, like 40 Hz and about 10–40 vesicle cycles per second: such estimates imply that local presynaptic mitochondria produce about 10 trillion ATPs every second directly at the bouton through oxidative phosphorylation, as the supply from somatic mitochondria is not sufficient (Underwood et al., 2020). Optical studies using synaptically targeted ATeam ATP biosensors show fast usage of synaptic ATP: on electrical stimulation there is rapid utilization and ~1 s recovery by the combination of oxidative phosphorylation in mitochondria and glycolysis (Rangaraju et al., 2014).

An equally significant metabolic hurdle faced by postsynaptic dendritic spines relates to protein synthesis. De novo synthesis of synaptic proteins underpins LTP-induced structural plasticity. A single dendritic spine large enough to support morphologic plasticity will undergo a stereotypical reaction to stimulation. These include entry of calcium, mitochondrial uptake of calcium, activation of dehydrogenases, generation of NADH, and local production of ATP to support de novo protein synthesis to enlarge the spine. This protein-synthetic plasticity support reaction is entirely local and can be stimulated within about 30 μm of the stimulated spine (Rangaraju et al., 2019). Complete depletion of the local dendritic mitochondrial pool eliminates translation and morphological plasticity. Local mitochondrial resource availability is therefore an essential and irreplaceable requirement for synaptic plasticity (Rangaraju et al., 2019).

6.2 Tau-induced mislocalisation and the synapse starvation paradigm

Kopeikina et al. (2011) demonstrated that in the rTg4510 model of tauopathy, mitochondrial content in neuronal processes decreases with age.

This decline occurs under conditions of tau overexpression and reflects a tau-dependent phenotype. Paradoxically, mitochondria show increased accumulation in the soma of Purkinje neurons despite being visibly eliminated from degenerating processes. Postmortem hippocampal neurons from AD subjects with significant tau pathology show more than a 50 % reduction in mitochondrial density. This reduction is observed in both proximal and distal dendritic trees compared to age-matched controls. Depletion of mitochondria at synapses is synergistic with decreased functionality. Mitochondria are not only fewer in number at the synapse, but also those that do reach it have signs of cristae degeneration and reduced matrix density. These structural features are compatible with bioenergetic failure and lack of significant energy supply to the synapse.

The correlation of higher phospho-tau in rat hippocampus to mitochondrial dysfunction and cognitive impairment was established by Kandimalla et al. (2018) with the use of a double-hemisphere stereotaxic injection model to demonstrate this. One month post-injection, a characteristic set of phenotypical hallmarks was clearly evident. These included loss of dendritic spines, decreased levels of spinogenic synaptic proteins, and fragmentation and dysfunction of mitochondria. Impairment of spatial cognition was also observed as part of this phenotypic profile (Kandimalla et al., 2018).

6.3 The PAD-mediated axonal transport inhibition mechanism

The molecular machinery by which pathogenic tau inhibits kinesin-1 transporter function has been established as a signalling-based mechanism. Kanaan et al. (2011) demonstrated that this is not a steric hindrance or physical blockade of tau on microtubules. Pathogenic tau does not directly occlude kinesin-1-binding domains on tubulin. Instead, it causes tau to undergo structural refolding that renders the otherwise hidden Phosphatase-Activating Domain (PAD) accessible. PAD is contained within the N-terminal 1–18 amino acid residues of the tau molecule. Once exposed on hyperphosphorylated tau at the AT8 epitope, or on caspase-cleaved tau, PAD activates protein phosphatase 1 (PP1). PP1 in turn dephosphorylates and primes GSK3 for activation (Kanaan et al., 2011). Active GSK3 subsequently phosphorylates the kinesin light chains, weakening cargo interactions in the carboxyl terminus of these subunits. This causes release of axonal cargo-containing kinesin-1 from anterograde transport tracks. This PAD-mediated mechanism is specific for anterograde transport and spares retrograde dynein-based cargo transport. Within affected tauopathy neurons, this results in a depletion of mitochondria in distal axonal compartments, alongside accumulation in the somatic region.

6.4 Synergistic amyloid-Tau toxicity at synaptic mitochondria

In the AD brain, tau and amyloid-beta target functionally distinct mitochondrial components synergistically rather than independently. In triple transgenic AD mice, mitochondrial membrane potential was preferentially compromised in animals carrying all three mutations rather than one or two alone (Rhein et al., 2009). This effect was supra-additive, exceeding the sum of individual contributions from each mutation. Tau inhibits NADH oxidation at Complex I of the electron transport chain. Amyloid-beta directly inhibits cytochrome oxidase subunit II of Complex IV through direct protein interaction (Eckert et al., 2011). Amyloid-beta appears to do so by disrupting mitochondria-associated endoplasmic reticulum membranes. Together, these effects create pro-oxidative entry and exit blockades at inter-complex electron transfer sites. Superoxide production is thereby significantly increased at these vulnerable transfer points (Rhein et al., 2009).

7. Glial cell contributions and neuroinflammation

7.1 Microglial metabolic reprogramming and Tau propagation

The neuronal aspect of AD dysfunction is just one aspect of pathogenesis; glial cell metabolism specifically that of the microglia and the astrocytcs are dramatically reprogrammed and serve to exacerbate the tau pathogenesis. These glial cells actively drive the pathology forward: in response to stimuli promoting neurodegeneration, microglia shift their phenotype from a house-keeping/surveillance state, to the so called disease-associated microglia (DAM) state. This change is dominated by TREM2-APOE signaling. Along with these phenotypic shifts, the genes related to lipid metabolism, mitochondrial activity, and phagocytosis undergo significant changes at a transcriptional level (Keren-Shaul et al., 2017; Krasemann et al., 2017).

Under normal physiology, microglia are highly dependent upon OXPHOS for energy demanding activities such as surveillance and clearing. Basal oxygen consumption is 95 plus or minus 15 pmol O2 per minute per 10^6 cells. Viability of the mitochondria are crucial for functional states of the microglia. TSPO, which is on the outer mitochondrial membrane is important for determining ATP production levels. When TSPO is deficient, oxygen consumption and ATP production are drastically reduced 65 % and 55 %, respectively and phagocytic activity is reduced 70 %.

At the DAM stage there is an extremely quick metabolic shift to aerobic glycolysis (a Warburg-like reprogramming), driven by increases in glycolytic drivers: hexokinase 2 (HK2) at roughly 3.5-fold; and pyruvate dehydrogenase kinase 1 (PDK1) at roughly 2.8-fold. This causes a shift of pyruvate away from mitochondrial oxidation resulting in the production of ATP which is very rapid but energetically inefficient (roughly 18 times less output per glucose molecule) with a reduction in proteostatic function (with about a 40 % decrease in amyloid-beta clearance).

Both soluble and insoluble species of tau can be internalised by microglia; however, insoluble tau degradation by microglia is extremely inefficient. These accumulations lead to alkalinisation of the lysosome as the degradative function fails. In essence, dysfunctional, metabolically impaired microglia secrete seed competent, partially degraded tau via exosomal release to trans-synaptically spread it and have an active role as a vector of tau spread. depletion of microglia and inhibition of exosome biogenesis led to a significant decrease in tau spread (Asai et al., 2015; Tong et al., 2022).

7.2 Neuroinflammation-metabolism crosstalk

Mitochondrial damage in AD causes a robust activation of innate immunity by releasing damage-associated molecular patterns (DAMPs). cytosolic mitochondrial DNA is one of the most crucial of these signals, which activates both cGAS-STING (cyclic GMP-AMP synthase-stimulator of interferon genes) efficiently to stimulate TBK1/IRF3-dependent production of interferons as well as NF-kB-dependent pro-inflammatory cytokine release (West et al., 2015). Additionally, mitochondrial DAMPs (along with Aß and hyperphosphorylated tau) trigger activation of the NLRP3 inflammasome, which activates caspase-1, triggering maturation of IL-1 ß (Li & Gong, 2025). Genetic inactivation of NLRP3 or caspase-1 in APP/PS1 mice leads to strong neuroprotection and M2 polarization (neuroprotective phenotype) of microglia (Heneka et al., 2013).

Such inflammatory conditions have an impact on neuronal metabolism through a network of different mechanisms: Firstly TNF-alpha upregulates the stress kinases JNK and IKK, which activate an inhibitory serine phosphorylation of IRS-1; Ser307 phosphorylation is 4.8-fold, and this action compromises PI3K-Akt signalling and establishes the neuronal insulin resistance. As a consequence of insulin resistance GSK3 beta activation is increased at 2.6-fold which consequently leads to hyperphosphorylation of tau at 3.2-fold: This creates a positive feedback loop between glial inflammation and neuronal function.

Astrocytes are similarly reprogrammed under chronic inflammatory conditions. IL-1 beta and TNF-alpha suppress GLUT1 expression via NF-kB-dependent transcriptional repression. Reductions of approximately 50–70 % are observed in the hippocampus and approximately 38–48 % in the cortex. This limits glucose availability to neurons and compounds existing metabolic deficits. Concurrent downregulation of the glutamate transporter GLT-1/EAAT2 at approximately 60 % reduction elevates extracellular glutamate concentrations to 5–12 micromolar. Concentrations at this level are sufficient to drive NMDA receptor overactivation and mitochondrial calcium overload. This culminates in opening of the mPTP and excitotoxic injury to neurons.

Sustained microglial activation further induces a neurotoxic A1 astrocyte phenotype through IL-1 beta, TNF-alpha, and complement component C1q (Liddelow et al., 2017). This switch involves high induction of complement factors such as C3 (~12 fold) and C1q (~8 fold). This transition is coupled with loss of astrocytic neuroprotection and over-activation of complement-mediated synaptic elimination, exceeding the tau-dependent synapse loss (Liddelow et al., 2017).

Additionally to neurons and astrocytes, oligodendrocytes are highly metabolically vulnerable in AD. The normal provision of ~30–40 % of axoplasmic ATP from oligodendrocytes to the myelinated axon by MCT1–mediated transfer of lactate, is impaired by an ~35 % reduction of white matter MCT1 levels in AD, where the carrier itself becomes oxidized (by 4-HNE adducts), resulting in an ~60 % decrease of its activity. Also, a ~50 % decrease of neuronal mitochondrial Complex IV activity impairs ATP production further. Resulting in an axoplasmic deficit down to ~30 % or lower than normal at distal axons in AD models (Lee et al., 2012; Wang et al., 2026; Looser et al., 2024), thereby amplifying defects of neuronal axonal mitochondrial trafficking and accelerating neurodegeneration (Rhein et al., 2009).

Together, these pathways produce a tightly linked neuroinflammatory-metabolic axis in AD, in which mitochondrial failure causes inflammation and inflammation, in turn, fuels metabolic dysfunction. These phenomena provide a vicious cycle that enhances tau pathology and neuronal death across affected brain regions.

8. Therapeutic implications and emerging strategies

8.1 Anti-Tau monoclonal antibodies: Mechanisms, clinical trial results, and lessons

The most advanced in terms of tau targeted immunotherapies are anti-tau monoclonal antibodies. These aim to trap extracellular tau seeds thus limiting the substrate for trans-synaptic spread. In some contexts, they also promote selective intracellular uptake of pathogenic tau epitopes via Fc receptor-mediated endocytosis (Leyns & Holtzman, 2017). This reduces intracellular tau burden in vulnerable neurons (Castillo-Casaña et al., 2025). Semorinemab is an Fc-silent humanised IgG4 monoclonal antibody targeting the tau N-terminus at amino acids 6–23. It yielded informative findings in the LAURIET Phase 2 trial (Monteiro et al., 2023). Peripheral target engagement was clear, with plasma tau concentrations rising more than 20-fold from baseline. CSF p-tau181 and p-tau217 levels were also significantly attenuated. A statistically significant 42.2 % slowing of ADAS-Cog11 decline was observed without measurable reduction in tau PET-defined NFT burden. This dissociation indicates that soluble tau clearance was achieved while fibrillar NFT reduction was not. These findings are consistent with an oligomeric tau toxicity model of neurodegeneration (Monteiro et al., 2023).

8.2 Antisense oligonucleotides and small molecule Tau aggregation inhibitors

Antisense oligonucleotides (ASOs) targeting MAPT mRNA offer a mechanistically broad therapeutic strategy. They reduce intracellular tau at its source, including both soluble and aggregated species. This contrasts with approaches that only limit extracellular tau interactions. MAPTRx (BIIB080) is an intrathecally administered gapmer that targets MAPT mRNA for RNase H1-mediated cleavage. It demonstrated dose-dependent and sustained reductions in CSF total tau of up to 50 % in a Phase 1b trial. Reductions in p-tau181 of up to 64 % and in p-tau217 of up to 69 % were also observed. The trial was a randomised placebo-controlled study involving 46 patients with mild AD (Mummery et al., 2023). These effects persisted for more than 24 weeks after the final dose. Preclinical support for this approach is likewise strong. Across multiple tauopathy mouse models, tau reduction prevents neuronal loss and reverses pathological tau deposition (DeVos et al., 2017). It also restores mitochondrial transport and morphology, lowers reactive oxygen species, and improves synaptic mitochondrial density.

8.3 Mitochondria-targeted therapeutics: From cardiolipin protection to mitophagy enhancement

Elamipretide, also known as SS-31 or MTP-131, is the most clinically advanced mitochondria-targeted protectant currently available. It is a water-soluble aromatic-cationic tetrapeptide with the sequence D-Arg-Dmt-Lys-Phe-NH2. It crosses the outer mitochondrial membrane and accumulates at the inner membrane through electrostatic interaction with cardiolipin head groups (Szeto, 2014). Its functions related to cardiolipin can be categorized into three basic mechanisms: preventing the formation of cytochrome c/cardiolipin peroxidase complex and therefore reducing the rate of cardiolipin peroxidation; sustaining cardiolipin-rich domains necessary for respiratory supercomplex assembly and channeling of electron transfer; maintaining cardiolipin-dependent proton buffering and enhancement of coupling efficiency when ATP production is high. In preclinical AD models, elamipretide reduces mitochondrial superoxide production, restores membrane potential, preserves ATP synthase dimer assembly and cristae architecture, and improves spatial memory performance (Manczak et al., 2010; Zhang et al., 2024).

Urolithin A, a bioactive metabolite which is a gut microbiome derived metabolite from ellagitannie rich foods (perilla, pomegranate, strawberry and walnut etc) has been proposed as a promising mitophagy inducing drug candidate (D'Amico et al., 2021; Fang et al., 2019). Urolithin A prevents PINK1 cleavage at the outer mitochondrial membrane and promotes activation of the full PINK1-Parkin mitophagy pathway. In human neuronal models and in APP/PS1 and 3xTg-AD mice, it reduces tau hyperphosphorylation and amyloid-beta accumulation (Fang et al., 2019). It also restores maximal oxygen consumption rate and improves spatial and recognition memory deficits. A Phase I study in healthy middle-aged adults reported that oral urolithin A was safe and achieved bioavailable plasma levels associated with mitophagy biomarker engagement. It also improved markers of muscle mitochondrial biogenesis and grip strength (D'Amico et al., 2021).

A third strategy involves NAD+ precursor supplementation using nicotinamide mononucleotide (NMN) or nicotinamide riboside (NR). This approach aims to counter tauopathy-associated mitochondrial dysfunction (Lautrup et al., 2019). NAD+ is required for the activity of sirtuins such as SIRT1 and SIRT3. These deacetylate and activate PGC-1 alpha, a master regulator of mitochondrial biogenesis. They also modulate multiple mitochondrial respiratory-chain proteins. NAD+ decline therefore directly contributes to reduced mitochondrial ATP output. In AD mouse models, NMN and NR restore impaired NAD+ levels and promote PGC-1 alpha-dependent mitochondrial biogenesis. They also activate neuronal mitophagy via SIRT3-dependent LC3 deacetylation, reduce amyloid-beta pathology and tau phosphorylation, and improve cognition (Fang et al., 2019; Gong et al., 2013).

8.4 Targeting mitochondrial dynamics: Restoring the fission-fusion balance

Pathological tau-driven Drp1 hyperactivation causes excessive mitochondrial fragmentation in affected neurons. Two major drug classes have been explored to address this in preclinical AD models. The first is mdivi-1, a quinazolinone inhibitor of Drp1 GTPase activity. It blocks Drp1 self-assembly into oligomeric constriction rings (Joshi et al., 2018). The second is the P110 peptide, which selectively blocks Drp1-FIS1 interaction without suppressing basal Drp1 activity or other receptor interactions. In AD model neurons, P110 selectively suppresses pathological fission while preserving physiologically required fission events. It maintains membrane potential and ATP production, reduces ROS, and improves synaptic density and function (Joshi et al., 2018). The HDAC inhibitor valproate can also upregulate mitofusin expression at low doses. It does so via increased histone acetylation at MFN1 and MFN2 promoters. This reduces mitochondrial fragmentation and restores cristae density in tauopathy models.

8.5 VDAC1 modulation and Tau-mitochondria interface targeting

The tau-VDAC1 interaction is an attractive targeting point in AD research. Pharmacological approaches that reduce the availability of VDAC1, as well as genetic approaches have proven effective in various preclinical AD models. VDAC1-interacting peptides, such as olesoxime-related compounds, a range of VDAC1-derived VDAC1 binder-inhibitor-tracker

(VBIT) series molecules compete with tau/amyloid-β binding to VDAC1. This prevents channel closure, but maintains essential metabolite exchange (Shoshan-Barmatz et al., 2018; Yang et al., 2024). An entirely mechanistically different approach involves reducing the relative abundance of VDAC1 in tau transgenic backgrounds, such that there remains a functional population of open channels, but less can be bound by tau. This enhances mitochondrial metabolite exchange flux and increases mitophagy flux, while maintaining mitochondrial morphology, synaptic protein levels and learning and memory performance (Vijayan et al., 2022).

8.6 Combination and multi-target strategies: Rational therapeutic architecture

The scope of evidence presented in this chapter shows that the tau-mediated failure of neuronal bioenergetics involves several interacting mechanisms that work in synergy. Accordingly, the disease is not likely to be cured by a single therapy alone (Eckert et al., 2011; Rhein et al., 2009). Such complexity rationalizes combination therapies that involves treatments that are mechanistically synergistic rather than overlapping (Hansson et al., 2023). An elegant and complementary drug design includes a combination of ASO-mediated reduction of tau and increase of mitophagy. Whereas ASO-mediated reduction of tau involves directly attacking the source of mitochondrial toxicity, urolithin A- or NAD+ precursor-mediated improvement of mitophagy clears and eliminates accumulated mitochondria which can perpetuate toxicity even followingtau reduction (DeVos et al., 2017; Fang et al., 2019).

8.7 Emerging mitochondrial biomarkers for early detection and treatment monitoring

A mitochondrial marker panel within CSF and plasma that are mechanistically linked to tau-mitochondria interactions has recently been discovered (Castillo-Casaña et al., 2025). This suggests that CSF/plasma mitochondrial biomarkers will serve as a valuable source for disease detection and treatment assessment in AD (Di Lorenzo et al., 2025). Cell-free mitochondrial DNA (cf-mtDNA) can be found in CSF and plasma, and corresponds to accumulated damage to mitochondria. This may be released into circulation following mPTP opening, outer mitochondrial membrane ruptures, or in mitochondria carrying exosomes. Increased CSF cf-mtDNA levels in AD are strongly associated with tau pathology and cognitive deficit, indicating its promise as an AD neurodegeneration marker.

Neuron-derived extracellular vesicles (neuroexosomes) provide a way of minimally invasively analyzing mitochondrial protein integrity. Informative markers include Complex I subunits NDUFB8 and NDUFS3 (targets of tau-mediated disruption), SDHB as a control for overall mitochondrial content, and VDAC1 as an indicator of outer mitochondrial membrane turnover. These proteins can be quantified in peripheral blood after affinity-based immunoprecipitation of neuron-derived vesicles, providing a peripheral readout of neuronal mitochondrial status. CSF levels of ATP synthase subunits ATP5B and ATP5O are reduced by approximately 30–40 % in AD, indicating mitochondrial protein loss and impaired oxidative phosphorylation. Restoration of these proteins (≥20 % increase) has been proposed as a pharmacodynamic marker of response, particularly for mitochondria-targeted interventions such as elamipretide.

Mitochondria encoded peptides such as humanin and MOTS-c were upregulated 2–3-fold in AD CSF; this is almost certainly due to a compensatory mechanism for mitochondrial impairment; therefore, levels would presumably fall after successful treatment, hence they are likely to be dynamic stress-response markers. Markers for oxidative damage were also upregulated in AD, namely 8-OHdG and 4-HNE protein adducts both within plasma and CSF in a disease-specific manner and there were positive correlations with in vivo estimates of respiratory-chain function using platelet bioenergetics, supporting their roles as systemic markers for systemic mitochondrial oxidative damage. These three methods present a complex picture of the mitochondrial system in AD, ranging from structural damage, through protein integrity, compensatory signalling, and finally evidence of oxidative stress, presenting a clear future for their application in early diagnosis, patient categorization and monitoring therapy over time in AD.

9. Conclusion and future perspectives

9.1 A hierarchical model of Tau-mitochondrial pathogenesis

The evidence supports tau as the major mediators of bioenergetic dysregulation in AD. It does so via multiple mechanismally distinct but convergent biopathways that go far beyond its microtubule association and assembly. These interactions constitute a hierarchy of pathway-stages that occurs over decades and connect early, subclinical metabolic defect to late, irreversible neurodegeneration. In the earliest stages, there appears to be concomitant, concurrent increases in amyloid burden and cerebral glucose hypometabolism that is detectable by longitudinal FDG-PET imaging nearly 22 years before clinically objective symptoms manifest (Gordon et al., 2018; Mosconi et al., 2008). This energy deficit impairs flux through the hexosamine biosynthetic pathway and reduces tau O-GlcNAcylation, thereby relieving a major inhibitory brake on tau phosphorylation, and allowing tau to become hyperphosphorylated in what appears to be a pathogenic, feed-forward loops. Although this early, tremendous bioenergetic pressure, neurons may yet maintain some degree of functional reserve.

At prodromal stages, hyperphosphorylated tau gains aberrant access to mitochondria and interacts directly with core bioenergetic proteins. The proteins that tau interact with include VDAC1, ANT and the subunits of ATP synthase. These interactions compromise ATP production and limit the exchange of ATP and ADP across the mitochondrial membrane (Atlante et al., 2008). Tau also sequesters Parkin, disrupts mitophagy, allowing accumulation of dysfunctional mitochondria (Cummins et al., 2019), with subsequent excessive ROS levels triggering tau kinases, causing yet more tau hyperphosphorylation, perpetuating the cycle of disease (Melov et al., 2007; Tracy et al., 2022). At the terminal stage there is widespread mitochondrial fragmentation and failure of axonal mitochondrial transport. The subsequent synaptic energy failure, mPTP induction, lysosomal failure and glial metabolic failure, contribute to the neuronal apoptosis that occurs, defining a threshold past which neuronal loss is inevitable. This hierarchical cascade has tau at the nexus of the progressive bioenergetic failure, placing its initial metabolic defect prior to mitochondrial failure and proteostatic loss, thus implying distinct windows of therapeutic opportunity during stages one (early metabolic rescue), two (mid-stage mitochondrial rescue) and three (late-stage neuroprotection) (Cao et al., 2026).

9.2 Research gaps and future directions

Even though much has been achieved, still the tau-mitochondria axis is underexplored, and some important gaps in understanding specific mechanisms need to be filled. An important yet unresolved issue remains isoform-specific and strain-specific mitochondrial toxicity. Different tauopathies are associated with different tau conformations and isoform compositions that are expected to determine their mitochondrial binding profiles. AD tau involves mixed 3R and 4R forms with C-shaped filaments and a seeding lag of approximately 3–4 h, with an EC50 of approximately 25 nM. PSP tau is 4R-predominant, forms narrow filaments, and shows a seeding lag of approximately 12–16 h with oligodendrocyte tropism. Pick's disease tau is 3R-predominant, forms Pick body filaments, and shows neuronal tropism. Hyperphosphorylated 4R tau also shows stronger binding to mitochondrial proteins such as VDAC1. The KD for 4R tau is around 45 nM and the KD for 3R tau is around 180 nM. This implies that 4R-predominant tauopathies are more vulnerable regarding mitochondria liability.

This raises the possibility that such differences could underpin the observed disease-specific vulnerability patterns, such as the more evident oligodendrocyte mitochondrial dysfunction (Narine & Colognato, 2022; Schoenfeld et al., 2010; Meyer & Rinholm, 2021) and rapid white-matter energy failure in 4R tauopathies such as PSP and corticobasal degeneration as opposed to AD, as well as the different regional degeneration patterns; however, these inconsistencies are still insufficiently well characterised and systematic testing is required in isogenic model systems that recapitulate disease relevant tau isoform burdens at physiologically relevant levels (DeVos et al., 2017; Mummery et al., 2023).

A second major problem still unresolved is reversibility of mitochondrial damage. It is possible that short-term suppression of tau may not be able to restore bioenergetic health of mitochondria once bioenergetics has become compromised long-term and, by extension, whether there is a pathological tipping point beyond which bioenergetic deficits translate into structural collapse and thus irreversible mitochondrial damage. Critical to addressing this question is its impact on therapeutic windows. Moreover, the extent to which tau- or mitochondria-targeted therapies will result in disease modification cannot be properly appraised without resolution of this question. Fortunately new techniques, open up new areas on which to focus these sorts of questions. Human-induced pluripotent stem cell- (iPSC-) derived neurons, three dimensional cerebral organoids, and super-resolution live-cell systems provide state-of-the-art tools that can be used to address questions pertaining to the interactions between tau and mitochondria on a scale that reflects human biology on a lucent and mechanistic scale: from early bioenergetic stress to structural mitochondrial collapse. Creating a fine distal-to-proximal cause-and-effect timeline of interactions that lead to bioenergetic collapse will ultimately be crucial for collaboration between basic and translational research, which is again pivotal in the

Table 1 Eight mechanisms of tau-mediated mitochondrial dysfunction.

Mechanism	Molecular target	Tau domain involved	Binding affinity	Functional consequence	Magnitude of impairment	Reversibility	Therapeutic target	Citations
Import channel blockade	TOMM40/ TIMM23	N-terminal projection domain (residues 1–150)	KD 85 nM (phospho-tau) vs 220 nM (non-phospho)	Mitochondrial protein import reduced 50–70 %; PINK1 import blocked triggering unintended mitophagy signalling	Severe	Partial (requires tau reduction)	tau N-terminal peptide inhibitors	Tracy et al. (2022), Needs et al. (2023), Amadoro et al., (2014)
ETC complex inhibition	Complex I (NDUFS3), III (cyt bc1), IV (COX-IV), V (ATP5B/ ATP5O)	MTBD; proline-rich domain	Complex I KD ~120 nM; Complex V subunits pulldown enhanced 12.5-fold in AD	Complex I activity −45–55 %; Complex IV −38–50 %; ATP synthase −40–55 %; total ATP −35–40 %	Severe	Uncertain requires mitochon-drial rescue	Elamipretide; respiratory complex stabilisers	Torres et al. (2021), Drummond et al. (2020), Rhein et al. (2009), Amadoro et al. (2014)
ANT competitive inhibition	Adenine nucleotide translocase 1 (ANT1/ SLC25A4)	N-terminal peptide residues 26–44	Ki = 180 nM (competitive)	ADP/ATP exchange −45–65 %; State 3 respiration −55–70 %; ATP production–export uncoupling	Severe	Partially reversible	Tau N-terminal targeting	Atlante et al. (2008), Amadoro et al. (2014)

VDAC1 pore closure	Voltage-dependent anion channel 1	Proline-rich domain (K163, R171, K174, R178); MTBD	KD 45–180 nM depending on phosphorylation; ~8 tau per VDAC1 oligomer	ADP/ATP flux −60–75 %; cytosolic ATP depleted 38–45 % despite intact mitochondrial synthesis; open probability 0.82 → 0.35	Severe	Partially reversible; VDAC1 heterozygous KO protective	VDAC1 stabilisers; tau-VDAC interaction disruptors	Manczak and Reddy (2012b), Vijayan et al. (2022), Smilansky et al. (2015)
Drp1 activation (fission)	Dynamin-related protein 1 (Drp1 GTPase)	Ser396-phosphorylated tau (PHF-1 epitope)	KD ~95 nM; AD brain co-IP efficiency 8.5-fold higher than controls	Drp1 GTPase activity 2.8-fold increase; fission rate 0.8 → 2.6 events/mitochondrion/hr; mean length 2.8 → 0.6 μm	High	Reversible with Drp1 inhibition (Mdivi-1)	Mdivi-1; P110 Drp1 inhibitory peptide	Manczak and Reddy (2012a), Torres et al. (2022), DuBoff et al. (2012)
OPA1/mitofusin inhibition (fusion)	OPA1 (IMM fusion); Mfn1/Mfn2 (OMM fusion)	Multiple domains; interaction via AD brain mitochondrial fractions	OPA1 HMW complexes 68 → 28 % of total; Mfn2 GTPase kcat 0.85 → 0.35 min^{-1}	Fusion events 0.6 → 0.15/mitochondrion/hr; fusion-incompetent OPA1 accumulates; irreversible fragmentation if sustained	High	Partially — OPA1 overexpression rescues morphology but not depolarisation	OPA1 activators; mitofusin agonists (phenelzine analogues)	Hu et al. (2016), Amadoro et al. (2014), Pérez et al. (2018), Pérez et al. (2024)

(continued)

Table 1 Eight mechanisms of tau-mediated mitochondrial dysfunction. *(cont'd)*

Mechanism	Molecular target	Tau domain involved	Binding affinity	Functional consequence	Magnitude of impairment	Reversibility	Therapeutic target	Citations
Parkin sequestration (mitophagy block)	Parkin RING1 domain (residues 238–293)	N-terminal projection domain (residues 1–150); K163, R171, K174	KD = 110 nM; PLA puncta 45–68/cell vs 3 in controls	Mitophagy efficiency 88 → 5–48 % (P301L abolishes completely); dysfunctional mitochondria accumulate; PINK1 orphaned → pro-death JNK/p38 signalling	Severe	Partial rescue with Parkin overexpression; ΔN150 construct fully rescues	Parkin activators; ubiquitin pathway enhancers; urolithin A	Cummins et al., (2019), Hu et al. (2016), Fang et al. (2019)
mPTP opening/ apoptosis	Cyclophilin D (CypD); ANT; BAX/ BAK	TauC3 (caspase-3 cleaved at Asp421)	TauC3-CypD KD = 75 nM vs 220 nM full-length tau; TauC3-BAX KD = 120 nM	Calcium threshold for mPTP 180 → 45 μM (4-fold sensitisation); cytochrome c release 80–90 %; caspase-3 activation 4–5-fold; cell death 72–75 %	Catastrophic	CypD-KO provides partial rescue (28 % vs 75 % death); BAX-independent component persists	cyclosporin A; CypD inhibitors; Bcl-2 modulators	Pérez et al. (2024), Quintanilla et al. (2009), Tapia-Monsalves et al. (2023)

MTBD, microtubule-binding domain; OMM/IMM, outer/inner mitochondrial membrane; ETC, electron transport chain; HMW, high molecular weight; PLA, proximity ligation assay.

Table 2 Therapeutic strategies targeting the tau–mitochondria axis across ten agent classes spanning antibodies, ASOs, aggregation inhibitors, mitochondrial protectants, and metabolic modulators.

Therapeutic class	Agent	Primary target	Mechanism of action	Key preclinical finding	Clinical stage	Clinical outcome	Key limitation (s)	Optimal disease stage	Citation (s)
Anti-tau mAb	Semorinemab	Extracellular tau N-terminus (aa 6–23)	Neutralises extracellular oligomers; Fc-mediated microglial clearance; peripheral sink	>95 % CSF free tau suppression; extracellular tau plaque reduction	Phase 2 (completed)	No clinical benefit despite >95 % target engagement; p-tau217 and NfL unchanged	Cannot access intraneuronal tau (~80–90 % of toxic species); disease stage too advanced; spreading not dominant mechanism at Braak IV+	Preclinical/ early (Braak I–II)	Ayalon et al. (2021), Monteiro et al. (2023)
Anti-tau mAb	Tilavonemab	Extracellular aggregated tau (aa 25–30)	Neutralises extracellular aggregates; blocks trans-synaptic spreading	CSF tau reduction 50–70 %	Phase 2 (completed)	No benefit in PSP or early AD	Same limitations as semorinemab; extracellular target insufficient	Presymptomatic	Höglinger et al. (2021)

(continued)

Table 2 Therapeutic strategies targeting the tau–mitochondria axis across ten agent classes spanning antibodies, ASOs, aggregation inhibitors, mitochondrial protectants, and metabolic modulators. *(cont'd)*

Therapeutic class	Agent	Primary target	Mechanism of action	Key preclinical finding	Clinical stage	Clinical outcome	Key limitation (s)	Optimal disease stage	Citation (s)
Antisense oligonucleotide	MAPTRx (BIIB080)	MAPT mRNA (intron 9)	RNase H1-mediated pre-mRNA degradation; reduces total intracellular tau production	>50 % tau reduction → cognitive and neuropathological benefit in transgenic models	Phase 1b (completed)	40–50 % sustained CSF tau reduction; acceptable safety; NfL stable	Intrathecal delivery required; reversibility of mitochondrial damage unclear post-tau reduction	Prodromal/early AD	Mummery et al. (2023)
Tau aggregation inhibitor	Hydromethylthionine (LMTX)	Tau β-sheet assembly; disulfide bonds (Cys291–Cys322)	β-strand intercalation disrupting H-bonding; reduces disulfide crosslinks; enhances autophagy	Disaggregates PHFs in vitro; autophagy upregulation	Phase 3 (completed)	No ITT benefit; monotherapy subgroup significant (42 % less ADAS-Cog decline, 38 % less brain atrophy)	Pharmacological interaction with AChEIs/memantine masks efficacy; modest BBB penetration at therapeutic doses	Mild–moderate; treatment-naïve patients	Gauthier et al. (2016), Wischik et al. (2019)

GSK3β kinase inhibitor	Tideglusib	GSK3β (non-ATP-competitive, allosteric)	Reduces tau hyperphosphorylation at Ser396/404, Thr231; prevents GSK3β-mediated priming	45% reduced phospho-tau; 35 % improved spatial memory in P301L mice	Phase 2 (completed)	No efficacy; insufficient CNS penetration (brain:plasma 0.3:1); only 20–30% GSK3β inhibition achieved	Inadequate BBB penetrance; need ≥70 % inhibition for efficacy	Early AD (if CNS penetration solved)	Lovestone et al. (2015)
Cardiolipin-targeting peptide	Elamipretide (SS–31)	Cardiolipin (IMM); cristae architecture	Stabilises cristae; enhances ETC supercomplex assembly; prevents cytochrome c release; protects mPTP	45 % reduced ROS; 60 % increased ATP; cristae density restored; cognitive improvement in APP/PS1 mice	Phase 2 (other indications)	Safe in mitochondrial myopathy/ HFpEF; no completed AD trial; excellent CNS penetration (brain:plasma 0.6–0.8)	No AD-specific Phase 2/3 data yet	Prodromal–early AD (before irreversible mitochondrial damage)	Calkins and Reddy (2011), Manczak et al. (2010)

(continued)

Table 2 Therapeutic strategies targeting the tau–mitochondria axis across ten agent classes spanning antibodies, ASOs, aggregation inhibitors, mitochondrial protectants, and metabolic modulators. (*cont'd*)

Therapeutic class	Agent	Primary target	Mechanism of action	Key preclinical finding	Clinical stage	Clinical outcome	Key limitation (s)	Optimal disease stage	Citation (s)
Mitophagy enhancer	Urolithin A	PINK1 stabilisation; NAD^+-SIRT1-PGC−1α axis	Induces selective mitophagy; restores mitochondrial biogenesis; activates FOXO3 autophagy genes	55 % reduced fragmentation; 3.2-fold mitophagic flux increase; 40 % phospho-tau reduction in mTau-HT22 cells	Phase 1 healthy elderly (completed)	Safe; plasma levels ≥EC50; upregulates mitochondrial biogenesis genes in muscle; improves exercise endurance	No AD cognitive endpoint trial yet; gut microbiota variability in conversion from ellagitannins	Preclinical–prodromal	Fang et al. (2019), D'Amico et al. (2021)
NAD^+ precursor	Nicotinamide riboside (NR)/NMN	NAD^+ salvage pathway; SIRT1/SIRT3	Elevates brain NAD^+; activates mitophagy; enhances synaptic plasticity; reduces phospho-tau	30–40 % phospho-tau reduction; 45 % spine density increase; 35–50 % cognitive improvement in P301S mice	Phase 2 in MCI/early AD (ongoing)	Interim data awaited (2025–2026); safe in healthy adults; brain NAD^+ elevation confirmed in preclinical models	BBB penetration of NMN debated; NAD^+ elevation in brain may be modest	Preclinical–prodromal (metabolic rescue)	Fang et al. (2019), Trammell et al. (2016)

AMPK activator/ metabolic modulator	Metformin	Complex I (mild); LKB1-AMPK axis	AMPK activation → PGC-1α upregulation → mitochondrial biogenesis; mTOR inhibition enhances autophagy; NF-κB suppression	Mitochondrial mass increase; reduced tau phosphorylation; anti-inflammatory effects	Phase 2 prevention (MAP trial ongoing)	Epidemiological: 20–35 % reduced AD incidence in diabetics; RCT results expected 2026	Evidence primarily observational (confounding likely); theoretical Complex I inhibition concern at high doses	Preclinical prevention; metabolically at-risk individuals	Markowicz-Piasecka et al. (2017), Luchsinger et al. (2025)
Ketogenic/ metabolic intervention	Exogenous ketones/ MCT/ ketogenic diet	PDH bypass; HDAC inhibition; NLRP3 blockade	Provides alternative fuel bypassing PDH; 25 % higher ATP/ O_2; HDAC inhibition upregulates PGC-1α; NLRP3 suppression	Cognitive improvement in AD mouse models; reduced neuroinflammation; enhanced mitochondrial biogenesis	Phase 2 pilot (completed)	4.2-point ADAS-Cog improvement acutely in MCI; increased cerebral ketone uptake on PET; preliminary 12-week cognitive benefit	Adherence challenging (30–50 % dropout); GI side effects; multiple daily doses needed for sustained therapeutic ketosis	MCI–early AD (metabolic rescue regardless of amyloid/ tau status)	Taylor et al. (2018), Henderson et al. (2009)

ITT, intention-to-treat; AChEI, acetylcholinesterase inhibitor; PSP, progressive supranuclear palsy; HFpEF, heart failure with preserved ejection fraction; MCT, medium-chain triglyceride; PDH, pyruvate dehydrogenase; NfL, neurofilament light chain.

Table 3 Hierarchical disease staging model with biomarker and therapeutic correlates across preclinical, prodromal/MCI, and dementia tiers.

Parameter	Tier 1 preclinical AD *(Years 0–10 before symptom onset)*	Tier 2 prodromal / MCI *(Years 10–15; symptomatic transition)*	Tier 3 dementia (Mild–Severe) *(Years 15–20+; established disease)*	Citation (s)
Braak stage	0–II (transentorhinal; tau confined to entorhinal cortex layer II)	III–IV (limbic; tau spreading to hippocampus, amygdala, basal forebrain)	V–VI (neocortical; tau in association and primary cortices)	Kfoury et al. (2012), Braak and Braak (1991)
Cognitive status	Cognitively normal (MMSE 28–30); subtle executive deficits detectable only on demanding cognitive stress tests	MCI or mild dementia (MMSE 20–28); episodic memory impairment; instrumental ADL partially preserved	Moderate–severe dementia (MMSE <20); global cognitive and functional impairment; dependent ADLs	Sperling et al. (2011)
Plasma biomarkers	p-tau217: >0.5 pg/ml (rising); GFAP: 150–200 pg/ml (1.5–2-fold elevated); NfL: <15 pg/ml (near-normal); p-tau181: borderline elevated	p-tau217: 2–8 pg/ml (substantially elevated); GFAP: 180–250 pg/ml; NfL: 15–30 pg/ml (moderate; >15%/yr increase = rapid progressor); p-tau181: clearly elevated	p-tau217: >8 pg/ml; GFAP: >250 pg/ml; NfL: >30 pg/ml (severe neurodegeneration); all tau markers markedly elevated	Ashton et al. (2022), Benussi et al. (2026), Palmqvist et al. (2023)

CSF biomarkers	Aβ42/40 ratio: decreasing (amyloid positivity); p-tau181: mildly elevated; t-tau: near-normal; 8-oxo-dG: borderline elevated	Aβ42/40: clearly reduced; p-tau181/217: substantially elevated; t-tau: 400–600 pg/ml; 8-oxo-dG: 2–2.5-fold above normal; ATP synthase subunits: 20–30 % reduced	All amyloid/tau markers markedly abnormal; t-tau: >600 pg/ml; 8-oxo-dG: 2.5-fold above normal; ATP synthase subunits: 30–40 % reduced; Humanin/MOTS-c: 2–3-fold elevated (compensatory)	Hampel et al. (2021), Janelidze et al. (2021)
Imaging findings	Amyloid PET: positive (SUVR > 1.17); Tau PET (^{18}F-MK−6240): Braak I–II pattern; entorhinal SUVR 1.3–1.5; FDG-PET: 10–20 % hypometabolism in PCC/precuneus; MRI: minimal atrophy (< 1 %/yr whole brain)	Amyloid PET: high positive; Tau PET: Braak III–IV; temporal SUVR 1.5–2.5; FDG-PET: 18–25 % hypometabolism in PCC, temporal, parietal; MRI: hippocampal atrophy 1–2 %/yr; medial temporal thinning	Amyloid PET: high or declining if treated; Tau PET: Braak V–VI; SUVR 3.5–6.0 in temporal/frontal; FDG-PET: 25–30 % hypometabolism; widespread; MRI: whole-brain atrophy >1.5–2 %/yr; ventricular expansion	Reiman et al. (2004); Buckley et al. (2019)

(continued)

Table 3 Hierarchical disease staging model with biomarker and therapeutic correlates across preclinical, prodromal/MCI, and dementia tiers. *(cont'd)*

Parameter	Tier 1 preclinical AD *(Years 0–10 before symptom onset)*	Tier 2 prodromal / MCI *(Years 10–15; symptomatic transition)*	Tier 3 dementia (Mild–Severe) *(Years 15–20+; established disease)*	Citation (s)
Dominant pathogenic mechanisms	Amyloid accumulation → GLUT1/3 downregulation (45–62 % reduction); reduced O-GlcNAcylation → compensatory tau hyperphosphorylation; early mitochondrial ROS elevation (1.5–2-fold) without major ATP depletion; amyloid-initiated tau seeding in entorhinal cortex; astrocyte GFAP elevation (reactive but compensated)	Tau mitochondrial localisation (4–8% → 15–25 % of neuronal tau); VDAC1/ANT inhibition → ATP reduction 25–40 %; Parkin sequestration → mitophagy 40–100 % impaired; ROS–kinase feedforward loop (GSK3β 2.3-fold; JNK 3.8-fold); calcium dysregulation; MAM contact sites reduced 60 %; axonal transport deficits (anterograde velocity −52 %); NLRP3 activation; IL-1β/TNFα elevation; insulin resistance	Mitochondrial fragmentation (length 0.6–0.9 μm); distal axon depletion −45–55 %; mPTP opening → cytochrome c release → caspase-3 activation 4–5-fold; lysosomal dysfunction (pH 4.6 → 5.4); cathepsin activity collapse; synaptic ATP 35–45 % of normal; synapse loss >50 % in hippocampus; glial metabolic support collapse (astrocyte lactate −60–75 %; oligodendrocyte MCT1 −35 %); cGAS-STING activation; IFN-α/β 15–35-fold elevated	Reddy (2011), Torres et al. (2021), Yao et al. (2009), Manczak and Reddy (2012a), Cummins et al. (2019), Pérez et al. (2024)

Neuronal loss estimate	<10–15 % in entorhinal cortex; minimal elsewhere; synaptic function largely intact	15–40 % in entorhinal cortex; 15–30 % in hippocampal CA1; partially compensated by synaptic plasticity	>60 % in entorhinal cortex; >50 % in hippocampal CA1; beyond salvageable threshold in vulnerable regions	Braak and Braak (1991), Koffie et al. (2011), Reddy (2011)
Primary therapeutic goal	**Prevention:** Prevent bioenergetic decline and tau accumulation before damage is established	**Disease modification:** Arrest amplification loop; rescue mitochondrial function; preserve remaining neurons	**Neuroprotection + symptom management:** Maximise function of surviving neurons; slow irreversible progression	Hampel et al. (2021)
Tau-directed therapies	Tau ASO (MAPTRx/NIO752): begin early to prevent accumulation; sustained ≥50 % tau reduction target; Anti-tau antibodies (if amyloid-positive): block initial trans-synaptic spreading at Braak I–II; GSK3β inhibitors: reduce compensatory hyperphosphorylation	Tau ASO: primary disease-modifying agent (40–50 % CSF tau reduction demonstrated); Anti-tau antibodies: limited benefit if Braak III–IV established; potentially combined with ASO; LMTX (monotherapy if AChEI-naïve): tau aggregation inhibition + autophagy enhancement; PP2A activators (sodium selenate): complement kinase inhibition	Tau ASO: may slow progression but limited by existing neuronal loss; Disease-modifying benefit marginal if >60 % neuronal loss; symptom-oriented approach preferred; AChEIs (donepezil, rivastigmine) + memantine: standard of care for symptomatic benefit	Niewiadomska et al. (2021), Lasagna-Reeves et al. (2011)

(continued)

Table 3 Hierarchical disease staging model with biomarker and therapeutic correlates across preclinical, prodromal/MCI, and dementia tiers. (*cont'd*)

Parameter	Tier 1 preclinical AD *(Years 0–10 before symptom onset)*	Tier 2 prodromal / MCI *(Years 10–15; symptomatic transition)*	Tier 3 dementia (Mild–Severe) *(Years 15–20+; established disease)*	Citation (s)
Mitochondria-directed therapies	Urolithin A (500–1000 mg/day): maintain mitophagic flux before impairment; PINK1 stabilisation; NAD^+ precursors (NR/NMN 500 mg/day): preserve brain NAD^+ and SIRT1 activity; Elamipretide (if available): cardiolipin stabilisation preventing early cristae deterioration	Elamipretide (primary mitochondrial agent): cristae restoration; ETC supercomplex stabilisation; mPTP protection; Urolithin A: restore impaired mitophagy (Parkin sequestration partially overcome); NAD^+ precursors: SIRT3 activation; mtDNA repair; synaptic plasticity support; Drp1 inhibitors (experimental): counteract tau-driven fragmentation	Elamipretide: protect residual mitochondria from further mPTP opening and cytochrome c release; NAD^+ precursors: support remaining bioenergetic capacity; Mitochondrial targeted antioxidants (MitoQ): reduce ROS burden in surviving neurons	Eckert et al. (2011), Manczak and Reddy (2012a)

Metabolic support therapies	Ketogenic diet/exogenous ketones (β-HB 1–3 mM): bypass early GLUT3 deficiency; maintain neuronal ATP; Metformin: AMPK activation; mitochondrial biogenesis; anti-inflammatory; prevention in at-risk individuals	MCT supplementation: hepatic ketogenesis; acute cognitive support; Exogenous ketones (β-HB mineral salts/esters): bypass PDH blockade; 4.2-point ADAS-Cog improvement in MCI trials; Ketogenic diet: NLRP3 suppression; autophagy enhancement; anti-inflammatory; Metformin: insulin sensitisation; GSK3β suppression via AMPK	GLUT3 restoration strategies (experimental); Exogenous ketones: most tractable metabolic support; bypasses all upstream metabolic blocks; MCT supplementation: feasible in care settings; tolerated at moderate doses; Nutritional optimisation: ensure adequate glucose and ketone substrate availability	Bonnechère et al. (2026), Chiang et al. (2016), Bonnechère et al. (2025), Yao et al. (2009)
Anti-inflammatory therapies	NLRP3 inhibitors (MCC950, experimental): prevent early inflammasome priming by sub-threshold mtDNA release; Lifestyle interventions: exercise (reduces GFAP, NfL); Mediterranean diet; sleep optimisation	NLRP3 inhibitors: reduce IL-1β; improve insulin signalling; slow tau phosphorylation; TREM2 agonists: enhance microglial phagocytosis of tau aggregates; Minocycline (clinical validation): reduces IL-1β/TNFα; improves insulin signalling; reduces AT8-tau burden	cGAS-STING inhibitors (experimental): reduce type I IFN-driven neurodegeneration; Anti-inflammatory support: NSAIDs limited by GI/CV risk; selective COX-2 inhibitors considered; TSPO agonists: enhance microglial ATP; restore phagocytic clearance of debris	Underwood et al. (2020), Li and Gong (2025), Scarmeas et al. (2009)

(continued)

Table 3 Hierarchical disease staging model with biomarker and therapeutic correlates across preclinical, prodromal/MCI, and dementia tiers. (*cont'd*)

Parameter	Tier 1 preclinical AD *(Years 0–10 before symptom onset)*	Tier 2 prodromal / MCI *(Years 10–15; symptomatic transition)*	Tier 3 dementia (Mild–Severe) *(Years 15–20+; established disease)*	Citation (s)
Combination priority strategy	**Anti-amyloid + Tau ASO + NAD^+ precursor + Ketones:** Address upstream amyloid + prevent tau accumulation + maintain metabolic reserve	**Tau ASO + Elamipretide + Urolithin A + Ketones + NLRP3 inhibitor:** Reduce tau production + rescue mitochondria + restore mitophagy + metabolic support + anti-inflammation	**AChEI/memantine + Elamipretide + Ketones + Cognitive rehabilitation:** Maximise remaining neuronal function; comfort-oriented disease management	Eckert et al. (2011), Rhein et al. (2009)
Expected therapeutic benefit	Highest neuronal populations largely intact; amplification loop not yet established; 5-year progression to MCI reduced 45–60 % in biomarker-positive individuals with early intervention (estimated)	Moderate 35–70 % neuronal preservation possible with combination strategies; synergistic neuroprotection (70 % neuronal survival with tau reduction + mitochondrial rescue vs 35–45 % monotherapy)	Limited irreversible neuronal loss in key regions; goal shifts from disease modification to quality-of-life preservation and slowing residual decline	Sims et al. (2023), van Dyck et al. (2023)

Key trial design considerations	Enrichment by p-tau217 >0.5 pg/ml + GFAP >150 pg/ml + amyloid PET positive; MMSE 28–30; primary endpoint: time to MCI or tau PET SUVR progression; 36–48-month duration minimum	Enrichment by tau PET Braak III–IV; MMSE 20–28; NfL 15–30 pg/ml; composite endpoint (PACC5 + CDR-SB + NfL trajectory); adaptive factorial platform trial design for combinations	Standard CDR-SB and ADCS-ADL endpoints; MMSE <20; focus on safety; quality-of-life instruments; caregiver burden measures	Sperling et al. (2011), Ashton et al. (2022)

PCC, posterior cingulate cortex; MCI, mild cognitive impairment; ADL, activities of daily living; AChEI, acetylcholinesterase inhibitor; ASO, antisense oligonucleotide; β-HB, β-hydroxybutyrate; MCT, medium-chain triglyceride; PDH, pyruvate dehydrogenase; MAM, mitochondria-associated ER membrane; PACC5, Preclinical Alzheimer Cognitive.

development of stage-matched intervention. Such knowledge will guide whether treatment should prioritise early metabolic rescue, mid-stage mitochondrial protection, or late-stage neuroprotection.

9.3 Integration and outlook

This chapter frames tau as the central effector of bioenergetic failure and defines the staged temporal framework of tau-mitochondrial pathogenesis and disease. It thus provides the scientific foundation and conceptual basis for next-generation therapeutic design. Such therapies will need to be stage-specific, multi-modal, biomarker-guided approaches. In the author‘s view, the most promising approach is the strategic combination of precision tau-lowering strategies with mitochondria-targeted interventions. ASOs are capable of around 50–70 % tau reduction, with acceptable safety profiles (Mummery et al., 2023). Mitochondria-targeted approaches include elamipretide to protect cardiolipinand cristae, urolithin A to promote mitophagy, and NAD+ precursors to promote mitochondrial QC (Fang et al., 2019; Szeto, 2014). Dietary or pharmaceutical ketosis can be tailored to potentially compensate for consistent mitochondrial energy shortfalls in affected neurons. Applied in bio-marker-positive early stage populations, identified via blood screening, these integrated strategies can, the author believes, provide a rational approach toward truly disease-modifying therapy. Today‘s emerging subtleties of molecular science, from exquisite science of protein-protein interactions to dynamic imaging of subcellular processes, crystallize this opportunity for rational therapeutic design; one that substantially outstrips the outdated and sclerotic protocol of amyloid-centric drug development in AD. Tables 1, 2, 3.

References

Acín-Pérez, R., Fernández-Silva, P., Peleato, M. L., Pérez-Martos, A., & Enriquez, J. A. (2008). Respiratory active mitochondrial supercomplexes. *Molecular Cell, 32*(4), 529–539.

Alavi Naini, S. M., & Soussi-Yanicostas, N. (2015). Tau hyperphosphorylation and oxidative stress, a critical vicious circle in neurodegenerative tauopathies? *Oxidative Medicine and Cellular Longevity, 2015*, 151979.

Alhadidy, M. M., & Kanaan, N. M. (2024). Biochemical approaches to assess the impact of post-translational modifications on pathogenic tau conformations using recombinant protein. *Biochemical Society Transactions, 52*, 301–318.

Aliu Opeyemi Yakubu, G. I. A., Chukwuma Emmanuel Nwaze, T. A. B., Temilade Oluwatoyosi Adeniyi, K. A., & Acheampong Benjamin Oladele, O. S. (2026). Expanding landscape of tau pathology in neurological disorders: A narrative review guiding diagnostic development and targeted therapeutics. *NeuroMarkers,* 100167. ISSN 2950-5887. https://doi.org/10.1016/j.neumar.2026.100167.

Alquezar, C., Arya, S., & Kao, A. W. (2022). Oxidative stress and tauopathies. *Antioxidants, 11*, 1421.

Alzheimer's Association (2023). 2023 Alzheimer's disease facts and figures. *Alzheimer's & Dementia, 19*(4), 1598–1695.

Amadoro, G., Corsetti, V., Atlante, A., Di Michele, G., Florenzano, F., Bobba, A., ... Calissano, P. (2014). AD-linked, toxic NH2 human tau affects the quality control of mitochondria in neurons. *Neurobiology of Disease, 62*, 489–507.

Asai, H., Ikezu, S., Tsunoda, S., Medalla, M., Luebke, J., Haydar, T., ... Ikezu, T. (2015). Depletion of microglia and inhibition of exosome synthesis halt tau propagation. *Nature Neuroscience, 18*(11), 1584–1593.

Ashton, N. J., Janelidze, S., Mattsson-Carlgren, N., Binette, A. P., Strandberg, O., Brum, W. S., ... Hansson, O. (2022). Differential roles of Aβ42/40, p-tau231 and p-tau217 for Alzheimer's trial selection and disease monitoring. *Nature Medicine, 28*, 2555–2562.

Atlante, A., Amadoro, G., Bobba, A., de Bari, L., Corsetti, V., Pappalardo, G., ... Passarella, S. (2008). A peptide containing residues 26–44 of tau protein impairs mitochondrial oxidative phosphorylation acting at the level of the adenine nucleotide translocator. *Biochimica et Biophysica Acta, 1777*(10), 1289–1300.

Attwell, D., & Laughlin, S. B. (2001). An energy budget for signaling in the grey matter of the brain. *Journal of Cerebral Blood Flow & Metabolism, 21*, 1133–1145.

Ayalon, G., Lee, S.-H., Adolfsson, O., Foo-Atkins, C., Atwal, J. K., Blendstrup, M., Booler, H., Bravo, J., Brendza, R., Brunstein, F., Chan, R., Chandra, P., et al. (2021). Antibody semorinemab reduces tau pathology in atransgenic mouse model and engages tau in patients with Alzheimer's disease. *Science Translational Medicine, 13*(593), eabb2639. pubmed.ncbi.nlm.nih+1.

Ban, T., Ishihara, T., Kohno, H., Saita, S., Ichimura, A., Maenaka, K., ... Ishihara, N. (2017). Molecular basis of selective mitochondrial fusion by heterotypic action between OPA1 and cardiolipin. *Nature Cell Biology, 19*(6), 856–863.

Barsoum, M. J., Yuan, H., Gerencser, A. A., Liot, G., Kushnareva, Y., Gräber, S., ... Bossy-Wetzel, E. (2006). Nitricoxide-induced mitochondrial fission is regulated by dynamin-related GTPases in neurons. *The EMBO Journal, 25*(16), 3900–3911. https://doi.org/10.1038/sj.emboj.7601253. Epub 2006 Jul 27. PMID: 16874299; PMCID: PMC1553198.

Barthélemy, N. R., Horie, K., Sato, C., & Bateman, R. J. (2020). Blood plasma phosphorylated-tau isoforms track CNS change in Alzheimer's disease. *Journal of Experimental Medicine, 217*(11), e20200861. https://doi.org/10.1084/jem.20200861. PMID: 32725127; PMCID: PMC7596823.

Benussi, A., Michelutti, M., Lombardo, T. M. I., Toffoletto, B., Palacino, F., Cenacchi, V., ... Manganotti, P. (2026). Diagnostic performance of plasma pTau217/Aβ42 ratio and a three-zone threshold model for Alzheimer's disease. *Neurobiology of Aging, 159*, 60–68.

Blum, T. B., Hahn, A., Meier, T., Davies, K. M., & Kühlbrandt, W. (2019). Dimers of mitochondrial ATP synthase induce membrane curvature and self-assemble into rows. *Proceedings of the National Academy of Sciences, 116*(10), 4250–4255.

Bonnechère, B., Stephens, E. B., Boileau, A. C., Ducker, M., & Stubbs, B. J. (2025). The effect of exogenous ketone bodies on cognition in patients with mild cognitive impairment, Alzheimer's disease and in healthy adults: A systematicreview and meta-analysis. *medRxiv [Preprint]*. https://doi.org/10.1101/2025.09.17.25335999. PMID: 41001501; PMCID: PMC12458514.

Bonnechère, B, Stephens, E. B., Boileau, AC, Ducker, M, & Stubbs, BJ (2026). The effect of exogenous ketone bodies on cognition across health and disease: a systematic review and meta-analysis. *Frontiers in Nutrition, 13*, 1802531. https://doi.org/10.3389/fnut.2026.1802531.

Braak, H., & Braak, E. (1991). Neuropathological stageing of Alzheimer-related changes. *Acta Neuropathologica, 82*(4), 239–259.

Buchholz, S., & Zempel, H. (2024 May). The six brain-specific TAU isoforms and their role in Alzheimer's disease and related neurodegenerative dementia syndromes. *Alzheimer's & Dementia, 20*(5), 3606–3628. https://doi.org/10.1002/alz.13784 Epub 2024 Mar 31. PMID: 38556838; PMCID: PMC11095451.

Buckley, R. F., Mormino, E. C., Rabin, J. S., et al. (2019). Sex differences in the association of global amyloid and regional tau deposition measured by positron emission tomography in clinically normal older adults. *JAMA Neurology, 76*, 542–551.

Calkins, M. J., & Reddy, P. H. (2011). Amyloid beta impairs mitochondrial anterograde transport and degenerates synapses in Alzheimer's disease neurons. *Biochimica et Biophysica Acta, 1812*, 507–513.

Cao, H., Liang, J., Dong, X., Xia, Z., Luo, X., & Liu, B. (2026). Targeting the astrocytic metabolic cascade in Alzheimer's disease: mechanisms, challenges and opportunities. *Frontiers in Aging Neuroscience, 18*, 1767811. https://doi.org/10.3389/fnagi.2026.1767811 PMID: 41858787; PMCID: PMC12996062.

Castillo-Casaña, Y., Arias, C., & Coria, R. (2025). Tau–mitochondria interactions in neurodegeneration: Mechanisms and therapeutic potential. *Cellular and Molecular Neurobiology, 46*, 3.

Chen, H., & Chan, D. C. (2006). Critical dependence of neurons on mitochondrial dynamics. *Current Opinion in Cell Biology, 18*(4), 453–459.

Cheng, Y., & Bai, F. (2018). The association of tau with mitochondrial dysfunction in Alzheimer's disease. *Frontiers in Neuroscience, 12*, 163.

Chiang, M. C., Cheng, Y. C., Chen, S. J., Yen, C. H., & Huang, R. N. (2016). Metformin activation of AMPK-dependent pathways is neuroprotective in human neural stem cells against Amyloid-beta-induced mitochondrial dysfunction. *Experimental Cell Research, 347*(2), 322–331. https://doi.org/10.1016/j.yexcr.2016.08.013 Epub 2016 Aug 21. PMID: 27554603.

Cieri, D., Vicario, M., Vallese, F., D'Orsi, B., Berto, P., Grinzato, A., ... Calì, T. (2018). Tau localises within mitochondrial sub-compartments and its caspase cleavage affects ER-mitochondria interactions and cellular Ca2+ handling. *Biochimica et Biophysica Acta - Molecular Basis of Disease, 1864*, 3247–3256.

Cohen, T. J., Guo, J. L., Hurtado, D. E., Kwong, L. K., Mills, I. P., Trojanowski, J. Q., & Lee, V. M. Y. (2011). The acetylation of tau inhibits its function and promotes pathological tau aggregation. *Nature Communications, 2*, 252.

Cummins, N., Tweedie, A., Zuryn, S., Bertran-Gonzalez, J., & Götz, J. (2019). Disease-associated tau impairs mitophagy by inhibiting Parkin translocation to mitochondria. *The EMBO Journal, 38*, e99360.

D'Amico, D., Andreux, P. A., Valdés, P., Singh, A., Rinsch, C., & Auwerx, J. (2021). Impact of the natural compound urolithin A on health, disease, and aging. *Trends in Molecular Medicine, 27*, 687–699.

Denton, R. M. (2009). Calcium in the regulation of intramitochondrial enzymes. *Biochimica et Biophysica Acta, 1787*, 1309–1316.

DeVos, S. L., Miller, R. L., Schoch, K. M., Holmes, B. B., Bhatt, N. P., Myrick, E., ... Miller, T. M. (2017). Tau reduction prevents neuronal loss and reverses pathological tau deposition and seeding in mice with tauopathy. *Science Translational Medicine, 9*(374), eaag0481.

Di Lorenzo, R., Zecca, C., Chimienti, G., Latronico, T., Liuzzi, G. M., Pesce, V., ... Lezza, S. (2025). Reliable new biomarkers of mitochondrial oxidative stress and neuroinflammation in CSF and plasma from Alzheimer's disease patients. *International Journal of Molecular Sciences, 26*, 7792.

Drummond, E., Pires, G., MacMurray, C., Askenazi, M., Nayak, S., Bourdon, M., ... Wisniewski, T. (2020). Phosphorylated tau interactome in the human Alzheimer's disease brain. *Brain, 143*(9), 2803–2817.

DuBoff, B., Götz, J., & Feany, M. B. (2012 Aug 23). Tau promotes neurodegeneration via DRP1 mislocalization in vivo. *Neuron, 75*(4), 618–632. https://doi.org/10.1016/j.neuron.2012.06.026 PMID: 22920254; PMCID: PMC3428596.

Eckert, A., Hauptmann, S., Scherping, I., Rhein, V., Muller-Spahn, F., Gotz, J., & Muller, W. E. (2011). The pathology of Alzheimer's disease: Mitochondrial dysfunction. *Alzheimer's Research & Therapy, 3*, 15.

Eckert, A., Nisbet, R., Grimm, A., & Götz, J. (2014). March separate, strike together—Role of phosphorylated TAU in mitochondrial dysfunction in Alzheimer's disease. *Biochimica et Biophysica Acta, 1842*, 1258–1266.

Fang, E. F., Hou, Y., Palikaras, K., Adriaanse, B. A., Kerr, J. S., Yang, B., ... Bohr, V. A. (2019). Mitophagy inhibits amyloid-β and tau pathology and reverses cognitive deficits in models of Alzheimer's disease. *Nature Neuroscience, 22*(3), 401–412.

Fernandes, T., Costa, M., Pereira, C., & Sant'Ana, M. (2023). The roles of mitochondria-associated membranes in regulating cellular processes. *Cellular and Molecular Neurobiology, 43*, 1323–1338.

Ferrer, I., Aguiló García, M., Carmona, M., Andrés-Benito, P., Torrejón-Escribano, B., Garcia-Esparcia, P., & Del Rio, J. A. (2019 May28). Involvement of oligodendrocytes in tau seeding and spreading in tauopathies. *Frontiers in Aging Neuroscience, 11*, 112. https://doi.org/10.3389/fnagi.2019.00112 PMID: 31191295; PMCID: PMC6546889.

Fiorese, C. J., Schulz, A. M., Lin, Y. F., Rosin, N., Pellegrino, M. W., & Haynes, C. M. (2016). The transcription factor ATF5 mediates a mammalian mitochondrial UPR. *Current Biology, 26*, 2037–2043.

Frey, T. G., & Mannella, C. A. (2000). The internal structure of mitochondria. *Trends in Biochemical Sciences, 25*(7), 319–324.

Gamblin, T. C., Chen, F., Zambrano, A., Abraha, A., Lagalwar, S., Guillozet, A. L., ... Binder, L. I. (2003). Caspase cleavage of tau: Linking amyloid and neurofibrillary tangles in Alzheimer's disease. *Proceedings of the National Academy of Sciences, 100*, 10032–10037.

Gauthier, S., Feldman, H. H., Schneider, L. S., et al. (2016). Efficacy and safety of tau-aggregation inhibitor therapy in patients with mild or moderate Alzheimer's disease. *Lancet, 388*, 2873–2884.

Gerson, J., Castillo-Carranza, D. L., Sengupta, U., Bodani, R., Prough, D. S., DeWitt, D. S., Hawkins, B. E., & Kayed, R. (2016). Tau oligomers derived from traumatic brain injury cause cognitive impairment and accelerate onset of pathology in htau mice. *Journal of Neurotrauma, 33*(22), 2034–2043. https://doi.org/10.1089/neu.2015.4262. Epub 2016 Apr 22. PMID: 26729399; PMCID: PMC5116695.

Gilkerson, R. W., Selker, J. M., & Capaldi, R. A. (2003). The cristal membrane of mitochondria is the principal site of oxidative phosphorylation. *FEBS Letters, 546*(2–3), 355–358.

Glancy, B., & Balaban, R. S. (2012). Role of mitochondrial Ca2+ in the regulation of cellular energetics. *Biochemistry, 51*, 2959–2973.

Goedert, M., Spillantini, M. G., Jakes, R., Rutherford, D., & Crowther, R. A. (1989). Multiple isoforms of human microtubule-associated protein tau: Sequences and localization in neurofibrillary tangles of Alzheimer's disease. *Neuron, 3*, 519–526.

Gong, C. X., Liu, F., Grundke-Iqbal, I., & Iqbal, K. (2005). Post-translational modifications of tau protein in Alzheimer's disease. *Journal of Neural Transmission, 112*, 813–838.

Gong, B., Pan, Y., Vempati, P., Zhao, W., Knable, L., Ho, L., ... Pasinetti, G. M. (2013). Nicotinamide riboside restores cognition through an upregulation of proliferator-activated receptor-γ coactivator 1αregulated β-secretase 1 degradation and mitochondrial gene expression in Alzheimer's mouse models. *Neurobiology of Aging, 34*(6), 1581–1588. https://doi.org/10.1016/j.neurobiolaging.2012.12.005. Epub 2013 Jan 9. PMID: 23312803; PMCID: PMC3632303.

Goode, Bruce L., Chau, Miu, Denis, Paul E., & Feinstein, Stuart C. (2000). Structural and functional differences between 3-repeat and 4-repeat Tau Isoforms: Implications for normal Tau function and the onsetof neurodegenerative disease*. ISSN 0021-9258 *Journal of Biological Chemistry, 275*(49), 38182–38189. https://doi.org/10.1074/jbc.M007489200.

Gordon, B. A., Blazey, T. M., Su, Y., Hari-Raj, A., Dincer, A., Flores, S., ... Benzinger, T. L. (2018). Spatial patterns of neuroimaging biomarker change in individuals from families with autosomal dominant Alzheimer's disease. *Lancet Neurology, 17*(3), 241–250.

Hampel, H., Hardy, J., Blennow, K., Chen, C., Perry, G., Kim, S. H., ... Vergallo, A. (2021). The amyloid-β pathway in Alzheimer's disease. *Molecular Psychiatry, 26*, 5481–5503.

Hansson, O., Blennow, K., Zetterberg, H., & Dage, J. (2023). Blood biomarkers for Alzheimer's disease in clinical practice and trials. *Nature Aging, 3*(5), 506–519.

Henderson, S. T., Vogel, J. L., Barr, L. J., Garvin, F., Jones, J. J., & Costantini, L. C. (2009). Study of the ketogenic agent AC-1202 in mild to moderate Alzheimer's disease. *Nutrition & Metabolism, 6*, 31.

Heneka, M. T., Kummer, M. P., Stutz, A., Delekate, A., Schwartz, S., Saecker, A., ... Latz, E. (2013). NLRP3 is activated in Alzheimer's disease and contributes to pathology in APP/PS1 mice. *Nature, 493*(7434), 674–678.

Höglinger, G. U., Litvan, I., Mendonca, N., Wang, D., Zheng, H., ... Rendenbach-Mueller, B. (2021). Safety and efficacy of tilavonemab in progressive supranuclear palsy: a phase 2, randomised, placebo-controlled trial. *Lancet Neurology, 20*(3), 182–192.

Hu, Y., Li, X. C., Wang, Z. H., Luo, Y., Zhang, X., Liu, X. P., ... Liu, G. P. (2016). Tau accumulation impairs mitophagy via increasing mitochondrial membrane potential and reducing mitochondrial Parkin. *Oncotarget*, 7, 17356–17368.

Isei, M. O., Girardi, P. A., Rodwell-Bullock, J., Nehrke, K., & Johnson, G. V. W. (2025). Site-specific phosphorylation of tau impacts mitochondrial function and response to stressors. *Journal of Neurochemistry. Advance Online Publication.*

Janelidze, S., Teunissen, C. E., Zetterberg, H., Allué, J. A., Sarasa, L., Eichenlaub, U., ... Hansson, O. (2021). Head-to-head comparison of 8 plasma amyloid-β 42/40 assays in Alzheimer disease. *JAMA Neurology, 78*, 1375–1382.

Jin, S. M., Lazarou, M., Wang, C., Kane, L. A., Narendra, D. P., & Youle, R. J. (2010). Mitochondrial membrane potential regulates PINK1 import and proteolytic destabilization by PARL. *The Journal of Cell Biology, 191*(5), 933–942.

Joshi, A. U., Saw, N. L., Shamloo, M., & Mochly-Rosen, D. (2018). Drp1/Fis1 interaction mediates mitochondrial dysfunction, bioenergetic failure and cognitive decline in Alzheimer's disease. *Oncotarget, 9*, 6128–6143.

Kanaan, N. M., Morfini, G. A., LaPointe, N. E., Pigino, G. F., Patterson, K. R., Song, Y., ... Binder, L. I. (2011). Pathogenic forms of tau inhibit kinesin-dependent axonal transport through a mechanism involving activation of axonal phosphotransferases. *Journal of Neuroscience, 31*(27), 9858–9868.

Kandimalla, R., Manczak, M., Yin, X., Wang, R., & Reddy, P. H. (2018). Hippocampal phosphorylated tau induced cognitive decline, dendritic spine loss and mitochondrial abnormalities in a mouse model of Alzheimer's disease. *Human Molecular Genetics, 27*(1), 30–40.

Kane, L. A., Lazarou, M., Fogel, A. I., Li, Y., Yamano, K., Sarraf, S. A., Banerjee, S., & Youle, R. J. (2014). PINK1 phosphorylates ubiquitin to activate Parkin E3 ubiquitin ligase activity. *Journal of Cell Biology, 205*(2), 143–153. https://doi.org/10.1083/jcb.201402104. Epub 2014 Apr 21. PMID: 24751536; PMCID: PMC4003245.

Karaa, A., Bertini, E., Carelli, V., Cohen, B. H., Enns, G. M., Falk, M. J., ... Mancuso, M. (2023). MMPOWER-3 trial investigators. efficacy and safety of elamipretidein individuals with primary mitochondrial myopathy: The MMPOWER-3 randomized clinical trial. *Neurology, 101*(3), e238–e252. https://doi.org/10.1212/WNL.0000000000207402. Epub 2023 Jun 2. PMID: 37268435; PMCID: PMC10382259.

Kazlauskaite, A., Kondapalli, C., Gourlay, R., Campbell, D. G., Ritorto, M. S., Hofmann, K., ... Muqit, M. M. (2014). Parkin is activated by PINK1-dependent phosphorylation of ubiquitin at Ser65. *Biochemical Journal, 460*(1), 127–139. https://doi.org/10.1042/BJ20140334. PMID: 24660806; PMCID: PMC4000136.

Keren-Shaul, H., Spinrad, A., Weiner, A., Matcovitch-Natan, O., Dvir-Szternfeld, R., Ulland, T. K., & Amit, I. (2017). A unique microglia type associated with restricting development of Alzheimer's disease. *Cell, 169*(7), 1276–1290 e17.

Kfoury, N., Holmes, B. B., Jiang, H., Holtzman, D. M., & Diamond, M. I. (2012). Trans-cellular propagation of Tau aggregation by fibrillar species. *Journal of Biological Chemistry, 287*(23), 19440–19451. https://doi.org/10.1074/jbc.M112.346072. Epub 2012 Mar 29. PMID: 22461630; PMCID: PMC3365982.

Koffie, R. M., Hyman, B. T., & Spires-Jones, T. L. (2011). Alzheimer's disease: Synapses gone cold. *Molecular Neurodegeneration, 6*(1), 63. https://doi.org/10.1186/1750-1326-6-63. PMID: 21871088; PMCID: PMC3178498.

Kopeikina, K. J., Carlson, G. A., Pitstick, R., Ludvigson, A. E., Peters, A., Luebke, J. I., ... Spires-Jones, T. L. (2011). Tau accumulation causes mitochondrial distribution deficits in neurons in a mouse model of tauopathy and in human Alzheimer's disease brain. *The American Journal of Pathology, 179*(4), 2071–2082.

Köpke, E., Tung, Y. C., Shaikh, S., Alonso, A. C., Iqbal, K., & Grundke-Iqbal, I. (1993). Microtubule-associated protein tau. Abnormal phosphorylation of a non-paired helical filament pool in Alzheimer disease. *Journal of Biological Chemistry, 268*(32), 24374–24384 PMID: 8226987.

Kozlov, A. V., Cammarata, D., & Szeto, H. H. (2024). Mitochondrial oxidative stress in neurodegeneration. *Cellular and Molecular Neurobiology, 46*(3), 1–22.

Krasemann, S., Madore, C., Cialic, R., Baufeld, C., Calcagno, N., El Fatimy, R., ... Butovsky, O. (2017). The TREM2-APOE pathway drives the transcriptional phenotype of dysfunctional microglia in neurodegenerative diseases. *Immunity, 47*(3), 566–581.e9.

Lapuente-Brun, E., Moreno-Loshuertos, R., Acín-Pérez, R., Latorre-Pellicer, A., Colás, C., Balsa, E., ... Enríquez, J. A. (2013). Supercomplex assembly determines electron flux in the mitochondrial electron transport chain. *Science (New York, N. Y.), 340*(6139), 1567–1570.

Lasagna-Reeves, C. A., Castillo-Carranza, D. L., Jackson, G. R., & Kayed, R. (2011). Tau oligomers as potential targets for immunotherapy for Alzheimer's disease and tauopathies. *Current Alzheimer Research, 8*(6), 659–665. https://doi.org/10.2174/156720511796717177. PMID: 21605039.

Lasagna-Reeves, C. A., Castillo-Carranza, D. L., Sengupta, U., Troncoso, J., Jackson, G. R., & Kayed, R. (2012). Identification of oligomers at early stages of tau aggregation in Alzheimer's disease. *FASEB Journal, 26*, 1946–1959.

Lautrup, S., Sinclair, D. A., Mattson, M. P., & Fang, E. F. (2019). NAD+ in brain aging and neurodegenerative disorders. *Cell Metabolism, 30*(4), 630–655.

Lee, Y., Morrison, B. M., Li, Y., Lengacher, S., Farah, M. H., Hoffman, P. N., ... Rothstein, J. D. (2012). Oligodendroglia metabolically support axons and contribute to neurodegeneration. *Nature, 487*(7408), 443–448. https://doi.org/10.1038/nature11314. PMID: 22801498; PMCID: PMC3408792.

Lee, Y., Stevens, D. A., Kang, S. U., Jiang, H., Lee, Y. I., Ko, H. S., ... Dawson, T. M. (2017). PINK1 primes parkin-mediated ubiquitination of paris in dopaminergic neuronal survival. *Cell Reports, 18*(4), 918–932. https://doi.org/10.1016/j.celrep.2016.12.090. PMID: 28122242; PMCID: PMC5312976.

Lee, G., Thangavel, R., Sharma, V. M., Litersky, J. M., Bhaskar, K., Fang, S. M., ... Ksiezak-Reding, H. (2004). Phosphorylation of tau by fyn: Implications for Alzheimer's disease. *The Journal of Neuroscience, 24*(9), 2304–2312. https://doi.org/10.1523/JNEUROSCI.4162-03.2004. PMID: 14999081; PMCID: PMC6730442.

Leyns, C. E. G., & Holtzman, D. M. (2017). Glial contributions to neurodegeneration in tauopathies. *Molecular Neurodegeneration, 12*, 50.

Li, Z, & Gong, C (2025). NLRP3 inflammasome in Alzheimer's disease: molecular mechanisms and emerging therapies. *Frontiers in Immunology, 16*, 1583886. https://doi.org/10.3389/fimmu.2025.1583886.

Li, Z., & Gong, C. (2025). NLRP3 inflammasome in Alzheimer's disease: molecular mechanisms and emerging therapies. *Frontiers in Immunology, 16*, 1583886. https://doi.org/10.3389/fimmu.2025.1583886. PMID: 40260242; PMCID: PMC12009708.

Li, T., & Paudel, H. K. (2006). Glycogen synthase kinase 3β phosphorylates Alzheimer's disease-specific Ser396 of microtubule-associated protein tau by a sequential mechanism. *Biochemistry, 45*(10), 3125–3133.

Liddelow, S. A., Guttenplan, K. A., Clarke, L. E., Bennett, F. C., Bohlen, C. J., Schirmer, L., ... Barres, B. A. (2017). Neurotoxic reactive astrocytes are induced by activated microglia. *Nature, 541*, 481–487.

Liu, F., Grundke-Iqbal, I., Iqbal, K., & Gong, C. X. (2005). Contributions of protein phosphatases PP1, PP2A, PP2B and PP5 to the regulation of tau phosphorylation. *European Journal of Neuroscience, 22*(7), 1942–1950.

Liu, J., He, X., Zheng, S., Zhu, A., & Wang, J. (2022). The mitochondrial unfolded protein response: A novel protective pathway targeting cardiomyocytes. *Oxidative Medicine and Cellular Longevity, 2022*, 6430342. https://doi.org/10.1155/2022/6430342. PMID: 36187338; PMCID: PMC9519344.

Liu, F., Iqbal, K., Grundke-Iqbal, I., Hart, G. W., & Gong, C. X. (2004). O-GlcNAcylation regulates phosphorylation of tau: A mechanism involved in Alzheimer's disease. *Proceedings of the National Academy of Sciences, 101*, 10804–10809.

Liu, F., Shi, J., Tanimukai, H., Gu, J., Gu, J., Grundke-Iqbal, I., ... Gong, C. X. (2009). Reduced O-GlcNAcylation links lower brain glucose metabolism and tau pathology in Alzheimer's disease. *Brain, 132*, 1820–1832.

Looser, Z. J., Faik, Z., Ravotto, L., Zanker, H. S., Jung, R. B., Werner, H. B., ... Saab, A. S. (2024). Oligodendrocyte-axon metabolic coupling is mediated by extracellular K+ andmaintains axonal health. *Nature Neuroscience, 27*(3), 433–448. https://doi.org/10.1038/s41593-023-01558-3. Epub 2024 Jan 24. PMID: 38267524; PMCID: PMC10917689.

Lovell, M. A., Xiong, S., Xie, C., Davies, P., & Markesbery, W. R. (2004). Induction of hyperphosphorylated tau in primary rat cortical neuron cultures mediated by oxidative stress and glycogen synthase kinase-3. discussion 673-81 *Journal of Alzheimer's Disease, 6*(6), 659–671. https://doi.org/10.3233/jad-2004-6610.

Lovestone, S., Boada, M., Dubois, B., Hüll, M., Rinne, J. O., Huppertz, H. J., ... del Ser, T. (2015). ARGO investigators. A phase II trial of tideglusib in Alzheimer's disease. *Journal of Alzheimer's Disease, 45*(1), 75–88. https://doi.org/10.3233/JAD-141959. PMID: 25537011.

Luchsinger, J. A., & Teresi, J. A. (2025). Protocol for a randomized phase II/III double-blind placebo-controlled trial to evaluate the safety and efficacy of extended-release metformin in amnestic mild cognitive impairment: Metformin in Alzheimer Dementia Prevention (MAP). Alzheimer Disease and Associated. *Disorders, 39*(2), 123–133.

MacAskill, A. F., Rinholm, J. E., Twelvetrees, A. E., Arancibia-Carcamo, I. L., Muir, J., Fransson, A., ... Kittler, J. T. (2009). Miro1 is a calcium sensor for glutamate receptor-dependent localization of mitochondria at synapses. *Neuron, 61*(4), 541–555.

Manczak, M., Mao, P., Calkins, M. J., Cornea, A., Bhatt, N. P., Kumar, S., ... Reddy, P. H. (2010). Mitochondria-targeted antioxidants protect against amyloid-beta toxicity in Alzheimer's disease neurons. *Journal of Alzheimer's Disease, 20*(.2), S609–S631.

Manczak, M., & Reddy, P. H. (2012a). Abnormal interaction between the mitochondrial fission protein Drp1 and hyperphosphorylated tau in Alzheimer's disease neurons. *Human Molecular Genetics, 21*, 2538–2547.

Manczak, M., & Reddy, P. H. (2012b). Abnormal interaction of VDAC1 with amyloid beta and phosphorylated tau causes mitochondrial dysfunction in Alzheimer's disease. *Human Molecular Genetics, 21*, 5131–5146.

Markowicz-Piasecka, M., Sikora, J., Szydłowska, A., Skupień, A., Mikiciuk-Olasik, E., & Huttunen, K. M. (2017). Metformin - a future therapy for neurodegenerative diseases: Theme: Drug discovery, development and delivery in Alzheimer's disease Guest Editor: Davide Brambilla. *Pharmaceutical Research, 34*(12), 2614–2627. https://doi.org/10.1007/s11095-017-2199-y. Epub 2017 Jun 6. PMID: 28589443; PMCID: PMC5736777.

Melber, A., & Haynes, C. M. (2018). UPRmt regulation and output: A stress response mediated by mitochondrial-nuclear communication. *Cell Research, 28*, 281–295.

Melov, S., Adlard, P. A., Morten, K., Johnson, F., Golden, T. R., Hinerfeld, D., ... Bush, A. I. (2007). Mitochondrial oxidative stress causes hyperphosphorylation of tau. *PLoS One, 2*(6), e536.

Meyer, N., & Rinholm, J. E. (2021). Mitochondria in myelinating oligodendrocytes: Slow and out of breath? *Metabolites, 11*(6), 359. https://doi.org/10.3390/metabo11060359. PMID: 34198810; PMCID: PMC8226700.

Mielke, M. M., Aggarwal, N. T., Vila-Castelar, C., Agarwal, P., Arenaza-Urquijo, E. M., Brett, B., ... Babulal, G. M. (2022). Consideration of sex and gender in Alzheimer's disease and related disorders from a global perspective. *Alzheimer's & Dementia, 18*(12), 2707–2724.

Milà-Alomà, M., Ashton, N. J., Shekari, M., et al. (2022). Plasma p-tau231 and p-tau217 as state markers of amyloid-β pathology in preclinical Alzheimer's disease. *Nature Medicine, 28*, 1797–1801. https://doi.org/10.1038/s41591-022-01925-w.

Mileykovskaya, E., & Dowhan, W. (2014). Cardiolipin-dependent formation of mitochondrial respiratory supercomplexes. *Chemistry and Physics of Lipids, 179*, 42–48.

Mink, J. W., Blumenschine, R. J., & Adams, D. B. (1981). Ratio of central nervous system to body metabolism in vertebrates. *American Journal of Physiology, 241*, R203–R212.

Mishra, P., Carelli, V., Manfredi, G., & Chan, D. C. (2014). Proteolytic cleavage of Opa1 stimulates mitochondrial inner membrane fusion and couples fusion to oxidative phosphorylation. *Cell Metabolism, 19*(4), 630–641.

Monteiro, C., Toth, B., Bhaskaran, S., & Ling, P. (2023). Efficacy and safety of semorinemab in mild-to-moderate Alzheimer disease: The Lauriet randomized clinical trial. *Neurology, 101*, e1391–e1401.

Mosconi, L., Pupi, A., & De Leon, M. J. (2008). Brain glucose hypometabolism and oxidative stress in preclinical Alzheimer's disease. *Annals of the New York Academy of Sciences, 1147*, 180–195.

Mummery, C. J., Börjesson-Hanson, A., Blackburn, D. J., et al. (2023). Tau-targeting antisense oligonucleotide MAPTRx in mild Alzheimer's disease: A phase 1b, randomized, placebo-controlled trial. *Nature Medicine, 29*, 1437–1447.

Narendra, D. P., & Youle, R. J. (2024). The role of PINK1-Parkin in mitochondrial quality control. *Nature Cell Biology, 26*(10), 1639–1651. https://doi.org/10.1038/s41556-024-01513-9. Epub 2024 Oct 2. PMID: 39358449.

Narine, M, & Colognato, H (2022). Current insights into oligodendrocyte metabolism and its power to sculpt the myelin landscape. *Frontiers in Cellular Neuroscience, 16*, 892968. https://doi.org/10.3389/fncel.2022.892968.

Navarro, E., & Montesinos, J. (2025). Mitochondria-associated endoplasmic reticulum membranes in microglia: One contact site to rule them all. *Contact (Geneva, Switzerland), 8*, 1–16.

Needs, H. I., Wilkinson, K. A., Henley, J. M., & Collinson, I. (2023). Aggregation-prone Tau impairs mitochondrial import, which affects organelle morphology and neuronal complexity. *Journal of Cell Science, 136*(13), jcs260993.pubmed.ncbi.nlm.nih+1.

Niewiadomska, G., Niewiadomski, W., Steczkowska, M., & Gasiorowska, A. (2021). Tau oligomers neurotoxicity. *Life (Basel), 11*(1), 28. https://doi.org/10.3390/life11010028. PMID: 33418848; PMCID: PMC7824853.

Nixon, R. A., & Rubinsztein, D. C. (2024). Mechanisms of autophagy–lysosome dysfunction in neurodegenerative diseases. *Nature Reviews. Molecular Cell Biology, 25*, 223–245.

Palmqvist, S., Stomrud, E., Cullen, N., Janelidze, S., Manuilova, E., Jethwa, A., ... Hansson, O. (2023). An accurate fully automated panel of plasma biomarkers for Alzheimer's disease. *Alzheimer's & Dementia, 19*, 1204–1215.

Paradies, G., Paradies, V., Ruggiero, F. M., & Petrosillo, G. (2019). Role of cardiolipin in mitochondrial function and dynamics in health and diseases. *Cells, 8*(7), 728.

Pérez, M. J., Ibarra-García-Padilla, R., Tang, M., Porter, G. A., Jr., Johnson, G. V. W., & Quintanilla, R. A. (2024). Caspase-3 cleaved tau impairs mitochondrial function through the opening of the mitochondrial permeability transition pore. *Biochimica et Biophysica Acta - Molecular Basis of Disease, 1870*(1), 166898.

Pérez, M. J., Vergara-Pulgar, K., Jara, C., Cabezas-Opazo, F., & Quintanilla, R. A. (2018). Caspase-cleaved tau impairs mitochondrial dynamics in Alzheimer's disease. *Molecular Neurobiology, 55*, 1–15.

Petrosillo, G., Matera, M., Casanova, G., Ruggiero, F. M., & Paradies, G. (2008). Mitochondrial dysfunction in rat brain with aging: Involvement of complex I, reactive oxygen species and cardiolipin. *Neurochemistry International, 53*(5), 126–131.

Petrucelli, L., Dickson, D., Kehoe, K., Taylor, J., Snyder, H., Grover, A., ... Hutton, M. (2004). CHIP and Hsp70 regulate tau ubiquitination, degradation and aggregation. *Human Molecular Genetics, 13*(7), 703–714. https://doi.org/10.1093/hmg/ddh083. Epub 2004 Feb 12. PMID: 14962978.

Puff, N., D'Avanzo, N., & Khalifat, N. (2021). Mitochondrial cristae: A model of the relationship between structure and function. *Membranes, 11*(7), 465.

Qin, R., Zhao, H., Gao, H., & Liu, H. (2025). Global trends in Alzheimer's disease and other dementias: A comprehensive analysis. *PLoS One, 20*(12), e0338018.

Quintanilla, R. A., Matthews-Roberson, T. A., Dolan, P. J., & Johnson, G. V. (2009). Caspase-cleaved tau expression induces mitochondrial dysfunction in immortalized cortical neurons. *Journal of Biological Chemistry, 284*, 18754–18766.

Rangaraju, V., Calloway, N., & Ryan, T. A. (2014). Activity-driven local ATP synthesis is required for synaptic function. *Cell, 156*(4), 825–835.

Rangaraju, V., Lauterbach, M., & Schuman, E. M. (2019). Spatially stable mitochondrial compartments fuel local translation during plasticity. *Cell, 176*(1–2), 73–84.e15.

Reddy, P. H. (2011). Abnormal tau, mitochondrial dysfunction, impaired axonal transport of mitochondria, and synaptic deprivation in Alzheimer's disease. *Brain Research, 1415*, 136–148. https://doi.org/10.1016/j.brainres.2011.07.052. Epub2011 Jul 31. PMID: 21872849; PMCID: PMC3176990.

Reiman, E. M., Chen, K., Alexander, G. E., Caselli, R. J., Bandy, D., Osborne, D., ... Hardy, J. (2004). Functional brain abnormalities in young adults at genetic risk for late-onset Alzheimer's dementia. *Proceedings of the National Academy of Sciences, 101*(1), 284–289.

Rhein, V., Song, X., Wiesner, A., Ittner, L. M., Baysang, G., Meier, F., ... Eckert, A. (2009). Amyloid-β and tau synergistically impair the oxidative phosphorylation system in triple transgenic Alzheimer's disease mice. *Proceedings of the National Academy of Sciences, 106*, 20057–20062.

Rieger, B., Arroum, T., Borowski, M. T., Villalta, J., & Busch, K. B. (2021). Mitochondrial F1FO ATP synthase determines the local proton motive force at cristae rims. *EMBO Reports, 22*, e52727.

Rissman, R. A., Poon, W. W., Blurton-Jones, M., Oddo, S., Torp, R., Vitek, M. P., ... Cotman, C. W. (2004). Caspase-cleavage of tau is an early event in Alzheimer disease tangle pathology. *Journal of Clinical Investigation, 114*(1), 121–130.

Rossi, M. J., & Pekkurnaz, G. (2019 Aug). Powerhouse of the mind: mitochondrial plasticity at the synapse. *Current Opinion in Neurobiology, 57*, 149–155. https://doi.org/10.1016/j.conb.2019.02.001 Epub 2019 Mar 12. PMID: 30875521; PMCID: PMC6629504.

Santos, R. X., Correia, S. C., Wang, X., Perry, G., Smith, M. A., Moreira, P. I., & Zhu, X. (2012). Nuclear and mitochondrial DNA oxidation in Alzheimer's disease. *Free Radical Research, 46*, 565–576.

Scarmeas, N., Luchsinger, J. A., Schupf, N., Brickman, A. M., Cosentino, S., Tang, M. X., & Stern, Y. (2009). Physical activity, diet, and risk of Alzheimer disease. *JAMA, 302*(6), 627–637. https://doi.org/10.1001/jama.2009.1144. PMID: 19671904; PMCID: PMC2765045.

Schlame, M., & Ren, M. (2009). The role of cardiolipin in the structural organization of mitochondrial membranes. *Biochimica et Biophysica Acta (BBA)-Biomembranes, 1788*(10), 2080–2083.

Schoenfeld, R., Wong, A., Silva, J., Li, M., Itoh, A., Horiuchi, M., Itoh, T., Pleasure, D., & Cortopassi, G. (2010). Oligodendroglial differentiation induces mitochondrial genes and inhibition of mitochondrial function represses oligodendroglialdifferentiation. *Mitochondrion, 10*(2), 143–150. https://doi.org/10.1016/j.mito.2009.12.141. Epub 2009 Dec 22. PMID: 20005986; PMCID: PMC3119038.

Schweers, O., Mandelkow, E. M., Biernat, J., & Mandelkow, E. (1995). Oxidation of cysteine-322 in the repeat domain of microtubule-associated protein tau controls the in vitro assembly of paired helical filaments. *Proceedings of the National Academy of Sciences of the United States of America, 92*(18), 8463–8467. https://doi.org/10.1073/pnas.92.18.8463. PMID: 7667312; PMCID: PMC41177.

Seirafi, M., Kozlov, G., & Gehring, K. (2015). Parkin structure and function. *The FEBS Journal, 282*(11), 2076–2088. https://doi.org/10.1111/febs.13249. Epub 2015 Mar 16. PMID: 25712550; PMCID: PMC4672691.

Shoshan-Barmatz, V., Nahon-Crystal, E., Shteinfer-Kuzmine, A., & Gupta, R. (2018). VDAC1, mitochondrial dysfunction, and Alzheimer's disease. *Pharmacological Research, 131*, 87–101. https://doi.org/10.1016/j.phrs.2018.03.010. Epub 2018 Mar15. PMID: 29551631.

Sims, J. R., Zimmer, J. A., Evans, C. D., Lu, M., Ardayfio, P., Sparks, J. D., ... Skovronsky, D. M. (2023). Donanemab in early symptomatic Alzheimer disease. *JAMA: the Journal of the American Medical Association, 330*, 512–527.

Smilansky, A., Dangoor, L., Nakdimon, I., Ben-Hail, D., Mizrachi, D., & Shoshan-Barmatz, V. (2015). The voltage-dependent anion channel 1 mediates amyloid β toxicity and represents a potential target for Alzheimer disease therapy. *Journal of Biological Chemistry, 290*(52), 30670–30683.

Sperling, R. A., Aisen, P. S., Beckett, L. A., Bennett, D. A., Craft, S., Fagan, A. M., ... Phelps, C. H. (2011). Toward defining the preclinical stages of Alzheimer's disease. *Alzheimer's & Dementia, 7*, 280–292.

Strauss, M., Hofhaus, G., Schröder, R. R., & Kühlbrandt, W. (2008). Dimer ribbons of ATP synthase shape the inner mitochondrial membrane. *The EMBO Journal, 27*(7), 1154–1160.

Sultana, R., Perluigi, M., & Butterfield, D. A. (2013). Lipid peroxidation triggers neurodegeneration: A redox proteomics view into the Alzheimer disease brain. *Free Radical Biology and Medicine, 62*, 157–169.

Szeto, H. H. (2014). First-in-class cardiolipin-protective compound as a therapeutic agent to restore mitochondrial bioenergetics. *British Journal of Pharmacology, 171*, 2029–2050.

Tamagno, E., Guglielmotto, M., Vasciaveo, V., & Tabaton, M. (2021). Oxidative stress and beta amyloid in Alzheimer's disease. *Antioxidants, 10*, 1479.

Tapia-Monsalves, C., Olesen, M. A., Villavicencio-Tejo, F., Quintanilla, R. A., & Cyclosporine, A. (2023). CsA) prevents synaptic impairment caused by truncated tau by caspase-3. *Molecular and Cellular Neuroscience, 125*, 103861. https://doi.org/10.1016/j.mcn.2023.103861. Epub 2023 May 12. PMID: 37182572.

Taylor, M. K., Sullivan, D. K., Mahnken, J. D., Burns, J. M., & Swerdlow, R. H. (2018). Feasibility and efficacy data from a ketogenic diet intervention in Alzheimer's disease. *Alzheimer's & Dementia: Translational Research & Clinical Interventions, 4*, 28–36.

Tong, B. C., Wu, A. J., Huang, A. S., Bhaskaran, S., & Bhattacharya, A. (2022). Tetrandrine ameliorates tauopathy and memory impairment by rescuing TPC2-mediated lysosomal alkalinization. *Journal of Biomedical Science, 29*, 85.

Torres, A. K., Jara, C., Olesen, M. A., & Tapia-Rojas, C. (2021). Pathologically phosphorylated tau at S396/404 (PHF-1) is accumulated inside of hippocampal synaptic mitochondria of aged wild-type mice. *Scientific Reports, 11*, 4448.

Torres, A. K., Rivera, B. I., Polanco, C. M., Jara, C., & Tapia-Rojas, C. (2022). Phosphorylated tau as a toxic agent in synaptic mitochondria: Implications in aging and Alzheimer's disease. *Neural Regeneration Research, 17*(8), 1645–1651.

Tracy, T. E., Madero-Pérez, J., Swaney, D. L., Chang, T. S., Moritz, M., Konrad, C., ... Gan, L. (2022). Tau interactome maps synaptic and mitochondrial processes associated with neurodegeneration. *Cell, 185*(4), 712–728 e14.

Trammell, S. A. J., Schmidt, M. S., Weidemann, B. J., Redpath, P., Jaksch, F., Dellinger, R. W., ... Brenner, C. (2016). Nicotinamide riboside is uniquely and orally bioavailable in mice and humans. *Nature Communications, 7*, 12948.

Underwood, E. L., Redell, J. B., Hood, K. N., Maynard, M. E., Hylin, M., Waxham, M. N., ... Dash, P. K. (2020). Enhanced presynaptic mitochondrial energy production is required for memory formation. *Scientific Reports, 10*, 3628.

van Dyck, C. H., Swanson, C. J., Aisen, P., Bateman, R. J., Chen, C., Gee, M., ... Bhatt, N. (2023). Lecanemab in early Alzheimer's disease. *New England Journal of Medicine, 388*, 9–21.

Vijayan, M., Alvir, R. V., Bunquin, L. E., Pradeepkiran, J. A., & Reddy, P. H. (2022). A partial reduction of VDAC1 enhances mitophagy, autophagy, synaptic activities in a transgenic Tau mouse model. *Aging Cell, 21*(8), e13663.

Wang, Y., Feng, B., Hong, R., Qiu, D., Zhao, J., & Zhao, L. (2026). Glial lactate metabolism and transport in Alzheimer's disease. *Frontiers in Aging Neuroscience, 18*, 1731089. https://doi.org/10.3389/fnagi.2026.1731089. PMID: 41930368; PMCID: PMC13038885.

Wang, X., & Schwarz, T. L. (2009). The mechanism of Ca2+-dependent regulation of kinesin-mediated mitochondrial motility. *Cell, 136*(1), 163–174.

West, A. P., Khoury-Hanold, W., Staron, M., Tal, M. C., Pineda, C. M., Lang, S. M., ... Bhaskaran, S. (2015). Mitochondrial DNA stress primes the antiviral innate immune response. *Nature, 520*, 553–557.

Wischik, C. M., Bentham, P., Gauthier, S., Miller, S., Kook, K., Schelter, B. O., ... Murray, A. D., et al. (2019). Concentration-dependent activity of hydromethylthionine on cognitive decline and brain atrophy in mild to moderate Alzheimer's disease. *Journal of Alzheimer's Disease, 72*(3), 931–944.

Wojsiat, J., Zoltowska, K. M., Laskowska-Kaszub, K., & Wojda, U. (2018). Oxidant/antioxidant imbalance in Alzheimer's disease: Therapeutic and diagnostic prospects. *Oxidative Medicine and Cellular Longevity, 2018*, 6435861.

Wu, J., Hao, S., Sun, X.-R., Zhang, H., Li, H., Zhao, H., ... Li, K. (2017). Elamipretide (SS-31) ameliorates isoflurane-induced long-term impairments of mitochondrial morphogenesis and cognition in developing rats. *Frontiers in Cellular Neuroscience, 11*, 119. https://doi.org/10.3389/fncel.2017.00119.

Yang, Y., Jia, X., Yang, X., Wang, J., Fang, Y., Ying, X., Zhang, M., Wei, J., & Pan, Y. (2024). Targeting VDAC: A potential therapeutic approach for mitochondrial dysfunction in Alzheimer's disease. *Brain Research, 1835*, 148920. https://doi.org/10.1016/j.brainres.2024.148920. Epub 2024 Apr 8. PMID: 38599511.

Yao, J., Irwin, R. W., Zhao, L., Nilsen, J., Hamilton, R. T., & Brinton, R. D. (2009). Mitochondrial bioenergetic deficit precedes Alzheimer's pathology in female mouse model of Alzheimer's disease. *Proceedings of the National Academy of Sciences of the United States of America, 106*(34), 14670–14675. https://doi.org/10.1073/pnas.0903563106. Epub 2009 Aug 10. PMID: 19667196; PMCID: PMC2732886.

Yuzwa, S. A., Shan, X., Macauley, M. S., Clark, T., Skorobogatko, Y., Vosseller, K., & Vocadlo, D. J. (2012). Increasing O-GlcNAc slows neurodegeneration and stabilizes tau against aggregation. *Nature Chemical Biology, 8*(4), 393–399. https://doi.org/10.1038/nchembio.797. PMID: 22366723.

Zhang, Y. M., Wang, Y. T., Wei, R. M., Li, X. Y., Luo, B. L., Zhang, J. Y., ... Chen, G. H. (2024). Mitochondrial antioxidant elamipretide improves learning and memory impairment induced by chronic sleep deprivation inmice. *Brain and Behavior, 14*(5), e3508. https://doi.org/10.1002/brb3.3508. PMID: 38688894; PMCID: PMC11061203.

CHAPTER ELEVEN

Brain lipid dysregulation as a driver of energy metabolism failure in Alzheimer's disease

Nitin Yadav[a,*], Jyotirmoy Banerjee[b], Manisha Tiwari[a], and Aparna Dixit[a,*]

[a]Dr. B.R. Ambedkar Center for Biomedical Research, University of Delhi, Delhi, India
[b]Department of Biophysics, All India Institute of Medical Sciences, Delhi, India
*Corresponding author. e-mail address: nitinyadav.hrc@gmail.com; aparnabanerjeedixit@gmail.com

Contents

International Review of Neurobiology, Volume 188
ISSN 0074-7742, https://doi.org/10.1016/bs.irn.2026.02.004

Abstract

Alzheimer's Disease (AD) is becoming more widely recognized as a condition of brain energy metabolism, wherein lipid dysregulation plays a crucial, although unrecognized, role. In addition to functioning as structural elements of neuronal membranes, lipids such as phospholipids, sphingolipids, and cholesterol play a crucial role in regulating mitochondrial bioenergetics, synaptic activity, and membrane-associated signaling pathways. In AD, modifications in lipid composition, distribution, and turnover compromise membrane fluidity, disturb mitochondrial dynamics, and obstruct lipid-mediated transport of energy substrates. These alterations intensify oxidative stress, impair glucose and ketone utilization, and stimulate neuroinflammatory pathways that further diminish metabolic capacity. Progress in lipidomics has uncovered disease-specific lipid signatures, providing a fresh understanding of the relationship between lipid homeostasis and neuronal energy loss. Despite extensive focus on amyloid and tau, lipid-mediated bioenergetic failure remains underrepresented in integrative AD models; this chapter addresses this gap and consolidates existing information connecting lipid modifications to metabolic dysfunction in AD, emphasizing molecular pathways and prospective treatment strategies aimed at lipid metabolism to re-establish bioenergetic equilibrium.

1. Introduction

Alzheimer's Disease (AD) is the most common type of dementia, affecting about 55 million people around the world. By 2050, the number of people with AD is expected to triple (Nichols et al., 2022). Historically, investigations have concentrated on the "amyloid cascade hypothesis" and the accumulation of neurofibrillary tangles resulting from hyperphosphorylated tau protein as the primary attributes of the disease (Nasb et al., 2024). While these proteinopathies are essential for diagnosis and pathophysiology, they do not fully clarify the complexities of AD. Clinical trials focused on the clearance of amyloid-β (Aβ) have consistently failed to produce substantial cognitive improvements, while tau-targeted therapies are still in the early stages of development (López-García et al., 2025; Viola & Klein, 2015). These disappointments have led to the understanding that AD is not merely a protein-centric disorder, but rather one arising from multifactorial disruptions in cellular metabolism, synaptic networks, and systemic physiology (Pasqualetti et al., 2022; Yin, 2023). Energy metabolism failure has become a consistent and early aberration in AD among these factors. Neuroimaging studies suggest that hypometabolism, marked by reduced glucose uptake, may transpire decades before the manifestation of clinical symptoms. This suggests that bioenergetic deficiencies are not a result of amyloid and tau accumulation but may act as a primary catalyst for neuronal susceptibility (Raut et al., 2023; Wang et al., 2020; Zilberter & Zilberter, 2017).

The brain makes up about 2 % of the body's total weight and uses about 20 % of the energy it has when it is not working. Most of this energy is needed to help synapses work, keep ion gradients stable, and send signals between neurons. Glucose is the main source of energy, and it is broken down into smaller pieces by glycolysis and oxidative phosphorylation in neuronal mitochondria. In the absence of glucose, ketone bodies (β-hydroxybutyrate and acetoacetate) may function as partial alternative fuels (Bonvento & Bolaños, 2021; Goyal et al., 2014). The brain is not just an organ that uses glucose; it is also one of the body's tissues with the most lipids. More than half of its dry weight is made up of lipids, which do more than just provide structural support. They actively participate in mitochondrial function, synaptic vesicle dynamics, receptor signaling, and the transport of energy substrates. Therefore, it is necessary to examine brain energy metabolism in conjunction with lipid biology (Bruce et al., 2017; Hussain et al., 2019).

1.1 Lipids as metabolic regulators

Phospholipids, sphingolipids, sterols (mostly cholesterol), glycerolipids, and Polyunsaturated fatty acids (PUFAs) are all types of lipids that are found in the brain. Each class has its own structural and functional contributions. Phospholipids maintain membrane fluidity and are crucial for the structural integrity of mitochondrial cristae. Sphingolipids serve as signaling molecules that regulate apoptosis and autophagy, while gangliosides influence synaptic plasticity. Cholesterol maintains lipid raft domains, promoting synaptic receptor clustering and neurotransmission. PUFAs, such as docosahexaenoic acid (DHA), are essential for synaptic function and anti-inflammatory signaling (Farooqui & Farooqui, 2024; Hussain et al., 2019). Changing these lipids affects both cell membranes and how well bioenergetic activities work, like taking in glucose and making ATP. Lipid dysregulation does not occur in isolation; it directly disrupts energy metabolism (Ademowo et al., 2017; Tian et al., 2023; Tracey et al., 2018).

Historically, lipid dysregulation in AD was identified through postmortem studies that demonstrated reduced PUFA levels and altered membrane phospholipids. Lipidomics, an advanced branch of metabolomics focused on extensive lipid profiling, has recently identified disease-specific lipid signatures in the blood, cerebrospinal fluid (CSF), and brain tissue of

AD patients (Akyol et al., 2021; He et al., 2025; Yin, 2023). These encompass the diminution of plasmalogens (ether phospholipids vital for oxidative defense) (Senanayake & Goodenowe, 2019). Elevated ceramide concentrations correlate with disease advancement (Mielke et al., 2012). Altered cholesterol esters and compromised cholesterol transport, particularly in individuals with the APOE4 genotype and reduced levels of DHA-containing phosphatidylcholines, are associated with cognitive decline (Raulin et al., 2022). These findings confirm that lipid abnormalities are not merely incidental but essential factors in the pathogenesis of AD.

1.2 Lipid dysregulation and mitochondrial dysfunction

Because neuronal ATP generation is overwhelmingly mitochondrial, lipid-mediated mitochondrial instability represents a central axis of metabolic failure in AD. A key feature of AD is mitochondrial dysfunction, which is shown by less ATP production, more reactive oxygen species (ROS) formation, and a broken fission-fusion balance. Lipids are necessary for keeping mitochondria healthy. For example, Cardiolipin, a phospholipid that is only found in mitochondria, is very important for making super-complexes of the electron transport chain (ETC). Loss of cardiolipin makes mitochondrial respiration less stable (Brothwell et al., 2025; Paradies et al., 2014, 2019; Bao et al., 2024). PUFAs in mitochondrial membranes control how fluid, permeable, and able to transport substrates they are (Morita et al., 2025). As a result, changes in lipid metabolism directly affect mitochondrial bioenergetics, which is what causes the hypometabolism seen in the brains of people with AD. Fig. 1 summarizes how lipid dys-regulation disrupts neuronal–astrocytic metabolic coupling at the synapse.

As discussed in later sections, lipid-derived inflammatory mediators further amplify mitochondrial and synaptic dysfunction. Inflammation is a defining feature of AD, with lipids being a key component of this process. Arachidonic acid, derived from membrane phospholipids, acts as the precursor for pro-inflammatory eicosanoids, such as prostaglandins and leukotrienes. Excessive production of these lipid mediators activates microglia and astrocytes, perpetuating a cycle of inflammation and oxidative stress. The concurrent depletion of pro-resolving lipid mediators, including resolvins and protectins derived from DHA, deprives the brain of critical anti-inflammatory signals (Biernacki & Skrzydlewska, 2025; Chiurchiù et al., 2018; Ortiz-Placín et al., 2023; Yui et al., n.d.). This dual role shows how lipid dysregulation causes both energy failure and neuroinflammatory cascades to get worse at the same time.

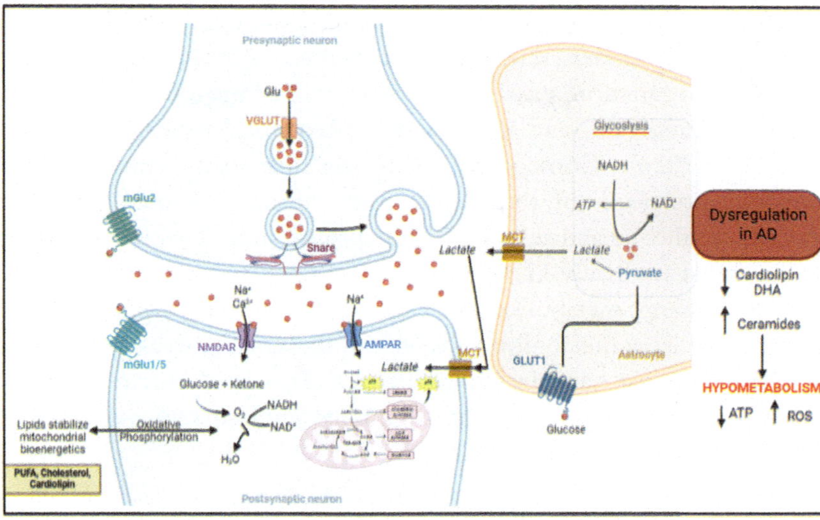

Fig. 1 Lipid–energy metabolism coupling at the tripartite synapse in AD. Schematic showing neuronal–astrocytic metabolic coupling supporting synaptic transmission, where astrocyte-derived lactate fuels neuronal mitochondrial ATP production. Lipids (PUFAs/DHA, cholesterol, cardiolipin) stabilize mitochondrial bioenergetics, while their dysregulation marked by reduced cardiolipin and DHA and increased ceramides leads to impaired oxidative phosphorylation, elevated ROS, and synaptic hypometabolism.

1.3 APOE4 and lipid metabolism

Apolipoprotein E (APOE) is the main lipid transporter in the brain. The APOE4 allele is the most important genetic risk factor for sporadic AD. APOE4 interferes with the movement of cholesterol and phospholipids between astrocytes and neurons, making synapses more likely to break down and mitochondria less likely to work properly. Compared to APOE3, APOE4 carriers exhibit diminished lipid recycling and lipid raft stability, leading to metabolic decline and amyloid buildup (Belaidi et al., 2025; Troutwine et al., 2022; Yang et al., 2023). This association underscores the complex relationship between genetic predisposition to AD and lipid regulation proficiency.

1.4 Shifting the paradigm: AD as a lipid-metabolic disorder

These findings collectively advocate for a paradigm shift in AD research, suggesting that the condition should be regarded not only as a proteinopathy but also as a disorder of lipid and energy metabolism. Lipid alterations may precede and intensify the detrimental effects of amyloid and tau, rather than being mere subsequent consequences. Lipid dysregulation indeed hinders the utilization of glucose and ketones, disrupts mitochondrial function, exacerbates oxidative and inflammatory damage, alters Aβ aggregation, and modifies tau phosphorylation (Golde, 2016; Liu et al., 2024). Consequently, investigating lipid pathways offers innovative prospects for therapeutic intervention, potentially restoring metabolic resilience and slowing disease progression. This chapter brings together what is already known about how lipid dysregulation and problems with energy metabolism work together in AD. By emphasizing the interaction of lipid metabolism, we seek to illuminate novel pathways for diagnosis, prognosis, and treatment in AD. This lipid-metabolic framework offers a unifying explanation for why amyloid removal alone has failed to restore cognitive function.

2. Brain lipid composition and energetic functions

Lipids are important for controlling the structure of membranes, signaling, synaptic activity, and mitochondrial energy metabolism (Bruce et al., 2017). The dysregulation of these systems directly contributes to the metabolic impairments observed in AD.

2.1 Phospholipids

Phospholipids are the main type of lipid in the brain. They are arranged in bilayers that make up the membranes of both neurons and mitochondria. The main subclasses are phosphatidylcholine (PC), phosphatidylethanolamine (PE), phosphatidylserine (PS), phosphatidylinositol (PI), and cardiolipin (CL) (Choudhary et al., 2024). About 30–35 % of the phospholipids in the brain are PC. It keeps the membrane intact, changes the shape, and gives choline for making acetylcholine. Reductions in PC-linked DHA are consistently associated with cognitive decline in AD patients (Huang et al., 2025). PE influences membrane curvature, a critical factor for synaptic vesicle budding and fusion. The PC:PE ratio checks the health of the mitochondrial membrane. A higher ratio, like the one seen in AD, messes up how mitochondria move (Calzada et al., 2016). PS, which is usually found in high amounts in the inner leaflet of membranes, shows that PS is being released from cells, which means that apoptosis is happening. Reduced PS levels have been noted in the brains of individuals with AD, possibly impairing synaptic communication (Kim et al., 2014). PI is a precursor to phosphoinositides, which control how cells send signals and move glucose transporters around. Altered phosphoinositide metabolism disrupts insulin signaling, worsening the traits of "type 3 diabetes" linked to AD (López-Gambero et al., 2020).

Phospholipids not only offer structural support but also directly regulate mitochondrial bioenergetics and neuronal metabolism.

2.2 Sphingolipids: Regulators of life and death

Sphingomyelin, ceramides, sphingosine, and complex gangliosides are all types of sphingolipids. Their combined roles in providing structural support and sending signals make them necessary for neurons to stay alive. Sphingomyelin, which is found a lot in myelin sheaths, keeps axonal conduction going. Alterations in sphingomyelin concentrations affect the localization of synaptic receptors within lipid rafts (Olsen & Færgeman, 2017). People with AD have a lot of ceramides in their brains. These are often called "lipid second messengers of stress." These effects converge on mitochondrial dysfunction, as detailed in Section 3. Elevated plasma ceramides predict cognitive decline and the progression from mild cognitive impairment (MCI) to AD (Pan et al., 2024). Sphingosine-1-phosphate (S1P) regulates neuronal survival, angiogenesis, and inflammatory responses. Lower levels of S1P in AD are linked to worse cognitive performance (Bigi et al., 2023).

Gangliosides (GM1, GD1a, GT1b) are glycosphingolipids that regulate synaptic plasticity and neurotrophic signaling. In AD, GM1 is diminished, compromising neuroprotection, while abnormal ganglioside-Aβ interactions facilitate aggregation into harmful oligomers (Cebecauer et al., 2017; Guo, 2023). The sphingolipid pathway demonstrates how lipid imbalances concurrently induce metabolic stress, apoptosis, and amyloid disease.

2.3 Cholesterol and sterol metabolism

Cholesterol makes up about 25 % of the lipids in the brain, and it is not evenly spread out. It is plentiful in myelin and synaptic membranes; however, it is carefully controlled because neurons cannot rely on peripheral cholesterol due to the blood-brain barrier. Astrocytes produce cholesterol and transport it to neurons via apolipoproteins, particularly APOE (Orth & Bellosta, 2012). Cholesterol strengthens lipid rafts, which are small areas that are important for receptor aggregation and neurotransmitter communication. Cholesterol dysregulation alters raft composition, affecting NMDA receptor and insulin receptor signaling

(Warda et al., 2025). APOE3 redistributes cholesterol well, but APOE4 is structurally unstable, which makes lipid transport less effective, synaptogenesis slower, and amyloidogenic processing of APP faster (Hauser et al., 2011; Raulin et al., 2022). Cholesterol metabolites, such as 24S-hydroxycholesterol, regulate cholesterol turnover. Their dysregulation in AD correlates with neuronal degeneration and mitochondrial distress. So, cholesterol is not just a building block; it also helps with signaling and moving things around.

2.4 PUFAs

PUFAs, particularly omega-3 (DHA, EPA) and omega-6 (arachidonic acid), are highly concentrated in neuronal membranes. Their flexible chains keep membranes fluid, which changes how proteins move and how receptors work. DHA makes up about 30 % of the phospholipid acyl chains in synaptic membranes. It helps synaptogenesis happen, makes long-term potentiation (LTP) easier, and makes anti-inflammatory mediators like resolvins and protectins. DHA deficiency in AD is consistently linked to memory decline. Eicosanoids derived from arachidonic acid (ARA) exacerbate neuroinflammation. In AD, too much activation of phospholipase A2 leads to more arachidonic acid being released, which makes inflammatory pathways worse. A changed ratio of ω-3 to ω-6 changes metabolism in a way that makes pro-inflammatory signals more likely to happen. Modern diets lacking in ω-3 fatty acids may exacerbate this imbalance, increasing the risk of AD. PUFAs serve as metabolic regulators, influencing inflammation, neural plasticity, and mitochondrial respiration (Gu et al., 2024; Tian et al., 2025).

2.5 Glycerolipids and triglycerides

Neurons seldom employ triglycerides for energy storage; however, glycerolipids function in signaling and serve as intermediates in phospholipid synthesis. Recent findings suggest altered glycerolipid metabolism in AD brains, possibly indicating impaired phospholipid turnover. Higher plasma triglycerides are linked to more amyloid buildup, which means that lipid metabolism in the body plays a role in brain disease (Watanabe et al., 2021).

Phospholipids and cholesterol control the fluidity of the membrane, which in turn controls where GLUT-1/3 transporters are located. Disruption of lipid rafts undermines transporter stability. Monocarboxylate transporters (MCTs) that are built into membranes depend on the lipid environment. When PUFAs are low, membranes become stiff, which makes it harder for medium-chain triglycerides to move around. Cardiolipin protects metabolic enzymes like the pyruvate dehydrogenase complex, which links lipid integrity to glucose oxidation directly. Lipid homeostasis is necessary for neurons to use glucose and ketones in the best way possible.

3. Lipid–mitochondria crosstalk in AD

Mitochondria are necessary for neurons to survive and function because the brain needs a lot of energy. These organelles mainly make ATP through oxidative phosphorylation (OXPHOS), help with calcium buffering, control apoptosis, and make metabolic intermediates that are necessary for making neurotransmitters. Still, lipids are a big part of how they work. They not only give mitochondrial membranes structural integrity, but they also take part in metabolic and signaling pathways. In AD, disturbances in brain lipid metabolism destabilize mitochondrial structure and impede energy production, thereby enhancing neuronal vulnerability. The intimate interplay between lipids and mitochondria represents a crucial pathogenic axis linking bioenergetic dysfunction to neurodegeneration (Clemente-Suárez et al., 2023).

Mitochondria are different from other organelles because they need certain lipids. The outer mitochondrial membrane (OMM) has a lot of PC, PE, and cholesterol, while the inner mitochondrial membrane (IMM) has a lot of cardiolipin and less cholesterol. This distribution ensures the right curvature of the membrane, the anchoring of proteins, and the efficiency of the ETC. Lipidomic studies in AD reveal significant alterations in lipid pools, particularly CL, PE, and sphingolipids, profoundly impacting mitochondrial bioenergetics. Cardiolipin oxidation by ROS destabilizes respiratory chain supercomplexes, reduces electron flow, and increases electron leakage, thereby sustaining a harmful cycle of oxidative stress (Falabella et al., 2021).

A key early sign of AD is a change from using mitochondrial respiration to using glycolysis for energy. This bioenergetic reprogramming goes along with changes to lipids that make it harder to use substrates. Lower levels of PUFAs, especially DHA, make membranes less fluid, which makes it harder for glucose and ketone substances to get through. At the same time, cholesterol buildup in mitochondrial membranes changes how permeable they are and messes up the structure of the cristae, which is important for oxidative phosphorylation. These changes make it harder for cells to make ATP and make neurons more sensitive to metabolic stress (Grimm et al., 2013).

Cardiolipin dysfunction is a pivotal factor in lipid-mitochondria interaction in AD. Under normal circumstances, cardiolipin attaches cytochrome c to the inner mitochondrial membrane, which helps electrons move through the electron transport chain. When cardiolipin is oxidized or depleted, it releases cytochrome c into the space between membranes, which starts the process of apoptosis. This phenomenon is well-documented in both experimental AD models and human brain tissue, emphasizing cardiolipin's dual role in bioenergetics and the regulation of apoptosis (Monteiro-Cardoso et al., 2014). Moreover, cardiolipin interacts with amyloid-β (Aβ), and its oxidation facilitates Aβ aggregation at mitochondrial membranes, exacerbating mitochondrial toxicity (Fuentes & Morcillo, 2024).

Ceramides make mitochondrial function even worse in AD. Higher levels of ceramide make the OMM more permeable by forming channels that help release apoptogenic proteins, such as apoptosis-inducing factor (AIF). Furthermore, ceramides interfere with mitochondrial dynamics, promoting excessive fission, a hallmark of neuronal pathology in AD. This imbalance causes synaptic loss by dysregulating the transport of mitochondria along axons and the supply of energy to synaptic terminals (Jazvinšćak Jembrek et al., 2015).

Cholesterol dysregulation is equally detrimental. Cholesterol is essential for maintaining the stability of OMM; however, its excessive accumulation leads to mitochondrial rigidity and interferes with protein import through the translocase of the outer membrane (TOM) complex. In AD, carriers of APOE4 demonstrate heightened mitochondrial cholesterol accumulation, correlating with compromised ETC function and increased oxidative damage. This method shows a reliable molecular link between genetic risk factors and mitochondrial dysfunction in AD (Goicoechea et al., 2023).

PUFAs, particularly DHA, are crucial for protecting mitochondrial membranes from oxidative stress. Phospholipids containing DHA reduce lipid peroxidation and enhance anti-inflammatory signaling. In the brains of individuals with AD, diminished DHA levels are associated with elevated lipid peroxidation markers, suggesting that insufficient PUFA protection aggravates mitochondrial oxidative injury. Conversely, DHA supplementation has shown efficacy in reinstating mitochondrial function in animal models by improving membrane fluidity and augmenting antioxidant defenses (Ng et al., 2005).

The mitochondria-associated membranes (MAMs) are specialized contact points between the endoplasmic reticulum and mitochondria. They are a key place where lipids and mitochondria interact, and this interaction fails to operate properly in AD. MAMs control the movement of lipids, calcium signaling, and the bioenergetics of mitochondria. The pathogenesis of AD is associated with increased mitochondrial-associated membrane activity, driven by the accumulation of presenilin and amyloid-beta, leading to impaired lipid exchange, excessive ceramide synthesis, and calcium overload within mitochondria. The hyperactive MAM phenotype exacerbates mitochondrial fragmentation and promotes neuronal apoptosis (Yu et al., 2021).

Experimental studies on transgenic AD mouse models and human postmortem brain analyses indicate that persistent lipid dysregulation occurs prior to significant mitochondrial dysfunction and neuronal degeneration. Hippocampal lipidomic profiling in 3xTg-AD mice revealed a reduction of cardiolipin and an increase of ceramide prior to cognitive decline, suggesting a potential causal role in disease progression (Kao et al., 2020). Clinical studies support this assertion, demonstrating that alterations in plasma lipid profiles (e.g., elevated ceramides, reduced DHA) correlate with mitochondrial dysfunction and cognitive deterioration in AD patients (Fig. 2).

The functional consequences of lipid-mitochondria disruption extend beyond typical energy metabolism. When the ETC becomes unstable, it makes too much mitochondrial ROS, which causes a lot of oxidative damage to DNA, proteins, and lipids, making metabolic problems worse. At the synapse, less ATP makes it harder for vesicles to recycle and neurotransmitters to move, which causes the first cognitive problems. Additionally, mitochondrial apoptotic pathways activated by lipid imbalances accelerate neuronal degeneration in vulnerable regions of the brain, such as the hippocampus and cortex (Misrani et al., 2021).

The interplay between lipids and mitochondria is a significant mechanism in the pathophysiology of AD. Problems with cardiolipin integrity, ceramide buildup, cholesterol balance, and PUFA depletion all hurt the bioenergetics and structural integrity of mitochondria. These disturbances not only threaten neuronal viability but also intensify amyloid and tau pathology, establishing a self-perpetuating pathogenic cycle. The acknowledgement of lipid–mitochondria interactions as a driver of AD progression provides a robust foundation for treatment strategies focused on lipid metabolism to preserve mitochondrial function.

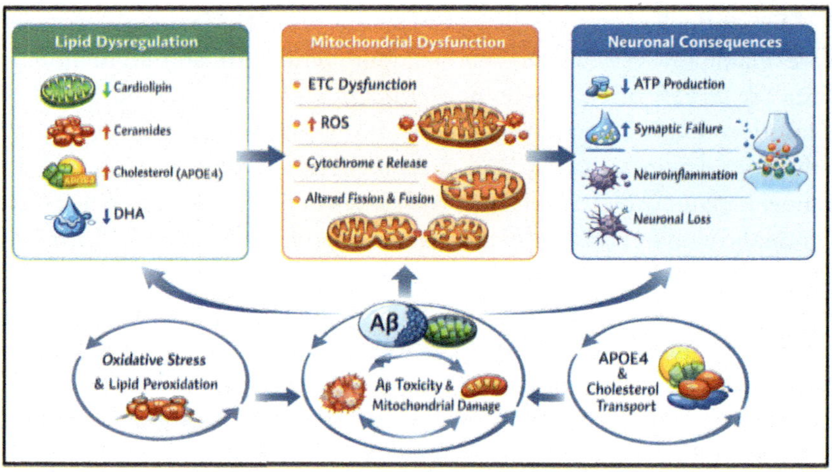

Fig. 2 Dysregulation of key lipid classes including cardiolipin, ceramides, cholesterol, and PUFA destabilizes mitochondrial membranes, impairs ETC, and increases oxidative stress. These mitochondrial defects reduce ATP availability, compromise synaptic transmission and plasticity, and amplify neuroinflammatory signaling, creating a self-perpetuating pathogenic cycle that drives cognitive decline in AD.

4. Lipidomic profiling in AD: Signatures of metabolic decline

The emergence of lipidomics as a powerful systems biology technique has altered our understanding of the pathogenesis of AD. It allows for comprehensive profiling of lipid classes, species, and their dynamics within the brain and peripheral systems, unlike classical biochemistry, which often examines individual metabolites in isolation. Lipidomic research has shown that disruptions in lipid homeostasis in AD precede, coexist with, and exacerbate metabolic dysfunction, rather than being mere epiphenomena of neurodegeneration. This section looks at the lipidomic evidence for AD, focusing on how lipid signatures change in plasma, CSF, brain tissue, and other biofluids.

The lipidome is naturally diverse, comprising glycerophospholipids, sphingolipids, sterols, triglycerides, and PUFAs, each fulfilling distinct structural functions and bioenergetic roles. Lipidomics in AD research provides crucial insights into the correlation between alterations in lipid composition and distribution and their impact on mitochondrial bioenergetics, membrane fluidity, and synaptic plasticity. Lipidomics reveals disease-specific "fingerprints" that may serve as biomarkers, enabling early detection and patient stratification (Köfeler et al., 2012).

One of the first uses of lipidomics in AD showed that PC and PE levels were much lower in cortical tissues. This was linked to problems with synapses. Recent high-throughput lipidomic analyses indicate that sphingolipids and ceramides are consistently elevated in AD, signifying the activation of neuroinflammatory and apoptotic pathways. These data collectively demonstrate that lipidomic alterations are fundamental to the pathophysiology of AD, rather than being mere secondary effects (Chiurchiù et al., 2022; Hachem et al., 2024).

4.1 Brain lipidomic alterations in AD

Post-mortem analyses of brains affected by AD have disclosed considerable disturbances in multiple lipid categories. These changes often happen in places where there are a lot of amyloid plaques and neurofibrillary tangles, like the hippocampus, temporal cortex, and posterior cingulate cortex (Kumar et al., 2021). PC and PE, two major phospholipids in neuronal membranes, are consistently reduced in the brains of individuals with AD. Notably, reductions in PC and PE correlate with mitochondrial dysfunction, as they play a crucial role in maintaining cristae architecture and energy synthesis (Blusztajn & Slack, 2023).

Elevated ceramide levels represent one of the most consistent lipidomic changes observed in AD. A number of studies have shown that higher levels of ceramides are linked to higher levels of amyloid-beta (Aβ). This suggests that ceramides increase Aβ production and Aβ activates sphingomyelinase activity (Novgorodov et al., 2008).

AD messes up how cholesterol is broken down in the body. High cholesterol levels in neuronal membranes change the composition of lipid rafts, which speeds up the amyloidogenic processing of APP. Oxysterols, oxygenated derivatives of cholesterol such as 24S-hydroxycholesterol, are elevated in the early stages of AD and may signify increased cholesterol turnover due to neurodegeneration (Di Paolo & Kim, 2011). Individuals with AD exhibit a substantial reduction in DHA, an omega-3 fatty acid crucial for synaptic plasticity. This depletion diminishes synaptic signaling and increases susceptibility to lipid peroxidation. Arachidonic acid (AA), an omega-6 polyunsaturated fatty acid with pro-inflammatory properties, is often elevated, signifying the activation of neuroinflammatory pathways (Thomas et al., 2015).

4.2 Peripheral lipidomic signatures: CSF, plasma, and serum

Since obtaining samples from brain tissue is so invasive, a lot of work has gone into finding lipid biomarkers in biofluids that are easier to access.

Lipidomic analysis of cerebrospinal fluid has shown that PC, PE, and sphingomyelin levels have gone down, while ceramides and lysophosphatidylcholine (LPC) levels have gone up. LPC, which is made by phospholipase A2, acts as a pro-inflammatory mediator and a lipid transporter. Elevated concentrations of ceramides in cerebrospinal fluid are significant indicators of cognitive decline and the progression from mild cognitive impairment (MCI) to AD (Byeon et al., 2021).

Peripheral lipidomic profiling has produced numerous promising biomarker candidates for AD. A specific plasma lipid panel characterized by reduced levels of PC (16:0/20:5, 18:0/22:6) and increased ceramides has demonstrated predictive accuracy for the progression from MCI to AD up to three years in advance. Abnormal triglyceride and diglyceride levels, signifying systemic metabolic dysfunction, are frequently noted in AD cohorts (He et al., 2025).

Serum sphingolipids and ceramides have demonstrated significant elucidatory potential. High levels of ceramide in the blood are linked to the shrinking of the hippocampus and a decline in cognitive function. A lipidomic analysis of serum has revealed sex-specific patterns, suggesting that women are more susceptible to ceramide-mediated pathology, which aligns with the elevated prevalence of AD in females (Mielke et al., 2012).

The strength of lipidomic profiling is that it can tell the difference between different stages of a disease. In preclinical and prodromal AD, there are clear changes in phosphatidylcholines that contain DHA and ceramides (Liu et al., 2021). Lipid abnormalities often manifest prior to detectable cognitive decline, suggesting their potential utility as early diagnostic markers. In mild-to-moderate AD, lipidomic changes become more pronounced, marked by significant depletion of membrane phospholipids and an elevation in pro-inflammatory lipid mediators. In the later stages of the disease, lipid profiles show more variety, which means that a lot of neurons have died and there are metabolic effects that follow (Chua et al., 2023; Ren et al., 2025).

4.3 Linking lipidomics to mitochondrial dysfunction

Changes in lipids are closely related to how mitochondria make energy. Lower levels of cardiolipin, a phospholipid that only exists in mitochondria, make oxidative phosphorylation and ATP production less effective. In the same way, ceramide buildup in mitochondrial membranes makes the membranes more permeable, makes it easier for cytochrome c to leave the cell, and speeds up cell death. By recording these lipidomic changes, scientists learn more about the bioenergetic problems that are common in AD (Gasanoff et al., 2021). Lipidomic research has clarified the impact of genetic risk factors for AD, specifically the APOE genotype, on lipid metabolism. Individuals with the APOE ε4 allele, characterized by the highest genetic risk, exhibit distinctive plasma lipid profiles characterized by reduced PC-DHA species and elevated ceramides. The lipidomic variations clarify the molecular mechanisms underlying the heightened susceptibility of APOE ε4 patients to impaired synaptic repair and mitochondrial dysfunction (Almeida et al., 2024).

4.4 Lipidomic approaches to therapeutics and precision medicine

Findings from lipidomics have prompted the exploration of therapeutic strategies targeting lipid metabolism. There are a lot of different ways that are being looked into right now:

1. Adding omega-3 to the diet PUFA- Research has been done on restoring DHA levels through dietary supplements or drugs, and the results have been mixed. Lipidomic analysis might help find groups of patients who respond well to PUFA treatment.
2. Modulation of sphingolipids- Inhibitors of sphingomyelinase, the enzyme that makes ceramide, are being tested to stop ceramide from building up and causing neurotoxicity.
3. Pharmacological agents aimed at cholesterol modulation- Statins and liver X receptor (LXR) agonists, which control how cholesterol is broken down, are being used to treat AD. Lipidomics makes it possible to watch how they work in vivo.
4. Lipids aimed at mitochondria- New treatments include giving cardiolipin-stabilizing agents or synthetic lipid analogs to make mitochondria stronger.
5. Targeted therapeutics- Combining lipidomic profiles with genetic, transcriptomic, and proteomic data allows for the creation of personalized treatments to restore lipid and energy metabolism for each patient.

5. Lipid-induced oxidative stress and neuroinflammation in AD

Oxidative stress and neuroinflammation are essential pathogenic features of AD, and both processes are intricately linked to lipid dysregulation. Problems with lipid metabolism make neurons more likely to build up ROS, cause mitochondria to stop working, and keep inflammatory pathways active. Recent findings suggest that lipid-induced oxidative stress and neuroinflammation establish a self-perpetuating cycle that accelerates synaptic degradation and cognitive decline in AD (Cerasuolo et al., 2024).

PUFAs, notably DHA and AA, are particularly susceptible to free radical attack due to their numerous double bonds. In the brains of people with AD, higher levels of ROS oxidize lipids, which leads to the creation of reactive aldehydes like malondialdehyde (MDA) and 4-hydroxynonenal (4-HNE). These substances change proteins by forming covalent bonds with them, making membranes less fluid, and making ion channels work less. This mechanism not only impedes mitochondrial ATP synthesis but also triggers apoptotic signaling, resulting in neuronal death (Petrovic et al., 2020).

Ceramides, bioactive sphingolipids elevated in AD, modulate oxidative stress via mitochondrial and NADPH oxidase pathways. High levels of ceramide make the ETC less effective, which leads to more ROS being made. Ceramides also stop antioxidant enzymes like superoxide dismutase from working, which makes the balance shift toward pro-oxidant conditions. Elevated ceramides stimulate protein phosphatase 2A, promoting tau hyperphosphorylation, which, alongside oxidative stress, worsens neurofibrillary disease (Jazvinšćak Jembrek et al., 2015).

Cholesterol buildup in lipid rafts changes how APP is processed, which leads to amyloidogenic cleavage and Aβ aggregation. Aβ peptides cause the production of ROS by interacting with metal ions and mitochondrial membranes. Oxidation of cholesterol creates oxysterols, such as 7-ketocholesterol, which damage mitochondria and cause microglial inflammatory responses. As a result, dysregulation of cholesterol not only worsens oxidative stress but also has a direct link to amyloid pathology (Rudajev & Novotny, 2022).

Lipid peroxidation increases oxidative stress, and PUFA make signaling molecules that change how inflammation works. AA generates pro-inflammatory eicosanoids (prostaglandins, leukotrienes), whereas DHA and eicosapentaenoic acid (EPA) produce specialized pro-resolving mediators (SPMs), such as resolvins, protectins, and maresins. In AD, reduced DHA levels impede SPM production, undermining the resolution of inflammation. The difference between pro-inflammatory and pro-resolving lipid mediators keeps microglial activation and cytokine release going (Oppedisano et al., 2020). Mitochondria serve as both generators and receptors of lipid-induced oxidative stress. Oxidized cardiolipin makes ETC complexes less stable, while ceramides make the outer membrane of the mitochondria more permeable. Cholesterol accumulation hardens mitochondrial membranes, making it harder for metabolites to move in and out. All of these changes to lipids make ROS leak more and weaken mitochondrial antioxidant defenses, creating a harmful cycle of oxidative damage and energy loss.

Microglial responses are directly affected by lipid imbalance. Oxidized phospholipids and cholesterol derivatives act as danger-associated molecular patterns (DAMPs) by turning on toll-like receptors (TLRs) and scavenger receptors on microglia. Ceramides and oxysterols augment NF-κB activation, promoting the secretion of TNF-α, IL-1β, and IL-6. Extended exposure to these lipid-derived stimuli sustains microglial activation, converting cells into a neurotoxic phenotype that damages synapses and exacerbates the pathogenesis of AD (Erridge et al., 2008).

Astrocytes, crucial regulators of cerebral lipid homeostasis, are also involved in inflammatory pathways in AD. When stressed, astrocytes release AA and prostaglandins, which makes neuroinflammation worse. Impaired astrocytic lipid metabolism diminishes cholesterol and PUFA transport to neurons, exacerbating synaptic dysfunction. Astrocytes take up oxidized lipids, which makes inflammatory pathways even more active and makes microglial responses worse (Calì et al., 2024). Aβ aggregates both originate from and intensify lipid-induced oxidative damage. Aβ interacts with lipid rafts, promotes cholesterol oxidation, and induces reactive oxygen species generation in mitochondria. Oxidized lipids, therefore, enhance Aβ aggregation and toxicity. This reciprocal relationship sustains oxidative stress and triggers inflammatory responses through microglial activation. Lipid dysregulation, oxidative stress, and amyloid pathology form a convergent pathogenesis axis in AD (Rangachari et al., 2018).

Addressing oxidative damage and inflammation caused by lipids shows promise as a way to treat these problems. Antioxidants like coenzyme Q10 and mitoTEMPO that stabilize membrane lipids lower peroxidation. Blocking ceramide synthase to change how sphingolipids are broken down lowers ROS and tau pathology. Adding PUFA, especially DHA, back into the diet makes membranes stronger and helps SPM-mediated inflammation go away faster. Statins and oxysterol-targeting agents control how cholesterol is broken down and lower the amount of damage caused by free radicals.

6. Lipid dysregulation and synaptic dysfunction in AD

Synaptic loss is the most significant predictor of cognitive deterioration in AD, beyond the impact of amyloid plaque or tau tangle accumulation. Lipids are essential for synaptic architecture and functionality, constituting the bilayer matrix of neuronal membranes, orchestrating receptor distribution, and facilitating synaptic communication via lipid-derived messengers. Alterations in lipid composition and metabolism significantly impact synaptic vesicle dynamics, receptor trafficking, and neurotransmitter release. Recent data suggest that lipid dysregulation is a fundamental contributor to synaptic vulnerability in AD, connecting molecular pathology to cognitive decline (Subramanian et al., 2020).

Cholesterol, sphingolipids, and PUFAs, such as DHA, are all found in large amounts in synaptic membranes. These lipids keep the membrane's shape and fluidity, which are important for vesicle fusion and receptor movement. Lipid rafts are small areas in the synaptic membrane that separate signaling molecules and neurotransmitter receptors. Dysregulation of lipid composition compromises raft integrity, impedes receptor localization, and disrupts signal transmission. In AD, decreased DHA, altered PS, and abnormal cholesterol distribution disrupt synaptic structure, hindering neuronal communication (Vallés & Barrantes, 2022). Cholesterol is necessary for synaptogenesis and vesicular recycling. Cholesterol regulates synaptic vesicle fusion proteins, such as synaptophysin and synaptobrevin, at presynaptic terminals, thereby promoting efficient neurotransmitter release. Cholesterol influences the aggregation of glutamate receptors, particularly NMDA and AMPA receptors, at the postsynaptic level, which are crucial for LTP and memory formation. In AD, dysregulation of cholesterol disrupts these mechanisms. Higher cholesterol levels speed up the amyloidogenic processing of APP in lipid rafts. On the other hand, removing cholesterol slows down receptor clustering and makes synaptic plasticity harder. Both an excess and a deficiency of cholesterol disrupt synaptic signaling, underscoring its dual role in the pathophysiology of AD (Thiele et al., 2000). DHA and AA are necessary for synapse function. DHA makes membranes more flexible, helps neurotrophic signaling, and encourages the growth of dendritic spines. It helps AMPA receptors get into postsynaptic membranes, which makes excitatory transmission stronger. On the other hand, AA is a precursor to pro-inflammatory eicosanoids, which can interfere with synaptic communication if there are too many of them. In AD brains, DHA levels are consistently reduced, while AA metabolism is elevated, facilitating the production of inflammatory lipid mediators. The resulting imbalance results in impaired LTP, reduced synaptic connections, and memory deficits. Dietary supplementation to restore DHA levels has shown improvement in synaptic resilience in animal models; however, practical applicability remains variable (Tanaka et al., 2012).

Sphingolipids affect the arrangement of postsynaptic receptors and the exocytosis of synaptic vesicles. When someone has AD, too much ceramide builds up and damages synaptic homeostasis. Ceramides make membranes stiffer, slow down vesicle transport, and stop the release of neurotransmitters. Additionally, ceramide-induced activation of protein phosphatase 2A promotes tau hyperphosphorylation, thereby indirectly interfering with the microtubule-dependent transport of synaptic components. The effects of both of them together make synapses less effective, linking problems with lipid metabolism to tau-mediated neurodegeneration (Young et al., 2012). PS and PE are necessary for vesicle docking and the release of neurotransmitters. Without PS, vesicles can't fuse, and without PE, the curvature needed for endocytosis is lost. Phosphatidylinositol 4,5-bisphosphate (PIP2) acts as a signaling lipid that changes how synaptic vesicles move. In AD, reduced PS levels and altered PI 4,5-bisphosphate dynamics impede vesicle trafficking and neurotransmission. These impairments result in synaptic exhaustion and reduced network connectivity, which are associated with memory deficits in patients (Varga et al., 2020).

Aβ peptides specifically interact with synaptic lipid rafts, disrupting receptor aggregation and membrane excitability. Aβ oligomers get through membranes that are high in PUFA, which causes lipid peroxidation and weakens the membrane, which makes synaptic communication less stable. Aβ reduces the surface expression of NMDA receptors at postsynaptic sites, impeding calcium-dependent signaling crucial for memory formation. Aβ disrupts lipid rafts, altering APP processing and creating a feedback loop of amyloidogenesis and synaptic dysfunction. The convergence of amyloid pathology and lipid dysregulation represents a critical factor in the etiology of AD (Midorikawa et al., 2024). Lipid disturbances affect structural components and modulate neurotransmitter systems. A lack of cholesterol makes it harder for acetylcholine to be released, which makes the cholinergic problems that are common in AD even worse. DHA depletion interferes with dopaminergic and serotonergic signaling, essential for cognitive and emotional regulation. The accumulation of ceramide disrupts GABAergic equilibrium, resulting in the hyperexcitability observed in AD.

Lipid dysregulation affects neurotransmitter balance throughout the system, which makes the synaptic networks that support cognition even weaker (Shin et al., 2024).

Lipid-induced inflammatory signals also affect synapses. Pro-inflammatory eicosanoids derived from arachidonic acid stimulate microglia at synaptic sites, initiating synaptic pruning and elimination. On the other hand, certain pro-resolving mediators (SPMs) that come from DHA help repair synapses and protect against excitotoxicity. In AD, reduced synthesis of specialized pro-resolving mediators undermines synaptic resilience, allowing uncontrolled inflammatory signaling to cause synaptic degeneration. This emphasizes the importance of lipid-derived mediators in maintaining the balance between synaptic loss and repair (Soteros & Sia, 2022). ApoE, the main lipid transporter in the brain, is very important for synaptic health because it brings cholesterol and PUFA to neurons. The ApoE4 isoform, the most important genetic risk factor for late-onset AD, has made lipid transport less effective, which means that there is less lipid available at synapses. Individuals with the ApoE4 allele demonstrate diminished dendritic spine density and altered receptor trafficking, which is associated with accelerated cognitive decline. On the other hand, ApoE2 protects neurons by making lipid transport more efficient and making synapses more resilient. This isoform-specific lipid transport underscores the essential role of lipid metabolism in synaptic health (Belaidi et al., 2025).

The cumulative effect of lipid disturbances leads to a slow decline in cognitive function. Changes in cholesterol levels, lower DHA levels, ceramide buildup, and changes in the movement of phospholipids lead to synaptic weakening, which changes the connections in hippocampal and cortical networks. These changes are what cause the memory problems that are common in people with AD. Functional imaging studies reveal hypoconnectivity in the default mode and hippocampal networks among AD patients, abnormalities that correlate with lipid metabolic disturbances. Lipid dysregulation creates a mechanistic link between molecular disease and cognitive impairment (Wu et al., 2023).

7. Therapeutic targeting of lipid metabolism in AD

7.1 Lipid-directed pharmacological interventions

7.1.1 Statins and cholesterol modulators

Cholesterol plays a dual role in the pathogenesis of AD. While essential for synaptic function and membrane fluidity, excessive cholesterol accumulation promotes the amyloidogenic processing of APP and disrupts mitochondrial dynamics. Statins, which block HMG-CoA reductase, have been studied a lot for their ability to lower the risk of AD by changing how cholesterol is broken down (Rudajev & Novotny, 2022). Observational studies suggest a reduced prevalence of dementia among long-term statin users; however, randomized clinical trials (RCTs) have yielded conflicting outcomes. For example, trials involving simvastatin and atorvastatin demonstrated marginal improvements in cognitive outcomes in moderate AD, whereas other studies revealed no significant benefits. The extensive range likely reflects the timing of intervention, fluctuating blood-brain barrier permeability, and the complex role of cholesterol in neuronal signaling (Westphal Filho et al., 2025).

In addition to statins, liver X receptor (LXR) and retinoid X receptor (RXR) agonists, which regulate cholesterol efflux through APOE pathways, have demonstrated promise in preclinical models by reducing amyloid accumulation and enhancing synaptic plasticity. However, their translation has been limited by systemic lipid adverse effects, particularly hypertriglyceridemia (Liang et al., 2004).

7.1.2 Sphingolipid and ceramide modulators

Ceramides, essential sphingolipids increased in the brains and CSF of AD patients, impair insulin signaling, augment oxidative stress, and promote neuronal death. Strategies targeting ceramide metabolism involve inhibitors of sphingomyelinase (SMase), the enzyme that synthesizes ceramide from sphingomyelin. Preclinical inhibition of acid sphingomyelinase diminished amyloid pathology, attenuated microglial activation, and improved cognitive function in murine models (Kalkman & Smigielski, 2025).

Fingolimod, an FDA-approved sphingosine-1-phosphate (S1P) receptor modulator for multiple sclerosis, is under investigation for AD due to its dual role in reducing neuroinflammation and modulating ceramide/S1P balance. Initial studies suggest improvements in synaptic plasticity and memory within AD models; however, controlled human trials have not yet been performed (Pournajaf et al., 2022).

7.1.3 Mitochondrial lipid targets

Lipids in the mitochondrial membrane, especially cardiolipin, are important for putting together the respiratory chain and making ATP. In AD, cardiolipin gets oxidized, which breaks the ETC and releases cytochrome c. Elamipretide (SS-31), a peptide that targets mitochondria, stabilizes cardiolipin and keeps bioenergetics stable when oxidative stress is present.

Initial clinical investigations in mitochondrial disorders demonstrate improved exercise tolerance and reduced oxidative biomarkers, indicating translational applicability for AD (Falabella et al., 2021).

Similarly, PUFA supplementation, particularly DHA, may restore mitochondrial lipid composition and improve neural resilience. While multiple randomized controlled trials involving AD patients demonstrated cognitive stability with DHA-rich diets, other studies reported no effect, underscoring the importance of early intervention, genetic predisposition (e.g., APOE genotype), and adherence (Quinn et al., 2010).

7.2 Nutritional and lifestyle-based approaches

7.2.1 Ketogenic diets and medium-chain triglycerides (MCTs)

A sign of AD is low glucose metabolism, and ketone bodies can be used as a different source of energy. Lipid-based ketogenic therapies aim to bypass impaired glucose metabolism and enhance mitochondrial ATP production. Clinical trials involving MCTs, which rapidly convert to ketones, demonstrate improved memory performance in individuals with moderate cognitive impairment (MCI) and early AD, particularly among APOE ε4 non-carriers (Dewsbury et al., 2021).

Formulations such as AC-1202 (Axona), a caprylic triglyceride medical food, exhibited cognitive advantages in Phase II trials; however, larger studies yielded inconsistent efficacy outcomes. Ketogenic diets reduce neuroinflammation and amyloid accumulation in animal models; however, their long-term sustainability and cardiovascular implications in older populations remain subjects of concern (Bohnen et al., 2023).

7.2.2 Omega-3 PUFAs

Omega-3 fatty acids, especially DHA and eicosapentaenoic acid (EPA), control how fluid the membranes are, help synaptogenesis, and make neuroprotective lipid mediators like resolvins and protectins. Many RCTs have looked at omega-3 supplementation in AD, and the results range from no noticeable effect to some cognitive improvement. Subgroup analyses demonstrate increased efficacy when administered during prodromal phases or in combination with antioxidants such as vitamin E. The OmegAD study showed that moderate AD patients who took DHA/EPA supplements for six months had a slower decline in their cognitive abilities (Dyall, 2015). New ways to make omega-3 fatty acids more bioavailable and help them get through the blood-brain barrier are being created. These include phospholipid-anchored omega-3 fatty acids and nanoparticle delivery methods.

7.2.3 Phospholipid replacement therapy

PS and PC, which are lower in neurons with AD, are important for recycling synaptic vesicles and releasing neurotransmitters. Dietary PS supplementation has demonstrated modest cognitive improvements in elderly individuals experiencing memory deficits. Lipid replacement treatment (LRT), which uses oral phospholipid formulations to fix mitochondrial membrane damage, has been helpful for people with chronic fatigue syndromes and is now being studied for neurodegenerative diseases (Nicolson & Ash, 2014).

7.3 Targeting APOE and lipid transport pathways

APOE is the most important genetic risk factor for late-onset AD. The ε4 allele is linked to problems with lipid transport, less effective synaptic repair, and changes in the composition of mitochondrial lipids. Therapeutic strategies targeting APOE encompass APOE mimetic peptides that enhance lipid efflux and neuroprotection, gene editing methodologies to modulate APOE isoform expression, and receptor-targeted therapies, including LDL receptor overexpression, to optimize lipid clearance and promote synaptic repair. Clinical research is still in its early stages; however, APOE-targeted strategies represent a precision medicine approach that could transform lipid-based therapies for AD (Liu et al., 2013).

7.4 Lipidomics and biomarker-guided therapies

Advancements in mass spectrometry-based lipidomics have enabled the identification of disease-specific lipid signatures in plasma and CSF. Elevated ceramides, reduced plasmalogens, and altered cholesterol esters are commonly linked to cognitive decline. The utilization of lipidomic indicators in clinical trials may enhance patient classification, expedite diagnosis, and enable real-time evaluation of therapeutic response (Lattau et al., 2024).

Baseline plasma DHA levels have been demonstrated to predict the response to omega-3 supplementation, highlighting the necessity for biomarker-guided precision nutrition. The ratios of ceramide and sphingomyelin may also help find people who would benefit the most from sphingolipid-targeted drugs (Casas et al., 2025).

7.5 Challenges and future perspectives

Even though there is a lot of evidence that lipid dysregulation is linked to the pathophysiology of AD, it is still hard to find effective treatments. AD is a biologically heterogeneous condition marked by lipid irregularities

Table 1 Major brain lipid classes, their bioenergetic roles, and alterations in AD.

Lipid class	Physiological role in brain energy metabolism	Alterations in AD	Consequences for bioenergetics & synaptic function	Therapeutic implications	References
Phospholipids (PC, PE, PS)	Maintain neuronal membrane integrity; regulate synaptic vesicle fusion and receptor trafficking; support mitochondrial cristae structure	Reduced PC and PE; altered PS distribution	Impaired membrane fluidity, disrupted synaptic transmission, reduced mitochondrial efficiency	Phospholipid replacement therapy; dietary choline and PS supplementation	Choudhary et al. (2024)
Cardiolipin	Anchors electron transport chain (ETC) complexes; stabilizes mitochondrial supercomplexes; supports ATP synthesis	Oxidation and depletion of cardiolipin in hippocampus and cortex	ETC destabilization, increased ROS, reduced ATP, activation of apoptotic pathways	Cardiolipin stabilizers (SS-31/elamipretide); mitochondrial antioxidants	Monteiro-Cardoso et al. (2014)
Sphingolipids/ Ceramides	Regulate cell signaling, apoptosis, and membrane microdomains	Elevated ceramides in brain, CSF, and plasma	Mitochondrial dysfunction, increased oxidative stress, enhanced neuroinflammation	Sphingomyelinase inhibitors; S1P receptor modulators (FTY720)	van Kruining et al. (2020)

Cholesterol	Maintains lipid raft integrity; regulates synaptic receptor clustering and APP processing	Altered cholesterol trafficking; increased cholesterol in lipid rafts, especially in APOE4 carriers	Enhanced amyloidogenic APP processing; impaired synaptic plasticity	Statins; LXR/RXR agonists; APOE-targeted therapies	Rudajev and Novotny, (2022)
PUFAs (DHA, AA)	Maintain membrane fluidity; regulate synaptic plasticity; generate lipid mediators controlling inflammation	Reduced DHA; dysregulated AA metabolism	Impaired synaptic signaling; exaggerated neuroinflammation; reduced resolution of inflammation	Omega-3 PUFA supplementation; pro-resolving lipid mediators	Morita et al. (2025)
Glycerolipids (DG, TG)	Serve as precursors for membrane lipids; DG acts as second messenger in synaptic signaling	Altered DG and TG species in cortex and hippocampus	Disrupted PKC signaling; impaired synaptic plasticity and energy balance	Targeting lipid turnover pathways; metabolic modulation	Liu et al. (2024)
Ketone-related Lipids/MCTs	Provide alternative energy substrates during glucose hypometabolism	Compensatory reliance in early AD	Partial rescue of ATP production; improved synaptic energetics	Ketogenic diets; MCT supplementation	Lin et al. (2021)

associated with amyloid-β accumulation, tau pathology, vascular dysfunction, and compromised energy metabolism. This indicates that lipid-targeted interventions are unlikely to be effective as standalone therapies and may require strategic combinations. Lipid disturbances, including altered ceramides, phospholipids, and PUFAs, can be detected decades before the onset of clinical symptoms. This suggests that treatment efficacy will be largely dependent on interventions during the preclinical or prodromal phases, rather than after the disease has presented. The distribution of drugs across the blood-brain barrier presents a substantial challenge, as numerous lipid-modulating medications exhibit insufficient penetration into the central nervous system, necessitating advanced delivery systems such as lipid nanoparticles and intranasal techniques (Chan et al., 2012). APOE genotype affects how well treatments work. For example, APOE4 carriers have different lipidomic profiles and respond differently to lipid-based therapies. This shows that we need precision methods that take genotype into account. In the end, long-term changes to lipid pathways, especially those that affect cholesterol production or nuclear lipid receptors, raise concerns about safety and systemic metabolic side effects. To address these issues, it is imperative to employ integrative multi-omics approaches that combine lipidomics with genetic and metabolic profiling to identify patient-specific lipid vulnerabilities and guide targeted therapies. Ultimately, combinatorial strategies that incorporate metabolic substrates, mitochondrial stabilizers, and anti-inflammatory lipid mediators may represent the most viable pathway for disease modification.

8. Conclusion

AD is now recognized not only as a pathology of protein aggregation but also as a condition characterized by impaired brain energy metabolism, significantly influenced by lipid dysregulation. Lipids serve as active regulators of mitochondrial integrity, synaptic transmission, and neuroinflammatory signaling rather than only passive structural components. The amalgamation of lipidomics, mitochondrial biology, and translational therapies heralds a transformative change in AD research. By focusing on lipid homeostasis, future therapies may restore neuronal energy metabolism and maybe postpone or avert cognitive deterioration. Ultimately, moving beyond descriptive lipidomic investigations to mechanistic and therapeutic frameworks will be essential for converting our comprehension of AD into practical therapies (Table 1).

Declaration of interests

The authors have no relevant affiliations or financial involvement with any organisation or entity with a financial interest in or financial conflict with the subject matter or materials discussed in the manuscript. This includes employment, consultancies, honoraria, stock ownership or options, expert testimony, grants or patents received or pending, or royalties.

References

Ademowo, O. S., Dias, H. K. I., Burton, D. G. A., & Griffiths, H. R. (2017). Lipid (per) oxidation in mitochondria: An emerging target in the ageing process? *Biogerontology, 18*(6), 859–879. https://doi.org/10.1007/s10522-017-9710-z.

Akyol, S., Ugur, Z., Yilmaz, A., Ustun, I., Gorti, S. K. K., Oh, K., ... Graham, S. F. (2021). Lipid profiling of AD brain highlights enrichment in glycerol(phospho)lipid, and sphingolipid metabolism. *Cells, 10*(10), 2591. https://doi.org/10.3390/cells10102591.

Almeida, F. C., Patra, K., Giannisis, A., Niesnerova, A., Nandakumar, R., Ellis, E., ... Nielsen, H. M. (2024). APOE genotype dictates lipidomic signatures in primary human hepatocytes. *Journal of Lipid Research, 65*(2), https://doi.org/10.1016/j.jlr.2024.100498.

Bao, H.-N., Yin, J., Wang, L.-Y., Wang, R.-H., Huang, L.-Q., Chen, Y.-L., ... Li, J. (2024). Aberrant accumulation of ceramides in mitochondria triggers cell death by inducing autophagy in Arabidopsis. *Journal of Experimental Botany, 75*(5), 1314–1330. https://doi.org/10.1093/jxb/erad456.

Belaidi, A. A., Bush, A. I., & Ayton, S. (2025). Apolipoprotein E in AD: Molecular insights and therapeutic opportunities. *Molecular Neurodegeneration, 20*(1), 47. https://doi.org/10.1186/s13024-025-00843-y.

Biernacki, M., & Skrzydlewska, E. (2025). Metabolic pathways of eicosanoids—Derivatives of arachidonic acid and their significance in skin. *Cellular & Molecular Biology Letters, 30*(1), 7. https://doi.org/10.1186/s11658-025-00685-y.

Bigi, A., Cascella, R., Fani, G., Bernacchioni, C., Cencetti, F., Bruni, P., ... Cecchi, C. (2023). Sphingosine 1-phosphate attenuates neuronal dysfunction induced by amyloid-β oligomers through endocytic internalization of NMDA receptors. *The FEBS Journal, 290*(1), 112–133. https://doi.org/10.1111/febs.16579.

Blusztajn, J. K., & Slack, B. E. (2023). Accelerated breakdown of phosphatidylcholine and phosphatidylethanolamine is a predominant brain metabolic defect in AD. *Journal of AD, 93*(4), 1285–1289. https://doi.org/10.3233/JAD-230061.

Bohnen, J. L. B., Albin, R. L., & Bohnen, N. I. (2023). Ketogenic interventions in mild cognitive impairment, AD, and Parkinson's disease: A systematic review and critical appraisal. *Frontiers in Neurology, 14*. https://doi.org/10.3389/fneur.2023.1123290.

Bonvento, G., & Bolaños, J. P. (2021). Astrocyte-neuron metabolic cooperation shapes brain activity. *Cell Metabolism, 33*(8), 1546–1564. https://doi.org/10.1016/j.cmet.2021.07.006.

Brothwell, M. J., Cao, G., Maschek, J. A., Poss, A. M., Peterlin, A. D., Wang, L., ... Funai, K. (2025). Cardiolipin deficiency disrupts electron transport chain to drive steatohepatitis. *eLife, 14*. https://doi.org/10.7554/eLife.106976.1.

Bruce, K. D., Zsombok, A., & Eckel, R. H. (2017). Lipid processing in the brain: A key regulator of systemic metabolism. *Frontiers in Endocrinology, 8*. https://doi.org/10.3389/fendo.2017.00060.

Byeon, S. K., Madugundu, A. K., Jain, A. P., Bhat, F. A., Jung, J. H., Renuse, S., ... Pandey, A. (2021). Cerebrospinal fluid lipidomics for biomarkers of AD. *Molecular Omics, 17*(3), 454–463. https://doi.org/10.1039/D0MO00186D.

Calì, C., Cantando, I., Veloz Castillo, M. F., Gonzalez, L., & Bezzi, P. (2024). Metabolic reprogramming of astrocytes in pathological conditions: Implications for neurodegenerative diseases. *International Journal of Molecular Sciences, 25*(16), 8922. https://doi.org/10.3390/ijms25168922.

Calzada, E., Onguka, O., & Claypool, S. M. (2016). Chapter Two—Phosphatidylethanolamine metabolism in health and disease. In K. W. Jeon (Vol. Ed.), *International review of cell and molecular biology: 321,* (pp. 29–88). Academic Press. https://doi.org/10.1016/bs.ircmb.2015.10.001.

Casas, R., Sánchez-García, N. D., Estruch, R., & López-Yerena, A. (2025). Impact of dietary interventions on the human plasma and lipoprotein lipidome. *Metabolites, 15*(9), 602. https://doi.org/10.3390/metabo15090602.

Cebecauer, M., Hof, M., & Amaro, M. (2017). Impact of GM1 on membrane-mediated aggregation/oligomerization of β-amyloid: Unifying view. *Biophysical Journal, 113*(6), 1194–1199. https://doi.org/10.1016/j.bpj.2017.03.009.

Cerasuolo, M., Di Meo, I., Auriemma, M. C., Paolisso, G., Papa, M., & Rizzo, M. R. (2024). Exploring the dynamic changes of brain lipids, lipid rafts, and lipid droplets in aging and AD. *Biomolecules, 14*(11), 1362. https://doi.org/10.3390/biom14111362.

Chan, R. B., Oliveira, T. G., Cortes, E. P., Honig, L. S., Duff, K. E., Small, S. A., ... Paolo, G. D. (2012). Comparative lipidomic analysis of mouse and human brain with Alzheimer disease. *Journal of Biological Chemistry, 287*(4), 2678–2688. https://doi.org/10.1074/jbc.M111.274142.

Chiurchiù, V., Leuti, A., & Maccarrone, M. (2018). Bioactive lipids and chronic inflammation: Managing the fire within. *Frontiers in Immunology, 9.* https://doi.org/10.3389/fimmu.2018.00038.

Chiurchiù, V., Tiberi, M., Matteocci, A., Fazio, F., Siffeti, H., Saracini, S., ... Sancesario, G. (2022). Lipidomics of bioactive lipids in Alzheimer's and Parkinson's diseases: Where are we? *International Journal of Molecular Sciences, 23*(11), 6235. https://doi.org/10.3390/ijms23116235.

Choudhary, R. C., Kuschner, C. E., Kazmi, J., Mcdevitt, L., Espin, B. B., Essaihi, M., ... Kim, J. (2024). The role of phospholipid alterations in mitochondrial and brain dysfunction after cardiac arrest. *International Journal of Molecular Sciences, 25*(9), 4645. https://doi.org/10.3390/ijms25094645.

Chua, X. Y., Torta, F., Chong, J. R., Venketasubramanian, N., Hilal, S., Wenk, M. R., ... Lai, M. K. P. (2023). Lipidomics profiling reveals distinct patterns of plasma sphingolipid alterations in AD and vascular dementia. *Alzheimer's Research & Therapy, 15*(1), 214. https://doi.org/10.1186/s13195-023-01359-7.

Clemente-Suárez, V. J., Redondo-Flórez, L., Beltrán-Velasco, A. I., Ramos-Campo, D. J., Belinchón-deMiguel, P., Martinez-Guardado, I., ... Tornero-Aguilera, J. F. (2023). Mitochondria and brain disease: A comprehensive review of pathological mechanisms and therapeutic opportunities. *Biomedicines, 11*(9), 2488. https://doi.org/10.3390/biomedicines11092488.

Dewsbury, L. S., Lim, C. K., & Steiner, G. Z. (2021). The efficacy of ketogenic therapies in the clinical management of people with neurodegenerative disease: A systematic review. *Advances in Nutrition, 12*(4), 1571–1593. https://doi.org/10.1093/advances/nmaa180.

Di Paolo, G., & Kim, T.-W. (2011). Linking lipids to AD: Cholesterol and beyond. *Nature Reviews. Neuroscience, 12*(5), 284–296. https://doi.org/10.1038/nrn3012.

Dyall, S. C. (2015). Long-chain omega-3 fatty acids and the brain: A review of the independent and shared effects of EPA, DPA and DHA. *Frontiers in Aging Neuroscience, 7.* https://doi.org/10.3389/fnagi.2015.00052.

Erridge, C., Kennedy, S., Spickett, C. M., & Webb, D. J. (2008). Oxidized phospholipid inhibition of toll-like receptor (TLR) signaling is restricted to TLR2 and TLR4:

ROLES FOR CD14, LPS-BINDING PROTEIN, AND MD2 as targets for specificity of inhibition*. *Journal of Biological Chemistry, 283*(36), 24748–24759. https://doi.org/10.1074/jbc.M800352200.

Falabella, M., Vernon, H. J., Hanna, M. G., Claypool, S. M., & Pitceathly, R. D. S. (2021). Cardiolipin, mitochondria, and neurological disease. *Trends in Endocrinology & Metabolism, 32*(4), 224–237. https://doi.org/10.1016/j.tem.2021.01.006.

Farooqui, A. A., & Farooqui, T. (2024). Metabolic reprogramming of astrocytes in pathological conditions: Implications for neurodegenerative diseases. *International Journal of Molecular Sciences, 25*(19), 10672. https://doi.org/10.3390/ijms251910672.

Fuentes, J. M., & Morcillo, P. (2024). The role of cardiolipin in mitochondrial function and neurodegenerative diseases. *Cells, 13*(7), 609. https://doi.org/10.3390/cells13070609.

Gasanoff, E. S., Yaguzhinsky, L. S., & Garab, G. (2021). Cardiolipin, non-bilayer structures and mitochondrial bioenergetics: Relevance to cardiovascular disease. *Cells, 10*(7), 1721. https://doi.org/10.3390/cells10071721.

Goicoechea, L., Conde de la Rosa, L., Torres, S., García-Ruiz, C., & Fernández-Checa, J. C. (2023). Mitochondrial cholesterol: Metabolism and impact on redox biology and disease. *Redox Biology, 61*, 102643. https://doi.org/10.1016/j.redox.2023.102643.

Golde, T. E. (2016). Overcoming translational barriers impeding development of AD modifying therapies. *Journal of Neurochemistry, 139*(S2), 224–236. https://doi.org/10.1111/jnc.13583.

Goyal, M. S., Hawrylycz, M., Miller, J. A., Snyder, A. Z., & Raichle, M. E. (2014). Aerobic glycolysis in the human brain is associated with development and neotenous gene expression. *Cell Metabolism, 19*(1), 49. https://doi.org/10.1016/j.cmet.2013.11.020.

Grimm, M. O. W., Zimmer, V. C., Lehmann, J., Grimm, H. S., & Hartmann, T. (2013). The impact of cholesterol, DHA, and sphingolipids on AD. *BioMed Research International, 2013*(1), 814390. https://doi.org/10.1155/2013/814390.

Gu, C., Philipsen, M. H., & Ewing, A. G. (2024). Omega-3 and -6 fatty acids alter the membrane lipid composition and vesicle size to regulate exocytosis and storage of catecholamines. *ACS Chemical Neuroscience, 15*(4), 816–826. https://doi.org/10.1021/acschemneuro.3c00741.

Guo, Z. (2023). Ganglioside GM1 and the central nervous system. *International Journal of Molecular Sciences, 24*(11), 9558. https://doi.org/10.3390/ijms24119558.

Hachem, M., Ahmmed, M. K., & Nacir-Delord, H. (2024). Phospholipidomics in clinical trials for brain disorders: Advancing our understanding and therapeutic potentials. *Molecular Neurobiology, 61*(6), 3272–3295. https://doi.org/10.1007/s12035-023-03793-y.

Hauser, P. S., Narayanaswami, V., & Ryan, R. O. (2011). Apolipoprotein E: From lipid transport to neurobiology. *Progress in Lipid Research, 50*(1), 62–74. https://doi.org/10.1016/j.plipres.2010.09.001.

He, D., Bao, Y., Yuan, S., Wang, Y., Cai, J., Xu, F., ... Zheng, Q. (2025). Integrated plasma and red blood cell membrane lipidomics analysis unveils novel biomarker panel for AD. *Alzheimer's Research & Therapy, 17*(1), 178. https://doi.org/10.1186/s13195-025-01830-7.

He, S., Xu, Z., & Han, X. (2025). Lipidome disruption in AD brain: Detection, pathological mechanisms, and therapeutic implications. *Molecular Neurodegeneration, 20*(1), 11. https://doi.org/10.1186/s13024-025-00803-6.

Huang, S.-Y., Yang, Z.-J., Cheng, J., Li, H.-Y., Chen, S., Huang, Z.-H., ... Lan, Q.-Y. (2025). Choline alleviates cognitive impairment in sleep-deprived young mice via reducing neuroinflammation and altering phospholipidomic profile. *Redox Biology, 81*, 103578. https://doi.org/10.1016/j.redox.2025.103578.

Hussain, G., Wang, J., Rasul, A., Anwar, H., Imran, A., Qasim, M., ... Sun, T. (2019). Role of cholesterol and sphingolipids in brain development and neurological diseases. *Lipids in Health and Disease, 18*(1), 26. https://doi.org/10.1186/s12944-019-0965-z.

Jazvinšćak Jembrek, M., Hof, P. R., & Šimić, G. (2015). Ceramides in AD: Key mediators of neuronal apoptosis induced by oxidative stress and Aβ accumulation. *Oxidative Medicine and Cellular Longevity, 2015*(1), 346783. https://doi.org/10.1155/2015/346783.

Kalkman, H. O., & Smigielski, L. (2025). Ceramides may play a central role in the pathogenesis of AD: A review of evidence and horizons for discovery. *Molecular Neurobiology, 62*(11), 14424–14441. https://doi.org/10.1007/s12035-025-04989-0.

Kao, Y.-C., Ho, P.-C., Tu, Y.-K., Jou, I.-M., & Tsai, K.-J. (2020). Lipids and AD. *International Journal of Molecular Sciences, 21*(4), 1505. https://doi.org/10.3390/ijms21041505.

Kim, H.-Y., Huang, B. X., & Spector, A. A. (2014). Phosphatidylserine in the brain: Metabolism and function. *Progress in Lipid Research, 56*, 1–18. https://doi.org/10.1016/j.plipres.2014.06.002.

Köfeler, H. C., Fauland, A., Rechberger, G. N., & Trötzmüller, M. (2012). Mass spectrometry based lipidomics: An overview of technological platforms. *Metabolites, 2*(1), 19–38. https://doi.org/10.3390/metabo2010019.

Kumar, A., Nemeroff, C. B., Cooper, J. J., Widge, A., Rodriguez, C., Carpenter, L., & McDonald, W. M. (2021). Amyloid and Tau in AD: Biomarkers or molecular targets for therapy? Are we shooting the messenger? *American Journal of Psychiatry, 178*(11), 1014–1025. https://doi.org/10.1176/appi.ajp.2021.19080873.

Lattau, S. S. J., Borsch, L.-M., auf dem Brinke, K., Klose, C., Vinhoven, L., Nietert, M., & Fitzner, D. (2024). Plasma lipidomic profiling using mass spectrometry for multiple sclerosis diagnosis and disease activity stratification (LipidMS). *International Journal of Molecular Sciences, 25*(5), 2483. https://doi.org/10.3390/ijms25052483.

Liang, Y., Lin, S., Beyer, T. P., Zhang, Y., Wu, X., Bales, K. R., ... Paul, S. M. (2004). A liver X receptor and retinoid X receptor heterodimer mediates apolipoprotein E expression, secretion and cholesterol homeostasis in astrocytes. *Journal of Neurochemistry, 88*(3), 623–634. https://doi.org/10.1111/j.1471-4159.2004.02183.x.

Lin, T.-Y., Liu, H.-W., & Hung, T.-M. (2021). The ketogenic effect of medium-chain triacylglycerides. *Frontiers in Nutrition, 8*. https://doi.org/10.3389/fnut.2021.747284.

Liu, C.-C., Kanekiyo, T., Xu, H., & Bu, G. (2013). Apolipoprotein E and Alzheimer disease: Risk, mechanisms and therapy. *Nature Reviews Neurology, 9*(2), 106–118. https://doi.org/10.1038/nrneurol.2012.263.

Liu, E., Zhang, Y., & Wang, J.-Z. (2024). Updates in AD: From basic research to diagnosis and therapies. *Translational Neurodegeneration, 13*(1), 45. https://doi.org/10.1186/s40035-024-00432-x.

Liu, Y., Thalamuthu, A., Mather, K. A., Crawford, J., Ulanova, M., Wong, M. W. K., ... Braidy, N. (2021). Plasma lipidome is dysregulated in AD and is associated with disease risk genes. *Translational Psychiatry, 11*(1), 344. https://doi.org/10.1038/s41398-021-01362-2.

Liu, Y., Yang, Z., Zhou, X., Li, Z., & Hideki, N. (2024). Diacylglycerol kinases and its role in lipid metabolism and related diseases. *International Journal of Molecular Sciences, 25*(23), 13207. https://doi.org/10.3390/ijms252313207.

López-Gambero, A. J., Sanjuan, C., Serrano-Castro, P. J., Suárez, J., & Rodríguez de Fonseca, F. (2020). The biomedical uses of inositols: A nutraceutical approach to metabolic dysfunction in aging and neurodegenerative diseases. *Biomedicines, 8*(9), 295. https://doi.org/10.3390/biomedicines8090295.

López-García, P., Tejero-Ojeda, M. M., Vaquero, M. E., & Carrión-Vázquez, M. (2025). Current amyloid inhibitors: Therapeutic applications and nanomaterial-based innovations. *Progress in Neurobiology, 247*, 102734. https://doi.org/10.1016/j.pneurobio.2025.102734.

Midorikawa, R., Wakazono, Y., & Takamiya, K. (2024). Aβ peptide enhances GluA1 internalization via lipid rafts in Alzheimer's-related hippocampal LTP dysfunction. *Journal of Cell Science, 137*(8), jcs261281. https://doi.org/10.1242/jcs.261281.

Mielke, M. M., Bandaru, V. V. R., Haughey, N. J., Xia, J., Fried, L. P., Yasar, S., ... Carlson, M. C. (2012). Serum ceramides increase the risk of Alzheimer disease. *Neurology, 79*(7), 633–641. https://doi.org/10.1212/WNL.0b013e318264e380.

Misrani, A., Tabassum, S., & Yang, L. (2021). Mitochondrial dysfunction and oxidative stress in AD. *Frontiers in Aging Neuroscience, 13*, 617588. https://doi.org/10.3389/fnagi.2021.617588.

Monteiro-Cardoso, V. F., Oliveira, M. M., Melo, T., Domingues, M. R. M., Moreira, P. I., Ferreiro, E., ... Videira, R. A. (2014). Cardiolipin profile changes are associated to the early synaptic mitochondrial dysfunction in AD. *Journal of AD, 43*(4), 1375–1392. https://doi.org/10.3233/JAD-141002.

Morita, S., Kondo, T., Tokuda, H., Kaneda, Y., Izumo, T., Nakao, Y., & Inoue, H. (2025). PUFAs in lipid membranes regulate human neuronal function and amyloid-β production. *iScience, 28*(6), https://doi.org/10.1016/j.isci.2025.112557.

Nasb, M., Tao, W., & Chen, N. (2024). AD puzzle: Delving into pathogenesis hypotheses. *Aging and Disease, 15*(1), 43–73. https://doi.org/10.14336/AD.2023.0608.

Ng, Y., Barhoumi, R., Tjalkens, R. B., Fan, Y.-Y., Kolar, S., Wang, N., ... Chapkin, R. S. (2005). The role of DHAin mediating mitochondrial membrane lipid oxidation and apoptosis in colonocytes. *Carcinogenesis, 26*(11), 1914–1921. https://doi.org/10.1093/carcin/bgi163.

Nichols, E., Steinmetz, J. D., Vollset, S. E., Fukutaki, K., Chalek, J., Abd-Allah, F., ... Vos, T. (2022). Estimation of the global prevalence of dementia in 2019 and forecasted prevalence in 2050: An analysis for the global burden of disease study 2019. *The Lancet Public Health*, 7(2), e105–e125. https://doi.org/10.1016/S2468-2667(21)00249-8.

Nicolson, G. L., & Ash, M. E. (2014). Lipid replacement therapy: A natural medicine approach to replacing damaged lipids in cellular membranes and organelles and restoring function. *Biochimica et Biophysica Acta (BBA) - Biomembranes, 1838*(6), 1657–1679. https://doi.org/10.1016/j.bbamem.2013.11.010.

Novgorodov, S. A., Gudz, T. I., & Obeid, L. M. (2008). Long-chain ceramide is a potent inhibitor of the mitochondrial permeability transition pore. *The Journal of Biological Chemistry, 283*(36), 24707–24717. https://doi.org/10.1074/jbc.M801810200.

Olsen, A. S. B., & Færgeman, N. J. (2017). Sphingolipids: Membrane microdomains in brain development, function and neurological diseases. *Open Biology*, 7(5), 170069. https://doi.org/10.1098/rsob.170069.

Oppedisano, F., Macrì, R., Gliozzi, M., Musolino, V., Carresi, C., Maiuolo, J., ... Mollace, V. (2020). The anti-inflammatory and antioxidant properties of n-3 PUFAs: Their role in cardiovascular protection. *Biomedicines, 8*(9), 306. https://doi.org/10.3390/biomedicines8090306.

Orth, M., & Bellosta, S. (2012). Cholesterol: Its regulation and role in central nervous system disorders. *Cholesterol, 2012*(1), 292598. https://doi.org/10.1155/2012/292598.

Ortiz-Placín, C., Castillejo-Rufo, A., Estarás, M., & González, A. (2023). Membrane lipid derivatives: Roles of arachidonic acid and its metabolites in pancreatic physiology and pathophysiology. *Molecules (Basel, Switzerland), 28*(11), 4316. https://doi.org/10.3390/molecules28114316.

Pan, Y., Li, J., Lin, P., Wan, L., Qu, Y., Cao, L., & Wang, L. (2024). A review of the mechanisms of abnormal ceramide metabolism in type 2 diabetes mellitus, AD, and their co-morbidities. *Frontiers in Pharmacology, 15*. https://doi.org/10.3389/fphar.2024.1348410.

Paradies, G., Paradies, V., De Benedictis, V., Ruggiero, F. M., & Petrosillo, G. (2014). Functional role of cardiolipin in mitochondrial bioenergetics. *Biochimica et Biophysica Acta (BBA) - Bioenergetics, 1837*(4), 408–417. https://doi.org/10.1016/j.bbabio.2013.10.006.

Paradies, G., Paradies, V., Ruggiero, F. M., & Petrosillo, G. (2019). Role of cardiolipin in mitochondrial function and dynamics in health and disease: Molecular and pharmacological aspects. *Cells, 8*(7), 728. https://doi.org/10.3390/cells8070728.

Pasqualetti, G., Thayanandan, T., & Edison, P. (2022). Influence of genetic and cardiometabolic risk factors in AD. *Ageing Research Reviews, 81*, 101723. https://doi.org/10.1016/j.arr.2022.101723.

Petrovic, S., Arsic, A., Ristic-Medic, D., Cvetkovic, Z., & Vucic, V. (2020). Lipid peroxidation and antioxidant supplementation in neurodegenerative diseases: A review of human studies. *Antioxidants, 9*(11), 1128. https://doi.org/10.3390/antiox9111128.

Pournajaf, S., Dargahi, L., Javan, M., & Pourgholami, M. H. (2022). Molecular pharmacology and novel potential therapeutic applications of fingolimod. *Frontiers in Pharmacology, 13*. https://doi.org/10.3389/fphar.2022.807639.

Quinn, J. F., Raman, R., Thomas, R. G., Yurko-Mauro, K., Nelson, E. B., Van Dyck, C., ... Aisen, P. S. (2010). DHAsupplementation and cognitive decline in Alzheimer disease: A randomized trial. *JAMA: the Journal of the American Medical Association, 304*(17), 1903–1911. https://doi.org/10.1001/jama.2010.1510.

Rangachari, V., Dean, D. N., Rana, P., Vaidya, A., & Ghosh, P. (2018). Cause and consequence of Aβ – Lipid interactions in Alzheimer disease pathogenesis. *Biochimica et Biophysica Acta (BBA) - Biomembranes, 1860*(9), 1652–1662. https://doi.org/10.1016/j.bbamem.2018.03.004.

Raulin, A.-C., Doss, S. V., Trottier, Z. A., Ikezu, T. C., Bu, G., & Liu, C.-C. (2022). ApoE in AD: Pathophysiology and therapeutic strategies. *Molecular Neurodegeneration, 17*(1), 72. https://doi.org/10.1186/s13024-022-00574-4.

Raut, S., Bhalerao, A., Powers, M., Gonzalez, M., Mancuso, S., & Cucullo, L. (2023). Hypometabolism, AD, and possible therapeutic targets: An overview. *Cells, 12*(16), 2019. https://doi.org/10.3390/cells12162019.

Ren, F., Wei, J., Chen, Q., Hu, M., Yu, L., Mi, J., ... Wu, A. (2025). Artificial intelligence-driven multi-omics approaches in AD: Progress, challenges, and future directions. *Acta Pharmaceutica Sinica B*. https://doi.org/10.1016/j.apsb.2025.07.030.

Rudajev, V., & Novotny, J. (2022). Cholesterol as a key player in amyloid β-mediated toxicity in AD. *Frontiers in Molecular Neuroscience, 15*. https://doi.org/10.3389/fnmol.2022.937056.

Senanayake, V., & Goodenowe, D. B. (2019). Plasmalogen deficiency and neuropathology in AD: Causation or coincidence? *Alzheimer's & Dementia: Translational Research & Clinical Interventions, 5*(1), 524–532. https://doi.org/10.1016/j.trci.2019.08.003.

Shin, K. C., Ali Moussa, H. Y., & Park, Y. (2024). Cholesterol imbalance and neurotransmission defects in neurodegeneration. *Experimental & Molecular Medicine, 56*(8), 1685–1690. https://doi.org/10.1038/s12276-024-01273-4.

Soteros, B. M., & Sia, G. M. (2022). Complement and microglia dependent synapse elimination in brain development. *WIREs Mechanisms of Disease, 14*(3), e1545. https://doi.org/10.1002/wsbm.1545.

Subramanian, J., Savage, J. C., & Tremblay, M.-È. (2020). Synaptic loss in AD: Mechanistic insights provided by two-photon in vivo imaging of transgenic mouse models. *Frontiers in Cellular Neuroscience, 14*. https://doi.org/10.3389/fncel.2020.592607.

Tanaka, K., Farooqui, A. A., Siddiqi, N. J., Alhomida, A. S., & Ong, W.-Y. (2012). Effects of DHAon neurotransmission. *Biomolecules & Therapeutics, 20*(2), 152–157. https://doi.org/10.4062/biomolther.2012.20.2.152.

Thiele, C., Hannah, M. J., Fahrenholz, F., & Huttner, W. B. (2000). Cholesterol binds to synaptophysin and is required for biogenesis of synaptic vesicles. *Nature Cell Biology, 2*(1), 42–49. https://doi.org/10.1038/71366.

Thomas, J., Thomas, C. J., Radcliffe, J., & Itsiopoulos, C. (2015). Omega-3 fatty acids in early prevention of inflammatory neurodegenerative disease: A focus on AD. *BioMed Research International, 2015*(1), 172801. https://doi.org/10.1155/2015/172801.

Tian, H., Zhao, X., Zhang, Y., & Xia, Z. (2023). Abnormalities of glucose and lipid metabolism in myocardial ischemia-reperfusion injury. *Biomedicine & Pharmacotherapy, 163*, 114827. https://doi.org/10.1016/j.biopha.2023.114827.

Tian, J., Zhang, Y., & Zhao, X. (2025). The effects and mechanisms of n-3 and n-6 PUFAs in the central nervous system. *Cellular and Molecular Neurobiology, 45*(1), 25. https://doi.org/10.1007/s10571-025-01543-3.

Tracey, T. J., Steyn, F. J., Wolvetang, E. J., & Ngo, S. T. (2018). Neuronal lipid metabolism: Multiple pathways driving functional outcomes in health and disease. *Frontiers in Molecular Neuroscience, 11*. https://doi.org/10.3389/fnmol.2018.00010.

Troutwine, B. R., Hamid, L., Lysaker, C. R., Strope, T. A., & Wilkins, H. M. (2022). Apolipoprotein E and AD. *Acta Pharmaceutica Sinica B, 12*(2), 496–510. https://doi.org/10.1016/j.apsb.2021.10.002.

Vallés, A. S., & Barrantes, F. J. (2022). The synaptic lipidome in health and disease. *Biochimica et Biophysica Acta (BBA) - Biomembranes, 1864*(11), 184033. https://doi.org/10.1016/j.bbamem.2022.184033.

van Kruining, D., Luo, Q., van Echten-Deckert, G., Mielke, M. M., Bowman, A., Ellis, S., ... Martinez-Martinez, P. (2020). Sphingolipids as prognostic biomarkers of neurodegeneration, neuroinflammation, and psychiatric diseases and their emerging role in lipidomic investigation methods. *Advanced Drug Delivery Reviews, 159*, 232–244. https://doi.org/10.1016/j.addr.2020.04.009.

Varga, K., Jiang, Z.-J., & Gong, L.-W. (2020). Phosphatidylserine is critical for vesicle fission during clathrin-mediated endocytosis. *Journal of Neurochemistry, 152*(1), 48–60. https://doi.org/10.1111/jnc.14886.

Viola, K. L., & Klein, W. L. (2015). Amyloid β oligomers in AD pathogenesis, treatment, and diagnosis. *Acta Neuropathologica, 129*(2), 183–206. https://doi.org/10.1007/s00401-015-1386-3.

Wang, W., Zhao, F., Ma, X., Perry, G., & Zhu, X. (2020). Mitochondria dysfunction in the pathogenesis of AD: Recent advances. *Molecular Neurodegeneration, 15*(1), 30. https://doi.org/10.1186/s13024-020-00376-6.

Warda, M., Tekin, S., Gamal, M., Khafaga, N., Çelebi, F., & Tarantino, G. (2025). Lipid rafts: Novel therapeutic targets for metabolic, neurodegenerative, oncological, and cardiovascular diseases. *Lipids in Health and Disease, 24*(1), 147. https://doi.org/10.1186/s12944-025-02563-0.

Watanabe, Y., Kasuga, K., Tokutake, T., Kitamura, K., Ikeuchi, T., & Nakamura, K. (2021). Alterations in glycerolipid and fatty acid metabolic pathways in AD identified by urinary metabolic profiling: A pilot study. *Frontiers in Neurology, 12*. https://doi.org/10.3389/fneur.2021.719159.

Westphal Filho, F. L., Moss Lopes, P. R., Menegaz de Almeida, A., Sano, V. K. T., Tamashiro, F. M., Gonçalves, O. R., ... Silveira Feitoza, P. V. (2025). Statin use and dementia risk: A systematic review and updated meta-analysis. *Alzheimer's & Dementia: Translational Research & Clinical Interventions, 11*(1), e70039. https://doi.org/10.1002/trc2.70039.

Wu, S., Liu, X., Yang, H., Ma, W., & Qin, Z. (2023). The effect of lipid metabolism on age-associated cognitive decline: Lessons learned from model organisms and human. *IBRO Neuroscience Reports, 15*, 165–169. https://doi.org/10.1016/j.ibneur.2023.08.2194.

Yang, L. G., March, Z. M., Stephenson, R. A., & Narayan, P. S. (2023). Apolipoprotein E in lipid metabolism and neurodegenerative disease. *Trends in Endocrinology & Metabolism, 34*(8), 430–445. https://doi.org/10.1016/j.tem.2023.05.002.

Yin, F. (2023). Lipid metabolism and AD: Clinical evidence, mechanistic link and therapeutic promise. *The FEBS Journal, 290*(6), 1420–1453. https://doi.org/10.1111/febs.16344.

Young, S. A., Mina, J. G., Denny, P. W., & Smith, T. K. (2012). Sphingolipid and ceramide homeostasis: Potential therapeutic targets. *Biochemistry Research International, 2012*(1), 248135. https://doi.org/10.1155/2012/248135.

Yu, W., Jin, H., & Huang, Y. (2021). Mitochondria-associated membranes (MAMs): A potential therapeutic target for treating AD. *Clinical Science, 135*(1), 109–126. https://doi.org/10.1042/CS20200844.

Yui, K., Imataka, G., Nakamura, H., Ohara, N., & Naito, Y. (n.d.). Eicosanoids derived from arachidonic acid and their family prostaglandins and cyclooxygenase in psychiatric disorders. *Current Neuropharmacology, 13*(6), 776–785. https://doi.org/10.2174/1570159×13666151102103305.

Zilberter, Y., & Zilberter, M. (2017). The vicious circle of hypometabolism in neurodegenerative diseases: Ways and mechanisms of metabolic correction. *Journal of Neuroscience Research, 95*(11), 2217–2235. https://doi.org/10.1002/jnr.24064.

CHAPTER TWELVE

Metabolic dysregulation in Alzheimer's disease: Mechanisms, markers, and therapeutic prospects

Radhika Mittal[a], Priya[a], Jyotirmoy Banerjee[b], and Aparna Dixit[a,*]
[a]Dr. B R Ambedkar Centre for Biomedical Research, University of Delhi, Delhi, India
[b]Department of Biophysics, All India Institute of Medical Sciences, Delhi, India
*Corresponding author. e-mail address: aparnabanerjeedixit@gmail.com

Contents

International Review of Neurobiology, Volume 188
ISSN 0074-7742, https://doi.org/10.1016/bs.irn.2026.05.011

Abstract

Metabolic dysregulation has emerged as a crucial pathogenic factor that contributes in progression of Alzheimer's disease (AD), and is often found to precede classical AD's pathologies, the amyloid-β accumulation and hyperphosphorylated tau proteinopathies. The key metabolic underpinnings associated with Alzheimer's disease (AD) includes cerebral glucose hypometabolism, insulin resistance, mitochondrial dysfunction, altered lipid metabolism, vascular and systemic metabolic impairments, disrupted amino acid and nitrogen metabolism driving secondary metabolic disturbances. The chapter outlines the current evidences on dysregulated metabolic processes, and highlights emerging metabolic biomarkers that are identified through advanced neuroimaging modalities, plasma/cerebrospinal fluid (CSF) profiles, lipidomic signatures and markers of mitochondrial impairment, underscoring their diagnostic and prognostic potential. Furthermore, the chapter discusses about the therapeutic prospects targeting metabolic pathways, addressing current challenges in development of therapeutic strategies, reinforcing the need for integrative and precision-based interventions for early diagnosis and disease modifying therapeutic strategies in AD.

1. Introduction

Alzheimer's disease is one of the complex neurodegenerative disorders which is characterised by progressive memory loss and cognitive dysfunction such as language and reasoning impairments, visual-spatial abilities and when severe, interferes with one's day to day activities. It is one of the leading causes of dementia worldwide, and the count of individuals suffering from AD has been manifested to reach to 78 million by 2030 and 139 million by 2050 (WHO, 2021). The increased count of people affected with AD imposes significant healthcare and socio-economic burden. Even though there is extensive research ongoing, yet current available therapies provide partial relief symptomatically and does not ensure permanent cure from the disease, thereby fails in mitigating disease progression, creating an alarming need for development of specific disease-modifying interventions (Zhang et al., 2024).

Conventionally, Alzheimer's disease pathology has been defined by amyloid-beta (Aβ) accumulation, hyperphosphorylation of tau protein, leading to plaque deposition and neurofibrillary tangles (NFTs) driving neuroinflammation and neurodegeneration (Cummings, 2021). While these hallmarks are central to AD pathology yet repeated failure of amyloid-tau targeted therapies clinically has led to a shift in AD research paradigm towards other integrative disease models (Li et al., 2021).

Evidence from advanced neuroimaging studies have revealed that amyloid and tau pathologies does not necessarily correlate with the cognitive decline, and these amyloid-tau hypotheses alone could not explain the heterogenous clinical manifestations observed across AD patients (Ottoy et al., 2025). Metabolic dysfunctions has emerged as one of the critical, yet often underappreciated component of disease pathogenesis (Kang et al., 2017). The systemic metabolic dysfunction involves impaired brain glucose utilisation, insulin resistance, altered lipid metabolism, mitochondrial dysfunction and oxidative stress, collectively contributes to neurodegeneration (Song et al., 2021).

One of the critical observations in relation with the metabolic paradigm in AD has been recognised with consistent hypometabolism of brain glucose detected years before the clinical onset of AD symptoms, as is being assessed with the help of neuroimaging techniques such as FDG-PET imaging (Calsolaro & Edison, 2016; Mosconi, 2013a). The early detection of metabolic dysfunction has attributed as a more sensitive and reliable biomarker of cognitive decline than amyloid deposition alone, suggesting potential fundamental role of altered metabolism in AD disease progression. The temporal precedence of metabolic abnormalities over pathological hallmarks presents energy metabolism as a primary mechanism in disease pathogenesis (Counts et al., 2017a; Bai & Yuan, 2023).

AD has been characterised as a multifactorial metabolic disorder where the various interconnected bioenergetic systems are concurrently compromised. The metabolic dysfunction in AD constitutes alteration in the multiple biological processes including intracellular energy production to disrupted glucose homeostasis. An integrative approach for targeting these connected metabolic nodes would pave the way for novel and potentially effective therapeutics rather than single target interventions. This review based on current evidence deciphers metabolic dysfunction as not just a downstream consequence of AD but as a fundamental factor that influences synaptic dysfunction and cognitive decline. It now becomes imperative to understand the metabolic alterations occurring in systemic and nervous system and their influence on disease initiation and progression. Identifying metabolic biomarkers for early detection and new therapeutic strategies targeting specific molecules altered in metabolic pathways represents an emerging frontier for development of disease modifying interventions in AD.

2. Metabolic dysregulation as a hallmark of AD

Even though brain comprises of only 2 % of total body weight, yet the energy demand of brain is exceptionally high, accounting for about 20–25 % of total energy production, making it more vulnerable to alterations associated with metabolic dysfunction (Camandola & Mattson, 2017). The high energy requirement is for the primary functions of the brain including neurotransmission, synaptic plasticity and memory formation and these processes depends on tightly regulated energy coupling reaction occurring between neurons and astrocytes. The neuron-astrocytic lactate shuttle system, where lactate produced by astrocytes is exported to support the mitochondrial respiration of the neurons. This interaction among the energy system compartments exemplifies its functional importance for maintenance of cognitive outcomes (Bélanger et al., 2011). The dysregulation associated with this integrated metabolic system directly imposes threat to neuronal survival and synaptic integrity (Watts et al., 2018).

The disturbance in metabolic system has emerged as early biomarkers of AD pathology, preceding conventional pathological hallmarks such as cognitive decline and amyloid-tau hypotheses by years or even decades. Longitudinal neuroimaging studies utilising FDG-PET has very well shown glucose hypometabolism progressively occurring in healthy individuals, and years later, they developed AD, highlighting metabolic dysfunction as a prodromal indicator (Mosconi et al., 2008a). Impaired cerebral glucose metabolism has been consistently observed in individuals at high risk of AD and this hypometabolic state is implicated with subsequent cognitive decline and correlates with early pathological changes, suggesting energy metabolism an initial event in AD pathophysiology (Mosconi et al., 2008b; Chen & Zhong, 2013). The reduction in glucose uptake and utilisation in glucose hypometabolic state leads to the altered activity of glucose transporters (e.g., GLUT1, GLUT3) and impaired insulin signaling pathways in brain, resulting in compromised substrate availability for ATP production (Simpson et al., 1994; Hölscher, 2019).

In addition, mitochondrial dysfunction occurring in presymptomatic stages of AD is also an early indicator of AD pathology that has emerge to potentially establish a foundation for subsequent neuropathological cascades (Swerdlow, 2018). Mitochondria is an essential organelle that is responsible for cell bioenergetics, regulates cellular calcium homeostasis, reactive oxygen species production, and apoptosis (Wang et al., 2020a). Multiple evidences demonstrated mitochondrial dysfunction in AD pathology such as altered activity of electron transport chain system, particularly complex IV (cytochrome c oxidase) and ATP synthase, with significant reduction in their activity in brains, platelets, and in peripheral cells of patients with AD, and in animal models (Du et al., 2010). These functional alterations lead to structural anomalies, affected mitochondrial dynamics (fission/fusion imbalance) (Wang et al., 2009), and altered mitochondrial morphology such as disrupted cristae architecture, which collectively results in energy shortfall across neurons and glia which not only impairs neuronal function but also fosters oxidative stress, synaptic dysfunction, and vulnerability to amyloid- and tau-mediated toxicity (Fernández-Checa et al., 2010). Hence, the alterations occurring in the metabolic system in early stage influences the pathological outcomes of the disease, highlighting energy dysfunction as one of the primary causes influencing the AD progression.

Several mechanistic studies delineate the close interplay between metabolic disturbances and neuropathological hallmarks of AD. Reduced astrocytic lactate production has been correlated with altered synaptic plasticity and long-term potentiation and is also implicated with memory formation (Suzuki et al., 2011; Wu et al., 2023). Furthermore, glucose hypometabolism has been known to disrupt energy dependent proteostasis, which leads to accumulation of misfolded amyloid-β protein and hyperphosphorylation of tau protein. This pathogenic cascade works bidirectionally creating a vicious cycle where dysfunction of metabolic processes leads to pathological protein accumulation, further exacerbating metabolic dysfunction (DuBoff et al., 2012). The initial compensatory metabolic responses such as shift towards alternate fuel substrates, are adaptive mechanisms, but with sustained activation they turn detrimental, causing deceleration in the bioenergetic system and network instability, culminating in cognitive failure (Constantino et al., 2025). Moreover, sex-specific alterations disproportionately shapes the vulnerability for the disease, wherein chronological transitions occurring during midlife, particularly in women, exhibiting distinct metabolic phenotypes substantially contributes to AD risk (Ding et al., 2013). Understanding these heterogenous phenotypes becomes important for development of precise therapeutic interventions, specifically targeting the individual's distinct vulnerabilities underlying disease trajectories.

3. Key metabolic pathways altered in Alzheimer's disease: Mechanisms and dysregulation

3.1 Glucose metabolism and insulin resistance

Glucose, being a primary energy source, is an exclusive substrate for energy production that is required by brain for neuronal survival, synaptic transmission and plasticity. The glucose metabolism is tightly regulated by a couple of processes including glycolysis, tricarboxylic acid (TCA) cycle, and mitochondrial oxidative phosphorylation, which are coordinated in a way that generates ATP, supporting action potential generation and propagation, neurotransmitter recycling and ion gradient across membranes (Zhang et al., 2021). It has been observed that in Alzheimer's disease, the fine-tuned metabolic network becomes progressively disrupted, leading to a state of cerebral glucose hypometabolism, which is known to be the earliest and most consistent pathophysiological phenomena of the disorder (Mosconi et al., 2008b; Adlimoghaddam et al., 2019). The decreased expression of glucose transporters and reduction in glucose utilisation in brain is directly correlated with cognitive decline, suggesting metabolic disruption as an active cause of disease progression and not just a consequence (Harr et al., 1995; Liu et al., 2008).

Altered glucose metabolism has now been recognised as a reliable and sensitive biomarker for predicting cognition, represented as a critical hallmark of AD. Positron emission tomography using 18F-fluorodeoxyglucose (FDG-PET) revealed distinct regions of glucose hypometabolism in brain, particularly tempo-parietal and frontal cortical regions (Hirono et al., 1998; Förster et al., 2012). Various studies suggest the temporal progression of metabolic alterations as a predictive marker for disease such as few studies reported amyloid and tau biomarkers as early predictors, while glucose hypometabolism is dominant in later stages (Hammond et al., 2020). Such stage dependent alterations implied the independent action as well as interconnected relation of metabolic dysfunction with neurodegeneration, presenting it not just as a downstream consequence of amyloid tau dysfunction but also primary driver of disease pathology.

Insulin resistance as hypothesised in Type 3 Diabetes has emerged as an integral component of Alzheimer's disease pathology, where metabolic dysfunction is driven by impaired insulin signaling in brain (Ferreira et al., 2018; Chapple et al., 2025; Andrade et al., n.d.). The insulin resistance may develop years before its clinical manifestations, and is often accompanied by reduced insulin availability in CSF. The insulin deficiency in CSF is known to affect downstream signaling cascades such as PI3K-Akt pathway, which plays an instrumental role in regulating synaptic neurotransmission, neuronal plasticity, and thereby memory formation (Gabbouj et al., 2019). The impaired expression and functional availability of primary cerebral glucose transporters mainly GLUT1 and GLUT3 is further influenced by disrupted insulin signaling in brain, exacerbating deficits in uptake and utilisation of cerebral glucose (Simpson et al., 1994). This cerebral glucose hypometabolism compromises with neural functions and synaptic integrity, progressively leading to cognitive decline. Interestingly, the altered cerebral metabolism and pathological hallmarks associated with AD are linked bidirectionally where one reinforces other, contributing to disease progression. Impaired insulin signaling leads to amyloidogenic processing of amyloid precursor protein (APP), promoting amyloid-beta (Aβ) accumulation (Wang et al., 2010; Chao et al., 2016). On the other hand, soluble Aβ and tau oligomers affects insulin receptor signaling via distinct mechanisms, including receptor internalisation and aberrant phosphorylation of insulin receptor substrates (Zhao et al., 2008; Ma et al., 2009). This pathogenic feed-forward loop worsens the disease severity by accelerating protein aggregation which further exacerbates metabolic dysfunction. Another factor that is in closely associated with cerebral glucose hypometabolism is thiamine diphosphate (TDP), which is an essential cofactor of key glucose metabolising enzymes, particularly pyruvate dehydrogenase and α-ketoglutarate dehydrogenase. The reduced availability of TDP in brain have been implicated with cerebral glucose hypometabolic patterns observed in FDG-PET, and is associated with cognitive decline (Sang et al., 2018). Collectively, these findings suggest the potential involvement of distinct interconnected pathways that drives metabolic dysfunction in Alzheimer's disease.

The distinct alterations in the glucose metabolism in early and later stage AD reveals significant pathological nuances that are implicated with specific therapeutic timeline and precise target selection. Region specific glucose hypermetabolism has been observed in early-stage AD, often deciphered to be a compensatory response to initial metabolic stress and network dysfunction (Martín-Saladich et al., 2025). But with sustained disease progression, these adaptive mechanisms turn exhaustive and ultimately leads to cerebral hypometabolism with increased neurodegeneration (Galli et al., 2025). It now becomes imperative to understand the temporal alterations that are crucial to Alzheimer's pathology, so as to develop precise, stage-specific and target specific therapeutic strategies for maximal efficacy. Several neuroimaging techniques utilising FDG-PET, Magnetic resonance imaging

(MRI), functional MRI (fMRI) and tau-PET have revealed glucose hypometabolism in normal cognition but amyloid positive individuals, and the altered glucose metabolism has emerged years before the first symptom onset in AD, suggesting coherence or precedence of metabolic dysfunction with amyloid-tau pathology of Alzheimer's disease (de Leon et al., 2001; Mosconi et al., 2008b), exemplifying metabolic dysfunction as a crucial pathogenic mechanism, highlighting it as a primary target for disease modifying strategies. Glucose hypometabolism, insulin resistance and cofactor deficiency associated with the AD pathology altogether highlights that metabolic dysfunction significantly contributes to AD progression independently, yet in association with classical amyloid-tau pathologies (Fig. 1).

3.2 Mitochondrial dysfunction and bioenergetics failure

Mitochondrial dysfunction represents another fundamental feature of AD pathology, characterised by mitochondrial DNA damage, imbalance in fusion-fission cycle, and altered axonal transport (Cabezas-Opazo et al., 2015).

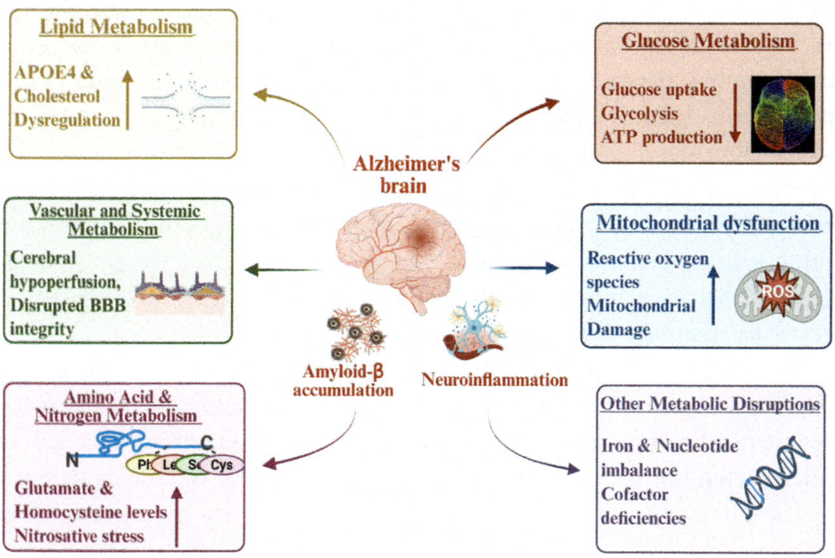

Fig. 1 Key metabolic pathways altered in Alzheimer's disease (AD). Impaired glucose and lipid metabolism, amino acid and nitrogen imbalance, and mitochondrial dysfunction reduce ATP production and increase oxidative stress. Secondary disturbances in iron, carbon, nucleotide, and cofactor metabolism promote oxidative damage, ferroptosis, and amyloid-β aggregation. Together, these interconnected metabolic impairments drive energy deficit, neuroinflammation, and neurodegeneration in AD.

These alterations result in reduced ATP production and increased ROS generation, contributing to bioenergetic failure and oxidative stress mediated neuronal damage (Nabi et al., 2022). Altered mitochondrial dynamics avoids compensatory events through fusion/fission cycle, that usually aids in restoring membrane potential and ATP synthesis capacity (Mani et al., 2021). The mitochondrial autophagy is essential for removal of dysfunctional organelles, if impaired, further contributes to oxidative damage (Goudarzi et al., 2021; Du et al., 2017). This collapse in energy levels directly affects neural processes including maintenance of synaptic transmission and neuroplasticity, thereby mitochondrial quality control is crucial for regulating functionality of synapses (Cai & Tammineni, 2017). It has also been studied that at molecular level, the loss of protein sensitive to oligomycin at F1-F0 subunit of ATP synthase (Complex V), coupled with amyloid-beta, has been observed to cause impairment in assembly of ATP synthase leading to altered catalytic function, thereby leads to energy deficit implicated in AD pathology (Antoniel et al., 2014; Giorgio et al., 2019).

Another complex of electron transport chain is Complex I (NADH reductase or ubiquinone oxidoreductase), one of the most complex enzymes that has been found to express pronounce disruption in AD brain tissues. It constitutes 44 distinct polypeptides, a multi-subunit respiratory enzyme, known to be vulnerable to assembly defects and age associated degeneration. Altered activity of Complex I significantly decrease the ATP production efficiency, and further promotes ROS production (Gao et al., 2025). It has also been noted that mutations in gene encoding mitochondrial ribosomal proteins, particularly PTCD1 (pentatricopeptide repeat protein 1) has been observed to be in higher number in AD patients. These variants further contributing to impairment of Complex I assembly, reduce 16S rRNA levels in mitochondria, thereby leading to altered efficiency of oxidative phosphorylation process, linking bioenergetic failure in AD with genetic predisposition. Under metabolic stress, cells possessing AD associated PTCD1 mutations exhibited energy deficiency, a pathophysiological state seen in ageing neurons (Fleck et al., 2019).

Other than alteration in enzyme complexes, the subsequent remodeling pattern in mitochondria, the fusion and fission, has found to have impaired balance. Excessive mitochondrial fission produces fragmented organelles with complementary defects of selective autophagy, leading to accumulation of defective organelles inside the cells, elevating oxidative stress (Pradeepkiran & Reddy, 2020; Liu et al., 2025). This dysregulation further depletes cellular bioenergetic reserves worsening progressive metabolic crisis. The disrupted

OXPHOS along with oxidative stress creates destructive pathogenic mechanisms, where dysfunctional electron transport with excessive ROS also compromises with anti-oxidant defense capabilities which are ATP dependent (Briston & Hicks, 2018; Zong et al., 2024). The increase in ROS production further destroys other molecules such as lipids, proteins and mtDNA (mitochondrial DNA), exacerbating initial damage into a destructive loop (Bokov et al., 2004; Giorgi et al., 2018). In the cases of genetic transcendents of AD, mutations in presenilin-1 (PS1) amplify the energy crisis by destroying membranes of endoplasmic reticulum and mitochondria (MAMs) and impaired calcium signaling, altering ATP depending calcium transport, thereby disrupting ATP and calcium homeostasis (Han et al., 2021; Area-Gomez et al., 2009). With age, neurons lose the capacity to generate new mitochondrial protein and is unable to restore its enzymatic function due to accumulation of mtDNA mutations, hence is another cause of energy deficit (Kukreja et al., 2014) (Table 1).

Table 1 Key mitochondrial functional parameters altered in Alzheimer's disease.

Mitochondrial parameters	AD associated alterations	References
ROS Production	Elevated ROS levels due to impaired electron transport chain activity.	Misrani et al. (2021)
Mitochondrial Membrane Potential	Depolarisation of inner mitochondrial membrane	Kerr et al. (2017)
Complex I-V activity	Progressive impairment in electron transport chain enzyme function, including NADH dehydrogenase and cytochrome c reductase.	Maurer et al. (2000), Wang et al. (2020b)
Calcium homeostasis	Dysregulated mitochondrial calcium uptake and impaired storage capacity.	Jadiya et al. (2019)
Mitophagy capacity	Reduced clearance of damaged mitochondria	Zeng et al. (2022), Park et al. (2020)
Mitochondrial DNA integrity	Accumulation of point mutations, deletions, and oxidative damage to mtDNA	Hoekstra et al. (2016)

3.3 Lipid metabolism and cholesterol homeostasis

CNS is one of the lipid-rich organs constituting about 25 % of total body cholesterol (Wuerch & Yong, 2025). Among metabolic alterations, lipid dyshomeostasis have emerged to play a crucial role in the pathogenesis of AD, accelerating amyloid tau proteinopathy. The lipid alterations have been recognised among the first neuropathological agents by Alois Alzheimer in 1907 (Alzheimer et al., 1995). Apolipoprotein E (ApoE), a lipid carrier known to pose significant lipid related genetic risk factor in AD pathology. Its primary function is to regulate the transport, packaging and distribution of lipids across all body parts (Budny et al., 2025). ApoE4 allele have demonstrated high risk for late onset AD, exhibiting dysregulated lipid metabolism and reduced efficiency in cholesterol transport (Corder et al., 1993). ApoE4 carriers promotes amyloid beta accumulation, altering lipid binding capacity compared to ApoE2 and ApoE3 (Kanekiyo et al., 2014). These abnormalities cause reduction in cholesterol transport to neurons, leading to impaired lipid metabolism, further compromising with neuronal membrane integrity and impaired amyloid-β clearance, thereby accelerating neurodegeneration. The lipid distribution in astrocytes is also disrupted by ApoE4, which further promotes oxidative stress and neuroinflammatory responses (Yang et al., 2023a). The genetic predisposition of lipids is linked with amyloid dependent and independent pathways, implicating altered lipid homeostasis with AD pathology.

The altered levels of sphingolipids, ceramides, phospholipids and many other bioactive lipids, have emerged to be crucial pathological hallmarks of Alzheimer's disease progression (Li et al., 2025). Importantly, elevated levels of ceramide in brains and plasma along with compromised sphingosine metabolism have been observed, and is known to contribute primarily in synaptic dysfunction and protein aggregation (Chakrabarti et al., n.d.). The high ceramide levels and cognitive dysfunction are strongly correlated, and mainly affects mitochondrial functions, primarily because of the stabilisation of β-secretase (BACE1) and altered lipid composition of the membrane (Pan et al., 2024). Notably, ceramides also influence electron transport chain complex function and activity, mainly complex I and III, further lead to increased ROS levels and ATP depletion (Villena et al., 2008). Moreover, in AD brains, there is increased sphingomyelin conversion into ceramides, reflecting altered membrane integrity and lipid raft composition, eventually forming a loop which further promotes amyloidogenic accumulation and synaptic dysfunction (Kalkman & Smigielski, 2025; Cao et al., 2024a; He et al., 2010). These disruptions in lipid metabolism during AD progression suggests their utility as crucial early biomarkers in AD.

Cholesterol, an important class of lipid family, plays vital role in regulating processes such as synaptogenesis, neurotransmission and membrane fluidity (Pfrieger, 2003; Guo et al., 2024). Cholesterol influences amyloid precursor protein (APP), as with its increased neuronal accumulation, APP localisation occurs within lipid rafts, driving Aβ production. These lipid rafts microdomains regulates β-secretase and γ-secretase binding, eventually aids in APP cleavage and Aβ pathogenesis (Xiu et al., 2006; Cao et al., 2024b). Moreover, the oxidised derivatives of cholesterol, such as oxysterol, regulates low density lipoprotein (LDL) receptor expression and sterol regulatory element binding protein (Choi & Finlay, 2020). The impaired cholesterol metabolism also affects the bile acid production, influencing gut microbiota composition, and further reduces secondary bile acids production (Kriaa et al., 2019). These alterations imposed detrimental effects on CNS metabolism through multiple pathways, such as bacterial LPS translocation across compromised intestinal barrier and altered TGR5 and FXR (farnesoid X) receptor signaling, promoting neuroinflammation (Wang et al., 2020; Balakrishnan et al., 2025). Notably, peripheral alterations in lipid metabolism directly influences systemic inflammation and blood brain barrier (BBB) integrity, thereby modulating neuroinflammatory responses exacerbating neuronal damage (Van Dyken & Lacoste, 2018a; Ge et al., 2025). Several epidemiological and experimental studies suggested a mechanistic link between hypercholesterolemia, dyslipidemia and metabolic syndrome with increased AD risk.

3.4 Vascular and systemic metabolic contributions

Impaired systemic metabolism and vascular dysfunctions represent one of the crucial elements of metabolic dysregulation in AD, yet not much recognised. Cerebral hypoperfusion emerges as early and critical metabolic impairments in Alzheimer's pathology, which is known to disrupt the biomolecules (such as glucose, oxygen and lipids) transportation to the brain regions demanding high energy (Love & Miners, 2016). Inadequate availability of substrates leads to energy decline in neuronal populations, mainly in hippocampal and prefrontal cortical regions, which are crucial for memory formation and cognition. Furthermore, hypoperfusion in brain induces oxidative stress and mitochondrial dysfunction, promoting amyloid-β protein accumulation and phosphorylation of tau (Lv et al., 2025).

Another vascular dysfunction includes impaired blood-brain barrier (BBB) integrity where permittance of selective biomolecules across brain is affected and transport of essential metabolites and vital energy substrates becomes severely compromised, disrupting cerebral metabolism. Due to structural and functional breakdown of BBB, peripheral metabolites, inflammatory mediators, and lipoproteins abruptly crosses brain parenchyma (Huang et al., 2020; Yin et al., 2026). Glucose and lipid homeostasis is disrupted due to this metabolic leakage, causing neuron-astrocytic energy coupling impairment, attenuating amyloid-β clearance mechanisms. Cerebral amyloid angiopathy (CAA), a pathological condition where amyloid-β gets accumulated within blood vessels, also contributes to structural and functional abnormalities in cerebral vasculature, affecting the interaction between pericytes and endothelial cells, leading to reduced capacity of capillary perfusion, enhancing leakage of plasma components and toxic substances to brain (Deng et al., 2022). The sustained dysfunction of the endothelium causes reduction in cerebral perfusion, exacerbating energy substrate deficits, creating metabolic crisis in the ageing brain (Vargas-George & Dave, 2022).

The impairments in the systemic, vascular and cerebral metabolism represents a web of multiple interconnected pathologies including dyslipidemia, type 2 diabetes mellitus (T2DM), obesity, hypertension, and the synergistic effect of these alterations have substantial contribution in elevating AD risk, mainly disrupting neurovascular function and synaptic integrity (Song et al., 2025). The perturbed adipokine signaling associated with obesity, hyperglycemia, chronic inflammation and Type 2 Diabetes Mellitus, is known to influence endothelial dysfunction, disrupts BBB, impairs glucose uptake and metabolism, thereby modulates neuroinflammatory responses (Van Dyken & Lacoste, 2018b). Such alterations in peripheral tissues induces systemic inflammatory milieu, breaching BBB and amplifying neuroinflammation with disease progression.

3.5 Amino Acid–Nitrogen imbalance and secondary metabolic dysfunction in AD

Alteration in the excitatory and inhibitory amino acids, notably glutamate, GABA, and serine, are a result of disturbances in the amino acid and nitrogen metabolism. These perturbations mainly affect neurotransmission across synapses, which leads to excitotoxicity and synaptic dysfunction (Sears & Hewett, 2021). The increase in glutamate levels induces

Table 2 Comparison of altered metabolic pathways in Alzheimer's disease.

Metabolic pathway	Physiological function	Metabolic deficit in AD	Functional consequences	References
Glucose Metabolism	Glycolysis and oxidative phosphorylation provide fuel for neuronal functions; GLUT1/GLUT3 mediate active glucose uptake	Impaired glucose uptake in posterior cingulate and precuneus regions; altered expression of glucose transporter; Brain insulin resistance	Reduced ATP generation; synaptic dysfunction, early cognitive decline	Szablewski (2021), Amin et al. (2023)
Lipid Metabolism	APOE facilitates cholesterol transport and neuronal membrane maintenance; membrane lipid composition supports receptor function	APOE4 exhibits reduced lipid transport efficiency; impaired cholesterol efflux; altered fatty acid composition; elevated lipid peroxidation	Increased amyloid-β accumulation; compromised membrane integrity; impaired protein trafficking	Yang et al. (2023b), Chen et al. (2025)
Mitochondrial Functions	Oxidative phosphorylation generates ATP; mitochondrial dynamics (fusion/fission) maintain organellar quality; efficient axonal transport delivers mitochondria to synaptic terminals	mtDNA mutations; impaired fusion-fission dynamics; defective mitochondrial autophagy; compromised axonal transport	Reduced ATP production; increased reactive oxygen species accumulation; bioenergetic collapse	Zhu et al. (2013), Flannery and Trushina (2019)
Vascular and Systemic Metabolism	Maintained cerebral blood flow, energy homeostasis	Cerebral hypoperfusion; impaired BBB integrity; Cerebral amyloid angiopathy (CAA)	Energy deficit, neuroinflammation, synaptic dysfunction, and impaired clearance	Murray et al. (2011)

Amino acid and Nitrogen metabolism	Glutamate and D-serine homeostasis regulate excitatory and inhibitory neurotransmission; normal homocysteine levels support methylation reactions	Impaired glutamate clearance; reduced D-serine production; elevated homocysteine; defective urea cycle function	Excitotoxicity and calcium overload; increased oxidative stress and neuroinflammation	Nasiri et al. (2025a), Al-Thani et al. (2025b)
Secondary Metabolites	Iron-dependent enzymes support oxidative phosphorylation and antioxidant defense; intact one-carbon metabolism supports methylation and trans-sulfuration pathways	Iron accumulation in selective brain regions; nucleotide synthesis defects; altered one-carbon metabolism	Enhanced Fenton-catalyzed oxidative stress; impaired DNA repair; defective antioxidant defense	Chen et al. (2025), Lionaki et al., 2022

calcium load and excitotoxic signaling in susceptible neurons driving network reorganisation whereas low levels of serine alter the N-methyl-D-aspartate (NMDA) receptor function essential for synaptic transmission (Castillo-Vazquez et al., 2024). In tandem, alterations in nitrogen metabolism including impaired urea cycle, causes accumulation of urea and ammonia in brain, augmenting neurotoxicity and cognitive decline in AD (Al-Thani et al., 2025a).

Other than primary metabolic pathway, alterations in levels of certain secondary metabolites such as iron, carbon, nucleotide and cofactors, also influence the cellular bioenergetic mechanisms and epigenetics (Yassine et al., 2023). Accumulation of iron in brain regions results in oxidative stress, and promotes ferroptosis, mitochondrial dysfunction and amyloid-β aggregation (Majerníková et al., 2024). The alterations in nucleotide metabolism induces impaired DNA synthesis and repair and compromised genome integrity, thereby accelerates the process of neurodegeneration (Wang et al., 2021). The reduced enzyme efficiency due to low levels of metabolic cofactors such as NAD^+, FAD and thiamine foster metabolic failure and cognitive decline (Ramamoorthy et al., 2022; Cobilinschi et al., 2024). The disruptions in secondary metabolites levels creates a permissive environment for primary pathogenic alterations including metabolic dysfunction mediating neuroinflammation and cognitive impairment with the progression of AD.

Collectively, these findings suggests that Alzheimer's is not solely centred to amyloid-β/tau pathology, but is a system level metabolic disorder in which aberrant metabolic activity of key pathways exacerbates synaptic dysfunction, neuroinflammation and cognitive decline (Table 2). Modulating these metabolic nodes offers promising avenues for biomarker discovery and therapeutic interventions.

4. Emerging biomarkers of metabolic dysfunction in AD

4.1 Imaging-based metabolic biomarkers

With the advances in neuroimaging, biofluid profiling and metabolomics, the identification of emerging biomarkers reflecting metabolic alterations in Alzheimer's disease brings promise for early diagnosis, disease stratification and therapeutic surveillance. Among neuroimaging techniques, FDG-PET is one of the most employed tools utilised in assessment of glucose metabolism in specific AD brain regions, particularly posterior cingulate cortex, precuneus, and tempo-parietal regions (Mattoli et al., 2023; Bouter et al., 2019). The hypometabolic state of cerebral glucose metabolism has been persistently observed in AD brain, often preceding cognitive decline and amyloid tau pathology, suggesting impaired glucose uptake and utilisation and synaptic dysfunction in affected neuronal populations (Blázquez et al., 2022). FDG-PET has been discovered as a sensitive marker for energy deficit state of neurons in prodromal and clinical AD (Ou et al., 2019). Beyond glucose metabolism, recent advanced have expanded metabolic biomarker toolkit, such as Magnetic Resonance Spectroscopy (MRS), which assess the metabolites in vivo, depicting lactate levels, alteration in glutamate-glutamine levels, altogether reflecting the neuronal integrity, glial activation and altered energy metabolism in AD (Mullins et al., 2018). 11C-acetate PET is another novel approach to assess reactive astrocytes in CNS through elevated levels of acetate uptake by monocarboxylate transporter-1 protein. The increased acetate uptake is correlated with reactive astrogliosis and aberrant GABA production, leading to inhibitory neurotransmission and decreased GLUT-3 expression linking astrocytic activation with the impaired glucose metabolism. Various reports have demonstrated significant 11C-acetate uptake in hippocampus, temporoparietal neocortex and entorhinal cortex in AD patients in comparison with healthy controls (Duong et al., 2021). By integrating multimodal imaging techniques, correlating FDG-PET signals with MRI and tau-PET findings, an early and accurate diagnosis of disease stage and associated functional pathological outcomes can be made as these modalities collectively captures altered metabolic state, structural anomalies, and neurodegeneration.

4.2 Plasma and cerebrospinal fluid metabolic biomarkers

A broad spectrum of metabolic biomarkers has been recognised through metabolomic profiling of plasma and CSF that depicts the altered biomolecules in pathways that are fundamental to AD pathogenesis (Trushina et al., 2013). Altered CSF insulin levels with impaired insulin receptor expression and adipokines levels including leptin and adiponectin, lactate and ketone bodies in CSF and plasma are the biomarkers of disrupted glucose and energy homeostasis, suggesting a correlation with cognitive decline, synaptic dysfunction and neurodegeneration (Garcia-Garcia et al., 2024). A shift in energy substrate utilisation towards other alternative bio-substrates has also been depicted, reflecting mitochondrial deficits and adaptive metabolic reorganisation. Moreover, the metabolites of TKP pathway, particularly kynurenic acid and quinolinic acid levels have been found to be altered in CSF and plasma with increased indoleamine-2,3-dioxygenase 1 (IDO 1) expression in astrocytes, highlighting them as crucial biomarkers, linking neuroinflammation and excitotoxicity with impaired metabolism in AD (Jacobs et al., 2019; González-Sánchez et al., 2020; Chatterjee et al., 2018). In CSF, altered products of phosphocreatine and high levels of creatinine have been correlated with compromised BBB integrity, and its altered permeability, suggesting tight regulation between altered metabolism and vascular dysfunctions in AD (Johansona et al., 2018).

4.3 Lipidomic biomarkers and sex-specific metabolic signatures

Lipidomic profiling aids in assessing specific alterations in lipid classes that are crucial for AD risk and progression. There are several lipid classes that are discovered to be dysregulated prominently, including ceramides, sphingomyelin and phospholipids, and these alterations are indicative of disrupted synaptic vesicle dynamics and membrane instability (He et al., 2025). These dysregulated lipid species have emerged to be persistent in AD pathogenesis and have been implicated with cognitive decline (Wang et al., 2024). Stage dependent alterations have also been examined with specific lipid moiety where phosphatidylcholines (PC) and their derivatives lysoPC exhibits distinct expression, differentiating AD patients from healthy controls (Whiley et al., 2014). Sex specific metabolic signatures have also emerged as crucial stratification tool for predicting disease stage and respective treatment response (Chang et al., 2023). Altered amino acid levels and oxidative stress markers have been revealed from metabolomics studies and these alterations could be detected years before AD symptoms onset, and hence can be employed as early predictive biomarkers of AD (Xie et al., 2021; Tiwari et al., 2025). Integration of multi-omics approaches have further revealed altered lipid and amino acid metabolism in APOE4 carriers as compared to non-carriers, establishing a robust correlation between metabolic dysfunction and cognitive decline and further highlighting metabolic dysregulation as a primary contributing factor in AD progression and not just a downstream consequence (Ren et al., 2025a; Dong et al., 2024).

4.4 Mitochondrial dysfunction biomarkers

The biomarkers of mitochondrial dysfunction in Alzheimer's disease are indicative of disrupted redox balance, bioenergetic failure and impaired oxidative metabolism. The identification of these biomarkers has now become accessible through plasma derived extracellular vesicles (EV based biomarkers) which allows non-invasive assessment of bioenergetic processes (Tam et al., 2024). Various research studies have revealed altered expression levels of TCA cycle intermediates and β-oxidation products in CSF and plasma both, suggesting dysregulated utilisation of mitochondrial substrate (Wen et al., 2025). Further, reduced activity of electron transport chain (ETC) complexes, including complex I, IV and ATP synthase, with increased ROS production and oxidative stress markers such as lipid peroxidation products have been found in AD patients (Kriebel et al., 2020). The reduction in ETC complex is about 25–35 %, which is indicative of high bioenergetic failure of nervous system (Adav et al., 2019; Yao et al., 2021a). Integrating EV based biomarkers and neuroimaging findings, including PET imaging with complex I tracer have exhibited distinct regional expression in mitochondrial complex I and its reduced levels have been correlated with cognitive decline in early-stage AD (Chen et al., 2025; Liu et al., 2025) The decreased expression of NAA (N-acetylaspartate) assessed through MRI further strengthens the evidence of mitochondrial dysfunction (Demougeot et al., 2001). Collectively, these biomarkers aids in delineating the key processes associated with metabolic dysfunction in AD and potentially offer translational dimension for disease modifying interventions and targeted therapeutic strategies (Table 3).

The integration of imaging modalities (such as FDG-PET and structural MRI) and plasma metabolomic profiles with CSF biomarkers facilitates longitudinal diagnostic approaches with enhanced specificity and sensitivity of about 90 % for recognising amyloid-positive individuals (Fiandaca et al., 2014; Wang et al., 2025). The temporal expression of these biomarkers often precedes cognitive symptoms by 5–10 years, representing them as crucial tool for evaluation of disease risk during prodromal stages of Alzheimer's disease (Counts et al., 2017b). Multimodal metabolic biomarker assessment substantially potentiates the prediction of MCI (mild-cognitive impairment) to AD conversion, delineating the importance of systemic metabolic assessment across multiple biological processes.

Table 3 Comparative analysis of metabolic biomarkers.

Metabolic parameter	Analyte/ biomarker type	Biological compartment	Clinical utility	Stage relevance	References
Glucose Metabolism	FDG-PET standardised uptake value ratio	Brain (in vivo imaging)	Early detection, disease stage stratification, prognostic indicator of cognitive decline	All stages, dominant predictor in late-stage AD	Alhasan et al. (2025)
Insulin Signaling	CSF insulin levels, PI3K-Akt pathway markers	CSF	Reflects brain insulin resistance; independent of systemic metabolism	Early stages; prodromal MCI	Geijselaers et al. (n.d.)
TCA cycle intermediates	α-ketoglutarate, succinate, malate	Plasma and CSF	Mitochondrial dysfunction; bioenergetic reserve	Progressive MCI to mild AD	Mohan and Kumar, (2025)
Amino Acids	Glutamate, aspartate	Plasma	Protein metabolism and neurotransmitter function	Early clinical stages, utility in preclinical	Nasiri et al. (2025b)
Ceramides and Sphingomyelins	SM, ceramide species	Plasma and brain tissues	Core lipid dysregulation	Preclinical, prodromal and clinical stages	Peña-Bautista et al. (2018), Casas-Fernández et al. (2022)
Electron Transport Chain complexes	Complex I, III, IV and V levels	Plasma neuron-derived extracellular vesicles (NDEVs)	Primarily assesses neuronal mitochondrial function	Early AD, sensitive to disease progression	Badhwar and Haqqani (n.d.), Yao et al. (2021b)
Multi-omics approaches	Combined FDG-PET, lipid and ETC markers	Integrated multi-compartments	Comprehensive assessment of metabolic dysfunction, improves accuracy	All stages; patient stratification	Venkatesh et al. (2025)

5. Therapeutic prospects: Targeting metabolic dysregulation

With the emergence of metabolic dysregulation as a central hallmark of Alzheimer's pathology, the focus for the development of therapeutics targeting brain and systemic metabolism have gained attention. Amyloid-tau targeting therapeutic interventions solely does not promise the cure of AD, rather integrated targeting of altered molecules responsible for metabolism dysfunction would aid in improving brain energetics, redox balance, lipid homeostasis, further resolve the molecular vulnerabilities of the pathogenic or ageing brain.

5.1 Modulation of glucose metabolism

Based on converging evidences of cerebral glucose hypometabolism and brain insulin resistance, pharmacological interventions modulating glucose metabolism has emerged as one of the fundamental therapeutic approaches in treatment of AD. Insulin sensitizers such as Metformin, a widely used antidiabetic drug and GLP-1 receptor agonists aids in promoting glucose utilisation, and concomitantly activating neuroprotective signaling mechanisms such as PI3K-Akt and AMPK pathways (Ma et al., 2022; Nuamnaichati et al., 2020). Other than managing peripheral glucose levels, the usage of these drugs has exhibited improved cerebral glucose metabolism and better cognitive outcomes in both preclinical models and clinical trials (Urkon et al., 2025). It has also been observed that these drugs further aids in mitigating neuroinflammation and also contributes in modulating classical amyloid tau pathology, highlighting their therapeutic potential. Insulin supplementation through intranasal mode of delivery offers unique advantage of bypassing BBB, thereby its direct delivery to the brain, preserving synaptic plasticity and augmenting cognitive function, as reported during preclinical and clinical trials (Wong et al., 2024; Hanson & Frey, 2008). Further dietary interventions, including ketogenic and low-sugar diets, supports brain energy system by providing ketone bodies as a substitute fuel, compensating for disrupted glucose homeostasis (Shahpasand et al., 2024).

5.2 Mitochondrial-targeted therapeutic strategies

Mitochondrial directed pharmacological interventions exemplified another critical approach to restore cellular bioenergetic balance and synaptic function in AD. Antioxidants such as N-acetylcysteine and α-lipoic acid, coupled with coenzyme supplementation reduces oxidative stress and enhances electron transport chain function (Elbaky et al., 2018). Another drug, Nilotinib, a selective tyrosine kinase inhibitor, initially developed for leukemia, have been shown to promote mitochondrial biogenesis and metabolic recovery via mediating PGC-1α signaling pathway, thereby improving mitochondrial respiration efficacy in astrocytes in AD models (Adlimoghaddam et al., 2021; Srivastava et al., 2024). This further emphasise on repurposing of existing drugs for novel therapeutic targets. By restoring energy production and minimising oxidative burden, these interventions seek to preserve synaptic integrity with attempt to retard the neurodegenerative cascades (Kakoti et al., 2022).

5.3 Lipid modulators and APOE-related therapies

Lipid targeted interventions and APOE-related therapies aim to maintain cholesterol and lipid homeostasis, ensuring membrane stability, amyloid processing and attenuating neuroinflammation. Statins, a known cholesterol lowering agent, inhibits HMG-CoA reductase, restores lipid levels and aids in mitigating neuroinflammatory responses, promoting synaptic function (Fracassi et al., 2019). Moreover, Apolipoprotein E targeted therapies are centred to provide personalised treatment, emphasising on neutralising APOE4 liabilities or elevating LDL receptor activity to ensure lipid clearance restoring cholesterol homeostasis (Yamazaki et al., 2016; Williams et al., 2020). Omega-3 polyunsaturated fatty acids have been discovered for their primary role in regulating neuronal membrane integrity, and influencing anti-inflammatory and neuroprotective responses through specialised pathways (Dighriri et al., n.d.; Layé et al., 2018).

5.4 Nutritional and metabolic intervention strategies

Introducing ketogenic diet and restricting calorie intake promotes metabolic shift from glucose to ketone bodies as alternate cerebral energy substrate which further enhances the mitochondrial efficacy and restores the NADH/NAD^+ balance, thereby supports electron transport chain function (Qu et al., 2021; Zhu et al., 2022). One of the most abundant ketone body, beta-hydroxybutyrate, induce Nrf2 mediated antioxidant effects and concomitantly supresses activation of leucine-rich-repeat-containing protein 3 (NLRP3) inflammasome (Youm et al., 2015). Moreover, calorie restriction promotes autophagy and stimulate sirtuins, responsible for protein quality control maintenance and ensures metabolic flexibility (Russo et al., 2025). The non-pharmacological based interventions include targeting of multiple pathogenic mechanisms while ensuring systemic safety.

5.5 Emerging and repurposed therapeutic modalities

Novel therapeutic interventions primarily focus on improving gut-brain axis, restoring metabolic functions, with complementary benefit of attenuating neuroinflammation and neurodegeneration. The effects of microbial modulation by probiotics, prebiotics, and postbiotics supplementation are widely studied in preclinical and clinical trials (Liu et al., 2022; Ji et al., 2023). The repurposing of existing pharmaceuticals such as IDO1 inhibitors, initially developed as oncologic drug, have been successfully repurposed for neurodegenerative disorders (Gopukumar et al., 2025; Savonije et al., 2023). In AD, it has been observed that restoring astrocytic glycolysis and lactate production ameliorates metabolic dysfunction. Specific targeting of metabolites, such as NAD^+ boosters, offer new treatment avenues, by conferring critical modulations occurring in bioenergetic systems at the molecular level (Xie et al., 2020).

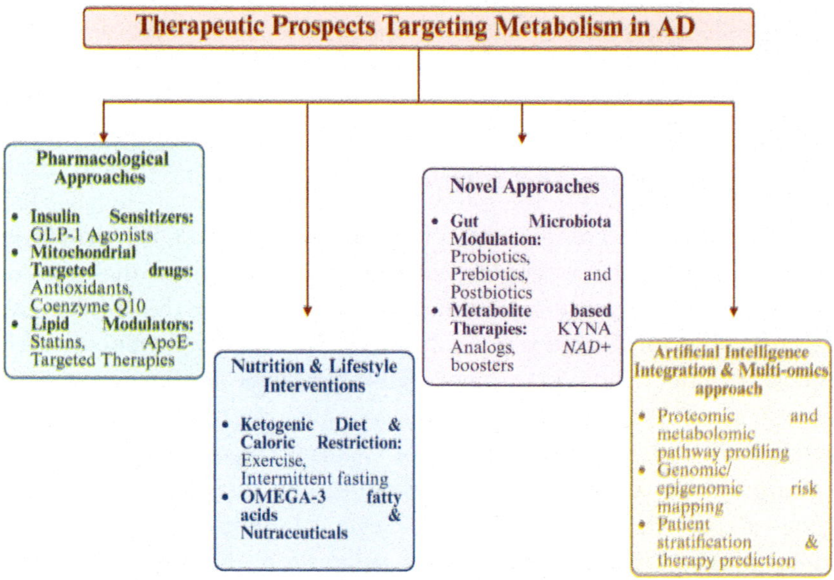

Fig. 2 Metabolism-targeted therapeutic strategies in Alzheimer's disease. Pharmacological interventions target glucose, lipid, and mitochondrial metabolism to restore bioenergetic homeostasis and nutritional/lifestyle interventions support metabolic flexibility. Novel strategies include gut microbiota modulation and metabolite-based therapies. Integration of multi-omics datasets with Artificial intelligence approaches enables systems-level mapping of metabolic alterations, facilitating biomarker discovery, patient stratification, and precision metabolic interventions in AD.

5.6 Therapeutic integration and patient stratification

Due to pathophysiological heterogeneity in metabolic dysfunction in Alzheimer's disease, it becomes imperative to employ patient stratification approach for optimal therapeutic treatment. Assessment of predominant factor influencing the AD pathogenesis is essential so as to provide effective treatment. Patients exhibiting insulin resistance, characterised by glucose hypometabolism, would get preferential benefit from GLP-1 agonists or intranasal insulin therapy. Whereas, the individual featuring mitochondrial bioenergetic collapse becomes ideal candidates for NAD^+ boosters and tyrosine kinase inhibition related interventions. ApoE-targeted interventions would aid individuals characterising lipid dysregulation such as upregulated ceramide levels and reduced APOE4 mediated cholesterol efflux (Ralhan et al., 2025). Altered gut metabolite levels influencing neuroinflammatory markers may respond optimally to gut-microbiota modulating approaches. Combination therapies such as intranasal insulin delivery in tandem with ketogenic diet would be targeting multiple nodes of metabolic dysregulation, exerting synergistic effects on preserving cognitive health and mitigating neuroinflammation (Fig. 2).

6. Challenges and future directions

The intricate complexities of metabolic networks in Alzheimer's disease, coupled with multiple compensatory mechanisms that activates dynamically in response to metabolic alterations, are the fundamental reason behind ineffective translation of single pathway therapeutics into clinical practice. The brain metabolism exhibits interconnected and adaptive networks, where perturbations in glucose, lipid and amino acid metabolism leads to their substitution with alternate energy substrates, limiting efficacy of single metabolic targets for sustained disease modifying effects (Zhou & Jiang, 2025). Several longitudinal metabolomics studies revealed that with the disease progression, the utilisation of substrates for energy production shifts dynamically, and every patient exhibits unique trajectory of timeline of metabolic adaptation/alterations (Marella et al., 2025; Yu et al., 2023). Also, the temporal heterogeneity presents complications for biomarker interpretation as these metabolic alterations indicate disease stage and compensatory processes undergoing but do not depict disease severity, intervention timings, since for specific metabolic targeting, therapeutic window is patient specific and is often narrow.

Such complexities present the need for intervention strategies that not only emphasise on inhibitory action but is an integrated system-level approach that substantially restores the metabolic function.

Other than pathological complexities, patient heterogeneity imposes second major challenge, where individual disease susceptibility and trajectory is impacted by one's physiology, age, biological sex, genetic constitution, and systemic comorbidities. The two individuals may reflect distinct pathological outcomes, such as one exhibiting insulin resistance may have preserved mitochondrial function, whereas another may demonstrate primary bioenergetic failure but stable glucose metabolism. Personalised medicine provides an escape for one size fit all concept for therapeutics, demonstrating precise assessment through integrated multi-omics profiling, depicting individual's metabolic signatures. These integrative strategies aids in patient stratification into their respective subtypes, reflecting the most benefitted targeted therapeutic interventions.

Sex differences among the population exhibits another critical dimension of distinct pathological outcomes. Where, women demonstrate greater vulnerability to age related disorders due to the hormonal influence associated with menopause, coupled with dysfunctions associated with altered estrogen levels (Turek & Gąsior, 2023). Currently available biomarkers lack sex-specific identification of disease progression, highlighting a gap among the gender specific therapeutic advancements (Lopez-Lee et al., 2024). Such sex-specific pathological alterations present the need for development of sex-stratified biomarkers, which accurately captures mechanistic differences among the individuals of opposite sexes.

The translational gap persisting between the therapeutic outcomes of metabolic interventions in preclinical models and their frequent failure in clinical trials remains a major challenge. The modulators of metabolic processes showing significant improvement in cognition in transgenic rodent models, often fails to achieve cognitive benefits in humans. Several factors may attribute in such translational failure, including inefficient BBB penetration of the drug, unsuitable pharmacokinetic profiles leading to subtherapeutic CNS concentrations, also most importantly, the inability of animal models replicate the biological complexities of the humans, along with their distinct physiological patterns and associated compensatory mechanisms, that are key characteristics of homo-sapiens.

To bridge this translational gap, refinement of preclinical models that would better capture complex metabolic networks along with improved clinical trial designs integrating patient specific interventions, may enhance treatment responsiveness. Future diagnostic and therapeutic development strategies must include multimodal biomarker identification by integrating characteristic metabolomics and lipidomics signatures along with advanced imaging modalities and blood-based biomarkers, to facilitate development of diagnostic and prognostic algorithms. Artificial Intelligence and machine learning offer new diagnostic avenues utilising biomarker datasets, identifying individual-specific metabolic traits that are responsible for bioenergetic impairment, altering neuronal function, enabling computational prediction for treatment responsiveness and cognitive outcomes (Ren et al., 2025b). Longitudinal biomarker analysis highlights a potential shift in therapeutic paradigm towards early interventions, due to the metabolic flexibility offered in initial stage of disease progression, contributing to effective therapeutic outcome. The real time biomarker monitoring, with long term metabolic dysfunction assessments, shares a promise to modulate static diagnostics and treatments of Alzheimer's disease into dynamic metabolic examination with personalised medicine approaches, thereby improving the likelihood of developing disease modifying therapies.

References

Adav, S. S., Park, J. E., & Sze, S. K. (2019). Quantitative profiling brain proteomes revealed mitochondrial dysfunction in Alzheimer's disease. *Molecular Brain, 12*(1), 8. https://doi.org/10.1186/s13041-019-0430-y.

Adlimoghaddam, A., Odero, G. G., Glazner, G., Turner, R. S., & Albensi, B. C. (2021). Nilotinib improves bioenergetic profiling in brain astroglia in the 3xTg mouse model of Alzheimer's disease. *Aging and Disease, 12*(2), 441–465. https://doi.org/10.14336/AD.2020.0910.

Adlimoghaddam, A., Snow, W. M., Stortz, G., Perez, C., Djordjevic, J., Goertzen, A. L., ... Albensi, B. C. (2019). Regional hypometabolism in the 3xTg mouse model of Alzheimer's disease. *Neurobiology of Disease, 127*, 264–277. https://doi.org/10.1016/j.nbd.2019.03.008.

Alhasan, M. S., Alhasan, A. S., Milburn, J., Khalil, M., Almaghraby, A., Alharthi, O., ... Azzam, A. Y. (2025). FDG-PET brain glucose hypometabolism predicts Alzheimer's disease progression pathways in cognitively normal adults: A longitudinal competing risks modeling. *Metabolism Open, 28*, 100400. https://doi.org/10.1016/j.metop.2025.100400.

Al-Thani, N. A., Stewart, G. S., & Costello, D. A. (2025a). The role of the urea cycle in the Alzheimer's disease brain. *Journal of Neurochemistry, 169*(3), e70033. https://doi.org/10.1111/jnc.70033.

Alzheimer, A., Stelzmann, R. A., Schnitzlein, H. N., & Murtagh, F. R. (1995). An english translation of Alzheimer's 1907 paper, "Uber eine eigenartige Erkankung der Hirnrinde". *Clinical Anatomy, 8*(6), 429–431. https://doi.org/10.1002/ca.980080612.

Amin, A. M., Mostafa, H., & Khojah, H. M. J. (2023). Insulin resistance in Alzheimer's disease: The genetics and metabolomics links. *Clinica Chimica Acta, 539*, 215–236. https://doi.org/10.1016/j.cca.2022.12.016.

Lionaki, E., Ploumi, C., & Tavernarakis, N. (2022). One-Carbon Metabolism: Pulling the Strings behind Aging and Neurodegeneration. *Cells, 11*(2), 214. https://doi.org/10.3390/cells11020214.

Andrade, L. J. de O., de Oliveira, L. M., Bittencourt, A. M. V., Lourenço, L. G. de C., & de Oliveira, G. C. M. (n.d.). Brain insulin resistance and Alzheimer's disease: A systematic review. *Dementia & Neuropsychologia, 18*, e20230032. https://doi.org/10.1590/1980–5764-DN-2023–0032.

Antoniel, M., Giorgio, V., Fogolari, F., Glick, G. D., Bernardi, P., & Lippe, G. (2014). The oligomycin-sensitivity conferring protein of mitochondrial ATP synthase: Emerging new roles in mitochondrial pathophysiology. *International Journal of Molecular Sciences, 15*(5), 7513–7536. https://doi.org/10.3390/ijms15057513.

Area-Gomez, E., de Groof, A. J. C., Boldogh, I., Bird, T. D., Gibson, G. E., Koehler, C. M., ... Schon, E. A. (2009). Presenilins are enriched in endoplasmic reticulum membranes associated with mitochondria. *The American Journal of Pathology, 175*(5), 1810–1816. https://doi.org/10.2353/ajpath.2009.090219.

Badhwar, A., & Haqqani, A. S. (n.d.). Biomarker potential of brain-secreted extracellular vesicles in blood in Alzheimer's disease. https://doi.org/10.1002/dad2.12001.

Bai, C., & Yuan, J. (2023). Metabolic biomarkers for Alzheimer's disease: A mini review. *Metabolism and Translational Medicine, 1*. https://doi.org/10.54844/mtm.2023.0409.

Balakrishnan, J., Kannan, S., Arivazhagan, G., Ramachandran, N., & Sundarasamy, V. G. (2025). Understanding the interconnected roles of gut microbiota and metabolomic profiles in alcoholic liver disease pathophysiology and their potential for innovative treatment strategies. *Medicine in Microecology, 26*, 100154. https://doi.org/10.1016/j.medmic.2025.100154.

Bélanger, M., Allaman, I., & Magistretti, P. J. (2011). Brain energy metabolism: Focus on astrocyte-neuron metabolic cooperation. *Cell Metabolism, 14*(6), 724–738. https://doi.org/10.1016/j.cmet.2011.08.016.

Blázquez, E., Hurtado-Carneiro, V., LeBaut-Ayuso, Y., Velázquez, E., García-García, L., Gómez-Oliver, F., ... Pozo, M. Á. (2022). Significance of brain glucose hypometabolism, altered insulin signal transduction, and insulin resistance in several neurological diseases. *Frontiers in Endocrinology, 13*, 873301. https://doi.org/10.3389/fendo.2022.873301.

Bokov, A., Chaudhuri, A., & Richardson, A. (2004). The role of oxidative damage and stress in aging. *Mechanisms of Ageing and Development, 125*(10–11), 811–826. https://doi.org/10.1016/j.mad.2004.07.009.

Bouter, C., Henniges, P., Franke, T. N., Irwin, C., Sahlmann, C. O., Sichler, M. E., ... Bouter, Y. (2019). 18F-FDG-PET detects drastic changes in brain metabolism in the Tg4–42 model of Alzheimer's disease. *Frontiers in Aging Neuroscience, 10*, 425. https://doi.org/10.3389/fnagi.2018.00425.

Briston, T., & Hicks, A. R. (2018). Mitochondrial dysfunction and neurodegenerative proteinopathies: Mechanisms and prospects for therapeutic intervention. *Biochemical Society Transactions, 46*(4), 829–842. https://doi.org/10.1042/BST20180025.

Budny, V., Ruminot, I., Wybitul, M., Treyer, V., Barros, L. F., & Tackenberg, C. (2025). Fueling the brain—The role of apolipoprotein E in brain energy metabolism and its implications for Alzheimer's disease. *Translational Psychiatry, 15*(1), 316. https://doi.org/10.1038/s41398-025-03550-w.

Cabezas-Opazo, F. A., Vergara-Pulgar, K., Pérez, M. J., Jara, C., Osorio-Fuentealba, C., & Quintanilla, R. A. (2015). Mitochondrial dysfunction contributes to the pathogenesis of Alzheimer's disease. *Oxidative Medicine and Cellular Longevity, 2015*, 509654. https://doi.org/10.1155/2015/509654.

Cai, Q., & Tammineni, P. (2017). Mitochondrial aspects of synaptic dysfunction in Alzheimer's disease. *Journal of Alzheimer's Disease: JAD, 57*(4), 1087–1103. https://doi.org/10.3233/JAD-160726.

Calsolaro, V., & Edison, P. (2016). Alterations in glucose metabolism in Alzheimer's disease. *Recent Patents on Endocrine, Metabolic & Immune Drug Discovery, 10*(1), 31–39. https://doi.org/10.2174/1872214810666160615102809.

Camandola, S., & Mattson, M. P. (2017). Brain metabolism in health, aging, and neurodegeneration. *The EMBO Journal, 36*(11), 1474–1492. https://doi.org/10.15252/embj.201695810.

Cao, Y., Zhao, L.-W., Chen, Z.-X., & Li, S.-H. (2024a). New insights in lipid metabolism: Potential therapeutic targets for the treatment of Alzheimer's disease. *Frontiers in Neuroscience, 18*. https://doi.org/10.3389/fnins.2024.1430465.

Casas-Fernández, E., Peña-Bautista, C., Baquero, M., & Cháfer-Pericás, C. (2022). Lipids as early and minimally invasive biomarkers for Alzheimer's disease. *Current Neuropharmacology, 20*(8), 1613–1631. https://doi.org/10.2174/1570159X19666211102150955.

Castillo-Vazquez, S. K., Massieu, L., Rincón-Heredia, R., García-de la Torre, P., Quiroz-Baez, R., Gomez-Verjan, J. C., & Rivero-Segura, N. A. (2024). Glutamatergic neurotransmission in aging and neurodegenerative diseases: A potential target to improve cognitive impairment in aging. *Archives of Medical Research, 55*(6), 103039. https://doi.org/10.1016/j.arcmed.2024.103039.

Chakrabarti, S. S., Bir, A., Poddar, J., Sinha, M., Ganguly, A., & Chakrabarti, S. (n.d.). Ceramide and sphingosine-1-phosphate in cell death pathways: Relevance to the Pathogenesis of Alzheimer's disease. *Current Alzheimer Research, 13*(11), 1232–1248. https://doi.org/10.2174/1567205013666160603004239.

Chang, R., Trushina, E., Zhu, K., Zaidi, S. S. A., Lau, B. M., Kueider-Paisley, A., ... Kaddurah-Daouk, R. (2023). Predictive metabolic networks reveal sex and APOE genotype-specific metabolic signatures and drivers for precision medicine in Alzheimer disease. *Alzheimer's & Dementia: The Journal of the Alzheimer's Association, 19*(2), 518–531. https://doi.org/10.1002/alz.12675.

Chao, A.-C., Lee, T.-C., Juo, S.-H. H., & Yang, D.-I. (2016). Hyperglycemia increases the production of amyloid beta-peptide leading to decreased endothelial tight junction. *CNS Neuroscience & Therapeutics, 22*(4), 291–297. https://doi.org/10.1111/cns.12503.

Chapple, B., Bayliss, E., Woodfin, S., Smith, M., Winter, J., & Moore, W. (2025). Type 3 diabetes: Linking insulin resistance to cognitive decline. *Diseases, 13*(11), https://doi.org/10.3390/diseases13110359.

Chatterjee, P., Goozee, K., Lim, C. K., James, I., Shen, K., Jacobs, K. R., ... Martins, R. N. (2018). Alterations in serum kynurenine pathway metabolites in individuals with high neocortical amyloid-β load: A pilot study. *Scientific Reports, 8*(1), 8008. https://doi.org/10.1038/s41598-018-25968-7.

Chen, J., Tian, C., Xiong, X., Yang, Y., & Zhang, J. (2025). Extracellular vesicles: New horizons in neurodegeneration. *eBioMedicine, 113*, 105605. https://doi.org/10.1016/j.ebiom.2025.105605.

Chen, L., Shen, Q., Liu, Y., Zhang, Y., Sun, L., Ma, X., ... Xie, J. (2025). Homeostasis and metabolism of iron and other metal ions in neurodegenerative diseases. *Signal Transduction and Targeted Therapy, 10*(1), 31. https://doi.org/10.1038/s41392-024-02071-0.

Chen, Y., Jin, H., Chen, J., Li, J., Găman, M.-A., & Zou, Z. (2025). The multifaceted roles of apolipoprotein E4 in Alzheimer's disease pathology and potential therapeutic strategies. *Cell Death Discovery, 11*, 312. https://doi.org/10.1038/s41420-025-02600-y.

Chen, Z., & Zhong, C. (2013). Decoding Alzheimer's disease from perturbed cerebral glucose metabolism: Implications for diagnostic and therapeutic strategies. *Progress in Neurobiology, 108*, 21–43. https://doi.org/10.1016/j.pneurobio.2013.06.004.

Choi, C., & Finlay, D. K. (2020). Diverse immunoregulatory roles of Oxysterols—the oxidized cholesterol metabolites. *Metabolites, 10*(10), 384. https://doi.org/10.3390/metabo10100384.

Cobilinschi, C., Andrei, C.-A., Grinţescu, I. M., & Mirea, L. (2024). Metabolic failure due to thiamine deficiency during critical illness. *Current Opinion in Clinical Nutrition and Metabolic Care, 27*(2), 155–162. https://doi.org/10.1097/MCO.0000000000001017.

Constantino, N. J., Ashley, C. C., & Macauley, S. L. (2025). The energetic collapse of the Alzheimer's brain: Metabolic inflexibility across cells and networks. *Journal of Neurochemistry, 169*(11), e70294. https://doi.org/10.1111/jnc.70294.

Corder, E. H., Saunders, A. M., Strittmatter, W. J., Schmechel, D. E., Gaskell, P. C., Small, G. W., ... Pericak-Vance, M. A. (1993). Gene dose of apolipoprotein E type 4 allele and the risk of Alzheimer's disease in late onset families. *Science (New York, N. Y.), 261*(5123), 921–923. https://doi.org/10.1126/science.8346443.

Counts, S. E., Ikonomovic, M. D., Mercado, N., Vega, I. E., & Mufson, E. J. (2017a). Biomarkers for the early detection and progression of Alzheimer's disease. *Neurotherapeutics: the Journal of the American Society for Experimental NeuroTherapeutics, 14*(1), 35–53. https://doi.org/10.1007/s13311-016-0481-z.

Cummings, J. (2021). New approaches to symptomatic treatments for Alzheimer's disease. *Molecular Neurodegeneration, 16*, 2. https://doi.org/10.1186/s13024-021-00424-9.

de Leon, M. J., Convit, A., Wolf, O. T., Tarshish, C. Y., DeSanti, S., Rusinek, H., ... Fowler, J. (2001). Prediction of cognitive decline in normal elderly subjects with 2-[(18) F]fluoro-2-deoxy-D-glucose/poitron-emission tomography (FDG/PET). *Proceedings of the National Academy of Sciences of the United States of America, 98*(19), 10966–10971. https://doi.org/10.1073/pnas.191044198.

Demougeot, C., Garnier, P., Mossiat, C., Bertrand, N., Giroud, M., Beley, A., & Marie, C. (2001). N-Acetylaspartate, a marker of both cellular dysfunction and neuronal loss: Its relevance to studies of acute brain injury. *Journal of Neurochemistry, 77*, 408–415. https://doi.org/10.1046/j.1471-4159.2001.00285.x.

Deng, W., Guo, S., van Veluw, S. J., Yu, Z., Chan, S. J., Takase, H., ... Bacskai, B. J. (2022). Effects of cerebral amyloid angiopathy on the brain vasculome. *Aging Cell, 21*(8), e13503. https://doi.org/10.1111/acel.13503.

Dighriri, I. M., Alsubaie, A. M., Hakami, F. M., Hamithi, D. M., Alshekh, M. M., Khobrani, F. A., ... Tawhari, M. Q. (n.d.). Effects of omega-3 polyunsaturated fatty acids on brain functions: A systematic review. *Cureus, 14*(10), e30091. https://doi.org/10.7759/cureus.30091.

Ding, F., Yao, J., Rettberg, J. R., Chen, S., & Brinton, R. D. (2013). Early decline in glucose transport and metabolism precedes shift to ketogenic system in female aging and Alzheimer's mouse brain: Implication for bioenergetic intervention. *PLoS One, 8*(11), e79977. https://doi.org/10.1371/journal.pone.0079977.

Dong, Y., Song, X., Wang, X., Wang, S., & He, Z. (2024). The early diagnosis of Alzheimer's disease: Blood-based panel biomarker discovery by proteomics and metabolomics. *CNS Neuroscience & Therapeutics, 30*(11), e70060. https://doi.org/10.1111/cns.70060.

Du, F., Yu, Q., Yan, S., Hu, G., Lue, L.-F., Walker, D. G., ... Yan, S. S. (2017). PINK1 signalling rescues amyloid pathology and mitochondrial dysfunction in Alzheimer's disease. *Brain: A Journal of Neurology, 140*(12), 3233–3251. https://doi.org/10.1093/brain/awx258.

Du, H., Guo, L., Yan, S., Sosunov, A. A., McKhann, G. M., & Yan, S. S. (2010). Early deficits in synaptic mitochondria in an Alzheimer's disease mouse model. *Proceedings of the National Academy of Sciences of the United States of America, 107*(43), 18670–18675. https://doi.org/10.1073/pnas.1006586107.

DuBoff, B., Götz, J., & Feany, M. B. (2012). Tau promotes neurodegeneration via DRP1 mislocalization in vivo. *Neuron, 75*(4), 618–632. https://doi.org/10.1016/j.neuron.2012.06.026.

Duong, M. T., Chen, Y. J., Doot, R. K., Young, A. J., Lee, H., Cai, J., ... Nasrallah, I. M. (2021). Astrocyte activation imaging with 11C-acetate and amyloid PET in mild cognitive impairment due to Alzheimer pathology. *Nuclear Medicine Communications, 42*(11), 1261–1269. https://doi.org/10.1097/MNM.0000000000001460.

Elbaky, N. A. A., El-Orabi, N. F., Fadda, L. M., Abd-Elkader, O. H., & Ali, H. M. (2018). Role of N-acetylcysteine and coenzyme Q10 in the amelioration of myocardial energy expenditure and oxidative stress, induced by carbon tetrachloride intoxication in rats. *Dose-Response, 16*(3), https://doi.org/10.1177/1559325818790158.

Fernández-Checa, J. C., Fernández, A., Morales, A., Marí, M., García-Ruiz, C., & Colell, A. (2010). Oxidative stress and altered mitochondrial function in neurodegenerative diseases: Lessons from mouse models. *CNS & Neurological Disorders Drug Targets, 9*(4), 439–454. https://doi.org/10.2174/187152710791556113.

Ferreira, L. S. S., Fernandes, C. S., Vieira, M. N. N., & De Felice, F. G. (2018). Insulin resistance in Alzheimer's disease. *Frontiers in Neuroscience, 12*, 830. https://doi.org/10.3389/fnins.2018.00830.

Fiandaca, M. S., Mapstone, M. E., Cheema, A. K., & Federoff, H. J. (2014). The critical need for defining preclinical biomarkers in Alzheimer's disease. *Alzheimer's & Dementia, Military Risk Supplement, 10*(3, ement), S196–S212. https://doi.org/10.1016/j.jalz.2014.04.015.

Flannery, P. J., & Trushina, E. (2019). Mitochondrial dynamics and transport in Alzheimer's disease. *Molecular and Cellular Neuroscience, 98*, 109–120. https://doi.org/10.1016/j.mcn.2019.06.009.

Fleck, D., Phu, L., Verschueren, E., Hinkle, T., Reichelt, M., Bhangale, T., ... Bingol, B. (2019). PTCD1 is required for mitochondrial oxidative-phosphorylation: Possible genetic association with Alzheimer's disease. *The Journal of Neuroscience, 39*(24), 4636–4656. https://doi.org/10.1523/JNEUROSCI.0116-19.2019.

Förster, S., Grimmer, T., Miederer, I., Henriksen, G., Yousefi, B. H., Graner, P., ... Drzezga, A. (2012). Regional expansion of hypometabolism in Alzheimer's disease follows amyloid deposition with temporal delay. *Biological Psychiatry, 71*(9), 792–797. https://doi.org/10.1016/j.biopsych.2011.04.023.

Fracassi, A., Marangoni, M., Rosso, P., Pallottini, V., Fioramonti, M., Siteni, S., & Segatto, M. (2019). Statins and the brain: More than lipid lowering agents? *Current Neuropharmacology, 17*(1), 59–83. https://doi.org/10.2174/1570159X15666170703101816.

Gabbouj, S., Ryhänen, S., Marttinen, M., Wittrahm, R., Takalo, M., Kemppainen, S., ... Natunen, T. (2019). Altered insulin signaling in Alzheimer's disease brain – special emphasis on PI3K-Akt pathway. *Frontiers in Neuroscience, 13*, 629. https://doi.org/10.3389/fnins.2019.00629.

Galli, A., Inglese, M., Presotto, L., Malito, R., Di, X., Toschi, N., ... Caminiti, S. P. (2025). Glucose metabolism in hyper-connected regions predicts neurodegeneration and speed of conversion in Alzheimer's disease. *European Journal of Nuclear Medicine and Molecular Imaging, 52*(12), 4639–4651. https://doi.org/10.1007/s00259-025-07379-9.

Gao, H., Jensen, K., Nesbitt, J., Ostroot, M., Cary, G. A., Wiley, J., ... Trushina, E. (2025). Mitochondrial complex I deficiency induces Alzheimer's disease-like signatures that are reversible by targeted therapy. *Alzheimer's & Dementia: The Journal of the Alzheimer's Association, 21*(8), e70519. https://doi.org/10.1002/alz.70519.

Garcia-Garcia, I., Kamal, F., Donica, O., Dadar, M., & Alzheimer's Disease Neuroimaging Initiative (2024). Plasma levels of adipokines and insulin are associated with markers of brain atrophy and cognitive decline in the spectrum of Alzheimer's Disease. *Progress in Neuro-psychopharmacology & Biological Psychiatry, 134*, 111077. https://doi.org/10.1016/j.pnpbp.2024.111077.

Ge, Z., Hu, Y., Kan, W., Li, L., Xu, J., Zhang, Y., ... Du, J. (2025). Lipid metabolic dysregulation-induced neuroinflammation in the pathophysiology of major depressive disorder. *Frontiers in Immunology, 16*. https://doi.org/10.3389/fimmu.2025.1625087.

Geijselaers, S. L. C., Aalten, P., Ramakers, I. H. G. B., De Deyn, P. P., Heijboer, A. C., Koek, H. L., ...Biessels, G. J. (n.d.). Association of cerebrospinal fluid (CSF) insulin with cognitive performance and CSF biomarkers of Alzheimer's disease. *Journal of Alzheimer's Disease, 61*(1), 309–320. https://doi.org/10.3233/JAD-170522.

Giorgi, C., Marchi, S., Simoes, I. C. M., Ren, Z., Morciano, G., Perrone, M., ... Wieckowski, M. R. (2018). Mitochondria and reactive oxygen species in aging and age-related diseases. *International Review of Cell and Molecular Biology, 340*, 209–344. https://doi.org/10.1016/bs.ircmb.2018.05.006.

Giorgio, V., Fogolari, F., Lippe, G., & Bernardi, P. (2019). OSCP subunit of mitochondrial ATP synthase: Role in regulation of enzyme function and of its transition to a pore. *British Journal of Pharmacology, 176*(22), 4247–4257. https://doi.org/10.1111/bph.14513.

González-Sánchez, M., Jiménez, J., Narváez, A., Antequera, D., Llamas-Velasco, S., Martín, A. H.-S., ... Carro, E. (2020). Kynurenic acid levels are increased in the CSF of Alzheimer's disease patients. *Biomolecules, 10*(4), https://doi.org/10.3390/biom10040571.

Gopukumar, S. T., Dwivedi, D., Gowri, K., Karpakavalli, M., Nair, C., Singh, D., & Das, U. (2025). Repurposing oncology drugs for Alzheimer's: Multi-omic convergence targeting cell-cycle, proteostasis, immunometabolism, and senescence. *Medicine in Drug Discovery, 28*, 100237. https://doi.org/10.1016/j.medidd.2025.100237.

Goudarzi, S., Hosseini, A., Abdollahi, M., & Haghi-Aminjan, H. (2021). Insights into parkin-mediated mitophagy in Alzheimer's disease: A systematic review. *Frontiers in Aging Neuroscience, 13*, 674071. https://doi.org/10.3389/fnagi.2021.674071.

Guo, J., Chen, S., Zhang, Y., Liu, J., Jiang, L., Hu, L., ... Chen, X. (2024). Cholesterol metabolism: Physiological regulation and diseases. *MedComm, 5*(2), e476. https://doi.org/10.1002/mco2.476.

Hammond, T. C., Xing, X., Wang, C., Ma, D., Nho, K., Crane, P. K., ... Lin, A.-L. (2020). β-Amyloid and tau drive early Alzheimer's disease decline while glucose hypometabolism drives late decline. *Communications Biology, 3*, 352. https://doi.org/10.1038/s42003-020-1079-x.

Han, J., Park, H., Maharana, C., Gwon, A.-R., Park, J., Baek, S. H., ... Jo, D.-G. (2021). Alzheimer's disease-causing presenilin-1 mutations have deleterious effects on mitochondrial function. *Theranostics, 11*(18), 8855–8873. https://doi.org/10.7150/thno.59776.

Hanson, L. R., & Frey, W. H. (2008). Intranasal delivery bypasses the blood-brain barrier to target therapeutic agents to the central nervous system and treat neurodegenerative disease. *BMC Neuroscience, 9*(3), S5. https://doi.org/10.1186/1471-2202-9-S3-S5.

Harr, S. D., Simonian, N. A., & Hyman, B. T. (1995). Functional alterations in Alzheimer's disease: Decreased glucose transporter 3 immunoreactivity in the perforant pathway terminal zone. *Journal of Neuropathology and Experimental Neurology, 54*(1), 38–41.

He, S., Xu, Z., & Han, X. (2025). Lipidome disruption in Alzheimer's disease brain: Detection, pathological mechanisms, and therapeutic implications. *Molecular Neurodegeneration, 20*(1), 11. https://doi.org/10.1186/s13024-025-00803-6.

He, X., Huang, Y., Li, B., Gong, C.-X., & Schuchman, E. H. (2010). Deregulation of sphingolipid metabolism in Alzheimer's disease. *Neurobiology of Aging, 31*(3), 398–408. https://doi.org/10.1016/j.neurobiolaging.2008.05.010.

Hirono, N., Mori, E., Ishii, K., Ikejiri, Y., Imamura, T., Shimomura, T., ... Sasaki, M. (1998). Frontal lobe hypometabolism and depression in Alzheimer's disease. *Neurology, 50*(2), 380–383. https://doi.org/10.1212/wnl.50.2.380.

Hoekstra, J. G., Hipp, M. J., Montine, T. J., & Kennedy, S. R. (2016). Mitochondrial DNA mutations increase in early stage Alzheimer's disease and are inconsistent with oxidative damage. *Annals of Neurology, 80*(2), 301–306. https://doi.org/10.1002/ana.24709.

Hölscher, C. (2019). Insulin signaling impairment in the brain as a risk factor in Alzheimer's disease. *Frontiers in Aging Neuroscience, 11*, 88. https://doi.org/10.3389/fnagi.2019.00088.

Huang, X., Hussain, B., & Chang, J. (2020). Peripheral inflammation and blood–brain barrier disruption: Effects and mechanisms. *CNS Neuroscience & Therapeutics, 27*(1), 36–47. https://doi.org/10.1111/cns.13569.

Jacobs, K. R., Lim, C. K., Blennow, K., Zetterberg, H., Chatterjee, P., Martins, R. N., ... Lovejoy, D. B. (2019). Correlation between plasma and CSF concentrations of kynurenine pathway metabolites in Alzheimer's disease and relationship to amyloid-β and tau. *Neurobiology of Aging, 80*, 11–20. https://doi.org/10.1016/j.neurobiolaging.2019.03.015.

Jadiya, P., Kolmetzky, D. W., Tomar, D., Di Meco, A., Lombardi, A. A., Lambert, J. P., ... Elrod, J. W. (2019). Impaired mitochondrial calcium efflux contributes to disease progression in models of Alzheimer's disease. *Nature Communications, 10*, 3885. https://doi.org/10.1038/s41467-019-11813-6.

Ji, J., Jin, W., Liu, S., Jiao, Z., & Li, X. (2023). Probiotics, prebiotics, and postbiotics in health and disease. *MedComm, 4*(6), e420. https://doi.org/10.1002/mco2.420.

Johansona, C., Stopa, E., Daiello, L., Monte, S., Keane, M., & Ott, B. (2018). Disrupted blood-CSF barrier to urea and creatinine in mild cognitive impairment and Alzheimer's disease. *Journal of Alzheimer's Disease & Parkinsonism, 08*. https://doi.org/10.4172/2161-0460.1000435.

Kakoti, B. B., Bezbaruah, R., & Ahmed, N. (2022). Therapeutic drug repositioning with special emphasis on neurodegenerative diseases: Threats and issues. *Frontiers in Pharmacology, 13*, 1007315. https://doi.org/10.3389/fphar.2022.1007315.

Kalkman, H. O., & Smigielski, L. (2025). Ceramides may play a central role in the pathogenesis of Alzheimer's disease: A review of evidence and horizons for discovery. *Molecular Neurobiology, 62*(11), 14424–14441. https://doi.org/10.1007/s12035-025-04989-0.

Kanekiyo, T., Xu, H., & Bu, G. (2014). ApoE and Aβ in Alzheimer's disease: Accidental encounters or partners? *Neuron, 81*(4), 740–754. https://doi.org/10.1016/j.neuron.2014.01.045.

Kang, S., Lee, Y., & Lee, J. E. (2017). Metabolism-centric overview of the pathogenesis of Alzheimer's disease. *Yonsei Medical Journal, 58*(3), 479–488. https://doi.org/10.3349/ymj.2017.58.3.479.

Kerr, J. S., Adriaanse, B. A., Greig, N. H., Mattson, M. P., Cader, M. Z., Bohr, V. A., & Fang, E. F. (2017). Mitophagy and Alzheimer's disease: Cellular and molecular mechanisms. *Trends in Neurosciences, 40*(3), 151–166. https://doi.org/10.1016/j.tins.2017.01.002.

Kriaa, A., Bourgin, M., Potiron, A., Mkaouar, H., Jablaoui, A., Gérard, P., ... Rhimi, M. (2019). Microbial impact on cholesterol and bile acid metabolism: Current status and future prospects. *Journal of Lipid Research, 60*(2), 323–332. https://doi.org/10.1194/jlr.R088989.

Kriebel, M., Ebel, J., Battke, F., Griesbach, S., & Volkmer, H. (2020). Interference with complex IV as a model of age-related decline in synaptic connectivity. *Frontiers in Molecular Neuroscience, 13*, 43. https://doi.org/10.3389/fnmol.2020.00043.

Kukreja, L., Kujoth, G. C., Prolla, T. A., Van Leuven, F., & Vassar, R. (2014). Increased mtDNA mutations with aging promotes amyloid accumulation and brain atrophy in the APP/Ld transgenic mouse model of Alzheimer's disease. *Molecular Neurodegeneration, 9*, 16. https://doi.org/10.1186/1750-1326-9-16.

Layé, S., Nadjar, A., Joffre, C., & Bazinet, R. P. (2018). Anti-inflammatory effects of omega-3 fatty acids in the brain: Physiological mechanisms and relevance to pharmacology. *Pharmacological Reviews, 70*(1), 12–38. https://doi.org/10.1124/pr.117.014092.

Li, D.-D., Zheng, C.-Q., Zhang, F., & Shi, J.-S. (2021). Potential neuroprotection by Dendrobium nobile Lindl alkaloid in Alzheimer's disease models. *Neural Regeneration Research, 17*(5), 972–977. https://doi.org/10.4103/1673-5374.324824.

Li, T., He, H., Zhang, E., Hu, F., Wang, Z., Xu, J., ... Peng, B. (2025). The physiological and pathological effects of sphingolipid metabolism and signaling in the central nervous system. *Brain Pathology, 36*(1), e70033. https://doi.org/10.1111/bpa.70033.

Liu, C., Meng, Y., Liu, R., Wang, Z., & Zhao, H. (2025). Pathological mechanisms and molecular imaging advances in Alzheimer's disease. *Clinical Interventions in Aging, 20*, 1583–1603. https://doi.org/10.2147/CIA.S534015.

Liu, T., Rong, Z., Li, J., Wu, H., & Wei, J. (2025). Three-dimensional interactive network: Mitochondrial-metabolic-calcium homeostasis driving Alzheimer's disease. *Genes & Diseases*101846. https://doi.org/10.1016/j.gendis.2025.101846.

Liu, Y., Liu, F., Iqbal, K., Grundke-Iqbal, I., & Gong, C.-X. (2008). Decreased glucose transporters correlate to abnormal hyperphosphorylation of tau in Alzheimer disease. *FEBS Letters, 582*(2), 359–364. https://doi.org/10.1016/j.febslet.2007.12.035.

Liu, Y., Wang, J., & Wu, C. (2022). Modulation of gut microbiota and immune system by probiotics, pre-biotics, and post-biotics. *Frontiers in Nutrition, 8*. https://doi.org/10.3389/fnut.2021.634897.

Lopez-Lee, C., Torres, E. R. S., Carling, G., & Gan, L. (2024). Mechanisms of sex differences in Alzheimer's disease. *Neuron, 112*(8), 1208–1221. https://doi.org/10.1016/j.neuron.2024.01.024.

Love, S., & Miners, J. S. (2016). Cerebral hypoperfusion and the energy deficit in Alzheimer's disease. *Brain Pathology, 26*(5), 607–617. https://doi.org/10.1111/bpa.12401.

Lv, H., Liu, Y., Yang, X., Xu, X., Zhou, J., & Yu, W. (2025). Research progress on the role of oxidative stress in the pathogenesis of vascular dementia and its treatment. *Journal of Stroke and Cerebrovascular Diseases, 34*(12), 108475. https://doi.org/10.1016/j.jstrokecerebrovasdis.2025.108475.

Ma, Q.-L., Yang, F., Rosario, E. R., Ubeda, O. J., Beech, W., Gant, D. J., ... Cole, G. M. (2009). Beta-amyloid oligomers induce phosphorylation of tau and inactivation of insulin receptor substrate via c-Jun N-terminal kinase signaling: Suppression by omega-3 fatty acids and curcumin. *The Journal of Neuroscience: The Official Journal of the Society for Neuroscience, 29*(28), 9078–9089. https://doi.org/10.1523/JNEUROSCI.1071-09.2009.

Ma, T., Tian, X., Zhang, B., Li, M., Wang, Y., Yang, C., ... Lin, S.-C. (2022). Low-dose metformin targets the lysosomal AMPK pathway through PEN2. *Nature, 603*(7899), 159–165. https://doi.org/10.1038/s41586-022-04431-8.

Majerníková, N., Marmolejo-Garza, A., Salinas, C. S., Luu, M. D. A., Zhang, Y., Trombetta-Lima, M., ... Dolga, A. M. (2024). The link between amyloid β and ferroptosis pathway in Alzheimer's disease progression. *Cell Death & Disease, 15*(10), 782. https://doi.org/10.1038/s41419-024-07152-0.

Mani, S., Swargiary, G., Singh, M., Agarwal, S., Dey, A., Ojha, S., & Jha, N. K. (2021). Mitochondrial defects: An emerging theranostic avenue towards Alzheimer's associated dysregulations. *Life Sciences, 285*, 119985. https://doi.org/10.1016/j.lfs.2021.119985.

Marella, B., Weinisch, P., Bless, J. J., Risacher, S. L., Blach, C., Karu, N., ... Arnold, M. (2025). A seven-year longitudinal study of the Alzheimer's disease blood metabolome. *medRxiv*. https://doi.org/10.64898/2025.12.05.25341709.

Martín-Saladich, Q., Pareto, D., Simó, R., Ciudin, A., Aparicio, C., Hammawa, K., ... Herance, J. R. (2025). Brain [18F]FDG uptake patterns in type 2 diabetes: New phenotypes relating to biomarkers of cognitive impairment. *Brain Communications*, 7(3), fcaf213. https://doi.org/10.1093/braincomms/fcaf213.

Mattoli, M. V., Cocciolillo, F., Chiacchiaretta, P., Dotta, F., Trevisi, G., Carrarini, C., ... Bonanni, L. (2023). Combined 18F-FDG PET-CT markers in dementia with Lewy bodies. *Alzheimer's & Dementia, 15*(4), e12515. https://doi.org/10.1002/dad2.12515.

Maurer, I., Zierz, S., & Möller, H. J. (2000). A selective defect of cytochrome c oxidase is present in brain of Alzheimer disease patients. *Neurobiology of Aging, 21*(3), 455–462. https://doi.org/10.1016/s0197-4580(00)00112-3.

Misrani, A., Tabassum, S., & Yang, L. (2021). Mitochondrial dysfunction and oxidative stress in Alzheimer's disease. *Frontiers in Aging Neuroscience, 13*, 617588. https://doi.org/10.3389/fnagi.2021.617588.

Mohan, G. S., & Kumar, R. (2025). Impairment of tricarboxylic acid cycle (TCA) cycle in Alzheimer's disease: Mechanisms, implications, and potential therapies. *Aging and Disease, 16*(5), 2553. https://doi.org/10.14336/AD.2025.0472.

Mosconi, L. (2013). Glucose metabolism in normal aging and Alzheimer's disease: Methodological and physiological considerations for PET studies. *Clinical and Translational Imaging: Reviews in Nuclear Medicine and Molecular Imaging, 1*(4), https://doi.org/10.1007/s40336-013-0026-y.

Mosconi, L., De Santi, S., Li, J., Tsui, W. H., Li, Y., Boppana, M., ... de Leon, M. J. (2008a). Hippocampal hypometabolism predicts cognitive decline from normal aging. *Neurobiology of Aging, 29*(5), 676–692. https://doi.org/10.1016/j.neurobiolaging.2006.12.008.

Mosconi, L., Pupi, A., & De Leon, M. J. (2008b). Brain glucose hypometabolism and oxidative stress in preclinical Alzheimer's disease. *Annals of the New York Academy of Sciences, 1147*, 180–195. https://doi.org/10.1196/annals.1427.007.

Mullins, R., Reiter, D., & Kapogiannis, D. (2018). Magnetic resonance spectroscopy reveals abnormalities of glucose metabolism in the Alzheimer's brain. *Annals of Clinical and Translational Neurology, 5*(3), 262–272. https://doi.org/10.1002/acn3.530.

Murray, I. V. J., Proza, J. F., Sohrabji, F., & Lawler, J. M. (2011). Vascular and metabolic dysfunction in Alzheimer's disease: A review. *Experimental Biology and Medicine, 236*(7), 772–782. https://doi.org/10.1258/ebm.2011.010355.

Nabi, S. U., Khan, A., Siddiqui, E. M., Rehman, M. U., Alshahrani, S., Arafah, A., ... Shen, B. (2022). Mechanisms of mitochondrial malfunction in Alzheimer's disease: New therapeutic hope. *Oxidative Medicine and Cellular Longevity, 2022*(1), 4759963. https://doi.org/10.1155/2022/4759963.

Nasiri, H., Nemati, R., Orangi, K., Mansouri, M., Saranjam, A., Esmizadeh, F., ... Sadi, S. (2025a). Serum amino acid levels are associated with brain metabolism in patients across the Alzheimer's disease continuum. *IBRO Neuroscience Reports*. https://doi.org/10.1016/j.ibneur.2025.09.007.

Nuamnaichati, N., Mangmool, S., Chattipakorn, N., & Parichatikanond, W. (2020). Stimulation of GLP-1 receptor inhibits methylglyoxal-induced mitochondrial dysfunctions in H9c2 cardiomyoblasts: Potential role of Epac/PI3K/Akt pathway. *Frontiers in Pharmacology, 11*, 805. https://doi.org/10.3389/fphar.2020.00805.

Ottoy, J., Owsicki, N., Bilgel, M., Binette, A. P., Salvadó, G., Kang, M. S., ... Betthauser, T. (2025). Recent advances in neuroimaging of Alzheimer's disease and related dementias. *Alzheimer's & Dementia, 21*(9), e70648. https://doi.org/10.1002/alz.70648.

Ou, Y.-N., Xu, W., Li, J.-Q., Guo, Y., Cui, M., Chen, K.-L., ... Yu, J.-T. (2019). FDG-PET as an independent biomarker for Alzheimer's biological diagnosis: A longitudinal study. *Alzheimer's Research & Therapy, 11*, 57. https://doi.org/10.1186/s13195-019-0512-1.

Pan, Y., Li, J., Lin, P., Wan, L., Qu, Y., Cao, L., & Wang, L. (2024). A review of the mechanisms of abnormal ceramide metabolism in type 2 diabetes mellitus, Alzheimer's disease, and their co-morbidities. *Frontiers in Pharmacology, 15*. https://doi.org/10.3389/fphar.2024.1348410.

Park, H., Kang, J.-H., & Lee, S. (2020). Autophagy in neurodegenerative diseases: A hunter for aggregates. *International Journal of Molecular Sciences, 21*(9), 3369. https://doi.org/10.3390/ijms21093369.

Peña-Bautista, C., Vigor, C., Galano, J.-M., Oger, C., Durand, T., Ferrer, I., ... Cháfer-Pericás, C. (2018). Plasma lipid peroxidation biomarkers for early and non-invasive Alzheimer disease detection. *Free Radical Biology & Medicine, 124*, 388–394. https://doi.org/10.1016/j.freeradbiomed.2018.06.038.

Pfrieger, F. W. (2003). Role of cholesterol in synapse formation and function. *Biochimica et Biophysica Acta, 1610*(2), 271–280. https://doi.org/10.1016/s0005-2736(03)00024-5.

Pradeepkiran, J. A., & Reddy, P. H. (2020). Defective mitophagy in Alzheimer's disease. *Ageing Research Reviews, 64*, 101191. https://doi.org/10.1016/j.arr.2020.101191.

Qu, C., Keijer, J., Adjobo-Hermans, M. J. W., van de Wal, M., Schirris, T., van Karnebeek, C., ... Koopman, W. J. H. (2021). The ketogenic diet as a therapeutic intervention strategy in mitochondrial disease. *The International Journal of Biochemistry & Cell Biology, 138*, 106050. https://doi.org/10.1016/j.biocel.2021.106050.

Ralhan, I., Do, A. D., Bae, J.-Y., Feringa, F. M., Cai, W., Chang, J., ... Ioannou, M. S. (2025). Protective ApoE variants support neuronal function by effluxing oxidized phospholipids. *Neuron, 6273*(25), 00847-5. https://doi.org/10.1016/j.neuron.2025.10.040.

Ramamoorthy, K., Yoshimura, R., Al-Juburi, S., Anandam, K. Y., Kapadia, R., Alachkar, A., ... Said, H. M. (2022). Alzheimer's disease is associated with disruption in thiamin transport physiology: A potential role for neuroinflammation. *Neurobiology of Disease, 171*, 105799. https://doi.org/10.1016/j.nbd.2022.105799.

Ren, F., Wei, J., Chen, Q., Hu, M., Yu, L., Mi, J., ... Wu, A. (2025a). Artificial intelligence-driven multi-omics approaches in Alzheimer's disease: Progress, challenges, and future directions. *Acta Pharmaceutica Sinica B, 15*(9), 4327–4385. https://doi.org/10.1016/j.apsb.2025.07.030.

Russo, L., Babboni, S., Andreassi, M. G., Daher, J., Canale, P., Del Turco, S., & Basta, G. (2025). Treating metabolic dysregulation and senescence by caloric restriction: Killing two birds with one stone? *Antioxidants, 14*(1), 99. https://doi.org/10.3390/antiox14010099.

Sang, S., Pan, X., Chen, Z., Zeng, F., Pan, S., Liu, H., ... Zhong, C. (2018). Thiamine diphosphate reduction strongly correlates with brain glucose hypometabolism in Alzheimer's disease, whereas amyloid deposition does not. *Alzheimer's Research & Therapy, 10*(1), 26. https://doi.org/10.1186/s13195-018-0354-2.

Savonije, K., Meek, A., & Weaver, D. F. (2023). Indoleamine 2,3-dioxygenase as a therapeutic target for Alzheimer's disease and geriatric depression. *Brain Sciences, 13*(6), https://doi.org/10.3390/brainsci13060852.

Sears, S. M., & Hewett, S. J. (2021). Influence of glutamate and GABA transport on brain excitatory/inhibitory balance. *Experimental Biology and Medicine, 246*(9), 1069–1083. https://doi.org/10.1177/1535370221989263.

Shahpasand, S., Khatami, S. H., Ehtiati, S., Alehossein, P., Salmani, F., Toutounchi, A. H., ... Karima, S. (2024). Therapeutic potential of the ketogenic diet: A metabolic switch with implications for neurological disorders, the gut-brain axis, and cardiovascular diseases. *The Journal of Nutritional Biochemistry, 132*, 109693. https://doi.org/10.1016/j.jnutbio.2024.109693.

Simpson, I. A., Chundu, K. R., Davies-Hill, T., Honer, W. G., & Davies, P. (1994). Decreased concentrations of GLUT1 and GLUT3 glucose transporters in the brains of patients with Alzheimer's disease. *Annals of Neurology, 35*(5), 546–551. https://doi.org/10.1002/ana.410350507.

Song, D., Li, Y., Yang, L.-L., Luo, Y.-X., & Yao, X.-Q. (2025). Bridging systemic metabolic dysfunction and Alzheimer's disease: The liver interface. *Molecular Neurodegeneration, 20*(1), 61. https://doi.org/10.1186/s13024-025-00849-6.

Song, T., Song, X., Zhu, C., Patrick, R., Skurla, M., Santangelo, I., ... Du, F. (2021). Mitochondrial dysfunction, oxidative stress, neuroinflammation, and metabolic alterations in the progression of Alzheimer's disease: A meta-analysis of in vivo magnetic resonance spectroscopy studies. *Ageing Research Reviews, 72*, 101503. https://doi.org/10.1016/j.arr.2021.101503.

Srivastava, A., Renna, H. A., Johnson, M., Sheehan, K., Ahmed, S., Palaia, T., ... Reiss, A. B. (2024). Nilotinib as a prospective treatment for Alzheimer's disease: Effect on proteins involved in neurodegeneration and neuronal homeostasis. *Life (Chicago, Ill.: 1978), 14*(10), https://doi.org/10.3390/life14101241.

Suzuki, A., Stern, S. A., Bozdagi, O., Huntley, G. W., Walker, R. H., Magistretti, P. J., & Alberini, C. M. (2011). Astrocyte-neuron lactate transport is required for long-term memory formation. *Cell, 144*(5), 810–823. https://doi.org/10.1016/j.cell.2011.02.018.

Swerdlow, R. H. (2018). Mitochondria and mitochondrial cascades in Alzheimer's disease. *Journal of Alzheimer's Disease: JAD, 62*(3), 1403–1416. https://doi.org/10.3233/JAD-170585.

Szablewski, L. (2021). Brain glucose transporters: Role in pathogenesis and potential targets for the treatment of Alzheimer's disease. *International Journal of Molecular Sciences, 22*(15), 8142. https://doi.org/10.3390/ijms22158142.

Tam, S., Wear, D., Morrone, C. D., & Yu, W. H. (2024). The complexity of extracellular vesicles: Bridging the gap between cellular communication and neuropathology. *Journal of Neurochemistry, 168*(9), 2391–2422. https://doi.org/10.1111/jnc.16108.

Tiwari, P., Gupta, A., Kaushik, M., Yadav, A., Anjali, A., Dwivedi, R., ... Dada, R. (2025). Comprehensive metabolomics profiling reveals novel biomarkers and pathways for early detection of Alzheimer's disease. *Brain Communications,* 7(6), fcaf410. https://doi.org/10.1093/braincomms/fcaf410.

Trushina, E., Dutta, T., Persson, X.-M. T., Mielke, M. M., & Petersen, R. C. (2013). Identification of altered metabolic pathways in plasma and CSF in mild cognitive impairment and Alzheimer's disease using metabolomics. *PLoS One, 8*(5), e63644. https://doi.org/10.1371/journal.pone.0063644.

Turek, J., & Gąsior, Ł. (2023). Estrogen fluctuations during the menopausal transition are a risk factor for depressive disorders. *Pharmacological Reports, 75*(1), 32–43. https://doi.org/10.1007/s43440-022-00444-2.

Urkon, M., Ferencz, E., Szász, J. A., Szabo, M. I. M., Orbán-Kis, K., Szatmári, S., & Nagy, E. E. (2025). Antidiabetic GLP-1 receptor agonists have neuroprotective properties in experimental animal models of Alzheimer's disease. *Pharmaceuticals, 18*(5), 614. https://doi.org/10.3390/ph18050614.

Van Dyken, P., & Lacoste, B. (2018a). Impact of metabolic syndrome on neuroinflammation and the blood–brain barrier. *Frontiers in Neuroscience, 12*, 930. https://doi.org/10.3389/fnins.2018.00930.

Vargas-George, S., & Dave, K. R. (2022). Models of cerebral amyloid angiopathy-related intracerebral hemorrhage. *Brain Hemorrhages, 3*(4), 189–199. https://doi.org/10.1016/j.hest.2022.05.001.

Venkatesh, R., Cardone, K. M., Bradford, Y., Moore, A. K., Kumar, R., Moore, J. H., ... Ritchie, M. D. (2025). Integrative multi-omics approaches identify molecular pathways and improve Alzheimer's disease risk prediction. *medRxiv: The Preprint Server for Health Sciences*. https://doi.org/10.1101/2025.05.31.25328688.

Villena, J., Henriquez, M., Torres, V., Moraga, F., Díaz-Elizondo, J., Arredondo, C., ... Quest, A. F. G. (2008). Ceramide-induced formation of ROS and ATP depletion trigger necrosis in lymphoid cells. *Free Radical Biology and Medicine, 44*(6), 1146–1160. https://doi.org/10.1016/j.freeradbiomed.2007.12.017.

Wang, H., Lautrup, S., Caponio, D., Zhang, J., & Fang, E. F. (2021). DNA damage-induced neurodegeneration in accelerated ageing and Alzheimer's disease. *International Journal of Molecular Sciences, 22*(13), 6748. https://doi.org/10.3390/ijms22136748.

Wang, R., Peng, S., Zhu, J., Xu, Y., Wang, M., Zhang, L., ... Liu, R. (2025). Innovations in Alzheimer's disease diagnostic technologies: Clinical prospects of novel biomarkers, multimodal integration, and non-invasive detection. *Frontiers in Neurology, 16*. https://doi.org/10.3389/fneur.2025.1651708.

Wang, S.-Z., Yu, Y.-J., & Adeli, K. (2020). Role of gut microbiota in neuroendocrine regulation of carbohydrate and lipid metabolism via the microbiota-gut-brain-liver axis. *Microorganisms, 8*(4), 527. https://doi.org/10.3390/microorganisms8040527.

Wang, W., Zhao, F., Ma, X., Perry, G., & Zhu, X. (2020a). Mitochondria dysfunction in the pathogenesis of Alzheimer's disease: Recent advances. *Molecular Neurodegeneration, 15*, 30. https://doi.org/10.1186/s13024-020-00376-6.

Wang, X., Li, H., Sheng, Y., He, B., Liu, Z., Li, W., ... Meng, X. (2024). The function of sphingolipids in different pathogenesis of Alzheimer's disease: A comprehensive review. *Biomedicine & Pharmacotherapy, 171*, 116071. https://doi.org/10.1016/j.biopha.2023.116071.

Wang, X., Su, B., Lee, H., Li, X., Perry, G., Smith, M. A., & Zhu, X. (2009). Impaired balance of mitochondrial fission and fusion in Alzheimer's disease. *The Journal of Neuroscience: The Official Journal of the Society for Neuroscience, 29*(28), 9090–9103. https://doi.org/10.1523/JNEUROSCI.1357-09.2009.

Wang, X., Zheng, W., Xie, J.-W., Wang, T., Wang, S.-L., Teng, W.-P., & Wang, Z.-Y. (2010). Insulin deficiency exacerbates cerebral amyloidosis and behavioral deficits in an Alzheimer transgenic mouse model. *Molecular Neurodegeneration, 5*, 46. https://doi.org/10.1186/1750-1326-5-46.

Watts, M. E., Pocock, R., & Claudianos, C. (2018). Brain energy and oxygen metabolism: Emerging role in normal function and disease. *Frontiers in Molecular Neuroscience, 11*. https://doi.org/10.3389/fnmol.2018.00216.

Wen, H., Deng, H., Li, B., Chen, J., Zhu, J., Zhang, X., ... Zhou, Y. (2025). Mitochondrial diseases: From molecular mechanisms to therapeutic advances. *Signal Transduction and Targeted Therapy, 10*(1), 9. https://doi.org/10.1038/s41392-024-02044-3.

Whiley, L., Sen, A., Heaton, J., Proitsi, P., García-Gómez, D., Leung, R., ... Legido-Quigley, C. (2014). Evidence of altered phosphatidylcholine metabolism in Alzheimer's disease. *Neurobiology of Aging, 35*(2), 271–278. https://doi.org/10.1016/j.neurobiolaging.2013.08.001.

Williams, T., Borchelt, D. R., & Chakrabarty, P. (2020). Therapeutic approaches targeting Apolipoprotein E function in Alzheimer's disease. *Molecular Neurodegeneration, 15*, 8. https://doi.org/10.1186/s13024-020-0358-9.

Wong, C. Y. J., Baldelli, A., Hoyos, C. M., Tietz, O., Ong, H. X., & Traini, D. (2024). Insulin delivery to the brain via the nasal route: Unraveling the potential for Alzheimer's disease therapy. *Drug Delivery and Translational Research, 14*(7), 1776–1793. https://doi.org/10.1007/s13346-024-01558-1.

Wu, A., Lee, D., & Xiong, W.-C. (2023). Lactate metabolism, signaling, and function in brain development, synaptic plasticity, angiogenesis, and neurodegenerative diseases. *International Journal of Molecular Sciences, 24*(17), 13398. https://doi.org/10.3390/ijms241713398.

Wuerch, E. C., & Yong, V. W. (2025). Cholesterol in the CNS: Functions, recycling and remyelination. *Journal of Neuroinflammation, 22*, 180. https://doi.org/10.1186/s12974-025-03490-8.

Xie, K., Qin, Q., Long, Z., Yang, Y., Peng, C., Xi, C., ... Wang, M. (2021). High-throughput metabolomics for discovering potential biomarkers and identifying metabolic mechanisms in aging and Alzheimer's disease. *Frontiers in Cell and Developmental Biology, 9*. https://doi.org/10.3389/fcell.2021.602887.

Xie, N., Zhang, L., Gao, W., Huang, C., Huber, P. E., Zhou, X., ... Zou, B. (2020). NAD+ metabolism: Pathophysiologic mechanisms and therapeutic potential. *Signal Transduction and Targeted Therapy, 5*, 227. https://doi.org/10.1038/s41392-020-00311-7.

Xiu, J., Nordberg, A., Qi, X., & Guan, Z.-Z. (2006). Influence of cholesterol and lovastatin on alpha-form of secreted amyloid precursor protein and expression of alpha7 nicotinic receptor on astrocytes. *Neurochemistry International, 49*(5), 459–465. https://doi.org/10.1016/j.neuint.2006.03.007.

Yamazaki, Y., Painter, M. M., Bu, G., & Kanekiyo, T. (2016). Apolipoprotein E as a therapeutic target in Alzheimer's disease: A review of basic research and clinical evidence. *CNS Drugs, 30*(9), 773–789. https://doi.org/10.1007/s40263-016-0361-4.

Yang, L. G., March, Z. M., Stephenson, R. A., & Narayan, P. S. (2023a). Apolipoprotein E in lipid metabolism and neurodegenerative disease. *Trends in Endocrinology & Metabolism, 34*(8), 430–445. https://doi.org/10.1016/j.tem.2023.05.002.

Yao, P. J., Eren, E., Goetzl, E. J., & Kapogiannis, D. (2021a). Mitochondrial electron transport chain protein abnormalities detected in plasma extracellular vesicles in Alzheimer's disease. *Biomedicines, 9*(11), 1587. https://doi.org/10.3390/biomedicines9111587.

Yassine, H. N., Self, W., Kerman, B. E., Santoni, G., Navalpur Shanmugam, N., Abdullah, L., ... Bowman, G. L. (2023). Nutritional metabolism and cerebral bioenergetics in Alzheimer's disease and related dementias. *Alzheimer's & Dementia: The Journal of the Alzheimer's Association, 19*(3), 1041–1066. https://doi.org/10.1002/alz.12845.

Yin, H., Lu, Z., Deng, Y., Tian, X., & Gong, Q. (2026). Decoding blood-brain barrier dysfunction in Alzheimer's disease: Innovations and challenges in multimodal MRI and PET imaging biomarkers. *Ageing Research Reviews, 114*, 102952. https://doi.org/10.1016/j.arr.2025.102952.

Youm, Y.-H., Nguyen, K. Y., Grant, R. W., Goldberg, E. L., Bodogai, M., Kim, D., ... Dixit, V. D. (2015). Ketone body β-hydroxybutyrate blocks the NLRP3 inflammasome-mediated inflammatory disease. *Nature Medicine, 21*(3), 263–269. https://doi.org/10.1038/nm.3804.

Yu, W., Chen, L., Li, X., Han, T., Yang, Y., Hu, C., ... Lü, Y. (2023). Alteration of metabolic profiles during the progression of Alzheimer's disease. *Brain Sciences, 13*(10), 1459. https://doi.org/10.3390/brainsci13101459.

Zeng, K., Yu, X., Mahaman, Y. A. R., Wang, J.-Z., Liu, R., Li, Y., & Wang, X. (2022). Defective mitophagy and the etiopathogenesis of Alzheimer's disease. *Translational Neurodegeneration, 11*, 32. https://doi.org/10.1186/s40035-022-00305-1.

Zhang, J., Zhang, Y., Wang, J., Xia, Y., Zhang, J., & Chen, L. (2024). Recent advances in Alzheimer's disease: Mechanisms, clinical trials and new drug development strategies. *Signal Transduction and Targeted Therapy, 9*(1), 211. https://doi.org/10.1038/s41392-024-01911-3.

Zhang, S., Lachance, B. B., Mattson, M. P., & Jia, X. (2021). Glucose metabolic crosstalk and regulation in brain function and diseases. *Progress in Neurobiology, 204*, 102089. https://doi.org/10.1016/j.pneurobio.2021.102089.

Zhao, W.-Q., De Felice, F. G., Fernandez, S., Chen, H., Lambert, M. P., Quon, M. J., ... Klein, W. L. (2008). Amyloid beta oligomers induce impairment of neuronal insulin receptors. *FASEB Journal: Official Publication of the Federation of American Societies for Experimental Biology, 22*(1), 246–260. https://doi.org/10.1096/fj.06-7703com.

Zhou, J., & Jiang, C. (2025). Targeting multilayered metabolic networks in brain diseases: Emerging perspectives on nanodelivery strategies. *Advanced Science, 12*(41), e03645. https://doi.org/10.1002/advs.202503645.

Zhu, H., Bi, D., Zhang, Y., Kong, C., Du, J., Wu, X., ... Qin, H. (2022). Ketogenic diet for human diseases: The underlying mechanisms and potential for clinical implementations. *Signal Transduction and Targeted Therapy*, 7(1), 11. https://doi.org/10.1038/s41392-021-00831-w.

Zhu, X., Perry, G., Smith, M. A., & Wang, X. (2013). Abnormal mitochondrial dynamics in the pathogenesis of Alzheimer's disease. *Journal of Alzheimer's Disease: JAD, 33*(0 1), S253–S262. https://doi.org/10.3233/JAD-2012-129005.

Zong, Y., Li, H., Liao, P., Chen, L., Pan, Y., Zheng, Y., ... Gao, J. (2024). Mitochondrial dysfunction: Mechanisms and advances in therapy. *Signal Transduction and Targeted Therapy, 9*(1), 124. https://doi.org/10.1038/s41392-024-01839-8.

At the other extreme of the balance of power, when the state controls the plutocracy, the latter cannot afford to resolve disputes abroad because it will risk sanctions by the sovereign. Outsourcing disputes undercuts the ruler's authority: it sends signals to both domestic and foreign audiences that state institutions cannot be trusted. Given the nature of state-controlled economies, any foreign legal battle is likely to implicate the state, as well. A plutocrat trying to use a foreign court will quickly increase the threat from the sovereign, who could cut off bank loans, trump up corruption charges, or even resort to political assassination. Countries similar to China and Saudi Arabia generally fit this pattern.

But when the state and independent plutocrats have relatively equal power resources, at the middle of the balance of power, outsourcing disputes becomes most likely. A sovereign will view the independent plutocrats as potential threats. They could try to fund an opposition party or a coup to ensure that plutocrats fall more in line with their interests or simply increase their economic rents. This dynamic could make a leader increasingly concerned with their tenure and creates clear incentives to bring the business elite into line. On the other side of this coin, business elites are aware that their very presence acts as a threat to the sovereign, making the plutocrats concerned with the possibility that the sovereign will attempt to bring them down.

This commitment problem can lead to political conflict between the state and individual plutocrats as each party seeks control. The losers of the political fight are then forced to exit. When plutocrats are locked out of their home system, the only way to recoup their economic damages is by using foreign legal systems. Liberal institutions then become an extraterritorial mechanism for perceived economic justice.

In a world of mobile capital, losers could still use the wealth that they have abroad to continue posing a political threat. One could continue funding opposition parties or disinformation campaigns from abroad. The winners of the political battle are then frequently forced to use foreign courts because extraterritorial disputes become the only option to seize their opponent's offshore wealth. Foreign courts are turned into a tool for autocrats to curb political opposition and delegitimize their rivals. Russia in the early years of Vladimir Putin's reign and Kazakhstan in the mid-2000s typify these political dynamics.

This book demonstrates that the balance of power between the state and the plutocracy dictates when and why parties outsource domestic disputes. The transparency associated with transnational law deters plutocrats from outsourcing disputes in both the lawless world of state capture and in the coercive context of state control. But when the state and the plutocracy have similar power resources, the divergence in interests leads to political clashes. The losers are left with little choice but to go abroad in search of new protection. In the past, the winners of such elite battles expropriate their rivals' assets and imprison their opponents, but with the rise of offshore finance, losers now abscond with many of their resources. This makes the winners use foreign courts as they search for ways to ensure that their rivals cannot come back fighting.

The argument does not explain the logic behind every single extraterritorial dispute. In some instances, plutocrats may simply be forum shopping for beneficial case law, and those functional factors create a baseline set of incentives to use courts in liberal jurisdictions. Nonetheless, these dynamics give transnational legal institutions a new set of authority over emerging markets while they simultaneously risk becoming the launderers of illiberal politics.

Extraterritorial Disputes and the Contradictions of the Liberal Order

At its core, this book is about a set of actors—emerging market plutocrats—whose role in global politics dramatically expanded with international economic integration but largely escaped academic analysis (Freund 2016; Hägel 2020). With the rise of the so-called BRIC (Brazil Russia, India, and China) countries, a record number of individuals were lifted out of poverty, and yet the elite of the elite thrived. When *Forbes* began collecting data on billionaires in 1987, less than thirty could be identified in emerging markets. By 2015, the number had risen more than twentyfold, to over six hundred. The super-wealthy have always flexed their financial, and thereby political, muscles. But events like the 2016 US election or the Brexit vote illustrate that they do not just operate within their home countries. Many work to shape politics in their national allies and adversaries alike.

Globalization has been the key to the development of this truly mobile class. The gains are often stashed in offshore bank accounts in a range of jurisdictions, from Singapore to Nevis, and in the real estate of safe havens like London and New York (Fernandez et al. 2016; Davis and Williams 2017). A number of International Political Economy (IPE) scholars have done critical work drawing attention to the rise of tax havens and the spread of offshore finance (Zucman 2015; Sharman 2006; Palan 2006), but undergirding many of these studies is the notion that these sites exist to increase economic gains. While tax avoidance is certainly a crucial logic of using foreign holding companies and complex ownership chains, extraterritorial dispute resolution illustrates the political gains: tax havens not only hide wealth but also provide transnational property rights. This highlights a need to analyze how the rules and structure of the global economy empower oligarchs and kleptocrats: the IPE of plutocracy.

Documenting the rise of extraterritorial dispute resolution and examining its strategic drivers form my initial attempt to kick-start that project. The cross-border spillover of elite conflict that extraterritorial disputes represent is only possible because of the transnational legal market (Dezalay and Garth 1998; Lowe 2002; Bookman 2019). Its broad reach challenges several core political science theories by illustrating the way strategic resources exist at multiple levels of analysis (Newman and Posner 2018; Dancy and Michel 2016; Eilstrup-Sangiovanni and Sharman 2021). Many scholars view elites as the key actors in pushing for more liberal and, with it, more democratic reform. The general premise of these approaches is that the strengthening of institutions aids elites because it creates a more structured playing field that limits predation by their rivals and, most important, the state (North and Weingast 1989). But now plutocrats can organize their commercial affairs in a way that allows them to buy access to these protections without putting in the costly work of political reform (Sharafutdinova and Dawisha 2017; Pistor 2019; Milanovic 2016). The incentives to fight for liberal institutions are then likely to have diminished, implying a potentially adverse relationship between economic interdependence and institutional development.

But when states are the initiators of extraterritorial disputes, they arbitrage the legitimacy of the transnational legal market for domestic political gains. A rich tradition of sociolegal scholarship demonstrates that ostensibly impartial legal systems may become vehicles of power consolidation (Galanter 1974; Ginsburg and Moustafa 2008). While many scholars have scrutinized the passage of new laws as tools for entrenching existing power inequalities (Goluboff 2009; Moustafa 2007), others have examined how court cases themselves can be leveraged for political ends (Stern and O'Brien 2012; Pereira 2005; Gillman 2002). Because plutocrats now hide and protect their wealth abroad, the perceived independence of British and US institutions are precisely what make them such valuable weapons in contemporary power struggles. From the United States (Teles 2008) to Kenya (Shen-Bayh 2018), to Singapore (Rajah 2012), scholars have documented how domestic courts can be strategically used to counter elite opposition. The extraterritorial disputes I analyze signify these dynamics unfolding transnationally under commercial pretenses, with the need and effectiveness of these strategies magnified by offshore finance.

The broad jurisdictional rules of global commerce may then be curtailing the prospects for liberal reform by providing plutocrats an insurance mechanism when they lose at home while these same rules nonetheless permit autocratic actors to achieve their illiberal ends. But that does not change the fact that these rules exist primarily to promote international economic exchange and have been highly successful in increasing cross-border lending and foreign investment. How to balance these competing effects raises important normative questions for proponents of both liberalism and globalization because it is the predictability and openness of the system—its core tenets—that are leading to its exploitation.

Examining these emerging legal dynamics and the contradictions they symbolize fits within a broader movement in political science to analyze strategic decision making and coalition building that unfolds across borders. While comparative scholarship often treats politics within countries as isolated from each other, IPE research frequently takes a bottom-up approach, moving from the domestic level to the international. Extraterritorial dispute resolution instead highlights a need to analyze how conflicts and institutions in one country shape the strategy and objectives of actors in other jurisdictions. There has been substantial research exploring such transnational connections, but most work in this vein focuses on intra-European or North-South connections (Simmons 2001; Gilardi 2005). The crux of my argument is that the balance of power between the state and the plutocracy conditions when and why such reverberations stem from emerging market politics. The theoretical point is that the interaction of transnational institutions and offshore finance has the potential to upend conventional understandings of domestic political bargains and changes the sites and tools of the fight when those bargains fail.

Methodology and Sources

This book is built on observations of the dynamics of extraterritorial dispute resolutions starting in the mid-1990s. When I started this project roughly eight years ago, I knew some of the details of the Berezovsky-Abramovich affair. I knew the former had been de facto labeled an enemy of Vladimir Putin's state. Handing over billions of dollars to the state's rival, as Abramovich would have been forced to do if he wanted to avoid a heated public battle, seemed beyond acceptable behavior in a rapidly personalizing regime. In other words, these cases didn't seem like they could be simple matters of forum shopping for convenient case law.

With that knowledge, I initially expected the intra-elite disputes in London to be an outgrowth of fleeing plutocrats needing new legal options, that is, political fights flowing across borders nearly as easily as the money laundered through London real estate. Always intrigued by the status games played by the superwealthy (growing up in Dubai, I got to see those duels firsthand), I thought there was probably some kind of sociological game at play. But when I went to collect the data on the range of extraterritorial fights in London, I was struck by two patterns. First, the number of outsourced battles involved plutocrats from only a handful of countries, with Russia and Kazakhstan leading the charge. Countries with similar legal and cultural histories that experienced even more frequent political turnover, such as Ukraine or Georgia, barely saw conflicts hosted in London or via ISDS. And it wasn't just plutocrats fighting each other; the state was regularly getting involved: state-affiliated enterprises voluntarily subjecting themselves to the courts of another country to take on their own citizens.

To figure out why outsourcing authority was popular with parties from only some weakly institutionalized settings, I drew on insights from comparative political economy's emphasis on the stability of state-business bargains (Evans 2012; Lindblom 1982; Rithmire 2023), law and society's recognition that even independent judicial systems can become tools of power consolidation (Moustafa 2007; Galanter 1974; Goluboff 2009), and offshore finance scholarship's recent emphasis on tax havens as sites of institutional arbitrage (Logvinenko 2019; Binder 2019). Engaging with this research, reading case law, and diving into the commercial history of Russia led me to reason that the cases taking place in London, Paris, and The Hague were fundamentally about domestic political struggles that are chronically triggered by the shocks associated with the macroeconomic aspects of globalization.

The bulk of this book focuses on Russia for both normative and inferential reasons. Through the 2000s, Russia experienced an unrivaled economic boom ushered in by the commodity super-cycle. Many name that growth as the primary driver behind Russians coming to dominate the luxury real estate market of London, the revenue sources of multinational banks, and thereby becoming lucrative sources of income for solicitors and barristers. Russia simultaneously underwent a radical return to authoritarianism, an exodus of much of the incumbent oligarchy, and the development of a new band of kleptocrats tethered to the state. The Kremlin was fortifying, and Russia was soon deemed a revisionist power when it first invaded Crimea, sought to interfere in the democratic processes across the United States and the European Union, and then fully renounced itself from international society as it rode another oil spike to invade Ukraine.

While countless books and articles have been written on the evolution of the Putin regime, the transnational dimensions of power consolidation are rarely brought front and center. As my project evolved, I realized my research question was shifting from the general causal determinants of outsourcing disputes to a narrower but arguably more important question of understanding how financial globalization and offshore finance influenced plutocrat-state conflict in Russia. Most scholars agree that 2000 to 2004 was a period of competition, but since then, Putin has had clear control over the plutocracy (Markus 2017a;

Sakwa 2014). I do not dispute that that five-year span is the clearest period of competition. Where I differ is that Russia efficiently moved into a *fully consolidated* era of state control after the Yukos affair in 2004. As I detail, the state's two primary plutocratic threats, Berezovsky and Mikhail Khodorkovsky, continued to challenge the Russian state through transnational means and most concretely through their extraterritorial dispute strategies. The combination of offshored wealth and the potential to refill their coffers via foreign courts ensured that they could continue posing a threat.

In other words, I agree with the consensus that Putin's unforeseen popularity, the spike in oil prices, and the backing of the security services moved structural political conditions into competition starting in 2000. But the process to move toward state control was more prolonged given the transnational nature of wealth and therefore law. It was only once these threats were largely defeated through foreign legal processes and the remaining plutocrats were shown the lengths to which the state would go to extinguish those threats that we saw a full consolidation. The individuals involved in the major turning points in recent Russian history are the focal points of the extraterritorial disputes that were resolved in London and through international arbitration venues. The political shifts that the cases both represent and cemented are the ones that set the stage for Russian revanchism.

Russia still has many plutocrats who earn their money through private means that could still pose a threat to the state if they were able to coordinate effectively. This large presence creates persistent potential for challenges to the state, so the occasional cases that we saw following the big battles with Berezovsky and Khodorkovsky need to be read as attempts by the state to keep these actors in check. These individuals regularly make public proclamations pledging their loyalty to the state to mitigate the threats that they know their wealth could pose.

From a research design vantage point, focusing on Russia provides several benefits. A within-case approach allows me to hold a number of potential explanatory variables constant, diminishing the likelihood that any latent cultural or historical factors are explaining the patterns of extraterritorial fights (Seawright and Gerring 2008; George et al. 2005). In addition, Russian plutocrat–state relations are one of the rare structures to switch completely from a situation where the plutocrats were able to command the political system to one where it's now difficult to argue there are still true oligarchs remaining.

The final decision to initiate or settle an extraterritorial litigation is taken by some of the wealthiest and most secretive individuals in the world. Systematic academic access to these plutocrats and state officials is almost impossible. I circumvented that obstacle by focusing on the other set of individuals deeply involved in the process of weighing the merits of an extraterritorial dispute: the British legal establishment. I focus on fights in the United Kingdom because its courts are the biggest players in the transnational legal market and, until the war in Ukraine, the jurisdiction acted as the ultimate safe haven for Russian wealth. I conducted sixty extended interviews with solicitors and barristers involved in the majority of intra-Russian disputes heard in front of the British courts. I identified them by analyzing public court documents and the Chambers rankings of legal professionals with just under a 40 percent response rate.[18]

I extend the analysis by applying my argument to the other countries that have had the most controversial and lucrative extraterritorial disputes in London and via the ISDS system. The cases were chosen to confirm further how a change from state capture toward competition can lead to the outsourcing of authority. Shocks that trigger competitive environments even when the state is initially in control can lead to extraterritorial disputes. Chapter 5 then details how Erdoğan's rise in Turkey mirrored several of the dynamics of Putin's ascent to power, leading to multiple offshore clashes between the Turkish state and the Uzan family. The analysis next illustrates how losses to state resources and public legitimacy in Angola and, conversely, the sudden enrichment of Nazarbayev's challengers in Kazakhstan provoked clashes that caused the state to leverage foreign courts in a battle to chase down their political opponents.

Plan for the Book

Chapter 1 sets up the puzzle and political economy of outsourcing domestic commercial claims. I highlight the functional benefits of using transnational legal institutions before I detail the potential costs that revolve around fears of disclosure and cross-examination. Next, the chapter lays out a typology of plutocrats in emerging markets that helps identify the nature of the threats to their wealth. I use that typology in conjunction with the logic of the balance of power between the state and the plutocracy to illustrate how plutocrats weigh the benefits of the rule of law against the costs of transparency, which then dictates the outsourcing of authority to foreign courts.

In chapter 2, I begin the analysis of business dispute resolution by Russian plutocrats. More specifically, the chapter illustrates how the use of foreign courts was not an option for the emerging oligarchy given the corruption that defines state capture. I begin by detailing the state of the courts and commercial law during the time and provide important background details on the rise of the Russian plutocracy through the privatization process and the control of Boris Yeltsin's administration. I show at a macro level how informal methods rather than foreign courts were the preferred means for resolving disputes before investigating the rise of two of the most important oligarchs of the 1990s and 2000s: Boris Berezovsky and Mikhail Khodorkovsky. I detail the private disputes involving companies that could logically trigger extraterritorial disputes, as well as the reciprocal relationship between the oligarchs and the state that were endemic to their individual growth. The latter likely prevented the use of London and New York courts despite the fact that Berezovsky and Khodorkovsky were quick to place assets abroad and embed themselves in transnational legal networks.

The next chapter begins by delineating the conditions that made Putin take the reins of the Russian state. It pays particular attention to the way his electoral victory shifted the balance of power from one of capture to one of competition. I provide macro evidence on how the change ushered in the use of London courts and international arbitration venues to resolve quintessentially Russian disputes. I continue the within-case individual comparisons by showing how the threat that Berezovsky and Khodorkovsky posed to the state led to their removal from the political game. I detail Berezovsky's relationship with his former business partner Roman Abramovich and show that the threat to the state prevented Abramovich from settling a $5.6 billion dispute in the London commercial court. Khodorkovsky's business partners waged a more than decade-long battle against the Russian state to recoup their losses from the expropriation of Yukos; again, their exits and Khodorkovsky's imprisonment curbed the transparency costs. The fact that this was a political rather than a commercial battle incentivized a variety of extraterritorial disputes.

In chapter 4, I illustrate how the state strikes back. It first discusses how cases involving SOEs in London are often driven by political motives. It then analyzes these dynamics by examining how various Russian SOEs were used to target both Berezovsky and Khodorkovsky as they continued to pose threats to the state even after conventionally losing the domestic political game. I generalize the findings from chapters 3 and 4 by using a quantitative billionaire-level analysis of the determinants behind the outsourcing authority. In other words, I show how domestic courts in major economic powers are new sites for political contestation for battles stemming from emerging markets. I show how those who want to remain on the Russian state's good side are unlikely to use transnational legal institutions because of potential retribution from the state. At a macro level, I illustrate how and why SOEs and subservient tycoons rarely take their fights to foreign courts.

I broaden the empirical scope of the project in chapter 5 to illustrate the value of the argument beyond the Russian case. It analyzes three countries that have experienced similar shifts in the balance of power that then reverberated into foreign courts: Kazakhstan; Turkey; and, most recently, Angola. These cases help establish that shifts toward competition, independent of where the balance of power starts, drive the outsourcing of authority. They cement how this is a game caused by and in search of offshored wealth.

Most of the conclusion explores the normative and theoretical implications of extraterritorial dispute resolution with an emphasis on how the liberal premises of these strategies effectively undercut the stated aims of the international economic order.

1
LIBERALISM AS A LAST RESORT

Business in emerging markets is plagued by a fundamental political dilemma: any state that is capable of enforcing property and contracting rights is also capable of stripping firms and individuals of their wealth (Weingast 1995). Without predictable protection mechanisms, trade and production grind to a halt and economic development is choked off (Acemoglu and Robinson 2013; Frye 2004). In other words, predatory behavior by state actors or by one's business partners needs to be rooted out by a centralized but not excessively powerful authority. The provision of the rule of law is then seen as the key to economic development (Glaeser et al. 2004; Aron 2000; Rodrik 2008).

Elite businesses have historically driven the strengthening of such liberal institutions (North et al. 2009; Amsden 2010; Amsden et al. 2012). They can use their economic clout to pressure the sovereign for reform that will be beneficial to all merchants, not just the already powerful few. Political development is seen as being in their long-term interest because it ensures that they will be able to keep their current spoils with plenty of room to grow, unencumbered by excessive state coercion.

While these dilemmas still affect most would-be capitalists in the developing world, globalization provides a select few with a transnational solution to their institutional needs. By writing their contracts in line with foreign laws, or filing against their opponent's offshore holdings, plutocrats, their private companies, and state-owned enterprises can secure property and adjudicate disputes through foreign courts. Plutocrats can even use their offshore companies to initiate arbitrations against their own government.

For three complementary reasons, extraterritorial dispute resolution is a game primarily involving plutocrats.

First, filing cases abroad can regularly involve millions in legal costs, making plutocrats or the state the only actors with adequate resources to take on a domestic opponent abroad. In one extraterritorial litigation discussed in chapter 4, *Berezovsky v. Abramovich*, more than 100 million pounds were spent just on lawyers (Sky News 2012). In the extraterritorial arbitration between Yukos and the Russian state, the private actors asked for more than $50 million worth of compensations for their legal expenses (Brauch 2014). This barely accounts for the millions more spent on private investigators and public relations professionals.

Second, the value of using foreign courts is conditioned by the size of a claim. The higher the claim, the greater the importance of a predictable, reputable dispute resolution process. Plutocrats will be most likely to have cases worth millions, or even billions, which makes spending money on extraterritorial disputes worthwhile. Conversely, plutocrats will be the ones likely to be defendants in such cases.

Third, and most important, even if a claimant wins, they need to be able to enforce the judgment. Enforcing against an opponent's assets at home would defeat the purpose of initially using London or New York because they would then have to deal with the home judiciary. For the case to make financial sense, the opponent would then need generally to have assets located abroad that can be claimed. Only plutocrats are likely to have large, international holdings worth the size of a multimillion-dollar claim that can eventually be seized. Extreme globalized wealth is thus the necessary condition for actors to engage in extraterritorial disputes.

Operating in weakly institutionalized environments, emerging market plutocrats are engaged in a constant battle to keep that wealth. They fear not only the state but also fellow plutocrats ready to raid their companies (Markus 2015; Gans-Morse 2012). Comparative scholarship expects economic elites to leverage political connections and extralegal means to guard their property, but the transnational nature of today's plutocracy provides them with multiple avenues for legal protection. They have the economic means and the functional needs for extraterritorial litigation against private parties and extraterritorial arbitration against their home governments. But as we saw in the introduction, outsourcing authority is not the norm with all plutocrats, even when they lack rule of law at home.

To resolve this puzzle, scholars must factor in the political costs of extraterritorial dispute resolution. Using a liberal institution forces plutocrats to reveal the details of their business and wealth, which could provide rivals with more ammunition in a commercial or political fight. Plutocrats then need to weigh the benefits of the rule of law against how it might affect future threats. The *balance of power between the state and the plutocracy* in an individual's home jurisdiction shapes the severity of these threats, which explains why extraterritorial dispute resolution is more popular with parties from some countries.

I develop this argument by drawing on a range of social science debates on the meaning of the law and on the nature of conflicts between capitalists and the state, and through my examination of the evolution of the plutocracy in Russia and elsewhere. I draw on the interviews I conducted to further justify some of the building blocks that are generally unrecognized in academic debates. I start by presenting the costs and benefits that plutocrats face when deciding whether to use foreign legal systems to resolve domestic disputes. I then define what I mean by "the balance of power," which is the primary explanatory factor underpinning my argument, illustrating different power arrangements between the state and the plutocracy. Next, I develop a typology of plutocrats. I show that how an individual makes their money—whether via private capital or through the state—combines with the balance of power to shape the kinds of threats their wealth is likely to augur. I then use these ideas, the balance of power and my typology of plutocrat wealth, to explain when and why plutocrats are likely to outsource legal authority to foreign courts. At the extremes of the balance of power, when the balance of power is clearly in favor of either the state or the plutocracy, disputes are rarely outsourced. Instead, when plutocrats and the state are in competition over control of their home system—when they have relatively equal power resources—political conflicts unfold under the guise of commerce through foreign courts.

The Stakes of Extraterritorial Dispute Resolution

Benefits of Using Foreign Courts

Previous scholarship notes that the primary benefit to using transnational institutions to settle domestic disputes is access to strong rule of law, where claimants and defendants will be treated equally (Dammann and Hansmann 2008; Sharafutdinova and Dawisha 2017). Opponents will be unable to bribe a judge, and their shared sovereign will be unable to interfere in proceedings. As one British lawyer told me, "Impartiality is the key."[1]

But if a claimant is more politically connected than their rival, weak rule of law may be beneficial as they generally know that the case will turn out in their favor. For example, Ang and Jia (2014) illustrate that businesses with stronger ties to Chinese Communist Party officials are more likely to take advantage of the domestic judicial system. In the absence of a strong state presence like China's, the wealthier player would almost certainly win a bidding war to bribe a judge. When in conflict with a more powerful player, the weaker plutocrat then has considerable incentives to seek foreign protection.

Although weaker parties have greater incentives to go abroad, the resources and flexibility of those courts may still incentivize (more) powerful actors to outsource legal battles. Why would a plutocrat tied to the state, whose judiciary could promise an outcome in their favor, go abroad? Why would the state ever choose extraterritorial conflict when it can guarantee a win at home? In short, transnational legal institutions are often just more efficient in their proceedings, saving parties valuable time and money. In other words, even if they can easily secure a victory at home, the price of a bribe and the length of proceedings could end up costing more than going abroad. With extraterritorial litigation, plutocrats would also be dealing with decades of predictable, established law that may provide additional advantages that the domestic institutional structure lacks. The United Kingdom, for instance, recognizes oral contracts, which are regularly used to cement deals in the developing world despite many judiciaries dismissing their validity. As one lawyer put it, when describing the resource differentials between London and an emerging economy's legal system: "We are Manchester United, they [courts in emerging markets] are Cardiff City."[2]

Transnational legal institutions also often have several interim measures and enforcement options that emerging economies lack. Worldwide freezing orders, a de facto method of private sanctions that prevents parties from accessing or moving their money, can be used to make an opponent's life impossible. This can lead to quicker settlement or payment. In the words of an individual involved in numerous extraterritorial disputes:

> Another big attraction is the freezing orders and injunctions . . . you can get one without having to fight a trial and it's often enough to get someone to settle . . . it's the first dossier that you bring to a freezing order that impresses the judge . . . that's what does the damage.[3]

Legal decisions that come from reputable institutions can then also be enforced in the host jurisdiction or in numerous other countries. The International Centre for Settlement of Investment Disputes and the New York convention ensure that arbitration awards won by private parties can be enforced in more than 150 countries (Efrat 2016). Barring major discrepancies between domestic laws, UK commercial judgments can be enforced in all the major economic hubs of the global economy. Because of treaties dating back to the 1920s, London judgments can be used to access wealth in most of the Commonwealth countries and territories, including the likes of Mauritius and Singapore, which are central sites of plutocratic wealth protection (Pinsent Masons 2013). The United Kingdom's reciprocal treaty arrangements with the European Union allow British judgments to be enforced in Cyprus, Malta, and the Netherlands, locations that are essential to the corporate shell game. The Lugano Convention ensures enforcement in Switzerland, which remains a central node for elite private banking (Tevendale et al. 2021). In sum, the United Kingdom's judgments can be enforced in nineteen of the top twenty offshore financial centers identified by the Tax Justice Network.[4] In contrast, Russia and China have only a handful of such reciprocal arrangements. Such an enforcement option is particularly important if one's rival has stashed their most prized assets abroad, which is now common with plutocrats in both liberal and illiberal countries (Alstadsæter et al. 2017).

Risks to Using Foreign Courts

When a plutocrat brings a case against a conational, they are not only subjecting themselves to foreign adjudicators but also to the laws governing disclosure and potential cross-examination. Such transparency is generally seen as a deterrent to legal proceedings for all businesses: no firm wants to reveal the inner workings of its organization or products for fear that it could give competitors an edge or reveal damaging information that could hurt firm value. With plutocrats in the developing world, additional layers of risk stem from the way they amass their wealth and how they manage to stay on top.

Examination Costs

To reach plutocrat status in an emerging market, it is highly likely that an individual has engaged in corrupt and likely criminal acts along the path to prosperity. Journalistic and academic accounts of business-government relations across numerous weakly institutionalized states highlight tax evasion, bribery, and even contract killings as common tactics that companies and individuals use to navigate an environment without a clear rule of law (Klebnikov 2000; Cuadros 2016; Freund 2016). As Kimberly Kay Hoang puts it in her remarkable ethnography of the superrich in emerging markets, plutocrats are constantly "playing in the gray." One of her interviewees described business success as a function of knowing "how to finesse the very boundary between legal and illegal activity . . . but with the tacit protection of officials who are often lubricated by gifts of bribes" (Hoang 2022, xiv). With business in emerging markets, "there is no neat dichotomy between legal and illegal, licit and illicit. The illegal and illicit facilitate legal and licit market making" (Hoang 2022, 10).

When involved in an extraterritorial dispute, plutocrats, their business partners, and their executives will be subject to questioning by some of the world's most experienced and astute lawyers. And if the dispute takes the form of extraterritorial litigation, it will all be public record. A counsel to numerous Eastern European oligarchs highlighted how Hoang's conception of "playing in the gray" is a regular deterrent to plutocrats whenever weighing the use of London's courts. "They [the barristers] are asking one thing but are thinking five steps ahead . . . you don't know what's going to come out. Some will pay [a settlement] just to avoid that."[5] Revealing the details of one's corrupt practices could have immediate negative economic consequences because business partners may retreat to avoid getting embroiled in a scandal. The information revealed during an examination could also leave parties subject to additional proceedings at home or abroad. And this has consequences, even if these are open secrets in the business community:

> It's one thing to say "everyone knows" but it's another thing to expose to public scrutiny and exposure . . . it depends on the legal consequences. Does it give rise to liabilities—civil or criminal? Here or at home? Can others come after you? Can it help the other side to defend the claim? Does it affect the judge's view of your credibility?[6]

Disclosure Costs

The costs go beyond what has happened in the past and directly influence and expose current business practices. As research illustrates (Haberly and Wójcik 2014, 2022; Alstadsæter et al. 2017), plutocrats across the developing world regularly use offshore havens to structure their investments. This is not necessarily illegal, and it helps explain why places like Mauritius, the British Virgin Islands, and Cyprus are huge foreign investors in places like India, China, and Russia, respectively (Linsi and Mügge 2019). The general logic for why plutocrats use these complex contracts is seen as straightforward: they are trying to avoid paying more in taxes (Xiao 2004; Ledyaeva et al. 2015). Almost 10 percent of the wealth of a country—whether developed democracies, hybrid regimes, or outright authoritarian jurisdictions—is hidden offshore (Alstadsæter et al. 2017).

Research also illustrates that offshore havens play a far more important, political role when the rule of law is weak (Sharman 2012; Bayer et al. 2020). They provide additional safety nets for plutocrats in case they come under political pressure from the state or economic pressure by a private rival, including giving them access to extraterritorial arbitration. For example, Mukhtar Ablyazov, who was accused of stealing almost $10 billion, used more than a thousand shell companies to guard the wealth he accumulated while running BTA Bank in Kazakhstan (Cooley and Heathershaw 2017; Nougayrède 2016). He primarily kept paper records that were stashed away in a giant yellow storage unit in London (Burgis 2017, 104–5). Plutocrats regularly go to such great lengths to prevent being exposed. Another lawyer illustrated that these measures build on informal domestic practices; his Ukrainian clients rarely sign their contracts themselves, giving them an additional layer of protection.[7]

By legally placing ownership of the assets abroad or obscuring ownership more generally by keeping assets with stewards, rivals are unlikely to know where to look when taking on a plutocrat. Participating in legal proceedings, however, will force a plutocrat to disclose what exactly their assets are and how exactly they are structured, leaving them exposed to new political challenges.[8] One lawyer explained the use of offshore structures bluntly: "The secrecy and the protection of assets . . . it's an instrument of survival."[9]

The gains of equality and efficiency in using extraterritorial dispute resolution thus need to be weighed against these examination and disclosure costs. The broader political environment in which plutocrats operate end up determining the severity of those costs.

The Balance of Power Between the State and the Plutocracy

In many democracies, politicians are expected to be highly dependent on plutocrats. They control capital and the money that is necessary to finance a campaign and keep an economy growing (Lindblom 1982). Alternatively, scholars have theorized the role of developmental states and the necessary embedded autonomy (Evans 2012) or highly autocratic countries where capital is completely subservient to the state (Rithmire 2023). In other words, institutional development and economic performance are often explained by pointing to the distribution of political resources between the business class and the state because the distribution creates an ordered set of interactions between the two entities.

This conceptualization of plutocratic-state relations relies on a power resources approach. Much of the seminal work on business power and oligarchy uses this premise as its starting point (Korpi 1985; Winters 2011; Fairfield 2015). What counts as a political resource varies by specific country settings, but at least five resources matter across contexts and can be held or acquired by the state and by plutocrats. The most important, across time and space, is coercive capacity: Who has control over the means and methods of violence? Most would agree that this is the primary and exclusive prerogative of the state, and that control can be channeled toward both direct and indirect expropriations of economic actors. If that control is seen as legitimate and effective, official and formal positions, such as the head of a state or bureaucratic organization, should be seen as an essential part of power.

The state is often viewed as having mobilization power. With authority stemming from the official positions, state actors can often lead individuals, change preferences, and develop networks that can be channeled toward both protecting and strengthening the state. In addition, the dominant governing ideology can sway in favor of the state, be it a function of religious, political, or even economic beliefs. Muslim monarchies get their power partly from this lineage, communist ideology buttressed many states during the Cold War, while most states today rely on using their potential to generate growth as the engine of their ideological legitimacy (Guriev and Treisman 2019).

The burgeoning literature on business power has identified how plutocrats often use each of these power resources to guard and protect their wealth. In weak states, plutocrats regularly acquire the means of coercion by directly hiring private security or building up their own mini-armies to protect their most lucrative economic assets (Sonin 2003). Official positions are not the purview of only the state; plutocrats regularly run for office and gain traditional state authority by occupying positions such as member of parliament, congressional representative, even head of state through democratic processes (Szakonyi 2018; Hägel 2020; Krcmaric et al 2023). Plutocrats can attain mobilizing power by hiring people to behave like civil society, be it to protest for economic or political change (Radnitz 2017; Werner 2012). Meanwhile, the dominant ideological movement of the past thirty years—neoliberalism—placed the economic and social value of plutocrats at its center (Freeland 2012; Slobodian 2018). The main point is that the most important political resources in a society are monopolized and are shared occasionally between the state and the plutocracy.

Varieties of Plutocrats

International relations scholars regularly define the 1990s–2010s as an era of American unipolarity or the neoliberal age, but it could just as easily be described as a time of global plutocracy. In 1987, when *Forbes* started ranking billionaires, they identified less than thirty across the developing world. Three decades later, the number had grown more than twentyfold, to over six hundred.[10] Although trade liberalization and global supply chains have lifted a record number of people out of poverty, the gains have primarily gone to those at the very top. In line with Winters (2011), I define "plutocrats" as actors who control massive concentrations of money that can be used to build and guard their wealth and thereby their political and social positions.[11] As Winters argues, all the power resources mentioned above matter for political outcomes, but money has been the most resistant to grander social changes. All plutocrats are imbued with substantial instrumental power via their wealth and, as Fairfield (2015) argues, they can use that wealth to acquire other power resources. The real power of money is its fungibility.

But where plutocrats get their money is likely to shape both their interests and the autonomy to exercise their power. Much of the literature and public debates on the super-wealthy focus on, unsurprisingly, the size of their fortunes. When analyses get into how plutocrats make their money, scholars and journalists distinguish between the industries that the individuals operate in, often focusing on asset mobility, or whether they have concentrated or diversified holdings (Markus and Charnysh 2017; Paniagua 2018). This misses how diverging sources of revenue have altered the rise of plutocrats in the past few decades.

Most plutocrats make their money through their private companies. But even as neoliberal logics pervaded global governance and governments auctioned off many of their most valuable resources, state-dictated rules continued to structure the economic playing field, and frequently the state is the literal means to make money (Stone et al. 2022; Babic et al. 2020). Positions at the top of effective state-owned enterprises, and the potential they hold for individuals to siphon off gains or extort side payments, is crucial to how many individuals in Russia and China get and stay rich (Cooley and Sharman 2017; Dawisha 2015). These countries still have productive private sectors, and the latter has seen the rise of a mammoth tech scene, with some of the most desired and lucrative intellectual property. Plutocrats inevitably care not just about their wealth but also maintaining and growing their income streams. Increasing incomes not only feed into growing their wealth but also provide them with substantial social status and potentially other power resources given the legitimacy that can come from running important and efficient firms. Their revealed preferences indicate as much: if plutocrats in emerging markets, where money can put a target on your back, only cared about their wealth, they would all cash out, put their money in the S&P500, and leave.

Terms like "plutocrat," "oligarch," "kleptocrat," "robber baron," "magnate," "tycoon," and "chief" are all used synonymously when discussing the super-rich, but to understand the wealthy, these terms should be differentiated. By bringing together the concept of the balance of power between the state and the plutocracy and the source of a plutocrat's income, the threats that plutocrats face to their wealth and the potential threat they pose to the state can be described more accurately. This conceptualization sheds light on when they are

TABLE 1.1. Varieties of plutocrats

	BALANCE OF POWER	
INCOME SOURCE	**PLUTOCRAT-DOMINATED**	**STATE-DOMINATED**
Via private capital	Oligarchs	Tycoons
Via state capital	Cronies	Kleptocrats

most likely to exercise the instrumental power of their wealth and if they will then decide to outsource their disputes to foreign courtrooms. In other words, by knowing the country context in which an individual operates and the source of the individual's wealth, scholars may identify what they are likely afraid of and how they are likely to use their money.

Table 1.1 lays out the typology combining the two common variants of the balance of power that appear stable and present in the literature on business-government relations: one where the state dominates power resources and one where the plutocrats dominate power resources. The other axis is constituted by how plutocrats make their money: either as a function of their own private resources or via the state.

As noted above, plutocrats are individuals with substantial amounts of wealth that can be deployed for economic and political ends; they all have instrumental power. Most countries, with the spread of free-market reform and increasing loss of state resources and legitimacy, largely see these actors monopolize the power resources of the country. Given the way their economies operate, capital often has structural power that tilts policymaking and policymaker preferences toward appeasing the elite business class. That allows for individuals that have their own substantial economic resources not only to dominate the market but also to dominate the political terrain regardless of whether they occupy official positions. These are true oligarchs in the classic sense of a governing few.

"Oligarch" is a term usually used to describe all Russians with super-wealth, but the term originally gained public prominence to describe the dozen or so individuals that came to conquer the 1990s market and political systems. It no longer describes the plutocrats in contemporary Russia. But reorienting the phrasing clarifies the positions of a host of individuals across countries and contextualizes their autonomy vis-à-vis the state. Chile, Ukraine, and India also fit the same pattern, where the state is relatively weak in terms of the power balance compared to the super-wealthy; the super-wealthy is made up of individuals that have their own private capital at the root of their incomes. These actors all face or faced very few state-directed threats to their wealth. They face few obstacles to using their power resources. They fear other dominant oligarchs operating in their industries more than the state. Sometimes their conflicts unfold through the market or through the political process when they support rival parties for control over relatively weak governments.

But even in these plutocrat-dominated societies, governments still have some rulemaking authority and some of its own government resources that can produce a set of cronies, as seen in the last column of table 1.1. At the very least, positions of state power have some baseline legitimacy or provide access to information on regulatory processes that can be useful toward improving one's economic position. These positions have also historically served as important gatekeepers to the broader global economy. They may have weak and ineffectual state-owned enterprises that can still provide some lucrative contracts, as is the case with the development of the smaller band of cronies around Theodore Obiang in Equatorial Guinea or under Sani Abacha in Nigeria. These individuals, given that they operate in situations where the oligarchs are still likely to have more power, will be unable to shift the structural status quo. The threat to their wealth is someone stepping in to take over the contracts or state positions they hold, that is, when oligarchs often aim to buttress their positions. Their preferences toward political change are likely to be mixed: strengthening the state could lead to more rents, but a stronger state could also clamp down on the corruption that fuels their wealth or simply seeks to keep the spoils for itself.

The value and autonomy of wealth sources flips in societies where the state dominates the power resources. Because money is the most important and fungible of the power resources, only states that are likely to have some direct control over their economies are likely to fall into this category of table 1.1. That means the most effective and viable path to wealth is through the state. Similar forms of clientelism exist as in the plutocrat-dominated settings but on a grander scale. In this set of cases, state enterprises tend to be larger; running them or servicing their needs becomes the way the state rewards allies and fortifies its control. This moves the analysis to the kleptocrats. The "silovarchs" coming from the security services in Russia and the individuals Vladimir Putin knew from his time in St. Petersburg like Igor Sechin and the Rotenbergs, who are currently sanctioned by the United Kingdom,[12] are prototypical examples of such kleptocrats (Treisman 2008). Russia is not unique in this form of governance by thieves; individuals in a host of authoritarian regimes like Saudi Arabia and Qatar and, as detailed in chapter 5, Angola under Dos Santos operate in a similar fashion. Given that the state is their dominant source of income, the state is unlikely to be wary of kleptocrats ever mounting a substantial challenge. A shift away from the status quo would mean a loss to their primary mechanisms of wealth generation; they know which side of their bread is buttered.

Kleptocrats are also then unlikely to pose a threat to the state because the state knows they are dependent on it for revenues and their plutocratic status.

Finally, in table 1.1 there are tycoons, who make their money through their own private initiatives but operate in conditions where the state dominates the balance of power. Even in the most state-dominated economies today, there is usually still some private enterprise. While some of these exist to service the state, many others earn revenue by selling goods and services to individual consumers or other private businesses. Plutocrats that come up through this process can still make lots of money—most of the well-known billionaires in China on the *Forbes* list fit this pattern—yet theirs is likely the most precarious predicament.

Tycoons may have tacit agreements with the state to stay out of the traditional political process, but they still have instrumental power from their wealth. They may even regularly channel that power to influence government regulations or stifle other private competitors. But their wealth also poses a potential threat to the state because a lack of clear reliance on the state for their income indicates that they may want to eventually shift the power balance. At the same time, kleptocrats will be eyeing their enterprises to take them over for themselves because this may be condoned by the state. Tycoons then play a constant balancing act of trying to avoid political conflict while still growing and protecting their income streams. And if they get too big, they could be cut back down to size. Given that they are in state-dominated settings, they will not be able to counter that process easily. Individuals like Jack Ma of AliBaba and Pony Ma of Tencent represent these individuals most clearly, but these actors operate in many country settings, including Russia, today. Those who were once oligarchs in the 1990s and somehow still persist in the country have been relegated to the travails of a tycoon, while the general growth of the country's economy in the 2000s and the 2010s has benefited a host of other predominantly private sector billionaires who came up under and abide by the rules of the game set out by the Putin regime.

Where do politicians fit in? It depends on whether the individual politician derives their income through their state position and the prevalent balance of power arrangement. Putin is instructive here. He started off as just a normal politician, one whose popularity and increasing state resources during the oil boom put the state on a path to challenge the incumbent oligarchs. Over time, Putin has transformed from a politician into a kleptocrat, at least according to reputable academic work on the topic (Dawisha 2015), because much of his hypothesized money has come from his ability to take cuts from different state contracts and for payments to ensure that certain companies can stay private. Putin of the early 2000s was a politician but is now also firmly a kleptocrat.

Yet there is still a substantial state with popular authority, a robust military, the rule *by* law, and a massive bureaucracy all working toward state goals.

In contrast, the economic circumstances meant that Boris Yeltsin was never able to build up his wealth to the point where his own resources would make him a serious threat. He wasn't even able to acquire crony status. Yeltsin came up through the political apparatus in a time of limited private enterprise. Other more contemporary examples of individuals operating in capture-like settings, such as Silvio Berlusconi in Italy or Michael Bloomberg in the United States, instead represent more prototypical examples of oligarchs whose wealth was not acquired through the political sphere but instead through their own private enterprises prior to taking office.

These are ideal typical forms of plutocrats. As with any ideal type, the lines blur in the real world. Many oligarchs of 1990s Russia displayed crony-like behavior to get their start, while many kleptocrats end up diversifying into more private forms of business. The purpose of the typology is to update the standard heuristics used to describe the super-wealthy given the varied forms of state-society relations that are present in emerging markets. The value of the typology is in delineating the types of threats that different plutocrats are likely to experience as a function of the equilibrium balance of power and their wealth source. The latter also matters for the propensity to become involved in extraterritorial disputes because those who depend on the state are unlikely to be considered meaningful threats worth taking on by the state. When these ideal types break down or become muddled is when political conflict emerges, leading to the outsourcing of authority to foreign courts.

How the Balance of Power Influences (Extraterritorial) Dispute Resolution

Legal scholars have long recognized that lawsuits are a form of bargaining failure (Macaulay 1963; Mnookin and Kornhauser 1979). Formal dispute resolution through state-provided institutions needs to be considered the final step of a broader dispute resolution process that usually begins with informal negotiations and ends in the courtroom only when all other options have been exhausted (Hendley et al. 2000; Albiston et al. 2014). In this situation, parties that litigate have recognized substantial grievances that collapse the bargaining space (Felstiner et al. 1980). Courtrooms more generally present a challenge for plutocrats in emerging markets because a plutocrat in a weakly institutionalized setting has been involved in some corrupt transactions that could be revealed over the course of a public legal process. As the typology in table 1.1 indicates, different types of plutocrats are likely to have engaged in their own specific forms of corruption. Yuen Yuen Ang (2020) argues that kleptocrats and cronies are often involved in "grand theft" from public purses because they rely outright on state positions or state officials for their incomes. Oligarchs and tycoons rely more heavily on so-called access corruption, where they reward state officials. For the former, they likely do so to keep the government reliant on them; for the latter, it is often the only way for them to survive. Regardless, all plutocrats have incentives to avoid revealing these corrupt practices, and so they undergird their wealth generation and protection strategies.

Plutocrats face two primary threats to their wealth: private threats from other plutocrats and public threats from the state. By conditioning the potential transparency costs associated with using transnational legal institutions, the balance of power between the state and the plutocracy structures the nature of the threats to plutocratic wealth.

While I conceptualize the balance of power as a spectrum, it can be broken down into three ideal types. At one extreme, where the plutocrats have all the power resources and the situation is dominated by oligarchs, society is stuck in state capture, where the government is fully reliant on the business class. At the other extreme, the state has all the resources, leaving plutocrats subservient to sovereign authority in the mold of kleptocrats described above. I label this scenario "state control."

But the world regularly goes through cycles where power balances shift. There are frequent waves of nationalization and privatization, as well as declines in the legitimacy of different governance models. To understand those shifts, the analytic focus needs to be on the middle of the distribution, with a relatively equal power balance between the state and the plutocracy. This is where I identify conditions of state competition, in which some plutocrats will be subservient to the state, and others have their own, independent sources of wealth that the state will want to acquire. Table 1.2 summarizes the key logics of each ideal type, with their relationship to the likelihood of extraterritorial dispute resolution. The table also includes countries that most clearly represent each ideal type.

It is also necessary to clarify that, under conditions of state competition, two strategies instigate extraterritorial disputes. First, fleeing plutocrats may use transnational legal institutions in a defensive manner, as a mechanism to regain their lost wealth. Second, the state, via its commercial enterprises, turns foreign courts into an offensive tool to delegitimize a fleeing plutocrat and attempt to access or drain the latter's offshore wealth. Previous literature on extraterritorial litigation has focused on the former (Sharafutdinova and Dawisha 2017; Dammann and Hansmann 2008). The next three subsections delineate why

TABLE 1.2. Balance of power ideal types and the likelihood of extraterritorial dispute resolution (EDR)

IDEAL TYPE	COMMERCIAL INCENTIVES FOR EDR	POLITICAL INCENTIVES FOR EDR	ALTERNATIVE(S) TO EDR	BARGAINING OUTCOME(S)	REPRESENTATIVE CASES
State capture: plutocracy primarily constituted by oligarchs	High: access to expertise and rule of law	Low: transparency of extraterritorial litigation would undermine beneficial home status quo for plutocrats	Informal private mediation; extralegal means	Settlements; unlikely to use foreign courts because it increases private threats to plutocrat's wealth	1990s Russia, Ukraine, South Africa
State control: plutocracy primarily constituted by kleptocrats	High: access to expertise and rule of law	Low: transparency would undermine state and risk retribution	Informal state mediation	Settlements; unlikely to use foreign courts because it increases public threats to a plutocrat's wealth	Contemporary Russia, Saudi Arabia, China
State competition: income streams of plutocrats is clear but conflict arises because they want to become oligarchs and fear becoming tycoons	High: access to expertise and rule of law	High: (1) fleeing plutocrats can regain their lost fortunes; (2) the state can capture an opponent's offshore wealth	None	Extraterritorial disputes; settling risks enriching political rivals	Early Putin years in Russia, mid-2000s Kazakhstan, contemporary Angola

outsourcing of authority at the extremes is rare and why extraterritorial disputes are more common when a country has a relatively equal balance of power between these two sets of actors.

State Capture Generally Leads Plutocrats to Avoid Foreign Courts to Prevent Unraveling a Beneficial System

In a system where the plutocracy has all the valuable political resources, the state will be completely reliant on the business class to keep it in power. Plutocrats are then mostly oligarchs, with their own private resources, who have virtually no concerns that the state will step in to expropriate their assets or implement regulations that may hinder their profit margins. That is, they do not face any public threats and instead are focused on private threats from rival oligarchs. The fact that the state lacks any capacity in relation to the plutocracy would create incentives to, and allow for, oligarchs to use extralegal means to resolve their conflicts (Volkov 2016).

When oligarchs dominate the political scene in emerging markets, they are operating in access-based corruption systems, frequently referred to as situations of state capture. If faced with a commercial conflict where the size of the claim far exceeds the legal costs (i.e., one worth taking to a foreign court), the claimant will then inevitably be going up against a fellow oligarch or one of a handful of cronies that is likely to exist through state resources. By going to London or New York, they will gain access to independent and expert judges that are usually lacking in emerging markets. If the claimant wins, they will have a reputable judgment that they can use to seize an opponent's assets almost anywhere in the world, including in tax havens that all plutocrats use to hide and guard their wealth.

But oligarch business networks are usually dense (Cárdenas 2016; Haselmann et al. 2018). Defendants are likely to know some details of the claimant's corrupt endeavors. Airing these details in a public courtroom, especially abroad, could quickly lead to business partners cutting their ties as the latter will want to avoid being caught up in a scandal. With the globalization of wealth and income portfolios, oligarchs could further risk being cut off from international markets. Even if these are well known acts of corruption among the political and economic elite, rumors are not the same as the highly reputable British courts calling you a fraudster. Although several multinational banks have been caught up in corruption scandals (Woll 2023), they have serious compliance checks in place that will likely force them to cut their ties, while any public companies owned by the oligarch will take a massive hit (Verdier 2023). Consider the brutal destruction of the Adani Group's share price in January 2023. After short sellers Hindenburg Research shed light on the company's use of offshore vehicles to allegedly manipulate their share price, more than $150 billion of market value evaporated (Kamdar 2024). As this book goes to press, nearly two years later, the share price is yet to fully recover despite Adani's consistent denial of the accusations (Reed 2024). If a small, self-interested hedge fund can have such an effect, a neutral indictment from a British court is likely to have substantially more impact.

At the same time, the literal acts of corruption could incentivize more legal actions. This could come from the home government that sees it as an opportunity to regain some power at home (assuming it is not heavily implicated), but it could also come from foreign sources. The United States, until the second Trump Administration, took an active role in policing global corruption through the

Foreign Corrupt Practices Act (FCPA), initiating over 250 since the early 2000s (Tomashevskiy 2021). New details coming to light in a reputable court could push the Department of Justice to begin investigating the oligarch because the US bureaucracy is able to use the broad reach of the dollar and US capital markets to claim jurisdiction. Lorenzo Crippa and I show that oligarchs become far more likely to move their money offshore when the United States initiates an FCPA case in the plutocrat's home jurisdiction, illustrating that oligarchs both fear and take precautions against anticorruption endeavors (Crippa and Kalyanpur 2024). The downside of a fair legal process, which extraterritorial disputes promise, is that you are also subject to the transparency of liberal procedure that could then instigate future threats to one's wealth. That risk is likely to cut both ways, affecting both defendants and claimants.

Even when faced with a major dispute where animosity is running high, oligarchs are fundamentally winning in the lawless world of state capture. The lack of strong institutions likely has even helped them consolidate their commercial empires (Sonin 2003). Unraveling the status quo could cost them their businesses and take away their rents, providing strong incentives for corrupt parties to avoid the (foreign) courtroom. The way disclosure and examination costs can increase private threats then facilitates settlement. These informational externalities deter the use of transnational legal institutions, tipping the scales in favor of using informal or extralegal measures and allowing the class to continue exercising indirect control outside the spotlight. This logic applies to both oligarchs and cronies because, in these contexts, the preexisting system is working to their benefit.

I document this process throughout chapter 2 of this book. The plutocrats controlling the 1990s Russian political terrain were true oligarchs whose wealth allowed them to dictate who was formally in charge of the state. As they built and fortified their income streams, they were quick to protect their money through offshore vehicles and embed themselves in the legal networks necessary to outsource the highly contentious disputes that characterized the particularly violent form of capitalism that emerged. But the corruption and coercion that characterized their business empires ensured that they avoided the foreign spotlight because, as one lawyer told me, "If you've got skeletons in your closet they will also be found out."[13] The oligarchs sought instead to resolve their disputes through informal, often extralegal means.

State Control Generally Leads Plutocrats to Settle to Avoid Sanctions by the Sovereign

On the other end of the spectrum, where the state has all the valuable political resources, there are almost no private threats to plutocrats because rivals are unlikely to take up predatory behavior without state approval. Instead, plutocrats are focused on the public threat from the state. Rulers can take an active role in managing intra-plutocracy conflicts, especially if the court system is weak. Assuming the sovereign is independent, it can often act as an impartial adjudicator between even otherwise wealthy or powerful oligarchs.

In a state-controlled society, there are typically two types of plutocrats, as described in the previous section. Kleptocrats rely on the state for their revenues, often through a clientelist exchange. The few tycoons have their own income streams and are often forced into an extortion-style arrangement with the state so that they can be allowed to remain nominally private. Both types of plutocrats will be unable to bring a claim if they want to keep their income because disclosing the nature of their dealings not only risks airing their own business secrets but also corruption by the state. If a claimant has managed to get to the top while somehow avoiding corrupt behavior, forcing an opponent to disclose the nature of commercial deals will still likely run the risk of implicating the state. The costs from a public legal process are then extremely high.

Taking the case abroad in a state-controlled environment implies that the state will have sufficient domestic tools to strike back. What comes out in cross-examination could be used as a basis for enforcing corruption measures, or the state could simply resort to political assassination. At the same time, disclosing the nature of one's holdings during a trial would then direct the state on how to execute an expropriation efficiently. Coercion will be accomplished even more effectively by the sovereign with the aid of the rest of the business class that can shut out a rebel plutocrat. This leads to the resolution of commercial conflict outside the foreign spotlight. In other words, power creates a basis for authority over the plutocratic class, curtailing the likelihood of extraterritorial dispute resolution.

Chapter 4 details the importance of focusing on the prospects of state retribution in dictating the outsourcing of authority. Honing in on how state-owned enterprises in Russia resolve their commercial conflicts and how plutocrats have actively avoided the foreign spotlight in recent years, it illustrates that, in state-controlled societies, extraterritorial dispute resolution can put plutocrats' economic and political fortunes in jeopardy. As the state strengthened under Putin, we've seen the development of a host of kleptocrats that rely on the state for their income. There are still several tycoons present in the country that could benefit from acquiring the rule of law abroad. But that would risk the state taking actions that would harm the companies that generate their income or, in more extreme cases, put the tycoon in personal danger. As one lawyer put it, today, the first thought for a Russian plutocrat considering extraterritorial dispute resolution is always, "Are you going to piss off Putin?"[14]

State Competition Leads to a Commitment Problem, Political Battles, Then Transnationalization

Extraterritorial disputes become most likely in the middle of the balance-of-power spectrum, when a country has a relatively strong state presence that is unable to completely control a similarly powerful plutocracy. The power struggles between the state and plutocrats places the property rights regime in flux. The nature of the power balance then leads to political conflicts that unfold under commercial guise in foreign venues.

When the state and independent plutocrats have relatively equal power, a sovereign views the independent plutocrats with their own revenue streams as a potential threat. Independent plutocrats could try to fund opposition parties to ensure that rules will fall more in line with their interests or to increase their potential rents. This dynamic could make a leader increasingly concerned with their tenure and creates clear incentives to bring the business elite into line. In contrast, would-be oligarchs are aware that their very presence acts as a threat to the sovereign, making the plutocrats concerned with the potential that the sovereign will attempt to bring them down.

Plutocrats generally prefer to avoid conflict with the state because it maximizes their ability to maintain their wealth. Some will likely attempt to align themselves directly with the sovereign. But given the lack of guarantees that they will not be targeted after giving up some of their resources, a plutocrat should be willing to do so only up to the point where they can still pose a credible threat to the sovereign. Conventional solutions to resolving commitment problems, such as increasing checks and balances or strengthening the courts, would push the balance of power further in favor of the plutocracy undermining the state's fundamental need to assert control.

As the state wants to move toward control, and the plutocrats with their own revenue want to shift toward capture to become oligarchs, bargaining solutions become unclear. Instead, clashes between the state and select rivals that attempt to gain political control are likely to emerge. Collective action among those in a plutocratic class is often difficult to sustain, especially because the opposing coalition usually has its own divisions and conflicts (Svolik 2009). This creates an opening for the state: the sovereign can take on specific plutocrats, rather than the class as a whole, to dispose of future threats preemptively (Albertus and Menaldo 2012). Other than eliminating a political challenger, an attack can illustrate to the rest of the independently wealthy who exactly is in charge. Successful targeting may then diminish the commitment problem: as they seek to maintain their income streams, plutocrats who transform into tycoons will generally refrain from taking actions that could make them the next example.

Why Plutocrats Will File Their Cases

Such distributional conflict is historically common. But in a globalized world of capital mobility and offshore finance, many of the targets can flee with some of their wealth intact. The losers of the power struggle (formerly oligarchs if the balance was in capture form before the conflict, or formerly tycoons if the prior form was state control) will then attempt to regain their losses through foreign courts. This not only gives plutocrats the option for monetary compensation but could also lead to legitimacy gains. For example, more attention is inevitably drawn to the case when it is filed in London or New York than if they try to settle, use domestic courts, or even use private arbitration. While such publicity can backfire when the plutocrat is an oligarch, having lost their domestic position takes off the corruption-induced shackles. Even if the claimant were to lose the case, global attention and potentially revealing an opponent's secret(s) could act as a roadblock against further predation.

Extraterritorial litigation is particularly effective when the claimant can freeze an opponent's assets over the years that these disputes take to come to a final verdict. Settling these cases becomes increasingly difficult regardless of whether the claimant has a conflict with a party subordinate or is independent of the state. Enriching someone who the sovereign identifies as a threat will quickly put a plutocrat's commercial interests in jeopardy.

When Putin took over the helm of the Russian state, buttressed by unexpected popularity and a surge in oil prices, the balance of power shifted from one of capture to competition. It triggered a commitment problem with the incumbent plutocracy, who still viewed themselves as oligarchs and sought to maintain that status. Conflicts emerged as Putin attempted to resolve the disequilibrium created by competition and to move it to a situation of control. The losers from those clashes—the likes of Boris Berezovsky and Mikhail Khodorkovsky—were then forced to try to regain their wealth through transnational institutions, as documented in chapter 3.

Why the State Will Be Forced to Use Foreign Courts

Leveraging liberal institutions like courts in countries with a strong rule of law is nonetheless a double-edged sword. The sovereign, filing through its own commercial entities or via would-be kleptocrats or tycoons, may choose to go after the plutocrats with waning power in foreign courts to ensure that they are unable to pose a future threat. Krcmaric and Escribà-Folch (2023) illustrate why states fear, and transnationally coerce, elites that have absconded: exiled leaders increase the likelihood of both coups and civil society protests. While their potential return casts a perpetual shadow over outcomes, they can take indirect steps to shape the political sphere at home. They may fund opposition parties, continue to provide motivation for their sympathizers back home, and feed them critical information about the government (Krcmaric and Escribà-Folch 2023, 407). They can also garner the support of the state's international rivals, promising to make better partners at home (Krcmaric and Escribà-Folch 2023, 408). The main point is that political elites can pose a serious threat to regime stability even when they flee their home country.

With some exceptions (Adamson 2013; Salehyan 2007, 2011), scholars have generally missed the fact that these same mechanisms of influence can be wielded by even those who are not part of the formal government apparatus yet possess material and instrumental power. The biggest threat that exiled plutocrats pose to the state comes from funding opposition groups, be they parties or armed insurgencies. Beyond financing the opposition, exiled plutocrats can garner international support and legitimacy that place pressure on their home state. Such strategies are likely to be most effective when major powers have an interest in the country, be it for geopolitical or economic reasons, and when a new leader is seeking to redevelop their own power base. Plutocrats who were once seen as part of the regime also frequently have information that can be damaging to the state. Their wealth often gives them access to knowledge on the corruption and backroom deals that are used to buttress state cohesion. Their revelations are arguably more credible than those brought up by former politicians because their motives can appear less self-serving by virtue of not having been part of the formal governing coalition. In sum, the private threat remains, but it is now transnational.

Trying a plutocrat in their domestic court for past corruption or criminality, while serving important symbolic purposes, rarely gives the state access to foreign resources. Courts in safe havens are likely to view such judgments as politically motivated, thus preventing enforcement. But the sovereign and its backers will also have the option to pursue a trial in London or New York based on the person or the assets being held abroad.[15] Taking this costly measure, which costs millions, risks revealing the state's secrets and could foment a heightened threat if the plutocrat wins, and it further illustrates to other potential rivals the lengths to which the sovereign will go to dispose of a rival to maintain control.

Even as Putin began to reassert state authority, the globalized wealth of today's plutocrats forced the Russian state to engage with foreign courts to execute its coercive imperatives effectively. Just as there was a wave of disputes initiated by former oligarchs seeking to regain their instrumental power, the Russian state filed a series of claims in foreign courtrooms. Given that the money and often people were now stationed abroad, the Russian state had full legal rights to do so. Chapter 4 will show that, as the state sought to acquire the offshore wealth, it was able to further delegitimize their rivals and illustrate to any future challengers how costly and contentious exercising the exit options that capital mobility allegedly provides can prove to be. These dynamics will be further illustrated in chapter 5 through analysis of extraterritorial conflicts stemming from Kazakhstan, Turkey, and Angola.

Implications

The development of markets relies on effective institutions that minimize predation by solving commitment and coordination problems. With increased global integration, the prospects for such conflict have dramatically increased. International relations scholars generally examine the international institutions and organizations that step in to fill the void of authority by creating nominally neutral third-party adjudication. But much of the global economy is governed by the domestic institutions of major economic powers, whose judicial independence and expertise have made them the natural sites for resolving transnational disputes. While this solves governance problems, it also creates new opportunities.

When oligarchs own the state, they are often concerned with avoiding publicizing their corrupt deals. This leads few to choose foreign courts to settle their disputes. Fear of upsetting the status quo by increasing private threats leads to resolution outside the (foreign) courtroom. When the state owns the plutocracy constituted by tycoons and kleptocrats, the consequences of undermining the state's authority removes the incentives for plutocrats to use foreign courts. Fear of retribution keeps business at home. Transparency costs limit the use of foreign courts, but in these contexts, who the plutocrat fears differs.

Disequilibrium exists when a relatively powerful state and wealthy plutocrats with their own income streams are both present. The outcome often leads to clashes, and the second-order consequences of settling make extraterritorial disputes likely. Using foreign courts becomes the last resort for now fleeing plutocrats, while the state-subservient actors subject themselves to liberal rules to achieve illiberal ends. When the fleeing plutocrat is cut off or wronged by an individual with independent income streams who is trying to stay on the state's side, an individual who is now fundamentally a tycoon, conflicts are most likely to unfold through foreign courtrooms via extraterritorial litigation. When the conflict can be attributed to a kleptocrat or state actor, the jurisdictional rules will force plutocrats to file investor-state dispute settlement (ISDS) claims against the state directly. But the state will be able to strike back through initiating its own extraterritorial litigation. The transnational legal market interacts with offshore finance to globalize intra-elite conflict.

2

CRACKS IN THE OLIGARCHY

> The very essence of the state—authority to set the rules of the game—was simply privatized by the new capitalism. The sequence was unmistakable: the wave of money came first, starting with early opportunities to sell oil and computers for super profits. The easy money was followed later by privatization of gigantic factories and natural resource treasures. Money and property invariably brought competition and conflict. And conflict needed a place to settle its dispute, but since the rules were still not drawn—the law not enforced, the courts not effective—the new money and property interests created their own rules outside the law, using bribery and corruption, using violence and coercion, all of which could be easily purchased. The cycle was complete: money ruled.
>
> David Hoffman (2011, 235)

Russia in the 1990s is the chaotic archetype of state capture. After the collapse of the Soviet Union, the new Russian state implemented a radical social experiment: the mass introduction of capitalism and democracy in a country that lacked economic or political freedoms for most of the twentieth century. From a functionalist perspective, the newly minted capitalists were ideal candidates to take their domestic economic disputes abroad. Although the courts improved in terms of efficiency and independence throughout the 1990s, they remained arenas ripe for corruption and confusion. In 1992, Russia saw the formation of a new model of resolving commercial conflicts through the *arbitrazh* courts with a second rewrite of the procedure codes in 1995 (Hendley 1998, 94–95). The latter brought in a host of changes that ought to encourage both the use and efficiency of formal dispute resolution, including moving toward single judges (rather than the traditional panel of three) deciding outcomes (Hendley 1998, 100) and a substantial shift toward more adversarial forms of engagement that placed the onus of proving cases on litigants rather than judges (Hendley 1998, 105). Contrary to some of the dramatized visions of coercion and private violence, the courts became part of the fabric of the average business in the 1990s (Hendley et al. 2000; Frye 2004).

But substantial budget shortfalls throughout much of the time period stymied reforms and facilities (Solomon 2008, 28). Enforcement was meagre and turned into the Achilles' heel of the system (Hendley 2004, 47), all while judicial expertise was still in its infancy (Burger 2004, 29). Effective legal systems rely on social legitimation. As Hendley (1998, 126) astutely summarized, "compliance stems from a combination of respect and fear . . . a basic respect for the institution of the court which presumably flows from the legitimacy of the state of which the court is a part. . . . For the most part, present-day Russian economic actors neither fear nor respect arbitrazh courts." Such a view of the legal establishment was likely only heightened for the new oligarchs, who were wielding political capital to build and protect commercial empires. Their business environment was awash with various forms of coercion and fraud, many of which were not yet formally classified as crimes in Russia but would have certainly been illegal under English law (Klebnikov 2000, 3).

The developing Russian plutocracy was quick to integrate into the international financial and legal networks that would give them the tools necessary to take disputes abroad. Many began stashing assets in offshore havens and property markets the moment the empire collapsed,[1] while law firms from the city of London were quick to set up shop in Moscow, recognizing the commercial void they could fill. In other words, dominant theories of extraterritorial dispute resolution would expect Russian plutocrats of the 1990s to storm foreign judiciaries to solve important commercial functions. But according to the British and Irish Legal Information Institute (BAILII), only one intra-Russian dispute was adjudicated before the British courts between 1995 and 2000, and no extraterritorial arbitrations were filed by de facto Russians against their home state.[2] In contrast, there were a number of traditional transnational litigations involving Russian parties in both the United States and the United Kingdom.[3]

Boris Yeltsin, who led the country through the transition, delegated authority over much of the economy to the "young reformers" (Kotkin 2008; Klebnikov 2000; Shleifer and Treisman 2001). A group of classical liberals, led by economists Anatoly Chubais and Yegor Gaidar, were ready to spare no social or political cost to give Russia a functioning market economy. They first released the Soviet price control system, which quickly led to hyperinflation, eating away the value of money and, with it, ordinary people's savings (Klebnikov 2000, 5). Foreign traders were subsidized, allowing them to reap huge arbitrage opportunities by selling natural resources on international markets (Klebnikov 2000, 5). The initial privatization attempts via a voucher system, which granted all Russian citizens shares in the major state-owned enterprises, spiraled out of control because the overwhelming majority of people were unfamiliar with concepts of equity and private ownership (Klebnikov 2000, 5). There were too many rumors of people trading vouchers that ended up being worth thousands for just a bottle of vodka. The reformers used negative interest rates to boost the creation of banking conglomerates that were already making huge profits by handling state accounts. The primary privatization scheme, the loans-for-shares program, turned into a colossal, corruption-fueled process that enriched the burgeoning bankers (Shleifer and Treisman 2001; Goldman 2003).

The hope was that creating a vibrant capitalist class would safeguard democratic institutions, setting the stage for economic development and Russia's integration into the post–Cold War, liberal order (Freeland 2005). There were certainly some successes in creating the treasured market economy: while less than 10 percent of the labor force was employed by the private sector in 1991, close to 67 percent earned their wages from private firms by 1995 (Freeland 2005, 66). But the result was a perversion, a Weingastian nightmare. With a weak, confused state running the country, the market for goods and the market for coercion integrated. The threat of violence was a regular shadow over business transactions and became a recurrent method to resolve economic conflicts (Belokurova 2014; Gans-Morse 2017; Volkov 2016).

The democratic institutions were the ones that suffered most. The loans-for-shares program, as detailed later in the chapter, was engineered to permanently bind the economically wealthy with the Yeltsin regime. Trading campaign funding for state property was seen as the only viable method to prevent a return to communism. In the words of Russian journalist Mikhael Berger, it was "'Corruption for the sake of Democracy'" (Freeland 2005, 67). The result was the oft-repeated, if somewhat exaggerated (Treisman 2001), statistic about wealth concentration under Yeltsin: some seven oligarchs apparently controlled close to 50 percent of the economy by the end of the decade. This radical inequality laid the groundwork for a politics dominated by the plutocracy, who took both law and policy into their own hands.

Russia in the 1990s then saw virtually no intra-oligarch disputes fought through transnational legal institutions despite a clear functional need for the rule of law. This chapter illustrates how the structural conditions and specifically the corrupt nature of state capture prevent legal outsourcing. It starts by delineating the development of the Russian plutocracy, highlighting how and why it is a clear example of state capture. The second section describes how violence and coercion were used to resolve disputes outside the foreign spotlight at a macro level. Next, it illustrates the growth and dispute resolution strategies of two oligarchs that came to define the world of state capture: Boris Berezovsky and Mikhail Khodorkovsky. Rather than using places like London to adjudicate conflicts, plutocrats sought to engage in private ordering and backroom deals. The final section of the chapter dives into such attempts and offenses by the key oligarchs while they exerted control over the state.

Plutocratic Capture

Corruption as Wealth Creation

Chrystia Freeland, now famous as the former deputy prime minister of Canada, covered Russia for the *Financial Times* throughout the 1990s. She recognized that the key difference between communism and capitalism is not about ownership but instead organization: "Western business schools tell their students that the secret of success is to know your customers. In the late Soviet Union, it was know your appartchiks" (Freeland 2005, 111). After the imperial collapse, the importance of government connections only amplified (Ledeneva 2006; Safavian et al. 2001; Varese 1997). With the economy in tatters and the rules of private property still developing, the best bet at amassing a fortune was raiding government coffers. Many would establish independent businesses and take risks setting up new trading ventures, but the one feature that united the individuals that would dominate the 1990s market for commerce and politics was the ability to see the state as a commercial opportunity. They needed to be cronies before they could turn into oligarchs.

The first common strategy was to set up a bank, take advantage of government connections to get some big state accounts, and then use that as a capital base. As Khodorkovsky summarized, "'Basically, we took money from the state, gave it to enterprises, and then took money from state enterprises and returned it to the state'" (Hoffman 2011, 124). While such arbitrage can hardly be critiqued as a profit-generating business model, and is in some ways crucial to a developing banking system, the second common strategy involved some unconventional practices. The soon-to-be oligarchs would seize management of a state-owned enterprise and then find creative ways to redirect the revenue streams into their own private bank accounts. More often than not, these accounts were held offshore as a mechanism for both tax evasion and political protection (Hoffman 2011; Sixsmith 2010).

While the country underwent a host of economic and political changes in the 1990s, the importance of government connections remained like a crude communist addiction (Darden 2008; Ledeneva 2001). But the new superrich weren't complaining and, in their more reflective moments, they hardly hid the facts. As media mogul Vladimir Gusinsky told Freeland: "I cannot say I am an absolutely honest man, an example for everyone. . . . Nor can any person who survived in this country before 1985, or who built great things after 1985. We all have things which we would not like to tell our children. As we say in Russia, the lucky child suckles on two mothers. There are people with whom you can establish relations and receive a lot" (Freeland 2005, 140–41).

Loans for Shares

But the graft of the early 1990s paled in comparison to what ensued between 1995 and 1997. The Russian state, led by Yeltsin, had worked admirably to try to shore up the economy, but with a ballooning deficit and billions in international debt, they needed something drastic to ensure that they would not fall off a monetary cliff (Desai 2005; Gaidar 1997; McFaul 1995). Vladimir Potanin devised the ultimate quid pro quo (Freeland 2005, 166). One of the few eventual oligarchs to come from a highly ranked communist family background, Potanin was raking in millions through his financial activities centered around Oneximbank (Hoffman 2011, 304–5). But his sights were set much higher. He wanted not just revenue streams but permanent ownership over some of Russia's most lucrative natural resources, specifically, Norilsk Nickel, a bastion of the Soviet Union that produced billions in revenue a year with enough smelting power to move global markets. To get his hands on the prize, he came up with a notorious scheme known as the loans-for-shares program, which still today constitutes the biggest transfer of public wealth into private hands.

The Yeltsin government needed money (Shleifer and Treisman 2001), and the only individuals who had the kind of cash needed to stave off creditors were the emerging plutocrats. The plan was for the plutocrats to loan the government money using equity in major state-owned enterprises as collateral. None of the plutocrats involved expected to be paid back. Instead, the assumption was that the government would default on these domestic loans, and the plutocrats would gain control of Russia's most valuable economic resources at knockdown prices (Klebnikov 2000, 197–205).

Until this point, the Russian superrich were constantly at each other throats, both metaphorically and physically. But loans-for-shares was one of the first instances in which they had collective class interests: only the wealthiest would be able to partake in the heist. They were quick to take advantage of the feeble government that was ready to sell the long-run future of the Russian economy to solve its short-run insolvency issues. Rather than get in each other's way, they collectively agreed to a division of the richest assets among the coterie of cronies (Freeland 2005, 166).

At the height of the commodity super-cycle in the 2000s, some of the companies would eventually be worth almost one hundred times more than the prices paid during privatization. The plutocrats fortified their control of the economy, but the young reformers and Yeltsin did not completely sell out without a price. The liberals structured the program so that the first stage—the loans—would occur in 1995, but the takeovers would only be complete in 1997. This was central to the regime's political success with the presidential elections arriving in between, in 1996. Alexey Ulyukaev, a young reformer, summarized, "We gave them just one of the two keys [to the property]. They would receive the second key only after the election" (Freeland 2005, 173). It was a safety net or the devil's bargain, depending on your vantage point, that the liberals made to safeguard the market economy. With the economy broken and many citizens yearning for a sense of order, Yeltsin's reelection chances were slim. To reap their gains and prevent the risk of expropriation, the plutocrats needed to vanquish a familiar foe: the resurgent Communist Party.

Davos

The World Economic Forum at Davos is the central playground for the global plutocracy. Each year, the bulk of the guests take private helicopters into the resort nestled away in the Swiss Alps to discuss pressing topics like climate change and inequality, and all the irony is lost. In 1996, Davos was the site of true collective action by Russia's emerging economic elite. The events will quickly dispel any doubts about how effectively the plutocrats ended up controlling Russian politics.

Heading into Davos, the Yeltsin regime was at an all-time low. Polls had the leader's popularity in the single digits, while Gennady Zyuganov was leading a resurgent communist charge. The likes of Berezovsky and Potanin had been too busy working out the details of loans-for-shares to appreciate the challenge in front of them, but Davos was a rude wake-up call. Rather than discuss Yeltsin or the prospects for foreign investment in Russia, all international eyes were on Zyuganov, and he knew how to work the crowd. "Even that stern arbiter of capitalist correctness, the *Wall Street Journal*, admiringly described Zyuganov's Davos performance as a combination of 'a bulldozer's power and a prize-fighter's balance' and said he 'bedazzled the West.'" (Freeland 2005, 183). Hedge fund giant George Soros told Khodorkovsky that they better ride their jets out of Russia in time or they would risk losing much more (Freeland 2005, 184).

Boris Berezovsky led the charge to prevent that outcome. While at Davos, he first went to see Vladimir Gusinsky who, as the next section details, was his biggest economic rival. Gusinsky recognized the shared, collective threat: they quickly buried the hatchet (Hoffman 2011, 327). The next step was to bring the rest of the major economic players together, and by the end of Davos, the group of seven—Berezovsky, Gusinsky, Khodorkovsky, Potanin, Vinogradov, Smolensky, and Friedman-Aven—were united to keep Yeltsin in power. They were soon being referred to as the *semibankrishcina*: a play on the word *semiboyarschina* used to describe the period of Russian history where seven boyars governed the country (Freeland 2005, 187).

The type of war chest that the plutocrats possessed would have allowed them to buy an election in virtually any emerging democracy. On July 3, 1996, Yeltsin won his campaign, defeating Zyuganov by almost three times the number of votes. A few weeks earlier, Zyuganov's popularity had been three times that of Yeltsin's (Freeland 2005, 227). It ensured that individuals like Khodorkovsky would be able to fortify their economic gains, and there was little question left about who had the political power. As Russian parliamentarian Grogroy Yavlinsky summarized, "'The government that was formed was without any clear ideology. It was neither red, nor white, nor green. It was based solely on personal greed. You got a system that was corporatist, oligarchic, and based on monopolized property rights and semicriminal relationships'" (Klebnikov 2000, 253).

Guns over Gavels

The previous section summarized the political and financial wizardry that created a foundation for the Russian plutocracy. These events led Joel Hellman (1998) to coin the term "state capture" while he studied the transition process across Eastern Europe. The Russian experience illustrates that, under such conditions, the transparency costs of extraterritorial dispute resolution become prohibitively high because the risks of exposing corrupt behavior could unravel a beneficial domestic status quo. Instead, extralegal means are the dominant form of business and dispute resolution among plutocrats.

The young reformers had imagined that the market forces driving a rapid shift to capitalism would leave little place for corrupt behavior. But by putting the market ahead of institution building, the opposite unfolded. As scholars like Vadim Volkov and journalists like Paul Klebnikov have documented, a crushing tax code coupled with a weak state incentivized firms to do business off the books and become sucked into the morass of organized crime acting as contract enforcers.

The state felt helpless. "'I feel like a person sitting in a pile of shit with his hands and legs tied—he can smell everything, he can see everything, but he can't do a damn thing,' grumbled one Moscow detective" (Klebnikov 2000, 34). As has often occurred in countries lacking rule of law, the void was filled by private actors. But not the benign type that libertarian theorists hope will develop and form when markets grow. Instead *krysha* took over (Varese 2001; Gans-Morse 2012). The Russian phrase directly translates to "roof" and implies a person or business that has protection—political, economic or coercive—from someone higher up. One would need to take action to illustrate it, but as with all deterrence, the best *krysha* never requires a gun to be shot (Mezrich 2016, 23).

Ben Mezrich summarized its importance: "In Russia, the walls didn't hold up the roof; the roof kept the walls from falling in. Without a strong roof, no matter how lavish your house, it would eventually come down" (Mezrich 2016, 28).

According to a report by the Ministry of Internal Affairs, close to 40 percent of private business, 60 percent of state-owned companies, and 85 percent of banks were connected to organized crime (Klebnikov 2000, 29). The statistics illustrate that mob ties and coercion were not restricted to ordinary businesspeople but instead formed a key part of the management and dispute resolution practices of the emerging plutocrats. While discussing the new plutocrats, one London-based lawyer who spent significant time in Russia said, "The '90s disputes were settled with Kalashnikovs."[4] Underlying his assessment was that the plutocrats hardly had a choice given the way business and the economy operated. Another lawyer noted that the rule of law was always going to be secondary given the structural political conditions: "We're talking about a period of historical and political instability, which meant that everything was up for grabs. They didn't have time for the courts . . . and part of it is just how they got the assets . . . you didn't have to litigate."[5]

The notion that the corruption and criminality characterizing state capture prevented the use of foreign courts was echoed by a number of different lawyers I interviewed. The way business, politics, and crime merged ensured that plutocrats would be unable to use outsource authority. The costs of transparency would be too high because "if you sue here [in London] you open yourself up to a counterclaim . . . if you've got skeletons in your closet they will also be found out."[6] Some directly pinpointed the costs from revealing the corruption: "Look the anecdotal data suggests that people were using other, non-legal means for their disputes . . . there was a lot of money to be made and just like anywhere there is money the criminals start to get involved . . . and then they facilitate the settlements . . . you couldn't go to court when that was a factor."[7] Others explicitly referenced a lack of state authority in diminishing the need for extraterritorial litigation: "Frankly it was too early and lawless [to come to London] . . . they could find more expedient, coercive ways to get things done . . . they did not feel trapped by the state in the same way."[8]

To summarize the chapter thus far, in the lawless world of state capture, criminality and corruption were regular features of Russian business. This was true for both everyday entrepreneurs and the Russian plutocracy. Such activity diminished the use of courts, both domestic and foreign, by the oligarchs because of the potentially high costs of transparency. Coercion became the standard means of resolving commercial conflicts. While the previous two sections provided a macro view of these developments, the next two sections provide a micro perspective by focusing on two of the primary plutocrats of the 1990s.

The Center of Attention

The Rise

After earning a PhD in applied mathematics from the prestigious Moscow State University, Boris Markovich Berezovsky was singularly focused on bringing glory to the Soviet Union. While running a lab at the Academy of Sciences for nearly a quarter of a decade, he set his sights on the peak of academic achievement. All that changed with perestroika. As with all the eventual oligarchs, his start came through leveraging his government connections. He teamed with state-owned car manufacturer Avtovaz, one of his clients at the academy (Klebnikov 2000, 53). While most Russians began to fear the passage of time because hyperinflation corroded their savings, Berezovsky figured out how to profit from it. He regularly took up cars on consignment and paid back the debts, months and sometimes even years later, to let inflation corrode the value of the de facto loans. Logovaz, Berezovsky, made 600 percent profit on nearly every car it sold (Mezrich 2016, 14–15).

Finding these arbitrage opportunities became key to Berezovsky's success as he ventured into a variety of industries. Not only did he become a banker, he also created the biggest financial Ponzi scheme of the era that netted him millions and even gained him comparisons to a notorious future US president (Klebnikov 2000, 138–39). Next, he borrowed a page from his rivals, like Potanin and Gusinsky, by seizing control of the state-owned airline Aeroflot. Given that the airline was dealing with huge amounts of foreign currency and lacked the real bureaucratic capacity to handle the accounts, Berezovsky and his lieutenants stepped in. They were accused by Swiss and later Russian authorities of diverting the revenue streams from Aeroflot into their Swiss bank accounts at lightning speed (Freeland 2005, 137).

But to achieve real oligarch status, Berezovsky recognized the need to be more brazen. He needed not only to manage Kremlin property but also to manage the Kremlin. While Berezovsky repeatedly tried to make inroads through Yeltsin's security chief, Aleksandr Korzhakov, his breakthrough came via recognizing another business opportunity. "Yeltsin, fresh from the bloody battle with parliament in October, had written his second book of memoirs, *Zapiski Prezidenta* (*Notes of a President*). Berezovsky, who had formed a useful friendship with Valentin Yumashev—the former journalist and trusted Yeltsin family friend who had ghosted the president's first book of memoirs—offered to publish the work" (Freeland 2005, 131). While sales of the book were unimpressive, it could not have hurt that Berezovsky managed to present an outright positive picture by placing money, nominally from book sales, in Yeltsin's offshore accounts (Klebnikov 2000, 118).

Berezovsky's fortunes rose immediately, according Korzakhov, as detailed by his unofficial biographer and former *Forbes* journalist Paul Klebnikov (2000, 119):

> "Through his access to Yeltsin, Berezovsky appropriated for himself a unique and leading role among the so-called oligarchs," Korzkhakov remembers. "In other words, he became the favorite of the president's family. Berezovsky was constantly demonstrating his special relations with the Yeltsin family; he reminded people of his closeness to the president on any pretext. High government officials feared refusing Berezovsky, who insisted on specially favorable conditions for his own business."

This exclusive treatment was still not enough to satisfy Berezovsky's voracious appetite for control. With Yeltsin's image struggling and his ability to shape the public limited, Berezovsky went after the airwaves. Leveraging his good graces, he convinced Boris Yeltsin to hand over control of ORT, commonly referred to as Channel 1, the large state-owned TV channel. While there were lucrative advertising contracts attached to the station, which Berezovsky would eventually try to monopolize for himself, he was open about his larger ambitions. He told Klebnikov in 1996, "I did not intend to make money with [Channel 1]. . . . I regarded it as an important source of political influence. All the subsequent events—especially the presidential elections—have confirmed that" (Klebnikov 2000, 147). He is referring to the way he ran Yeltsin's campaign and got him reelected following Davos.

Nonetheless, Berezovsky still needed to ensure that ORT would not sink all his profits, and Roman Abramovich gave him an easy way to cover costs. An orphan from Siberia, Abramovich was exporting everything from toy dolls to oil, quickly marching toward plutocrat status (Midgley and Hutchins 2011, 37–42). He met Berezovsky on a fellow oligarch's yacht and proposed that the two of them get together to capitalize on the privatization frenzy. Abramovich wanted to combine oil producer Noyabrskneftegaz and the Omsk refinery to create a new major oil company, Sibneft (Klebnikov 2000, 195). Abramovich would take over the business side, but to make it all possible, he needed Berezovsky to handle the political process. Berezovsky easily delivered, and the two snatched up the company for just over $100 million in 1997.[9] Abramovich eventually sold the company for $13 billion roughly a decade later (Ostrovsky 2005).

From the beginning, Berezovsky realized that the bandit capitalism he was running was always at risk of ending. He took a variety of steps to ensure that he would be maximizing profits while also creating a host of safety nets. Key to the latter was his ability to leverage the laws and rules of foreign jurisdictions. While setting up Logovaz, his car trading endeavor, he learned that a joint venture with a foreign entity would minimize his tax bill and allow him to keep his revenues hidden away. Paul Klebnikov detailed some of this early complexity and explains that 1992 was the start of a long and lucrative relationship between Berezovsky and offshore finance. Forus Services S.A., a Swiss-based company, was nominally a joint venture with commodities trader Andre & Cie but quickly turned into an offshore holding company for a series of Berezovsky's business affairs, including Obyedinyonny Bank, which financed much of his endeavors with Avtovaz and Aeroflot (Klebnikov 2000, 89).

Logovaz, Aeroflot, ORT, Sibneft, and their offshore connections would eventually lay the groundwork for Berezovsky's extraterritorial disputes. They formed the foundation of his economic and political success but eventually became a mix of political culpability and opportunity. Despite having the resources and the ability to access foreign courts in the 1990s, Berezovsky's corruption and business practices made extralegal means the preferred way to resolve economic disputes with fellow Russian businesspeople.

The Conflicts

Where Berezovsky went, controversy and even bullets followed. But when he got his start in the car industry, it was already amid a heated battle between rival Chechen and Slavic gangs. While Badri Patarkatsishvili, Berezovsky's long-term business partner and adopted brother, was the chef intermediary with the mob bosses, Berezovsky was regularly swept up in the violence.[10] Over the course of a year, his car dealership experienced three different armed attacks, leading Berezovsky to flee abroad for much of the end of 1993 and the start of 1994. Soon after he returned, the door of his apartment had a grenade rigged to it (Klebnikov 2000, 35).

The car industry was one of the reasons why Moscow detectives found themselves feeling completely helpless. By 1997, when the police were finally able to investigate Berezovsky's partner, Avtovaz, more thoroughly, they found evidence linking the company to sixty-five contract killings (Klebnikov 2000, 92). Berezovsky had to count himself lucky on numerous occasions. On June 7, 1994, he entered his chauffeured car outside the Logovaz headquarters, but the car made it only a few blocks. An Opel Sedan full of explosives had been left idly waiting for Berezovsky to pass and was detonated remotely. Berezovsky would suffer severe burns and barely made it out alive. His bodyguard ended up losing an eye. Berezovsky retreated to Switzerland, while the headquarters of his private Obyedinyonny Bank was bombed not long after. Who exactly ordered the hit was lost in the chaotic world of the 1990s, but many assumed it had to be one of the rival gangs trying to force Berezovsky out of the car dealership market. One suspect was Sergei Timofeev, an apparent Moscow mobster whose bank had recently been in a loan-related dispute with Logovaz. Four months after the attack on Berezovsky, Timofeev was found dead in a car in central Moscow (Blake 2019, 45). But Berezovsky eventually came to believe that the TV magnate Vladimir Gusinsky was behind the attack (Klebnikov 2000, 28), despite no concrete evidence.

Gusinsky was a former theater director turned oligarch who enriched himself by forming an alliance with Moscow mayor Yuri Luzhkov (Freeland 2005, 140–41). His Most Group pioneered many of the tactics that became part of the political and economic strategy for the oligarchs. To a certain degree, Berezovsky owed his success to Gusinsky. Berezovsky was quick to follow whenever Gusinsky implemented a new commercial strategy, be it joint ventures with foreign companies, starting a bank, or becoming a media mogul. Gusinsky took the first step; Berezovsky followed (Klebnikov 2000, 148).

Thus, Gusinsky and Berezovsky were constantly vying for control over the same slice of the plutocracy pie. Berezovsky had little choice but to retaliate once he returned from Switzerland. Rather than go through any courts, he went to General Korzhakov. The head of Yeltsin's security services claims that on numerous occasions, Berezovsky tried to convince him to organize the assassination of Gusinsky (Klebnikov 2000, 153). Berezovsky continued to claim that Gusinsky was a threat to Yeltsin's work and all the efforts the three of them had put in to reforming the country.

While these pleas failed, Berezovsky did manage to gain some retribution while the two oligarchs were battling for control over Aeroflot. Going directly to Yeltsin, Berezovsky kept arguing that Gusinsky was aligning with Luzhkov to take Yeltsin on (Freeland 2005, 151). Matters only got worse for the oligarch when NTV's coverage of the ongoing battles in Chechnya grew more negative. On December 2, 1994, Gusinsky found himself being followed by a group of men dressed in black, their faces covered, and carrying automatic rifles. They chased his car from his home into the parking lot of his office building, *Novy Arbat*. Gusinsky tried to get his friends in the Moscow police to help. Sitting in his penthouse office, Gusinsky watched for hours as the group roughed up his bodyguards in the parking lot. At one point, when Gusinsky's driver refused to get out of the car, one of the men in black placed a grenade on top of the car (Klebnikov 2000, 156).

Gusinsky called in journalists, hoping cameras might dissuade his attackers, but it only made matters worse. A second group of men dressed in the same nondescript attire, but with a larger set of guns, showed up and forced his bodyguards to lie face flat in the snow for hours. The men were sent by Korzhakov to do Berezovsky's bidding. The incident became known as the faces in the snow fiasco, and Gusinsky fled to London to regroup.[11] If he needed to, Berezovsky could have then filed a case against his media rival abroad, but Berezovsky had already gotten the better of him through his extralegal measures.

All this stands in stark contrast to how the member of the Yeltsin "family" worked to protect his image. Klebnikov was the head of *Forbes* Russia and began to publish stories about Berezovsky's growing economic and political influence. When a bombshell story on Berezosky was filed by *Forbes*, Berezovsky did not hesitate to go abroad for retribution (Klebnikov 2000, 1). He filed a libel suit against the magazine in the London courts, illustrating that if he really needed to, Berezovsky could have found a way to turn his domestic disputes into extraterritorial litigations. But he didn't need to, nor could he politically afford to, once he captured the state. Berezovsky's political dexterity was unrivaled.

The Chosen One

The Rise

As a child, Mikhail Khodorkovsky set his sights on becoming a soviet factory manager. "He was so focused, so young, on that one goal that even in kindergarten the other children nicknamed him 'director,' the Soviet equivalent of referring to a kid as 'CEO'" (Freeland 2005, 114). He joined the *komsomol*, the youth wing of the Communist Party, as a teen and was quickly viewed as the future of a changing party. David Hoffman reports that, on numerous occasions, party bosses put their hands on Khodorkovsy's shoulder, telling him that he would become their experimental capitalist. In Hoffman's words, he "was anointed" (Hoffman 2011, 101), and looking back at his success, he did not disappoint.

While pursuing his degree in chemistry, Khodorkovsky worked his way up to the deputy head of the institute's *komsomol* committee (Freeland 2005, 116). He used this position expertly to take advantage of the new opportunities created by perestroika. First, along with well-connected classmate Alexei Golubovich, he set up a private café through the institute (Sixsmith 2010, 16). Claiming they were creating services to improve scientific research, they began to run information technology (IT) services and sell computer parts for various Soviet ministries. It was unbelievably lucrative for the time. By 1987, they were working with most Soviet departments, and their employee ranks had risen into the hundreds (Hollingsworth and Lansley 2009, 214–15). The use of these cooperatives was crucial to the rise of many Russian plutocrats.

The capital and acumen they gained from their computer business left Khodorkovsky and his associates perfectly poised to set up Bank Menatep when Mikhail Gorbachev liberalized the rules around banking licenses. Their success, and the fact that Khodorkovsky was living up to the capitalist hype, led to their group becoming key advisers during the transition from the Yeltsin administration to Gorbachev's. This created an even stronger position as they entered the lawless 1990s, where government contracts provide the foundations for their growth (Freeland 2005, 117).

Borrowing a page from Gusinsky and Potanin became the root of their success. In what Freeland called "the real Klondike" (Freeland 2005, 117), Khodokovsky and company started managing the government's finances, including everything from basic servicing of former Soviet bank accounts to interdepartment transfers. At one point, the bank considered opening a US-style investment bank to reap percentages off the gains from the new company trying to raise capital, but members of Menatep openly admitted that would defy commercial logic when it was so efficient to woo a bureaucrat simply to hand over parts of their government budget (Hoffman 2011, 23)

According to political scientist Richard Sakwa (2014, 10), "The relationship was largely one-way, with Menatep shaping government policy, while the nascent state had few levers with which to exert influence in the other direction." The company became such effective government operators that they created their own lobbying firm to act on behalf of its various clients. When the young reformers began the voucher phase of privatization, it was Menatep that stepped in, acting as buyer of last resort in addition to snapping up some serious bargains. Their portfolio grew to include "textile mills, chemicals, metallurgy, glass, food processing, wood pulp and paper, fertilizers, and oil. It was a risky process because no one was sure what conditions these industries were in—the Soviet way was never to open the books and to guard profit and loss figures as if they were national secrets—but Khodorkovsky and his partners took the risk" (Sixsmith 2010, 31). Menatep slowly came to embody the diversified holding group that would come to characterize much of Russian capitalism (Johnson 1997).

By 1995, Khodorkovsky's economic and political resources already placed him in the elite of Russian society, inevitably making him one of the biggest beneficiaries of loans-for-shares. He put in the work lobbying the Yeltsin administration and Chubais, securing the guarantee that no foreigners would be involved (Hoffman 2011, 313). As a banking entity, his company became responsible for the auction of the entire episode's biggest prize: Yukos. The giant oil company had close to 2 percent of the world's oil reserves under its control (Hollingsworth and Lansley 2009, 48). Khodorkovsky and Menatep wound up paying $309 million for 78 percent of the company (Hollingsworth and Lansley 2009, 48). By 1997, it was trading at thirty times that value (Hollingsworth and Lansley 2009, 48). The process was not as smooth as many had expected, with rival oligarch groups putting in serious bids, but they were largely disqualified on arbitrary technical grounds (Hoffman 2011, 317).

As the discussion in this chapter shows, government connections and political patronage were essential to Khodorkovsky's early growth and eventual status as the richest man in Russia. These were facts that he and his business partners, like Leonid Nevzlin, recognized. Khodorkovsky noted that his rise under the Soviet Union would not have been possible without his connections. In his words, "It wasn't the money but the patronage. At the time, you had to have political sponsorship" (Hoffman 2011, 9).

But more than arguably any other eventual oligarch, he perfected these methods. According to one preeminent Russian journalist of the 1990s, "He was one of the first to understand the advantages of investments in government officials. . . . Receptions for high-level guests at the bank's country cottages on Rublevskoye Highway . . . would be returned in income a thousandfold" (Hoffman 2011, 124). While Berezovsky targeted the top, Khodorkovsky recognized the value of a diversified political portfolio. The group believed they could survive through any administration because they had actively structured the company "to parallel the government. Khodorkovsky and his lieutenants all had direct lobbying links to key ministers, and their deputies to the deputy ministers" (Hoffman 2011, 232).

Nonetheless, Khodorkovsky and Menatep took several steps to safeguard their wealth. They started advertising to the public in an apparent attempt to rake in new revenue, yet they had little desire for new customers, with the government accounts taking up their time and resources. The goal was to create a constituency as an additional form of political insurance (Hoffman 2011, 123). But in line with the other oligarchs, the meat of the protection strategy was offshore finance. There are reports that as early as 1990, Menatep was helping Soviet directors stash their assets abroad. Gibraltar, the British Virgin Islands, Cyprus, and the Isle of Man all became key intermediaries in these everyday transactions (Hoffman 2011, 125). As David Hoffman details, the webs of complexity associated with corporate tax evasion in the twenty-first century were effectively at work in Khodorkovsky's empire. They ranged from shell companies in places like Gibraltar and Switzerland for Menatep and several offshore outposts to soak up and redistribute Yukos's foreign sales. Khodorkovsky even bought up 20 percent of Balmet Group, an asset manager in Switzerland, that specialized in tax planning (Hoffman 2011, 447).

Khodorkovsky perfected a host of political and economic strategies that made him one of the biggest beneficiaries of state capture. These were essential resources in his rise to the top of the Russian *Forbes* list. This type of wealth accumulation was inevitably accompanied by a host of business conflicts. Despite his links to the international financial and legal system, however, he avoided the extraterritorial spotlight.

The Conflicts

After Yukos was expropriated by the Russian state in the early 2000s, Mikhail Khodorkovsky became a symbol for the liberal opposition against the Putin regime. Many Russian observers still find this a startling, ironic shift. As Bill Browder, the US investor who has himself become a strident political force against Putin, put it (Hollingsworth and Lansley 2009, 209):

> This wasn't Mother Teresa versus Mike Tyson. There were two big tough guys at it with each other and one of them won. Putin won, Khodorkovsky lost, but it was a fair fight. He certainly didn't deserve the pity of some type of political prisoner because all the things he did basically disallow him from claiming that.[12]

Over the course of consolidating Yukos and building up Menatep, Khodorkovsky's group encountered a number of different economic conflicts and opposition. A recurring theme was that these individuals would be bought off, find themselves coerced through kompromat, and occasionally even go missing. Or key documents or paper trails would suddenly disappear. "For example, a truck carrying 607 boxes of Menatep Bank documents mysteriously plunged into the Dubna River on May 24, 1999. That was how things were done in Russia, a state where the rule of law had yet to be established and enforcing the securities law was a distant dream" (Hoffman 2011, 452). An analyst of the oil sector summarized as follows (Sixsmith 2010, 49):

> Basically, Yukos quite ruthlessly and effectively squeezed out minority shareholders from its various subsidiaries during the consolidation process. They used to call it greenmailing. . . . There is no doubt that large amounts of blood were spilt. This is all absolutely off the record, but there are all kinds of tales—and untangling fact from friction is of course very difficult—but I think most people would accept that this went on sometimes at a local level. My impression of how the murders were planned is that they were not planned from above—somebody like Khodorkovsky was not actually giving a direct order.

During the investigations of Yukos in the 2000s, the Putin regime brought up dozens of deaths linked to the company and its staff. In one case against the head of Yukos's security services, Russian prosecutors argued that two assassins, a husband and wife team, were sent to murder a business rival who had allegedly crossed the Yukos management team over a business deal (Sixsmith 2010, 93). The murder was attempted but failed. Aleksei Pichugin, the security head, is then alleged to have had the assassins murdered, although no bodies and little corroborating evidence were uncovered (Sixsmith 2010, 93). Martin Sixsmith, author of one of the most detailed accounts of the entire Yukos affair, provides compelling evidence that this particular case could not have unfolded in the way the Russian authorities argued but that Yukos's reputation and business dealings created the room for such political maneuvering.[13]

Other investigations involving Russian politicians continued to cast a darker shadow over Yukos. The murder of Nefeteyugansk's mayor, Vladimir Petukhov, is arguably the most controversial. Shot dead on Khodorkovsky's thirty-fifth birthday in 1998, Petukhov had been embroiled in a long-running battle with Yukos over its tax payments (Sakwa 2014, 117). The company accounted for some 70 percent of the oil town's revenues and had engaged in a variety of different payment mechanisms under the previous mayor, including using property, promissory notes, and even delivering money directly to the government agencies to, in theory, minimize corruption (Sakwa 2014, 117–19). Petukhov engaged in a variety of tactics to claw back money from Yukos, including an eight-day hunger strike. His conviction is hardly surprising; part of why he ran for mayor was to pressure Yukos to pay back his private company that had become a large creditor of the local Yukos subsidiary.

Following his death, more than 300,000 residents of the town marched against Yukos and almost stormed its local corporate offices (Sakwa 2014, 119). The workers were already frustrated with Yukos having cut wages and jobs in the region as a function of macroeconomic turmoil, and they had little doubt that Khodorkovsky was behind the killing. Petukhov's wife would eventually testify, stating, "I can think of no one else but Khodorkovsky who could have done it. There was no one else" (Sakwa 2014, 121). Yukos argued that the killing was actually against their interests because it would interfere in an ongoing arbitration that they expected would resolve their tax issues with Nefteyugansk, and the reputational costs would be substantial (Sakwa 2014, 120). Following the testimony by Petukhov's wife, the main Yukos shareholders declined to comment, while representatives for Nevzlin stated, "If they accuse Khodorkovsky of [murder] tomorrow, it won't be because they think he committed any crime, it's that they've received orders from the government that it's on their political agenda to charge him" (Siegel 2008).

The same assassin that allegedly carried out the Petukhov attack was also at the center of a clash between Yevgeny Rybin and Yukos (Sakwa 2014, 121–22).

The head of an Austrian company's Russia operations, Rybin survived two attempts on his life as he tried to recoup losses incurred when his company was collateral damage in a takeover by Yukos in 1997. The assassin's confession, which became a key feature of the trials against Yukos, noted that the orders had come directly from Nevzlin, who at the time sat on Yukos's board of directors (Sakwa 2014, 115).

Given that Khodorkovsky's extraterritorial fights were directly with the Russian government, it's worth focusing on how he handled conflicts with government officials at the center of power. We have already discussed his ability to capture bureaucrats and politicians, but such skill also extended to the security services, who are generally considered Putin's biggest backers. As Hollingsworth and Lansley (2009, 215) report, the close ties were almost an official position:

> Testimony to the US Senate Banking and Financial Services Committee in 1999 . . . described a "KGB money-laundering operation with stolen funds that were passed through Khodorkovsky's Menatep Bank as a KGB-controlled front firm." In effect, the committee was told the bank was being used by Kremlin insiders and top KGB officials as their paymaster. The reward for the oligarch was a large pot of gold, which one source described as "a platinum parachute."

Menatep did eventually clash, however, with the Russian bureaucracy in 1999. After the collapse of the ruble and the economic crash in 1998, Yukos amassed millions in debt that it had no intention of paying back. Khodorkovsky, by now well-versed in using offshore structures to avoid paying his taxes, came up with an audacious plan to move the entire Yukos company offshore. The goal was then to leave the creditors and minority shareholders with a hollowed shell of a company while he and his directors turned Russia's biggest company inside out (Hoffman 2011, 453). It led to a series of major scandals and fights with shareholders, most infamously with Kenneth Dart, the US plutocrat (Hoffman 2011, 448–51).

Dmitri Vasiliev, the chair of the Russian Securities Commission, was infuriated as Yukos insulted any notion of minority shareholder rights. He was adamant that the company would not get away with it, and, after consulting with Chubais, announced a full investigation. Vasiliev was facing an uphill battle by going against one of the most powerful companies in Russia, particularly while his ministry was indebted to some of Yukos's public relations consultants (Hoffman 2011, 453–54). His office lacked substantial powers beyond blocking the registration and issuance of publicly traded equity. He initially tried to outsource the investigation by encouraging the minority shareholders to file suits against Khodorkovsky. Dart followed through across six different offshore courts (Hoffman 2011, 452). There were no major, related Russian extraterritorial litigations.

While this only worked to block some of the initial transfers, Vasiliev then looked to bring in enforcement from the rest of the Russian government. But the Fuel and Energy Ministry and a corrupt State Tax Service did nothing. The chair then raised the ante by announcing a criminal investigation. Yukos shot back quickly. First, it started to publicize that Vasiliev was choosing a famous US speculator over Yukos. Following the relatively tame announcement of the criminal investigation, Yukos filed a case in the Russian courts accusing Vasiliev of slander; under Russian law, this was a criminal offense (Hoffman 2011, 455). The action was quickly interpreted as a warning shot, and a handful of days later, a despondent Vasiliev resigned from his post, knowing his office couldn't compete with Yukos's coffers (Hoffman 2011, 455–56). In the end, the creditors settled, and Khodorkovsky kept complete control of the company. In the captured 1990s, Khodorkovsky could always find a way to resolve his disputes without risking the exposure of Menatep's or Yukos's inner workings.

Limits to Private Ordering

The violence and coercion used to resolve disputes took a substantial toll on Russia's rising plutocracy. In a captured world, the plutocrats were left with little choice but to mitigate the costs themselves. There were several attempts to lay down an informal set of rules or establish a form of private ordering to facilitate dispute resolution and bring order to their chaotic system. The Club on Sparrow Hills is arguably the most notable attempt. In 1994, at the insistence of Vasily Shaknovsky, an aide to Yuri Luzhkov, who witnessed the bloodshed firsthand in Moscow, a broad band of plutocrats assembled in a sprawling home in the historic Moscow residential neighborhood. The hill boasted a number of prestigious institutes and at one point was home to the Moscow film industry (Hoffman 2011, 271). As a location, the villa would have fit right into any mafia reboot.

Khodorkovsky, Berezovsky, Potanin, Smolensky, and Vinogradov were all at the first meeting of the club, which took place in September 1994. They were joined by a host of other players with growing influence, like Mikhail Friedman, whose Alfa Group was quickly turning him into a major economic force. Shakhanovsky had ambitious goals. He wanted *bolshoi capital* to recognize its collective interests, to start working together for the sake of the country rather than shooting each other (Hoffman 2011, 272). But the plutocrats did not trust each other, and before any major political projects, like the 1996 election, could come together, they needed to agree on what the actual rules were. The plutocrats even drafted a charter, vowing not to attack each other, be it physically or through the airwaves (Hoffman 2011, 272). The kompromat that the ex-KGB officers were hired to collect was now supposed to be off limits. Vinogradov told Hoffman that he thinks these collective rules were broken the minute the first meeting ended, yet the group continued to meet formally on a regular basis for a full year, and again for crisis cleanup after that (Hoffman 2011, 273). The way these rules were broken forcefully came into public view only two years later when almost all members of the club tried to reap the spoils of a paralyzed, recently reelected Yeltsin.

Bankers War

Igor Malashenko pointed out one of the essential features of state capture in Russia: "It's funny when people talk about the seven bankers ruling the country; they *hate* each other. They have conflicting interests. When they sit together, around the same table, you can feel the tension in the air" (Hoffman 2011, 376). While the plutocrats tried to establish rules to govern themselves, indiscretions would inevitably arise when real money was at stake, leading to feuds that none could take abroad given the corruption propping up their fiefdoms. As Freeland notes, the fight over the airwaves of Russia that unfolded in 1997 was almost logical from the vantage point of the informal system: Gusinsky worked on the election campaign, never got a big return during loans for shares, and now the debts were due (Freeland 2005, 272).

Privatizing the telephone giant Svyazinvest was never going to be easy. Past attempts had failed because the directors of the disjointed network all commanded their own little areas, while the security services viewed the wires as an essential part of their territory. But Vladimir Gusinsky put in the work, lobbying the necessary directors for over a year before he took his proposal to the head of the property ministry, Alfred Kokh. Gusinsky knew this was a complicated, high-capital operation so he brought in some juggernauts: Friedman's Alfa Group and Credit Suisse First Boston were ready to finance any deal, and the Spanish telecom giant Telefonica was onboard to help run the enterprise (Hoffman 2011, 375).

But given the structural conditions at the time, Gusinsky knew it was only a matter of time before another oligarch tried to get in his way. It turned out to be Vladimir Potanin, who relinquished his government post to get back in the commercial action (Freeland 2005, 264–65). Potanin built up his own impressive backers, including the investors Boris Jordan and George Soros, and Gusinsky was furious. The entire financial setup for a deal this big involved a web of offshore companies in places like Cyprus and the British Virgin Islands (Hoffman 2011, 378).

With tempers running high, the Kremlin tried to broker a deal to split the telephone company down the middle, but neither plutocrat was interested. The state was simply not strong enough, in a captured world, to force a settlement. Potanin offered to pay off Gusinsky with a deal in the hundreds of millions, stating that his consortium would even give Gusinsky the reins over the company, admitting that the former theater director was the only one who actually understood the business (Hoffman 2011, 379). It was immediately rejected, with Gusinsky next threatening to reveal the corruption that he viewed as the foundation of Potanin's war chest (Hoffman 2011, 379–80). As the date of the auction neared, Gusinsky still thought he was owed the company for his work on the reelection. Overseeing the auction was another clear sign to him that the company was his for the taking, a logical extrapolation from the loans-for-shares phase. Potanin tried to broker one last deal. The two of them, with Berezovsky tagging along as he tried to live up to his Godfather anointment by Klebnikov, went to see Anatoly Chubais while he was on holiday in the south of France.

Chubais's reputation had been damaged badly by loans-for-shares. Spurred on by the true liberal reformer Boris Nemtsov, he was adamant that they were going to disband the system of state capture, and he flat out refused to deal with Gusinsky. The auction would proceed. To almost everyone's shock, Potanin and his group emerged victorious, with a $1.87 billion bid that topped Gusinsky's offer of $1.71 billion (Hoffman 2011, 384). The transaction was the biggest privatization return for the Russian state, and it appeared that the government was finally growing out of its plutocrat shackles.

Gusinsky was in shock and furious. Berezovsky was even more alarmed because the political system he helped build looked like it might be changing (Klebnikov 2000, 270):

> "You cannot play according to one set of rules and then decide that from four o'clock tomorrow morning you are establishing a new set of rules," Berezovsky declared. "Chubais is trying to break the government and big capital. But big capital is the real support of the government."

The results heightened the paranoia that Gusinsky had been feeling in the previous weeks: that Kokh and Chubais were in Potanin's pocket. Gusinsky and Berezovsky then teamed up to see what allegations would stick. Rather than going to a domestic or foreign court to contest the privatization, they waged a war through the media. The battle then turned to kompromat. Journalists started learning details about Kokh's tax returns. He had filed a $100,000 advance for an apparently detailed book of the privatization process. The payment had come through an organization with ties to Potanin's Oneximbank (Hoffman 2011, 391). Over the coming weeks, a shaky video showing Kokh and Potanin in a hot tub in the Côte d'Azur soon after the Svyazinvest privatization surfaced. Next the two media moguls went after Chubais. They attacked both his dignity and his competence. Chubais only made matters worse for himself when he noted in an interview that he too was writing a book on privatization. He and his team received a total of $450,000 in advance (Hoffman 2011, 393). The story that it was a bribe for the privatization continued to dominate airwaves, and Chubais was eventually relieved from his government post. Kokh went into the private sector and lost his credibility (Klebnikov 2000, 265).

Svyazinvest was enough to wreck the emerging political leadership and handicap a number of their plutocrats. Chubais complained of constant surveillance, of his team and even his family, and soon proclaimed that all that was left was for him to be shot (Hoffman 2011, 396). While the plutocrats tried to govern themselves, the economic stakes and the lack of real authority led to an all-out war between the most powerful private players. This happened only a few months after all involved were united in their pursuit to reelect Yeltsin. Even though the individuals could have found multiple potential logistics to have a case heard in a foreign court, the dispute instead played out through the domestic press. One of the biggest advantages of this approach is that it was asymmetric: the de facto claimants, Berezovsky and Gusinsky ran the media. They were never subject to any cross examination and were left to stretch the truth as a weapon against their rivals through their privately owned airwaves.

Lack of rule of law and limited commercial law are considered the key drivers of extraterritorial dispute resolution. With the collapse of the Soviet Union, Russia in the 1990s epitomized the need for such functional forum shopping. In this chapter, I showed that, while this need persisted, the nature of business and politics under a captured state mitigated the use of foreign courts; corruption and coercion were far too common to put a case in the foreign spotlight.

The plutocrats that ran the country began building their wealth through the currency of government connections. They were quick to stash their assets abroad and avoid paying taxes by integrating into the offshore financial and legal networks that would allow them to take cases abroad. Many fortified their positions as the elite of the elite, turning from cronies to oligarchs, through the loans-for-shares program that should not only be considered the economic sale of the century but also the *political* sale of the century. It left the Yeltsin government and the young reformers even more indebted to the handful of individuals who came together at a Swiss resort to ensure a beneficial oligarchic system.

The degree of wealth concentration was astounding. According to work done by Boone and Rodionov, by the end of the decade, twelve companies had revenues greater than the entire federal budget, with eight oligarch groups accounting for some 85 percent of the revenues of the biggest sixty-four companies (Hollingsworth and Lansley 2009, 59). Almost all these individuals seemed to have strong links to the criminal underworld, which served as their original mechanism for dispute resolution. With rare exceptions, like Yeltsin's reelection, the relationships between the economic wealthy were adversarial and dominated by paranoia. Chrystia Freeland wrote, "The rich have always been different from you and me, but in Russia the biggest difference was that they were more scared, they were always looking for the emergency exits" (Freeland 2005, 106). Few could blame them given the frequency with which their lives were threatened.

The examples of Berezovsky and Khodorkovsky detail how individuals can come to dominate the political scene and turn it into even greater economic gains. They also indicate how violence and corruption were their own routes to resolving disputes even when individuals could gain access to foreign courts. Berezovsky repeatedly tried to get rid of his eventual partner Vladimir Gusinsky, while crimes plague Yukos's past even as they now serve as a symbol against the Putin regime. Lacking real state-dictated rules, the plutocrats inevitably tried to set up their own informal system of governance, but the plutocrats were ever ready to undercut the institutions for individual economic gain.

3

EXTRATERRITORIAL DISPUTES AS PLUTOCRATIC INSURANCE

> London proved the perfect playground for superrich Russians on the run from [Vladimir] Putin's regime. Its booming banks and skyrocketing property market gave them a safe place to stash the money they had looted during the smash-and-grab post-Communist era, while its opulent hotels, luxury department stores, and star-studded nightclubs made for appealing places to spend it. England was a land where both an ill-gotten fortune and a tarnished reputation could be laundered to look as white as a sheet in a flash. Its world-class lawyers and accountants were on hand to help siphon cash safely out of Moscow and into respectable-looking UK companies via opaque offshore structures. Its estate agents were ready to hand over the keys to the country's most prestigious addresses without asking too many questions, and its lacquered PR [public relations] gurus flocked to polish away any lingering reputational taint from the mucky business of getting rich in Russia.
>
> Heidi Blake (2019, 14)

Soon after Boris Yeltsin left office in 1999, there was a sharp increase in Russian extraterritorial litigations. While there were no major Russian extraterritorial judgments in London through the 1990s, there were over thirty-five disputes at the London Commercial Court and the Chancery Division over the next decade and half.[1] There were another seven extraterritorial arbitrations at major international tribunals. So prevalent were the oligarch and state-owned company battles that apparently "at one point there was even a Russian rate," at the major British law firms.[2] These disputes came from a range of sectors, with just over half consisting of conflicts stemming from finance, real estate, and oil and gas, with plutocrat versus plutocrat battles making up roughly 70 percent of the disputes. More than 60 percent of the conflicts arose from corporate governance– or fraud-related claims. This wave occurred despite general improvement in the functional push factors within Russia—rule of law, judicial expertise, legal certainty—widely considered to drive the outsourcing.

The start of the Vladimir Putin years saw the regime substantially invest in improving the legal environment. The Plan for the Improvement of the Courts saw 44 billion rubles budgeted to improve the courts (Solomon 2008, 28), while in 2003, judges' salaries were increased by 40 percent (Burger 2004, 33). A host of measures were also undertaken to improve the transparency of the system, with the development of websites and databases to make most legal decisions easily available and further foster a stronger degree of predictability (Solomon 2008, 29). The efficiency of the courts far outpaced those in places like London and New York, with almost 95 percent of cases at the *arbitrazh* courts meeting their time requirements (Solomon 2008, 31). By 2008, survey evidence indicates that 89 percent of managers believed the courts would protect their rights in a fight against another business, with the state instead seen as the primary property threat (Frye 2010). The use of domestic courts by Russian business surged with the *arbitrazh* courts hearing more than 1 million cases a year (Gans-Morse 2017).

As Timothy Frye (2010, 94) summarized in his review of the state of the Russian legal system, "Russian courts do not work badly in run-of-the-mill disputes between private firms but are much less effective in politically sensitive or high-profile cases involving large stakes." While his analysis suggests that business could get justice at home, it also helps explain the ensuing wave of extraterritorial conflicts. Most disputes that found their way to London involved not just commerce but politics: "It all began with Putin."[3]

As this book goes to press, Putin has been in power for two and a half decades. When he was elected, few insiders foresaw the way he would reshape both Russian politics and Russian society. He was supposed to help continue the burgeoning system of capture, but he was buoyed by public popularity and resurgent oil prices. He was thus armed with the tools to take on the incumbent oligarchy (Colton and McFaul 2003; Rutland 2000). He hardly hid his ambitions once he officially came into power. Putin told Chrystia Freeland (2005, 326): "Those who have made money shouldn't control society . . . the oligarchs must not, and have no right to influence government decisions. If somebody doesn't like this, if somebody got used to anarchy, I am sorry, but they will have to agree with the new rules."

His rise then ushered in a shift from state capture to a world that I label "state competition," where a relatively powerful state and an independently wealthy plutocracy both exist. I argue that this gives rise to a commitment problem because the state and the plutocracy cannot trust each other, which results in political clashes. Some of these clashes then play out as economic conflicts through transnational legal institutions. Given the size of the claims in many of the disputes, these cases have the potential to rearrange who has control within the home political system. Even if no money is forced to change hands, the public process can expose the corruption and coercion that frequently characterizes both commerce and politics in emerging markets. The primary purpose of this chapter is to illustrate how these dynamics unfold.

The chapter generally follows the same structure as the previous one on state capture. I begin by delineating how Putin came to power and how his rise led to state competition. The second section provides a macro view of why intra-Russian cases began to arrive in places like London. I rely heavily on the interviews I conducted with sixty corporate lawyers who worked with Russian clients. The third and fourth sections continue the within-case analysis at the micro level by diving deeper into the political and economic conflicts of Boris Berezovsky and Mikhail Khodorkovsky. They illustrate how the threat they posed to the Russian state led to clashes between the plutocrats and the Kremlin. Foreign courts became the only way to get their money back, especially because they were unencumbered by the potential costs of transparency: they were in exile (Berezovsky) and in prison (Khodorkovsky).

From Capture to Contestation

In line with most dictators, Putin's rise has become like folklore.[4] A schoolyard bully growing up, Putin found discipline and salvation through the noble sport of judo. In 1975, he joined the KGB at the height of the Cold War with the expectation that such service would bring him much needed status and stability. A relatively unremarkable recruit, he did not turn into one of the spies he admired as a child. Instead, he found himself stationed in a run-of-the-mill outpost in East Germany as the Soviet Union disintegrated. When the KGB eventually tried to remove Yeltsin during a coup in 1991, an oft-repeated quote is that Putin immediately resigned from the security services. He firmly stood on the side of the democrats.

On returning to Russia, Putin went home to St. Petersburg and joined his former professor Anatoly Sobchak, who was serving as mayor of the city. He held a series of positions during his time with Sobchak, including head of external relations in charge of foreign investment and developing new commercial ventures. The mayor's office was notoriously corrupt, and Putin would find himself regularly ensnared in crises. Within Putin's first year, there was an investigation into Putin's dealings that allegedly included siphoning off almost $100 million for his cabal (Dawisha 2015).

Once Sobchak lost the mayor's office, Putin moved onto the Moscow political scene. Over various stints in the Yeltsin government, he built up a reputation as a by-the-book, direct, loyal lieutenant who would find a way to get difficult jobs done, almost like a no-nonsense fixer. His labor came to fruition; in 1998 he was named the head of the Federal Security Service, the FSB (*Federalnaya Sluzhba Bezopastnosti*), which took the KGB's place. The move was a result of Yeltsin and the family searching for someone they believed would keep the administration rather than the security services at heart.

The cycle repeated only a few months later. Following the economic crash of 1998 and the ever-growing inequality, Yeltsin's popularity was at another low point. Term limited, the family needed to start thinking about an exit strategy that would provide them safety from prosecution and ensure the continuity of a system that would still enrich them. They needed someone who could win popular support by appearing to be independent of the status quo while remaining a malleable, controllable political figure. They needed, as Masha Gessen (2013) titled her Putin biography, *The Man Without a Face*. Yeltsin was going through prime minister after prime minister, and it was during these heated weeks that Boris Berezovsky apparently suggested Putin would be the ideal fill-in. After a few conversations with Yeltsin, on August 9, 1999, Putin was chosen as the de facto successor.

But making the move from prime minister to president was not going to be straightforward. Putin was chosen partly because he was an unknown, but that meant that he had no name recognition as a politician. Unlike in 1996, there was no communist threat that Berezovsky and the family could leverage. The oligarchs were split on how the system needed to evolve to suit their needs. Vladimir Gusinsky cut loose and teamed with longtime patron Yuri Luzhkov, who had been running Moscow for close to a decade (Klebnikov 2000, 298). The stage was set for another heated Berezovsky-Gusinsky battle, with numerous other oligarchs aligning themselves with the two different factions.

But Putin and the Yeltsin family were lent a major electoral boost from one of the city's greatest tragedies. Over the first two weeks of September, a series of apartment explosions across different Moscow neighborhoods led to 293 deaths and over 1,000 injuries (Satter 2016). Moscow was yet again in a state of chaos and paranoia, with the government quickly blaming Chechen terrorists. Putin, as prime minister, took over and, through a message of law, order, and revenge, became the face of the fight back.

During this episode, Berezovsky felt most vindicated over his decision to lobby for the KGB agent. Gessen recounts one of her interactions with a Moscow veteran journalist that illustrates some of the signals Berezovsky was ignoring (Gessen 2013, 32): "Berezovsky would keep calling me and asking, 'Isn't he fucking amazing?,' she [Natalia Gevorkyan, a celebrated reporter at *Moscow News*] told me years later. I would say, 'Borya your problem is, you have never known a KGB colonel. He is not fucking amazing. He is perfectly ordinary.'"

Overseeing the second Chechen war and repeatedly exclaiming its barbarity, Putin quickly ascended in the polls. The United Russia Party that was formed de facto to support the prime minister took home a majority in December's Duma elections (Colton and McFaul 2000). The origins of the apartment bombing are still hotly debated, with a large body of credible evidence indicating that the FSB was responsible for the attack (Eckel 2019).

The rally-around-the-flag effect was not enough to put the Yeltsin family at ease. To cement Putin's position, Yeltsin shocked Russia by announcing his abdication of the presidency on New Year's Eve. It made Putin acting president. Given the rules of the Russian constitution, the election was soon moved forward, blindsiding the other contenders. Putin was officially elected as president in March, winning comfortably with 53 percent of the vote.

Putin was more popular than even Berezovsky imagined. The economy had begun to rebound, buoyed by a recovery in oil and natural resource prices.[5] Putin also had the support base of the *siloviki*, the security services, that almost no Russian leader had enjoyed in decades. In other words, the timing put Putin on a path to reassert the Russian state's authority over the incumbent oligarchy that was managing the economic and political system. As Sakwa summarized (2014, 44), "A new model of political economy was to emerge in which the notion of an 'oligarch' as a politically independent entrepreneur was to become anachronistic. Putin's attempts to restructure Russian political economy signaled a sea change in the legal environment."

The Russian state was quickly put into gear. The first target was Vladimir Gusinsky. Not only had Gusinsky sided with Luzhkov and his running mate Yevgeny Primakov prior to the election, but his NTV media station had also started looking into events that Putin did not want investigated. Shortly after the apartment bombings, NTV was the only station to thoroughly investigate the rumors surrounding the FSB's role in the apartment terror attack. They even suggested that Putin's hand was at play. Within a few weeks of the new administration, Gusinsky was under investigation for embezzlement.[6]

Gusinsky was swiftly thrown behind bars and given an ultimatum: give up control over NTV or stay in prison. Gusinsky was left with little choice but to sell Media Most to Gazprom for $300 million (Lambroschini 2000). The most popular news channel and one of the biggest banking groups was going to be transferred to state control at a valuation that would resemble the stealing during the loans-for-shares scheme described in chapter 2, which, ironically Gusinsky did not even participate in. It was loans for shares to loans for freedom.

The battle was not as smooth as Putin and his colleagues had hoped. They faced resistance from both domestic and transnational sources. As Gessen (2013, 164) wrote:

> Soon enough, this kind of takeover of private business large and small would become commonplace. But the system Boris Yeltsin had left behind was not quite ready to accommodate "state racketeering." Yeltsin's successive governments had not succeeded in turning the Russian courts into a functioning justice system, but they had succeeded in planting the seeds of ambition in them. Now these courts, mostly at the lower levels, would refuse some of Gazprom's claims with one city court even throwing out the case against Gusinsky altogether. In the end, it took the state monopoly almost a year to gain control of Gusinsky's media empire.

Once he left Russia, fleeing to his mansion in Spain, Gusinsky also tried to repudiate the deal by claiming that it was done under duress. But Alfred Kokh, the new head of Gazprom's media arm that took over Media Most, spared no transnational punches. With the help of the prosecutor general's office, he obtained an Interpol red notice and sought extradition. Gusinsky was forced to relent and pay, but he did end up making a claim against the Russian state through the European Court of Human Rights (ECHR).

Gusinsky would serve as a warning for the rest of the plutocrats. On July 28, 2000, he invited all the leading businessmen for a meeting that is now infamous in Russia. Putin simply stated that the plutocrats were done with politics. They would be allowed to keep their ill-gotten gains as long as they did not cross into Putin's territory. One invitee said, "It was more like a gathering ordered by Don Corleone than a meeting summoned by a leader of the Western world" (Hollingsworth and Lansley 2009, 66). The resurgent state was trying to stave off the costly fights that stem from state competition.

Nonetheless, Putin continued to deny that he had anything to do with Gusinsky's legal troubles. Yet anytime he was questioned about the cases, he seemed to be aware of every detail. In David Hoffman's words, "He had his hands on the hangman's noose now being strung around Gusinsky's neck" (Hoffman 2011, 479). The new structure of business-government relations was neatly summarized by Putin himself during an interview with French newspaper *Le Figaro* (Mezrich 2016, 173):

> Generally, I don't think that the state and the businessmen are natural enemies. Rather, the state has a cudgel in its hands that you use to hit just once, but on the head. We haven't used this cudgel yet. We've just

> brandished it, which is enough to keep someone's attention. The day we get really angry, we will not hesitate to use it. It is inadmissible to blackmail the state. If necessary, we will destroy those instruments that allow this blackmail.

The changes in the economy and the simultaneous election of a new leader had turned Russia from a phase of plutocratic capture into state competition. The fight with Gusinsky was an attempt by the state to lay down a new set of rules. But given the power balance, many more conflicts catalyzed over the coming decade. The former theater director was a powerful actor, but his saga has now become a footnote compared to other plutocrat-state battles, particularly because many became public consumption when they unfolded in foreign courtrooms.

Why (Some Economic) Disputes Migrated Abroad

The early years of Putin's reign can then be understood as a fine balancing act. The Russian state increasingly grew in power but in the presence of a large, independently wealthy plutocracy. There was constant tension between who really had control. I argue that these dynamics led to political clashes as plutocrats and the state both tried to assert their authority. The losers who were forced to exit only really had one option to get their money back: taking the dispute abroad. Having exited the political game, their fear about the transparency costs diminished and, without equitable access to courts at home, extraterritorial dispute resolution became their last resort.

The majority of the interviews I conducted revealed stories that followed a similar path. When asked about why Russians end up bringing disputes to London, lawyers would rattle off a list of the functional factors: "It's the flexibility." "Our judiciary is unrivaled." "The Russian courts cannot compare." Although a handful of participants were adamant that the functional features were all that mattered, it would generally only be a few minutes before lawyers brought up the political structure and the plutocracy having a target on its back as primary features. One lawyer who worked on several major extraterritorial litigations said, "People here often have a snowy idea for why this happens like independent judiciary and quality of judges . . . but that's really a tiny aspect. . . . Necessity is the mother of invention."[7] In a similar vein, another lawyer told me to disregard the first fifteen minutes of our conversation: that is when they discussed the functional features. He noted, "These are well rehearsed conversations we have in the pub and in the boardroom."[8]

It Starts with Mr. P

The recurring theme was that once Yeltsin left office, life for the plutocrats became more politically tenuous. "It's only since Putin came to power that they [plutocrats/oligarchs] started to fear and seek safety . . . before Putin it was chaotic . . . after Putin it's more dangerous . . . people start thinking I need a plan B."[9] These new conditions, according to several lawyers, pushed many Russian elites to move to London, and most not by choice. In the words of a highly established lawyer in London, "It's the boom of the oligarchs, and then eventually moving away and having disputes . . . it's about the movement of the high-net-worth individuals."[10]

Precisely because most of these individuals were forced out or simply feared too much for their lives in Russia, foreign courts became their only reprieve. "In the 2000s you get a lot of people fleeing to London, and then a lot are here on asylum so they obviously can't get a hearing in Russia."[11] "For many, it's the first and last one [option]."[12] A frequent adviser to Russian corporates noted that the movement of people, triggered by state competition, led directly to the extraterritorial litigations: "I feel that probably a lot depends on the cost to the claimant and they were now living outside of Russia. They are refugees from Russia, they've escaped, and they were also using outside counsel. There are a number of cases initiated outside of Russia which should certainly come to Russia but land up in the UK."[13]

The competition explains the shift in litigation behavior between the 1990s and the 2000s: "Honestly, it's because they hadn't fallen out at that point. Then it's dominoes."[14] Another individual went even further, stating that the London battles would have been handled in a private, backroom fashion if the political structure had not shifted: "And then there's the change at the top . . . a lot of the cases have to do with what happened with Yeltsin and the traders that made off . . . these are arguments between two people who should have never had to do anything with the assets in the first place. . . . I really think that if Yeltsin had stayed in power . . . there would have been an internal system devised to resolve all this."[15]

One Piece of a Bigger Political Puzzle

A consistent point across the interviews was that extraterritorial disputes should not be viewed in isolation. The cases are often a function of a political clash and need to be understood as a struggle over control in Russia: "Litigation is only ever one aspect of a bigger more complex situation. It's always political . . . always layers of commercial decision, and the lawyers don't usually find out

. . . these are sophisticated, robust operators."[16] "Then there's the nature of the disputes. They are strategic. They are not about the person or the assets. It's got a bigger scope. . . . They [Russian clients] are some of the most strategic people. I know I'm not getting the full picture. There's the bit you know, the bit you don't know, and the bit you can't know.[17]

Extraterritorial dispute resolution is often the only option that losers of the clash have left. Foreign courts become the mechanism that they can use to try to regain economic and political power: "It's part of a game in a globalized world where the traditional players who don't have power but were once thought to have power."[18] "This [extraterritorial dispute resolution] happens when someone's back is against the wall . . . they are left with no choice . . . it's a way to fight back against what's happening to you at home."[19]

Because of these political stakes that are wrapped up with the balance of power, many of the intra-Russian disputes go to some kind of judgment. "Asking for a settlement, and this is not just for the Russians, can be seen as a sign of weakness. . . . I need to have him on the back foot before I am going to consider it."[20] Parties cannot be seen giving away money to an opponent of the state. Safeguarding economic and political positions prevents the types of out-of-court deals scholars would expect given the economic costs and liabilities of the disputes: "The political situation can help or hinder settlement. For smaller disputes it doesn't really matter but when it's larger with a lot at stake there will be some effect . . . some political influence . . . the political situation becomes relevant and could facilitate or not facilitate."[21] "When these begin, no one thinks that they are going to take it all the way unless there is really bad blood and political pressure."[22]

Berezovsky's Fall and Fight

The Political Clash

As much as Putin may have wanted, Berezovsky was not going to go down without a fight. Berezovsky viewed the takedown of Gusinsky, once one of his biggest rivals, with disdain and it must have felt like a sense of déja vu. Anatoly Chubais had tried to change the rules around privatization overnight, and Putin was now trying to rearrange the entire governance system within weeks of his election. Berezovsky believed he was the engineer behind the electoral victory. The whole point of picking Putin was to ensure that *Berezovsky's oligarchy* would continue. He had lived through multiple attempted mob hits, Gusinsky, the communists, and the bankers' war, so Berezovsky inevitably thought he could handle a no-name former KGB agent of his own apparent creation.

Putin repeatedly sent Berezovsky signals to illustrate that his reign over the Russian political scene was ending. But Berezovsky had control over arguably the most important political resource. Even if Putin had seized NTV for himself, Berezovsky still had ORT, at the time, a channel commanding a viewership greater than all of ABC, NBC, and CBS combined (Mezrich 2016, 53). As a novice politician, the beginning of Putin's reign was rocky, with a series of questionable decisions that Berezovsky ensured ORT would amplify at any given opportunity. This included castigating Putin for trying to usurp power by changing the constitution to allow him to fire governors and criticizing a botched rescue mission during an infamous hostage situation at a theater in Moscow. The goal was to show Putin that the state still needed Berezovsky's and thus big capital's support.

The signaling turned into a full-scale battle for control after the drowning of the *Kursk* submarine. A symbol of the Soviet navy was out on an exercise, its first in a decade, when it sunk. One hundred eighteen men were trapped on board for several days, all while Putin was on holiday at his dacha in Sochi (BBC News 2000). Rather than return early or even make a statement, Putin was quiet through the first few days. ORT went all out in its coverage blasting the president. The rest of the media followed suit and soon images of Putin hosting a dinner party surfaced while the 118 men were still stuck. The president initially refused the aid of the international community but finally gave in once he returned to Moscow. A Norwegian rescue mission reached the submarine but found every man dead. The coverage—lambasting the President's mishandling of the crisis and his feeble attempts to console the victims' families—dragged on throughout August, spearheaded by Berezovsky's team.

Shortly after, while Berezovsky was celebrating the way another tragedy appeared to turn in his favor, he was summoned to the Kremlin.[23] Expecting to meet with Putin directly, Berezovsky instead found himself in the office of his old protégé, Alexander Voloshin. Voloshin had stayed on as chief of staff when the office passed from Yeltsin to Putin. He made the purpose of the meeting clear from the beginning: Berezovsky was no longer going to be involved with ORT and its content. He would need to give up the channel, or he would be forced out.[24]

Berezovsky, of course, refused, but he didn't realize what he was really dealing with. This was no longer his Kremlin. Putin wanted to turn him from an oligarch into just another tycoon. The Russian state had its own media apparatus, the economy was growing, and the rest of the plutocrats were getting in line. The prosecutor general signaled that Berezovsky's dealings with Aeroflot, the charges which he squashed under Yeltsin, were going to be investigated again. But Berezovsky preemptively left Russia on October 30, 2000 (Mezrich

2016, 172). He eventually sold his shares in ORT to his former business partner Roman Abramovich, but only after his close associate Nikolai Glushkov was arrested. Abramovich promptly transferred the ORT shares directly into Kremlin control. As Heidi Blake, journalist and author of *From Russia with Blood*, summarized, "Within a matter of months, the man who had helped bring Putin to power had made himself the number one enemy of the Russian state" (Blake 2019, 15).

Berezovsky-Abramovich

The Partnership

Roman Abramovich is arguably Russia's most impressive rags-to-riches story. Now famous the world over as the former fabled owner of Chelsea Football Club, he started off as a mechanic before selling dolls and children's toys in the 1980s (Midgley and Hutchins 2011, 37). His first big break came when he managed to work his way through the bureaucracy to acquire an oil export license, which at the time amounted to "a license to print money" (Midgley and Hutchins 2011, 42). The wealth he began to accumulate put him on the outskirts of the real oligarchs like Berezovsky and Khodorkovsky, but he finally met Berezovsky off the coast of Antigua on the yacht of Alfa Bank's Pytor Aven. Abramovich was not going to squander the opportunity, quickly pitching the idea of combining the Noyabrsk oilfield and the Omsk refinery to create Sibneft, as discussed in chapter 2.

According to Ben Mezrich (2016), who chronicled their relationship in *Once Upon a Time in Russia*, Berezovsky was immediately impressed by Abramovich's ambition. With Berezovsky's eyes then set on making ORT his personal fiefdom, he needed the money. In keeping with his godfather moniker, Berezovsky said that he would require $30 million annually just to keep things running smoothly.[25] Abramovich had his *krysha*.

On his return to Moscow from Antigua, Berezovsky immediately set out to make the deal happen. He secured the rights to run the auction but still needed to get buy-in from the red directors running the oilfield and the refinery. Ivan Litsekvich, the head of Omsk, would not acquiesce, but then the following (Mezrich 2016, 73):

> In the sort of coincidence of timing that seemed to happen more and more often in modern Russia, at some point during the weeks the merger was being discussed, Litskevich had organized a late-night party for himself and an attractive young woman on the banks of the

> Irtysh River; halfway into the tryst, Litskevich decided to try to impress his date by jumping into the frozen waters in front of them—and almost instantly after hitting the 7-degree water, had a heart attack and drowned. His driver—one of the few witnesses to the event—died in a bar fight shortly afterward.

The actual auction was a farce even by the standards of the 1990s. It was alleged that the wrong address was put on the advertisement to prevent competitors showing up. Security guards even stopped people from entering, and those that managed to get in were ruled out on technicalities like a form that was supposed to be in yellow being filed in blue (Klebnikov 2000, 268). So Abramovich's company NFK took home Sibneft for a little over $100 million (Klebnikov 2000, 207). Neither Berezovsky or Badri Patarkatsishvili's names showed up on the papers. No one else was satisfied with the outcome, as Paul Klebnikov reported (2000, 269):

> "This was a comedy and it remains a comedy," snorts Alfred Kokh. "In fact the actions in acquiring Sibneft were so thinly disguised that the quote that comes to mind is Lebed's: 'Bereozvsky . . . is not satisfied with stealing—he wants everybody to see that he is stealing with impunity.'"

It was the start of an incredibly lucrative, mutually beneficial relationship. Abramovich would have literal suitcases filled with money delivered to his business partner's office on a regular basis (Mezrich 2016, 69–70). At the same time, Berezovsky took a liking to Abramovich and brought him into the Yeltsin family fold. Abramovich could regularly be found in the Kremlin residence and over time became particularly close to Yeltsin's powerful daughter, Tatyana, and her husband, former chief of staff Valentin Yumashev (Klebnikov 2000, 284–85).

But the two oligarchs were not going to rest on their laurels. A couple of years later, they made a play to join the ever-violent aluminum sector. The 1990s in Russia is regularly referred to as a time of the "aluminum wars" because rival gangs had control over major swaths of the industry (Mezrich 2016, 122). Bloodshed was common, but over time power had begun to consolidate into the hands of a few individuals.

Abramovich, Berezovsky, and their respective righthand men, Badri Patarkatsishvili and Eugene Shvidler, smelled opportunity. They needed to convince the rivals to stop shooting each other and start cooperating, and the best way to do that would, of course, be with money. Shvidler brought all the relevant parties together and supposedly locked them in a hotel room overnight to make a deal happen (Mezrich 2016, 162). By the next morning, Abramovich's political and financial clout had brought an end to the bloodiest episode in Russian capitalism at the cost of $560 million. A few years later, Rusal, the company that came out of the negotiations, would become the backbone of the global aluminum market, worth billions.

The Decoupling

While Berezovsky was buying the 1996 elections, Abramovich was already becoming wary of his partner's brash political antics (Mezrich 2016, 87). He suggested that Berezovsky back away from the company to ensure its growth, with the important caveat that the payments would keep coming along. His fears proved prescient when Putin and Berezovsky openly clashed. In the interim years, Abramovich became an effective political player in his own right. Beyond forming part of the powerful Yeltsin family clique, he recognized the winds were changing once Putin came to power and took several steps to ingratiate himself with the new president. He was one of the driving forces behind the creation of the United Russia Party that backed Putin to his eventual victory (Midgley and Hutchins 2011, 106). He was even instrumental in staffing the new administration, sitting down with every person who wanted to run a department or be part of the cabinet (Midgley and Hutchins 2011, 102).

In other words, Abramovich appears to have recognized that he could be perceived as a threat given his independent wealth, but he took decisive steps to show Putin whose side he was on. He was ready to be a "tycoon." With animosity reaching a tipping point, Abramovich became the natural interlocutor between the Russian state and Berezovsky as they negotiated the latter's exit from Russian economics and politics.[26] Abramovich skillfully managed Berezovsky's divestment from ORT, even if he was unable to secure Glushkov's release for weeks.[27]

But those negotiations would pale in comparison to breaking up their partnership. Putin wanted Berezovsky out of the Russian political and economic system, so Abramovich needed to end his arrangement with his longtime patron. While vacationing in the Alps, Badri, Berezovsky, and Abramovich sat down to negotiate a final exit. Abramovich's family was seated in the same café. The offer was remarkable.[28] Per Berezovsky, Abramovich would pay $1.3 billion to his two associates in exchange for control of their holdings. According to Mezrich (2016, 199): "The fact that Roman was willing to hand over this enormous lump sum, a historic amount by any consideration, was, in Badri's view, a testament to the younger man's respect and honor of their relationship, of what they had accomplished."

But to Berezovsky, this deal was testament to the fact that he was no longer relevant. He struggled to accept the offer but recognized that there was little else he could do. After some serious financial engineering, including bringing in a series of intermediaries to the transaction, the money was stashed away in Berezovsky's and Badri's offshore accounts (Hollingsworth and Lansley 2009, 96–101).

A couple of years later, Berezovsky was still frustrated with the events and felt cheated. He told Abramovich's biographers, Dominic Midgley and Chris Hutchins, the following (2011, 239): "I'm not happy because it was done under pressure, Abramovich told me that if I won't sell, Putin will destroy the company. . . . We owned fifty per cent . . . it was fifty per cent Abramovich and fifty per cent Badri and me. It surprises me to read that there are oil companies who try to buy Sibneft without understanding that Abramovich doesn't have it in a legal way because I had to sell it under pressure." When asked if he was planning to do something, he responded: "No, because I think now it's useless. Useless, because in Russia we don't have court [*sic*]. And I don't want to spend my time for nothing. But I want to stress that those who try to buy Sibneft have to understand that Abramovich is not the legal owner of Sibneft because he got the other fifty per cent from us using pressure."

The Case

On October 5, 2007, Berezovsky had set out to find a fancy new shirt, as he often did when trying to boost his mood. The past few years were not kind to him, but October 5 became an opportunity for redemption (Mezrich 2016, 233–37). When he left the Dolce & Gabbana store in West London, frustrated with the ill-fitting options, his bodyguards signaled that his target was in reach. Abramovich was spotted shopping a few doors down at Hermès. Berezovsky grabbed the nondescript envelope that had been sitting in the glove compartment of his Bentley for weeks, and with his bodyguards struggling to keep pace, stormed the designer shop on Sloane Street (Mezrich 2016, 236).

Abramovich had his own men stationed outside, and onlookers must have thought that the military trained professionals were about to come to blows. Berezovsky spotted the opportunity to squeeze through the commotion and handed the envelope to Abramovich (Mezrich 2016, 235). Abramovich refused to hold it and let it fall to the floor, as though that would somehow make it legally invalid. But after weeks of searching for his former partner, including trying to stalk him at Chelsea games at Stamford Bridge, Berezovsky had succeeded. Using the London courts, he was officially suing Abramovich for $5.6 billion.

Since Berezvosky fled Russia, he had fought off repeated attempts by the Russian government to extradite him despite gaining political asylum in the United Kingdom. He had thwarted attempts on his life, but many close to him had started to drop, including Alexander Litvinnenko, who was infamously poisoned using polonium.[29] Meanwhile, Abramovich's fortunes had never been better. He had become the global face of a modernizing Russia, and his relationship with Putin continued to grow stronger (Hollingsworth and Lansley 2009, 159). After a series of failed attempts to cash out some of his Sibneft earnings, he was handsomely rewarded by the Russian state. Abramovich sold Sibneft to the government for approximately $13 billion (Ostrovsky 2005). Berezovsky was furious; he believed it was still his company and that was his money.

So Berezovsky filed a case against Abramovich claiming his share of what the latter had earned from the Sibneft deal and from Abramovich's equity in Rusal. The crux of Berezovsky's legal claim was that he was the true owner of half of Abramovich's stake and that he had been coerced into giving up the equity that Abramovich had sold off. The use of coercion made the original deal invalid; hence, he was still owed the money.[30] He gained jurisdiction by serving Abramovich on British soil, and while Abramovich could have challenged the validity of an extraterritorial litigation by asserting Russia would be the appropriate forum, he relented. A challenge would have certainly brought up the fact that Berezovsky would be targeted the moment he stepped on Russian soil, so he would have almost certainly lost that challenge while embarrassing the Russian state through the process.

Few expected the case to go to trial, primarily because of the transparency costs Abramovich would incur. As one lawyer who works with the Russian elite told me, "It is surprising that Abramovich came out. That was Berezovsky's bet. No one would expect it, no one expected that he would let himself be cross-examined [given the nature of his dealings]."[31] According to another lawyer, Berezovsky was expecting a big payday to keep his mouth shut, similar to the deal struck in the Alps.

Instead, there were weeks of trial detailing the Wild West that was Russian capitalism in the 1990s. It involved some of the biggest players being called as witnesses, and Abramovich's team even convinced a barrister to delay his appointment to the supreme court of the United Kingdom so that he would have the finest representation. But this was more than just a battle between plutocrats: "When you cut to the heart *of Berezovsky v. Abramovich*, it is more than the dispute. It is *the* question about Soviet Russia. Who actually owns the property? Who actually has property rights?"[32]

The Stakes, and Why They Couldn't Settle

When Abramovich dissolved the partnership, his hope must have been that Berezovsky would leave politics to embrace his international socialite side fully. But in the interim years, Berezovsky was set on one goal. In the words of Andrei

Vassiliev, the editor of Russia's premier business newspaper *Kommersant*, "[Berezovsky] is not addicted to gambling or drugs, but to politics. He has fallen ill with politics" (Hollingsworth and Lansley 2009, 260). According to the authors of *Londongrad*, journalists Mark Hollingsworth and Stewart Lansley (2009, 254), "Most of his fortune became devoted to one cause: blacking Putin's name in the West and destabilizing him in his own country."

Berezovsky used a variety of different tactics to try to undermine the Russian government. His nonprofit, the International Foundation for Civil Liberties, became a central player in organizing protests across Russia well before paid protestors became a global fad or talking point. He took out ads in major institutions like the *New York Times* and the *Financial Times* criticizing Putin's approach to human rights (Hollingsworth and Lansley 2009, 275). But he truly made the fight transnational, targeting Putin's allies inside and outside his homeland:

> In his November 2004 speech to the Oxford University Russian Society he claimed that he had spent some $25 million in former Soviet states, all aimed at weakening the power of Russia. He also substantially underwrote the "Orange Revolution" in Ukraine in 2004–5, which led to the victory of Viktor Yushchenko over the Moscow-backed Viktor Yanukovich. It was a major setback for Putin. (Hollingsworth and Lansley 2009, 276)

These claims were further buttressed by Berezovsky, who argued that he was actively plotting a coup against Putin after apparently buying off members in his inner circle (Hollingsworth and Lansley 2009, 280; Blake 2019, 145). Berezovsky also started to fund investigations into the apartment bombings prior to the 2000 election. This type of effort was particularly difficult for Berezovsky because he viewed himself as the central player in Putin's rise. He told Masha Gessen almost a decade later that "it never even occurred to me that there was a parallel game to ours—that someone else was doing what they thought was right to get Putin elected. Now I am convinced that was exactly what was going on" (Gessen 2013, 42).

Given these very public acts of subversion, few had any doubts about what Berezovsky would do with the spoils from the case against Abramovich. As one British lawyer told me: "For Boris it was his way, his plan to expose his enemies. He enjoyed the wealth and all that comes with it . . . but he wanted the money for the power it would give him to further his causes in the former CIS [Commonwealth of Independent States] countries to destabilize the pro-Putin regimes . . . then eventually Putin."[33]

These high political stakes prevented the case from settling, leading to the extraterritorial litigation. Abramovich is a notoriously private person, and he would have had no desire to see his past business activities laid bare under a public spotlight. But he needed to stay on Putin's good side. This was the argument from several leading lawyers in London who work almost exclusively with Russian clients: "It could have settled because Abramovich would have wanted to avoid the publicity. It's possible that people in the Kremlin would have said to do that [avoid settling]. . . . He [Abramovich] would have had to run settling by the Kremlin."[34] "Of course, Putin wouldn't want him [Abramovich] to settle . . . he knows what Berezovsky would do with it."[35]

The fact that Berezovsky had been labeled an enemy of the state would also inevitably heighten the stakes, making settlement more difficult. He was an independent actor that the Kremlin had long stopped relying on or trusting: "Abramovich, he could not be seen settling with someone like Berezovsky . . . because the oligarchs have been divided into those for and against the state."[36]

There is a general consensus among the British lawyers that there was overt coordination between the Kremlin and Abramovich's camp. One lawyer said that he had it on good authority that Abramovich went directly to Putin and asked if he could settle, but when the ruler heard the size of the claim, he quickly refused.[37] Another lawyer recounted a rumor that Putin wanted the case to go forward so that he, via Abramovich, would have a chance to correct the record of misinformation Berezovsky had apparently been spreading over the previous few years.[38]

The key point is that politics, and the way a settlement might further entrench a threat to the state, prevented settlement from going forward: "Abramovich was not allowed to settle . . . it would have been a direct payment, public payment, to the enemy of Putin's regime . . . were he to lose he would have at least had an excuse [to pay]."[39]

The Outcome

Rather than a backroom deal, there was a multiweek trial held in court 26 at the new Rolls Building housing the London Commercial Court. Proceedings hinged on a series of oral agreements starting in 1995, when the idea to form Sibneft emerged. The evidence brought to bear was astounding and lived up to the litigation's price tag:

> There were 20 different sets of bundles, designated from A to T, with each set in itself often comprising further subsets of bundles, and numerous numbered individual files. For example: there were 9 subsets of what were euphemistically referred to as "Selected Disclosure" bundles, of which the principal subset of chronological documents

> itself ran to what in hard copy was 99 lever arch files; there were 13 files of witness statements, 36 files of expert evidence addressing Russian history and Russian law, and 57 files of expert valuation reports which, in the event, the court did not need to address. Most importantly, for the trial alone, there were approximately 4,000 pages of written submissions submitted by counsel.[40]

The opening was a shock, with Abramovich de facto admitting to past criminal behavior. He repeatedly stated that he would never have been able to acquire Sibneft without the political influence of Berezovsky and the latter's relationship with Yeltsin. But their deals never gave Berezovsky rights over the company, only de facto protection payments. Using a translator throughout his deposition, he argued that the $1.3 billion paid to Berezovsky and his business partner was a final payment to end their debilitating *krysha* arrangement.[41] But he painted himself as an honest man trying to survive in a broken system, who needed the aid of a real crook—Berezovsky—to survive. He was assessed ". . . to be a truthful, and, on the whole, reliable witness."[42]

While it appeared that Abramovich avoided smearing Berezovsky, the latter took every chance to question his younger protégé's intelligence. He regularly cited his own brilliance as a key source of his business success in the 1990s. Abramovich's version of events reaffirmed the centrality of Berezovsky's importance to the Yeltsin administration and the operation of state capture. As did Berezovsky's version:

> Q. I assume that you would accept that your Russian history expert knows about Russian history?
> A. Definitely he knows. Much less than me, but knows.
> Q. Would you look at paragraph 83 . . .
> A. Because I made the history; he just learned the history.[43]

Over the course of the proceedings, it soon came to light that Berezovsky had offered at least two witnesses 1 percent of his winnings, which had not been previously disclosed.[44] At the center of the case was a recording that Patarkatsishvili had made in secret, which Berezovsky had managed to acquire for 5 percent of the assumed winnings from a source that could not be named for fear of their safety. The transparency costs of the litigation process also came back to haunt. The libel lawsuit that Berezovsky had brought against *Forbes* years earlier was repeatedly brought up as a reason to doubt his claims because he had challenged many of the issues surrounding Sibneft's ownership structure that he now regarded as facts.[45] The net effect was that Justice Elizabeth Gloster "found Mr. Berezovsky an unimpressive, and inherently unreliable, witness, who regarded truth as a transitory, flexible concept, which could be moulded to suit his current purposes."[46]

The court recognized that it needed to avoid seeing the proceedings through "Western Eyes," indicating this was "a uniquely personal story, centred as it was on a relationship between two men, of different ages, both in their time extremely successful and powerful, and whose fortunes changed as the dynamic of their relationship changed."[47] But this was admittedly a case that relied less on interpretations of the law: "The case was one where, in the ultimate analysis, the court had to decide whether to believe Mr. Berezovsky or Mr. Abramovich." After both parties spent more than 50 million pounds on their legal teams, Berezovsky's claim was flatly rejected.[48]

Lawyers familiar with the proceedings definitely agreed that Berezovsky was lying. But they doubted that it was an overt tactic: "Berezovsky was much closer to the truth. I'm afraid that for Boris he didn't quite remember the real details . . . he was focused on the money just coming in."[49] Despite the spectacle of the trial, many were surprised by its outcome. According to several lawyers I spoke to, people on both sides believe that, based on the customs of the 1990s, Berezovsky was owed the money. It was how business was and often still is done in their home country. Shortly after the trial, Berezovsky gave a soul-searching interview where he expressed a wide range of emotions including guilt and remorse for the way he had acted over the past two decades. The next day he was found dead in the bathroom of his home on the outskirts of London.[50] There are rumors that he was killed, but many close to him view it as a suicide.[51] After the legal loss and embarrassment, Berezovsky saw no way back.

Around the World with Yukos

The Political Clash

After fighting off determined creditors and experiencing the real macroeconomic chaos of market forces, Mikhail Khodorkovsky found a new mission. He set out to reinvigorate Yukos and Menatep by importing Western business practices, hoping that his lead would reshape the rest of his Russian competition. He lured US and French executives from major foreign oil companies, started placing former foreign politicians on his board, while consultants from places like McKinsey became mainstays in the Yukos offices (Hollingsworth and Lansley 2009, 222). Khodorkovsky and the magnificent seven that started and managed Menatep together were rewarded handsomely. The rise in oil prices and the changes in corporate governance made Yukos's value skyrocket;

by the start of 2003, Khodorkovsky was the richest man in Russia, with a fortune estimated at $15 billion (BBC News 2013).

Khodorkovsky was not just a convert to the importance of corporate governance; he also came to believe in the value and necessity of social institutions to create a sustainable market economy. In his own words, “We started out as robber barons, but you have got to understand that rules were hazy in those days and we changed ourselves as we became familiar with what was expected of us. . . . I don’t think I actually violated the law, but it took me some time to realize that above and beyond the law there were also ethical laws” (Hollingsworth and Lansley 2009, 221).

Like his fellow plutocrats around the world, and like Berezovsky, he created a foundation to build up Russian civil society. *Otkrytaya Rossiya* (Open Russia) was expansive. Not only did Khodorkovsky fund internet cafés across rural provinces to begin connecting people, he also funded journalists and investigative training (Gessen 2013, 237). He donated close to $100 million to the prestigious Russian State University to build up their humanities program (Gessen 2013, 237). He became a vocal advocate for human rights and, in an attempt to open up the social and political space, bought the floundering weekly newspaper *Moskovsky Novosti* (*Moscow News*) (Hollingsworth and Lansley 2009, 244). While Western nonprofits were backing away from Russia as they came under political pressure, Khodorkovsky’s efforts only stepped up. By some estimates, he was directly supporting up to 80 percent of nongovernment organizations (NGOs) (Gessen 2013, 237). His ambitions appeared to be increasingly geared toward changing the way society functioned, and he pinned the blame on the institutions that were vital to the state’s effort to move toward conditions of control (Sixsmith 2010, 80):

> By creating an atmosphere of uncertainty and vulnerability, our legal system has done great damage to young entrepreneurs in our country. . . . These people need to feel independent and safe in conditions of democracy . . . the statist ideology that pervades our country has prompted many of them to emigrate. You can work out the damage this has done. If a 25-year-old entrepreneur decides to leave Russia because he has been insulted by a policeman or because the tax authorities have come and rubbed his nose in the dirt, the country loses millions in dollars in lost potential. We’ve just thrown it away. And a hundred thousand such entrepreneurs are leaving Russia every year!

To many observers it was difficult to fully discern Khodorkovsky’s motives, but to the Kremlin, his intentions were clear. Putin and his inner circle of former-KGB members and allies from his days in St. Petersburg saw the development of civil society as a threat and a potential weapon for Khodorkovsky. And it would be difficult to deny these conclusions given the way Khodorkovsky's political tactics were overtly developing. When he bought *Moscow News*, one of his first acts was to hire an outspoken Kremlin critic as editor (Hollingsworth and Lansley 2009, 244). The newspaper soon put out statements indicating that they were on a fight to regain their liberal credentials (Sixsmith 2010, 82).

It looked like he was taking a page out of Berezovsky's book. As early as 1998, Khodorkovsky publicly noted that he wanted to retire from business by the age of forty-five. "But in the spring of 2003, as Khodorkovsky, stepped up his civic and political activities, commentators began to point out that his 45th birthday happened to fall in 2008, the year Putin was constitutionally obliged to step down" (Freeland 2005, 327).

As documented in chapter 2, purchasing government officials was key to Khodorkovsky's success. This was common under conditions of state capture, but with the balance of power changing, the same activities risked becoming a liability. Putin had given multiple warning shots to the plutocrats and most had, at least on the surface, succumbed to the pressure. Khodorkovsky nonetheless doubled down on his nonmarket strategy in 2003 with the Duma elections approaching. He met with leading members of all the opposition parties and began to promise them sums of money that paled in comparison to those spent on lobbying in the 1990s. Some put Khodorkovsky's offers in the hundreds of millions (Sixsmith 2010, 52). It began to look like "Khodorkovsky was systematically buying up the Russian Parliament to make it his own private fiefdom" (Sixsmith 2010, 52), with an ambitious goal of owning up to one hundred different deputies that would be in Yukos's pocket (Sixsmith 2010, 52–53). Some of the events began to border on outrageous given the way the balance of power was shifting (Sixsmith 2010, 53):

> The Speaker of the Duma, Gennady Seleznyov, reported that one MP was interrupted by a mobile phone call in the middle of his speech: it was Khodorkovsky ringing him what to say! Seleznyov told Putin about the incident and the president flew into a rage. Seleznyov later said that when it came to legislation on the oil industry, it seemed to him that there are "250 Vladimir Dubovs" in the chamber (and fortunately for Yukos, Dubov was also the chairman of the Duma's Taxation Committee, so he could look after the company's interests there too . . .).

Tensions between Yukos and the Russian state began to mount, but Khodorkovsky continued to build what appeared to be a formidable political coalition. The tensions came into public view at one of Putin's plutocrat-only events that were otherwise becoming increasingly irregular, sullen affairs. When Khodorkovsky questioned the way state-owned Rosneft, directed by key Putin ally Igor Sechin, took over an oil company with limited regard for the process or the law, Putin did not bother to hide his disdain for the plutocrat (Sixsmith 2010, 62):

> "Let me tell you," he said, looking directly at Khodorkovsky, "that Rosneft bought Northern Oil because the state has a duty to increase the stocks of oil at its disposal. But some companies like Yukos have got themselves fantastic, excessive reserves of oil. I think the real question is: how did they get them? The ball is in your court!"
>
> Putin paused for a moment before turning the knife.
>
> "And another matter. I believe Yukos has got into a few problems with its tax affairs. You may claim you are dealing with those problems, but the question that needs to be addressed is why they arose in the first place . . ."

Soon after this event, a special unit staffed with fifty people was allegedly set up in the general prosecutor's office to find ways to charge Khodorkovsky.[52] The latter should have had little doubt that he was getting under Putin's skin and, given the already established record with Berezovsky and Gusinsky, he could become the next target for the state to illustrate its authority. But he also knew he was the biggest player outside the Kremlin, managing one of Russia's most important companies at a time when the Kremlin was trying to boost investment. He knew he had structural power (Zygar 2016, 48): "Yukos was the exclusive supplier of oil products in forty-two regions of Russia. If I'd stopped supplying petroleum products, the country would have ground to a halt. The regions would have revolted. Ambulances, fire engines, and all essential services would have stopped running after just three days."

Wealth should have been a protection, and tensions only heightened as rumors around Khodorkovsky wanting to cash out some of his Yukos shares to a foreign player began to swirl. Talks had been ongoing between Putin and Abramovich about reviving a deal between Sibneft and Yukos, but now international oil companies were being considered as potential bidders just at the time the Russian state wanted to regain the oil wealth it failed to maximize because of the previous decade's privatization (Sixsmith 2010, 132).

Despite increasing pressure and clearly living on the political edge, Khodorkovsky remained a calm and collected figure, but he did not fail to seek out some outlandish insurance (Hollingsworth and Lansley 2009, 229):

> In early 2003 he [Khodorkovsky] made secret contingency plans to avoid arrest. His business partner Leonid Nevzlin devised a plan whereby

oligarchs would live on a ship that sailed around the world in international waters. A Cyprus security company was commissioned and it bought a mini-*QE2* cruise liner for the operation. The 100-cabin ship was dismantled, refitted, and refurbished and its cabins reduced to twelve luxury suites. Named *Constellation*, the plan was for the ship to fly an international flag, in the manner of Liberia, and remain in international waters, thereby avoiding extradition and international arrest warrants. While Khodorkovsky was on board, other oligarchs would be flown in by helicopter and plenty of food and drink would be provided.

Yukos Versus Russia

The Dismantling

There would be no party on the *Constellation* because the full force of the Russian state soon struck the Yukos empire. On October 24, 2003, during an unscheduled grounding of his private jet at Nizhny Novgorod, Mikhail Khodorkovsky was arrested (Sixsmith 2010, 5–14). The landing happened while he was flying across Russia, meeting governors in what looked like another brazen attempt to strengthen his political base. He was also lobbying to prevent the arrest of fellow Yukos executive Vasily Shaknovsky. Khodorkovsky appears to have known exactly what was coming—he had preemptively shaved his head (Sixsmith 2010, 142).

In the previous few weeks, Yukos and its employees had found themselves in a number of heated situations. During the summer, Platon Lebedev, the man in charge of Menatep's finances and a member of Yukos's inner circle, had been arrested on fraud charges related to the privatization of a chemicals and fertilizer company. Aleksei Pichugin, the head of security services (discussed in chapter 2), was accused of a string of murders. Shakhnovsky, another core member of the management team, was accused of tax evasion (Sixsmith 2010, 139). The Russian state was getting closer to the man at the top, and the Kremlin apparently used intermediaries to tell Khodorkovsky that he would be left alone as long as he fled Russia (Hollingsworth and Lansley 2009, 230).

Over the course of the year, there was still plenty of confusion around who was driving the fight against Yukos. Many thought that the attack was the work of the *silovarchs* pressuring the tax and prosecutor's offices (Sixsmith 2010, 91–100), with Putin only acting as an indifferent bystander. Khodorkovsky's arrest—a huge statement—would inevitably have required Putin taking charge and dispelled any claims of plausible deniability.

A variety of charges were brought against Khodorkovsky, Lebedev, and the company, which all boiled down to tax evasion and defrauding the Russian government related to Yukos revenues (Sixsmith 2010, 147–52). For years, Menatep and Yukos had diverted their earnings through various subsidiaries to take advantage of domestic Russian tax havens. The practice was common in the oil industry and was even encouraged in the 1990s.[53] Over the coming years, numerous lawyers and employees of Yukos were harassed and prosecuted. Nonetheless, most of the key players, such as fellow founders Leonid Nevzlin and Mikhail Brudno, fled Russia to escape imprisonment. Every week that the saga dragged on, the charges continued to mount. By the end, Khodorkovsky and Yukos were accused of illegally siphoning off almost $28 billion, which would amount to net revenue from all Yukos oil sales for the entire five-year period before Khodorkovsky's arrest (Sixsmith 2010, 199).

According to Gessen (2012), the trial of Khodorkovsky and Lebedev was an opportunity squandered by the Russian authorities. A trial that took ten months involved enforcement of many laws that were retroactively changed to make Khodorkovsky look guilty and other tax avoidance charges that according to the Russian code were generally considered legal. Genrikh Padva, the esteemed advocate representing the Yukos officials, appeared to be in disbelief throughout but still had to engage every point and provide every rebuttal. Even if it would have no effect on the outcome, he needed the record for any future extraterritorial fights. No one was surprised by the nine-year prison sentence that both defendants received.[54]

In December 2004, what many still see as the primary reason for the legal assault on Yukos finally unfolded (Gessen 2012; Sakwa 2014; Sixsmith 2010). With the huge tax fine, the Russian authorities took it on themselves to auction some of the company's most valuable assets in what can now be seen as a deprivatization. Following an inverted script of the 1990s, the government managed to restrict the auction to two bidders: the state-owned Gazprom and a shell company Baikalfinans. As Sixsmith (2010, 176–77) details:

> No one had heard of them [Baikalfinans]; they had no previous record of financial transactions and their registered headquarters seemed to be a vodka bar in the provincial town of Tver. Astoundingly, Baikalfinans produced the winning bid: the other supposed contender, the state gas monopoly Gazprom, simply failed to make a counterbid. Baikalfinans became the owner of Yukos's crown jewels for $9.3 billion—money went immediately in the state's coffers as part payments towards Yukos's alleged "tax arrears." . . . The whole charade was explained a few days later, when the state oil company Rosneft announced it had

> bought Yuganskneftegaz from Baikalfinans for an undisclosed price. And by happy coincidence, the man who controlled Rosneft was . . . Igor Sechin.

If there are any doubts about the political nature of the entire episode, one need only consider the behavior of Khodorkovsky's fellow plutocrats. Yukos was charged with exploiting Russia's tax havens to boost his bottom line, but the structures they used were common across the entire oil and gas sector in Russia. Companies like TNK and Sibneft all had a web of onshore and offshore structures to minimize their government tax bill. Paul Klebnikov, author of *Godfather of the Kremlin*, brought up such a point hours before he was assassinated in Moscow (Midgley and Hutchins 2011, 275):

> "As regards all formal and informal accusations made against Yukos—non-payment of taxes, a lack of patriotism and political interests—Sibneft is much worse than Yukos. Yet Sibneft is flourishing and being supported by the Kremlin while Yukos is being broken up into little bits." . . . Klebnikov attributed this discrepancy to Abramovich's relationship with Putin. "I think one of them is simply the personal friend of the president," he said. "The second is simply an independent person. If the law is applied too severely to one oligarch, why is the same not done to another oligarch who has offended public morals much more seriously?"

Yukos shareholders would soon make a similar argument through foreign legal venues.[55] Within a few months, Putin and the Russian state managed to neuter a serious threat from an independent plutocrat and reap huge economic gains in the process. But the fight was far from over.

The Case(s)

Unlike with Berezovsky, there was no single extraterritorial litigation that defined Yukos's efforts to regain its wealth.[56] Instead, the company used its multijurisdictional structure to make a variety of claims both to defend the company and to challenge the Russian state. They were now generally free from transparency costs. Khodorkovsky and Lebedev were in jail, but the regime had made clear that they were ready to bend the truth during the Russian trials. The state appeared willing to make up charges so any substantive past wrongdoings would be trivial. The remaining founders and directors were abroad and saw no economic future left in Russia with the way they had been treated. The criminal charges that were filed against many of them coupled by the corrupt judiciary had all but guaranteed it.

The US courts first appeared as part of Yukos's strategy in 2004, when Yukos argued that it could take advantage of the United States' lenient Chapter 11 bankruptcy code (Hibbitts 2004). Based on a few million dollars in a Texas account and one of their directors now living there, Yukos was hoping to stop the Russian state's ability to sell off their assets and place them in receivership. While there appeared to be some leeway despite the weak domicile grounds, the Houston, Texas, court eventually rejected the claim. A few years later, Yukos shareholders filed a claim in the London courts to block Rosneft, the biggest beneficiary of the deprivatization, from publicly offering its shares. They argued that Rosneft had illegally seized the prime oil piece of the Yukos empire, *Yuganskeneftegaz*, so any public offering was the equivalent of selling shares in stolen property (Sixsmith 2010, 77).

While these extraterritorial defensive attempts failed, there have been a host of successes. On numerous occasions, the European Court of Human Rights has ruled in favor of Khodorkovsky for his imprisonment and the dismantling of Yukos. Shareholders were eventually awarded over $2 billion for their efforts (Davies et al. 2014). As these cases progressed, the Yukos holding company in the Netherlands became a site for a series of battles. In April 2009, a financial arm of Yukos, Yukos Capital, won an award of almost $400 million against Rosneft (Akin 2009). The claimants argued that Rosneft had failed to provide payment for outstanding debt from *Yuganskeneftegaz* that would now be Rosneft's liability. Although Rosneft immediately challenged the decision, the Dutch supreme court backed it. Rosneft was forced to pay but refused to hand over the additional $140 million in interest. Yukos then made claims in several courts to seize Rosneft assets worth that amount, and the British courts eventually froze over a half billion dollars' worth of Rosneft assets in the process (Vorbyova 2010).

But the big (now notorious) win came several years later. Using its offshore holding companies in Cyprus and the Isle of Man, the Yukos team used the Energy Charter Treaty (ECT) to bring an extraterritorial arbitration against the Russian state in the Permanent Court of Arbitration in The Hague. Several lawyers I spoke to expressed surprise that the venue accepted the claim in 2009 given the political stakes of the battle. The ECT was designed to protect foreign firms' investments in energy across a host of signatory countries, but Yukos and its core management team were Russian. It was the de jure way that it structured its investments, often used by the Kremlin to attack them in public, that they argued made them foreign investors.

The Russian state contested the ECT's jurisdiction based on several legal grounds. The most detailed came through accusing the claimants of twenty-eight different instances of illegal and bad faith conduct related to how they acquired Yukos, their abuse of the Cyprus-Russia double tax treaty and domestic tax havens, and the steps the company took to prevent the state from collecting taxes owed to it.[57] Any of these could have allowed the tribunal to reject the claim under the clean hands doctrine,[58] which prevents parties from initiating an arbitration when corrupt or criminal behavior is involved in the purchase of the agreement, but these appeals were all rejected.[59]

The ECT was ambiguous on the definition of "foreign" and expansively interpreted its own jurisdiction, so the case went forward. The entire ruling process took close to five years and involved nearly a dozen witnesses for both claimant and respondents. In the midst of Russia's 2014 invasion of Crimea, the court validated Yukos's claims: GML was awarded $50 billion, which was half of the $100 billion it had sought but still the largest judgment ever handed down in an investor-state dispute (DLA Piper 2016).

Why They Couldn't Settle

These extraterritorial disputes have been fought for over a decade, with tens of millions spent on legal fees by both sides. The overwhelming consensus is that the underlying cause of the attack was Khodorkovsky's independence: settling was not an option because the purpose of the entire endeavor was to eliminate a political threat to the resurgent Russian state. Regaining the oil company was an important part of the Russian state's political strategy. But it was the independence and the signaling value to the other plutocrats that made the entire episode inevitable. Mikhail Kasyanov, one of Putin's former prime ministers, recounted that Putin actually admitted to him that the tax charges were a political act. In a signed affidavit to the ECHR, Kasyanov said that Putin told him that Khodokorvsky had "crossed a line" by funding opposition parties, and in particular the communists, and that was why Yukos was being pursued (Sixsmith 2010, 298). "Finance minister Alexei Kudrin explained that the Yukos affair was inevitable, 'not in the personal sense, but in the sense of a clarification of the rules of the game'" (Miller 2018, 42).

The fact that Khodorkovsky had shown an independent streak was identified by several different key players as the root cause of the conflict and the reason the legal battles became acrimonious despite the high costs. Berezovsky directly pinpointed these dynamics while recounting the change in the political structure (Midgley and Hutchins 2011, 242):

> Again it's not a fight against journalists. It's not a fight against politicians. It's a fight against independent journalists, and independent politicians. And then came the logical third step. How can you control, let's say, political life and mass media if you don't have money

> under control? So the logical third step is to put under control independent business and by putting Khodorkovsky in jail he sent a clear message to all the business community: "I don't allow you to touch the political life of this country." And it's impossible because business, as we have discussed before, has a great interest in the political life of the country. Naturally.

Extraterritorial dispute resolution is frequently a mechanism for a losing plutocrat to get their money back after they have been politically targeted. There are gains to the political conflict that go beyond the specific case as a resurgent state tries to curb future challengers. Taking on the biggest plutocrat can potentially bring the rest in line. Chrystia Freeland (2005, 325), based on interviews with several political players, wrote this shortly after Khodorkovsky was arrested:

> Machismo should never be discounted as a powerful factor in Russian politics, but ultimately the struggle between Khodorkovsky and the prosecutors was about something even more crucial in modern-day Russia: state power and its limits. Wary of scaring off already skittish foreign investors, the Kremlin took some pains to insist the fight was not about revisiting Russia's controversial division of property in the 1990s. What the Kremlin did not say—but what everyone else in Russia understood—was that the conflict was part of the state's wider effort to control the political space ahead of the upcoming parliamentary and presidential elections. "Yukos is part of Putin's political game," the leader of one of Russia's liberal political parties told me. "He is against the independent position of anyone, including business. I think Yukos is important for Putin because he wants to show business that if you make money, you have to be loyal. If you are not loyal, you will be in prison."

Stanislav Belkovsky, a Kremlin adviser during the early years of the fight, took a similar view. He was seen as the ideological face of the *siloviki* and wrote a report titled *The State and the Oligarchy* that acted as a rationale for the fight with Khodorkovsky (Sixsmith 2010, 65). Although he later moderated his position, he saw the fight as a battle for the future of Russia (Sixsmith 2010, 68): "The main cause is the object fact of growing tension between big business and the national interests of Russia. But the catalyst for the Kremlin's reaction [in attacking Yukos] is the political ambition of Mikhail Khodorkovsky."

Nearly every lawyer that I discussed the case with highlighted the political dynamics and the changed balance of power as preventing any form of settlement. One lawyer who focuses on Russian disputes pinpointed the fact that this was a direct battle between the state and a defiant former oligarch, leaving settlement impossible: "This was even more sensitive than Berezovsky-Abramovich. It was to the death . . . it's so personal . . . all of the other disputes involved someone close to the Kremlin but this was with Putin himself."[60]

Another British lawyer who worked with numerous Russian clients called it a "a proxy war for Russian politics."[61] One individual went even further, calling a settlement "not part of political reality . . . it is out of the question."[62] The fact that Khodorkovsky was labeled an enemy, directly by Putin, meant that no one around Putin could even fathom bringing up the idea of settling the cases, even if it was the economically rational decision: "Do I want to enter P's [Putin's] office and say we are paying 1 billion to Khodorkovsky? It would be my end!"[63]

The political stakes were then the key reason why the costly disputes took place. Khodorkovsky and Yukos certainly knew as much. A lawyer said that this was one of the few instances where you would expect the case to go all the way. "[The Russian government was] never going to settle. Yukos knew that and a lot of this is now more for PR [public relations]."[64] The individuals involved on the GML side knew this was not going to stop without the power of foreign courts compelling the defendant Russian state and Rosneft.

Khodorkovsky was a threat, and a settlement could, and likely would, find its way to him even if he were in jail. Mikhail Brudno, one of the Yukos inner circle members, was expecting his boss to win the entire game before the political conflict that triggered the extraterritorial cases (Sixsmith 2010, 83): "They were sure that he was after their power and authority, that he would take their power away from them. And he would have done, if he had had enough time! He would have taken away their absolute power!"

As historian Christopher Miller (2018, 58), author of *Putinomics*, summarized, the entire affair was about a broader war toward regaining control:

> The management of the country's energy sector demonstrates again the priorities of Putinomics. When the state's authority was in question—when the Kremlin was unable to tax fully the oil industry, and when Khodorkovsky threatened Putin politically—the government's overriding priority was to reassert control. Jailing Khodorkovsky was controversial, but both statist conservatives and liberal technocrats supported the effort to re-establish central authority and rebuild government finances. Once the state's primacy was established, technocrats were given the space to use tax incentives to optimize production. In the gas sector where the state's predominance was never in doubt, the role of private players has increased—though all the non-Gazprom producers remain quick to pledge their loyalty to the Kremlin.

It was the macro battle between the once-oligarchs and the state, and the conditioning effects on the rest of the plutocracy, that Khodorkovsky himself pointed to as the core logic behind the Yukos Affair (Zygar 2016, 48): "Putin was behaving like a classic military commander. When I served as an army officer, we were taught not to yell at soldiers standing to attention. So when he noticed broad disaffection among big business, Putin located the command center [that is, he found the key person] and smashed it. After that, no one dared step out of line."

Outcomes

Nearly ten years after his initial imprisonment, Khodorkovsky was released by the Russian authorities. He quickly fled abroad. Lebedev was released a few weeks later. In the interim years, Khodorkovsky had turned into the symbol of the opposition. He published six books during his time in prison and continued to correspond with journalists and everyday dissidents, but the timing was important. The Yukos/GML team never let up their fight against the Kremlin, and the publicity created by the cases kept Khodorkovsky an international story. As Russia was getting ready to host the winter Olympics, Putin needed a win and an end to the Yukos affair. A deal was struck, and Khodorkovsky became yet another Russian exile, albeit with a bigger following and offshore fortune than most.

The Permanent Court of Arbitration in 2014 gave GML the biggest investor-state dispute settlement (ISDS) payout that would continue to shadow any future judgments. But the case came through the ECT, which Russia's Duma had never actually ratified. Two years later, a new set of lawyers representing the Russian state made the case that the ECT could not be considered valid per the Russian constitution. The Dutch appeals court agreed, and the initial decision was struck down (DLA Piper 2016). Nonetheless, the GML fleet of lawyers continued trying to enforce it around the world. On February 18, 2020, the Dutch appeals court ruled that Russia was still bound by the initial award of $50 billion (BBC News 2020). In a perfect twist of fate, the appeals court used a small section of a decision by the Russian supreme court legally justifying the 2014 invasion of Crimea as the basis to overturn the verdict. As Oliver Bullough neatly summarized, "Game back on" (Bullough 2020). Almost exactly four years later, when it initially looked on the verge of finally getting rid of its $50 billion bill, the Russian state lost another appeal (Reuters 2024). Just as this book was going to press, a final decision by the Dutch courts granted GML a $65 billion windfall.

As Putin came to power, riding the waves of public popularity and high oil prices, the balance of power in Russia shifted. The political system turned from one of state capture, where the plutocrats ran the show, to one of state competition, where a resurgent state and an independent plutocracy were both present (Sakwa 2014; Djankov 2015; Gel'Man 2008). As individual plutocrats and the state tried to assert their authority, a series of political clashes emerged. These fights explain the stark increases in Russian extraterritorial dispute resolution whereby plutocrats and state-owned enterprises became mainstays of foreign courts.

Berezovsky clashed with the Russian state when his major news channel ORT continued to be independent and continued to do his, rather than the Kremlin's, bidding. Berezovsky was one of the reasons Putin was even brought to power, but he became the regime's number one enemy. When his former business partner Abramovich sold off his stakes in Sibneft, it opened the door for Berezovsky to mount a comeback. A dispute worth over $5.6 billion, which gave us one of the only firsthand accounts of the chaos of state capture, could not settle. Handing over that kind of money to an enemy of the Russian state was out of the question.

Khodorkovsky was one of the biggest beneficiaries of state capture in the 1990s, and his oil and banking empire owed much of its success to his political dealings. After initially stepping back from politics when Putin came to power, the richest man in Russia made several moves that were seen as positioning himself and his company to mount a challenge against the Russian state. But Putin and the security services spared no cost or effort to take down Khodorkovsky and the Yukos empire. Despite Khodorkovsky's imprisonment, those closest to him fled the country and mounted a multiyear, multijurisdictional fight against the regime. It involved cases against Rosneft, who took over Yukos's most valuable assets, and the Russian state directly. The extraterritorial dynamics were inevitable because the point of the Kremlin attacking Yukos was to get rid of a political opponent. Implicitly admitting they were wrong by giving part of the money back would have only invigorated the threat.

For both Berezovsky and Khodorkovsky's camp, extraterritorial dispute resolution became a form of offshore protection. Liberal jurisdictions became the final way to fight back against a regime that tried to take them down. This chapter illustrated the first driver of the outsourcing of authority. At the height of state capture, there were almost no foreign court battles, but they became an essential dynamic of the losers from state competition to regain their money and legitimacy. The conflicts would have unfolded historically within the Russian jurisdiction, but with capital mobility and the rise of offshore finance, traditional elite battles spread across liberal and illiberal contexts.

4

EXTRATERRITORIAL DISPUTES AS STATE RETALIATION

> The right to property in Russia is entirely conditional upon the property owner's loyalty to the Russian government. The system is tending to evolve not in the direction of freedom and a postindustrial society, but rather back toward feudalism, when the sovereign distributed privileges and lands to his vassals and could take them away at any moment. The only difference is that, in today's Russia, the things that [Vladimir] Putin is distributing and taking away are not parcels of land, but gas and oil companies.
>
> Andrei Piontkovsky (2009, 52)[1]

As states liberalized their capital accounts through the 1990s, scholars and policymakers were fixated on globalization's effects on state autonomy. The ability of firms and individuals to move money across borders was expected to leave states at the whim of business, circumscribing policy options, as governments raced to the bottom in search of foreign capital (Strange 1996; Drezner 2001; Tonelson 2000). The way plutocrats are now able to use the transnational legal market to pick and choose the rules and courts that best suit their material needs, at a surface level, echoes these fears. Individuals can subvert their state's authority by simply changing a few words in their private contracts.

But a large body of research now shows that state authority, although certainly limited for some smaller countries (Rudra 2008), has continued to be a key feature of globalization (Drezner 2008). States flex their own market power to make firms and individuals abide by their preferred rules (Bradford 2012; Newman and Posner 2011; Bach 2010). They can even leverage the way firms are structured using complex global supply chains and multisite production networks to extend their reach beyond their borders (Crasnic et al. 2017). The dark(er) side of extraterritorial dispute resolution follows the same pattern.

As detailed in chapter 3, in a world of state competition, the losers of a political clash are given a new lease on life by using foreign courts to potentially get their money back. Arguably the most important mechanism to bring a case abroad is using the broad scope of personal jurisdiction that is usually applied in the United Kingdom and the United States. But states can also take advantage of such rules. If a dissident or opponent has moved abroad to seek protection, the latter will be deemed domiciled in the jurisdiction, opening the space for the state to engage its own form of lawfare (Kittrie 2016).

Using one's state-owned enterprises (SOEs) or subservient kleptocrats and tycoons to bring a case against a fleeing opponent fills in the second strategy, the boomerang, leading to the outsourcing of authority. Plutocrats can still mount a political challenge from abroad, funding parties or engaging in disinformation campaigns, particularly if the individuals have substantial wealth stashed in offshore and foreign jurisdictions. Extraterritorial litigation can provide the means for a regime to get its hands on that wealth, similar to the attempts to bring back the loot of a deposed kleptocrat (Sharman 2017). It can further help in discrediting the figure and signal to any future challengers the length to which the regime will go to take on the opposition. Roughly 30 percent of London's emerging market extraterritorial litigations involve a state or state-owned entity. A state-backed actor is part of almost half of the Russian extraterritorial litigations. During the mid-2000s, the Kremlin was forced to deal with the political uncertainty caused by an independently wealthy plutocracy. The British courts became the natural venue to continue the political fight with those who have fallen out with the regime.

Nonetheless, the Russian state has been able to consolidate substantial authority more recently. As Richard Sakwa (2014, 220) bluntly put it:

> In March 2000 Putin promised that if elected president, "Russia's oligarchs will be abolished as a class." Within five years a new system of state-business relations was created. Independent oligarchs like [Mikhail] Khodorkovsky ceased to exist and business became dependent on the state, while a shadow class of "beneficiaries" intercepted natural-resource rents and enjoyed a privileged relationship with the power system.

In other words, political dynamics have begun to resemble more of a state control–like setting where the plutocracy is predominantly constituted by the new winners: the kleptocrats, and the tycoons hanging on to their own private resources. This has put the Russian state in a position to govern the relations and resolve disputes between wealthy parties that have accepted or gained from the new status quo. Examining how disputes are mediated between those subservient (or those that need to be seen as subservient) to the Russian state gives us a window into conventional state control settings. Researchers like Dawisha (2015), and Soldatov and Borogan (2010) have detailed the way the Kremlin has used its resources to build up its own band of super-wealthy supporters, largely coming out of the security services and Putin's trusted cabal from his time in St. Petersburg. These are kleptocrats that are loyal to the state, who frequently and openly take political and commercial actions to further their interests. As Putin's tenure grows longer, more and more plutocrats begin to fall into this subservient category, especially as the nominally independently wealthy, the tycoons, seek to ensure that they are now aligned with the Kremlin (Djankov 2015).

The purpose of this chapter is twofold. First, it illustrates how and when the state strikes back against those that could pose a challenge from abroad. Second, the chapter documents how the Russian state manages disputes when in control, mitigating the likelihood of extraterritorial legal battles. I begin by providing a broad overview of why some of the Russian extraterritorial litigations that involve state actors are unable to settle. I then continue the within-case comparisons of Russian plutocrat dispute resolution strategies across the balance of power. The chapter dives into how the Putin regime faced off against Boris Berezovsky via the state-owned airline Aeroflot, and the following section details the fight waged by Rosneft across foreign courts as it tried to get its hands on Yukos's offshore wealth. Next, I present quantitative evidence that highlights the relationship between independent wealth and extraterritorial dispute resolution. The final section of the chapter provides a glimpse into the more conventional, state-controlled world of dispute resolution by examining a dispute between two tycoons who needed to stay on Putin's good side, preventing them from taking full advantage of the authority of foreign courts.

Why the State Goes Abroad

Just as Russian plutocrats became mainstays of the London courts in the 2000s, Russian SOEs became increasingly common claimants. From a functional viewpoint this may not be surprising: under Putin, the role of the state in the economy was growing, so there were sure to be more disputes for them to take part in. While this is no doubt true, it fails to consider what might be stopping these expensive cases from settling and why they need to be taken abroad, where they may risk high transparency costs. Diving into the details of the cases and discussing them with the lawyers involved illustrate that there are much larger stakes to these cases that go beyond a simple commercial claim. "There are a lot of people in the background [of a dispute] that are really the ones pulling the strings."[2]

The timing, the fact that Russian SOEs became major users of foreign courts in the 2000s, was no coincidence. It was driven by the balance of power as plutocrats began to clash with the Russian state and were then forced to flee. The state found new, extraterritorial ways to take on its opposition. As one lawyer neatly summarized, "It's usually a pattern of they come here and so they get sued here."[3] Plutocrats migrated abroad in search of safety, but the same liberal rules that make places like the United Kingdom valuable havens also create political opportunities. A lawyer who has worked on cases involving SOEs said, "It's very common to see a claimant that may not be based here but the defendant has come here to escape . . . the downsides of parking your ass here, in a relatively tolerant, safe place, is that you make yourself a target."[4]

The gains from the cases go well beyond any basic commercial sense because one of the primary drivers is to ensure that those who have run cannot pose a future threat. Having wealth in places like London or the British Virgin Islands puts plutocrats in a position to continue shaping politics back home, even if they have escaped Russia. The state needs to find ways to prevent that, so the cases are really "all about capital that is outside of Russia."[5] The political stakes inevitably make settlement rare, and the lawsuit is fundamentally about moving the balance of power further toward state control. As one lawyer put it, "If the state is involved they basically want to crush the people . . . and whoever might want to settle would be blamed for trying to help the fraudster . . . it's a power-play."[6]

Most state-initiated cases in London center around some form of fraud, which again feeds into the larger political games at play. Cases alleging fraud tend to be valid for a longer period of time. Given the way business is done in emerging markets, there is frequently some type of business activity or corruption in a fleeing plutocrat's past that could be considered fraudulent. As one lawyer working with numerous Russian clients told me, "Most cases involve some corruption that we were all aware of, but no one did anything about it until there is a reason to remember."[7]

Extraterritorial litigations are then frequently an option for the state once a plutocrat has been targeted. "These cases tend to come up when the economically powerful fall out . . . then you see accusations around fraud because it is easy to put in place."[8] But the mix of the political stakes and the nature of the allegations can further entrench parties, making settlement less likely: "But they [the cases] often include matters of dishonesty. If it's a breach of contract where I say you owe me a million, you say it's 200 thousand, we can deal. But the moment you allege dishonesty it's much harder . . . often there is a political angle, or at least the people involved perceive there is."[9] "Because of the nature of the allegations, people calling each other fraudsters, it's very hard to settle . . . and as soon as that starts, they also open criminal proceedings in Russia . . . these are politically motivated and very hard to stop."[10]

This type of entrenchment can still be valuable to the state. Although the primary aim of any of these cases may be to get the money of the fleeing or allegedly fraudulent plutocrat, there are additional publicity gains. There may be symbolic value to having an independent British court affirm the SOE, or the state's case, because it will delegitimize a potential opposition figure: "And back to the political aspects, they want to win and recover but they also want the political value of being shown to be right."[11] "Sometimes it's not only the case of the money, sometimes one part is just about the punishment. Some want the decision. They want the official position that the opponent was wrong."[12]

In this way, an extraterritorial litigation can have political value beyond even the specific opponent. There are additional signaling gains. Taking the costly measure of filing a case abroad shows any future political challengers and those that might still be independent at home the length to which the state will go to exhaust its political opposition. One lawyer, who regularly works with oligarchs across Eastern Europe, said, "Some cases do not settle because the state wants to send a message . . . if you are stealing from us, we will go after you . . . at least it's better than killing people."[13]

Chasing Berezovsky

As a member of "the family" that ran the Boris Yeltsin administration, Boris Berezovsky effectively captured the Russian state. Throughout the 1990s, he used a variety of extralegal means to acquire his wealth and resolve his disputes. This activity came back to haunt Berezovsky under Putin. Berezovsky was forced to flee Russia because Putin made it clear that he would not tolerate the independence and thereby political threat that Berezovsky was adamant on maintaining. The conflict was fundamentally driven by the balance of power, which now resembled a world of competition as opposed to the one of capture that Berezovsky had effectively engineered. Nonetheless, the proximate cause for Berezovsky leaving Russia was that he was expecting to be arrested for his dealings with state-owned airline Aeroflot.

Aeroflot's books were in tatters following the transition from communism, but it was a highly valuable company to any plutocrat looking to diversify and safeguard their wealth. The international nature of the airline business meant that Aeroflot was regularly dealing with valuable foreign currencies, giving the company plenty of legitimate reasons to be moving around money in foreign and offshore accounts. As detailed in chapter 2, Berezovsky borrowed a page from his fellow plutocrats by gaining control of the management of Aeroflot so that he could transfer some of those foreign funds into his own accounts.

Berezovsky had to fight off Vladimir Gusinsky for the management rights (Freeland 2005, 136), but the bitter dispute proved extremely valuable once Berezovsky's lieutenant Nikolai Glushkov formally started running the airline (Freeland 2005, 136–37):

> Swiss authorities would later allege that Aeroflot's hard currency receipts had been systematically misappropriated to the tune of at least $200m, through Andava and Forus [Berezovsky's entities], two Swiss-based companies which ostensibly provided the airline with financial services. But thanks to the Kremlin connection—after all, Aeroflot had become the Yeltsin family firm—inside Russia the airline had political Teflon: no matter what machinations investigations uncovered, they could never make them stick.

While there was an initial investigation in 1999, when Yeltsin was still in charge, Berezovsky effectively used kompromat to squash it (Klebnikov 2000, 287–91). But the charges were eventually brought up again after Berezovsky and Putin began to clash. The rumors of a new criminal investigation into Aeroflot were confirmed on November 1, 2000, by Deputy Prosecutor Vasily Kolmogorov (Hollingsworth and Lansley 2009, 90). This was within two days of Berezovsky boarding his private jet to flee Moscow permanently. His exit was not enough to protect his protégé Glushkov, who was arrested in relation to his role in the fraud allegations at Aeroflot.[14] He was eventually used as leverage by Putin when initially negotiating Berezovsky's divestment of his Russian assets.[15]

Taking the Chase Extraterritorial

When Roman Abramovich bought out Berezovsky, the Kremlin's hope was that the former godfather would turn into an afterthought of the 1990s. Berezovsky's incessant criticism of the Putin regime and his funding of nongovernment organizations (NGOs) and opposition parties instead made his personality loom large across contemporary Russian politics. This instigated the Kremlin's attempts to extradite Berezovsky from the United Kingdom. In addition to charges related to conspiracy to commit a coup, the Aeroflot fraud charges were one of the major focal points of the Russian government's case. But the outright animosity between the state and Berezovsky made the United Kingdom grant Berezovsky asylum in 2003 (BBC News 2003). The result soured UK-Russian relations and set the stage for another legal battle that would stretch over a decade.

As Berezovsky began mounting his claim against Abramovich, the prosecutor's office in Russia ramped up its Aeroflot cases. In November 2007, Berezovsky and Glushkov were tried for fraud with the individuals eventually sentenced to six years in absentia. "Moscow's Savelovsky Court convicted Berezovsky of fraud and found him jointly and severally liable with Mr. Glushkov to pay Aeroflot over 200 million rubles. The judgment took effect in February 2008, after Moscow City Court dismissed Berezovsky's appeal" (Burke 2014). In 2011, the Golovinsky Court approved adjusting Aeroflot's claim for inflation, making the total 2 billion rubles. Berezovsky refused to pay as he sat in London looking for ways to make more money, with his fortune taking a massive hit after the financial crisis. In the interim, Glushkov had managed to flee, joining his former employer in London.

With the defendants now stationed abroad, the state was left with little choice but to take its claims extraterritorial. Aeroflot filed a case in London seeking to enforce their domestic judgments against the Russian state's political opponents. Although initially rejected by the courts in 2012 (Brick Court Chambers 2012), Aeroflot appealed the case. Before any conclusion could be reached, Berezovsky was found dead in his apartment in March 2013.

Why They Couldn't Settle

A fraud case that most elites in Russia knew about in the 1990s dragged on through the 2000s and was eventually brought to the London courts. A couple of lawyers I spoke to viewed *Aeroflot v. Berezovsky* as nothing more than a standard fraud claim, where an individual was accused of illegally stealing funds and the wronged simply intended to recoup their money. But to many others there, the stakes were less financial and more political. As one lawyer summarized, "They were out to get him . . . and everyone knew. . . . It would have gone away if it wasn't a political opponent."[16] Aeroflot fully had that right given Berezovsky's new residence in London; as long as Aeroflot were genuinely after the money, they had a legitimate claim.

Although the way Berezovsky and Glushkov appeared to handle the Aeroflot accounts was morally questionable, they exploited loopholes in Russian law that several other plutocrats had also taken advantage of. One lawyer plainly described it as "tit for tat."[17] The timing of the case further suggests that the stakes went beyond a commercial motive. "It's strange, the timing. Any lawyer, if we are in England is going to tell you to move as quickly as possible but they waited. In that situation it's unusual that's for sure."[18] It appears to some that the cases were strategically timed to counter some of Berezovsky's political pressure. One lawyer was explicit in saying that the bringing of the judgment to London as Berezvosky brought his claim against Abramovich was likely a concerted effort to delegitimize Putin's opponent. It was "an opportunity to kick a man when he's down."[19]

Thus, the actual case and the claim were regarded as secondary considerations by several London-based lawyers. As one individual said, "The Russian state just wanted him under investigation . . . to attack him in any way possible."[20] Others pointed out that, with state-initiated cases, victory or settlement may not even be the goal. Even if the claim failed, they would succeed in questioning Berezovsky's legitimacy and force him to exert substantial time and money in a defense: "It was a way to put pressure on people who have run away from Russia. They were after the person's money . . . the worst case is that you drain their money through the legal fees."[21]

None of this is to say that Aeroflot did anything legally wrong. It was completely within its rights to bring a case to London with Berezovsky using the city and its huge army of professional services as its base. These are the rules of a liberal system.

Outcome

Once Berezovsky died, there was a general air of expectation that the case against him would soon be resolved; the trustees of the estate eventually settled for 82 million pounds, but Aeroflot will likely never see any of that money because the estate was said to be bankrupt, and Aeroflot was low on the creditor totem pole.[22] Aeroflot and the state were still pressing the case against Glushkov, Berezovsky's partner, who was also party to the claim and the fraud charges. Once Glushkov fled Russia, he continued to work as part of Berezovsky's extraterritorial cabal and had even given evidence against Abramovich during the London Commercial Court trial. Nonetheless, he was a relatively small-time player.

A handful of days after the poisoning of Sergei Skripal in March 2018, Glushkov was found strangled in his home in London (BBC News 2021). Less than a month after his death, the lawyers representing Aeroflot, according to Justice Vivien Rose, sent "intrusive and hectoring" questions to Glushkov's daughter and civil partner:

> The High Court is familiar with high-value litigation being hard fought with no quarter given on either side. But it is, fortunately, "out of the norm" for a litigant and its legal representatives to so lose sight of any basic standard of decent and compassionate behaviour as to send a letter in those terms. It is all the more shocking given that it seems very unlikely that at the time the letter was drafted and sent, Aeroflot had no inkling that the next day it would serve a notice of discontinuance, abandoning all claims against Mr Glushkov's estate.[23]

As Justice Rose notes, with the two political opponents, Berezovsky and Glushkov, out of the picture, the case was finally discontinued just before the scheduled trial (Herbert Smith Freehills 2018). In a witness statement in 2011, Glushkov said, "I believe that it was a direct result of my close relationship with Boris [Berezovsky] that I was targeted for persecution by the Russian government" (Bond et al. 2018). Justice Rose ordered Aeroflot to pay the legal costs.

Yukos's Foreign Coffers

Mikhail Khodorkovsky was the chosen communist to lead Russia in its search for an equitable market economy. His young roots in Komsomol became his route to building up a lucrative banking conglomerate. The wealth of connections put him in a place to rig the auction for what would become Russia's biggest oil company, Yukos. At the time, the company appears to have rarely been afraid to bend the law to resolve its disputes.

From the start, Bank Menatep and Yukos stashed their earnings in a web of domestic and foreign tax havens. The integration of Khodorkovsky and the magnificent seven's wealth into global financial structures reached its apex in 1997, when British lawyer Steve Curtis inverted the ownership of Yukos (Hoffman 2011, 448): the company that constituted almost 2 percent of the Russian government's tax revenue was made a de jure foreign entity when ownership rights were transferred into a variety of shell companies to protect Khodorkovsky and Yukos's inner circle.

By 2003, Khodokovsky emerged from an initial fight against the incumbent oligarchs. In fact, he was thriving at the top of Russia's *Forbes* list and was set to merge with Abramovich's Sibneft. But Khodorkovsky also clearly appeared to have become dissatisfied with the emerging move toward state control. He began funding a host of opposition parties. Russian civil society was fundamentally dependent on his charity, and he was even in talks to sell off parts of Yukos to major foreign oil companies. In other words, he was posing a serious threat to the Russian state, and the Kremlin reacted with what Paul B. Stephan (2013) called "expropriation by taxation."

For the most part, Yukos followed the standard practices of the Russian oil industry, routing various transactions through domestic subsidiaries to minimize its tax burden. These mechanisms were used as the basis to target Khodorkovsky and the company initially. Through its various machinations, the Russian state eventually charged the company with owing close to $28 billion. They had even secured a freezing order on domestic assets that prevented Yukos from cashing any of its securities and then insisted that payment needed to be completely in cash (Stephan 2013, 25). This left the company and its shareholders with almost no domestic room to maneuver, and the Russian state was not backing down.

The state was never afraid to illustrate its control of the situation; when a judge tried to remove the freezing order, he was summarily dismissed (Stephan 2013, 24). Following the trials of Khodorkovsky and Platon Lebedev, and with Yukos having no real options to pay the Russian government, the underlying logic behind the attack became clear. In 2004, Yukos's most valuable asset, the oil field *Yuganskneftegaz*, was auctioned off to the shell company Baikalfinans, which quickly turned around and sold the jewel to Rosneft. The Russian state not only managed to bankrupt their political opponent but also took his biggest revenue stream for itself. Yukos was eventually left with no option but to file for bankruptcy.

Taking the Case Extraterritorial

The sale of *Yuganskneftegaz* put a dent toward paying off Yukos's tax bill; the Russian government received just under $10 billion for an asset that was nonetheless valued closer to twice that (Stephan 2013, 25). Rosneft continued to file charges against the company domestically. As noted, Yukos was well versed in protecting its money through using subsidiaries and offshore jurisdictions, so there was money for Rosneft to seize—but not in Russia. Although Khodorkovsky and Lebedev were behind bars, the other major players in Yukos such as Leonid Nevzlin and Mikhail Brudno had safely secured passage abroad. They were armed with their offshore wealth, and who knew when and how Khodorkovsky might find his own way out. The Russian state still needed to neutralize the political threat, and it enlisted the help of a host of foreign courts in the process.

Extradition charges were filed in numerous jurisdictions, from the United Kingdom to Lithuania (Stephan 2013, 28). The Russian state even tried to get foreign authorities to help raid Yukos's offices abroad by filing petitions through the foreign courts in places like Switzerland (Stephan 2013, 28). But the biggest offensive play by Rosneft and the Kremlin was to take Yukos's Russian bankruptcy and use it as the basis for seizing the assets of the company's foreign subsidiaries. Some of these were essential to international commerce; others were more conventional offshore outfits facilitating tax avoidance (Stephan 2013). While there were several such attempts, two of the extraterritorial litigations stand out. The first involved the Dutch subsidiary Yukos Finance B.V.; a Dutch court eventually rejected the recognition of the bankruptcy (Stephan 2013, 31–33). This helped Yukos shareholders transfer a broad swath of the most valuable resources into a highly protected Dutch trust. The second now infamous case was filed against Yukos CIS in Armenia, a country that maintained close ties to the Kremlin and was quick to recognize the validity of the bankruptcy (Belton and Stott 2016). Yukos fought the charges for years with no apparent end in sight.

Why They Couldn't Settle

The extraterritorial battles waged by Rosneft and the Russian state appear to have had a clear purpose. Of course, the state wanted the money. But beyond that, the general consensus is that the Kremlin was out to break Yukos and its leaders, who were seen as political threats. Such a viewpoint becomes clear when you consider how some of the Yukos management tried to handle negotiations about the tax payment after Khodorkovsky was imprisoned. According to Bruce Misamore, one of the senior foreign employees of the company, they repeatedly tried to settle, going through different branches of the Kremlin. He told Martin Sixsmith (2010, 175), "They would lead us on . . . and we would think that we were very close to a settlement, but then someone would meet Putin and the whole thing would get dropped. That is why I absolutely 100 per cent believe that every major action in the destruction of Yukos was directed or personally approved by Vladimir Putin."

During the hearings for the arbitration claims against the state, Yukos employees argued that they went up to offering $21 billion, although there was no paper trail to confirm this.[24] The sum came close to matching some of the grandiose claims that the Russian state was pressing and would have surpassed the amount the Russian state has thus far been able to collect extraterritorially. According to the Yukos team, the negotiations ended when their lead negotiator was forced to leave Russia after being threatened with arrest.[25] The Kremlin appeared hellbent on ruining the company and taking out Khodorkovsky, but a settlement would have left the firm in the prior management's hands. As Misamore again said (Sixsmith 2010, 175–76), "Once we got close to a solution with other parties in the Kremlin, I believe that when they took it to Putin on the weekends or whatever he was the one who said, 'No we are going to destroy this company; no deals.'"

In the years after the first tax bill against Yukos, the shareholders claim they made roughly eighty attempts to settle.[26] Thus, the decade spent chasing down the company could have likely been avoided even if the tax claim was completely legitimate, but there were larger implications for the balance of power. The Russian state spent millions on foreign and domestic lawyers chasing down assets that Rosneft could seize. But that money meant little in the larger political scheme as the Russian state tried to move further toward a world of control. After all, the individuals in government were not the ones paying the costs. As one lawyer cynically put it, "They are not going to settle. Who is paying the bills? It's the people. It's the Russian state. They don't care about the legal fees. 1 million less here and there who cares. It's too distant from the Kremlin."[27] "It was to the death."[28] The Russian state did not feel encumbered by the economic costs. Even if the attempts failed, they would succeed in draining and tying up whatever money the Yukos exiles had left stashed away. The longer it went on, the generally better the outcome for the Russian state because even failure would, to a certain extent, further the political cause. One specialist familiar with the case said, "There were never any approaches [for settlement]. . . . [Igor] Sechin was hell bent . . . this was about the rise of Khodorkovsky and the challenge to authority."[29]

The extent to which the legal proceedings have become politicized has even created a nice informal precedent for future plutocrats who were chased out. Because the fight has become so openly about politics over commerce, any Russian trying to get asylum in the United Kingdom apparently aims to find anything that might connect them to Yukos to garner favorable treatment from the British courts.[30] Rosneft was exercising its rights, however, under how most liberal legal systems operate. The premise of a liberal system is that you are entitled to your day in court as long as you have some legitimate commercial motives.

Outcome

After a decade, dozens of trials, and tens of millions in legal fees, Rosneft's offensive relented. The state-owned oil company that was languishing before it acquired Yukos settled with the shareholders in early 2015 (Farchy 2015). While the exact sum was not disclosed, the agreement would allow the Yukos trusts in Russia to begin distributing some of the $400 million it had acquired through its own extraterritorial claims (Farchy 2015). The terms of the settlement meant that all Yukos- and Rosneft-related cases worldwide would be considered over, but it had no bearing on the GML cases against the Russian state or on any other general cases involving Yukos and other Russian parties.

The settlement came as a shock to a number of observers, but the logic became clearer about a year later. The *Financial Times* reported that Rosneft's lawyers had manipulated the legal proceedings in Armenia in their favor (Belton and Stott 2016). The newspaper had emails showing Rosneft lawyers telling senior members of the Armenian judiciary how they wanted five different trials to go and even dictated some of the rulings. A judge fled Armenia after refusing to put out a prewritten judgment that was given to him on a USB drive. He eventually testified in one of the Dutch proceedings related to Yukos CIS, where the emails were also revealed. As Judge Surik Ghazaryan put it, "A clear and unequivocal message was sent to any Armenian judge presiding in a Yukos-related case: obey the instructions that come from the top and rule in favour of Rosneft and against Yukos or face harsh consequences" (Belton and Stott 2016).

Three months later, the case was settled. According to a number of different lawyers I spoke to, the revelations meant that Rosneft was almost certainly going to lose, finally bringing them to the table. The evidence was eventually admitted in another Dutch case involving Yukos International and upped the stakes in the extraterritorial arbitration battle. Citing political motivations, in January 2019, the company's bankruptcy in Russia was formally recognized in the High Council (the Netherlands supreme court) (Reuters 2019).

These cases fit a broader pattern in which SOEs are sometimes used to mount an attack, through the courts, against the Kremlin's political rivals. One litigator said, "The same legal instrument in private hands, when deployed by the state, will and should be interpreted in a different way."[31] A lawyer who has worked on SOE-related disputes stated, "Many of these cases are really a campaign of harassment."[32] But they are campaigns that any truly liberal system needs to permit.

Economic Independence and Extraterritorial Dispute Resolution

The qualitative evidence from chapters 2 and 3 indicates that as the balance of power in Russia shifted toward competition, the independently wealthy became more likely to use transnational legal institutions to resolve their domestic commercial conflicts. They may initiate the claims or be forced into defending one. Berezovsky and Khodorkovsky were involved in both types of claims. Next, I provide quantitative, plutocrat-level evidence to generalize the findings further. Using a sample of the 164 Russians that appeared on the *Forbes* billionaires list between 1996 and 2015, I assess how the nature of an individual's wealth has an impact on the likelihood of that individual becoming involved in an extraterritorial litigation. The value of this analysis mirrors the logic of the qualitative cases (latent cultural factors and country-level institutional and economic factors are constant across all the individuals) and provides an additional layer of generalizability.

To proxy for whether the plutocrat was independently wealthy, I code whether the individual gained their major holdings through the 1990s privatization process or directly through their connections with one of privatization's winners. As chapter 2 details, privatization created the Russian plutocracy. The individuals were heavily involved during state capture and operated as the incumbent oligarchic class when Putin took over and shifted the balance of power toward competition. These former oligarchs should be more likely to be targeted by the Russian state and thereby have the political need to file a case abroad and be subject to extraterritorial disputes filed by Kremlin-aligned actors. Those with wealth derived from privatization who remain in Russia, such as Roman Abramovich, should also be more likely to be part of a case in London. Their fleeing business partners may file cases against them that are difficult to settle because they are de facto political battles under commercial pretenses.

I created the dataset by updating the Russian billionaire data collected by Treisman (2016), who coded the privatization status, net worth, and the sectoral and industry wealth sources for the individuals appearing on the 2005 and 2015 *Forbes* lists. Based on their online archives, I coded these variables for the additional seventy-six Russians who appear on the list in the time period of analysis but dropped off prior to 2015. If an individual's privatization status was unclear from *Forbes*, I surveyed the major academic and journalistic books about the time period and conducted in-depth reviews on the individuals and the relevant companies using the database Factiva.

Figure 4.1 compares the sectors in which plutocrats made their wealth during the privatization process versus the second wave of economic growth, in which individuals would be deemed tycoons or kleptocrats. A few patterns stand out. Fitting some of the common conceptions of the Russian economy, the hydrocarbons sector produced the most billionaires during privatization and continued to grow the size of the plutocracy through the second wave. Finance, however, has become a crucial source of revenue for Russia's plutocracy. The starkest differences between the two subclasses comes in the construction and real estate sectors, which are core to the second wave of generally state-directed economic activity.

The dependent variable for the analysis is involvement in extraterritorial dispute resolution, represented by a dummy variable, where 1 indicates if the plutocrat or a company in which they have a significant interest was involved in an extraterritorial litigation in either the London courts or through extraterritorial arbitration between the years 1996 and 2020. The variable was coded by cross-referencing each billionaire with all the intra-Russian cases filed at the two major commercial courts in London and the arbitrations publicly available through the UN Conference on Trade and Development (UNCTAD). To ensure accurate coding, I searched the individual and the company name through the Westlaw legal database. Forty of the 164 individuals were part of an extraterritorial litigation.

Figure 4.2 plots the year of the plutocrat's first appearance on the *Forbes* billionaires list against their average net worth during the years they appeared on the list. It further breaks down the relationship by whether the individual

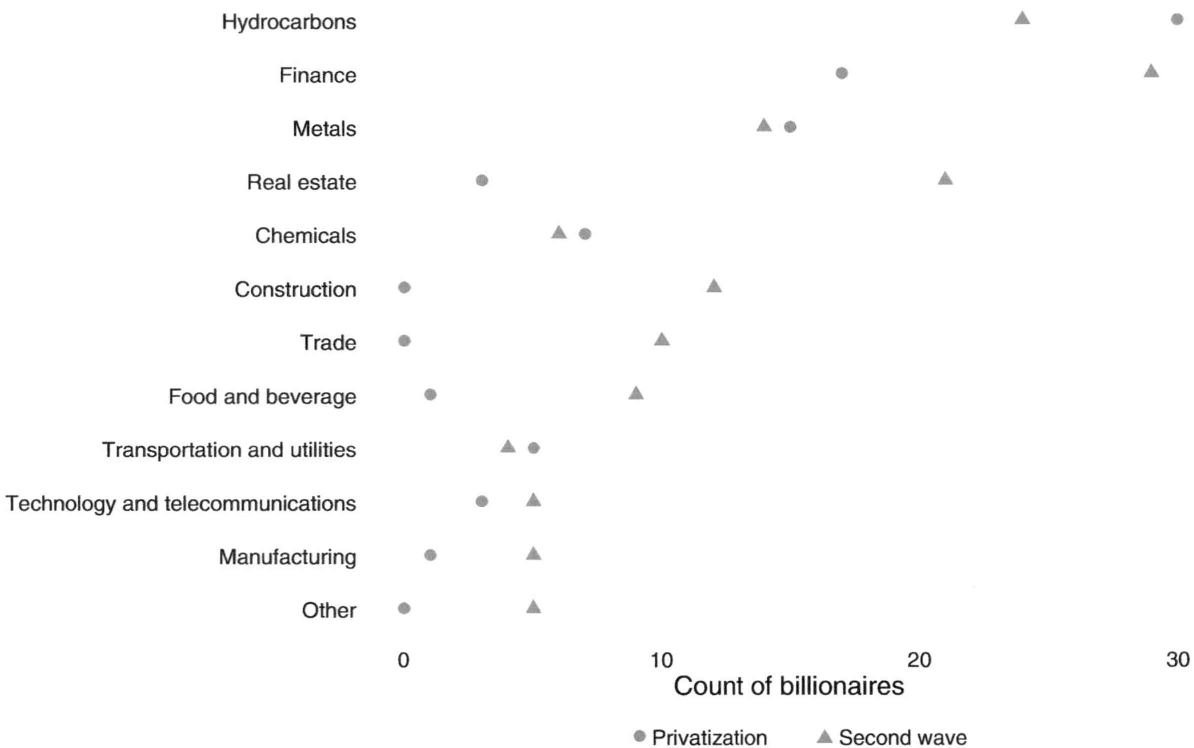

FIGURE 4.1. Billionaires by sector

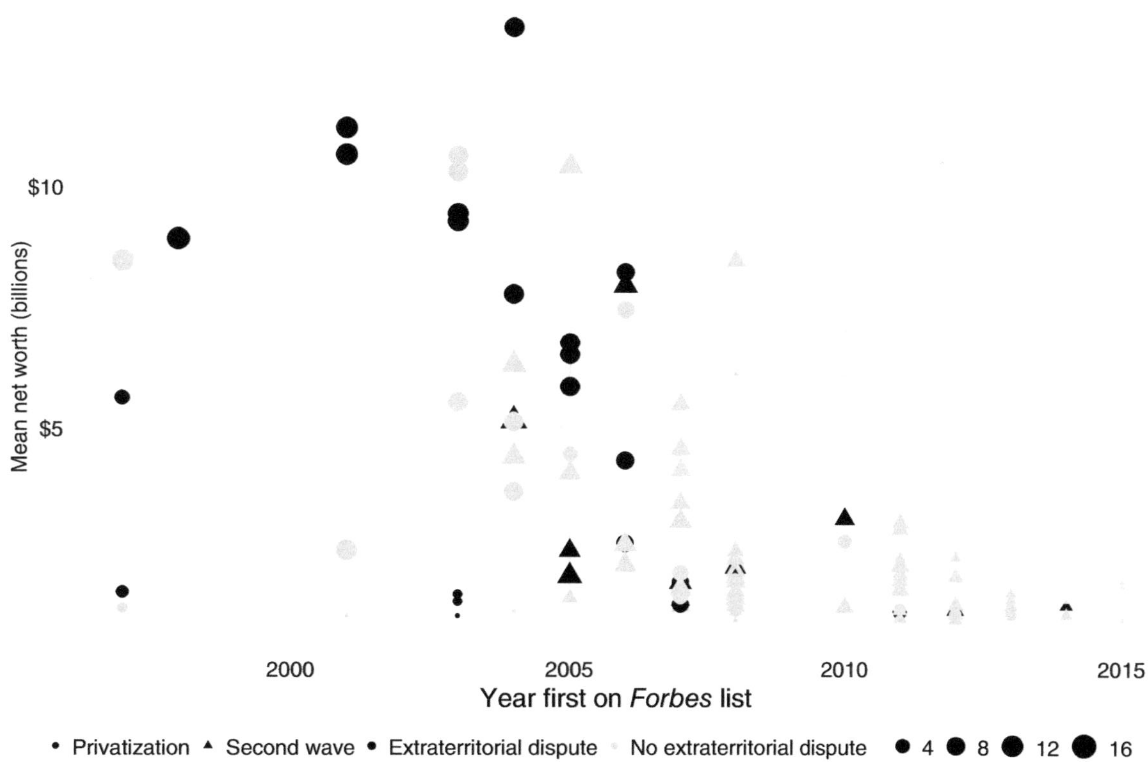

FIGURE 4.2. Year of first appearance on *Forbes* list versus mean net worth

was part of the 1990s privatization and whether the individual was part of a foreign court battle. The size of each data point represents the number of years they appear on the *Forbes* billionaire list. Fifty-nine of the 164, or just under 36 percent of the individuals, gained their wealth from privatization and are thereby classified as the incumbent oligarchs. The independently wealthy dominate the initial years of the Russian entries on the *Forbes* list, while they are also regularly more involved in extraterritorial disputes. Of the forty individuals involved in an extraterritorial case, twenty-seven were part of the privatization plutocrats, which suggests a positive relationship between the variables. By contrast, only thirteen of the 105 second-wave plutocrats were involved in an extraterritorial dispute.

I use a series of ordinary least squares regressions to assess the relationship between extraterritorial dispute resolution and the source of a billionaire's wealth. Standard errors are clustered by the sector of the plutocrat's primary wealth source, as coded by *Forbes*. I start by estimating the relationship between only the Dependent Variable (DV) and the explanatory variable. As the model to the far left of figure 4.3 shows, an individual who gained their wealth from privatization is more likely to be involved in an extraterritorial court battle. The result is significant at the 1 percent level.

Next, I add individual economic factors that are likely to increase the chances the individual was involved in a case that was worth taking abroad. Specifically, I include their mean net worth during the period they appear on the *Forbes* list and the number of years they appear. The more wealth they have and the longer they have that wealth should increase the odds of using foreign courts if only functional factors determine extraterritorial litigation. The main result remains positive and significant at the 1 percent level.

The type of assets that a plutocrat holds could also influence the state's relationship with the individual. In other words, I include additional political variables that might mitigate the relationship between privatization and extraterritorial litigation. First, I create a dummy variable for whether the plutocrat operates in mobile sectors of the economy. It is coded as 1 if they have holdings in technology and telecommunications or in banking and finance. These individuals are likely to have more credible threats of exit and therefore have relatively more power vis-à-vis the state compared to individuals with fixed holdings. I also include whether the individual has diversified sources of wealth, which I code as 1 when *Forbes* lists multiple sectoral sources of wealth for the individual. Prior research indicates that more diversified sources of revenue would also allow the plutocrat to have a more flexible political strategy (Markus and Charnysh 2017).

The relationship between privatization and extraterritorial dispute resolution remains positive and significant at the 1 percent level. These findings are

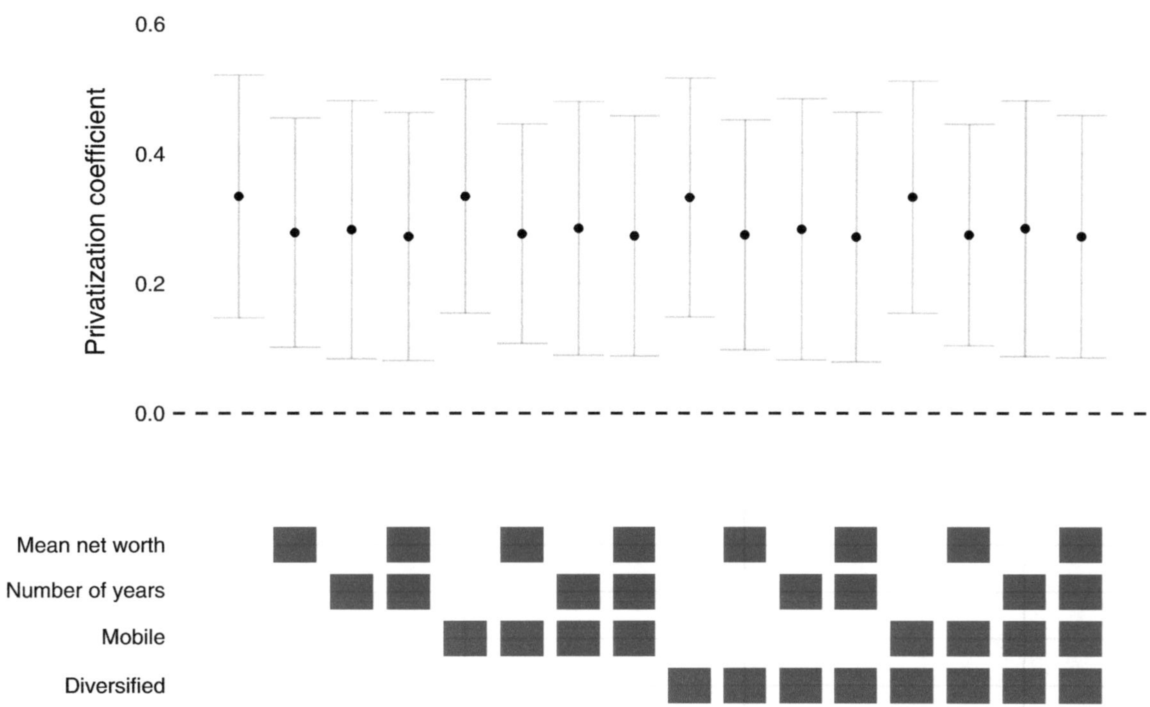

FIGURE 4.3. Privatization coefficient and attributes

robust to the use of an alternative measure of the explanatory and control variables; different sampling strategies, including restricting the sample to only those plutocrats who showed up on the *Forbes* list before 2008; and a narrower dependent variable that codes only for disputes that occurred between 1996 and 2015.

Figure 4.3 plots the coefficient for the privatization variable while accounting for the various controls discussed above. The bottom panel illustrates which controls are included for each estimate. The model with all four controls suggests that plutocrats who gained their wealth through privatization are approximately 28 percent more likely to be involved in an extraterritorial dispute. The only control variable that appears to have a meaningful impact on the privatization coefficient is the mean net worth of the individual. It could be that more wealth is associated with more disputes or that richer plutocrats are conceived of as greater political threats.

One should be cautious with overly generalizing from these findings because there could still be selection effects, for example, those involved in privatization were not random. Nonetheless, the strong, statistically significant correlation provides further evidence consistent with the qualitative examinations of the biggest extraterritorial disputes. As the balance of power in Russia became competitive, the incumbent oligarchs were forced more often to use transnational legal institutions to resolve their nominally domestic commercial conflicts.

State Control over Extraterritorial Dispute Resolution

As the resurgent Russian state took on the incumbent oligarchs, Kremlin-owned enterprises were regularly party to extraterritorial litigations. More often than not, they were used as an offensive tool against a rival to seize foreign assets. Despite the companies clearly being embedded in the transnational network for financial and legal services, SOEs did not litigate against each other in foreign courtrooms. There were no cases in the London courts between Kremlin-owned or Kremlin-aligned companies. This comes as little surprise to the lawyers who work with these companies: "You know in Russia we've got special phrase, we don't like to throw our garbage from the house outside of the house. If you've got two SOE companies most likely they would not go abroad."[33]

When Putin came to power, many of the state-owned companies were invigorated by rising oil and commodity prices. The SOEs, by definition, were subservient to the state. Over time, Putin built up his own band of plutocrats, the kleptocrats, and provided them with lucrative state contracts and plum positions toward the top of valuable government entities. They generally come out of the security services, creating a class that Daniel Treisman (2008) famously labeled the *silovarchs*. Karen Dawisha (2015, 2), author of *Putin's Kleptocracy*, bluntly summarized the situation: "With the return under Putin to state capitalism, the state nationalizes the risk but continues to privatize the rewards to those closest to the president in return for their loyalty."

As state power consolidated, even the independently wealthy took steps to try to align themselves with Putin. Following the first set of battles with the likes of Berezovsky and Khodorkovsky, the balance of power began moving toward state control, making those who wanted to stay on Putin's good side actively avoid steps that might signal disloyalty or undercut Kremlin interests. Thus, these tycoons still need to avoid the transparency costs associated with extraterritorial litigation. A lawyer with more than a decade of experience with Russian clients said that when weighing a case in London, "The first concern is always 'are you going to piss off Putin?'"[34]

Cases like *Berezovsky v. Abramovich* detailed some of the skeletons in the Russian government's closet. By coming to London, "[the plutocrats] would run the risk of embarrassing the country, of embarrassing the Russian Federation, and the President."[35] If they were to attempt such a case, there would almost certainly be political ramifications, potentially turning an economic dispute into a political problem: "It's the publicity angle. If you owe your position to the state, it would be naïve to think that this is the end of it."[36]

While there is still some contestation, the state needs to ensure that no plutocrat takes actions that could symbolize a lack of control. The state needs to build up an image of cohesion and will find ways to sanction those who undercut the effort by using foreign courts. When discussing why SOEs do not come to London for their own disputes, one lawyer brought up the fate of Vladimir Yakunin. "Yakunin's career followed Putin's for over two decades. When Putin began work at the Presidential Property Management Department, Yakunin was named the head of its office in the Northwest Region. When Putin became president Yakunin was appointed deputy minister of transport and given control over the nation's seaports. Later he was made chief of Russian Railways" (Miller 2018, 84). But he was dismissed from his post when news of his son acquiring a British passport surfaced. The British lawyer said, "If you are close to the Kremlin there is a great expectation that you buy into what they do. Using our courts would not fit this."[37] As another practitioner neatly summarized: "If they were both equally close [to Putin] it would not happen here, and if it did, they would cease to be close."[38]

The signals sent by an extraterritorial dispute and the potential costs of transparency then diminish the likelihood of using London to resolve disputes between parties close to or wanting to be close to the state. Instead, actors often use informal institutions and the power of the state to settle commercial conflicts. "The real dispute mechanism is the political connections . . . litigation is the tip of the iceberg . . . it's not really what's happening."[39] In other instances, when the stakes are particularly high, plutocrats and SOEs are forced to go directly to the authorities to have them act as the neutral arbiter. Avoiding the courtroom, be it domestic or foreign, will make sure to keep the dirty laundry hidden: "Those who are on the right side of the government will not come here. . . . They are likely to use the real power brokers."[40] "I don't really see them going to court. . . . I see them going to a boardroom at the Kremlin."[41]

The War for Norilsk Nickel

The decade-long duel between Vladimir Potanin and Oleg Deripaska illustrates how commercial disputes are resolved when the state is generally able to influence rival tycoons. Potanin was one of the original designers of the loans-for-shares program, and his reward was Norilsk Nickel. The company had the largest nickel reserves in the world and controlled one-fifth of global palladium supplies. Once Putin turned the balance of power toward competition, Potanin could have easily become a plutocrat to make an example of, but Potanin appears to have actively eschewed politics as the balance shifted. Putin's namesake was in some ways the ideal tycoon who would not touch the electric fence of electoral politics and could be called on to do the state's bidding. In 2010 Potanin announced that he would be donating half his wealth to charity; when Putin needed funds for his pet Olympics project in Sochi, Potanin put up $262 million for a resort through his holding company Interros (Buckley 2007).

Deripaska is now famous as the oligarch that former Donald Trump campaign manager Paul Manafort once worked for (Reuters 2018). Deripaska cut his teeth in the dangerous aluminum industry of the 1990s and, by the start of the millennium, he was positioned atop the aluminum conglomerate that Berezovsky and Abramovich had helped consolidate. In the coming years, he would become the primary shareholder of Rusal, which forms the backbone of the Russian, and the global, aluminum industry (Katz et al. 2020).

The meteoric rise did not satisfy Deripaska. He built a reputation for being a hard-nosed, ruthless capitalist. As Potanin and former Brooklyn Nets owner Mikhail Prokhorov began divvying up their control of Interros and Norilsk Nickel, Deripaska spotted an opportunity. He wanted to create a Russian, multicommodity conglomerate that would rival the likes of BHP Billiton. Acquiring Norilsk Nickel would be the next piece. It did not matter that there were few synergies between the two companies; Deripaska had decided the company was his target, and after months of heated negotiations, he convinced Prokohorov to sell in the spring of 2008 (Belton 2008a).

Rusal bought up 25 percent of the nickel giant while Potanin maintained just under 30 percent through his holding company Interros. Everyone knew Potanin was not going to let Deripaska take over. Shortly after Prokhorov's sale went through, Potanin and fellow magnate Alisher Uzmanov announced that they had plans to create a three-way merger that would trample Deripaska's ambitions (Robinson 2008). While the deal was stalling, Potanin continued the offensive strategy. At the next board meeting for Norilsk Nickel, he managed to take control of four of the nine board seats while Deripaska was left with three (Belton 2008d). By July, he managed to make himself the company's new chair while pushing out the outspoken CEO Denis Morozov and replacing him with loyalist Sergei Batekhin (Belton 2008d).

Shortly after, however, Vladimir Strzhalkovsky, former KGB officer and rumored Putin confidant, was lined up to become chief executive at the following board meeting (Belton 2008c). Potanin openly noted the move was an effort to speed up the deal involving Uzmanov, but whether this was forced by the government or was a political insurance effort is still unclear. Nonetheless, the fight for control was soon called to a halt because the 2008 financial crisis exhausted the fortunes of both Deripaska and Potanin. Interros and Basic Element, Deripaska's holding company, were heavily indebted and required government assistance via state banks VEB and VTB just to stay afloat (Belton 2008b). Former Kremlin chief of staff Alexander Voloshin was then brought in as chair to keep the peace (Gorst 2008). The Russian state was managing the plutocrats.

But the truce was relatively short lived. In the summer of 2010. Potanin managed to gain a majority of the seats on the Norilsk Nickel board once again. Interros had four while Rusal was left with three, and Vasily Titov, first deputy chief executive of VTB who appeared to have Interros's backing, was suddenly chair, and Voloshin was kicked out (Belton 2010a). Rusal alleged that the company's management, led by Strzhalkovsky, had colluded to ensure the outcome. By the end of July, Deripaska was threatening domestic and extraterritorial legal action (Belton 2010b).

The boardroom tactics appeared to have cost Potanin a $1 billion loan from a consortium of Western financial groups that did not want to get involved in another oligarch battle that looked to be to the death (Belton 2010c). Interros was on the verge of a London Court of International Arbitration case (Belton 2010e), and Russian prosecutors appeared to provide some backing for Deripaska's claims of violations of the voting rules (Belton 2010d). But the efforts were in vain: by November, Deripaska's own boardroom coup failed. Potanin then offered to buy out Rusal's stake at a massive premium, in a deal worth $12 billion that was immediately rejected (*Mining Journal* 2010).

Deripaska was not ready to deal and Potanin again went on the offensive: Norilsk Nickel offered a share buyback worth $3 billion for a little over 6 percent of the company (Belton 2010f). The shares would then go to the company's treasury that was aligned with Interros. The deal was coupled with an unexpected 8 percent sale of shares to commodities trader Trafigura, which also appeared to have come into Potanin's fold. Few details of the deal were released, and it was generally assumed that Trafigura were given a nice rate, well below market, to safeguard Potanin's control (Belton 2010f). Deripaska managed to get a temporary injunction on the transactions through the courts in Nevis, but it was eventually rejected (Belton 2011). He then set out to take on Interros through courts in the United States and the United Kingdom to expose the real details around who was pulling the company's strings. The litigations promised to reveal some of the most controversial parts of the company's practices.

With more legal challenges pending, Norilsk Nickel's management tried once again to resolve the dispute through a buyout, but Deripaska was not giving up on his multi-metal dream (Weaver 2011). So they went for the jugular. The company offered another $4.5 billion share buyback at a 6 percent premium on the market rate. The orders came flying in and, within a few months, management's shares of the company went up 17 percent (Weaver 2011). In the interim, however, Trafigura appeared to have purchased 8 percent of the shares, but it was eventually written down to less than 1 percent. No one came forward claiming the remaining 7 percent that would in theory be worth billions (Blas and Belton 2011). Nonetheless, management and Interros had already secured close to a governing majority, which would allow them to shut down Deripaska's goals. For years, he had been trying to increase the company's dividends to help him actually pay down the debt that Rusal took on when buying up its Norilsk Nickel shares.

The Settlement

Toward the end of 2012, four years after the war for nickel control began, there was still no light at the end of the tunnel. The Kremlin was growing frustrated with the public nature of the plutocrats' accusations. Another potential courtroom fiasco was set for December and was running parallel to a number of other extraterritorial plutocrat battles. But the financial and legal community was particularly giddy about this fight given the power and value of Norilsk Nickel. They should not have gotten their hopes up.

A handful of days before the first set of hearings were scheduled, the warring tycoons announced a settlement. Rather than either Deripaska or Potanin buying the other out, as both had repeatedly tried to do in the past, a third, familiar plutocrat was brought into the fold. Abramovich bought up just under 6 percent of the shares for $1.5 billion but was given 20 percent of the voting rights (Weaver 2012). In other words, he was put in a position to act as de facto mediator between the two rivals. The deal further included a fixed dividend policy for the coming years, surely to Deripaska's delight. The KGB executive Strzhalkovsky would also be gone, but Potanin would be taking over the role for himself. The board would become an even split of four pieces each for Interros and Rusal, while Abramovich and the independent directors would fill out the rest. The penalties for backtracking on the deal were staggering and could have reached up to $560 million for each of the three billionaires (Wagstyl 2012).[42]

Although some of the details are surprising, the settlement was what close observers were already expecting because these were plutocrats seeking to keep the Kremlin happy. Putin had intimated that he wanted this resolved quickly and quietly, so the plutocrats knew they needed to come to the table. As one London lawyer said, "The Kremlin does step in, it did step in, it's obvious, and it's public."[43] The general justification aligns with the potential costs of publicity as "the Kremlin doesn't want its dirty laundry, like who killed who and who bribed who, aired abroad."[44] A lawyer for numerous other plutocrats said that this was par for the course but that the success of these brokered deals inevitably relies on the degree of state control: "You always have attempts [by the state to broker] but it depends on how decisive. Depending on the level of interference, the higher up you get the more they settle, and the level of vulnerability. If all your assets are in Russia . . . or you've put all your eggs in one basket, you will listen."[45]

Potanin and Deripaska were both actors whose revenue was heavily focused on and tied up in Russia, and who had gone out of their way to ensure good relations with the government. They did not want to undermine the Kremlin. As another lawyer said, "If anyone has influence on all these players it's the state. No matter what anyone says, they all want to make peace with the state."[46] Although one individual I spoke to was absolutely adamant that there was no Kremlin involvement, and Potanin has also gone out of his way to say the same (Weaver and Clover 2013), there was a broad consensus on how the settlement was brokered. The state had to be involved: "There is no other logical explanation. And it made sense from a political perspective—you bring in this neutral trusted third party . . . this is a company important to Russia, it's at a time during the big oligarch cases. This was the beginning of the clampdown by the Kremlin [to stop cases going abroad]."[47] The fact that the state (implicitly) stepped in to prevent an extraterritorial case aligns with the model of a country moving toward state control: "It fits with the more general pattern of Putin as arbitrator, if necessary, be it between politicians or oligarchs."[48]

Postscript

Come 2018, a little over six years after the brokered deal, London finally witnessed a legal conflict between the plutocrats, but it was more of a footnote than a real battle in the war over Nornickel. By December 2018, the shareholder lockup had expired and Potanin soon offered to buy Abramovich's shares outright. In the preceding few months, Deripaska's political capital had taken a toll. He was embroiled in the Mueller investigation through his connections to Paul Manafort, and a female escort who accompanied Deripaska on a yachting vacation claimed to have tapes of him discussing Russia's role in hacking the 2016 American election (AP News 2018). It appears Potanin smelled blood and was ready to pounce. Abramovich, always the willing businessman, appeared ready to cash out. News of an impending legal battle knocked $2.5 billion off Norilsk Nickel's market cap (Foy 2018b).

Per the terms of the 2012 deal, a roulette-style shootout would need to decide who would gain complete control (Foy 2018c). It would be all or nothing. But the riskiness of an auction, coupled with new sanctions on Deripaska, halted the procedure in April 2018 (Foy 2018a). After Deripaska's initial injunction failed, there was a trial to determine the specific legality of a sale. It was narrow and did little to provide real clarity on the company or expose either party to true transparency costs. A British judge ruled in June that the sale could not actually go through, and Potanin was forced to return the 2 percent he had already picked up from Abramovich prior to any legal action (Foy and Croft 2018). Potanin appeared to be readying an appeal (Lyrchikova and Ivanova 2018). When confronted with a dispute in a now state-dominated Russian society, the choice for tycoons is clear. When weighing up a fight abroad, "The question is do I take the risk and upset Mr. P or do I forego the right?"[49]

When Putin was elected president, the plutocratic class was expecting business as usual. But after a wave of high oil prices and the popularity engineered by the plutocrats, the balance of power shifted from one of capture to competition. The commitment problem created by the relatively equal balance led to clashes between the state and select powerful oligarchs of the Yeltsin era. Those like Berezovsky who were forced to flee were able to use foreign courts to try to regain their economic wealth. Yet they had also run away with vast amounts of their profits stashed in foreign real estate and offshore bank accounts. This chapter has illustrated what the state can do in a world of mobile capital where its political opponents can continue to pose a transnational threat to its authority.

The state-owned airline Aeroflot was the proximate driver behind Berezovsky's exit from Russia. There is little double that Berezovsky had fleeced its accounts through the 1990s (Klebnikov 2000). The legality of his actions is definitely debatable, but he was also not the only plutocrat who used SOEs to bolster his bottom line when the state was captured. Berezovsky continued to cast a shadow over Russian politics through the 2000s, despite being stationed abroad. The Aeroflot charges were then brought to London, drowning Berezovsky in another complicated legal challenge.

Although Khodorkovsky was locked away after he and the state clashed, several of his closest business partners continued to pose problems for the Kremlin from abroad. They made their aims of not just regaining their wealth but also challenging Putin clear. Even auctioning off Yukos's most valuable assets only dented the outstanding payments owed to the Russian state following the tax charges the Russian state filed against the company. Rosneft, the state-owned oil company that was most empowered by the expropriation, was then able to bring cases against Yukos's subsidiaries to get its hands on the wealth that Khodorkovsky and his colleagues had moved abroad.

Both state-directed extraterritorial battles with Berezovsky and Khodorkovsky took more than a decade to resolve. Millions were spent on legal fees on both sides, but the process becomes logical when considering these cases as going beyond commercial gains. The choice to take cases abroad and not to settle would have ripple effects on the balance of power.

As SOEs became weaponized, they did not actually begin using foreign courts for their own disputes with Kremlin-aligned actors. With Putin consolidating his power, the balance began shifting toward state control, ensuring that SOEs and tycoons that want to stay on the inside refrain from publicly using foreign courts. The Kremlin's power resources make it the only authority able to resolve disputes effectively. Tycoons and kleptocrats know that they can expect to pay a political price for undermining the state if they publicly purchase liberalism abroad.

5

WEAPONIZED LEGAL DEPENDENCE

Extraterritorial dispute resolution is a result of financial globalization. It is substantially easier to initiate a claim against a fleeing plutocrat if they have stashed their money abroad and particularly if they find themselves spending time in havens like New York and London, where they can safely spend and display their wealth. Extraterritorial arbitrations are usually started by a plutocrat claiming to be a foreign investor in their own country via their offshore protection strategies. The hidden wealth and how it allows plutocrats to continue shaping politics at home compel the state to seek the legitimacy of foreign legal institutions to plunder both the reputations and coffers of their rivals. In other words, the phenomenon this book focuses on is another in a long list of pathologies plaguing the liberal international economic order.

Optimizing comparative advantage was the much belabored and celebrated goal of international economic integration. Decentralized systems would develop through that process, one whose economic gains and socializing effects would deter autocratic movements and, just as critically, mitigate the likelihood of great power conflict. The costs would just be too high. As Farrell and Newman (2019) identified in their classic article, those dreams were shattered by the most vocal progenitor of the liberal order, the United States. Instead of breaking down past concentrations of power, globalization saw them shift toward a centralized nodal form. The hegemon can claim jurisdiction over almost any international actor through the control of the dollar and has repeatedly used the choke points of the global financial system to wreak havoc on the economies of its adversaries. Today's world is one of weaponized interdependence that looks like it will only heat up as geopolitical tensions mount: through US-China competition, Russian belligerence, and continued efforts by numerous states to stay part of the so-called nonaligned.

While many expected globalization to diminish state power, the net result was a shift in the tool kit. That is most pronounced with US control of the global financial networks that Farrell and Newman theorize and that Haberly and Wójcek (2022) expertly document. The scope for such statecraft is larger than most analysts realize because global financial networks are, in the latter's terms, "sticky." They argue that analysts perpetually go through waves of expecting network power to diffuse only for it to grow more centralized because money rests on social trust. It gains its power from legitimacy, and legitimacy is finite and fickle. In their words: "Money itself—as nothing more than a representation of future expectations—is so inherently abstract and insubstantial. This creates a powerful countervailing imperative for the production of money to be rooted in relationships of trust that need to be earned over very long periods of time, and are moreover deeply co-constituted with the institutional 'legacy standards' through which they originally developed" (Haberly and Wójcik 2022, 17).

In other words, when financial innovation occurs, the creators need to borrow the credibility of existing influential actors. New products are co-opted by the banks and financial institutions with preexisting status in places like New York and London before eventually ensconcing the novelty through the laws of these jurisdictions. They become the legal backbone of economic exchange and the global financial networks that many thought would create benign prosperity before they were co-opted into the geopolitical tool kit.

The preceding chapters of the book illustrate that similar forces are at play with extraterritorial dispute resolution—but with three major distinctions. First, rather than seeing the financial system exploited, it is the courts that make that system function who turn into the locus of politics. Second, while foreign actors are usually the ones experiencing the coercion, here intra-elite (rather than interstate) conflicts are the logic driving weaponization. Third, this is not done by the United States or the European Union. Instead, their institutions become the venues through which statecraft is pursued by actors that are normally on the receiving end of "liberal" coercion. While these actions are done voluntarily by the initiators, they illustrate the dependence of renegade actors on incumbent financial and legal nodes. Given the dominant forms of global wealth protection and the rules and norms around how to seize that wealth, plutocrats instrumentalize the legitimacy of external institutions based in countries renowned for a strong rule of law to wage their political battles.

In the remainder of this chapter, I illustrate the value of the balance-of-power argument that I develop for understanding the outsourcing of disputes in other cases around the world. The analysis also sheds further light on state-plutocrat relations, elucidating how conflicts are likely to unfold in a financially globalized world still rooted around two financial and legal nodes. While this chapter does not offer the depth presented in chapters 2 through 4, the cases were explicitly chosen to illustrate the benefits and limits of my argument. I first analyze intra-Kazakh conflicts that have taken place in London. Kazakhstan is an important case to understand in its own right because the country's plutocrats are responsible for the second largest number of extraterritorial disputes. And the evolution of the balance of power inverts the Russian case. While the two countries have a host of shared historical and cultural experiences, Kazakhstan was fundamentally a world of state control in the 1990s before moving toward more competitive dynamics through the 2000s. My next two cases move outside the former Soviet Union context to ensure that these fights are not simply an outgrowth of the contested property rights regimes that stemmed from the collapse of the Soviet empire. I focus on the evolution of elite politics in Turkey and Angola. Both provide analytic leverage given that they have not only had major extraterritorial battles but also because they reached contested forms of the balance of power from opposite ends of the spectrum. Turkey moved from capture-style dynamics with the election of Recep Tayyip Erdoğan and the Justice and Development Party (*Adalet ve Kalkınma Partisi*, AKP) in the early 2000s, while Angola moved from state control once the country's decades-long dictator, José Eduardo Van-Dúnem dos Santos, lost legitimacy following the 2014 oil price collapse. These cases were thus selected for their ability to shed light on different dimensions of the broader argument of the book.

The Ever-Expanding Arm of the Kazakh State

When Nursultan Nazarbayev officially gave up the reins of power in Kazakhstan in 2019, he achieved a goal few autocrats live to boast about—survival. Ruler of the country for nearly three decades, initially appointed by Mikhail Gorbachev as the USSR was crumbling, the former steelworker turned politician appeared to have an iron grip on the country even amid years of speculation over his succession. Nazarbayev survived the death and resurgence of the country's main benefactor, the Russian state; a volatile political space; multiple coup attempts, including from his son-in-law; and even managed to convert the world's greatest financial crisis into a political opportunity (Burgis 2020). While the state was firmly in control through the 1990s, the economic growth and in particular the boom in the financial sector coinciding with the commodity super-cycle shifted power dynamics toward competition. The changing balance of power triggered a series of extraterritorial legal battles involving the individuals that sought to challenge Nazarbayev's grip over the resource-rich country.

Given its substantial reliance on the USSR, Kazakhstan was the last of the former members of the republic to declare independence. Nazarbayev fully recognized the need to liberalize the economy, opting for a Boris Yeltsin type of shock therapy treatment. It went through three initial phases: first hitting the smaller sectors like housing, then a voucher style mass privatization, before select companies were eventually auctioned off (Cummings 2005, 29). On the surface, this all looked like a recipe to place the state in a subservient position to the band of plutocrats that would form from the kick-starting of a market economy, but Nazarbayev took great pains to ensure that the state would remain in charge. First, a host of international companies were brought into the mix to leverage the abundant and diversified natural resources that the country is blessed (some would say cursed) by; the likes of Shell and Chevron had become major players in the economy by the early part of the 1990s (Peyrouse 2012, 349–50). Unlike Yeltsin's experience, the actual transfer of wealth was not done to a band of people who were already multi-millionaires. Instead, Nazarbayev was able to channel substantial wealth to his own political backers and family members (Peyrouse 2012, 358): more patronage than political survival.

All this was only possible because Kazakhstan's political structure remained largely intact after the fall of the USSR. Democracy was nominally brought into place, but Nazarbayev successfully fought to keep a strong presidency and security services integrated (Cummings 2005). As the economic volatility of the early years continued, bringing the specter of political challenge, the state continued removing any liberal shackles. The constitutional court used a quirk in the electoral map to rule the entire 1994 parliamentary election illegal, fortifying Nazarbayev's position, while a new constitution was adopted to further entrench the state's position (Cummings 2005, 26). Greater investigative authority was moved into the offices controlled by the president, and the first signs of democratic political debate were squashed when emerging media outlets like *Karavan* were bought up by individuals subservient to the state (Cummings 2005, 28). When Akezhan Kazhegeldin, prime minister from 1994 to 1997, looked like he might oppose the regime, he was quickly charged with money laundering and soon fled the country in political exile (Cummings 2005, 28).

From Control to Contestation

The 1990s were then largely a success for the Kazakhstan state. Nazarbayev consolidated political power, built up a steady band of kleptocrats loyal to the state largely coming from the president's family and inner circle, and the country became one of the fastest growing economies in the region. But that success also bred the grounds for threat. The economic gains meant that those closest to the state reap rewards, but so did all individuals who had been able to get their hands on a slice of the growing pie. A new tycoon class emerged that, by 2000, looked to be uniting as their interests increasingly diverged from the state. As Junisbai and Junisbai (2005, 386) put it:

> President Nazarbaev's greatest achievement—Kazakhstan's ambitious economic liberalization carried out since independence—appears to have had an unintended side effect. Under conditions of rapid economic transformation, the tried and true Soviet solution to maintaining the political status quo—the monolithic national elite—began to shatter as the elite actors' economic interests diversified and ultimately came into conflict.

The establishment of the Democratic Choice of Kazakhstan (DCK) was a shock to the heretofore state-controlled system. Unlike prior minor political challenges, the union was coming from a range of powerful actors that would, on the surface, fall into standard inner-circle elites representing a range of business and technocratic backgrounds coming together. Their goals were ambitious and nominally fit with much of modernization theory (Junisbai and Junisbai 2005). Fear of a growing state was leading the tycoons to find new means of securing their wealth, and democracy looked like it was their best bet. Its founders further admitted that they feared how the deinstitutionalization was allowing for their assets to be taken away, while many were implicitly barred from the latest wave of privatizations of the oil and gas sector that could have fortified both their income and wealth. Unlike earlier opposition figures, who had come from the ranks of the intelligentsia, key DCK founders had independent resources with which to finance their political activities (Junisbai and Junisbai 2005, 380).

This wake-up call did not go unanswered. The DCK founding members were swiftly attacked. Those who were part of the government apparatus were rapidly replaced. Meanwhile, the two biggest proponents of the party, Galymzhan Zhakiyanov and Mukhtar Ablyazov, found themselves charged with tax evasion (Junisbai and Junisbai 2005, 380). The DCK was soon deemed an illegal organization. But with its uprising, "the authorities became aware that the oligarchs possess enough revenues to guarantee them political autonomy, which, a decade after independence, could be used against the president as the strong man of the country and against his family as an economic actor" (Peyrouse 2012, 353). It set the stage for multibillion-dollar political clashes that would be forced to unfold in foreign courts.

The Rise of Mr. Ablyazov

Ablyazov is the man most associated with the DCK. He got his financial start in the late 1980s like many other eventual plutocrats: by wheeling and dealing whatever electronics he could get his hands on. He started a private equity shop in the early 1990s that became the basis for his acquisition of Bank TuranAlem (BTA) during the voucher phase of privatization in 1998 (Burgis 2020, 66–67). He was recruited into the administration as an energy minister, but that was not enough cover in a political terrain that was quickly becoming contested when Kazakhstan's economy lifted off. He knew his independent revenue sources would become a threat, so DCK became a failed vehicle toward protection. He knew because apparently Nazarbayev told him as much while DCK was getting its start. Journalist Tom Burgis (2020, 67), author of *Kleptopia*, writes of this relationship:

> He alone stood up to the president, denouncing the promotion of Nazarbayev's relatives and the creeping establishment of what he called a "clan-ocracy." When he refused to continue in the cabinet, Nazarbayev demanded to see him. "You don't respect your president." He ranted, "You don't respect me as a person, you're not loyal to me." He calmed down and asked again: "Come back and work for me." Again Ablyazov refused. "Well, in that case, you're going to have to give me a chunk." They would be partners. The president would take half of BTA. . . . It would be a guarantee against disloyalty. Ablyazov stood firm: no. He knew change was coming. He wanted it, and found that others did too.

Ablyazov was thrown behind bars and Nazarbayev thought he was able to loot his assets completely. But Ablyazov had become as familiar as any would-be oligarch with the intricacies and machinations of offshore finance and wealth protection. He transferred most of his shares to his business partner, Yerzhan Tatishev, through a maze of offshore entities and allowed his partner's wife to hold many of the shares (Lillis 2022, 44). Tatishev would die in a freak hunting accident in 2004, by which time Ablyazov had talked himself out of Kazakh prison and nominally out of its political scene. Illustrating that the plutocrats still maintained some power even amid the DCK crackdown, they were able to lobby for Ablyazov's early release as long as he promised to leave the political game (Lillis 2022, 43).

Ablyazov appeared to retreat to Moscow, where he took on the chairmanship of BTA. The bank's remarkable growth in the coming years reasserted his instrumental power. It became one of the biggest players in the region's finance. As Cooley and Heathershaw (2017, 57) summarize, "Between 2003 and 2007, the bank's outstanding loans grew by 1,100 per cent as it acquired an international reputation for aggressive investments that yielded eye-popping results. Western institutional investors including Royal Bank of Scotland, Barclays and HSBC poured about $10 billion into the Kazakh bank between 2003 and 2007." The bank borrowed another $15 billion. This looked like a political war chest to Nazarbayev, so when Ablyazov returned to his home country, he claimed there was another shakedown. Nazarbayev allegedly asked for another chunk in BTA, majority status, because as he reportedly told Ablyazov (Burgis 2020, 68): "I am afraid of you and afraid of your billions, that you can turn them against me, they are in the bank that you are controlling and that is why I need to control the bank with you."

The ammunition that Ablyazov was stockpiling was partly driven by the surge in global liquidity of the early and mid-2000s. How that economic upturn ended is now well known, and BTA looked like it was going to be another casualty of the bankruptcies and bailouts of the global financial crisis. In February 2009, Kazakhstan's sovereign wealth fund took over the bank after it looked to be on the verge of default, putting in a $1.4 billion balance sheet injection for an 80 percent stake in the company (Cooley and Heathershaw 2017, 60). A few years earlier, in the midst of the Kazakhgate scandal, the fund had been restructured, which allowed the president to have any final say on investments (Yessenova 2018). The incumbent BTA team was adamant that they were not at risk, and even as property prices were suffering, Ablyazov held firm. Roman Solodchenko, a member of the bank's executive board prior to the takeover, was vehement that they were in the black and that the government's actions had led to the company's downfall. In particular, the government changed the accounting rules for its subsidiaries and triggered international default payments, which diminished market confidence (Hug 2010, 12). Once the Kazakh state upped the bank's capital requirements, its fate was sealed (Hug 2010, 12).

With BTA now in state-hands there was a shift in the global accounting firms at the helm. EY was replaced by PricewaterhouseCoopers (PwC), who found a $10 billion blackhole in BTA's balance sheet (Burgis 2020, 105).

London Courts Will Make You Reveal Your Assets

Based on a transaction with a UK-domiciled shell company worth approximately $285 million, BTA established jurisdiction against Ablyazov and eventually a host of other defendants (Cooley and Heathershaw 2017, 61). The bank argued these individuals had aided Ablyazov's fraud that ranged between $6 billion and $10 billion. The first crucial legal step in any of these fraud-related cases, as discussed in chapter 2, is the acquisition of a worldwide freezing order. As the name suggests, the injunction prevents a defendant from accessing or moving money worth the value of fraud claim. BTA acted swiftly and received several favorable judgments that would legally force Ablyazov to detail all his offshore holdings (Nougayrède 2016, 413–15). Next, the bank sought a receivership order that would move legal authority over any companies, real or shell, associated with Ablyazov over to third-party professionals (Nougayrède 2016, 412–13). It then moved to guaranteeing mandatory disclosure orders over individuals that could have valuable information on Ablyazov's commercial dealings, including legal services providers in a range of offshore jurisdictions and Ablyazov's brother-in-law (Nougayrède 2016, 412–13).

The freezing orders proved to be the beginning of the end for Ablyazov. Over a series of hearings, BTA's representatives in London showed that the defendant repeatedly lied about his offshore wealth structures. They soon revealed literally more than five hundred entities where Ablyazov would be deemed the beneficial owner. In sum, the offshore empire appeared to have more than one thousand intricately interconnected companies, with Ablyazov's name rarely ever mentioned in the legal documentation; family members and beneficial owners reigned supreme (Nougayrède 2016, 411). Key to establishing these facts was a discovery of a storage facility in London that private investigators, hired by BTA, found by tracking a close associate of Ablyazov (Burgis 2017, 104–5). For the failure to disclose his assets and lying under oath in the process, Ablyazov was held in contempt of court and handed a twenty-two-month prison sentence, but he had managed to exit the jurisdiction just before the judgment came down (Cooley and Heathershaw 2017, 61).

When Your Wealth Structures Foil Your Rescue Plan

While Ablyazov was on the backfoot in London, he was not simply relenting. In testimony to the court, he argued that his offshore chicanery was necessary: the gist was that, when you do business in an environment like Kazakhstan, one where state arbitrariness is par for the course, when your wealth creates fear over comfort, these are the steps you must take. He also used one of these entities to initiate an extraterritorial arbitration against the Kazakh state for indirect expropriation, echoing the claims referenced by BTA board member Solodchenko above. The claim was worth $1.5 billion.

The case was initiated under the Kazakhstan-Netherlands bilateral investment treaty (BIT) and should have been relatively straightforward on jurisdictional grounds. The claimant was formally KT Asia BV which had acquired 9.96 percent of shares in BTA from two British Virgin Islands (BVI) companies. Ablyazov was honest that he was behind the claim. He even went on to say that he was the owner of these two BVI companies. The tribunal agreed that he should be deemed a foreign investor as a function of these holdings, but because he had transferred the shares to KT Asia BV well below market price, they did not constitute an actual investment that was necessary for a claim (Nougayrède 2016, 414). Ablyazov's counterargument was that these should be considered internal restructurings, but this contradicted the fact that the corporations were not legally in his name so not formally part of a single group in even a generous legal reading (Weiniger and Kantor 2015, 538). The proceeding did nonetheless go some way to establishing the fact that Ablyazov controlled 75.18 percent of BTA at the time of the nationalization, which was concealed from the Kazakh government by keeping any individual under the 10 percent necessary reporting threshold (Cooley and Heathershaw 2017).

English Courts as the Ultimate Condemnation

There have been more than seventy judgments issued in proceedings related to the Ablyazov affair. Few if any cases have had a greater impact on English law. A host of other related individuals, like Ablyazov's son-in-law Ilyas Khrapunov, have found themselves on the receiving end of a legal onslaught. Khrapunov has been accused of fronting Ablyazov's money in US courts while his father, Viktor Khrapunov, was charged with embezzling $300 million from the city of Almaty (Cooley and Heathershaw 2017, 75). The root of the claims in London is the alleged fraud that Ablyazov perpetrated against the bank that he had bought just over a decade prior to the renationalization. Ablyazov attempted to have the dispute dismissed on the grounds that it was politically motivated. While the courts recognized that the circumstances of the case were truly unlike any that they had previously presided over, the case was seen as legitimate for a host of reasons, including the legal fact that the bank was its own entity separate from Nazarbayev and the state, the fraud had occurred before the change in ownership, and it was legitimate to seek the fraudulent funds when BTA was in such dire financial conditions.[1] Even if getting rid of a opposition figure was collateral damage, it did not constitute an abuse:

> At first sight the elimination of Mr. Ablyazov as a political opponent would appear to be clearly illegitimate because it is far removed from the remedy which the law gives for misappropriation of assets and appears not to be "reasonably related to the provision of some form of redress" for Mr. Ablyazov's alleged wrong. However, the elimination of Mr. Ablyazov as a political opponent is said to be the consequence of undermining and damaging Mr. Ablyazov's reputation and facilitating the expropriation of Mr. Ablyazov's assets worldwide. If the Bank succeeds in its actions, which are essentially for fraud, Mr. Ablyazov's reputation is likely to be undermined and damaged and his assets are likely to be seized in order to execute the judgment. Thus those consequences cannot be an illegitimate purpose of the proceedings. If the likely effect of those consequences is to eliminate Mr. Ablyazov as a political opponent that consequence too cannot be an illegitimate purpose. As Simon Brown LJ said in *Broxton v McClelland* "a plaintiff is entitled to seek the defendant's financial ruin if that will be the consequence of properly prosecuting a legitimate claim." So for this reason also I do not consider that the proceedings in this court are an abuse of process.[2]

Having absconded from the jurisdiction, Ablyazov could barely mount a defense. By 2018, English courts handed down judgments worth nearly $5 billion against Ablyazov (Bland 2018). Even with these losses, Ablyazov was not deterred. He used the proceedings to try to shift the costs of transparency of the liberal process against the Kazakh state. He accused the president's son-in-law Timur Kulibayev of stealing state assets by selling them below market value and puppeteering control of the Kazakh sovereign wealth fund, the value of which had ballooned into the billions (Cooley and Heathershaw 2017, 65). Moving to France to try to escape a new set of government authorities, Ablyazov ended up in a foreign prison for 3.5 years before escaping extradition charges by the skin of his teeth—the French court was widely expected to send him home (Reuters 2016). In the interim, his wife and daughter were part of a dramatic diplomatic affair as Italy inadvertently sent them to their homeland (Gorst 2013). Despite these various machinations, Ablyazov remained active in shaping politics at home, and the threat to the state turned transnational. He admits to funding a host of opposition actors, including most clearly the Alga! Party, which operates as one of the closest descendants of the DCK movement (Lillis 2022).

According to Ablyazov's witness statements, prior to the legal actions in London, Nazarbayev had offered Ablyazov a final deal. If he were to hand over the bank's Russian assets plus an extra $50 million, they would consider easing off (Burgis 2020, 116). That would firmly shift the balance of power back from competition toward control. Ablyazov refused because he continued to see the battle as existential, rooted in needing to change how the structure of politics worked in the country. Instead, according to leaked documents, BTA spent nearly $500 million on lawyers to chase down Ablyazov's wealth and, with that, his instrumental power (Burgis 2020, 256). The bank has been able to seize only a small fraction of money compared to the judgments they won, but Ablyazov's name is now just as synonymous with fraud as it is with any real or feigned democratic aspirations.

Turkey's Coercive Coalition

Recep Tayyip Erdoğan is frequently mentioned in the same breath as Vladimir Vladimirovich Putin. Both ascended to power following major economic volatility that they tamed and turned into relatively prosperous growth. Both were viewed with a mix of concern and intrigue by foreign governments and investors as they battled against inflation. They steadily strengthened the authoritarian nature of their regimes and eventually spurned those foreign actors to lead the charge against the liberal order. While these comparisons are frequent in public and academic discourse, the legal tools of the international economic architecture also became sites of conflict as Erdoğan and his Justice and Development Party (AKP) clashed with the incumbent oligarchy. This section of the chapter details the shifting dynamics in Turkey from one of capture in the 1990s to a competitive balance of power between the state and the plutocracy, which then unfolded in a series of extraterritorial arbitrations against the Turkish state. While Erdoğan was able to win an important set of battles that curbed the political excesses of the super-wealthy, unlike in Russia, the country has broadly remained in competitive dynamics over the past two decades, leading these clashes to recur.

Capture with Turkish Characteristics

Turkey began its integration into the global economy in the early 1980s. Spearheaded by Prime Minister Turgut Özal, who is frequently considered a Margaret Thatcher or Ronald Reagan type of figure in the country's history, international barriers to trade and finance quickly eroded. The attempted growth model mirrored some of the rising East Asian Tigers, moving toward export-oriented industrialization after years of import substitution that had characterized the previous two decades and led to the initial buildup of a plutocratic class (Karadag 2010, 13–14). Turkey is, like many emerging markets today, synonymous with large, diversified, family-run companies sitting at the top of the economic hierarchy. Some that continue to dominate today's economy were already well established by the 1980s and directly benefited from many of the export incentives created at that time (Karadag 2010, 22). Even if Özal managed to command the political sphere, his developmental ambitions meant he had to favor the preexisting economic elite because they were the "only national champions available" for his party to support (Karadag 2010, 22).

Part of that political and economic success by the family companies stemmed from the fact that they had been forced to deal with a volatile state-structured environment for several years already. They needed to be able to turn different revenue streams on and off in order to maintain their instrumental power (Colpan 2010, 496). The dexterity of the wealthy families proved to be particularly effective through the 1990s, when the country experienced a series of major economic crises. The banking system in particular became a central vehicle for state and business actors to seek rents and patronage resources (Karadag 2010, 19). The full opening up of the capital account in 1989, arguably Özal's most lasting achievement, set the stage for the battles discussed later in this section: they kick-started a host of boom-bust cycles across growth and inflation driven primarily by global liquidity flowing into the country's banking system for the first time.

Foreign borrowing fueled the ambitions of the existing families while a host of plutocrats who got their start during the opening of the economy in the 1980s began to acquire essential power resources. The private airways, critical in any competitive developing democracy, were up for grabs. The first private TV station was set up by Cem Uzan working alongside Ahmet Özal, son of Turgut Özal (Karadag 2010, 20). Privatization, long promised, also finally began to proceed with many of the super-wealthy ready to snap up underperforming assets. Finding ways to garner political support from the plutocratic class was then becoming essential, especially because the political space embraced a host of new entrants, which often occurs in times of economic upheaval. This included the rise of Islamist political forces for the first time since the early 1980s. The power of the plutocratic class is highlighted by what happened following the election of the Workers Party, an Islamic powerhouse that foreshadowed Erdoğan's eventual rise, in 1997. The media apparatus was quickly turned on the government, particularly those outlets controlled by the Doğans, using Turkey's broadly secular principles as a cudgel. Within six months, the government headed by Necmetten Erbakan was removed by force and barred from running again by the courts (Silverman 2014, 139). As Roy Karadag (2010, 6) summarized in his comprehensive analysis of the evolution of business-government relations in Turkey:

> If we must interpret the post-1980 transformations, we can conclude that those reforms resulted in the emergence of what is here labeled "oligarchic capitalism." The fragmentation of the political arena, the end of corporatist social control, and the establishment of new, closed elite political business cartels that capture the state represent the crucial elements of oligarchic dynamics that have undermined state power and institutional trust.

The AKP's Triumph Shifts the Balance Toward Competition

In 2001, Turkey suffered another financial crisis, its third in under a decade. Devastating in terms of its impact for income distribution, with stories of state capture and corruption dynamics, the political class was in shambles (Öniş 2004, 3). By the 2002 parliamentary election, all the incumbents suffered devastating electoral defeats; not one managed to gain the 10 percent vote threshold that would have promised them a place in parliament (Öniş 2004, 4). Instead, the nominally new Justice and Development Party cofounded by Erdoğan as the de facto successor to the Islam-oriented Workers Party was able to take control, capturing 34.3 percent of the vote and 66 percent of the parliamentary seats (Karadag 2010, 22).

After the economic crisis, Turkey was again forced to turn to the International Monetary Fund (IMF). While much needed economic reforms had stalled under the state capture dynamics of the 1990s, the clear majority that the AKP had at its disposal ensured that they could withstand any pressure to move back to the recalcitrant practices in the banking sector that had particularly benefited the incumbent plutocrats (Karadag 2010, 26). In a rare case in which IMF bailouts actually engineered the promised outcomes, the AKP's legitimacy was bolstered by the early years of its economic growth. In the first quarter of 2004 (Öniş 2004, 12), the country's growth clocked in at 12 percent, foreshadowing Turkey's greatest period of sustained, relatively inflation-free development since the 1950s (Öniş 2012, 139).

The party's control of the state and the accompanying power resources were further strengthened by its broad coalition base. Most important for our purposes and unlike any of the successful governments of the previous two decades, the party had its own economic backers that were independent of the incumbent plutocratic state structure. While the powerful family businesses found ways to benefit from neoliberal reform, the policy changes and loose financing also initiated the growth of the Anatolian Tigers. Smaller in size, religiously inclined, and well organized through the Independent Industrialists and Businessmen Association (*Müstakil Sanayici ve İşadamları Derneği*, MUSIAD), the exporting firms from the interior of the country became part of the capital backbone for Erdoğan and the AKP, thus diminishing their reliance on the incumbent plutocracy (Öniş 2004, 6–7).

The combination of political legitimacy and the party's economic resources shifted the balance of power toward competition with the incumbent plutocracy. But the state was still being watched and under a host of international pressures. Still so reliant on foreign financing and with clear hopes of joining the European Union, the state was not able to bring the resources under its own control. Instead, it resorted to creating a class of aligned tycoons. A new wave of privatizations was ushered in, with the winners largely coming from the Anatolian block and the incumbent plutocrats mostly shut out (Esen and Gumuscu 2018, 351). Meanwhile, media suppression increased, the independence of the courts was diminished, and the regulatory institutions brought under party control (Esen and Gumuscu 2018). These changes triggered an increasingly prodemocratic response from the Turkish Industry and Business Association (*Türk Sanayicileri ve İş İnsanları Derneği*, TUSIAD), the group representing the largest six hundred firms in Turkey and who feared both loss of international financing and an increasingly powerful and arbitrary state that could cut off their revenue streams (Ozel 2013).

TUSIAD was right, and the Saving Deposits Insurance Fund (*Tasarruf Mevduatı Sigorta Fonu*, TMSF) quickly became central to that power consolidation process. Several diversified family groups were in trouble after the 2001 financial crisis (particularly with many of these groups having the core of their empires in the banking sector), which triggered a series of investigations that led to major resource transfers from former oligarchs to plutocrats subservient to the state. More than sixty companies were taken over from Dinç Bilgin's Medya Holding, another nine from Mehmet Emin Karamehmet's Çukurova Holding, and nearly forty from Askoy Holdings (Esen and Gumuscu 2018, 359). The bulk of these companies were then sold below market value to companies with close ties to the AKP (Esen and Gumuscu 2018). But these moves paled in comparison to the conflict between the Turkish state and the Uzan family.

The Rise and Takedown of the Uzans

By the 2001 crisis, the Uzans had come to epitomize the successful diversified plutocrats of Turkey, at least on the surface. With their wealth estimated at

$1.6 billion, they had competitive firms across a host of industries, from construction to banking, to sports (*The Economist* 2001). They were some of the biggest winners of the initial 1990s privatization wave, walking away with two substantial electricity companies, Cukurova Elektrik A.S. (ÇEAŞ) and Kepez Elektrik Turk Anonim Sirketi (Kepez) (*The Economist* 2001). Prior to that, they bought up Imar Bank, which turned into one of fastest growing firms in the economy after the country opened up its capital account. Undergirding their success, both economically and politically, was their vast media holdings, which included several TV stations, with Star serving as the company's crown jewel. Like many other emerging market plutocrats, they were no stranger to controversy. The government's Capital Markets Board had called for an investigation of ÇEAŞ in 1995 (Reuters 1995), but as is common in situations of capture, the allegations eventually dissipated. At the forefront of the company was Cem Uzan, son and heir apparent, whose playboy lifestyle included "a Boeing 747 aircraft, a long-range corporate-style jet, two helicopters and three yachts—one of which he used to entertain the Duchess of York in 1998" (Doran 2001).

The family's troubles began once the wreckage of the crisis was sifted through. The first big blow was dealt not by the Turkish state but instead by telecommunications multinationals Nokia and Motorola. They jointly accused the Uzans of reneging on a loan for equipment bought by Telsim, one of Uzan's companies and the second biggest telecommunications operator in the country. US District Court Judge Jed Rakoff found them guilty, with damages worth nearly $4 billion; he identified them as having "sought to advance and conceal their scheme through an almost endless series of lies, threats and chicanery" (McClam 2003). Cem Uzan's lifestyle was already impaired: he had been found in contempt of UK courts because he failed to show up when the companies attempted to freeze his and the family's assets (Reuters 2002), forcing him to avoid the jurisdiction where they had a host of properties or risk prison time. However, that did not stop Cem from mounting a political challenge in the 2002 parliamentary elections. Buoyed by the family's media properties, he led the Youth Party. Although it didn't form part of the coalition, the party saw its popularity steadily increase throughout 2002 and 2003 (Açikel 2003, 197).

While a $4 billion bill is substantial for any plutocrat, it only served as a harbinger for far more debilitating economic problems. ÇEAŞ and Kepez had already been found guilty of violating contracts with the Turkish state and were seized by government authorities (O'Brien 2001). Imar Bank, the family's main financial institution, was caught up in a fraud that the state argued amounted to nearly $6 billion and was soon placed under receivership by TMSF (Karadag 2010, 26). Those claims soon became the basis for some 219 companies affiliated with the Uzans to be taken over by the state (Esen and Gumuscu

2018, 219). While the family tried to fight back through their media holdings, they found themselves on the receiving end of a libel lawsuit spearheaded by Prime Minister Erdoğan (Associated Press Newswire 2003). Although many foreigners applauded these efforts by the state as important corruption crackdowns necessary to modernize the Turkish economy, others saw parallels to the Russian structural predicament. As one investor noted (Karadag 2010, 26):

> Well, you have seen what they have done to the Uzan family. Of course, he was implicated in fraudulent activities, but he was only targeted because he was head of the Youth Party, which is the same thing that happened to Khodorkovsky in Russia. There are a lot of people involved in the same things, but you are charged only when you're actively engaged in politics.

Refracting Extraterritorial Conflict

With the balance of power in state competition, the Uzans were caught in a conflict initiated by the Turkish state as it sought to reassert its authority over the plutocracy. They were forced to flee their home country. Like the plutocrats referenced in the previous section, they used their offshore positions to initiate claims against what they viewed as a political persecution. Fugitives in their home system, where the courts were increasingly dominated by state prerogatives, they tried every trick in the extraterritorial tool kit, attempting a range of different methods to gain jurisdiction. Given that their claims were directly against the state, investor-state dispute settlement (ISDS), rather than national courts, was their potential mechanism for reprieve. While they clearly structured their offshore wealth to minimize their tax burden and seek extraterritorial protections, they perpetually failed in their transnational legal endeavors.

In *Cementownia v. Turkey*, the reneging of the contracts between ÇEAŞ and the Turkish state was used as the basis to initiate a claim worth $4.6 billion. Cementownia, a Polish company, and the individuals at the front of the claim had received shares from the Uzans. In *Libananco v. Turkey*, a Cypriot shell company was the next option for the fleeing family, but this time they were after a whopping $10 billion. The Uzans explicitly stated that the purpose of their offshoring was to have protection and exit options given Cem's entry into the political sphere.[3] Over the course of the arbitrations, the claimants argued that they and their counsels were subject to surveillance and intimidation by the state, which led to their primary Turkish legal expert dropping out. Turkey retorted that there was some surveillance[4]—the government confirmed they had issued a domestic order to keep track of what were now de facto fugitives—but that this had nothing to do with arbitration because the order was issued in relation to the Imar Bank fraud.[5]

The Turkish state cited several legal principles for why the case needed to be thrown out, but arguably the most damning was the fact that the claimants in several cases against Turkey, all purportedly initiated by the Uzans, were launching their cases based on the ownership of the same set of shares. They could not all be owners because the sum of the shares equaled more than 100 percent.[6] The arbitrators noted that, during the hearings, the claimants presented four different versions for how Libananco came to acquire the shares.[7]

Ownership over Telsim, which had doomed the Uzans in their fight against Motorola and Nokia, was also used as the basis to start a $19 billion exterritorial arbitration. Turkey was quick to point to the New York judgments that condemned the Uzans during the proceedings.[8] The additional arbitration was brought by Jordanian-Dutch dual national, Sabah Fakes, who argued that the Turkish state owed him that sum given their alleged expropriation and resale of Telsim shares to Vodafone, the latter of which netted the government $4.5 billion. According to the claimant, he had come to own these shares legally after a middleman had put him in touch with Hakan Uzan, scion of the family: he had transferred legal ownership of 66.96 of Telsim to Fakes, but the beneficial owner was still Hakan Uzan as a result of an oral agreement.[9] The purpose of the transfer was to attract new buyers for the shares using the neutrality of Fakes as "bait."[10] The transfer occurred for the price of $3,800, with the expectation of a follow-up payment of $17.6 million that never materialized.[11]

As these legal machinations unfolded, the Uzans were on the run and eventually found themselves in France under political asylum (Gross et al. 2021). While they initiated a host of disputes through the shell companies and proxies referenced above, the Uzans next used their French residence as the basis to file a claim where they were finally the de jure claimants. Initiated under the multilateral Energy Charter Treaty (ECT), à la Khodorkovsky, the claim was for $2.5 billion and again referred to the electricity companies and the reneged contracts. At the center of the dispute was the claim that Cem Uzan was, at the time the investments were made, a UK permanent resident, qualifying him as a foreign investor. But the costs of transparency hurt Cem when the Turkish state responded by noting that he had argued he was not bound by British jurisdiction in that same time period while he was fighting the fraud claims from Motorola and Nokia.[12]

The Uzans lost on jurisdictional grounds in each of these disputes, but not before they spent tens of millions of dollars on starting them, leaving them on the hook for even more because, in each instance, they were told to pay the Turkish state's legal costs. These cases illustrate the lengths and depths to which plutocrats can try to use extraterritorial arbitrations to defend themselves, but they need some degree of effective planning. The unexpected turn in the balance of power from capture to competition, brought about by the 2001 financial crisis, appeared to catch the Uzans flat-footed.

The Turkish state managed to strip them of their revenue streams through domestic expropriations, and it had little need to go after their offshore wealth, not when the Uzans spent so much on their legal challenges and had a $4 billion judgment hanging over them from the Telsim-related fraud charges. The following years saw Erdoğan and the AKP continue to strengthen the state's control, but this was a perpetually uneasy balancing act: the privatizations and reforms needed to stabilize the economy left many of the plutocrats intact even if now they feared and lost battles against the state. The latter unfolded when another major media mogul Aydin Dogan faced tax charges and capitulated in 2009 (Silverman 2014, 151). But the growth of the plutocrats continued with many of the revenue streams still private, and volatility in the economy and thereby state legitimacy returned after the 2008 financial crisis. When a rift in the AKP coalition emerged with the Gulenist movement mounting its own challenge, another round of extraterritorial conflict began in the late 2010s. This time they involved businesspeople accused of aiding in the coup, especially Atin Ipek, chair of the Koza Ipek group (Plowman 2023). The Uzans returned to centerstage with Cem attempting a failed presidential bid in 2023. That only occurred after the family launched a $68 billion claim against Vodafone, Motorola, and Blackrock for benefiting from what they claim was fraudulent sales related to the assets they once controlled but were then auctioned off by the state (Gross et al. 2021).

Angolan Overreach

More than one million killed and more than a decade of gross domestic product (GDP) growth above 10 percent: these are the two statistics that define José Eduardo dos Santos's (JES) tenure at the helm of the Angolan state (Musacchio et al. 2010). The collapse of oil prices in 2014 triggered the unraveling of his system of control that, at its peak, mirrored the power of 1990s Nazarbayev. As in the Kazakh case, macroeconomic forces shifted the balance of power to state competition, instigating a series of extraterritorial fraud claims against the incumbent plutocrats with their own power resources.

First elected leader of the People's Movement for Liberation of Angola (*Movimento Popular de Libertação de Angola*, MPLA) in 1979, JES effectively navigated great power struggles in the country throughout the Cold War, despite the

Angolan civil war basically splitting the country into two different governance regimes. Dos Santos and the MPLA were quick to abandon their communist roots once the currents of the liberal order deemed it necessary in the 1990s, but the MPLA's onslaught against the National Union for Total Independence of Angola (*União Nacional para a Independência Total de Angola*, UNITA) Party continued. It only abated in 2002 when UNITA's leader Jonas Savimbi was murdered at the hands of the Angolan military (Musacchio et al. 2010, 5).

This was not like the numerous brokered deals to end civil war that defined international organizational efforts throughout the period; they had tried and failed after the Cold War ended, with MPLA's electoral victories perpetually disputed (Musacchio et al. 2010, 5). This was all-out victory, leaving JES and the MPLA in full command. The MPLA had roughly fifty times more oil revenue than UNITA when the war ended (de Oliveira 2015, 51). Nearly all that power advantage could be attributed to state oil company Sonangol, which was the backbone of both the war economy and the eventual growth miracle of the 2000s.

While Angola had once managed to develop a relatively diversified economy, the war economy was oil. The country regularly ranked behind only Saddam Hussein's Iraq as the most concentrated of economies. Because most of its deposits were physically located offshore, Sonangol thrived through the 1970s and 1980s despite the violence, lucratively aided by major American oil companies like Chevron, even as the United States formally supported UNITA (Friedman and Reinert 2020, 5). JES effectively commandeered these revenue streams for both military and personal gain. With the stated need for war-time exceptions to conventional governance, Sonangol was fenced off from the rest of the MPLA's party apparatus and left in the hands of the centralized state, with JES sitting at the top (de Oliveira 2015, 33).

By the time the economy began liberalizing, Sonangol, and thereby JES, were well positioned to capitalize. The company was restructured as a holding company in 1992, leading to an expansion of operations into a range of sectors, from real estate to construction and even catering (de Oliveira 2015, 35), all while fortifying its domination of the country's oil sector. That same year, JES amended the constitution to ensure that the president could not be prosecuted except for acts of treason and bribery, opening the doors to continue turning Sonangol's revenue streams into a kleptocracy machine (*Premium Times* 2020). The IMF infamously sounded the alarm, claiming there was a multibillion "Bermuda triangle"–style black hole in the state's finances between 1997 and 2002, which they implied was the doing of JES and his emerging band of kleptocrats (de Oliveira 2015, 39). The company's role continued to expand, however, including allotment of oil exploration and concession rights and greater autonomy over the profits that needed to be sent to official coffers. At times, it even managed the public debt (Musacchio et al. 2010, 7).

Once the civil war ended, Angola's fiscal situation was inevitably on shaky ground. Because the IMF requested substantial transparency for any potential rescue, the state ventured on a now well-worn path: it turned to China, with Sonangol's oil used as the basis for loans (Musacchio et al. 2010). But their fiscal needs would quickly change because the surge in global oil prices would once again make the country rich in cash. Focusing on consolidating its position across territory once occupied by its rivals, the state launched a series of welfare programs and a host of infrastructure projects to bring in political elites from outside its base in Luanda (de Oliveira 2015, 97). But the bulk of the money went to continuing JES's control of the state and in building up its own brand of plutocrats. Taking a script out of the 1990s Russian playbook, the MPLA repeatedly called for the creation of a "national bourgeoise" to safeguard the country's interests (Friedman and Reinert 2020, 7):

> As one senior MPLA cadre argued, "The country needs to develop the cosmopolitan elites it never had . . . which presupposes allowing them access to wealth and power. . . . It may not please you to hear it, but Angola needs rich, strong, and powerful families that can stand up to international competition. To have a republic of better fed poor people, all with the third grade, is not a viable model for us."

The mechanism for building that bourgeoise was connections to JES. This was clearly a state-controlled system. The only way to gain plutocrat status was to be a kleptocrat, and it launched the fortunes of a tight-knit cabal largely centered around the JES family (de Oliveira 2015, 130–40). Given this power structure, the lack of outright intra-elite conflict is unsurprising. For a country at war for most of his tenure, it is remarkable that no MPLA figure was ever killed under JES (de Oliveira 2015, 99). A culture of avoiding conflict among the ruling economic elites permeated. As one Angolan businessperson put it, they "divide up the market in ways that minimize friction" (de Oliveira 2015, 149). Courts, be they domestic or foreign, were not an option given the interlinks of the system with the state. According to Angolan economist Manuel Alvas da Rocha, "the MPLA created a web, bonds between politicians [so that no one will talk about the others]. Because if one of them talks, they are all gone. Because the relationships are umbilical" (de Oliveira 2015, 153). Sitting at the top, as de facto adjudicator, per a celebrated Angolan journalist, was JES: "He is not only brother, father, grandfather, leader or President, he is the master of existence, the 'godfather' of a supreme and unscrupulous clan which dominates Angola" (Verde 2022, 59).

From Godfather to Terminator

Angola under JES looked like a prototypical example of the resource curse: economically and thereby politically concentrated. This concentration left the power balance up to international forces. Once the commodity super-cycle hit its first bump, coupled with the transnational forces emerging with the Arab Spring in 2010, Angola experienced its first signs of sustained civil society–initiated protests. JES and the state responded with increased oppression. Angola had one of the largest armies, with one of the highest ratios of military members to citizens, and it was well-fed under the resource-rich regime. Some measures verged on the farcical; the state famously put seventeen students on trial for reading books on nonviolent resistance. Although oil prices soon recovered, they experienced another sharp crash in 2014, short-circuiting JES's control (OMCT 2015). In theory, Angola had begun to follow international best practices for such dire times. It had set up a sovereign wealth fund, but its investments were suffering severe losses as well (Verde 2022, chap. 2). The slush fund had dried up. Worse still, the backbone of the economy, Sonangol, had been just as poorly managed, leaving the state with nowhere to turn (Verde 2022, chap. 4).

Rumors of succession rapidly picked up pace. The MPLA was facing elections in a couple of years and couldn't risk losing its stronghold with both popular discontent and UNITA's base consolidating. None of that would be surprising given the dire living standards in the country. Inflation was ballooning; basic medicines and consumer staples were in short supply, while hospitals were both underfunded and overcrowded (Verde 2022, 88). Diversification became the hot topic, but an economy so completely mired in oil wealth and the travails of the resource curse meant that diversification was a pipedream, especially on the timeline needed to maintain state control (Verde 2022, 88). The MPLA soon effectively executed an "internal and discreet palace coup" (Verde 2022, 100). Manuel Vicente, who headed Sonangol for years under JES, was discredited during the political jockeying because of how poorly he had left the country's crown jewel. JES lost enough credibility that none of his family members, who at this point were occupying key positions across the state in addition to their own private, largely offshore wealth, could pass muster. In a move reminiscent of late 1990s Russia, the MPLA was seeking "an accommodating candidate who would make the transition smooth, and who respected the party" (Verde 2022, 99). In other words, this was still fundamentally seen as a state-dominated system, so the MPLA were after a "kind of back seat driver . . . someone bound by the structure that he could not stray from the previously laid path" (Verde 2022, 100). Defense Minister João Lourenço (JLo) was ushered in as vice president of the MPLA in August 2016 shortly before the election. Although he had once fallen out with JES, he "was not seen as a charismatic or transformative figure" (Friedman and Reinert 2020, 8).

What the MPLA and the rest of the elite establishment missed was how the balance of power was inevitably shifting toward state competition. Already rocked by the economic crisis, JES and his family, who commanded a vast fortune and were soon set to be largely outside the government, presented a serious threat to the functioning of the state. As Angolan political economy expert Ricardo Soares de Oliveira put it more than a year before JLo's ascent to power, "Their collective stake in the economy is so large that there is nothing left on a similar scale for newcomers to appropriate. Merely in order to consolidate his own patronage structure, a new head of state would have to redistribute privileges away from JES's core supporters" (de Oliveira 2015, 207). JES's final acts in power illustrated such fears. He strengthened the rules favoring the head of Sonangol, a position held by his daughter Isabel dos Santos, removing the few restrictions on the position's ability to dictate the company's governance. He further passed a series of decrees that would limit the executive's ability to staff key positions in government. And in one final act of kleptocracy, $500 million were transferred from the Bank of Angola to a London bank account under the guise of a guarantee for $30 billion in financing for the state (Verde 2022, 199). In other words, JES seems to have made a series of decisions to shift the balance of power toward capture once he was no longer nominally in charge of the state.

These actions set the stage for JLo to live up to the anticorruption rhetoric that had bolstered his electoral victory and thereby a host of extraterritorial disputes. He soon seized the real estate and media holdings of several key advisers to JES, although the grounds for these expropriations were never made fully clear (Verde 2022, 144). He fired José Filomeno, one of JES's sons, from his position as head of the sovereign wealth fund, and initiated cases through the London high courts to freeze the $500 million that were transferred from the national Bank of Angola. The state was trying to get its hand on its opponents' offshore war chest. But interpreting the case as a commercial one, which was necessary because of the way the case was filed, those efforts were eventually rebuked (Verde 2022, 137–38). Filomena could not escape the charges at home and was eventually convicted of criminal charges related to the looting of the state, which came with a prison sentence. But JES's other chosen child would not to be brought down as easily.

The dos Santos Boomerang

At the peak of her instrumental powers, Isabel dos Santos was worth more than $2 billion, with a diverse set of revenue streams that could make even the most successful Turkish families jealous (Dolan 2021). She was a major shareholder in two of the key financial players of the Angolan economy, including an outright majority in Banco Fomento de Angola (Pitcher and Rodrigues Sanches 2019). She was active across the real estate pipeline, including dealing in cement, luxury properties, and restaurants, and held a 25 percent stake in Angola's dominant telecommunications company Unitel (Pitcher and Rodrigues Sanches 2019). A joint venture with Brazilian and Portuguese investors, which would come back to haunt her, doled out $5 billion in dividends between 2006 and 2015 (*Premium Times* 2020). She influenced public opinion through her TV channel ZAP and controlled *Forbes*'s Angola edition: the same magazine that had crowned her a billionaire in the early 2010s and ran an exposé on the political sources of her wealth in 2013.

These were just her revenue streams in Angola. In an inverted colonial pattern, Portugal became increasingly dependent on Angolan wealth, with Isabel dos Santos at the center. She had substantial holdings in multimedia firm Zon and Banco Internacional de Credito (BIC) (Pitcher and Rodrigues Sanches 2019). In 2006, she acquired one-third of the shares in Galp Energia, Portugal's oil-related cash cow that had a turnover of €15 billion. Sonangol was central to these efforts, acting "both as a partner and as a financier to Isabel dos Santos. Sonangol acquired shares in Galp and assigned them to Isabel dos Santos under her new company Exem who would reimburse Sonangol with future dividends obtained from the very same Galp shares" (Verde 2022, 53). Dos Santos's involvement only became public in 2008. Her husband, Sindika Dokolo, was named to the board of directors of the front company that had acquired the shares. A couple of years later, he purchased struggling Swiss jewelry outfit de Grisogno that would go on to host lavish bashes, including at the Cannes film festival, with attendees like Leonardo DiCaprio of *Blood Diamond* and *Great Gatsby* fame (Forsythe et al. 2020).

When Isabel dos Santos was named as the new head of Sonangol, there was an uproar among the Angolan elite given the struggling economy and the MPLA's flailing political fate. She appears to have been central in ousting Vicente from the company and thereby crushing any of his potential political ambitions (Verde 2022). She had amassed substantial wealth through the state and a diversified flow of income streams that stretched beyond the need to loot directly. Despite her carefully groomed international reputation (she hired top brass public relations firms when she stepped into the spotlight starting in 2013, even giving headline presentations at the World Economic Forum and the London School of Economics), her dealings were catching up with her. Major international banks like Citibank and Barclays had abandoned deals with her in 2012 (*Premium Times* 2020). Soon after JLo came to power, dos Santos found herself sacked from her position in Sonangol and absconding abroad.

Extraterritorial Searches for the Missing Millions

At almost the moment the clocks dawned on 2020, the first blow in a multiyear, multijurisdictional war between the Angolan state and Isabel dos Santos was struck as the former sought to move from competitive dynamics back to a system of state control. Domestic courts froze dos Santos's assets, using an unpaid $75 million loan from Sonangol to her and her husband's Exem Energy BV as the basis, on top of an alleged $1.1 billion against a business associate (*Business Week* 2021). The ante was quickly raised when the couple were accused of money laundering and embezzlement of public funds. Dos Santos denied the allegations, labeling them "politically motivated" (*Reuters News* 2020). The state was given what appeared to be a decisive helping hand from the International Consortium of Investigative Journalists (ICIJ). Most famous for the Panama Papers, the ICIJ published the Luanda Leaks in January 2020, providing granular detail on dos Santos's offshore structures: more than four hundred companies across more than forty countries, with nearly one-quarter of those entities registered in traditionally defined offshore havens like Malta and Mauritius (*Premium Times* 2020). The couple again denied all wrongdoing and discussed the allegations as a "political persecution" (AllAfrica 2020).

There were nearly immediate, dire consequences for both dos Santos's income and wealth. EuroBic, a Portuguese bank that she had long held shares in, announced it was cutting ties with dos Santos, who was now going to seek a sale rumored to be worth $200 million (Dolan 2020). But that was nothing compared to the Portuguese courts announcing that they were freezing and, eventually seizing, all her assets in the former colonial power (BBCAP 2020). Dos Santos took each chance she was presented to argue that she was innocent, a private businessperson, who was being politically targeted. At one point, she argued that the cases needed to be dismissed because they were based on fabricated documents: the passport used in some of the evidence appeared to have a false signature based on actor Bruce Lee (Reuters 2020a). But there did appear to be some initial bargaining for a settlement that did not lead to any settled outcome (*CE NoticiasFinancieras* 2020b). Instead, Atlantic Ventures, which was tied to dos Santos, filed a claim against the Angolan state at the Chambre Arbitrale Internationale de Paris worth $850 million. The plutocrat attempted to use foreign courts to regain her losses from the political turmoil triggered by state competition. Atlantic Ventures sought damages for JLo's 2018 presidential decree that revoked concessions around the Port of Dande (*CE NoticiasFinancieras* 2020a). The tribunal refused jurisdiction, however, handing the Angolan state a victory (*CE NoticiasFinancieras* 2020a).

Foreign courts continued to attack dos Santos. In September 2020, a court in Amsterdam froze her stake in Exem Energy, the vehicle she had used to garner shares in Galp through the financing that appeared to have been provided by Sonangol. While this did not go unchallenged, it was another battle that dos Santos lost a year later. The court ruled that the half-billion-dollar stake needed to be transferred back to Sonangol and the Angolan state. After the hearing, the state-owned oil company's representatives publicized part of the judgment that stated that the shares in the company were "tainted by illegality, enabling Ms. Isabel dos Santos . . . to reap an extraordinary financial gain to the detriment of Sonangol and, consequently, of the State of Angola" (Reuters 2021). Dos Santos continued to deny being directly affiliated with the company, noting that the 40 percent stake in Esperanza Holding BV at dispute, now with a market value of nearly $700 million, were her late husband's property (Reuters 2021).

While the legal battles consistently dealt blows to dos Santos, she did not sit idly by, waiting for the law to do its work. She repeatedly stated that she saw this as a political battle, and it appears JLo saw it the same. Dos Santos used a range of arguments to strike back according to academic and activist Rui Santos Verde (Verde 2022, 160):

> Since then, it is easy to see that she has become his main enemy. Only she had the financial capacity, network and international articulation to face the president. And that is exactly what she has done, smoothly and discreetly, but with exhausting effects. In Angola, Isabel dos Santos shrewdly takes avail of the various economic blows that Lourenço [JLo] has suffered, sending the message that she is the largest private employer in the country, so she is the best equipped to create jobs. The same argument is referred to in order to defend herself from governmental procedures against her interests. According to her admirers, by attacking her the government is undermining Angolan private economic development. This argument has percolated in the country. I have heard several supporters of Lourenço speaking out that it is necessary to spare Isabel dos Santos, since she is the largest employer in the country. With this argument, Isabel dos Santos has earned a capital of sympathy that she did not have at the end of her father's term in September 2017. At the same time, through her communication agencies, Isabel dos Santos created a huge anti-Lourenço barrage on social media networks. From the president's golf swing failures in the Mangais golf course to the dissemination of technical reports critical of Lourenço's policy, everything found its way.

The political threat posed by Isabel dos Santos has persisted despite her physically exiting the country in early 2020. She has even shown up as a parallel state-style figure at important international events like the Russia-Africa summit that

JLo was also attending (Verde 2022). And so the fight for political control percolated on, even after the United States sanctioned dos Santos in 2021 while JLo was presiding over a once again contentious political scene back home. An Interpol Red Notice was issued for dos Santos in November 2022 (Reuters 2022), in the same month that she was quoted in Bloomberg as saying that she still had presidential aspirations (Almeida 2022). A few months later, Unitel, the telecommunications company that dos Santos held shares in, requested that the London courts name her as a direct defendant (Gemmell 2023). The claim dated back to issues first raised in 2015. The company's main shareholder had been Brazilian telecommunications company, Oi, but it had recently been nationalized via Sonangol (Reuters 2020b). When fresh criminal charges were filed by the Angolan state against dos Santos at the start of 2024, her lawyer responded by saying: "Isabel is presently engaged in court proceedings in various jurisdictions to clear her name. The aim of the Angolan authorities is to intimidate her from returning to Angola so she is unable to run for political office" (Chibelushi 2024).

When Putin came to power in the early 2000s, the balance of power between the state and the plutocracy switched from capture to competition. It triggered a series of extraterritorial legal battles that have bounced between London and a host of other offshore and international legal venues. The value of this chapter is to illustrate how the argument that I developed, centered around the potential costs of transparency and the need to acquire and delegitimize the wealth of one's opponents, explains the offshoring of elite conflict.

While Nazarbayev withstood the downfall of the Soviet Union and strengthened his grip over the political system in Kazakhstan in the 1990s, his success in engineering economic growth under a newly privatizing system switched the balance of power from one of control to one of competition. He won the early domestic political battles and repeatedly tried to bring the new tycoons under his thumb, but they wanted a more predictable, more capturable system. The ripples from the 2008 financial crisis presented an opportunity to counter these trends. BTA was nationalized, triggering one of the biggest ever fraud cases in British and international legal history against Nazarbayev's clearest political challenger, Ablyazov. The latter's attempts to regain his money failed through the ISDS mechanism, even though he exploited both legal and extralegal means to bring to light the corruption that he saw at the core of the Kazakh state. But it was his offshore dealings and his need to continue hiding his wealth that was the clearest cause of his legal downfall, leaving him on the hook for literally billions of dollars.

The Kazakhstan case shows that one-way changes to the balance of power are not the sole cause of outsourcing authority. The case study is an instance of moving from control to competition. Turkey provides an example mirroring more of the Russian dynamics. The increased political competition of the 1990s and its integration into global capital markets created a dynamic that strongly favored the incumbent plutocracy. The same forces—the vagaries of international capital mobility that induce economic crises—set the stage for a relegitimization of the Turkish state under Erdoğan and the AKP. The state then tried to move toward control and, in the process, took on a series of individuals with their own private resources that they sought to redistribute or simply turn into more kleptocratic actors. At the center of these battles was the Uzan family. Cem Uzan had attempted his own political challenge but failed in the elections that brought the AKP to power, even as he and his family were fighting fraud charges abroad. The Uzans tried every tool available to them in ISDS to reclaim their losses, but this overindulgence was one of the main reasons those attempts failed. Unlike in Russia, the state in Turkey was not able to clamp down fully on the plutocracy, that is, firmly turn it toward a system of control, given its international capital dependence and its own limited economic resources that it needed to redistribute to engineer the economic returns. State competition has instead been more of the norm, leading to extraterritorial legal battles returning in the late 2010s.

That same period saw João Lourenço take over the Angolan state, which for almost four decades had been under the control of José Eduardo dos Santos. Throughout JES's reign, oil wealth was the core to his and the Angolan state's power. There were no cronies or tycoons for most of his time, just a series of kleptocrats, most notably JES's own family members, whose wealth originated from their ties to the state. The kleptocrats internationalized both their wealth and income streams and turned into potential threats to the state once JLo took over after a massive decline in state revenues and legitimacy. With the system in disequilibrium caused by state competition, the state sought to take on the individuals that had the instrumental power resources to move the system toward capture. JES's final actions, increasing shackles on the state both through political and commercial laws, likely created an even greater need for JLo to tamp down his authority. The result was a series of extraterritorial legal battles against the dos Santos clan, most notably against Isabel dos Santos, whose international status and acclaim would quickly fade as a function of both the revelations of her offshore dealings through the Luanda Leaks and the foreign legal cases against her.

These three case studies illustrate how battles over political control become global as a function of offshore wealth. The intermediaries of these conflicts are the legal institutions that enable and enforce the structures of financial globalization. The analysis in this chapter showed that plutocrats and the state have a menu of transnational legal tools at their disposal. The balance of power influences when extraterritorial battles are likely to take place, but legal specifics dictate precisely where and how they end up fought. The rules of the system do frequently leave plutocrats wanting because they have often lost many of their disputes. But the state, echoing how great powers have used the choke points of the global economy for their own geopolitical statecraft, regularly weaponize global legal institutions to further their domestic agendas. Enforcing those victories and the state actually getting its hands on the wealth are always challenging, as BTA's experience highlights most clearly. States nonetheless can and have successfully discredited and delegitimized their opponents through arduous, high-profile, foreign legal conflict.

CONCLUSION

The Tragedy of Transnational Liberalism

Why do conationals sue each other using foreign courts? Extraterritorial dispute resolution should be driven by a search for the rule of law. Actors go to the United Kingdom and international tribunals for the fair and equal treatment that they are unlikely to receive at home, with the added promise of an efficient and expert process. While these factors no doubt shape the calculus of firms and individuals choosing to outsource their disputes, it fails to explain why plutocrats from only some emerging markets fight legal battles abroad. In other words, it fails to explain why Russian plutocrats regularly sue each other in London, yet South African and Saudi Arabian plutocrats rarely sue abroad despite having similarly weak legal institutions at home.

This book proposes a complementary, distributional explanation for extraterritorial dispute resolution. While the gains from fair treatment may create a baseline set of incentives, the potential costs of a liberal process are critical determinants for outsourcing authority.[1] The actors who have the functional need for courts in London or tribunals in The Hague and the economic means to go abroad are primarily emerging market plutocrats. To gain that wealth and status, these plutocrats have almost certainly engaged in corrupt, possibly criminal, activity that could easily come to light in a public, foreign legal process. An extraterritorial dispute could force plutocrats to reveal where they hide their assets, who their real business partners are, and even the intricacies of their political patronage. Divulging these business practices can easily invite new threats to a plutocrat's wealth, be it from fellow plutocrats, the state, and even international actors. Thus, individuals need to strategically weigh the gains of the rule of law against the costs of its transparency.

I argue that the balance of power between the state and the plutocracy guides plutocrat decision making. I conceptualize the balance of power as a spectrum. At one extreme is a world of state capture (Hellman 1998), where the plutocrats have the most valuable resources. In this scenario, the government is subservient and fundamentally replaceable, so the oligarchs are not concerned about the public threat from the state but instead are primarily focused on intraplutocracy battles in which a plutocrat is constantly concerned that a fellow plutocrat might step up to raid their company and take over their assets (Firestone 2008; Markus 2017a). When a major dispute arises, both parties to a lawsuit are likely individuals with corrupt pasts. Even if both actors despise each other, they are fundamentally winning under conditions of state capture. Potentially detailing their murky business histories or how they leverage or control state actors could bring the entire system into question. Those prospects create strong incentives to settle cases outside the foreign spotlight. Plutocrats are then left to resolve disputes themselves.

At the other extreme is a world of state control, where the state has all the most valuable resources. In this scenario, the private threat is diminished because no one will make a major economic or political move without the state's approval. Instead, kleptocrats and tycoons are focused on mitigating the threat from the state, which an extraterritorial dispute would quickly increase. Filing a case in London or New York would send a signal to international and domestic audiences that the wealthy do not trust their home institutions. Given the way many state-controlled economies work, a case would almost certainly implicate the state. The latter would then be left with no choice but to react; this fear of retribution prevents the super-wealthy from taking advantage of a more efficient, liberal process abroad, with the state instead often acting as informal mediator.

At the extremes of the balance of power, extraterritorial cases should rarely occur, but such cases are likely to occur in the middle of the balance of power, in a world of competition where I find plutocrats outsource authority. When the state and the plutocracy have relatively equal power, there is uncertainty around who really has control, with the state and the plutocracy both preferring a move toward their beneficial extreme. Whether it's a world dominated by winning oligarchs or subservient tycoons that is up for grabs, the state wants control while the plutocrats want capture. These opposing interests leave the state in fear that plutocrats will fund an opposition party or a coup to bring the power balance in their favor. At the same time, plutocrats are concerned that the state might expropriate their assets or trump up corruption charges to bring them into line. The divergence in interests leads to different actors trying to assert themselves. While the losers of these clashes would historically have been thrown behind bars, many are now able to flee. Once they've exited the economic and political game, the costs from transparency diminish, and the only means of recourse is to use foreign courts. They now have far less to lose from transparency, and they are locked out of their home institutions. Liberalism, through London or Paris or The Hague, becomes their last resort.

But the losers of these clashes do often run away with substantial wealth intact: foreign bank accounts and real estate holdings are central features of the average plutocrat's portfolio (Freund 2016; Hägel 2020). If they still want to mount a political challenge, they will then have an arsenal, so they could still fund an opposition party or a disinformation campaign from abroad. In other words, fleeing plutocrats remain a potential political threat that has turned transnational. So the only way for the winners of the clash to get rid of their opponents is to also use foreign courts. But the motive differs compared to plutocrat-initiated cases. Extraterritorial litigation can be an attempt by the state to get their hands on a plutocrat's offshore wealth. Even if the claim does not amount to a huge proportion of the plutocrat's wealth, it will drain their time and money in legal fees. A victory would not only come with material gains but also with increased legitimacy because a court in the United Kingdom or the United States could implicitly deem the political winner as legally right. It illustrates to any future challengers the lengths to which a winner will go to secure their power, even subjecting themselves to the possible costs from transparency. Extraterritorial dispute resolution is often the result of political conflict unraveling under commercial guise.

I developed this argument in relation to the Russian cases that, from a functionalist perspective, represent a clear empirical puzzle. In the 1990s, Russian plutocrats lacked an independent judiciary and developed commercial law, but they rarely if ever used transnational legal institutions to resolve their conflicts. Despite their domestic courts steadily improving in the 2000s, there was a huge upsurge in intra-Russian fights outsourced to foreign legal systems.

The 1990s saw a shift in the balance of power from an infamous form of state capture to a period of competition ushered in by the rise of Vladimir Putin (Dawisha 2015; Sakwa 2007; Myers 2015; Judah 2013). The within-case analysis allows me to hold a number of cultural and institutional features constant. Based on historical accounts and the sixty interviews I conducted with lawyers working with Russian plutocrats and corporations, I examine the macro-level drivers for conflict and adjudication across the balance of power. I further track the dispute resolution strategies of two plutocrats—Boris Berezovsky and Mikhail Khodorkovsky—over time, which should further provide confidence that no idiosyncratic, personal-level factors determine who uses the courts. While most scholars see 2004 as the end of plutocrat-state competition in Russia, that analysis misses the extraterritorial dynamics that were used by potential Putin challengers and by the Russian state to prevent any future return to conditions of capture. The offshore legal coercion is an indication of the prolonged, transnational process of Putin's power consolidation—and it worked. As these cases were evolving, Russia steadily moved toward a world of state control, so examining how disputes are resolved by plutocrats under Putin's control provides a cross-spectrum analysis. Those now wanting, or needing, to stay on the state's good side generally avoid foreign courtrooms.

In recent years, extraterritorial disputes involving Russia have decreased in number but Russians have not stopped engaging foreign courts. Conflicts between the Russian state and tycoons with their own resources still occasionally take place. But the larger trend is for disputes to turn geopolitical. Over the course of nearly a decade, Russia sued Ukraine in a claim worth $3 billion at the London high courts over outstanding bond payments. These inter-sovereign conflicts would normally be handled informally through the Paris Club, but the Putin regime was seeking a quick and efficient way to lend money to its client, Viktor Yanukovych. That resulted in a deal structured through private markets, with a series of debilitating clauses, including allowing Russia to force a default on the country if its debt–gross domestic product (GDP) ratio went above 60 percent. London was chosen as the seat of jurisdiction. The money was meant to prop up Yanukovych, but he was soon ousted during the Orange Revolution. So dire was the country's need to maintain international credibility that Ukraine handed over some of the initial payments, even though Russia had now invaded Crimea in 2014. The most boring aspect of the international order—contract law—was used to keep a rival down and indebted. The UK supreme court finally ruled against Russia a year after its full-scale invasion of Ukraine, and a host of Russian plutocrats are active in London legal proceedings to challenge sanctions against them related to the war.[2]

While great powers like the United States and Russia are seen as the countries most capable of weaponizing the rules of the liberal order, chapter 5 showed how plutocrats and state actors from a range of other countries have exploited the system. Once the 2008 global financial crisis hit Kazakhstan, the state deemed it necessary to nationalize Bank TuranAlem (BTA). Kazakhstan was a clear instance of state control in the 1990s and saw no extraterritorial disputes in that decade, but the commodities and financial boom of the 2000s turned the dynamics toward state competition. The newly state-owned entity, that is, BTA, initiated and won the biggest ever fraud case in British legal history against Nursultan Nazarbayev's political opponent, Mukhtar Ablyazov, and kick-started a host of other disputes against those close to the fleeing plutocrat. Ablyazov did try to get some of his money back through an extraterritorial arbitration, but his offshore dealings caught up with him. The extent he went to protect his wealth undercut his ability to claim being a legitimate foreign investor.

The Uzans complex offshore dealings also made them a threat and interfered with their ability to sue the Turkish state after more than two hundred of their companies were nationalized. In the 1990s, Turkey shared several similarities with Russia: economic volatility and state capture at the heart of a system accustomed to substantial state authority, which in Turkey was then reasserted once Recep Tayyip Erdoğan and the Justice and Development Party (AKP) rode a wave of discontent to power. Some of Erdoğan's first acts of power consolidation were to attempt to move toward state control by taking on the incumbent plutocrats. The Uzans tried multiple legal avenues to win back their money but perpetually lost, while the state had little need to chase their money because the Uzans were being challenged by Motorola and Nokia for fraud-related damages worth $4 billion. In an economy still trying to integrate globally, the state was forced to continue privatizing, building up a group whose members were more like tycoons than kleptocrats, so there was never complete state control. Instead, Turkey has gone through waves of intra-elite conflict that have rippled through foreign courts.

Turkey is not blessed with the same commodity resources that often make state control plausible and valuable, but Angola follows the opposite dynamic. One of the most oil-dependent economies in the world, Angola and its wealth became the basis for multidecade rule by José Eduardo dos Santos (JES). His iron grip quickly loosened with the oil collapse in 2014 and soon led to João Lourenço (JLo) replacing him at the top of the political hierarchy. JES had built up a regime that is often considered the epitome of kleptocracy, but on his way out, he attempted to weaken the state and move toward a system more like one of capture. But that shifted the power balance toward state competition, with the incumbent plutocrats now threatening a reforming state. Next, a host of extraterritorial fights broke out as JLo sought to take on the critical relics of JES's reign: the wealth and status of his daughter Isabel dos Santos.

My argument further helps us understand why several countries with large, globalized plutocracies and weak judicial institutions have generally operated outside the foreign spotlight. South Africa has recently taken the mantle of true state capture from 1990s Russia following the revelations detailing the links between the Gupta family and the Zuma administration (Mahajan 2019; Martin and Solomon 2016). But there have been no intra-elite conflicts in South Africa decided by British courts, despite the country experiencing substantial macroeconomic volatility throughout the 2010s. A more direct comparison can be made with Ukraine, which is frequently juxtaposed with Russia given their shared cultural and institutional histories. While the Russian state returned to power, Ukraine has generally been characterized by weak state authority and frequent clashes between rival plutocratic factions (Puglisi 2003; Pleines 2009). Ukrainian extraterritorial disputes involving different cliques of the country's plutocracy have occasionally occurred but far fewer compared to Russia, and far fewer ever make it to court according to the lawyers I spoke to, which allows Ukrainian elites to avoid the transparency costs. For example, the $2 billion battle pitting Victor Pinchuk against Ihor Kolomyski and Gennady Bogolyubov was settled the night before it was scheduled to have public hearings (Shirbon 2016). Lawyers regularly traced the fewer battles between Ukrainian plutocrats, compared to their Russian counterparts, to the differences in the balance of power: "With Ukraine there wasn't the same kind of state domination."[3]

At the other end of the spectrum, from a functional perspective, Saudi plutocrats should be some of the most likely litigants in London. Their kleptocrats and tycoons are highly integrated into London property and legal markets, and the state enjoys near monopoly control over the domestic legal system. Although there have been a handful of Saudi extraterritorial cases, the numbers pale in comparison to those from Russia or Kazakhstan. According to the lawyers I spoke to, the limited extraterritorial activity goes back to the lack of political conflict in settings of state control: "In Saudi and other Middle Eastern countries, they are smaller and less likely to fall out. They know where they sit."[4] Even though Saudi Arabia experienced a crackdown of the tycoon class once Mohammed Bin Salman came to power (the infamous Ritz-Carlton affair that turned the luxury hotel into a prison for some of the Saudi super-wealthy), no extraterritorial disputes were initiated. The plutocrats have not appeared to have ever sought or have had the material means to change the system. The system was defined by state control, and the crackdown was a reminder of who is in charge through an internal redistribution of rents.

The lack of cases from China compared to Russia can be explained by looking at the power balance: "Russia has a whole lot of central control, but also a whole wave of less connected people. . . . China has been much more monolithic, and you've only been able to go up through the party."[5] The dominance of the Chinese state and the dearth of outright political conflict that changes the transparency calculus are likely the biggest deterrents to using foreign courts: "People are more afraid of stepping out of line. . . . The collapse hasn't happened in China . . . the regime is still there. It [a wave of extraterritorial cases] requires a rupture."[6] In recent years there is little doubt that private capital has made substantial strides; a host of successful tycoons have formed private enterprises, but they have seen their economic fortunes deteriorate because the Chinese state seeks to keep them in check, most notably through the shackles put on the technology sector. These disputes have not boiled over into foreign courts because they never reached a disequilibrium of state competition (Xu 2022). Even through the years of technology growth, the party-state's control over the rest of the economy fortified, and there was little doubt about who was actually in charge. As this book goes to press, the first real signs of Chinese state decline look underway. A debt overhang, lack of aggregate demand, an aging population, and the risks of US-China decoupling are all beginning to fray the system. The legitimacy the state derives from economic growth is finally in question. If this pattern continues, the plutocratic class, largely unaccustomed to entering the political fold, would be incentivized to mount its first substantive challenge, which would then likely spill over into commercial courts.

More Authority, More Politics

The findings raise a number of theoretical and normative questions that academics and policymakers need to grapple with. They suggest that transnational legal institutions regularly act as intermediaries in the political conflicts of emerging markets. In terms of international relations theory, the disputes represent the creation of new hierarchies (Nedal and Nexon 2019; Lake 2009; Cooley 2005): the authority of often domestic, liberal institutions is used voluntarily to adjudicate the clashes between the most powerful actors that, implicitly or explicitly, govern foreign states.

While not every extraterritorial case involves a fleeing plutocrat or an authoritarian proxy (the goal of my argument is to explain the cross-country differences and differences over time in usage, not each specific case), the most legally and materially consequential disputes generally have substantial political stakes. When a plutocrat is thrown out, foreign courts become one of the only mechanisms for them to gain some form of redemption. It is easy *not* to feel any sympathy for a plutocrat who has lost: they are still fundamentally plutocrats with many of their commercial deals accompanied by heavy suspicion of corruption. But a key tenet of liberalism is that all individuals are treated equally under the law. Past behavior should have little bearing on the ability to initiate a legal claim. Plutocrats do not break laws to have their cases heard in London or even when they use their offshore entities to initiate extraterritorial arbitrations. If anything, they are practicing the very process that makes liberalism an effective governing ideology.

The stakes become even more complicated when one considers cases where the state appears to be exploiting liberal institutions to target its political opposition. Once again, the state, its entities, or its private proxies are at a basic level simply exercising their rights to gain jurisdiction and enforce claims abroad. In numerous cases that appear to be politically motivated, the underlying legal claim, such as an apparent fraud, seems to be backed by the facts. But these claims are usually issues that everyone in the home business community knew about and may even be regarded as accepted business practice in the home jurisdiction. The state may have even condoned the acts, but they may come to public light or become a legal matter when the political tides switch and someone with power has reason to weaponize it. Because the courts look only at the legal issues, as is their prerogative, they risk missing the forest for the trees, inadvertently allowing their legal powers to stifle political competition. In other words, liberal institutions can be used to achieve illiberal ends. Courts are and have always been political (Ginsburg and Moustafa 2008; Shapiro and Stone Sweet 2002; Voeten 2008), as one senior lawyer remarked when discussing these state-initiated cases, so it is hardly surprising that, in an economically integrated world, they might become used for strategic ends by foreign actors.

When I began the research for this manuscript, I assumed that these state-initiated claims undermined the legal system. But my discussion with British lawyers, some of whom have been on the losing end of these types of cases, made me realize the effects are complex, with no silver-bullet solutions. If the courts would just throw out cases that they deem political, judges would be asked to draw lines around claimant intentions and individual rights that would directly counter the system's liberal principles. The task becomes even more difficult when one considers that the apparently political cases operate in a much larger market where the legal services that are provided are crucial to the global economy. And what are the alternatives? To paraphrase a number of lawyers I spoke with, isn't it better that these plutocrats fight each other in court rather than using their AK-47s? That is the very purpose of the law.

Answering these questions is becoming increasingly important as a new competition for authority over the global economy emerges. London and New York have dominated the transnational legal market for decades, often making them the default and most efficient authorities for extraterritorial disputes. They have acquired these leading positions because of a mix of historical legacy, institutional strength, and sheer legal capacity. Yet they do not govern the global economy just to promote international trade and international finance. The authority is lucrative. Transnational dispute resolution is the bread and better for big law in New York and for the magic circle and silver circle firms of London.

But several challengers are now trying to break up the historic duopoly. Countries like the Netherlands, France, Singapore, and even Kazakhstan have set up their own commercial courts to tap into the profits (Bookman 2019; Erie 2020). These new setups often enjoy substantial state backing because the government hopes to foster a new export industry. English is often the language used for proceedings, and in many instances, retired British barristers are imported to provide both the legal prowess and the international legitimacy.

The overarching fear of the 1990s was that globalization and economic integration would lead to a race to the bottom in which countries would weaken their regulations to compete for capital (Drezner 2001; Strange 1996; Rudra 2008). While this phenomenon has played out in some areas, there is a general consensus that numerous states have maintained their autonomy (Drezner 2008; Newman and Posner 2011; Damro 2012). The proliferation of commercial courts inverts this logic. Rather than the subversion of standards, a race to the top has unfolded, with countries creating institutions that rest on liberal rules. That raises the question of why only some jurisdictions are entering this apparent push toward neutrality and who really stands to benefit. The costs still prohibit the average business or individual from using these developing institutions, so the new legal options are inevitably distributional. An increase in the number of courts is almost certainly going to lead to more domestic (political) conflicts turning transnational through instances of extraterritorial dispute resolution.

Is Liberalism a Zero-Sum Game?

The individuals who would have the economic means to take advantage of these new legal options are plutocrats. They will be able to afford the lawyers and the court fees and have opponents whose assets are abroad and ready to be seized, allowing them to sidestep their home legal system. According to several theories of political development, plutocrats were historically the actors fighting for liberal reform; they were fighting for property rights and for democracy in attempts to restrain a coercive state (North and Weingast 1989; Albertus and Menaldo 2014). But because of globalization and the transnational legal market that has grown to support it, these actors can now buy those protections. They write their contracts using British and American law as the governing law and structure their offshore holdings to gain access to extraterritorial arbitration. Although their businesses and their revenue streams are often physically located at home, their legal protections are transnational, which is another example of how globalization has undermined the coterminous relationship between authority and national borders (Kalyanpur and Newman 2019; Newman and Posner 2011).

The fights that plutocrats used to develop liberal institutions involved a mix of backroom bargaining and occasionally bloody rebellion. In other words, they were the hard, difficult work of political reform. But now that they can access the rule of law abroad, their incentives to fight for political development have likely diminished (Sharafutdinova and Dawisha 2017; Pistor 2019; Milanovic 2016). They do not need to organize themselves and pressure the state when the rules that they are after can be arbitraged. While these effects are difficult to disentangle empirically, the new incentives for plutocrats are relatively clear. Why should they risk their wealth through reform when they can protect it via offshore sites? Extraterritorial dispute resolution and the politics surrounding it suggest that the liberal institutions of developed democracies may be undercutting the development of liberalism in emerging markets. One of the goals of the international liberal order was to use the greater connectivity to socialize and transmit values like individual freedom (Ikenberry 2009; Johnston 2003), but because access to liberal institutions has a high price, it has differential, potentially contradictory effects.

A natural next step in the research agenda is to unravel these relationships through careful quantitative and qualitative work. Analysis by Ananyev (2018) suggests that the amount of offshore wealth held abroad correlates with a steady decline in a country's Polity score. These efforts could be extended by examining whether assets held abroad, which increases a plutocrat's legal protections, then also diminishes how frequently elites initiate coups. Although capital mobility and offshore finance have led to the rethinking of several theories based on the masses as the key actors behind democratization (Freeman and Quinn 2012; Acemoglu and Robinson 2005), elite-led frameworks clearly need such updating (Winters 1996).

While this book focuses on the public sources of dispute resolution, future work will need to incorporate the private options available to plutocrats. Instead of designating British or US courts as their authorities, plutocrats could have legal claims heard in private arbitration venues such as the London Court of International Arbitration or the International Court of Arbitration in Paris. The fact that they do not have public proceedings or publish judgments ensures that there are limited costs of transparency and thus makes it unlikely that my argument would explain these disputes. It's possible that a story about more standard forum shopping applies, but private arbitration could also provide states with a bargaining chip. If they do not want to succumb to whatever plutocrat pressure remains, or they simply do not want to keep themselves involved in managing big business, they could tell plutocrats to outsource their disputes to these private venues. It could provide the needed neutrality but with far more mitigated transparency costs.

Transnational Political Development

The feedback effects of extraterritorial disputes are not restricted to the countries whose elites have access to foreign courts. There are also substantial spillovers for the hosts because extraterritorial dispute resolution has the potential to create alternate pathways toward political development. Political economy scholars tend to focus on how big business, and interest groups more generally, can shape rules in their favor by capturing legislatures and bureaucrats (Frye 2002; Mattingly 2016; Grzymala-Busse 2008). But the legislative phase is only the first stage of legal development. Private actors take a direct role in influencing how the law is practiced by deciding which of their disputes are worth taking to court (Putnam 2016; Kahraman et al. 2020), which could then alter the precedents that guide the future strategies of both state and nonstate actors.

Although they are now decreasing in number, Russian disputes took place in London for more than a decade. They built up a huge body of case law in areas like bankruptcy, fraud, and even some parts of family law. These may now act as legal precedent, reshaping how British actors can engage and exchange with each other. Rather than the legislature or the citizens of a country updating English law, exterritorial litigations lead to the business practices and the politics of a handful of emerging markets rewriting the rules. These effects are not confined to the United Kingdom. When parties from different countries litigate in London, these Russian-generated precedents may apply. And because English law is the global gold standard, there are inevitably global consequences.

By fomenting new economic complementarities (Hall 2001; Hancké et al. 2007), the impact of these cases can move beyond the law. *Berezovsky v. Abramovich* involved numerous law firms and barristers, and each took home millions of pounds through the process. The major British law firms were engaged in a constant competition to ensure that they were first in line for the next multibillion-dollar extraterritorial litigation. Many created their own dedicated Russia or Commonwealth of Independent States (CIS) desk that often built on their own sectoral expertise, familiarizing themselves with the intricacies of the Russian legal code and contrasting it with the English counterpart. Others have become even more specialized, dedicating themselves to representing state-owned Russian banks that are chasing fraudsters in London. Each year, conferences like Russia Law Week in London or an English Law Week in Moscow fortified both legal and social connections.

The complexity of extraterritorial disputes creates a market for several other specialist services. Public relations (PR) firms saw a huge boom from Russian plutocrats trying to protect their images because the court of public opinion could always play a role in a plutocrat's long-term fate, especially when facing possible extradition. Many PR efforts are under way before any formal battles begin because plutocrats, like all business actors, try to use the media narrative to shape early settlements. In the same vein, specialized companies involving accountants, assets-tracing specialists, and numerous private security and private investigatory firms have grown while Russian oligarchs became mainstays of the London legal and social scenes. The Russian invasion of Ukraine and its accompanying sanctions regime are serving as dramatic hit for these industries. But the legal precedents will remain.

Economic integration has the ability to disrupt political bargains (Berger 2000; Gourevitch 1978; Farrell and Newman 2014). Actors can seek to create new coalitions across borders and leverage international rules to further their own domestic political agendas. This book showed that the process is well underway for extraterritorial rules and that it has an impact on the actors who demand new political resources as well as those that act as witting or unwitting suppliers. Scholars have long recognized that the existence of safe havens and offshore sites could change the incentives for elites (Palan 2002; Sharman 2011; Shaxson 2011), but how does providing those services change the way the broader institutions of a safe haven operate?

The International Political Economy of Plutocracy

Finally, this book focused on individuals. International relations scholarship often exhibits a fallacy of composition problem. The state is usually taken as the main unit of analysis, but with globalization and economic integration, non-state and substate actors have become empowered (Avant et al. 2010; Chwieroth 2007; Bartley 2007). International political economy (IPE) generally corrects for this by moving research to the firm level (Kim and Osgood 2019). Scholars theorize firm incentives and test hypotheses using firm-level data (Abdelal 2015; Baccini et al. 2017; Wilf 2016). But the move to the firm level was incorporated without fully realizing the potential consequences of ownership (Braun 2015; Fichtner et al. 2017; Puente and Schneider 2020). Unlike in the developed world, the biggest firms in emerging markets are still run by a handful of plutocrats and their families. And as the Robert Mueller–Russia probe clearly brought to light, these plutocrats are integrated across economic and political systems. They park assets offshore; buy real estate in safe havens like New York; and now they even buy passports from places like Malta, Cyprus, and St. Kitts (Surak 2021; Mavelli 2018). In other words, the world appears to be governed by a transnational elite constituted by maybe two thousand people who have amassed great wealth and with it huge political power (Freeland 2012; Graz 2003; Hägel 2020).

What this book then brought to light is the need for a systematic research agenda that places the super-rich at the center: an IPE of plutocracy that examines how the structure and institutions of economic integration enforce and enable rule by the wealthy. Plutocrats have learned the tricks of the global economy and have learned how to exploit the gaps in the international liberal order. Extraterritorial dispute resolution illustrates that this mobility and integration now allow them to arbitrage institutions. They buy liberalism when politics deems necessary.

Acknowledgments

This book has been a labor of love, a labor of pain and paranoia, punctuated by moments of joy and disbelief. Like with any book, I have incurred countless debts turning the manuscript into something I can be proud of. Since my fieldwork several years ago, I am still in awe of the individuals that took time out of their busy lives to discuss extraterritorial litigation with a random PhD student, one who frankly did not know what he was getting himself into. I learned an incredible amount from these lawyers, not only on process and politics but also on the normative challenges of holding up a system that you believe in despite its flaws. My fieldwork was graciously supported by Georgetown's Graduate School of Arts and Sciences and Columbia's Harriman Institute. I also thank Cambridge University Press and the *Review of International Studies* for publishing my article, "Weaponised Legal Dependence," and allowing me to reprint parts of the paper across chapters 1, 4, and 5 of this book. In addition, parts of chapter 1 first appeared as "Twilight of the Oligarchs" in the *Review of International Political Economy* in 2026, which was published under a creative commons license. Thank you to the LSE and Taylor & Francis for promoting easily accessible work.

That said, no group of people have done more to influence this project than my mentors at Georgetown University. First and foremost, working with Abe Newman has been a revelation. I distinctly remember first pitching him the project in the kitchen at the Mortara Center. He gave me a smile that made me sure I was finally onto something after maybe thirty tries at picking a topic. He has been an unrelenting supporter, mentor, coauthor, and friend throughout this process. Kate McNamara has helped me grow in more ways than she knows. She is ready to debate, to critique, and to always find a way to help. One of the fortunately constant highlights of my life is getting a message from Kate to grab a coffee or a drink while she's visiting London. Erik Voeten continues to teach me intellectual dexterity and is a force of nature. James Vreeland quite literally got me started on this journey in 2009.

My time at Georgetown was made meaningful by several other faculty members, especially Diana Kapiszewski, Marko Klasnja, Daniel Nexon, and Irfan Nooruddin. I am lucky to have been part of a diverse cohort, collectively grinding through our problem sets. A special shoutout goes to Kristen Collins, who helped me clarify my interests from the get-go and continues to be a key part of my life. Many who were further along in their studies, in particular Lori Crasnic, Paula Ganga, and Dani Nedal, guided me through several messes and even cared to write things with me. Henry Farrell is a remarkable inspiration.

Mark Blyth and Michael Kennedy are the reasons I decided to go get a PhD in the first place, and I am grateful for what they saw in me (despite my frustrating undergraduate tendencies). I had no semblance of what an academic career could look like until they taught me that I was asking the right questions. Mark continues to keep me knocking on the right doors.

COVID-19 was weird but made far more manageable by my time at the Niehaus Center for Globalization and Governance. Helen Milner has been a trailblazer in our field, and I deeply appreciate her ability to bring eclectic and interesting postdoctoral scholars together. Making connections and friends over multiple Zoom meetings a week can be a surprisingly beautiful thing.

I joined the London School of Economics (LSE) in September 2021, and I am amazed at how it has surpassed my expectations. The International Relations Department is filled with faculty, students, and administrators that all show up. Our group is wonderful and wide-ranging, and I am fortunate to be a part of it. Special thanks to Dan Berliner and Milli Lake, who have created the foundations for both my professional and personal lives, constantly encouraging and pushing me on both my best and strangest ideas. They are my first port of call. Natalya Naqvi has turned into a close friend and interlocutor, and I regularly thank the heavens for her coming back to the LSE. Vicky Paniagua has been a constant throughout my time in London, where our coffee breaks and gigs never fail to make me feel lighter. Alex Hartman is an unrivaled source of intellectual provocation.

The broader political science community has inevitably and positively shaped this book. As I began the project, conversations with Alex Cooley, Julia Gray, Nate Jensen, Stan Markus, and Rachel Wellhausen were especially helpful. As I was making some headway, discussions with Vincent Arel-Bundock, Phil Ayoub, Sarah Bauerle Danzman, Andrea Binder, Jeff Colgan, Adam Dean, Stefan Eich, Marty Finnemore, Orfeo Fioretos, Tim Frye, Alex Gourevitch, Leslie Johns, Juliet Johnson, Filiz Kahraman, Peter Katzenstein, Thomas Oatley, Krzysztof Pelc, Thomas Pepinsky, Elliot Posner, Alex Reisinbichler, Bryn Rosenfeld, Mark Schwartz, Swati Srivastava, Taylor St. John, David Szakonyi, Todd Tucker, and Will Winecoff all pushed me in interesting directions. I am also thankful to audiences at International Studies Association 2019, American Political Science Association 2021, Brown, Cornell, George Washington, Purdue, Temple, Toronto, and University College London for all their insights.

Four individuals have taken me under their wings in unexpected and undeserving ways. Eric Helleiner co-organized a workshop where I first presented some of these ideas. Getting such encouragement from an International Political Economy giant, both at the start and at the tail end, of this manuscript was an absolute privilege. Lou Pauly has been a champion of mine before he even met me. Jason Sharman has taught me more about the offshore world than anyone. His broad-ranging interests and tenacious execution, while still being so wonderfully kind, is something I aspire toward. Igor Logvinenko is the rare (possibly only) person I've met who has shared the same intersection of academic and existential angst.

The intersection of mostly sad indie pop and interest in IPE is, to my knowledge, rather limited. Calvin Thrall and Lorenzo Crippa both entered my life in roundabout ways and have become such terrifically fun collaborators. I owe an especially large debt to Calvin, who first worked on some of the ISDS data presented in this book.

I decided to submit the manuscript to Cornell because I respect many of the books the press has published, and because I had always heard that their review process is generative. My time with them fully lived up to the hype. Jackie Teoh has been a phenomenal editor, going far beyond her remit to become someone I was able to workshop many inchoate ideas. My reviewers were a terrific blend of generous and critical. Parts of the book that I think provide the greatest value would simply not exist without their questioning. In the final stages of the process, I also got extremely useful advice from Martin Soames and Blaise Gaymer. I would not feel right about publishing this without their aid.

While it doesn't necessarily come through in the contents of the book, I could not have developed the ideas in this manuscript without reading Rawi Abdelal, Oliver Bullough, Albert Hirschman, Michel Kalecki, Charles Lindblom, Katharina Pistor, and Adam Tooze.

Arjun, Brett, Chelsea, Evan, John, Manu, Nikos, Raaj, Shilpa, Tom, Yuxi, and Wayne have all provided me the support and balance I needed to see this through. Julie was in this from the very start, and it is bittersweet that she wasn't there for the end.

This book was written over almost eight years. Thank you to the following entities for providing me company on the best and worst of writing days: Big Red Machine; Big Thief; James Blake; Bon Iver; Caroline Rose; Dijon; Kendrick Lamar; Khruangbin; Maribou State; Mk.Gee; and, at the most critical of stages, The War on Drugs.

Nothing I have done or do in my life would be possible without the never-ending love and support of my family. My sister's ability to expand her horizons and bring others in keeps impressing me. My parents are the rare cohesive unit who genuinely balance each other and use that to support those closest to them through any vagary life decides to throw their way.

Notes

INTRODUCTION

1. The use of the term "plutocrat" is more specific than the use of "elite" in comparative politics, which can refer to political, commercial, or cultural positions. Plutocrats can constitute a large part of *the* political elite (politicians can be plutocrats) but also often operate informally outside it. I define "plutocrats" in a manner similar to recent scholarship referenced above (Winters 2011; Markus and Charnysh 2017): individuals who control massive concentrations of money that can be used to build and guard their wealth and thereby their political and social positions.

2. *De Molestina & Ors v. Ponton & Ors*, [2002] EWHC 2413 (Comm).

3. *Zambia, Attorney General of Zambia for and On Behalf of v. Meer Care & Desai (a firm) & Ors*, [2007] EWHC 952 (Ch). *Islamic Republic of Pakistan v. Zardari & Ors*, [2006] EWHC 2411 (Comm).

4. *State of Brunei & Ors v. HRH Prince Jefri & Ors*, [2008] EWHC 1247 (Ch).

5. *Tajik Aluminium Plant v. Ermatov & Ors*, [2005] EWHC 2241 (Ch).

6. As legal scholars have long recognized, litigation should be seen as a last-ditch attempt to resolve disputes given the high monetary and informational costs. For a particularly cogent typology on different dispute resolution strategies, see Hendley et al. (2000, 630). On how dispute strategies interact, see Albiston et al. (2014).

7. *Amin Rasheed Shipping Corp v. Kuwait Insurance Co.*, [1983] 1 WLR 241 (HL) and *Spiliada Maritime Corp v. Cansulex Ltd*, [1986] AC 460 (HL) are the primary precedents to determine if the United Kingdom is considered an inappropriate forum. For full details on the United Kingdom's approach to conflict of laws, see Collins (2006), chaps. 11 and 12.

8. *Sawiris & Ors v. Marwan*, [2010] EWHC 89 (Comm).

9. *Rochester Resources Ltd & Ors v. Lebedev & Anor*, [2014] EWHC 2926 (Comm).

10. *North Shore Ventures Ltd v. Anstead Holdings Inc & Ors*, [2010] EWHC 1485 (Ch).

11. *Gorbunova v. Berezovsky (aka Platon Elenin) & Ors*, [2013] EWHC 76 (Ch).

12. *Pacific International Sports Clubs Ltd v. Soccer Marketing International Ltd & Ors*, [2009] EWHC 1839 (Ch).

13. *(1) Jsc Mezhdunarodniy Promyshlenniy Bank (2) State Corporation "Deposit Insurance Agency" v. Sergei Viktorovich Pugachev*, [2014] EWHC 3547 (Ch).

14. *Nigeria v. Santolina Investment Corp & Ors*, [2007] EWHC 437 (Ch).

15. *Boreh v. Republic of Djibouti & Ors*, [2015] EWHC 769 (Comm).

16. Some investment treaties do vary on what constitutes an "investment" and provide further clarification on whether such extraterritorial cases may be disallowed.

17. The figures represent the data from the London courts and the United Nations Commission on International Trade Law (UNCITRAL) described in the previous sections aggregated to the country level.

18. Some of the lawyers had been working with the same Russians for more than a decade, while others worked with a range of actors from plutocrats to SOEs. I also interviewed lawyers brought into cases for their specific areas of expertise such as fraud or asset tracing. Several of the interviewees work at some of the most historic, reputable firms and chambers in the city; others operate from newer partnerships that focus on

Eastern European corporate law and dispute resolution. In sum, I obtained data from the full range of enablers of extraterritorial dispute resolution to assess the weight of the economic and political drivers of outsourcing authority.

1. LIBERALISM AS A LAST RESORT

1. Lawyer19, January 2018. Given the high stakes nature of the disputes discussed in the book, I anonymize all quotes from interview participants. Lawyer identification numbers are randomized within each month of the interviews conducted.

2. Lawyer16, January 2018.

3. Lawyer40, February 2018.

4. Based on my calculations. The only exception is China.

5. Lawyer29, February 2018.

6. Lawyer55, July 2018.

7. Lawyer56, July 2018.

8. If they lie, they can be held in contempt of court and subject to years in prison, as happened to Mukhtar Ablyazov and Cem Uzan, whose political and legal plights will be discussed in chapter 6.

9. Lawyer47, March 2018.

10. Based on my calculations.

11. This book is heavily indebted to Winters's scholarship. We both focus on the tactics used by super-wealthy individuals to guard and protect their wealth. But my goal and Winters's goal differ. The purpose of his book is to define different forms of oligarchy rather than specific types of oligarchs, which is the form that Winters argues is the most important for contemporary politics is the development of civil oligarchies, where oligarchs have given up the means of coercion and delegated powers to the state in order to create a more predictable, tame setting. My argument focuses on contemporary politics and is valid only in places that have some semblance of state authority (even if this may be diminished vis-à-vis oligarchs), so I think it is appropriate to use his definition of "oligarchs" as my definition of plutocrats in line with more recent scholarship on the topic (Markus and Charnysh 2017, 1633). From a strict reading of Winters, you could say that he might see me creating a classification of oligarchs within civil oligarchies, but my book further differs from his by also focusing on the means through which actors gain their wealth: through the private sphere or through the state. The latter is largely absent from Winters's analysis as a method to derive wealth. One exception is in his logic of sultanistic oligarchies, where one individual steps forward to create order among the oligarchs, but his examples rely on individuals who had mostly already developed their wealth prior to taking office (Haji Mohammad Suharto in Indonesia, Ferdinand Marcos in the Philippines) rather than through the looting of the state that is key to developing wealth in my definition of kleptocrats and cronies. Another way to put it is that Winters analyzes oligarchies but does not believe every emerging market *has to be* an oligarchy. I assume that oligarch-like actors, what I deem as plutocrats (those with substantial material wealth that can be used to guard their positions) can exist independent of whether the regime as a whole should be classified as an oligarchy.

12. For details on the United Kingdom's sanctions against Igor Sechin, see Foreign, Commonwealth & Development Office, "Abramovich and Deripaska Among 7 Oligarchs Targeted in Estimated £15 Billion Sanction Hit," press release, UK Government, March 10, 2022, https://www.gov.uk/government/news/abramovich-and-deripaska-among-seven-oligarchs-targeted-in-estimated-15bn-sanction-hit. For details on the United Kingdom's sanctions against the Rotenbergs, see Spencer Woodman, "UK Targets Three Oligarchs and Five Russian Banks in First Tranche of New Sanctions," *International Consortium of Investigative Journalists*, February 22, 2022, https://www.icij.org/investigations/pandora

-papers/uk-targets-three-oligarchs-and-five-russian-banks-in-first-tranche-of-new-sanctions/. Comprehensive details on the United Kingdom's sanctions against Russia can be found at Foreign, Commonwealth & Development Office, "Russia Sanctions Regime: Industry and Specialist Guidance," UK Government, April 15, 2019, last updated June 27, 2025, https://www.gov.uk/government/collections/uk-sanctions-on-russia.

13. Lawyer40, February 2018.

14. Lawyer34, February 2018.

15. The recognition of foreign judgments is governed by principles of comity and reciprocity, coupled with a patchwork of bilateral treaties. Tax havens respect the judgments of courts in New York and London because of their reputations, so tax havens will be more likely to enforce their judgments over those from weakly institutionalized states. Judgments from London are particularly useful because many crown dependencies and territories, such as the British Virgin Islands and Guersney, act as sites for plutocrats to stash their wealth. See Michaels (2009) for further details.

2. CRACKS IN THE OLIGARCHY

1. For example, Chrystia Freeland (2005, 131) reports, "By the mid-1990s, red directors and *kooperators* were already salting away tens of millions of dollars in Swiss bank accounts; yet, like the rest of the Russian establishment, they had only the foggiest notion of how advanced capitalism worked."

2. Based on data described in chapter 1. This does not mean that plutocrats were not using English law or bargaining in the shadow of the English courts or arbitration. But they rarely went to judgment.

3. My analysis of WestLaw cases in the Chancery Division and commercial courts in London.

4. Lawyer17, January 2018.

5. Lawyer31, February 2018.

6. Lawyer40, February 2018.

7. Lawyer30, February 2018.

8. Lawyer25, February 2018.

9. *Berezovsky v. Abramovich* [2012] EWHC 2463 (Comm), 2012 WL 3963559 para. 228.

10. *Berezovsky v. Abramovich* [2012] EWHC 2463 (Comm), 2012 WL 3963559 para. 154.

11. The incident is covered in detail by a number of writers, including Freeland (2005), Klebnikov (2000), and Hoffman (2011).

12. Browder's tune has certainly changed since his own blacklisting by the Putin regime. For an informative, personal account of his rise, fall, and anticorruption efforts, see Browder (2015).

13. For a detailed examination of the facts, see Sakwa (2014, 114–17).

3. EXTRATERRITORIAL DISPUTES AS PLUTOCRATIC INSURANCE

1. Based on data on public judgments described in chapter 1.

2. Lawyer15, January 2018.

3. Lawyer23, January 2018.

4. This section summarizes the rise of Vladimir Putin and draws on a number of important academic and journalistic texts, including Dawisha (2015), Gessen (2013), Goldman (2008), Sakwa (2007), Mezrich (2016), Myers (2015), and Zygar (2016).

5. Between 1996 and 2006, Russian state revenue from oil increased tenfold, from $14 billion to $140 billion (Sakwa 2014, 33).

6. For detailed examinations of the media under Putin, see Gehlbach (2010), Kagarlitsky (2002), and Ostrovsky (2017).

7. Lawyer39, February 2018.
8. Lawyer15, January 2018.
9. Lawyer40, February 2018.
10. Lawyer54, July 2018.
11. Lawyer43, February 2018.
12. Lawyer54, July 2018.
13. Lawyer6, December 2017.
14. Lawyer49, July 2018.
15. Lawyer39, February 2018.
16. Lawyer40, February 2018.
17. Lawyer15, January 2018.
18. Lawyer20, January 2018.
19. Lawyer47, March 2018.
20. Lawyer58, July 2018.
21. Lawyer30, February 2018.
22. Lawyer47, March 2018.
23. *Berezovsky v. Abramovich* [2012] EWHC 2463 (Comm) at [595].
24. *Berezovsky v. Abramovich* [2012] EWHC 2463 (Comm) at [595].
25. *Berezovsky v. Abramovich* [2012] EWHC 2463 (Comm) at [269].
26. *Berezovsky v. Abramovich* [2012] EWHC 2463 (Comm) at [597].
27. For conflicting views on the ORT negotiations, see *Berezovsky v. Abramovich* [2012] EWHC 2463 (Comm) at [595–607].
28. For full details, see *Berezovsky v. Abramovich* [2012] EWHC 2463 (Comm), 2012, Sections 11 and 12.
29. For a detailed look at the suspected murders by the Russian government on British soil, and in particular the targeting of Berezovsky's inner circle, see Blake (2019).
30. For details on the claims, see *Berezovsky v. Abramovich* [2012] EWHC 2463 (Comm) at [6–7].
31. Lawyer17, January 2018.
32. Lawyer15, January 2018.
33. Lawyer51, July 2018.
34. Lawyer14, January 2018.
35. Lawyer47, March 2018.
36. Lawyer60, July 2018.
37. Lawyer40, February 2018.
38. Lawyer8, January 2018.
39. Lawyer21, January 2018.
40. *Berezovsky v. Abramovich* [2012] EWHC 2463 (Comm) at [34].
41. *Berezovsky v. Abramovich* [2012] EWHC 2463 (Comm).
42. *Berezovsky v. Abramovich* [2012] EWHC 2463 (Comm) at [123].
43. *Berezovsky v. Abramovich* [2012] EWHC 2463 (Comm) at [99 (iii)].
44. *Berezovsky v. Abramovich* [2012] EWHC 2463 (Comm) at [105].
45. *Berezovsky v. Abramovich* [2012] EWHC 2463 (Comm) at [453–455].
46. *Berezovsky v. Abramovich* [2012] EWHC 2463 (Comm) at [100].
47. *Berezovsky v. Abramovich* [2012] EWHC 2463 (Comm) at [38].
48. *Berezovsky v. Abramovich* [2012] EWHC 2463 (Comm) at [96].
49. Lawyer32, February 2018.
50. For details on the investigation into the circumstances surrounding Berezovsky's death, see Blake (2019, 273–77).
51. Based on several lawyer interviews.

52. *Hulley Enterprises Limited (Cyprus) v. The Russian Federation*, PCA Case No. AA 226 18 July 2014, 258–59.

53. For a detailed understanding of the legal aspects of the takeover, see Stephan (2013).

54. For a detailed examination of the trial period, see Sakwa (2014, 91–111).

55. *Hulley Enterprises Limited (Cyprus) v. The Russian Federation*, PCA Case No. AA 226 18 July 2014, 218.

56. Note that there are dozens of Yukos-related litigations that have happened over the past decade and a half. The extraterritorial arbitrations primarily involved members of the magnificent seven, while others were driven by minority shareholders. For clarity's sake, I only recap some of the most consequential legal episodes.

57. *Hulley Enterprises Limited (Cyprus) v. The Russian Federation*, PCA Case No. AA 226 18 July 2014, 408–23.

58. Note that the Energy Charter Treaty (ECT) does not explicitly contain a clean-hands-doctrine clause, but the tribunal interpreted it as a general principle of international investment law.

59. *Hulley Enterprises Limited (Cyprus) v. The Russian Federation*, PCA Case No. AA 226 18 July 2014, 435.

60. Lawyer14, January 2018.

61. Lawyer22, January 2018.

62. Lawyer5, December 2017.

63. Lawyer21, January 2018.

64. Lawyer10, January 2018.

4. EXTRATERRITORIAL DISPUTES AS STATE RETALIATION

1. First seen at Dawisha (2015, 335).

2. Lawyer18, January 2018.

3. Lawyer54, July 2018.

4. Lawyer39, February 2018.

5. Lawyer33, February 2018.

6. Lawyer53, July 2018.

7. Lawyer17, January 2018.

8. Lawyer13, January 2018.

9. Lawyer13, January 2018.

10. Lawyer45, February 2018.

11. Lawyer51, July 2018.

12. Lawyer4, December 2017.

13. Lawyer60, July 2018.

14. *Berezovsky v. Abramovich* [2012] EWHC 2463 (Comm) at [89].

15. *Berezovsky v. Abramovich* [2012] EWHC 2463 (Comm) at [89].

16. Lawyer21, January 2018.

17. Lawyer14, January 2018.

18. Lawyer30, February 2018.

19. Lawyer14, January 2018.

20. Lawyer32, February 2018.

21. Lawyer21, January 2018.

22. *PJSC Aerflot v. Leeds* [2018] EWHC 1735 (Ch) at [11].

23. *PJSC Aerflot v. Leeds* [2018] EWHC 1735 (Ch) at [87].

24. *Hulley Enterprises Limited (Cyprus) v. The Russian Federation*, PCA Case No. AA 226 18 July 2014, 307.

25. *Hulley Enterprises Limited (Cyprus) v. The Russian Federation*, PCA Case No. AA 226 18 July 2014, 307.

26. *Hulley Enterprises Limited (Cyprus) v. The Russian Federation*, PCA Case No. AA 226 18 July 2014, 307.
27. Lawyer21, January 2018.
28. Lawyer14, January 2018.
29. Lawyer54, July 2018.
30. Lawyer54, July 2018.
31. Lawyer26, January 2018.
32. Lawyer31, February 2018.
33. Lawyer11, January 2018.
34. Lawyer34, February 2018.
35. Lawyer25, January 2018.
36. Lawyer15, January 2018.
37. Lawyer14, January 2018.
38. Lawyer53, July 2018.
39. Lawyer47, March 2018.
40. Lawyer13, January 2018.
41. Lawyer54, July 2018.
42. Note that the potential fee fluctuates by share price, so the $560 million value corresponds to the share price at the time of the agreement.
43. Lawyer47, March 2018.
44. Lawyer50, July 2018.
45. Lawyer29, February 2018.
46. Lawyer33, February 2018.
47. Lawyer50, July 2018.
48. Lawyer31, February 2018.
49. Lawyer14, January 2018.

5. WEAPONIZED LEGAL DEPENDENCE

1. *JSC BTA Bank v. Mukhtar Ablyazov, Roman Solodchenko, Drey Associates Limited, Zhaksylyk Zharimbetov* [2011] EWHC 1136 (Comm) and *JSC BTA Bank v. Mukhtar Ablyazov and Others* [2011] EWHC 202 (Comm).
2. *BTA Bank v. Ablyazov* [2011] EWHC 1136 (Comm) at [53].
3. *Libananco Holdings Co. Limited v. Republic of Turkey*, ICSID case no. Arb/06/8, Decision on Preliminary Issues, Dated June 23, 2008, 17.
4. *Libananco Holdings Co. Limited v. Republic of Turkey*, ICSID case no. Arb/06/8, Decision on Preliminary Issues, Dated June 23, 2008, 4.
5. *Libananco Holdings Co. Limited v. Republic of Turkey*, ICSID case no. Arb/06/8, Decision on Preliminary Issues, Dated June 23, 2008, 18.
6. *Libananco Holdings Co. Limited v. Republic of Turkey*, ICSID case no. Arb/06/8, Award, Date of Dispatch to the Parties September 2, 2011, 17.
7. *Libananco Holdings Co. Limited v. Republic of Turkey*, ICSID case no. Arb/06/8, Award, Date of Dispatch to the Parties September 2, 2011, 11.
8. *Mr. Saba Fakes v. Republic of Turkey*, ICSID case no. Arb/07/20, Award, Date of Dispatch to the Parties July 14, 2010, 11.
9. *Mr. Saba Fakes v. Republic of Turkey*, ICSID case no. Arb/07/20, Award, Date of Dispatch to the Parties July 14, 2010, 17.
10. *Mr. Saba Fakes v. Republic of Turkey*, ICSID case no. Arb/07/20, Award, Date of Dispatch to the Parties July 14, 2010, 43.
11. *Mr. Saba Fakes v. Republic of Turkey*, ICSID case no. Arb/07/20, Award, Date of Dispatch to the Parties July 14, 2010, 44.
12. *Cem Cengiz Uzan v. Repulic of Turkey*, Arbitration V 2014/023, Award on Respondent's Bifurcated Preliminary Objection April 20, 2016, 35.

CONCLUSION

1. For a broader examination of the politics of transparency across regimes, see Berliner (2014) and Hollyer et al. (2018).

2. For a detailed examination of the Ukraine-Russia bond standoff, see Hess (2018) and Kalyanpur (2023).

3. Lawyer47, March 2018.

4. Lawyer13, January 2018.

5. Lawyer33, February 2018.

6. Lawyer31, February 2018.

References

Abdelal, Rawi. 2015. "The Multinational Firm and Geopolitics: Europe, Russian Energy, and Power." *Business and Politics*. http://www.degruyter.com/view/j/bap.ahead-of-print/bap-2014-0044/bap-2014-0044.xml.

Acemoglu, Daron, and James A. Robinson. 2005. *Economic Origins of Dictatorship and Democracy*. Cambridge University Press.

Acemoglu, Daron, and James A. Robinson. 2013. *Why Nations Fail: The Origins of Power, Prosperity, and Poverty*. Crown Business.

Açikel, Fethi. 2003. "Mapping the Turkish Political Landscape Through November 2002 Elections." *Journal of Southern Europe and the Balkans* 5 (2): 185–203. https://doi.org/10.1080/1461319032000097932.

Adamson, Fiona. 2013. "Mechanisms of Diaspora Mobilization and the Transnationalization of Civil War." *Transnational Dynamics of Civil War*, 63–88.

Akin, Melissa. 2009. "YUKOS Unit Wins $389 Mln Dutch Ruling vs Rosneft -Paper." *Reuters*, April 29. https://uk.reuters.com/article/russia-rosneft-yukos-idUKLT73368620090429.

Albertus, Michael, and Victor Menaldo. 2012. "If You're Against Them You're with Us: The Effect of Expropriation on Autocratic Survival." *Comparative Political Studies* 45 (8): 973–1003.

Albertus, Michael, and Victor Menaldo. 2014. "Gaming Democracy: Elite Dominance During Transition and the Prospects for Redistribution." *British Journal of Political Science* 44 (3): 575–603.

Albiston, Catherine R., Lauren B. Edelman, and Joy Milligan. 2014. "The Dispute Tree and the Legal Forest." *Annual Review of Law and Social Science* 10 (1): 105–31. https://doi.org/10.1146/annurev-lawsocsci-110413-030826.

AllAfrica. 2020. "How Africa's Richest Woman Exploited Family Ties, Shell Companies and Inside Deals to Build an Empire." *Premium Times*, January 20.

Almeida, Henrique. 2022. "Angola's Isabel dos Santos Considers Presidential Run from Exile." *Bloomberg*, November 25. https://www.bloomberg.com/news/articles/2022-11-25/isabel-dos-santos-says-from-exile-she-may-run-to-become-president-of-angola.

Alstadsæter, Annette, Niels Johannesen, and Gabriel Zucman. 2017. "Who Owns the Wealth in Tax Havens? Macro Evidence and Implications for Global Inequality." Working Paper 23805. National Bureau of Economic Research. https://doi.org/10.3386/w23805.

Amsden, Alice H. 2010. "Elites and Property Rights." UNU-Wider Working Paper No. 2010/109. https://www.wider.unu.edu/sites/default/files/wp2010-109.pdf.

Amsden, Alice H., Alisa DiCaprio, and James A. Robinson. 2012. *The Role of Elites in Economic Development*. Oxford University Press.

Ananyev, Maxim. 2018. "Political Economy of Offshore Finance." PhD diss., University of California, Los Angeles.

Ang, Yuen Yuen. 2020. *China's Gilded Age: The Paradox of Economic Boom and Vast Corruption*. Cambridge University Press.

Ang, Yuen Yuen, and Nan Jia. 2014. "Perverse Complementarity: Political Connections and the Use of Courts Among Private Firms in China." *Journal of Politics* 76 (2): 318–32.

AP News. 2018. "Escort Says Oligarch Now Has Tapes on Russia Interference." *AP News*, August 20. https://apnews.com/article/78eef0ecec49412b816ac5169d63279a.

Aron, Janine. 2000. "Growth and Institutions: A Review of the Evidence." *World Bank Research Observer* 15 (1): 99–135.

Associated Press Newswire. 2003. "Turkish Prime Minister Seeks Defamation Damages from Rival." *Associated Press*, June 25.

Avant, Deborah D., Martha Finnemore, and Susan K. Sell. 2010. *Who Governs the Globe?* Vol. 114. Cambridge University Press.

Aykut, Dilek, Apurva Sanghi, and Gina Kosmidou. 2017. *What to Do When Foreign Direct Investment Is Not Direct or Foreign: FDI Round Tripping*. The World Bank.

Babic, Milan, Javier Garcia-Bernardo, and Eelke M. Heemskerk. 2020. "The Rise of Transnational State Capital: State-Led Foreign Investment in the 21st Century." *Review of International Political Economy* 27 (3): 433–75. https://doi.org/10.1080/09692290.2019.1665084.

Baccini, Leonardo, Pablo M. Pinto, and Stephen Weymouth. 2017. "The Distributional Consequences of Preferential Trade Liberalization: Firm-Level Evidence." *International Organization* 71 (2): 373–95. https://doi.org/10.1017/S002081831700011X.

Bach, David. 2010. "Varieties of Cooperation: The Domestic Institutional Roots of Global Governance." *Review of International Studies* 36 (3): 561–89.

Bartley, Tim. 2007. "Institutional Emergence in an Era of Globalization: The Rise of Transnational Private Regulation of Labor and Environmental Conditions." *American Journal of Sociology* 113 (2): 297–351.

Bayer, Ralph-C., Roland Hodler, Paul A. Raschky, and Anthony Strittmatter. 2020, March. "Expropriations, Property Confiscations and New Offshore Entities: Evidence from the Panama Papers." *Journal of Economic Behavior & Organization* 171: 132–52. https://doi.org/10.1016/j.jebo.2020.01.002.

BBCAP. 2020. "Portuguese Court Orders Seizure of Isabel dos Santos' Assets." *BBC Monitoring Africa*, March 15.

BBC News. 2000. "Putin Admits 'Guilt' for Sub Disaster." *BBC News*, August 23. http://news.bbc.co.uk/1/hi/world/europe/893294.stm.

BBC News. 2003. "UK Grants Asylum to Russian Tycoon." *BBC News*, September 10. http://news.bbc.co.uk/1/hi/business/3097880.stm.

BBC News. 2013. "Profile: Mikhail Khodorkovsky." *BBC News*, December 22, sec. Europe. https://www.bbc.com/news/world-europe-12082222.

BBC News. 2020. "Dutch Court Backs $50bn Yukos Claim Against Russia." *BBC News*, February 18, sec. Europe. https://www.bbc.com/news/world-europe-51547011.

BBC News. 2021. "Nikolai Glushkov: Putin Critic 'Strangled in London Home by Third Party.'" *BBC News*, April 9, sec. London. https://www.bbc.com/news/uk-england-london-56695489.

Belokurova, G. 2014. "When Does Business Turn Violent? Elections and Business-Related Violence in Russia, 1995–2010." PhD diss., University of Wisconsin, Madison.

Belton, Catherine. 2008a. "Close to the Wind: Russia's Oligarchs." *Financial Times*, October 24, sec. Financial Times. https://www.ft.com/content/d96aa8ac-a1f9-11dd-a32f-000077b07658.

Belton, Catherine. 2008b. "Moscow Forces Truce in Battle for Norilsk." *Financial Times*, November 25. https://www.ft.com/content/c30b4566-baff-11dd-bc6c-0000779fd18c.

Belton, Catherine. 2008c. "Putin Ally Set to Take Helm at Norilsk Nickel." *Financial Times*, July 30. https://www.ft.com/content/dc92db1e-5dc3-11dd-8129-000077b07658.

Belton, Catherine. 2008d. "Russian Rivals Win Seats on Norilsk Nickel Board." *Financial Times*, June 30. https://www.ft.com/content/05811a96-46c6-11dd-876a-0000779fd2ac.

Belton, Catherine. 2010a. "Accusations Reignite Debate Over Norilsk." *Financial Times*, June 29. https://www.ft.com/content/8fd7d79c-8306-11df-8b15-00144feabdc0.

Belton, Catherine. 2010b. "Deripaska Steps up Battle with Norilsk." *Financial Times*, July 30. https://www.ft.com/content/bf8f7428-9c0f-11df-a7a4-00144feab49a.

Belton, Catherine. 2010c. "Pressure Intensifies on Nickel Tycoon." *Financial Times*, August 5. https://www.ft.com/content/bd528bb8-a0b6-11df-badd-00144feabdc0.

Belton, Catherine. 2010d. "Probe Uncovers 'Violations' at Norilsk AGM." *Financial Times*, September 15. https://www.ft.com/content/828c9930-c0bc-11df-94f9-00144feab49a.

Belton, Catherine. 2010e. "Rusal Takes Norilsk Dispute to Court." *Financial Times*, August 11. https://www.ft.com/content/0ad2a414-a570-11df-a5b7-00144feabdc0.

Belton, Catherine. 2010f. "UC Rusal Hits out at Norilsk $3bn Buy-Back." *Financial Times*, December 29. https://www.ft.com/content/e601008c-137d-11e0-a367-00144feabdc0.

Belton, Catherine. 2011. "UC Rusal Wins Ruling Against Norilsk Deals." *Financial Times*, February 8. https://www.ft.com/content/24ab8c06-32e3-11e0-9a61-00144feabdc0.

Belton, Catherine. 2020. *Putin's People: How the KGB Took Back Russia and Then Took on the West*. Farrar, Straus and Giroux.

Belton, Catherine, and Michael Stott. 2016. "Rosneft Lawyers Manipulated Yukos Rulings, Emails Suggest." *Financial Times*, November 28. https://www.ft.com/content/56650382-b495-11e6-ba85-95d1533d9a62.

Berger, Suzanne. 2000. "Globalization and Politics." *Annual Review of Political Science* 3 (1): 43–62.

Berliner, Daniel. 2014. "The Political Origins of Transparency." *Journal of Politics* 76 (2): 479–91.

Binder, Andrea. 2019. "All Exclusive: The Politics of Offshore Finance in Mexico." *Review of International Political Economy* 26 (2): 313–36. https://doi.org/10.1080/09692290.2019.1567571.

Blake, Heidi. 2019. *From Russia with Blood: The Kremlin's Ruthless Assassination Program and Vladimir Putin's Secret War on the West*. Mulholland.

Bland, Stephen M. 2018. "The Ablyazov Affair: 'Fraud on an Epic Scale.'" *The Diplomat*, February 23. https://thediplomat.com/2018/02/the-ablyazov-affair-fraud-on-an-epic-scale/.

Blas, Javier, and Catherine Belton. 2011. "New Twist in Riddle of Norilsk Stake." *Financial Times*, December 19. https://www.ft.com/content/dfe024ae-2a5f-11e1-9bdb-00144feabdc0.

Bond, David, Neil Buckley, and Jane Croft. 2018. "Murder Probe Launched into Russian Businessman." *Financial Times*, March 16. https://www.ft.com/content/0baeef76-2945-11e8-b27e-cc62a39d57a0.

Bookman, Pamela K. 2019. "The Adjudication Business." *Yale Journal of International Law* 45 (2): 227–83.

Bradford, Anu. 2012. "The Brussels Effect." *Northwestern University Law Review* 107: 1–68.

Brauch, Martin Dietrich. 2014. "Yukos v. Russia: Issues and Legal Reasoning Behind US $50 Billion Awards." *Investment Treaty News*, September 4. https://www.iisd.org/itn/2014/09/04/yukos-v-russia-issues-and-legal-reasoning-behind-us50-billion-awards/.

Braun, Benjamin. 2015. "From Performativity to Political Economy: Index Investing, ETFs and Asset Manager Capitalism." *New Political Economy* 21 (3): 257–73. https://doi.org/10.1080/13563467.2016.1094045.

Brick Court Chambers. 2012. "Berezovsky Granted Summary Judgment: Enforcement of Russian 'Indexation' Judgment Contrary to English Public Policy." *Brick Court Chambers*, October 30. https://www.brickcourt.co.uk/news/detail/berezovsky-granted-summary-judgment-enforcement-of-russian-indexation-judgment-contrary-to-english-public-policy.

Browder, Bill. 2015. *Red Notice: A True Story of High Finance, Murder, and One Man's Fight for Justice*. Simon and Schuster.

Buckley, Nick. 2007. "Putin's Lobbying Wins Sochi the Olympics." *Financial Times*, July 5. https://www.ft.com/content/fe4ad9cc-2ad0-11dc-85f9-000b5df10621.

Bullough, Oliver. 2020. "Oliver Bullough on Twitter: 'Game Back On. This Case Has Been Running for Years and—If the Russian Government Doesn't Find a Way to Overturn It Again—Resulting Court Cases Could Prove a Major Impediment to Russian State Companies Trying to Operate in the West." *Twitter*, February 18. https://twitter.com/OliverBullough/status/1229710419780128774.

Burger, Ethan S. 2004, Spring. "Corruption in the Russian Arbitrazh Courts: Will There Be Significant Progress in the Near Term." *International Lawyer* 38 (1): 15–34.

Burgis, Tom. 2017. "Spies, Lies and the Oligarch: Inside London's Booming Secrets Industry." *Financial Times*, September 28. https://www.ft.com/content/1411b1a0-a310-11e7-9e4f-7f5e6a7c98a2.

Burgis, Tom. 2020. *Kleptopia: How Dirty Money Is Conquering the World*. William Collins.

Burke, Ingrid. 2014. "London High Court Allows Aeroflot Appeal in Berezovsky Case." *RAPSI* (Russian Legal Information Agency), January 16. http://www.rapsinews.com/news/20140116/270462492.html.

Business Week. 2021. "How Isabel dos Santos Went Broke." *Business Week*, December 20.

Cárdenas, Julián. 2016. "Why Do Corporate Elites Form Cohesive Networks in Some Countries, and Do Not in Others? Cross-National Analysis of Corporate Elite Networks in Latin America." *International Sociology* 31 (3): 341–63. https://doi.org/10.1177/0268580916629965.

CE NoticiasFinancieras. 2020a. "Angolan State Wins Case Against Isabel dos Santos at Paris Court." *CE NoticiasFinancieras*, July 13.

CE NoticiasFinancieras. 2020b. "Isabel dos Santos Negotiates to Pay Back Money to Angola." *CE NoticiasFinancieras*, February 1.

Chibelushi, Wedaeli. 2024. "Isabel dos Santos: Angolan Billionaire Rejects Fresh Criminal Charges." *BBC News*, January 16, sec. Africa. https://www.bbc.com/news/world-africa-67998451.

Chwieroth, Jeffrey. 2007. "Neoliberal Economists and Capital Account Liberalization in Emerging Markets." *International Organization* 61 (2): 443–63.

Collins, Lawrence, ed. 2006. *Dicey, Morris and Collins on the Conflict of Laws*. Sweet and Maxwell. https://archive.org/details/diceymorriscolli0001unse/page/n9/mode/2up.

Colpan, Asli M. 2010. *Business Groups in Turkey*. Oxford University Press. https://doi.org/10.1093/oxfordhb/9780199552863.003.0017.

Colton, Timothy J., and Michael McFaul. 2000. "Reinventing Russia's Party of Power: 'Unity' and the 1999 Duma Election." *Post-Soviet Affairs* 16 (3): 201–24.

Colton, Timothy J., and Michael McFaul. 2003. *Popular Choice and Managed Democracy: The Russian Elections of 1999 and 2000*. Brookings Institution Press.

Cooley, Alexander A. 2005. *Logics of Hierarchy: The Organization of Empires, States, and Military Occupation*. Cornell University Press.

Cooley, Alexander A., and John Heathershaw. 2017. *Dictators Without Borders: Power and Money in Central Asia*. Yale University Press.

Cooley, Alexander A., and Jason Sharman. 2017. "Transnational Corruption and the Globalized Individual." *Perspectives on Politics* 15 (3): 732–53. https://doi.org/10.1017/S1537592717000937.

Crasnic, Lori, Nikhil Kalyanpur, and Abraham Newman. 2017. "Networked Liabilities: Transnational Authority in a World of Transnational Business." *European Journal of International Relations* 23 (4): 906–29.

Crippa, Lorenzo, and Nikhil Kalyanpur. 2024. "American Economic Coercion and Elite Re-Globalization." Working Paper. https://lorenzo-crippa.github.io/files/FCPA_offshore_wealth.pdf.

Cuadros, Alex. 2016. *Brazillionaires: Wealth, Power, Decadence, and Hope in an American Country*. Spiegel & Grau.
Cummings, Sally N. 2005. *Kazakhstan*. Tauris.
Dammann, Jens, and Henry Hansmann. 2008. "Globalizing Commercial Litigation." *Cornell Law Review* 94 (1): 1–71. https://papers.ssrn.com/sol3/papers.cfm?abstract_id=1373848.
Damro, Chad. 2012. "Market Power Europe." *Journal of European Public Policy* 19 (5): 682–99.
Dancy, Geoff, and Verónica Michel. 2016, August. "Human Rights Enforcement from Below: Private Actors and Prosecutorial Momentum in Latin America and Europe." *International Studies Quarterly* 60 (1): 173–88. https://doi.org/10.1111/isqu.12209.
Darden, Keith. 2008. "The Integrity of Corrupt States: Graft as an Informal State Institution." *Politics & Society* 36 (1): 35–59.
Davies, Megan, Vladimir Soldatkin, and Gilbert Reilhac. 2014. "European Court Rules Russia Must Pay Yukos Shareholders 1.9 Billion." *Reuters*, July 31. https://uk.reuters.com/article/uk-russia-yukos-echr-idUKKBN0G00QO20140731.
Davis, Aeron, and Karel Williams. 2017. "Introduction: Elites and Power After Financialization." *Theory, Culture & Society* 34 (5–6): 3–26.
Dawisha, Karen. 2015. *Putin's Kleptocracy: Who Owns Russia?* Simon and Schuster.
de Oliveira, Ricardo Soares. 2015. *Magnificent and Beggar Land: Angola Since the Civil War*. Oxford University Press.
Desai, Padma. 2005. "Russian Retrospectives on Reforms from Yeltsin to Putin." *Journal of Economic Perspectives* 19 (1): 87–106.
Dezalay, Yves, and Bryant G. Garth. 1998. *Dealing in Virtue: International Commercial Arbitration and the Construction of a Transnational Legal Order*. University of Chicago Press.
Djankov, Simeon. 2015. "Russia's Economy Under Putin: From Crony Capitalism to State Capitalism." Peterson Institute for International Economics.
DLA Piper. 2016. "A Case of Winning the Battle and Losing the War: *Yukos v Russian Federation*."
Dolan, Kerry A. 2020. "As Angola Accuses Billionaire Isabel dos Santos of Fraud, Her Empire Begins to Unravel." *Forbes*, January 23.
Dolan, Kerry A. 2021. "How Isabel dos Santos, Once Africa's Richest Woman, Went Broke." *Forbes*, January 22. https://www.forbes.com/sites/kerryadolan/2021/01/22/the-unmaking-of-a-billionaire-how-africas-richest-woman-went-broke/.
Doran, James. 2001. "Motorola Traces Luxury Assets in Loan Wrangle." *The Times*, July 23.
Drezner, Daniel W. 2001. "Globalization and Policy Convergence." *International Studies Review* 3 (1): 53–78.
Drezner, Daniel W. 2008. *All Politics Is Global: Explaining International Regulatory Regimes*. Princeton University Press.
Eckel, Mike. 2019. "Two Decades on, Smoldering Questions About the Russian President's Vault to Power." *Radio Free Europe/Radio Liberty*, August 7, sec. Russia. https://www.rferl.org/a/putin-russia-president-1999-chechnya-apartment-bombings/30097551.html.
Efrat, Asif. 2016. "Legal Traditions and Nonbinding Commitments: Evidence from the United Nations' Model Commercial Legislation." *International Studies Quarterly* 60 (4): 624–35.
Eilstrup-Sangiovanni, Mette, and J. C. Sharman. 2021. "Enforcers Beyond Borders: Transnational NGOs and the Enforcement of International Law." *Perspectives on Politics* 19 (1): 131–47.

Erie, Matthew S. 2020, September 17. "The New Legal Hubs: The Emergent Landscape of International Commercial Dispute Resolution." *Virginia Journal of International Law* 60 (2): 225–98.

Esen, Berk, and Sebnem Gumuscu. 2018. "Building a Competitive Authoritarian Regime: State–Business Relations in the AKP's Turkey." *Journal of Balkan and Near Eastern Studies* 20 (4): 349–72. https://doi.org/10.1080/19448953.2018.1385924.

Evans, Peter B. 2012. *Embedded Autonomy: States and Industrial Transformation.* Princeton University Press.

Fairfield, Tasha. 2015. *Private Wealth and Public Revenue.* Cambridge University Press.

Farchy, Jack. 2015. "Rosneft Settles Legal Dispute with Yukos Shareholders." *Financial Times,* April 1. https://www.ft.com/content/e95c3cd4-d87b-11e4-ba53-00144feab7de.

Farrell, Henry, and Abraham Newman. 2014. "The New Politics of Interdependence Cross-National Layering in Trans-Atlantic Regulatory Disputes." *Comparative Political Studies* 48 (4): 497–526. https://doi.org/10.1177/0010414014542330.

Farrell, Henry, and Abraham Newman. 2019. "Weaponized Interdependence." *International Security* 44 (1): 42–79.

Felstiner, William L. F., Richard L. Abel, and Austin Sarat. 1980. "The Emergence and Transformation of Disputes: Naming, Blaming, Claiming." *Law & Society Review* 15 (3/4): 631–54. https://doi.org/10.2307/3053505.

Fernandez, Rodrigo, Annelore Hofman, and Manuel B. Aalbers. 2016. "London and New York as a Safe Deposit Box for the Transnational Wealth Elite." *Environment and Planning A: Economy and Space* 48 (12): 2443–61.

Fichtner, Jan, Eelke M. Heemskerk, and Javier Garcia-Bernardo. 2017. "Hidden Power of the Big Three? Passive Index Funds, Re-Concentration of Corporate Ownership, and New Financial Risk." *Business and Politics* 19 (2): 298–326.

Firestone, Thomas. 2008. "Criminal Corporate Raiding in Russia." *International Lawyer* 42 (4): 1207–29.

Forsythe, Eliza, Lisa B. Kahn, Fabian Lange, and David Wiczer. 2020, September. "Labor Demand in the Time of COVID-19: Evidence from Vacancy Postings and UI Claims." *Journal of Public Economics* 189: 1–7. https://doi.org/10.1016/j.jpubeco.2020.104238.

Foy, Henry. 2018a. "Oligarchs Step Back from 'Shootout' for Control of Norilsk Nickel." *Financial Times,* April 13. https://www.ft.com/content/e6dabae8-3e3b-11e8-b7e0-52972418fec4.

Foy, Henry. 2018b. "Oligarch Vladimir Potanin on Money, Power and Putin." *Financial Times,* April 13. https://www.ft.com/content/a48fedf2-3da0-11e8-b7e0-52972418fec4.

Foy, Henry. 2018c. "Rusal Prepares for 'Shoot Out' to End Battle Over Norilsk." *Financial Times,* February 23. https://www.ft.com/content/252389d4-1866-11e8-9376-4a6390addb44.

Foy, Henry, and Jane Croft. 2018. "Deripaska Wins London Court Battle Over Norilsk Shares." *Financial Times,* June 27. https://www.ft.com/content/5113ec3e-7a20-11e8-bc55-50daf11b720d.

Freeland, Chrystia. 2005. *Sale of the Century: Russia's Wild Ride from Communism to Capitalism.* Abacus.

Freeland, Chrystia. 2012. *Plutocrats: The Rise of the New Global Super-Rich and the Fall of Everyone Else.* Penguin.

Freeman, John R., and Dennis P. Quinn. 2012. "The Economic Origins of Democracy Reconsidered." *American Political Science Review* 106 (1): 58–80.

Freund, Caroline. 2016. *Rich People Poor Countries: The Rise of Emerging-Market Tycoons and Their Mega Firms.* Peterson Institute for International Economics.

Friedman, Jeremy, and Sophus Reinert. 2020, December. "Angola Starts Now." Harvard Business School Case 719-007, January 2019. (Revised December 2020.)

Frye, Timothy. 2002. "Capture or Exchange? Business Lobbying in Russia." *Europe-Asia Studies* 54 (7): 1017–36.
Frye, Timothy. 2004. "Credible Commitment and Property Rights: Evidence from Russia." *American Political Science Review* 98 (3): 453–66.
Frye, Timothy. 2010. "Corruption and Rule of Law." In *Russia After the Global Crisis*, ed. Anders Aslund, Sergei Guriev, and Andrew Kuchins. Peterson Institute for International Economics.
Gaidar, Yegor. 1997. "The IMF and Russia." *American Economic Review* 87 (2): 13–16.
Galanter, Marc. 1974. "Why the 'Haves' Come Out Ahead: Speculations on the Limits of Legal Change." *Law & Society Review* 9 (1): 95–160.
Gans-Morse, Jordan. 2012. "Threats to Property Rights in Russia: From Private Coercion to State Aggression." *Post-Soviet Affairs* 28 (3): 263–95.
Gans-Morse, Jordan. 2017, May. "Demand for Law and the Security of Property Rights: The Case of Post-Soviet Russia." *American Political Science Review* 111 (2): 338–59.
Gehlbach, Scott. 2010. "Reflections on Putin and the Media." *Post-Soviet Affairs* 26 (1): 77–87.
Gel'Man, Vladimir. 2008. "Party Politics in Russia: From Competition to Hierarchy." *Europe-Asia Studies* 60 (6): 913–30.
Gemmell, Katharine. 2023. "Angola's Isabel dos Santos Pulled into Defaulted London Loan Dispute." *Bloomberg*, May 23. https://www.bloomberg.com/news/articles/2023-05-23/exiled-angolan-isabel-dos-santos-pulled-into-unitel-s-london-loan-default-spat.
George, Alexander L., Andrew Bennett, Sean M. Lynn-Jones, and Steven E. Miller. 2005. *Case Studies and Theory Development in the Social Sciences*. MIT Press.
Gessen, Masha. 2012. "The Wrath of Putin." *Vanity Fair*, April. https://www.vanityfair.com/news/politics/2012/04/vladimir-putin-mikhail-khodorkovsky-russia.
Gessen, Masha. 2013. *The Man Without a Face: The Unlikely Rise of Vladimir Putin*. Riverhead.
Gilardi, Fabrizio. 2005. "The Institutional Foundations of Regulatory Capitalism: The Diffusion of Independent Regulatory Agencies in Western Europe." *Annals of the American Academy of Political and Social Science* 598 (1): 84–101.
Gillman, Howard. 2002. "How Political Parties Can Use the Courts to Advance Their Agendas: Federal Courts in the United States, 1875–1891." *American Political Science Review* 96 (3): 511–24.
Ginsburg, Tom, and Tamir Moustafa, eds. 2008. *Rule by Law: The Politics of Courts in Authoritarian Regimes*. Cambridge University Press.
Glaeser, Edward L., Rafael La Porta, Florencio Lopez-de-Silanes, and Andrei Shleifer. 2004. "Do Institutions Cause Growth?" *Journal of Economic Growth* 9 (3): 271–303.
Goldman, Marshall I. 2003. *The Piratization of Russia: Russian Reform Goes Awry*. Routledge.
Goldman, Marshall I. 2008. *Petrostate: Putin, Power, and the New Russia*. Oxford University Press.
Goluboff, Risa L. 2009. *The Lost Promise of Civil Rights*. Harvard University Press.
Gorst, Isabel. 2008. "Moscow Tightens Grip on Norilsk." *Financial Times*, December 26. https://www.ft.com/content/764c308e-d37b-11dd-989e-000077b07658.
Gorst, Isabel. 2013. "Kazakhstan: Authorities Secure Control of Fugitive Banker's Wife and Daughter." *Financial Times*, June 3, sec. beyondbrics. https://www.ft.com/content/56b50a97-dcc2-3275-ab0c-5a7a1f962c77.
Gourevitch, Peter. 1978. "The Second Image Reversed: The International Sources of Domestic Politics." *International Organization* 32 (4): 881–912.
Gray, Julia. 2018. "Treaty Shopping and Unintended Consequences: BRICS in the International System." In *BRICS and the Global Economy*, ed. Soo Yeon Kim, 259–85.

https://www.worldscientific.com/doi/10.1142/9789811202308_0011. World Press Publishing.

Graz, Jean-Christophe. 2003. "How Powerful Are Transnational Elite Clubs? The Social Myth of the World Economic Forum." *New Political Economy* 8 (3): 321–40.

Gross, Anna, Victor Mallet, and Ayla Jean Yackley. 2021. "Uzan Family to Sue Groups Including Motorola and Vodafone for $68bn." *Financial Times*, July 16. https://www.ft.com/content/9debb4e0-c23a-4dd9-83f2-017659b0d6a2.

Grzymala-Busse, Anna. 2008. "Beyond Clientelism: Incumbent State Capture and State Formation." *Comparative Political Studies* 41 (4–5): 638–73.

Guriev, Sergei, and Daniel Treisman. 2019. "Informational Autocrats." *Journal of Economic Perspectives* 33 (4): 100–127. https://doi.org/10.1257/jep.33.4.100.

Haberly, Daniel, and Dariusz Wójcik. 2014. "Tax Havens and the Production of Offshore FDI: An Empirical Analysis." *Journal of Economic Geography* 15 (1): 75–101.

Haberly, Daniel, and Dariusz Wójcik. 2022. *Sticky Power: Global Financial Networks in the World Economy*. Oxford University Press.

Hägel, Peter. 2020. *Billionaires in World Politics*. Oxford University Press.

Hall, Peter A. 2001. *Varieties of Capitalism*. Wiley Online Library. http://onlinelibrary.wiley.com/doi/10.1002/9781118900772.etrds0377/full.

Hancké, Bob, Martin Rhodes, and Mark Thatcher. 2007. *Beyond Varieties of Capitalism: Conflict, Contradictions, and Complementarities in the European Economy*. Oxford University Press.

Haselmann, Rainer, David Schoenherr, and Vikrant Vig. 2018. "Rent Seeking in Elite Networks." *Journal of Political Economy* 126 (4): 1638–90. https://doi.org/10.1086/697742.

Hellman, Joel S. 1998. "Winners Take All: The Politics of Partial Reform in Postcommunist Transitions." *World Politics* 50 (2): 203–34.

Hendley, Kathryn. 1998, Winter. "Remaking an Institution: The Transition in Russia from State Arbitrazh to Arbitrazh Courts." *American Journal of Comparative Law* 46 (1): 93–127.

Hendley, Kathryn. 2004. "Enforcing Judgments in Russian Economic Courts." *Post-Soviet Affairs* 20 (1): 46–82. https://doi.org/10.1080/1060586X.2004.12005943.

Hendley, Kathryn, Peter Murrell, and Randi Ryterman. 2000. "Law, Relationships and Private Enforcement: Transactional Strategies of Russian Enterprises." *Europe-Asia Studies* 52 (4): 627–56.

Herbert Smith Freehills. 2018. "High Court Orders Indemnity Costs Following Discontinuance of Proceedings Alleging Serious Fraud." *Mondaq Business Briefing*, September 12.

Hess, Maximilian. 2018. "Bond of War: Russian Geo-Economics in Ukraine's Sovereign Debt Restructuring." *Foreign Policy Research Institute*, September 19. https://www.fpri.org/article/2018/09/bond-of-war-russian-geo-economics-in-ukraines-sovereign-debt-restructuring/.

Hibbitts, Bernard. 2004. "Yukos Files for US Bankruptcy Protection to Stop Russian Sell-Off." *Jurist News*, December 15. https://www.jurist.org/news/2004/12/yukos-files-for-us-bankruptcy/.

Hoang, Kimberly Kay. 2022. *Spiderweb Capitalism: How Global Elites Exploit Frontier Markets*. Princeton University Press.

Hoffman, David E. 2011. *The Oligarchs: Wealth and Power in the New Russia*. Hachette UK.

Hollingsworth, Mark, and Stewart Lansley. 2009. *Londongrad: From Russia with Cash: The Inside Story of the Oligarchs*. Fourth Estate London.

Hollyer, James R., B. Peter Rosendorff, and James Raymond Vreeland. 2018. *Transparency, Democracy, and Autocracy: Economic Transparency and Political (In)Stability*. Cambridge University Press.

Hryniuk, Olga. 2018. "The Enduring Saga of Tajikistan's TALCO Dispute." *CIS Arbitration Forum*, January 15. http://www.cisarbitration.com/2018/01/15/the-enduring-saga-of-tajikistans-talco-dispute/.

Hug, Adam. 2010. "Kazakhstan at a Crossroads: Governance, Corruption & International Investment." *Foreign Policy Centre*, September. https://fpc.org.uk/publications/kazakhstanatacrossroads/.

Ikenberry, G. John. 2009. "Liberal Internationalism 3.0: America and the Dilemmas of Liberal World Order." *Perspectives on Politics* 7 (1): 71–87.

Johnson, Juliet. 1997. "Russia's Emerging Financial-Industrial Groups." *Post-Soviet Affairs* 13 (4): 333–65.

Johnston, Alastair Iain. 2003. "Socialization in International Institutions." In *International Relations Theory and the Asia-Pacific*, ed. G. John Ikenberry and Michael Mastanduno. Columbia University Press.

Judah, Ben. 2013. *Fragile Empire: How Russia Fell in and out of Love with Vladimir Putin*. Yale University Press.

Junisbai, Barbara, and Azamat Junisbai. 2005. "The Democratic Choice of Kazakhstan: A Case Study in Economic Liberalization, Intraelite Cleavage, and Political Opposition." *Demokratizatsiya* 13 (3): 373–92. https://doi.org/10.3200/DEMO.13.3.373-392.

Kagarlitsky, Boris. 2002. *Russia Under Yeltsin and Putin: Neo-Liberal Autocracy*. Pluto.

Kahraman, Filiz, Nikhil Kalyanpur, and Abraham L. Newman. 2020. "Domestic Courts, Transnational Law, and International Order." *European Journal of International Relations* 26 (1, suppl): 184–208.

Kalyanpur, Nikhil. 2023. "An Illiberal Economic Order: Commitment Mechanisms Become Tools of Authoritarian Coercion." *Review of International Political Economy* 30 (4): 1238–54. https://doi.org/10.1080/09692290.2023.2211280.

Kalyanpur, Nikhil. 2025. "Weaponised Legal Dependence: How States Repress Their Globalised Oligarchs." *Review of International Studies*. https://doi.org/10.1017/S0260210525101216.

Kalyanpur, Nikhil. 2026. "Twilight of the Oligarchs." *Review of International Political Economy*. http://dx.doi.org/10.1080/09692290.2026.2627936.

Kalyanpur, Nikhil, and Abraham L. Newman. 2019, Winter. "Mobilizing Market Power: Jurisdictional Expansion as Economic Statecraft." *International Organization* 73 (1): 1–34.

Kamdar, Bansari. 2024. "Explainer: Adani vs. Hindenburg Research: What You Need to Know." *Reuters*, January 3, sec. India. https://www.reuters.com/world/india/adani-vs-hindenburg-research-what-you-need-know-2024-01-03/.

Karadag, Roy. 2010. "Neoliberal Restructuring in Turkey: From State to Oligarchic Capitalism." MPIfG Discussion Paper, No. 10/7, Max Planck Institute for the Study of Societies, Cologne.

Katz, Alan, Kitty Donaldson, and Stephanie Baker. 2020. "Deripaska's Rusal Role Was Flagged for U.S. Sanctions Review." *Bloomberg*, December 17. https://www.bloomberg.com/news/articles/2020-12-17/deripaska-s-rusal-role-spurred-europe-sanctions-warning-to-u-s.

Kim, In Song, and Iain Osgood. 2019. "Firms in Trade and Trade Politics." *Annual Review of Political Science* 22: 399–417.

Kittrie, Orde F. 2016. *Lawfare: Law as a Weapon of War*. Oxford University Press.

Klebnikov, Paul. 2000. *Godfather of the Kremlin: Boris Berezovsky and the Looting of Russia*. Harcourt.

Korpi, Walter. 1985. "Power Resources Approach vs. Action and Conflict: On Causal and Intentional Explanations in the Study of Power." *Sociological Theory* 3 (2): 31–45.

Kotkin, Stephen. 2008. *Armageddon Averted: The Soviet Collapse, 1970–2000*. Oxford University Press.

Krcmaric, Daniel, and Abel Escribà-Folch. 2023. "I'll Be Back? Exiled Leaders and Political Instability." *Journal of Conflict Resolution* 67 (2–3): 402–27. https://doi.org/10.1177/00220027221115622.

Krcmaric, Daniel, Stephen C. Nelson, and Andrew Roberts. 2023. "Billionaire Politicians: A Global Perspective." *Perspectives on Politics* 22: 357–71.

Lake, David A. 2009. *Hierarchy in International Relations*. Cornell University Press.

Lambroschini, Sophie. 2000. "Russia: Prosecutors Charge Media-MOST Magnate Gusinsky." *Radio Free Europe/Radio Liberty*, November 13, sec. Archive. https://www.rferl.org/a/1095124.html.

Ledeneva, Alena V. 2001. "Unwritten Rules." *How Russia Really Works, London*. http://cer-staging.thomas-paterson.co.uk/sites/default/files/publications/attachments/pdf/2011/e246_unwritten_rules-2203.pdf.

Ledeneva, Alena V. 2006. *How Russia Really Works: The Informal Practices That Shaped Post-Soviet Politics and Business*. Cornell University Press.

Ledyaeva, Svetlana, Päivi Karhunen, Riitta Kosonen, and John Whalley. 2015. "Offshore Foreign Direct Investment, Capital Round-Tripping, and Corruption: Empirical Analysis of Russian Regions." *Economic Geography* 91 (3): 305–41.

Lillis, Joanna. 2022. *Dark Shadows: Inside the Secret World of Kazakhstan*. Bloomsbury.

Lindblom, Charles E. 1982. "Politics and Markets." http://philpapers.org/rec/LINPAM-7.

Linsi, Lukas, and Daniel K. Mügge. 2019, June. "Globalization and the Growing Defects of International Economic Statistics." *Review of International Political Economy* 26 (3): 361–83. https://doi.org/10.1080/09692290.2018.1560353.

Logvinenko, Igor. 2019. "Before the Interests Are Invested: Disputes Over Asset Control and Equity Market Restrictions in Russia." *Review of International Political Economy* 26 (4): 695–721.

Lowe, A. Vaughan. 2002. *Transnational Legal Processes: Globalisation and Power Disparities*. Vol. 9. Cambridge University Press.

Lyrchikova, Anastasia, and Polina Ivanova. 2018. "Russia's Potanin to Appeal UK Court Decision in Battle of Billionaires." *Reuters*, November 27. https://www.reuters.com/article/us-russia-nornickel-rusal-court-idUSKCN1NW1VQ.

Macaulay, Stewart. 1963, February. "Non-Contractual Relations in Business: A Preliminary Study." *American Sociological Review* 28 (1): 55–67. https://doi.org/10.2307/2090458.

Mahajan, Karan. 2019. "How the Gupta Brothers Hijacked South Africa Using Bribes Instead of Bullets." *Vanity Fair*, March 3. https://www.vanityfair.com/news/2019/03/how-the-gupta-brothers-hijacked-south-africa-corruption-bribes.

Markus, Stanislav. 2015. *Property, Predation, and Protection*. Cambridge University Press.

Markus, Stanislav. 2017a. "The Atlas That Has Not Shrugged: Why Russia's Oligarchs Are an Unlikely Force for Change." *Daedalus* 146 (2): 101–12.

Markus, Stanislav. 2017b. "Oligarchs and Corruption in Putin's Russia: Of Sand Castles and Geopolitical Volunteering." *Georgetown Journal of International Affairs* 18 (2): 26–32.

Markus, Stanislav, and Volha Charnysh. 2017. "The Flexible Few: Oligarchs and Wealth Defense in Developing Democracies." *Comparative Political Studies* 50 (12): 1632–65. https://doi.org/10.1177/0010414016688000.

Martin, Michaela Elsbeth, and Hussein Solomon. 2016. "Understanding the Phenomenon of 'State Capture' in South Africa." *Southern African Peace and Security Studies* 5 (1): 21–35.

Mattingly, Daniel C. 2016. "Elite Capture." *World Politics* 68 (3): 383–412.

Mavelli, Luca. 2018. "Citizenship for Sale and the Neoliberal Political Economy of Belonging." *International Studies Quarterly* 62 (3): 482–93.

McClam, Erin. 2003. "Wealthy Turkish Family Loses Telcom Case." *AP Online*, July 31.

McFaul, Michael. 1995. "State Power, Institutional Change, and the Politics of Privatization in Russia." *World Politics* 47 (2): 210–43.

Mezrich, Ben. 2016. *Once Upon a Time in Russia: The Rise of the Oligarchs—A True Story of Ambition, Wealth, Betrayal, and Murder.* Simon and Schuster.

Michaels, Ralf. 2009. "Recognition and Enforcement of Foreign Judgments." In *The Max Planck Encyclopedia of Public International Law*, ed. Rüdiger Wolfrum. Oxford University Press.

Midgley, Dominic, and Chris Hutchins. 2011. *Abramovich: The Billionaire from Nowhere.* BookBaby.

Milanovic, Branko. 2016. *Global Inequality: A New Approach for the Age of Globalization.* Harvard University Press.

Miller, Chris. 2018. *Putinomics: Power and Money in Resurgent Russia.* University of North Carolina Press.

Mining Journal. 2010. "Rusal Rejects Norilsk's Buy-Back Offer." Internet Archive Wayback Machine, December 19. https://web.archive.org/web/20101219222841/http://www.mining-journal.com/finance/rusal-rejects-norilsks-buy-back-offer.

Mnookin, Robert H., and Lewis Kornhauser. 1979. "Bargaining in the Shadow of the Law: The Case of Divorce." *Yale Law Journal* 88 (5): 950–97.

Moustafa, Tamir. 2007. *The Struggle for Constitutional Power: Law, Politics, and Economic Development in Egypt.* Cambridge University Press.

Murat, Akser. 2018. "News Media Consolidation and Censorship in Turkey: From Liberal Ideals to Corporatist Realities." *Mediterranean Quarterly* 29 (3): 78–97.

Musacchio, Aldo, Eric Werker, and Jonathan Schlefer. 2010, September. "Angola and the Resource Curse." Harvard Business School Case 711-016.

Myers, Steven Lee. 2015. *The New Tsar: The Rise and Reign of Vladimir Putin.* Simon and Schuster.

Nedal, Dani K., and Daniel H. Nexon. 2019. "Anarchy and Authority: International Structure, the Balance of Power, and Hierarchy." *Journal of Global Security Studies* 4 (2): 169–89.

Newman, Abraham L., and Elliot Posner. 2011. "International Interdependence and Regulatory Power: Authority, Mobility, and Markets." *European Journal of International Relations* 17 (4): 589–610.

Newman, Abraham L., and Elliot Posner. 2018. *Voluntary Disruptions: International Soft Law, Finance, and Power.* Oxford University Press.

North, Douglass C., John Joseph Wallis, and Barry R. Weingast. 2009. *Violence and Social Orders: A Conceptual Framework for Interpreting Recorded Human History.* Cambridge University Press.

North, Douglass C., and Barry R. Weingast. 1989. "Constitutions and Commitment: The Evolution of Institutions Governing Public Choice in Seventeenth-Century England." *Journal of Economic History* 49 (4): 803–32.

Nougayrède, Delphine. 2013. "Outsourcing Law in Post-Soviet Russia." *Journal of Eurasian Law* 3 (6). https://papers.ssrn.com/sol3/papers.cfm?abstract_id=2433771.

Nougayrède, Delphine. 2016. "The Use of Offshore Companies in Emerging Market Economies: A Case Study." *Columbia Journal of European Law* 23: 401.

O'Brien, David. 2001. "AKP Facing Dispute with Business Dynasty: Row Over Turkish Electricity Firms Could Escalate" *Daily Star*, June 3.

OMCT. 2015. "Why the 'Angola 15' Trial Is a Parody of Justice?" OMCT. December 18. https://www.omct.org/en/resources/urgent-interventions/why-the-angola-15-trial-is-a-parody-of-justice.

Öniş, Ziya. 2004. "The Political Economy of Turkey's Justice and Development Party." *SSRN Electronic Journal.* https://doi.org/10.2139/ssrn.659463.

Öniş, Ziya. 2012. "The Triumph of Conservative Globalism: The Political Economy of the AKP Era." *Turkish Studies* 13 (2): 135–52. https://doi.org/10.1080/14683849.2012.685252.

Ostrovsky, Arkady. 2005. "Gazprom Buys Sibneft Stake for $13.1bn." *Financial Times*, September 28. https://www.ft.com/content/1aee02b4-3016-11da-ba9f-00000e2511c8.

Ostrovsky, Arkady. 2017. *The Invention of Russia: From Gorbachev's Freedom to Putin's War.* Penguin.

Ozel, Isik. 2013. "Is It None of Their Business? Business and Democratization, the Case of Turkey." *Democratization* 20 (6): 1081–1116. https://doi.org/10.1080/13510347.2012.674369.

Palan, Ronen. 2002. "Tax Havens and the Commercialization of State Sovereignty." *International Organization* 56 (1): 151–76. https://doi.org/10.1162/002081802753485160.

Palan, Ronen. 2006. *The Offshore World: Sovereign Markets, Virtual Places, and Nomad Millionaires.* Cornell University Press.

Paniagua, Victoria. 2018. "Protecting Capital: Economic Elites, Asset Portfolio Diversification, and the Politics of Distribution." PhD diss., Duke University.

Pereira, Anthony W. 2005. *Political (In)Justice: Authoritarianism and the Rule of Law in Brazil, Chile, and Argentina.* University of Pittsburgh Press.

Peyrouse, Sebastien. 2012. "The Kazakh Neopatrimonial Regime: Balancing Uncertainties Among the 'Family,' Oligarchs and Technocrats." *Demokratizatsiya* 20 (4): 345.

Pinsent Masons. 2013. "Enforcing England and Wales Judgments Abroad." Pinsent Masons, April 1. https://www.pinsentmasons.com/out-law/guides/enforcing-england-and-wales-judgments-abroad.

Piontkovsky, Andrei. 2009. "Reading Russia: The Dying Mutant." *Journal of Democracy* 20 (2): 52–55.

Pistor, Katharina. 2019. *The Code of Capital: How the Law Creates Wealth and Inequality.* Princeton University Press.

Pitcher, Anne, and Edalina Rodrigues Sanches. 2019, June. "The Paradox of Isabel dos Santos: State Capitalism, Dynastic Politics, and Gender Hostility in a Resource-Rich, Authoritarian Country." *Cahiers d'études Africaines* 234: 597–624. https://doi.org/10.4000/etudesafricaines.26276.

Pleines, Heiko. 2009. "The Political Role of the Oligarchs." In *Ukraine on Its Way to Europe: Interim Results of the Orange Revolution*, ed. I. Bredies, A. Unland, and V. Yaushik. Peter Lang Verlag.

Plowman, Hugo. 2023. "High Court Hands Down Another Judgment on Koza Ltd & Akin Ipek v Koza Altin." Mishcon de Reya LLP, August 10. https://www.mishcon.com/news/high-court-hands-down-another-judgment-on-koza-ltd-and-akin-ipek-v-koza-altin.

Premium Times. 2020. "How Africa's Richest Woman Exploited Family Ties, Shell Companies and Inside Deals to Build an Empire." *Premium Times*, January 19.

Puente, Ignacio, and Ben Ross Schneider. 2020. "Business and Development: How Organization, Ownership and Networks Matter." *Review of International Political Economy* 27 (6): 1354–77.

Puglisi, Rosaria. 2003. "The Rise of the Ukrainian Oligarchs." *Democratization* 10 (3): 99–123.

Putnam, Tonya L. 2016. *Courts Without Borders: Law, Politics, and US Extraterritoriality.* Cambridge University Press.

Radnitz, Scott. 2017. *Weapons of the Wealthy: Predatory Regimes and Elite-Led Protests in Central Asia.* Cornell University Press. https://doi.org/10.7591/9780801466175.

Rajah, Jothie. 2012. *Authoritarian Rule of Law: Legislation, Discourse and Legitimacy in Singapore.* Cambridge University Press.

Reed, John. 2024. "Gautam Adani Says 'Every Attack Makes Us Stronger' as He Faces US Charges." *Financial Times*, November 30, sec. Adani Group. https://www.ft.com/content/30852082-e487-42f6-82ce-c61c33bd7c8f.

Reuters. 1995. "Cukurova Elektrik Share Deals Raise Suspicions." *Reuters News*, December 6.

Reuters. 2002. "UK Court Sentences Turkish Media Baron to Jail." *Reuters News*, December 20.

Reuters. 2016. "French Court Cancels Kazakh Tycoon Ablyazov's Extradition to Russia." *Reuters*, December 9, sec. World. https://www.reuters.com/article/world/french-court-cancels-kazakh-tycoon-ablyazovs-extradition-to-russia-idUSKBN13Y2FB/.

Reuters. 2018. "Manafort Had $10 Million Loan from Russian Oligarch: Court Filing." *Reuters*, June 27, sec. World. https://www.reuters.com/article/world/manafort-had-10-million-loan-from-russian-oligarch-court-filing-idUSKBN1JN2YE/.

Reuters. 2019. "Dutch Supreme Court Rules That Russia Unlawfully Bankrupted Yukos." *Moscow Times*, January 18. https://www.themoscowtimes.com/2019/01/18/dutch-supreme-court-rules-that-russia-unlawfully-bankrupted-yukos-a64195.

Reuters. 2020a. "Angola's dos Santos Wants Graft Case Against Her Dropped, Citing 'Fake' Documents." *Reuters News*, May 13.

Reuters. 2020b. "Brazilian Telecom Oi Sells Its 25 Percent Stake in Unitel to Sonangol: Filing." *Reuters*, January 24, sec. Media and Television. https://www.reuters.comarticle/idUSE6N27M00V/.

Reuters. 2021. "Dutch Court Rules Against dos Santos in Oil Asset Case—Lawyers." *Reuters*, July 27. https://www.reuters.com/business/energy/dutch-court-rules-against-dos-santos-oil-asset-case-lawyers-2021-07-27/.

Reuters. 2022. "Interpol Confirms Red Notice for Angolan Billionaire Isabel dos Santos." *Reuters*, November 30. https://www.reuters.com/world/europe/interpol-confirms-red-notice-angolan-billionaire-isabel-dos-santos-2022-11-30/.

Reuters. 2024. "Dutch Court Denies Russia's Appeal Against $50 Bln Yukos Payout." *Reuters*, February 20, sec. Europe. https://www.reuters.com/world/europe/dutch-court-denies-russias-appeal-against-50-bln-yukos-payout-2024-02-20/.

Reuters News. 2020. "Billionaire dos Santos' Suitability as Shareholder in Question After Asset Freeze—Bank of Portugal." *Reuters*, January 2, 2020. https://www.reuters.com/article/business/finance/billionaire-dos-santos-suitability-as-shareholder-in-question-after-asset-freez-idUSKBN1Z11JO/.

Rithmire, Meg. 2023. *Precarious Ties: Business and the State in Authoritarian Asia*. Oxford University Press.

Robinson, Gwen. 2008. "Russian Tycoons Target Norilsk." *Financial Times*, May 29, sec. FT Alphaville. https://www.ft.com/content/1007b458-5d16-3bb6-856e-b16507732342.

Rodrik, Dani. 2008. *One Economics, Many Recipes: Globalization, Institutions, and Economic Growth*. Princeton University Press.

Rudra, Nita. 2008. *Globalization and the Race to the Bottom in Developing Countries: Who Really Gets Hurt?* Cambridge University Press.

Rutland, Peter. 2000. "Putin's Path to Power." *Post-Soviet Affairs* 16 (4): 313–54.

Safavian, Mehnaz S., Douglas H. Graham, and Claudio Gonzalez-Vega. 2001. "Corruption and Microenterprises in Russia." *World Development* 29 (7): 1215–24.

Sakwa, Richard. 2007. *Putin: Russia's Choice*. Routledge.

Sakwa, Richard. 2014. *Putin and the Oligarch: The Khodorkovsky-Yukos Affair*. IB Tauris.

Salehyan, Idean. 2007. "Transnational Rebels: Neighboring States as Sanctuary for Rebel Groups." *World Politics* 59 (2): 217–42.

Salehyan, Idean. 2011. *Rebels Without Borders: Transnational Insurgencies in World Politics*. Cornell University Press.

Satter, David. 2016. "Vladimir Putin & 1999 Russian Apartment-House Bombings—Was Putin Responsible?" *Hudson Institute*, August 17. https://www.hudson.org/national-security-defense/vladimir-putin-1999-russian-apartment-house-bombings-was-putin-responsible.

Seawright, Jason, and John Gerring. 2008. "Case Selection Techniques in Case Study Research: A Menu of Qualitative and Quantitative Options." *Political Research Quarterly* 61 (2): 294–308.

Shapiro, Martin, and Alec Stone Sweet. 2002. *On Law Politics and Judicialization*. Oxford University Press.

Sharafutdinova, Gulnaz, and Karen Dawisha. 2017. "The Escape from Institution-Building in a Globalized World: Lessons from Russia." *Perspectives on Politics* 15 (2): 361–78. http://kclpure.kcl.ac.uk/portal/files/49208144/Instit_Arbitrage_final_Dec.14_2015_1.docx.

Sharman, Jason Campbell. 2006. *Havens in a Storm: The Struggle for Global Tax Regulation*. Cornell University Press.

Sharman, Jason Campbell. 2011. *The Money Laundry: Regulating Criminal Finance in the Global Economy*. Cornell University Press.

Sharman, Jason Campbell. 2012. "Chinese Capital Flows and Offshore Financial Centers." *Pacific Review* 25 (3): 317–37.

Sharman, Jason Campbell. 2017. *The Despot's Guide to Wealth Management: On the International Campaign Against Grand Corruption*. Cornell University Press.

Shaxson, Nicholas. 2011. *Treasure Islands: Tax Havens and the Men Who Stole the World*. Random House.

Shen-Bayh, Fiona. 2018. "Strategies of Repression: Judicial and Extrajudicial Methods of Autocratic Survival." *World Politics* 70 (3): 321–57.

Shirbon, Estelle. 2016. "Ukrainian Oligarchs Settle $2-Billion Dispute Days Before London Court Hearing." *Reuters*, January 22. https://uk.reuters.com/article/uk-britain-ukraine-oligarchs-idUKKCN0V02O3.

Shleifer, Andrei, and Daniel Treisman. 2001. *Without a Map: Political Tactics and Economic Reform in Russia*. MIT Press.

Siegel, Matt. 2008, March. "Widow Puts Blame on Khodorkovsky." *Moscow Times*.

Silverman, Reuben. 2014. "Dogan Versus Erdogan: Business and Politics in AKP-Era Turkey." *Mediterranean Quarterly* 25 (2): 131–51. https://doi.org/10.1215/10474552-2685794.

Simmons, Beth A. 2001. "The International Politics of Harmonization: The Case of Capital Market Regulation." *International Organization* 55 (3): 589–620.

Sixsmith, Martin. 2010. *Putin's Oil: The Yukos Affair and the Struggle for Russia*. Bloomsbury.

Sky News. 2012. "Russian Oligarch Pays £35m in Legal Costs." *Sky News*, October 12. https://news.sky.com/story/russian-oligarch-pays-35m-in-legal-costs-10467372.

Slobodian, Quinn. 2018. *Globalists: The End of Empire and the Birth of Neoliberalism*. Harvard University Press.

Soldatov, Andrei, and Irina Borogan. 2010. *The New Nobility: The Restoration of Russia's Security State and the Enduring Legacy of the KGB*. PublicAffairs.

Solomon, Peter H. 2008, October. "Assessing the Courts in Russia: Parameters of Progress Under Putin." *International Journal for Court Administration* 1 (2): 26–32.

Sonin, Konstantin. 2003. "Why the Rich May Favor Poor Protection of Property Rights." *Journal of Comparative Economics* 31 (4): 715–31.

Stephan, Paul B. 2013. "Taxation and Expropriation: The Destruction of the Yukos Oil Empire." *Houston Journal of International Law* 35 (1): 1–51.

Stern, Rachel E., and Kevin J. O'Brien. 2012. "Politics at the Boundary: Mixed Signals and the Chinese State." *Modern China* 38 (2): 174–98.

Stone, Randall W., Yu Wang, and Shu Yu. 2022. “Chinese Power and the State-Owned Enterprise.” *International Organization* 76 (1): 229–50.

Strange, Susan. 1996. *The Retreat of the State: The Diffusion of Power in the World Economy.* Cambridge University Press.

Surak, Kristin. 2021. “Millionaire Mobility and the Sale of Citizenship.” *Journal of Ethnic and Migration Studies* 47 (1): 166–89.

Svolik, Milan W. 2009. “Power Sharing and Leadership Dynamics in Authoritarian Regimes.” *American Journal of Political Science* 53 (2): 477–94.

Szakonyi, David. 2018. “Businesspeople in Elected Office: Identifying Private Benefits from Firm-Level Returns.” *American Political Science Review* 112 (2): 322–38.

Teles, Steven M. 2008. *The Rise of the Conservative Legal Movement: The Battle for Control of the Law.* Princeton University Press.

Tevendale, Craig, Hannah Ambrose, Ergin Mizrahi, and Özlem Çavuş Gerçek. 2021. “Enforcement of Court Judgments in the UK, the EU and Turkey After Brexit & Enforcement Strategy.” Herbert Smith Freehills Kramer, August 17. https://www.hsfkramer.com/en_US/insights/2021-08/enforcement-of-court-judgments-in-the-uk-the-eu-and-turkey-after-brexit-enforcement.

The Economist. 2001. “Facing Down the Uzan Clan.” *The Economist,* August 16. https://www.economist.com/business/2001/08/16/facing-down-the-uzan-clan.

Thrall, Calvin. 2021. “Spillover Effects in International Law: The Case of Tax Planning and Investor-State Dispute Settlement.” University of Austin at Texas. https://s18798.pcdn.co/gripe/wp-content/uploads/sites/18249/2021/02/taxplanning_v4.pdf.

Tomashevskiy, Andrey. 2021. “Economic Statecraft by Other Means: The Use and Abuse of Anti-Bribery Prosecution.” *International Studies Quarterly* 65 (2): 387–400. https://doi.org/10.1093/isq/sqab030.

Tonelson, Alan. 2000. *The Race to the Bottom: Why a Worldwide Worker Surplus and Uncontrolled Free Trade Are Sinking American Living Standards.* Westview.

Treisman, Daniel. 2001. *After the Deluge: Regional Crises and Political Consolidation in Russia.* University of Michigan Press.

Treisman, Daniel. 2008. “Putin’s Silovarchs.” *Orbis* 51 (1): 141–53.

Treisman, Daniel. 2016. “Russia’s Billionaires.” *American Economic Review* 106 (5): 236–41.

Van Os, Roos, and Roeline Knottnerus. 2011. “Dutch Bilateral Investment Treaties: A Gateway to ‘Treaty Shopping’ for Investment Protection by Multinational Companies.” Social Science Research Network (SSRN). https://papers.ssrn.com/sol3/papers.cfm?abstract_id=1961585.

Varese, Federico. 1997. “The Transition to the Market and Corruption in Post-Socialist Russia.” *Political Studies* 45 (3): 579–96.

Varese, Federico. 2001. *The Russian Mafia: Private Protection in a New Market Economy.* Oxford University Press.

Verde, Rui Santos. 2022. *Angola at the Crossroads: Between Kleptocracy and Development.* Bloomsbury.

Verdier, Pierre-Hugues. 2023. “Sanctions Overcompliance: What, Why, and Does It Matter?” *North Carolina Journal of International Law* 48: 471–98.

Voeten, Erik. 2008. “The Impartiality of International Judges: Evidence from the European Court of Human Rights.” *American Political Science Review* 102 (4): 417–33.

Volkov, Vadim. 2016. *Violent Entrepreneurs: The Use of Force in the Making of Russian Capitalism.* Cornell University Press.

Vorbyova, Toni. 2010. “Yukos Gets Security from Rosneft in Asset Freezing Case.” *Reuters,* April 8. https://www.reuters.com/article/rosneft-yukos-idUSLDE6370H320100408.

Wagstyl, Stefan. 2012. "Norilsk Nickel: Abramovich Steps in to Keep the Peace—but for How Long?" *Financial Times*, December 4. https://www.ft.com/content/51ea86cc-5286-341f-a324-4e9c296a10ea.

Weaver, Courtney. 2011. "Norilsk Offers to Buy Back Stake from Rusal." *Financial Times*, August 19. https://www.ft.com/content/033047b6-ca7c-11e0-94d0-00144feabdc0.

Weaver, Courtney. 2012. "Abramovich to Take Reduced Norilsk Stake." *Financial Times*, December 11. https://www.ft.com/content/b0cd0502-43ad-11e2-844c-00144feabdc0.

Weaver, Courtney, and Charles Clover. 2013. "Vladimir Potanin Plans Norilsk Nickel Overhaul." *Financial Times*, September 9. https://www.ft.com/content/78709aa8-18b3-11e3-83b9-00144feab7de.

Weingast, Barry R. 1995. "The Economic Role of Political Institutions: Market-Preserving Federalism and Economic Development." *Journal of Law, Economics, & Organization* 11 (1): 1–31.

Weiniger, Matthew, and Elizabeth Kantor. 2015. "KT Asia Investment Group BV v Republic of Kazakhstan: *Ratione Personae* and *Ratione Materiae*." *ICSID Review* 30 (3): 533–38. https://doi.org/10.1093/icsidreview/siv024.

Werner, Timothy. 2012. *Public Forces and Private Politics in American Big Business*. Cambridge University Press.

Wilf, Meredith. 2016. "Credibility and Distributional Effects of International Banking Regulations: Evidence from US Bank Stock Returns." *International Organization* 70 (4): 763.

Winters, Jeffrey Alan. 1996. *Power in Motion: Capital Mobility and the Indonesian State*. Cornell University Press.

Winters, Jeffrey Alan. 2011. *Oligarchy*. Cambridge University Press.

Woll, Cornelia. 2023. *Corporate Crime and Punishment: The Politics of Negotiated Justice in Global Markets*. Princeton University Press. https://www.torrossa.com/gs/resourceProxy?an=5559978&publisher=FZO137.

Xiao, Geng. 2004. *Round-Tripping Foreign Direct Investment and the People's Republic of China*. Vol. 58. Asian Development Bank Institute.

Xu, Yan. 2022, September. "Fragile Fortune: State Power and Concentrated Wealth in China." *Politics & Society* 51 (4): 463–92. https://doi.org/10.1177/00323292221124408.

Yessenova, Saulesh. 2018. "Chapter 14. The Political Economy of Oil Privatization in Post-Soviet Kazakhstan: Life Worlds of Oil and Gas." In *Subterranean Estates: Life Worlds of Oil and Gas*, ed. Hannah Appel, Arthur Mason, and Michael Watts. Cornell University Press. https://doi.org/10.7591/9780801455407-017.

Zucman, Gabriel. 2015. *The Hidden Wealth of Nations: The Scourge of Tax Havens*. University of Chicago Press.

Zygar, Mikhail. 2016. *All the Kremlin's Men: Inside the Court of Vladimir Putin*. PublicAffairs.

Index

Figures and tables are indicated by "f" and "t" following the page numbers.

www.ingramcontent.com/pod-product-compliance
Ingram Content Group UK Ltd.
Pitfield, Milton Keynes, MK11 3LW, UK
UKHW042002200726
473393UK00012B/38/J